Catalog of Vascular Plants of Arequipa, Peru

With Species Descriptions and Taxonomic Key

Edgar Heim

Publisher: BoD · Books on Demand GmbH, In de Tarpen 42, 22848 Norderstedt
Print: Libri Plureos GmbH, Friedensallee 273, 22763 Hamburg

ISBN: 978-3-7597-9358-4

Cite as:
Heim E (2024). Catalog of Vascular Plants of Arequipa, Peru. Books on Demand, Norderstedt (D).

Cover pictures: Austrocylindropuntia subulata
                    Adesmia miraflorensis
                    Hypseocharis pimpinellifolia

About the Author

Edgar Heim, born in Lima, Peru, studied food technology and dedicated more than 30 years of his career to this field as a manager and lecturer. Alongside his professional activities, he developed a deep personal interest in nature, biodiversity and ecology, with a particular passion for botany and travel. His travels have largely focused on South America and Southeast Asia, regions known for their rich biodiversity.

He lived in Peru for almost 20 years and now resides in Switzerland. Over the years, he has compiled a comprehensive database of plant and vegetation data, focusing on the flora of Peru, which serves as the foundation for this catalog.

In 2014, he published *"Flora of Arequipa, Peru: An Illustrated Field Guide"* followed by the *"Catalog of Ferns, Gymnosperms, and Flowering Plants of the Department of Arequipa, Peru"* in 2021. This current catalog represents a continuation of his dedicated research into the flora of southern Peru and underscores his commitment to documenting and preserving the plant biodiversity of this remarkable region.

Preface

This catalog aims to enhance understanding of the flora and vegetation of the department of Arequipa, Peru, and to serve as a practical resource for researchers, conservationists, and nature enthusiasts. By expanding our knowledge, we hope not only to inspire appreciation for Arequipa's natural heritage but also to underscore the importance of preserving its biodiversity.

Although ongoing research continues to reveal previously unknown species, the department of Arequipa remains botanically underexplored. This catalog provides an updated overview of plant species in the region, based on botanical literature, databases, herbarium specimens, and fieldwork across all provinces of Arequipa. It is based on the author's "Catalog of Ferns, Gymnosperms, and Flowering Plants of the Department of Arequipa, Peru", which has been completely revised and updated with additional information.

The catalog contains 1,350 wild-growing, non-cultivated ferns and flowering plants. Each species is presented with a brief morphological description, details of its ecological and geographic distribution, information on its human uses, and notes on its taxonomy. The appendix includes a comprehensive taxonomic key to all species discussed, as well as a list of commonly used synonyms.

Acknowledgments

I am deeply grateful for the assistance of botanists around the world. Special thanks go to those who generously answered my questions and shared their expertise: Hamilton Beltrán, Andrew Efremov, David Ferguson, Antonio Galán de Mera, Eric Gouda, Rafaël Govaerts, Marilu Huertas, Sandra Knapp, Yero Kuethe, Blanca León, Eliana Linares Perea, Ary Mailhos, Daniel Montesinos-Tubée, Jhon Muñuico, Nataly O'Leary, John Pruski, Victor Quipuscoa Silvestre, Ruth Ripley, Kai Schablewski, Hanno Schaefer, Julian Shaw, Oscar Vargas, Maximilian Weigend, Felipe Zapata, and all the many botanists on iNaturalist who provided valuable input on my observations.

I am equally grateful to Remijio Castro Huamani, a skilled guide who introduced me to the breathtaking landscapes of Arequipa, and to my wife, Ruth, whose support and companionship enriched most of my travels.

The database on which this work is based is updated periodically. It is available on the author's homepage.

Please feel free to reach out with any questions or to report any errors.

Edgar Heim

erhsgch@gmail.com
http://www.edgarsplantguide.com/

Content

Introduction

Aim

This catalog serves several purposes. First, it aims to improve our understanding of the flora and vegetation of the department of Arequipa, providing a valuable resource for researchers, conservationists, and nature enthusiasts. Second, it seeks to highlight the natural heritage of the region and promote greater awareness of the importance of conserving its rich plant life. Third, it provides a completely revised and expanded version of the author's previous Catalogue of Ferns, Gymnosperms, and Flowering Plants of the Department of Arequipa, Peru, incorporating new data and findings. Finally, it introduces a taxonomic key to help professionals and nature enthusiasts identify and appreciate Arequipa's diverse plant species.

State of Knowledge

Although ongoing research continues to reveal previously unknown species, the department of Arequipa remains botanically underexplored. This underscores the dynamic nature of the region's biodiversity and the need for continued study.

The following is a non-exhaustive, chronologically arranged list of key contributions to the study of Arequipa's flora and vegetation. Each work has added valuable insight into the complexity and richness of the region's plant life:

- Dillon (1997): Published a checklist of plants from Lomas de Mejía.
- Rodriguez Diaz (1998): Focused on the biodiversity of the Cuencas de Cotahuasi, with particular attention to medicinal plants.
- Quipuscoa & Huamantupa (2010): Cataloged the plants of the Reserva Nacional de Las Salinas y Aguada Blanca.
- Arce Condori (2010): Created an introductory booklet on the flora of Lagunas de Mejía.
- Dillon et al. (2011): Studied the vegetation of the Peruvian Lomas and compiled a checklist.
- Galán de Mera & Linares (2012): Published a phytosociological analysis of Arequipa's vegetation, supplemented in 2022.
- Montesinos & Mondragón (2013, 2016): Investigated the flora and vegetation of the Acarí River.
- Heim (2014): Produced an illustrated field guide to Arequipa's flora.
- Cano (2015): Focused her thesis on the ferns of Lomas de Atiquipa.
- Quipuscoa et al. (2016): Cataloged the plants of Lomas de Yuta.
- Talavera et al. (2017): Published an illustrated guide to the plants of Lomas de Atiquipa.
- Pauca & Quipuscoa (2017, 2020): Published taxonomic keys and studies on the cacti of Arequipa.
- Montesinos et al. (2019): Investigated the vegetation of the Chilli River in Arequipa.
- Montesinos & Zegarra (2019): Authored a guide to wild plants in urban and rural Arequipa.
- Heim (2021): Published the first edition of this catalog.
- Bedoya et al. (2023): Studied the genus Stevia (Asteraceae) in Arequipa.

In addition to these contributions, several publications on the flora and vegetation of neighboring departments have significantly enriched the regional understanding of southern Peru's flora and vegetation:

- León et al. (2004): Published a study on the flora of Tacna.
- Whaley et al. (2010): Produced a flora guide for Ica.
- Montesinos (2015): Compiled a guide to Moquegua's flora.
- Montesinos (2016): Conducted a detailed analysis of mountain vegetation in southern Peru, particularly the plant communities of Moquegua.

Building upon these foundational studies, this catalog aspires to be a comprehensive resource for botanists, conservationists, policymakers, and enthusiasts, facilitating informed decisions on species protection, environmental management, and future research initiatives. With continued research and exploration, we are getting closer to discovering the full botanical richness of Arequipa.

Materials and Methods

Data and Sources

The catalog was compiled using the following sources:

- Current botanical literature: Referenced in the Cited Literature chapter.
- Database records for Arequipa: Obtained from GBIF.org (as of December 2023), Tropicos.org (as of June 2023), and Herbariovaa.org (as of June 2023).
- Observations from iNaturalist.org: Records for Arequipa as of September 2024.
- Field observations by the author: Conducted during trips to all provinces of Arequipa. New species records for Arequipa were documented photographically and uploaded to iNaturalist.org.

For the purposes of this catalog, only wild-growing, non-cultivated species are included. Subspecific taxa are discussed within species descriptions rather than listed independently, allowing the user to focus on the primary species record.

Verification Criteria

Species occurrences were classified as certain when supported by at least one verified herbarium specimen or a credible, well-referenced source. This approach ensures that the presence of each species is grounded in verifiable evidence. Species mentioned without reference to herbarium specimens were classified as doubtful.

Digital herbarium specimens from repositories (B, CUZ, F, GA, K, L, MO, NY, and US) underwent partial verification. While this provided valuable data, the lack of digitalization at HSP and HUSA limited our ability to confirm some records.

Many CUZ and HUSA specimens were collected by anonymous collectors. Verification of digitally available anonymous specimens revealed some misidentifications. Consequently, anonymous specimens were classified as doubtful if they could not be verified.

Taxonomy and Nomenclature

The catalog's taxonomic framework is based on the Angiosperm Phylogeny Group (APG IV, 2016) for flowering plants and the Pteridophyte Phylogeny Group (PPG I, 2016) for ferns and related taxa. This ensures that classification and nomenclature reflect current scientific consensus. A phylogenetic tree of all plant families found in Arequipa is provided in Appendix 3.

The accepted scientific names primarily follow POWO (2024), with specific exceptions:

- Asteraceae, genus Baccharis: Classification follows Müller (2013).
- Oxalidaceae, genus Oxalis: Species synonymizations as listed by Hassler (2024), with updates from Shaw (2021, 2023).
- Pteridaceae, subfamily Cheilanthoidea: The proposed merger of the genera Argyrochosma, Astrolepis, Cheilanthes, Myriopteris, Notholaena, and Pellaea into Hemionitis is not followed, as it lacks broad acceptance.
- Verbenaceae: The classification follows O'Leary et al. (2021), O'Leary et al. (2016), O'Leary & Mulgura (2014), and Binder (2002).

Additional minor deviations from POWO (2024) are noted in the text.

The phylogenetic tree in Appendix 3 may assist with the identification of unknown species.

Description of Species

Species descriptions are based primarily on online scientific publications. In case where no publication was available, the descriptions rely on information from herbarium vouchers, preferably from type specimens.

Ecology

Wherever possible, ecological data were sourced from Brako & Zarucchi (1993) and Tropicos.org (2023), supplemented and adapted with information from recent research and field observations. The ecological data refer to the ecological behavior of the species in Peru and are not localized for the department of Arequipa.

Altitude ranges are categorized based on the levels established by Brako & Zarucchi (1993):

Coastal:	0–1000 m
Andean I:	500–1500 m
Andean II:	1500–3500 m
Andean III:	>3500 m

Distribution

The global distribution data were mainly derived from POWO (2024).

Notes on the Distribution of Species in Arequipa

Species in Arequipa were categorized as native or introduced according to Brako & Zarucchi (1993) and POWO (2024). Species have been further categorized as follows:

- Adventive species: Introduced species with a single recorded occurrence.
- Doubtful species: Species whose presence in Arequipa is uncertain due to insufficient evidence.
- Expected species: Species not yet officially recorded, but likely to be present based on their geographic range.
- Excluded species: Species previously mentioned for Arequipa but now excluded. The species list and reasons for exclusion are provided in Appendix 6.

Additional comments on the occurrence of the species in Arequipa have been included where necessary. The term voucher is used to include herbarium specimens and observational records.

Verified records and data from credible sources were used to identify the provinces in which each species occurs. The provinces of Arequipa are abbreviated as follows:

Arequipa	ARE
Camaná	CAM
Caravelí	CAR
Castilla	CAS
Caylloma	CAY
Condesuyos	CON
Islay	ISL
La Unión	UNI

Human Use of Plants

This catalog is not a comprehensive account of human uses of Arequipa's flora. Instead, it highlights selected uses supported by reliable sources, offering a basic understanding of certain species' cultural and economic importance. Information was drawn from expert reports and online sources. While many publications address plants' pharmacological and chemical properties, no in-depth review was conducted.

Synonyms

The 1,350 species listed in the catalog have over 10,000 synonyms, requiring a selective approach to their inclusion. The following criteria were used to streamline the list:

- Introduced species: Synonyms for introduced species were minimized, retaining only those that are still commonly used (e.g., Agave cordillerana, Centranthus ruber, Canna edulis).
- Eponymous synonyms: These were largely excluded as they are of limited practical value for field botany.
- Geographic epithets: Synonyms with epithets referring to locations outside of Arequipa were largely excluded.

This approach resulted in a curated list of 2,800 synonyms. The complete list for each species is available in POWO (2024).

Taxonomic Key to Species

The taxonomic key is an artificial key designed for accessibility. Specific taxonomic terms have been omitted to ensure that users with a basic knowledge of botany can use the key effectively. Wherever possible, observable characteristics, either visible with the naked eye or with a magnifying glass, have been included.

The key has been adapted from existing sources. In cases where no pre-existing keys were available or the information was contradictory, the key was developed based on species descriptions and/or type specimens.

Developing the key for the angiosperm families was a significant challenge. Several taxonomic guides were consulted and adapted to correspond to the plant families found in Arequipa. Despite these efforts, minor inconsistencies may remain in this section of the key.

The keys for the following genera are considered tentative and provisional due to inconsistent or incomplete information: Cinnagrostis (Poaceae), Dalea and Lupinus (Fabaceae), Nassella (Poaceae), Nolana (Solanaceae), and Nototriche (Malvaceae).

Catalog Structure

The species catalog is divided into two main sections: ferns and flowering plants. Within each group, families are arranged alphabetically, and the species within each family are listed alphabetically by their scientific names.

The appendices provide additional information, facilitating a deeper understanding of Arequipa's flora:

- Appendix 1 and 2: Alphabetical lists of genera, families, and orders of the species found in Arequipa.
- Appendix 3: A phylogenetic tree illustrating the relationships among the plant families in Arequipa.
- Appendix 4: A compilation of selective synonyms for the species documented in Arequipa.
- Appendix 5: A list of endemic species of Arequipa, arranged alphabetically by family.
- Appendix 6: A record of excluded species for Arequipa, organized alphabetically by species name.
- Appendix 7: A taxonomic key aligned with the structure of the species catalog for ease of use.

Arequipa and its Vegetation

Geography of Arequipa

Arequipa, one of Peru's 24 departments, lies between 14°40' S and 17°20' S and 70°50' W and 75°10' W. It is bordered by the Pacific Ocean to the west, the departments of Ica, Apurímac, and Ayacucho to the north, Cuzco and Puno to the east, and Moquegua to the south (Figure 1).

Figure 1: Location of the department of Arequipa in Peru and its 8 provinces. Source: https://www.tes.com/

Arequipa spans an area of 63,345 km² and has an estimated population of approximately 1.6 million, with around 1 million residing in the capital, the city of Arequipa. The economy is primarily driven by trade (about 15 % of GDP), agriculture (13 % of GDP), construction (about 11% of GDP), and mining (9 % of GDP).

Geology of Arequipa

The landscape of Arequipa is characterized by active volcanoes. In the north, towering volcanoes such as Nevado Solimana (6093 m), Coropuna (6425 m), and Ampato (6288 m) rise from the desert. The city of Arequipa itself is surrounded by three volcanoes: Chachani (6057 m), Misti (5822 m), and Picchu-Picchu (5664 m). Evidence of volcanic activity is visible in ash layers and lava flows exposed in road cuts. In addition, the region's distinctive sillar, a white volcanic rock, is used extensively for construction.

Despite its volcanic history, most of Arequipa's soils are sedimentary or alluvial. Around 150 million years ago, as the South American plate moved westward, it collided with the Nazca plate, creating a rift valley that was filled with eroded material. About 60 million years ago, the Andes began to rise, lifting these sediments into high plateaus and altering water systems. Rivers that once flowed into the Pacific were blocked by the rising Andes, forming inland lakes. Over time, some rivers, such as the Colca and Cotahuasi, carved deep canyons back to the Pacific, while others found new paths eastward to the Amazon Basin. Some systems, such as Lake Titicaca, became enclosed basins with no outlet to the ocean. Ancient lakes eventually filled with sediment, to form flat inter-Andean plains, such as the Pampa de Cañahuas.

Limited information is available on the reactivity of the soil in Arequipa. Galán de Mera and Linares (2012) report that the vegetation in the region is predominantly acidophilic. However, due to the volcanic and sedimentary origin of the soils, combined with reduced leaching in the arid climate, areas with neutral to alkaline soils are expected. A notable example is the Las Salinas salt lake, where the mined salt is rich in boron and lithium, indicating an alkaline environment.

For more detailed geological information, see the geological map of the Instituto Geológico Minero y Metalúrgico (INGEMMET, 2021) and the maps of SINIA (2024).

Phytogeography of Southern Peru

Peru is part of the Neotropical floristic kingdom, specifically within the Andean floristic region. According to Galán de Mera & Linares (2012), southern Peru is divided into five biogeographical provinces (Figure 2).

Using the biome classification of Walter & Breckle (1999), the desert provinces Limeño-Ariqueña and Atacama belong to the zonobiome III (subtropical desert vegetation), the Oruro-Arequipeña, Ancashino-Paceña and Urubambense provinces belong to the orobiome II (seasonal vegetation of tropical mountains), and the province of Madre de Dios is a transitional stage between zonobiome I and II, a gradual change from evergreen tropical evergreen forests to tropical seasonal green forests.

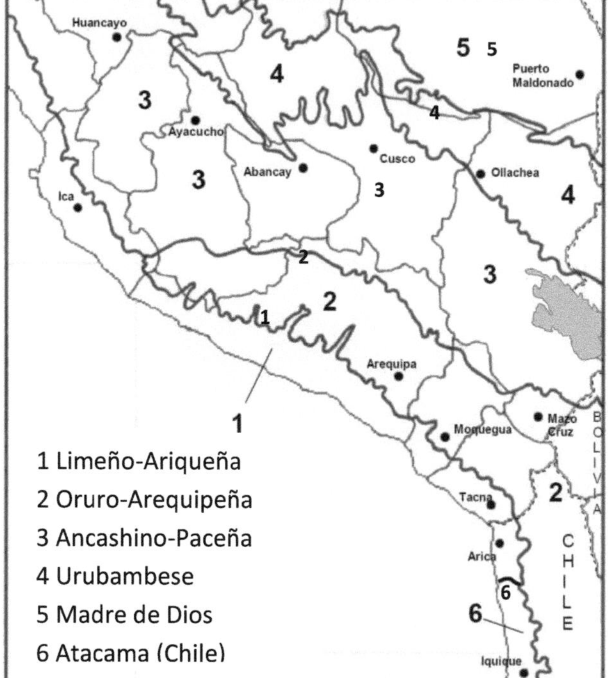

Figure 2: Phytogeographic provinces of southern Peru
Source: Galán de Mera & Linares (2012)

Climate and Vegetation

Arequipa's climate is strongly influenced by the Humboldt Current, a cold ocean current that flows north along the Pacific coast from southern Chile to central Peru. This current cools the sea air and prevents it from absorbing moisture, making onshore precipitation rare. Rainfall along the coast occurs only during El Niño events, which happens approximately every 7-10 years. Between such rains, the Atacama Desert is the driest place on earth, with an average rainfall of only 1 mm per year.

Temperature and precipitation vary along the transect from the Pacific coast to the inter-Andean valleys, as shown in Figure 3. Temperatures are relatively stable throughout the year but decrease with altitude. Intense solar radiation at higher elevations partially compensates for this cooling. Precipitation increases dramatically with altitude. The coast is extremely dry, while Arequipa receives rain in January and February. In Cuzco, the rainy season lasts seven months. Northeastern Arequipa probably has a climate similar to that of Cuzco, although long-term climate data are not available.

The Humboldt Current prevents rainfall but contributes to the formation of fog and clouds. During winter, the coastal plains are often covered in dense fog for months, resulting in precipitation at altitudes between 250 and 1000 m. This supports seasonal vegetation known as fog deserts, fog oases, or lomas.

Above 800-1000 m, the influence of fog decreases, and the high mountains are too far away to induce orographic rain. The 1000–2000 m zone is extremely dry and almost lifeless, called the abiotic zone.

Above 2500 m, the landscape is dominated by xeric scrub with many cacti. At 3000 m, the scrub becomes denser and the plant diversity increases. On the southern slopes, these scrublands give way to Polylepis forests, ecosystems of high biodiversity that extend up to 4500 m. On the northern slopes and in sunny plains, tolares (heath-like vegetation) dominates, while in drier areas Festuca chrysophylla grasslands outcompete shrubs.

Above 4500 m, conditions become harsher. Night frosts are common, even in summer, while daytime temperatures can exceed 40°C due to intense solar radiation. Plants at this altitude become smaller and form cushions for insulation. High evaporation, icy winds, and grazing pressure favor cushion plants and species with thorns, spines, or coarse leaves. Above 5000 m, the landscape becomes rocky, with only a few highly specialized plants. Above 5500 m there is permanent snow and ice.

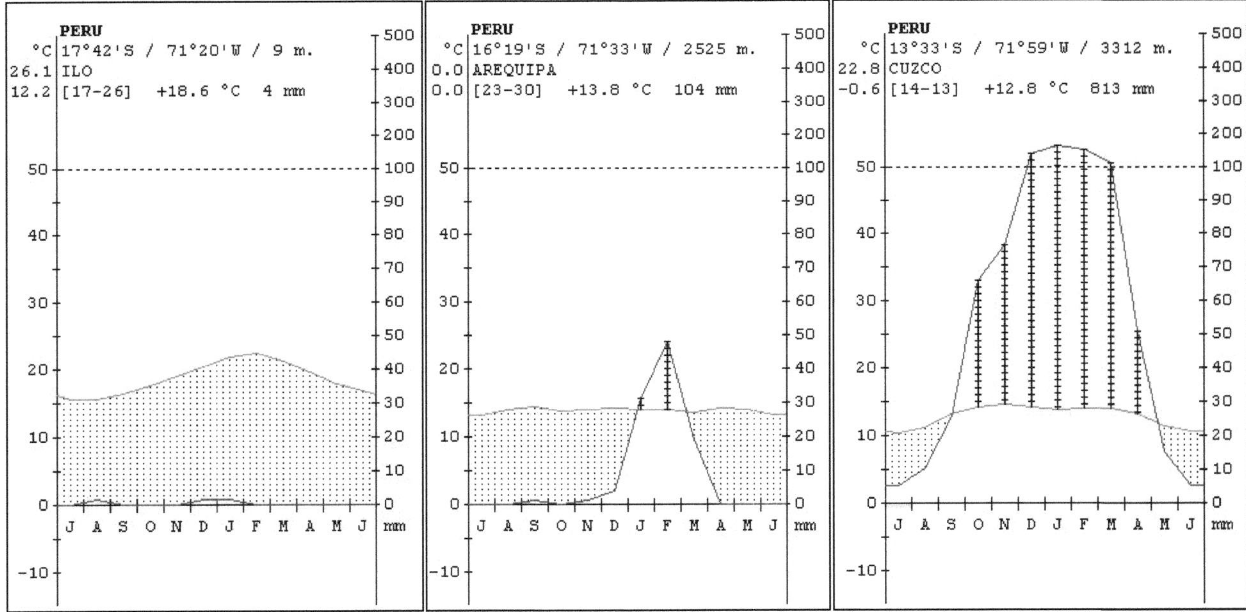

Figure 3: Climate diagrams of Ilo (Pacific Coast), Arequipa (2500 m), and Cuzco (3300 m)
Source: Rivas-Martínez & Rivas-Sáenz (2018)

In more humid regions of Arequipa, various azonal vegetation types thrive. Bofedales, or upland wetlands, form in humid areas at high altitudes. Depressions without drainage form seasonally dry salt lakes with halophilic vegetation dominated by algae. The original riverine vegetation of shrubs and small trees at mid elevations has been largely replaced by agriculture. Along the coast, the rivers create large wetlands, such as those in Camaná and Mejía, which stand out as green belts against the sandy desert.

The bioclimatic zones of southern Peru, as described by Galán de Mera & Linares (2012), are summarized in Table 1. These zones illustrate the transition from the Pacific coast to the highlands of Arequipa, along with the characteristic vegetation and vernacular names associated with each altitude range.

Bioclimate	Altitude (m)	Indicative plants	Vernacular name
Thermotropic hyperarid	0 - 1000	Tillandsia purpurea	Chala
Thermotropic ultrahyperarid	1000 - 2000	Abiotic desert without any plant growth	Yunga
Thermotropic arid	2000 - 2100	Neoraimondia arequipensis	Yunga
Mesotropic arid-semiarid	2100 - 3100	Armatocereus matucanensis Euphorbia apurimacensis	Yunga
Supratropic semiarid-dry	3100 - 3800	Cantua buxifolia Diplostephium meyenii Puya densiflora	Quechua
Orotropic dry-subhumid	3800 - 4800	Festuca chrysophylla Parastrephia quadrangularis	Puna
Cryotropic dry-subhumid	4800 - 6380	Nototriche obcuneata Werneria poposa	Janca

Table 1: Bioclimatic transect from the Pacific Ocean to the Arequipa highlands
Source: Galán de Mera et al. (2009)

Results and Discussion

The Flora of Arequipa in Figures

This study identifies 1,350 species of vascular plants growing wild in Arequipa. 1,187 species (86%) are native, while 163 species (14%) were introduced, mainly from Europe and North America, and are now widely distributed in South America. The figures include 99 species that have been reported for Arequipa, but whose presence is uncertain due to insufficient evidence (status: doubtful), and 32 species that have not yet been officially recorded, but are likely to be present due to their geographic range (status: expected).

| | native | endemic | | introduced | total | % introd./ | % endemic/native | |
	# species	# Peru	# Arequipa	# species	# species	total	Peru	Arequipa
Coastal	468	135	33	109	577	23%	28.8%	7.1%
Andean I	418	83	7	100	518	24%	19.9%	1.7%
Andean II	758	155	21	124	882	16%	20.4%	2.8%
Andean III	601	79	5	43	644	7%	13.1%	0.8%
Total	1187	287	60	163	1350	14%	24.2%	5.1%

Table 2: Altitudinal distribution of native, endemic and introduced vascular plants in Arequipa.

As shown in Table 2, the Coastal and Andean I zones contain a high proportion of introduced species, mainly from Mediterranean climates in Europe and North America. In these arid areas, the plants grow mainly in the valleys or lomas. Andean II zone (1500-3500 m) has the highest biodiversity, with introduced species coming mainly from temperate regions. Andean III zone (>3500 m) also has a surprisingly high level of biodiversity and shows the lowest proportion of introduced plants.

A quarter of the native species of Arequipa are Peruvian endemics and 5% of the native species grow only in Arequipa. The highest endemism is found in coastal areas.

Arequipa is home to species from 117 plant families. The most diverse families are Asteraceae (238 species), Poaceae (132 species), and Fabaceae (85 species). Table 3 shows that the 20 most diverse families account for about 75% of all species in the region. The family Asteraceae has the most endemic species (12), followed by Brassicaceae and Fabaceae with 8 endemic species each.

Family	# species	% species	# endemic species
Asteraceae	238	17.6%	12
Poaceae	132	9.8%	0
Fabaceae	85	6.3%	8
Solanaceae	80	5.9%	2
Malvaceae	66	4.9%	3
Brassicaceae	57	4.2%	8
Cactaceae	42	3.1%	2
Amaranthaceae	33	2.4%	1
Boraginaceae	32	2.4%	2
Caryophyllaceae	31	2.3%	0
Cyperaceae	25	1.9%	0
Apiaceae	24	1.8%	3
Convolvulaceae	22	1.6%	3
Lamiaceae	21	1.6%	0
Plantaginaceae	19	1.4%	0
Verbenaceae	18	1.3%	0
Pteridaceae	17	1.3%	0
Calceolariaceae	16	1.2%	0
Bromeliaceae	15	1.1%	3
Onagraceae	14	1.0%	1
all other (97 families)	363	26.9%	12
Total	1350	100%	60

Table 3: Number of species per family (without cultivated species)

Discussion of the Results

To assess whether the number of vascular species recorded in this study for Arequipa is realistic, a comparison with neighboring regions is helpful.

- Tacna: León et al. (2004) recorded 708 species of vascular plants, without distinguishing between native and introduced species. Of these, 92 species were classified as Peruvian endemics and 36 species were endemic to Tacna.
- Ica: Whaley et al. (2010) estimated 530 native plants in coastal areas (480 recorded), including 141 species endemic to Peru and S-Ecuador and 55 species endemic to coastal S-Peru.
- Moquegua: Montesinos (2015) listed 610 species in the Flora Moqueguana. However, this is not a complete inventory, and no additional information is available for this department.
- Chile: Flores Fuentes (2016) recorded 726 species for the Arica y Parinacota region, of these 90.8% were native. Squeo et al. (2008) recorded 1,099 species (native 980) for the Atacama region.

Brako & Zarucchi (1993) listed 788 flowering plants for the department of Arequipa (excluding ferns, subspecies, and varieties, and species later recognized as synonyms). Quipuscoa, Dillon, & Ortíz recorded 1,160 vascular plants for Arequipa at the XI Congreso Nacional de Botánica in Puno in 2006.

Barthlott, Lauer, and Placke (1996) divided the western slopes of the Andes in southern Peru into diversity zones and estimated vascular plant diversity ranging from less than 100 to 1,000-1,500 species per 10,000 km².

A realistic estimate for Arequipa's vascular plant species is 1,300-1,500 (including 1,100-1,300 native species). Higher estimates, e.g., more than 1,800 species as sometimes postulated, probably include cultivated plants. The number of introduced species may increase in the future due to the international exchange of goods.

The definition of a species often varies depending on the botanist describing it. Local botanists sometimes emphasize minor morphological differences, occasionally designating local biotypes or subspecies as separate species. This tendency is evident in Arequipa with genera such as Nolana (Solanaceae), Palaua and Nototriche (both Malvaceae), Cumulopuntia (Cactaceae), and Senecio (Asteraceae), among others. Phylogenetic analyses can help to clarify these classifications.

Arequipa's endemism is particularly remarkable along the coast, where 33 endemic species have been identified. Of these, 21 grow in the lomas, unique fog-fed ecosystems separated by vast deserts. These deserts act as barriers, promoting evolution and the formation of new species. However, some endemic species, such as Bomarea latifolia, Cistanthe weberbaueri, and Cuscuta hitchcockii, are taxonomically unresolved, known only from their type locality.

Rapid population growth, especially in coastal areas and the suburbs of the capital Arequipa, is exerting severe pressure on natural habitats. Key problems include uncontrolled urban expansion, poor land-use planning and inadequate waste management. The ecosystem of the lomas, in particular, require sustainable protection to ensure their long-term survival.

Protecting biodiversity is primarily the responsibility of national and regional governments. However, individuals can also play an important role by respecting nature and acting in an environmentally responsible manner in their daily life.

Species Catalog

Ferns and Allies (Pteridophyta)

Aspleniaceae

Asplenium gilliesii Hook.

Description: Epipetric, rarely terrestrial. Leaves densely cespitose, 1-pinnate, 8-20 x 1-1.5 cm. Petiole 0.2-0.4 mm thick, this and rachis flexuous and greenish or stramineous. Pinnae 7-20 pairs, margins broadly and deeply lobed and cuspidate.

Ecology: Andean II-III: 2700-4000 m. Among rocks, in dry meadows.

Distribution: Peru, Bolivia, NW-Argentina, N-Chile.

Arequipa: native. CAY.

Note: Tryon & Stolze (1989) note that A. gilliesii and A. peruvianum may simply be variants with intermediates. Both species may be conspecific.

Asplenium lorentzii Hieron.

Description: Terrestrial. Fronds erect or decumbent. Blade 1-pinnate, 16-30 x 1.7-2.7 cm. Rachis 4-10 cm long, reddish or greenish-brown, glabrous. Pinnae numerous, 10-14 mm long, subopposite or alternate, most of them rhombic, deeply dentate, lacking a distinct midrib. Veins distinct, slightly darker than the blade. Pinna base angle approx. 45°. Sori 4-6 on larger pinnae.

Ecology: Coastal, Andean I: 500-1200 m. Lomas.

Distribution: S-Peru, Bolivia, NW-Argentina.

Arequipa: native. CAR.

Asplenium monanthes L.

Description: Epipetric or terrestrial. Stalk short, erect, at the apex with filiform to narrowly deltoid, gray-brown to blackish scales. Blade 1 -pinnate, cespitose, 60 x 3 cm, linear, tapering to a pinnatifid apex and strongly reduced at the base. Rachis, glossy, deep brown, purple or blackish, essentially glabrous. Pinnae up to 1.5 cm long, sessile, quadrangular to subacute and falcate, inequilateral at base. Sori solitary or in pairs near the margins toward base.

Ecology: Andean I-III: 1000-4500 m. In forests and thickets, on and among rocks, in crevices of rock cliffs.

Distribution: Tropical and subtropical America and Africa.

Arequipa: native. CAR.

Asplenium peruvianum Desv.

Description: Epipetric, rarely terrestrial. Blade densely cespitose, 1-pinnate, 6-30 x 0.6-1.5 cm. Rachis 0.5-1 mm thick, stiff and usually reddish or grayish-brown.

Pinnae 12-30 pairs, rhomboid, ± as long as wide, margins entire to deeply dentate, rarely lobed. Pinna base angle >45°.

Ecology: Andean II-III: 2750-4500 m. In open forests or thickets, in rock crevices, on and among rocks.

Distribution: Andes from Venezuela to Argentina.

Arequipa: native. ARE, CAY.

Note: See note on A. gilliesii.

Asplenium praemorsum Sw.

Description: Plants epipetric, epiphytic, or sometimes terrestrial. Fronds erect or decumbent, stalks with abundant scales. Leaves 1- pinnate-pinnatisect, fasciculate, 12-60 x 3-14 cm. Pinnae 7-18 pairs, deltoid, pointed, short- stalked, divided into 4-6 cuneate pinnules almost to the rachis. Veins indistinct. Sori linear, dense.

Ecology: Andean I-II: 500-3500 m. Forests, rocky cliffs, on trunks, and branches of trees.

Distribution: Mexico to Bolivia and Argentina.

Arequipa: native. CAR.

Asplenium triphyllum C.Presl

Description: Plants epipetric, rarely terrestrial. Fronds lanceolate. Stalk small to stout, with linear scales, blackish. Pinnae numerous, subsessile to short-petiolate, subdistant to dense, most trifoliolate or 2-foliolate, or occasionally with another pair of lateral pinnules, terminal segments obovate or oblanceolate, entire to bifid.

Ecology: Andean II-III: 3000-4800 m. Among rocks and boulders, rarely in wet soil.

Distribution: Colombia to Argentina.

Arequipa: native. ARE, CAY.

Blechnaceae

Blechnum occidentale L.

Description: Stalk small, erect to usually decumbent, and short creeping. Fronds monomorphic. Blades pinnatisect to 1- pinnate at the base, the apex gradually reduced, the pinnae essentially flat, the basal ones truncate, semicordate to cordate. Sori on both sides of the main vein.

Ecology: Coastal, Andean I-II: 400-3500 m. Moist and shady places along streams, in thickets, between rocks.

Distribution: Tropical America.

Arequipa: native. CAR.

Cystopteridaceae

Cystopteris diaphana (Bory) Blasdell

Description: Stalk scales light to dark brown, usually rather broad. Fronds 10-50 cm long, the petiole usually glabrous. Blade 1-3-pinnate-pinnatisect, gradually reduced at the apex, glabrous, pinnae almost sessile to short- stalked. Veins free. Sori roundish, borne on the veins.

Ecology: Coastal, Andean I-III: 300-4500 m. Usually in rocky, locally moist places.

Distribution: Widely distributed all over the world, not in Australia.

Arequipa: native. ARE, CAR, CAY.

Note: Some authors consider this species to be a subspecies of C. fragilis.

Dennstaedtiaceae

Pteridium esculentum (G.Forst.) Cockayne

Description: Terrestrial. Stem slender, long-creeping, often branched, with trichomes. Frond spaced, to 7 m long, sometimes scandent, blade 2-pinnate-pinnatifid to 4-pinnate, pubescent to rarely glabrous, veins free. Sori marginal, covered by a well-modified marginal indusium.

Ecology: Coastal, Andean I-III: 400-3000 m. Pastures, thickets, rocky sites, and clearings.

Distribution: Central and South America, from Mexico to N-Argentina.

Arequipa: native. CAR.

Note: In Arequipa, Pteridium esculentum ssp. arachnoideum.

Dryopteridaceae

Polystichum orbiculatum (Desv.) Gay

Description: Frond 15-100 cm long. Rachis scales persistent, usually brown to dark brown near the base and brown to whitish beyond. Scales <1 cm long. Blade 2-3-pinnate, 2-25 cm wide. Rachis bearing a few to many brownish to whitish scales. Pinnae usually ascending, apex obtuse to subacute, pinnules flat, with revolute margins and apex.

Ecology: Andean II-III: 2700-4900 m. Rocky, shady places.

Distribution: Mexico to Bolivia.

Arequipa: native. ARE, CAY.

Equisetaceae

Equisetum bogotense Kunth

Description: 20-50 cm tall, 1-2 mm wide, usually bearing an apical strobilus. Sheaths with short, brown, papery teeth. Branches irregular, with 4 ridges. Strobilus stalked.

Ecology: Coastal, Andean I-III: 100-4200 m. Moist open places, stream banks, irrigation ditches.

Distribution: Costa Rica to Argentina and Chile.

Arequipa: native. ARE, CAR, CAY, ISL, UNI.

Equisetum xylochaetum Mett.

Description: Giant horsetails, usually 2-4 m tall. Stout stems, up to 4 cm in Ø. Sheaths of the main stem usually light brown, with long deciduous teeth. Branches in regular whorls, with 8-10 ridges.

Ecology: Coastal, Andean I-II: 100-3500 m. Wet open areas, irrigation ditches.

Distribution: S-Peru and N-Chile (Atacama).

Arequipa: native. ARE, CAM, CAR, CAS, ISL.

Note: E. xylochaetum was separated from E. giganteum L. based on the phylogenetic studies of Christenhusz et al. (2019).

Ophioglossaceae

Ophioglossum crotalophoroides Walter

Description: Stem conspicuously globose, fleshy, Ø 3-12 mm. Blades arise from a cavity in the upper part of the stem, sterile lamina ovate to deltoid. Fertile segment solitary, borne at the base of the sterile blade, exceeding the blade.

Ecology: Andean I-III, Amazonian: 600-4300 m. On moist, open grassy slopes.

Distribution: Tropical highlands of the Americas.

Arequipa: native, expected. No voucher, Tryon & Stolze (1989) mention it for Bolivia, Chile and Peru, Montesinos- Tubée (2015) mentions it for Moquegua, his observation in iNaturalist (2022) is very close to the border to Arequipa.

Note: An inconspicuous plant that is easily overlooked and therefore probably under-collected.

Polypodiaceae

Pleopeltis macrocarpa (Willd.) Kaulf.

Description: Stem creeping, epiphytic or lithophytic. Fronds 6-35 x 0.8-3 cm. Stalk 0.5-7 cm long. Blade narrowly elliptic, attenuate at base, coriaceous or subcoriaceous, scales castaneous, with whitish to tawny margin, erose or erose-ciliolate. Sori round to oval, up to 5 mm in Ø, in 2 rows between midvein and margins in the upper half of the lamina, exindusiate.

Ecology: Coastal, Andean I-II: 500-3000 m. Humid places, lomas, occasionally in rocky soil.

Distribution: Mexico to Chile and Argentina, also Africa and India.

Arequipa: native. CAR.

Pleopeltis pycnocarpa (C.Chr.) A.R.Sm.

Description: Stem creeping, fronds distant or approximate, scales adpressed or spreading. Blade pinnatifid to pinnatisect, to 15 cm long, scaly beneath, not reduced at the base. Pinna-segments sometimes pinnatifid or partly lobed, the margins notched, scaly beneath, usually fully adnate at the base. Veins not readily visible, free. Sori in 1 row between midvein and margin.

Ecology: Coastal, Andean I-III: 400-4100 m. Rock crevices, on rocky slopes, partly underneath rocks.

Distribution: Colombia to Argentina and Chile.

Arequipa: native. CAR.

Pteridaceae

Adiantum chilense Kaulf.

Description: Stem slender, rhizome creeping. Fronds 20-50 cm long, 3-pinnate, segments cuneate to flabellate, mostly suborbicular, 1.5 x 2 cm, sparsely hairy, veins ending in the sinuses of the lobes. False-indusial flaps oblong to crescentic, up to 2.5 mm long.

Ecology: Coastal, Andean I-III: 100-4200 m. Hillsides, open woods, thickets, lomas and rocky places.

Distribution: Mexico and the Caribbean to Argentina, tropical Africa.

Arequipa: native. ARE, CAR, CAS, CAY, CON, ISL.

Note: The species is listed under several binomials: A. thalictroides var. hirsutum, A. chilense var. hirsutum, and A. poiretii var. hirsutum. Huiet et al. (2018) propose a broader understanding of A. poiretii and list it as a pantropical species.

Adiantum subvolubile Mett.

Description: Stem slender. Fronds 20-60 cm long, rachis glabrous. Blade ± elongate-ovate, 2-or 3-pinnate in the center. Pinnae subsessile, segments glabrous. The basal pinnule undivided, overlaying the rachis, basal pinnae often reduced, not articulate, the color of the apex of the rachis passing into the base of the segment. False indusia few, roundish or reniform to lunate.

Ecology: Coastal, Andean I-II: 50-3300 m. Lomas, on rocks.

Distribution: Peru, Bolivia, Ecuador.

Arequipa: native. ARE, CAR, CAS, ISL.

Argyrochosma nivea (Poir.) Windham

Description: Fronds 10-30 cm tall, stalk shorter than the blade. Blade deltoid-lanceolate, 2-3-pinnate, pinnules ovate, coriaceous. Underside of pinnules farinose, upper side glabrous. Sporangia along the veins in the outer third of the pinnule.

Ecology: Andean I-III: 500-4000 m. Rocky hillsides and crevices of rocks.

Distribution: The Andes from Colombia south to Argentina.

Arequipa: native. ARE, CAR, CAS, CAY, UNI.

Astrolepis sinuata (Lag. ex Sw.) D.M.Benham & Windham

Description: Frond 45 x 7 cm, evergreen. Blade 1-pinnate, leathery, with star-shaped scales on the upper surface. Pinnae 20-30 pair, 3-6 lobes per pinna, margins slightly rolled down, veins obscure, free, forking. Sporangia scattered on underside of pinnule, no pseudoindusium.

Ecology: Andean II-III: 1000-3000 m. Crevices of rocks, rocky banks, and hillsides, lomas.

Distribution: SW-United States to Argentina.

Arequipa: native. ARE, CAR, CAY, CON, UNI.

Cheilanthes arequipensis (Maxon) R.M.Tryon & A.F.Tryon

Description: Fronds up to 8 cm tall. Blade deltoid-oblong, 2-pinnate, pinnules hardly lobed. Underside of pinnules densely covered with large, pale reddish-brown scales. Margins slightly revolute.

Ecology: Andean II: 2200-3300 m. Rocky places and crevices of rocks.

Distribution: Peru, N-Chile, Argentina.

Arequipa: native. ARE, CAS.

Cheilanthes fractifera R.M.Tryon

Description: Fronds 5-12 cm long, stalk ± grooved on upper side, with large whitish scales at the base. Blade and axis eglandular. Blade deltoid to broadly ovate, 2-pinnate. Pinna moderately whitish pubescent beneath, thinly pubescent above.

Ecology: Andean I: 1000-2000 m. Rocky hillsides.

Distribution: Peru.

Arequipa: native. ARE.

Cheilanthes incarum Maxon

Description: Fronds 12-38 cm long, scales linear to acicular, castaneous. Petiole 4-14 cm long, equaling or shorter than blade, scaly. Blade narrowly lanceolate, 2-pinnate-pinnatifid, basally tripinnate, scaly. Pinnae ovate-triangular. Pseudoindusium distinct, lunate, margins erose. Sori discrete, sporangia with 32 spores.

Ecology: Andean II-III: 2200-4200 m. Rocky slopes, dry ravines.

Distribution: Peru, Bolivia.

Arequipa: native. Not specified by province. Mentioned with specimen references by Rivera et al. (2024).

Cheilanthes mollis (Kunze) C.Presl

Description: Frond up to 30 cm long, stalk shorter than the blade, terete, sparsely to densely pubescent, rachis similar. Blade lanceolate, 2-3-pinnate. Pinnae beneath densely covered with whitish to ferruginous, stellate trichomes, terminal segments small, many suborbicular. Margins strongly revolute.

Ecology: Coastal: 0-1000 m. Lomas.

Distribution: S-Peru, N-Chile.

Arequipa: native. CAR.

Note: Ch. mollis, Ch. peruviana and Ch. pilosa are morphologically very similar.

Cheilanthes peruviana T.Moore

Description: Stems erect, stout, scales linear-ligulate, long-attenuate, margins entire, brown, concolorous. Fronds up to 30 cm long, stalk terete, scaly, stalk similar. Blade narrowly lanceolate, 2-3-pinnate, with some pinnules lobed. Pinna glabrous above, beneath densely covered with deep brown scales.

Ecology: Coastal, Andean I-III: 300-3900 m. In soil and on rocks of lomas, and exposed rocky slopes.

Distribution: Peru.

Arequipa: native. ARE, CAR, CAS, CAY.

Cheilanthes pilosa Goldm.

Description: Stem rather slender, short-creeping, scales lance subulate, dark brown, rather sclerotic. Fronds glutinous without dark yellow secretion. Stalk with rigid, shining scales. Blade herbaceous to chartaceous.

Ecology: Andean II: 1500-2500 m.

Distribution: Peru to Argentina and N-Chile.

Arequipa: native. ARE, CAY.

Cheilanthes pruinata Kaulf.

Description: Stem moderately stout, creeping, scales dark reddish-brown, concolorous, or with very narrow pale borders. Fronds 20-50 cm long, petiole terete, glutinous, glandular hairs with dark yellow secretion. Stalk terete, lamina linear, 2-3-pinnate-pinnatifid. Blade subcoriaceous. Pinnae deltoid, deciduously pubescent on both sides except along the axes and midveins beneath.

Ecology: Andean II-III: 2500-4400 m. Crevices or on ledges of cliffs and in rocky soil.

Distribution: Peru, Bolivia, N-Chile, Argentina.

Arequipa: native. ARE, CAS, CAY.

Myriopteris aurea (Poir.) Grusz & Windham

Description: Stem short creeping, knotted, scales lanceolate, entire, brownish, ± bicolorous. Fronds to 60 cm long, stalk <1/3 as long as blade. Blade linear-elliptic, long-attenuate at the base, 1-pinnate. Pinnae oblong with the lower surface covered with a dense tawny (white when young) tomentum of fine, matted trichomes.

Ecology: Andean I-III: 1200-3800 m. On soil banks, rocky slopes, shrubby hillsides, or cliffs.

Distribution: Southern North America to N-Chile and Argentina.

Arequipa: native. ARE, CAY, UNI.

Myriopteris myriophylla J.Sm.

Description: Short creeping with moderately stout stems. Fronds 15-40 cm long, 4-pinnate.

Ecology: Andean I-II: 1500-3500 m. Rocky soils, shrubland, cliffs.

Distribution: Mexico to Argentina.

Arequipa: native. ARE, CAR, CAS, CAY, UNI.

Notholaena sulphurea (Cav.) J.Sm.

Description: Stem short-creeping to nearly erect, scales lanceolate, attenuate, dark sclerotic with narrow brownish and deciduously glandular-ciliate margins. Fronds 20 cm, stalk much longer than the blade. Blade pentagonal or ± elongate, 2-pinnate the base, 1-pinnate above basal pinnae, terminal segments adnate, underside densely white or yellow-farinose. Sporangia on the vein end.

Ecology: Andean II: 1500-2000 m. Open rocky places.

Distribution: S-Mexico to Chile.

Arequipa: native. ARE, UNI.

Pellaea sagittata (Cav.) Link.

Description: Stem stout, compact, decumbent, scales straight or nearly so, concolorous, tan to rust-colored, lanceolate-triangular, usually cordate. Fronds 15-80 cm long, 1-2-pinnate, pinna ascending at acute angles to the stalk, pinna entire or of 3-numerous sagittate segments, long-petiolate.

Ecology: Andean I-II: 1000-3000 m. On dry banks, among rocks, and on stone walls.

Distribution: Mexico to Bolivia.

Arequipa: native. CAR, UNI.

Pellaea ternifolia (Cav.) Link

Description: Stem flexuous, elongate, decumbent, scales straight or falcate, bicolorous with a slender sclerotic stripe narrower than the borders. Fronds 4-50 cm long, stiff, erect, stalk and rachis plane on the adaxial surface or sulcate, purple to black. Blade linear to narrowly lanceolate, 1 -pinnate, the pinnae ternate or entire, sessile, or subsessile.

Ecology: Andean II-III: 1800-4600 m. Crevices of igneous rock or on Inca walls, in sun and semi-shade.

Distribution: SW-United States to Argentina.

Arequipa: native. ARE, CAR, CAY, UNI.

Pityrogramma trifoliata (L.) R.M.Tryon

Description: Distinguished by the entire to 1-pinnate pinnae, when pinnate usually with only three pinnules (trifoliate). Underside of pinnule with yellow or whitish farina.

Ecology: Coastal, Andean I-II: 0-2300 m. Open rocky ground, riverbanks, irrigation ditches.

Distribution: Tropical America, from S-USA to Argentina.

Arequipa: native. CAR.

Salviniaceae

Azolla filiculoides Lam.

Description: Plants pinnately branched, often 2-6 cm long. Leaves densely packed, imbricate, oblong to ovate, ca. 1 mm long. Emerging from the water surface (not floating planar).

Ecology: Coastal, Andean I-III: 0-4000 m. Calm waters.

Distribution: Native to South America and introduced all over the world.

Arequipa: native. ARE, CAR, CAY, ISL.

Use: It is grown for its nitrogen-fixing ability, which can be used to increase the growth rate of rice, or as green manure. In recent years, Azolla sp. has become important in wastewater treatment to remove metals and nitrogen.

Note: Evrard & Van Hove (2004) suggest that A. caroliniana and A. microphylla are synonyms of A. filiculoides. Metzgar et al. (2007) note a significant genetic difference between the two tribes. Here A. filiculoides is used in a broad sense.

Thelypteridaceae

Amauropelta glandulosolanosa (C.Chr.) Salino & T.E.Almeida

Description: Creeping stem. Fronds 2-pinnate, 20-120(-250) cm long. Pinnae and pinnule falcate, pinnula with 4-12 pairs of often forked, sunken veins.

Ecology: Andean II: 2500-4000 m. Rock crevices, moist and shaded places, along streams and irrigation ditches.

Distribution: S-Ecuador to Bolivia.

Arequipa: native. ARE, CAR, CAY.

Amauropelta rufa (Poir.) Salino & T.E.Almeida

Description: Closely related to A. glandulosolanosa but differs in the simple veins that are often raised beneath, the absence of deep red globose or pyriform glands along the midveins on the underside of pinna and pinnules, and the generally smaller indusia.

Ecology: Andean I-II: 1000-3200 m. Irrigation ditches, small streams, and disturbed sites.

Distribution: Ecuador, Peru and Bolivia.

Arequipa: native, doubtful. ARE. No specimen and no observation. Mentioned by Montesinos & Zegarra (2019).

Note: A. glandulosolanosa and A. rufa grow together, and it would not be surprising if they hybridize (Tryon & Stolze, 1989).

Woodsiaceae

Woodsia montevidensis Hieron.

Description: Stem light brown to dark brown, usually broadly scaled. Fronds 8-40 cm long, the stalk glabrous or often slightly scaly, especially at the base. Blade 1-pinnate-pinnatifid to pinnatisect, gradually reduced at apex, glandular- pubescent, pinnae sessile, the basal ones reduced. Veins free. Sori roundish, borne on a vein.

Ecology: Coastal, Andean I-III: 200-4500 m. Lomas, cliffs, rocky places, on Inca walls, grassy slopes, stream banks, forests.

Distribution: South America, from Colombia and Venezuela to Argentina and N-Chile.

Arequipa: native. ARE, CAY.

Flowering Plants (Gymnospermae and Angiospermae)

Acanthaceae

Dicliptera congesta Kunth

Description: Much-branched herb, up to 40 cm tall. Stems slender, pubescent, but not whitish. Leaves elliptic, 2-3x as long as wide, base and apex acute, margins entire, both sides slightly hairy. Inflorescence axillary and terminal often branched. Flowers pink, some pedicellate, others sessile.

Ecology: Andean II: 2000-3000 m. Shrublands.

Distribution: Peru.

Arequipa: native. CON.

Note: Humboldt's type specimen was first thought to be from Mexico, later revised to S-Peru.

Dicliptera tomentosa (Vahl) Nees

Description: Small annual herb. Stem robust, whitish-tomentose, scarcely branched. Leaves elliptic, almost as wide as long. Flowers sessile, axillary and terminal, corolla pink.

Ecology: Coastal, Andean I-II: 0-3000 m. Lomas, rocky slopes.

Distribution: Peru.

Arequipa: native. ARE, CAR, CAY, CON, ISL.

Dyschoriste repens (Nees) Kuntze

Description: Stem spreading, geniculate from a perennial base, densely pubescent, branched. Branches short, densely foliate. Leaves obovate 3 x 1 cm, apex obtuse to acute, base tapering, ciliate, entire, upper surface hirsute. Flowers axillary subtended by leafy bracts. Calyx 5-lobed, corolla 11-12 mm long, pubescent on the external surface, whitish-blue to violet.

Ecology: Coastal: 0-500 m. Lomas, rocky slopes.

Distribution: Peru, also mentioned for Venezuela.

Arequipa: native. CAR, ISL.

Aizoaceae

Mesembryanthemum cordifolium L.f.

Description: Succulent herb, prostrate. Leaves fleshy, cordate at base. Flowers bright pink to red.

Ecology: Coastal, Andean I-II: 0-2500 m. Cultivated and naturalized in disturbed areas.

Distribution: Native to South Africa.

Arequipa: introduced. ARE, CAM, CAR.

Use: Ornamental plant.

Mesembryanthemum crystallinum L.

Description: Prostrate, succulent herb, covered with large, glistening blister cells. Ground leaves triangular. Flowers white.

Ecology: Coastal, Andean I-II: 0-3000 m. It can tolerate nutritionally poor or saline soils.

Distribution: Native to Africa and S-Europe.

Arequipa: introduced. ARE, CAM, CAS.

Use: Edible leaves and seeds.

Sesuvium portulacastrum (L.) L.

Description: Prostrate, succulent, hairless herb, rooting at the nodes. Leaves opposite, oblong or oblanceolate, fleshy, rounded at apex, decurrent into the petiole below, the petiole base connate with that of the opposite leaf. Flowers solitary, green outside, pink, red, or purplish inside.

Ecology: Coastal: 0-1000 m. Beaches, lomas, tidal flats. Salt tolerant.

Distribution: Native to Africa, Asia, Australia and the Americas.

Arequipa: native. CAM, CAR, ISL.

Tetragonia crystallina L'Hér.

Description: Annual, up to 25 cm tall, glabrous. Leaves elliptic to ovate, sessile, apex acute. Flowers 1 per axil, petals yellow. Fruit obovoid, smooth sides and sharp angles, broadest above middle, 6-9 mm long, peduncles 3-8 mm long. The fruits may be somewhat flattened, appearing 3-angled.

Ecology: Coastal, Andean I-II: 0-2000 m. Lomas, rocky slopes.

Distribution: S-Peru.

Arequipa: native. CAR, ISL.

Tetragonia macrocarpa Phil.

Description: Annual, 5-10 cm tall, prostrate, glabrous. Leaves elliptic to oblong, sessile, apex obtuse, 6-25 x 2-8 mm, oldest 1 to >3 x as long as youngest. Flowers axillary, petals yellow. Fruit 3-7 mm long, glabrous to pilose, moderately tuberculate to usually longitudinally ridged.

Ecology: Coastal, Andean I: 0-500 m. On gravel and sand near coasts and inland.

Distribution: S-Peru, N-Chile.

Arequipa: native, doubtful. Not specified by province. Taylor (1994) restricts T. macrocarpa to N-Chile. Dillon et al. (2011) mention the species for the lomas of Peru.

Tetragonia microcarpa Phil.

Description: Annual, up to 30 cm tall, glabrous. Leaves narrowly elliptic to oblanceolate, sessile, apex obtuse, 3-25 x 0.5-9 mm, oldest 1 to 2 x as long as youngest. Flowers 1-3 per axil, petals yellow-green. Fruit 3-8 mm long, glabrous rhomboidal to ellipsoid, widest at middle, 3-4-angled.

Ecology: Coastal, Andean I-II: 50-3100 m. On sand, growing in dense clumps in runoff channels.

Distribution: S-Peru and N-Chile.

Arequipa: native. CAR, ISL.

Tetragonia ovata Phil.

Description: Annual, up to 20 cm tall, glabrous. Leaves elliptic to oblong, sessile, apex obtuse, 15-65 x 3-40 mm, oldest 1 to >3 x as long as youngest. Flowers axillary, petals yellow-green. Fruit ovoid, widest below middle, 4-5 mm long, peduncles 5-20 mm long.

Ecology: Coastal, Andean I-II: 500-2000 m. Grasslands, lomas.

Distribution: N-Chile.

Arequipa: native, doubtful. Not specified by province. The specimens from Arequipa were re-identified as T. crystallina by Taylor (1994). The same author restricts T. ovata to N-Chile. Dillon et al. (2011) mention it for the lomas of Peru.

Tetragonia pedunculata Phil.

Description: Annual, up to 12 cm tall, prostrate, densely villose. Leaves elliptic to oblong, sessile, apex obtuse, 6-30 x 3-8 mm. Flowers axillary, petals yellow. Fruit 9-10 mm long, tuberculate.

Ecology: Coastal: 0-500 m. Lomas, rocky slopes.

Distribution: S-Peru, N-Chile.

Arequipa: native, doubtful. Not specified by province. Several specimens. Taylor (1994) restricts T. ovata to N-Chile. Dillon et al. (2011) mention it for the lomas of Peru.

Tetragonia tetragonoides (Pall.) Kuntze

Description: Prostrate, sprawling annual, up to 2 m long. Leaves lanceolate to ovate or triangular-hastate, 2-8 cm wide. Flowers greenish-yellow. Fruit with distal horns.

Ecology: Coastal: 0-100 m. Cultivated, occasionally naturalized on coastal dunes and stony beaches.

Distribution: Native to New Zealand, introduced all over the world.

Arequipa: introduced, adventive. ARE. Observation in iNaturalist (2024) and mentioned by Brako & Zarucchi (1993).

Use: Grown for its edible leaves and as an ornamental plant for ground cover.

Tetragonia vestita I.M.Johnst.

Description: Annual, up to 20 cm tall, with straight trichomes, at least on young stems and fruit. Leaves elliptic or rhombic, 10-60 x 5-30 mm, 3-5 palmate veins, apex obtuse to acute. crystalline-papillose and ± villous. Flowers 1 per axil, yellowish. Fruit obovoid, 6-8 mm long.

Ecology: Coastal: 0-1000 m. Lomas, rocky slopes.

Distribution: S-Peru.

Arequipa: native. ARE, CAM, CAR, CAS, ISL.

Alstroemeriaceae

Alstroemeria chorillensis Herb.

Description: Tuberous perennial herb, erect, 15-60 cm tall. Main stem leafy. Leaves twisted, green, oblanceolate, 5- 7 x 0.6-1.2 cm, with wavy margins. Flowers broadly funnel-shaped, tepals 5 cm long, strongly zygomorphic, lilac or pinkish, purple-streaked. The pattern on the upper pair of tepals varies considerably.

Ecology: Coastal: 0-500 m. Lomas, rocky slopes.

Distribution: Peru, N-Chile.

Arequipa: native. CAM, CAR, ISL.

Note: Hofreiter & Rodriguez (2006) propose to synonymize A. lineatiflora Ruiz & Pav. and the Peruvian population of A. violaceae Phil. with A. chorillensis.

Bomarea dulcis (Hook.) Beauverd

Description: Perennial vine, 0.1-2 m high. The stems straight, often spiral. Leaves alternate, narrowly lanceolate, 3.5-7 x 0.2-1.5 cm, acute, green to bluish-green. Inflorescence pendulous, outer tepals pinkish-red, rarely yellow, with green tip, inner tepals pink, with green tip. Fruit an ovoid capsule.

Ecology: Andean II-III: 1500->4500 m. Rocky slopes, elfin forests, shrublands, disturbed areas, grasslands.

Distribution: Bolivia, N-Chile and Peru.

Arequipa: native. ARE, CAY.

Bomarea involucrosa (Herb.) Baker

Description: Erect herb, stem erect, 1-3 m tall, densely leafy throughout, recurved at top. Leaves linear, 5-20 x 0.5-3 cm (usually much narrower), slightly to strongly revolute, sharply pointed, coriaceous. Flowers compact, yellow- green, not spotted, 6-8 cm long.

Ecology: Andean II-III: 2500-4500 m. Rocky slopes.

Distribution: Peru and Bolivia.

Arequipa: native. ARE, CAY.

Bomarea latifolia (Ruiz & Pav.) Herb.

Description: Stem stout, glabrous. Leaves broadly oblong-ovate, 10-16 x 4-8 cm, strongly veined, glabrous on upper surface, pubescent on underside. Inflorescence a thyrse. Flowers 4 cm long, outer tepals pink with green tip, inner tepals slightly longer than outer, greenish-yellow with a green tip and a pink stripe.

Ecology: Coastal: 500-1000 m. Lomas.

Distribution: Peru.

Arequipa: native, endemic?. CAM, CAR, ISL.

Bomarea ovata (Cav.) Mirb.

Description: Suberect, twining, tuberous herb. Stem slender, glabrous. Leaves ovate to lanceolate-ovate, 3-8 x 0.5-4 cm, acuminate, membranous, upper surface glabrous, underside pubescent. Bracts like the leaves, usually much reduced, deciduous. Inflorescence an umbel in weak plant, a thyrse in strong plants. Outer tepals pink, inner tepals subequal to outer, usually narrower, yellow, green at the tip, with round dark spots.

Ecology: Coastal, Andean I-II: 100-3700 m. Disturbed areas, lomas, shrublands.

Distribution: N-Peru to W-Argentina, Bolivia.

Arequipa: native. ARE, CAR, CAY, UNI.

Amaranthaceae

Alternanthera albosquarrosa Suess.

Description: Like A. pubiflora but with larger flowers.

Ecology: Andean II: 1500-2500 m. Rocky slopes.

Distribution: Ecuador, Peru.

Arequipa: native. ARE, CAR.

Note: Probably a synonym of A. pubiflora. Flora of Peru (1936ff.) notes that A. albosquarrosa may differ from A. pubiflora in having larger flowers, but there appear to be intergrading forms.

Alternanthera albotomentosa Suess.

Description: Prostrate herb, perennial. Leaves opposite, shortly petiolate, furry tomentose. Flowers white, heads sessile.

Ecology: Andean II: 1500-2500 m. Rocky slopes.

Distribution: Peru.

Arequipa: native. CAR.

Note: Svenson (1946) synonymizes this species with A. halimifolia.

Alternanthera arequipensis Suess.

Description: Perennial herb, base woody. Leaves and stems adpressed pubescent. Leaves opposite. Heads at the apices of the lateral branches, partly solitary on peduncles of <12 mm, partly subsessile, flowers white.

Ecology: Andean II: 2000-2500 m. Open rocky slopes.

Distribution: S-Peru.

Arequipa: native. ARE, UNI.

Note: May be a synonym of A. pubiflora.

Alternanthera brasiliana (L.) Kuntze

Description: Spreading perennial herb, much-branched, geniculate, up to 4 m tall. Stem hairy. Leaves opposite, ovate, purple, at least the veins. Inflorescence pedunculate, flowers white.

Ecology: Coastal, Amazonian: 0-500 m. Riversides, disturbed areas. Cultivated and naturalized near villages.

Distribution: Native to Tropical America, introduced worldwide.

Arequipa: introduced, adventive. ISL.

Use: Ornamental plant.

Alternanthera caracasana Kunth

Description: Perennial herb with long, prostrate stems, rooting from the lower nodes. Stems densely villous.

Leaves opposite, longer than broad. Heads sessile, with tiny stiff white flowers.

Ecology: Coastal, Andean I-II: 500-3000 m. Lomas, grasslands, disturbed areas.

Distribution: Central and South America, naturalized in North America, Europe, Africa and Australia.

Arequipa: native. ARE, CAY, ISL, UNI.

Alternanthera halimifolia (Lam.) Standl. ex Pittier

Description: Prostrate herb, perennial, stem hairy, sometimes reddish. Leaves opposite, elliptic, shortly petiolate, furry gray-green. Heads globose, sessile, flowers brownish-white.

Ecology: Coastal, Andean I-II, Amazonian: 0-2500 m. Disturbed areas, lomas, riversides, rocky slopes.

Distribution: The Andes, from Panama to N-Chile.

Arequipa: native. ARE, CAR, CAY, ISL.

Alternanthera porrigens (Jacq.) Kuntze

Description: Suffrutescent, erect, much-branched. Stems densely pilose with adpressed with white hairs. Leaves opposite, shortly petiolate, ovate to elliptic, acute or acuminate, densely adpressed-pilose on both surfaces. Inflorescence a large panicle, heads purple or pink, rarely almost white.

Ecology: Coastal, Andean I-II, Amazonian: 0-3500 m. Disturbed areas, grasslands, lomas, rocky slopes, shrublands.

Distribution: Ecuador, Peru, Bolivia, introduced in Java.

Arequipa: native. ARE, CAR, ISL, UNI.

Use: Used medicinally as a cardiovascular agent.

Alternanthera pubiflora (Benth.) Kuntze

Description: Herb or subshrub. Leaves opposite, ovate and entire, short petiolate, sparsely to densely pilose. Heads in clusters of 3, lateral ones pedunculate, terminal ones sessile, flowers white.

Ecology: Coastal, Andean I-II, Amazonian: 0-3500 m. Dry disturbed areas, lomas, riversides, shrublands.

Distribution: Central America and W-South America.

Arequipa: native. ARE, CAM, CAR, ISL.

Note: Alternanthera mollendoana Suess. is synonymized here with A. pubiflora, as some authors do, but not POWO (2024). A. arequipensis could also be a synonym of A. pubiflora.

Alternanthera pungens Kunth

Description: Prostrate perennial herb, branched and rooted at the nodes, branches densely pilose, forming dense mats. Leaves opposite, orbicular or rhombic, 1.5-5 cm long, usually as broad as long (difference to A. caracasana), rounded and apiculate at the apex, adpressed-pilose when young, soon glabrate. Solitary flower heads, sepals 5-7 mm, sparsely villous, prickly.

Ecology: Andean I-II, Amazonian: 0-2500 m. Disturbed areas, roadsides, in settlements on trampled areas.

Distribution: Native of tropical America, now widespread in the tropics and subtropics worldwide.

Arequipa: native, doubtful. ARE, CAY. No specimen, 2 doubtful observations from Caylloma and prov. Arequipa (iNaturalist, 2024). It is likely to occur.

Use: The plant is a diuretic. A decoction is used internally to treat gonorrhea.

Amaranthus cruentus L.

Description: Annual herb, erect, up to 2 m tall, often reddish tinted throughout, stems stout, branched, angular, ± glabrous. Leaves spirally, simple, without stipules, long-petiolate, lamina broadly lanceolate to rhombic-ovate, 2-18 cm × 2-15 cm, attenuate or shortly cuneate at base, obtuse to subacute at apex, mucronate, entire, glabrous to sparsely pilose. Inflorescence consists of numerous cymes in axillary and terminal racemes, up to 45 cm long.

Ecology: Coastal, Andean I-II, Amazonian: 0-2500 m. Cultivated herb, disturbed areas.

Distribution: Native to South America, cultivated worldwide.

Arequipa: introduced. ARE, CAY, ISL. Several observations in iNaturalist (2024).

Use: Valued as a leafy vegetable.

Amaranthus dubius Mart. ex Thell.

Description: Slender, erect herb, up to 1 m tall, much-branched, glabrous. Inflorescence not showy, dull green. Bracteoles shorter than 2 mm, not exceeding the stigma branches in intact infructescence. Carpellate flowers with 5 sepals. Seeds dark brown to black.

Ecology: Coastal, Andean I, Amazonian: 0-1000 m. Riversides, grasslands, disturbed areas.

Distribution: Native to South America.

Arequipa: native. ISL.

Use: Valued as a leafy vegetable.

Amaranthus hybridus L.

Description: Erect or slightly spreading annual, up to 1.5 m tall, usually glabrous, stem often reddish. Inflorescence terminal, ± drooping. Bracts 3-4 mm long, spinescent, at least one of them clearly elongated. Perianth segments 5. Stamens 5. Ovary as long as perianth.

Ecology: Coastal, Andean I-II, Amazonian: 0-3200 m. Cultivated, disturbed areas.

Distribution: Native to N- and C- America, introduced worldwide.

Arequipa: introduced. ARE, CAR, ISL, UNI.

Use: Cultivated on a small scale for its edible leaves and seeds.

Amaranthus spinosus L.

Description: Stem erect, green or ± tinged purple, 0.3-1 m tall. Plants bearing 2 rigid, sharp-pointed spines 5-20 mm long at most nodes. Terminal spike with staminate flowers in apical portion and carpellate flowers near the base.

Ecology: Coastal, Andean I-II, Amazonian: 0-2500 m. Disturbed areas, grasslands, riversides.

Distribution: Native to tropical America, introduced worldwide.

Arequipa: native. ARE, CAM, ISL.

Use: Valued as a vegetable.

Amaranthus viridis L.

Description: Stems erect, 0.2-1 m tall. Inflorescence composed of a terminal panicle, 10-20 cm long, often with axillary spikes or glomeruli. Carpellate flowers with 3 sepals. Fruit indehiscent, rugose when dry, 1-1.6 mm long.

Ecology: Coastal, Andean I, Amazonian: 0-1300 m. Disturbed areas, grasslands.

Distribution: Origin possibly South America. Naturalized weed all over the world.

Arequipa: native. ARE, CAR, ISL.

Use: The leaves and stems are eaten like spinach and the seeds are used as a grain.

Atriplex espostoi Speg.

Description: Plants herbaceous or suffrutescent, coarse and stout, 50 cm tall, finely farinose or green and glabrous, sparsely branched. Leaves ± sessile, rounded-deltoid to broadly rhombic, base auriculate, margins ± coarsely dentate. Fruiting bracteoles fused only at base. Flowers yellowish.

Ecology: Coastal: 0-500 m. Lomas, rocky slopes.

Distribution: Peru.

Arequipa: native. ARE, CAM, CAR, ISL.

Atriplex imbricata (Moq.) D.Dietr.

Description: Densely tomentose shrubs, monoecious, 30-50 cm high. Leaves alternate, imbricate at apex, amplexicaul, triangular or ovate, with cordate base and rounded apex, 5-10 x 5-6 mm, covered with vesicular hairs. Male inflorescences in multiflorous glomeruli, 2 x 0.5 cm, terminal. Female inflorescences in few-flowered glomeruli, in the basal part of the branches.

Ecology: Coastal, Andean I-II, Amazonian: 0-3500 m. Lomas, rocky slopes.

Distribution: S-Peru, Bolivia, NW-Argentina and N-Chile.

Arequipa: native. ARE, CAR, CAY, ISL.

Atriplex myriophylla Phil.

Description: Small creeping, monoecious shrub, tufts up to 30 cm in ∅. Leaves alternate, orbicular, oblong or obovate-angular, entire, apex rounded or subacute, base decurrent, 2-15 x 1-5 mm. Male inflorescences in multiflorous glomeruli, apical, grouped in spikes. Female inflorescences in few-flowered glomeruli, in axils of lower leaves. Fruiting bractlets shortly pedicellate, rhombic, fused at the margins to the middle.

Ecology: Andean II-III: 2000-4400 m.

Distribution: S-Peru, Bolivia, NW-Argentina and N-Chile.

Arequipa: native, doubtful. ARE. No specimen, convincing observation from Socabaya in iNaturalist (2024).

Atriplex peruviana Moq.

Description: Small perennial shrub, often prostate and forming mats. Leaves alternate, 5-15 x 2-3 mm, oblong to elliptic, whitish, somehow succulent, sessile or short petiolate. Female inflorescences in few-flowered glomeruli, in axils of lower leaves.

Ecology: Coastal, Andean I-II: 500-3000 m. Lomas, riversides.

Distribution: Ecuador, Peru, Bolivia, Chile.

Arequipa: native. CAR.

Atriplex semibaccata R.Br.

Description: Prostrate perennial shrub. Leaves narrow-elliptic to elliptic, up to 3 cm long, the vesicular hairs usually denser on the lower surface, margins entire or sinuate, petiolate. Flowers axillary, solitary, or in small clusters. Fruiting bracteoles rhombic, 4-6 mm long, apex entire to 3-lobed, dry or succulent and red.

Ecology: Coastal, Andean I-II: 0-3000 m. Lomas, roadsides, disturbed areas.

Distribution: Native to Australia, naturalized in warmer climates.

Arequipa: introduced. ARE, ISL. Brako & Zarucchi (1993) mention it for coastal areas, own observation from Chiguata at 3000 m.

Atriplex suberecta I.Verd.

Description: Sprawling herb to 40 cm high, monoecious. Stem decumbent to ascending, base densely scaly. Leaves petiolate, diamond-shaped to oblanceolate, 1-4 cm long, apex acute to obtuse, margins coarsely sinuate-toothed. Flowers in axillary clusters. Fruiting bracteoles subglobose, acute, 3-4 mm long, with a short stipe-like base, appendages absent.

Ecology: Coastal, Andean I: 0-1500 m. Saline and degraded sites.

Distribution: Native to Australia, introduced to Europe, Africa and America.

Arequipa: introduced. ARE. Montesinos & Zegarra (2019) mention it for the first time for Peru in La Joya.

Chenopodiastrum murale (L.) S.Fuentes, Uotila & Borsch

Description: Annual herb, up to 70 cm tall with an erect, leafy stem, usually red or red-striped. Leaves oval to broadly triangular, margins dentate, smooth on the upper surface and powdery on the underside. Inflorescence a cluster of spherical buds.

Ecology: Coastal, Andean I-II: 0-2500 m. Disturbed areas, rocky slopes.

Distribution: Native to Europe, W-Asia and N-Africa, introduced worldwide.

Arequipa: introduced. ARE, CAR, CAY, ISL.

Use: Edible seeds and leaves.

Note: Renamed from Chenopodium murale due to phylogenetic research in 2012.

Chenopodium album L.

Description: Erect, much-branched annual herb, up to 150 cm tall. Stem often tinged reddish, gray mealy hairy. Leaves variable, even on the same plant, rhombic-ovate to lanceolate, gray-green, paler underneath, with entire to shallowly toothed margins. Inflorescence a large cluster of small, rounded heads. Flowers gray-green.

Ecology: Coastal, Andean I-III: 0-4000 m. Disturbed areas. In soils rich in nitrogen.

Distribution: Native to Europe, widely introduced elsewhere.

Arequipa: introduced. ARE, CAY, ISL, UNI.

Chenopodium petiolare Kunth

Description: Erect or spreading herb, slender, much-branched, pale, branches rather densely mealy. Leaves slender- petiolate, blades thin, densely farinose or sometimes glabrate, very variable in outline, small, usually ± deltoid and distinctly hastate-lobate at the base. Inflorescence large and open, much-branched, the spikes slender and much interrupted, leafless or with few reduced leaves.

Ecology: Coastal, Andean I-III: 0-4000 m. Desert, disturbed areas, lomas, riversides, rocky slopes.

Distribution: The Andes of South America.

Arequipa: native. ARE, CAM, CAR, CAY, ISL, UNI.

Use: Cultivated since pre-Inca times for its nutritious seeds.

Dysphania ambrosioides (L.) Mosyakin & Clemants

Description: Annual or short-lived perennial, growing up to 1.2 m tall, irregularly branched, sticky glandular, odorous (turpentine). Leaves oblong-lanceolate, up to 12 cm long. Flowers female and bisexual, in small clusters along spike- like branches within a leafy or leafless panicle.

Ecology: Coastal, Andean I-III, Amazonian: 0-4000 m. Disturbed areas.

Distribution: Native to Central and South America, introduced all over the world.

Arequipa: native. ARE, CAS, CAY, ISL, UNI.

Use: Used as a leaf vegetable and a tisane for its pungent flavor.

Dysphania incisa (Poir.) ined.

Description: Stems erect, up to 50 cm tall, sparsely hairy. Leaves aromatic, blade 4 cm long, base cuneate, margins pinnatifid or entire in distal leaves, apex acute to acuminate, mucronate, with sessile glands adaxial. Inflorescence an axillary borne compound cyme.

Ecology: Andean II: 2000-3500 m. Rocky slopes.

Distribution: Native to North and South America.

Arequipa: native. ARE.

Froelichia interrupta (L.) Moq.

Description: Perennial from a thick root, stems ascending or decumbent, <1 m high, white-lanate or sericeous. Leaves petiolate or the upper subsessile, the blades ovate-orbicular to oblong, 3-10 cm long, obtuse or acute, sericeous or floccose-tomentose beneath. Inflorescence lax and interrupted. Fruiting calyx deltoid, broadly winged laterally, the thin wings entire or crenulate.

Ecology: Coastal: 0-500 m. Rocky slopes, sandy places. Probably also present at higher elevations.

Distribution: Mexico to Peru and Paraguay.

Arequipa: native, doubtful. ISL. Specimens Mexia 4172 (MO) (not seen) and Stafford 938 (F) from Mollendo in the 1930s, no further vouchers since then.

Gomphrena meyeniana Walp.

Description: Perennial, 2-4 cm tall, with one or several stemless open rosettes. Leaves obovate to elliptic-spatulate, petiole pinkish, twice as long as the blade. Heads 1.5 cm across, flowers white.

Ecology: Andean III: 3500-4500 m.

Distribution: Peru, Chile, Bolivia and Argentina.

Arequipa: native. ARE, CAY.

Gomphrena umbellata J.Rémy

Description: Small annual, glabrous, umbellately branched, up to 7.5 cm tall. Leaves ± fleshy, spatulate, up to 1 cm long, obtuse, clasping at the base, radical leaves linear-spatulate and petiolate. Heads globose, dense, few- flowered, pedunculate, or almost sessile. Bractlets white, oblong-obovate, thin, scarious, glabrous, sepals 1.5 mm long, linear or subspatulate, obtuse or acute.

Ecology: Andean III: 3500-4000 m. Dry valleys, dry Puna.

Distribution: S-Peru, Bolivia, NW-Argentina, N-Chile.

Arequipa: native, doubtful. ARE. No specimen, mentioned for RN Salinas y Aguada Blanca (Quipuscoa & Huamantupa, 2010).

Guilleminea densa (Willd. ex Schult.) Moq.

Description: Stems numerous from a thick, vertical root, 25 cm long or less, densely lanate. Leaves on short, winged petioles, the blades elliptic to broadly oval, 3-15 mm long, obtuse or acute, abruptly contracted at the base, densely villous or lanate beneath. Flowers white, the glomeruli much shorter than the leaves.

Ecology: Andean I-III: 1000-4000 m.

Distribution: North and South America, naturalized in S-Africa and E-Australia.

Arequipa: native. CAR, CAY, UNI.

Salicornia cuscoensis Gutte & G.K.Müll. ex Freitag, M.Á.Alonso & M.B.Crespo

Description: Perennial, suffruticose, prostrate. Branches delicate, rooting at the basal nodes, forming dense, carpet- like, circular clumps. Stems 2-3(-6) cm high, 1-2 mm Ø, appearing like an annual. Leaves opposite, succulent, fused along the internodes. Inflorescences in dense terminal spikes. Flowers in groups of 3, each inserted in an individual cavity.

Ecology: Andean II-III: 2500-4500 m. Halophilic communities of high altitude.

Distribution: Peru.

Arequipa: native, expected. Alonso et al. (2017) separated this species from S. andina. No voucher. Likely to occur due to suitable habitats and distribution in Cuzco and Ayacucho.

Salicornia neei Lag.

Description: Erect to decumbent shrub, often greenish-red. Stem robust, woody, up to 80 (-150) cm tall, neither creeping nor rooting at nodes. Leaves succulent, connate, sheathing completely the internodes, apex rounded. Inflorescence many-flowered, a terminal spike with many short lateral branchlets.

Ecology: Coastal: 0-500 m. Saline and humid soils along the coast. Characteristic of the "Salicornial".

Distribution: Ecuador to Argentina, mainly along the Pacific coast, also on inland lowlands of Argentina.

Arequipa: native. ISL.

Note: Alonso & Crespo (2008) installed the species Sarcocornia neei, based on morphological characteristics.

Salsola kali L.

Description: Low herb, 5-50 cm tall, papillose to hispid or glabrous. Stems erect to ascending, branched from base. Leaves alternate, linear, 1-2 mm wide, fleshy, apex spiny. Inflorescence interrupted at maturity, usually 1-flower per axil of bract, bracts alternate, reflexed, not distinctly swollen at base, apex narrowing into subulate spine. Flowers with bracteoles free or becoming connate and adnate to perianth base.

Ecology: Coastal, Andean I: 0-1000 m. Beaches, salty places, obligatory halophilic.

Distribution: Native to Europe and N-Africa, introduced worldwide.

Arequipa: introduced. ARE. No specimen, one mention and observation from the salt lake of La Joya (Montesinos & Zegarra, 2019).

Suaeda foliosa Moq.

Description: Perennial, glabrous, much-branched, stout branches roughened by the persistent leaf bases of fallen leaves. Leaves mostly 5-8 mm long, glaucous, semi terete and fleshy, obtuse or acute. Flowers minute, green, solitary or in clusters of 3.

Ecology: Coastal. Andean I-III: 0-4000 m. Desert, lomas, saline wetlands.

Distribution: S-Peru, N-Chile, Bolivia.

Arequipa: native. ARE, CAR, ISL.

Amaryllidaceae

Clinanthus humilis (Herb.) Meerow

Description: Crocus-like habit, the very short (<10 cm) peduncle bearing a single erect orange-red flower. Leaves linear, only 2-4 cm long at flowering, developing to 30 cm long. Flowers 5-6 cm long.

Ecology: Andean II-III: 2500-4500 m. Grasslands, rocky slopes.

Distribution: Peru, Bolivia, N-Chile, W-Argentina.

Arequipa: native. ARE.

Clinanthus incarum (Kraenzl.) Meerow

Description: Bulb globose. Leaves before flowering up to 30 cm long, to 12 mm wide. Flowers 4 or 5, nodding, 6 cm long, gradually amplified to nearly 2 cm wide, cup not toothed, filaments very short, style and stamens subequal.

Ecology: Andean I-II: 500-2000 m. Lomas, grasslands.

Distribution: Peru.

Arequipa: native. CAM, CAR, ISL.

Nothoscordum andicola Kunth

Description: Bulbs obconical, 2-3 x 1-1.5 cm, scape 3-15 cm long. Leaves 1-4 mm wide. Umbels 5-10 flowered. Pedicels subequal, up to 1 cm long. Flowers pinkish-white, the obovate-elliptic tepals united below for a third of their length, style equal to the oblong ovary.

Ecology: Andean I-II: 1000-4500 m. Rocky slopes.

Distribution: Peru, Bolivia, N-Chile, W-Argentina.

Arequipa: native. ARE, CAM, ISL.

Nothoscordum bivalve (L.) Britton

Description: Bulb subglobose, often proliferous, scape 10-20 cm tall. Leaves 1-3 mm wide. Umbels of 3-6(-10) flowers. Pedicels unequal, up to 2 cm long. Flowers not fragrant, tepals white with brown or green veins, oblong- lanceolate, 1 cm long. Fruit depressed, globose.

Ecology: Coastal: 0-500 m. Lomas.

Distribution: Southern North America to Argentina and C-Chile.

Arequipa: native. CAR, ISL.

Nothoscordum gracile (Aiton) Stearn

Description: Bulb ovoid, ca. 1.5 cm in Ø, scape 25-70 cm high. Leaves 4-12 mm wide. Inflorescence an umbel of 10-20 flowers. Flowers fragrant, tepals white with pink midvein and greenish base, 9-15 x 4 mm. Capsule obovoid. No onion- or garlic-like smell.

Ecology: Andean II: 2000-3500 m. Riversides, sandy roadsides, woods and cultivated ground.

Distribution: Native to South and Central America, cultivated and naturalized worldwide in warm climates.

Arequipa: native. ARE.

Use: Ornamental plant.

Note: Often confused with Allium neapolitanum, Mediterranean, not occurring in South America.

Pyrolirion albicans Herb.

Description: Scapes and narrow leaves 10-20 cm high. Flowers white, very fragrant, sessile in the tight spathe, the tube 7 cm long, slightly widened only at the throat, the lobes 4 cm long.

Ecology: Coastal: 0-500 m. Lomas.

Distribution: S-Peru.

Arequipa: native. CAM, CAR, ISL.

Pyrolirion arvense (F.Dietr.) Erhardt, Götz & Seybold

Description: Perianth up to 11 cm across in the upper part, stamens long, no scent. Perianth tube 3 cm wide. Flowers yellow.

Ecology: Coastal: 0-500 m. Lomas.

Distribution: S-Peru.

Arequipa: native. CAM.

Note: Herbarium specimen Iltis 1546 (US) from Camaná is Pyrolirion arvense and not Chlidanthus fragrans (N. Cuba, pers. comm).

Anacardiaceae

Schinus molle L.

Description: Tree up to 15 m tall, with fissured bark. Branches pendulous, foliage weeping. Leaves imparipinnate, 25 cm long. Leaflets linear-lanceolate, 3-4 cm long. Inflorescence a lax panicle, pendulous. Flowers dioecious, green- yellow. Fruit a pink-red berry, 1-seeded.

Ecology: Coastal, Andean I-II, Amazonian: 0-3500 m. Rocky slopes.

Distribution: Peru, Bolivia, Chile and Argentina. Widespread ornamental, naturalized in semi-desert and Mediterranean climates.

Arequipa: native. ARE, CAR, CAY, ISL, UNI.

Use: The berries are traded as pink peppercorns. Used as a spice and medicinally to treat wounds, toothache and rheumatism.

Note: The relationship to Schinus areira L. (syn. S. molle var. areira L. (DC.)) is not clear. Hassler (2024) excludes S. areira for Peru, other authors mention both species for Peru. Here we exclude S. areira for Arequipa because we have no evidence of its occurrence, except for the specimen Williams 255 (US) in 1901.

Annonaceae

Annona cherimola Mill.

Description: Tree 3-10 m tall. Leaves ovate-lanceolate, 12-20 x 8 cm, persistent brownish tomentose underneath. Flowers fragrant, opposite a leaf, solitary or 2-3, outer petals 3 cm long, pale yellow with a purple spot at base, inner 3 petals small, reddish to purple. Fruit heart-shaped to globular, surface with U-shaped areoles or smooth, 10-15 cm in Ø, flesh white, seeds black, obovate.

Ecology: Coastal, Andean I-II: 0-3000 m. Cultivated, naturalized (or native?) in disturbed areas.

Distribution: Native to Ecuador, Peru and Bolivia, cultivated in frost-free areas.

Arequipa: introduced, adventive. ARE. Cultivated and introduced here and there.

Use: Edible fruits.

Apiaceae

Apium graveolens L.

Description: Erect biennial to 1 m high, with ribbed stems, aromatic, glabrous. Leaves usually 1-pinnate, upper leaves much smaller, leaflets 3-5, ovate-cuneate, to 5 cm long, each ± deeply 3-lobed, with crenate to toothed margins. Umbels 3-4 cm Ø, sessile or subsessile with 4-8 unequal rays. Petals white. Fruit broadly ovoid, 1.5-2.5 mm long, mericarps with 5 slender whitish ribs.

Ecology: Coastal, Andean I-II: 0-2500 m. Disturbed areas, riversides.

Distribution: Probably native to the Mediterranean, introduced worldwide.

Arequipa: introduced. ISL. Observations from Mejía in iNaturalist (2024).

Use: Cultivated worldwide as a vegetable or for its hypocotyl, which is eaten raw or cooked.

Arracacia peruviana (H.Wolff) Constance

Description: Slender, branching, 0.6-0.9 m tall, squamulose to scaberulous throughout, the stem base clothed with dry sheaths, from a branched taproot. Leaves ovate-lanceolate, 20-30 cm long, 2-pinnate, the leaflets ovate to lanceolate, acute, cuneate at base, the lower distinct and short-petiolate, the terminal sessile and confluent, 2-5 x 1- 4 cm, coarsely sinuate and mucronate-serrate. Inflorescence of alternate axillary peduncles, 7-15 cm long, Rays 4- 8 cm long, bractlets herbaceous, linear, exceeding flowers. Flowers reddish-brown.

Ecology: Andean II: 2500-3500 m. Rocky slopes, shrublands.

Distribution: Peru and Bolivia.

Arequipa: native, doubtful. ARE. No voucher, mentioned for Arequipa (Ruhm et al., 2022), could be a confusion with A. xanthorrhiza.

Arracacia xanthorrhiza Bancr.

Description: Plants stout, caulescent, 0.5-1.2 m tall, from a greatly swollen taproot. Leaves broadly ovate, 10-30 cm long and broad, 2-ternate or 2-pinnate, the leaflets ovate-lanceolate to ovate, 4-12 x 1.5-6.5 cm, acuminate, mucronate-serrate and coarsely incised or lobed. Petioles 8-45 cm long. Cauline leaves with narrow sheaths. Inflorescence a compound umbel. Rays 1-4 cm long, bractlets herbaceous, linear, not exceeding flowers.

Ecology: Andean II-III: 2000-3000 m. Rocky slopes, shrublands, disturbed areas.

Distribution: Native to Ecuador, Peru and Bolivia.

Arequipa: introduced, adventive. CAS, UNI. Cultivated and persistent near fields.

Use: Cultivated for thousands of years for its edible tubers.

Azorella compacta Phil.

Description: Evergreen perennial subshrub forming dense cushions up to 1 m tall and several meters in Ø. Leaves triangular, rosulate, apex obtuse. Umbels sessile, 2-4-flowered. Flowers greenish-white to pink. Fruits often with purple stripes.

Ecology: Andean III: 4000->4500 m. Grasslands, shrublands, indifferent to soil conditions.

Distribution: Native to the Andes of Peru, Bolivia, Argentina and Chile.

Arequipa: native. ARE, CAS, CAY, CON, UNI.

Use: Used as fuel and in traditional medicine.

Note: The growth rate is estimated at 1.5 cm per year and many Yaretas are >1000 years old.

Azorella diapensioides A.Gray

Description: Plants cespitose, forming hard, dense, convex green to blue-green cushions to 10 cm high. Branches clothed with persistent leaf sheaths. Leaves rosulate, the leaf blades linear, 4-10 x 1.5-2 mm, obtuse, apiculate to spinulose-tipped, decurrent, entire or rarely 1-3-lobed. Umbels sessile, several-flowered, the flowers pale yellow to greenish.

Ecology: Andean III: 3500->4500 m. Rocky slopes, grasslands.

Distribution: Peru, Bolivia.

Arequipa: native. ARE, CAY, CON.

Bowlesia lobata Ruiz & Pav.

Description: Perennial, vining herb, up to 60 cm long, branched, glabrous. Leaves opposite, weakly clustered toward the base, petiolate, thinly membranous, the basal orbicular-reniform, strongly cordate, 5-7-palmately lobed, the terminal lobe longer and entire, lobes incised to one-third of blade. Inflorescence of solitary or paired axillary peduncles bearing simple umbels, the umbels 1-3 (-5) -flowered, peduncles short. Petals white or purple-tingled.

Ecology: Andean II-III: 2500->4500 m. Grasslands, forests, livestock concentration areas.

Distribution: Ecuador, Peru, Bolivia W-Argentina, Chile.

Arequipa: native, doubtful. ARE. No specimen, mentioned for RN Salinas y Aguada Blanca (Quipuscoa & Huamantupa, 2010), observation form Chaclaya. An occurrence in Arequipa is likely, due to the distribution range of the species in Peru (Cuzco, Puno, Moquegua, Tacna).

Bowlesia palmata Ruiz & Pav.

Description: Annual, erect, prostrate or climbing, 0.3-1.2 m long, densely scabrous, the young leaves and inflorescence densely hirsutulous with stellate and simple hairs. Lower leaves alternate, the others opposite, petiolate, thinly membranous, the basal and cauline ovate to orbicular-reniform, 1.5-5 x 2-7 cm, cordate to truncate, palmately 3- or 5- parted nearly to base, leaflets pinnately parted. Umbel 1-4-flowered, sessile, axillary. Flowers greenish-yellow to white, involucre absent.

Ecology: Coastal, Andean I-II: 0-3000 m. Disturbed areas, lomas, rocky slopes.

Distribution: Ecuador, Peru, Bolivia.

Arequipa: native. CAR, ISL.

Bowlesia sodiroana H.Wolff

Description: Annual, prostrate or ascending, 5-60 cm long, branched. Stems rather stout, armed with usually 4-rayed processes. Lower leaves alternate, upper opposite, petiolate, membranous, the basal and lower cauline ovate to orbicular-reniform, 1-4 x 1.5-5.5 cm, cordate, palmately 3-7-lobed shallowly or well below middle. Umbel solitary, subsessile in leaf-axils, 1-5-flowered. Flowers greenish-white or purplish, involucre absent.

Ecology: Coastal, Andean I-III: 0-4000 m. Grasslands, lomas.

Distribution: Ecuador, Peru, Chile.

Arequipa: native. ARE, CAY, ISL.

Bowlesia tenella Meyen

Description: Small annual herb, prostrate to ascending, to 35 cm tall, branched, stems filiform, sparsely pubescent. Basal leaves alternate, petiolate, orbicular-reniform, 0.8-2.2 x 1-3 cm, cordate, palmately 5-7-lobed, lobes ovate, acute- obtuse, subequal, mostly entire, upper leaves smaller. Umbels 1-3-flowered, subsessile in leaf axils of the main stem or on filiform axillary shoots. Flowers yellowish-white.

Ecology: Andean III: 3500-4500 m. Rocky slopes.

Distribution: Peru, Bolivia.

Arequipa: native. ARE, CAY.

Bowlesia tropaeolifolia Gillet & Hook.

Description: Perennial, 5-120 cm tall, prostrate, or half-climbing, dichotomously branched. Young foliage and inflorescence white- or grayish-tomentose. Leaves alternate below, opposite above, orbicular-reniform, 0.5-3.5 × 0.7-4.5 cm, cordate, 5-9-palmately lobed, lobes subequal, incised ± to mid-blade. Flowers yellowish- or greenish- white.

Ecology: Andean II-III: 3000->4500 m. Grasslands, rocky slopes, shrublands.

Distribution: Peru, Bolivia, Argentina, Chile.

Arequipa: native. ARE. No voucher. Mentioned by Brako & Zarucchi (1993) for Arequipa. An occurrence in Arequipa is likely, due to the distribution range of the species in Peru (Cuzco, Puno, Moquegua, Tacna). Some specimens may be confused with B. lobata.

Chaerophyllum andicola (Kunth) K.F.Chung

Description: Very variable species, usually acaulescent, sometimes erect up to 50 cm tall. Leaves entire to 1-2-pinnate, white-tomentose, or grayish-hirsute to green and glabrate. Leaflets ovate to oblong, margins entire. Umbels 10-30-flowered, subtending bracts 6-10. Flowers white or pink.

Ecology: Andean II-III: 2500->4500 m. Grasslands, rocky slopes, shrublands.

Distribution: Central America to Argentina.

Arequipa: native. ARE.

Conium maculatum L.

Description: Herbaceous biennial, 1.5-2.5 m tall, glabrous. Stems smooth, green, hollow, usually spotted or streaked with red or purple at the base. Leaves 2-4-pinnate, finely divided and lacy, deltoid, up to 50 cm x 40 cm. Flowers small, white, clustered in umbels of 10-

15 cm in Ø. Crushed leaves and roots emit an unpleasant parsnip-like odor.

Ecology: Andean II-III: 2000-4000 m. Aquatic or terrestrial herb, disturbed areas, emergent, floating.

Distribution: Native to Europe, W-Asia and N-Africa, introduced worldwide.

Arequipa: introduced. CAY, ISL, UNI.

Note: Fresh Conium contains very poisonous alkaloids. The poison is reduced when dried.

Cyclospermum laciniatum (DC.) Constance

Description: Plants slender, annual, branching above, 20 cm high. Leaves oblong-ovate, blades 0.5-3 x 0.5-3 cm, ternate, then 1-2-pinnate, the segments linear, 4 x 1 mm, entire to incised-lobed, petioles 3.5 cm long. Umbel axillary, sessile, 2-3-rayed, rays unequal, 1 cm long. Flowers greenish-pinkish-white, fruit ovoid, 1.5-2 x 1-2 mm, glabrous, to rugulose or hispidulous.

Ecology: Coastal: 0-1000 m. Lomas, rocky slopes.

Distribution: Ecuador, Peru, Chile.

Arequipa: native. CAR, ISL.

Cyclospermum leptophyllum (Pers.) Sprague ex Britton & P.Wilson

Description: Annuals, alternately branched above, up to 60 cm tall. Leaves oblong-ovate, blades 3.5-10 x 3.5-8 cm, 3-4- pinnately decompound, leaflets linear to filiform, 1.5-7 x 0.5-1 mm, petioles 2.5-11 cm, sheath white scarious- margined. Cauline leaves ternate-pinnately decompound, leaflets filiform. Umbel compound, sessile, rays to 2 cm long, fruit ovoid, 1.2-3 mm long, 1.5-2 mm broad.

Ecology: Coastal, Andean I-III: 0-4000 m. Disturbed areas.

Distribution: Native to South America, introduced worldwide.

Arequipa: native. CAR, ISL.

Daucus montanus Humb. & Bonpl. ex Schult.

Description: Annual, 0.1-1 m, stems solitary, erect, simple or few-branched, papillate-hispid with retrorse hairs. Leaf- blades oblong, 3.5-11 x 2.5-5 cm, leaflets linear, 2-5 x 1 mm, acute, ± hispid, petioles 3-12 cm long. Umbel with 4-20 rays, unequal. Flowers whitish to red-purple. Fruit oblong, 3-6 x 2-3 mm, armed with bristles.

Ecology: Coastal, Andean I-III: 0-4500 m.

Distribution: Central America to Chile.

Arequipa: native. CAR, ISL.

Domeykoa amplexicaulis (H.Wolff) Mathias & Constance

Description: Slender, prostrate, or spreading annual herbs, the branches 10-100 cm long, striate. Leaves ± rosulate at the base, orbicular to obovate, 0.9-3.5 x 1.2-5 cm, cuneate, repandly spinulose-dentate. Umbels compact, 3-8 mm in Ø, with usually 15-40 yellowish-white flowers.

Ecology: Coastal: 0-1000 m.

Distribution: Peru.

Arequipa: native. CAR, ISL.

Domeykoa saniculifolia Mathias & Constance

Description: Rather slender, prostrate or ascending annual herb, branches 10-30 cm long, faintly ribbed. Leaves ± rosulate at the base, orbicular to obovate, 0.6-2 x 0.8-2.2 cm, cuneate, deeply 3- or 5-lobed, the lobes oblanceolate to obovate, coarsely spinulose-dentate, petiole 2-5 cm. long. Umbels compact, 5-7 mm in Ø, with 8-15 pink to purplish flowers.

Ecology: Coastal: 0-1000 m.

Distribution: Peru.

Arequipa: native. CAR.

Eremocharis ferreyrae Mathias & Constance

Description: Slender, woody-based perennial herbs, 30-70 cm tall, branches slender, ascending, terete, striate, not grooved or ribbed. Leaves numerous, solitary, 1.5-3 cm long and broad, ovate-deltoid, deeply tri-lobed, lobes ovate, 3-15 mm broad, the central a little larger than the 2 lateral, all spinose-dentate or -lobed, petiole linear, 3-5 cm long. Umbels compact, 4-7 mm in Ø, with 20-30 flowers. Flowers bright yellow to greenish-white. Fruit orbicular, 1.5 mm long and broad, sharply angled.

Ecology: Coastal: 0-500 m. Lomas.

Distribution: Peru.

Arequipa: native. CAR. Type specimen Ferreyra 6364 (MO) from Atico, Camaná, known only from the type locality.

Eremocharis hutchisonii Mathias & Constance

Description: Low shrub, 30-100 cm tall, intricately branched, branches short, stiff, ± spinescent, angled, prominently ribbed, the older ones with exfoliating bark. Leaves sparse, alternate, 2-3 x 1.5-2 cm, ovate, 1-2-ternate, primary lobes oblong-spatulate, 0.5-1.5 x 0.5 cm, divaricate, subequal. Petiole broadly linear, 1-3 cm long, gradually dilated toward base. Cauline leaves upwardly reduced, uppermost trilobed. Umbels lax, 1-1.5 cm in Ø, 8-20 flowers, bracts oblong, 1-2 mm long, flowers purplish-brown. Fruit sharply angled.

Ecology: Andean II: 3000-3500 m.

Distribution: Peru.

Arequipa: native. CAR.

Eremocharis piscoensis Mathias & Constance

Description: Slender shrub 1-2 m tall, branches slender, terete, ribbed. Leaves alternate, the basal up to 7 x 5 cm, broadly ovate, ternate, primary and ultimate divisions linear or filiform, acuminate, divergent, 0.3-3 cm x 1 mm, petiole linear, 1-2 cm long, flattened, somehow dilated. Umbels lax, 7-10 mm in Ø, 5-10 flowered. Flowers purplish- brown, fruit orbicular, 2 x 1.5-2 mm, sharply angled.

Ecology: Coastal, Andean I-II: 500-2000 m. Lomas.

Distribution: Peru.

Arequipa: native. CAM.

Foeniculum vulgare Mill.

Description: Robust, hairless, ± glaucous perennial herb, to 2 m tall. Stem finely ribbed. Leaves with prominent sheathing bases, 3-4-pinnate, finely divided into filiform ultimate segments, these not borne in one plane, aromatic. Umbels 4-8 cm in Ø, terminal and leaf-opposed, rays 10-30, ± glaucous. Petals bright yellow. Fruit 4-8 mm long.

Ecology: Coastal, Andean I-II: 0-2000 m.

Distribution: Mediterranean, introduced all over the world.

Arequipa: introduced. ARE.

Use: Culinary and medicinal uses.

Lilaeopsis occidentalis J.M.Coult. & Rose

Description: Perennial, submersed, stoloniferous aquatic plant. Leaves of septate linear phyllodes, giving a characteristic segmented appearance.

Ecology: Coastal, Andean I-III: 0-4000 m. Seasonally inundated areas, submerged, terra firme forests.

Distribution: Colombia to Chile and Argentina.

Arequipa: native. ARE, CAY.

Note: POWO (2024) synonymized L. macloviana with the North American L. occidentalis.

Spananthe peruviana (H.Wolff) G.K.Müll.

Description: Plants 7-40 cm high, dichotomously branched. Leaf-blades 1.2-3 cm long, 1.5-4 cm broad, acute, subcordate to truncate, dentate-crenate, sparingly setose on veins, petiole to 5 cm long. Peduncles 1-4.5 cm long, 1-3 in the axils, involucre of several scarious linear-lanceolate bracts, shorter than the white flowers. Fruit truncate- ovoid, 2-2.5 mm long, 0.5-2 mm broad.

Ecology: Coastal, Andean I-II: 0-2500 m. Lomas, rocky slopes.

Distribution: Peru.

Arequipa: native. ISL.

Visnaga daucoides Gaertn.

Description: Annual or biennial herbs growing from a taproot erect to a maximum height near 80 cm. Leaves up to 20 cm long and generally oval to triangular but dissected into many small linear to lanceolate segments. Inflorescence a compound umbel of white flowers like those of Daucus. Seeds compressed oval-shaped, less than 3 mm long with prominent ribs.

Ecology: Coastal, Andean I-II: 0-3000 m. Disturbed areas, lomas.

Distribution: Mediterranean, introduced elsewhere.

Arequipa: introduced. ISL.

Apocynaceae

Asclepias curassavica L.

Description: Perennial subshrubs up to 1 m tall, pale gray stems and milky sap. Leaves opposite lanceolate or oblong-lanceolate, acuminate or acute. Inflorescence in cymes with 10-20 flowers, purple to red with yellow to orange. Fusiform-shaped fruits. Flat seeds with silky hairs.

Ecology: Coastal, Andean I-II, Amazonian: 0-2500 m. Disturbed areas, grasslands, riversides.

Distribution: Native to Central and South America, introduced elsewhere.

Arequipa: native. ARE, CAM, CAR, CAY, ISL.

Use: Ornamental plant.

Funastrum clausum (Jacq.) Schltr.

Description: Twining or trailing vine, gray-green, glabrescent to sericeous. Leaves caducous, elliptic to oval, acuminate to mucronate, obtuse to shallowly cordate, 2-5 x 0.5-2.5 cm, membranous, with one or more glands on the midrib at the base, petiole 0.2-0.6 cm long. Inflorescences umbelliform, 7-15-flowered, peduncle stout, equaling the subjacent internode in thickness, 3.5-6 cm long. Corolla rotate-subcampanulate, greenish-yellow to white, tube 2.5 mm long.

Ecology: Andean I-II, Amazonian: 0-2000 m. Disturbed areas, desert, forests, riversides, seasonally inundated areas.

Distribution: Native to Central and South America.

Arequipa: native. CAR.

Gomphocarpus physocarpus E.Mey.

Description: Shrub 1-2 m tall, few stemmed, stem pubescent when young. Leaves narrowly lanceolate, 5-10 × 0.6- 1.5 cm, above sparsely pubescent, underside hairy along midvein, both ends tapering or acute. Peduncle 2-4.5 cm, short hairy. Pedicel 1.5-2.5 cm. Sepals lanceolate. Corolla white, 1.4-2 cm in Ø, lobes reflexed, margins densely bearded, without appendage. Fruit an ovoid to globose balloon, 6-8 × 2.5-5 cm, beakless, pericarp with soft prickles, glabrous when ripe.

Ecology: Coastal, Andean I-II: 0-3000 m. Disturbed areas, gardens.

Distribution: Native to Africa, naturalized in warmer climates.

Arequipa: introduced, adventive. ARE.

Use: Poisonous ornamental plant.

Jobinia formosa (N.E.Br.) Liede & Meve

Description: Twining vine with milky sap, slightly woody below. Leaves cordate to truncate at base. Inflorescence paniculate, calyx green, corolla pale green, reflexed, corona bright white, ± completely fused, plicate and emarginate.

Ecology: Andean II: 1500-3000 m. Rocky slopes, riversides.

Distribution: Ecuador, Peru.

Arequipa: native. ARE. Syntype from Arequipa, no other specimens. Observation from Cayma in iNaturalist (2024).

Note: The differentiation between J. formosa and J. tiarana is problematic. There are populations with lobed petals and cordate leaves.

Jobinia tiarata (Malme) Liede & Meve

Description: Twining vine with milky sap, slightly woody below. Leaves lanceolate, base truncate to acute. Inflorescence paniculate, calyx green, corolla pale green, reflexed, corona not plicate and only fused for about half its length and crowned by 5 long, slenderly lanceolate, free and erect lobules.

Ecology: Andean II: 2500-3000 m.

Distribution: Peru.

Arequipa: native. CAR, CAY, UNI.

Pentacyphus andinus (Ball) Liede

Description: Vine to suberect herb, ± twinning. Leaves lanceolate, base subcordate. Inflorescence few-flowered. Flowers cream-green tinged reddish, corolla incised to one third.

Ecology: Andean II-III: 1500->4500 m. Disturbed areas, forests, rocky slopes, shrublands.

Distribution: Peru.

Arequipa: native. ARE, CAS. No specimen but several mentions and observations.

Philibertia solanoides Kunth

Description: Plants suberect or low scramblers. Leaves lanceolate, cordate at base, veins palmate. Inflorescences strictly umbelliform with a stout, suberect peduncle recurved at the apex. Flowers yellow with brown-purple, corolla rotate-subcampanulate.

Ecology: Coastal, Andean I-III, Amazonian: 500->4500 m. Disturbed areas, grasslands, rocky slopes.

Distribution: Peru, Bolivia, Argentina.

Arequipa: native. ARE, CAR, CAS, CAY, UNI.

Vallesia glabra (Cav.) Link

Description: Shrub up to 3 m tall. Stem green, with blackish nodes. Leaves smooth, lanceolate. Flowers small, tubular, white. Fruit oval, whitish, pearly and translucent. Seeds oval, light brown.

Ecology: Coastal: 0-1000 m. Shrublands.

Distribution: Mexico to N-Argentina.

Arequipa: native. CAR, UNI.

Araceae

Colocasia esculenta (L.) Schott

Description: Evergreen, perennial. Leaves clustered on a tuberous rootstock, 0.4-2 m long, sagittate, not glossy. Petiole long, erect. Spathe white or yellowish-orange, constricted, spadix white.

Ecology: Coastal, Andean I, Amazonian: 0-1500 m. Cultivated and escaped here and there.

Distribution: Probably native to Malaysia. Cultivates for 5000 years and introduced worldwide in the subtropics and tropics.

Arequipa: introduced. CAY.

Use: The corm is a good source of non-allergenic starch for making baby food.

Lemna minuta Kunth

Description: The smallest species of Lemna. Fronds 0.8-3.5 mm long, elliptical, not translucent, with a central vein and only one root. Floating in one layer.

Ecology: Coastal, Andean I-II: 0-3000 m. Ponds, stagnant freshwater. Propagated by waterfowl.

Distribution: Native to North and South America, naturalized in Europe and Japan. Invasive weed in ponds and aquatic cultures.

Arequipa: native. ARE, CAY, ISL. No specimen. Some observations of L. minor s.l. for Arequipa in iNaturalist (2024), which are probably L. minuta.

Use: Due to its high protein content, duckweed is used as feed for cattle and fowl.

Zantedeschia aethiopica (L.) Spreng.

Description: Evergreen perennial to 1.5 m tall. Leaves leathery, glossy, dark green, sagittate, margins wavy, petiole up to 1 m. Spathe greenish-white, conical, spadix yellow. Fruits green or yellow.

Ecology: Coastal, Andean I: 0-1500 m. Forests, riparian habitats, freshwater wetlands, irrigation ditches.

Distribution: Native to southern Africa. Introduced in subtropical regions.

Arequipa: introduced, adventive. ARE.

Use: Ornamental plant.

Araliaceae

Hydrocotyle bonariensis Comm. ex Lam.

Description: Creeping perennial herb. Leaves peltate, orbicular, up to 12 cm in Ø. Inflorescence a basal, profusely branched umbel, flowers white.

Ecology: Coastal, Andean I-II: 0-3500 m. Disturbed, moist areas, in sandy soil near brackish lagoons.

Distribution: Native to America, alien to Australia and South Africa.

Arequipa: native. ARE, CAR, ISL.

Use: The roots are used medicinally to relieve toothache.

Note: The name comes from Bonaria in Buenos Aires. Formerly placed in Apiaceae.

Hydrocotyle ranunculoides L.f.

Description: Perennial herb. Stems horizontal, fleshy, rooting at the nodes. Leaves orbicular to reniform (not peltate), palmately lobed and crenate, petiole up to 35 cm long, 2-3 mm thick and fleshy. Flowers greenish-yellow.

Ecology: Coastal, Andean I-III, Amazonian: 0-4500 m. Shallow freshwater, moody soil, riversides.

Distribution: Native to America, introduced in Europe and Australia.

Arequipa: native. ARE, CAR, ISL.

Note: Herbarium specimen M.Dillon 8841 (F) from Atiquipa is mislabeled as H. alchemilloides (leaves parted to the middle, 5 lobes, long petiole).

Hydrocotyle umbellata L.

Description: Glabrous herb, usually aquatic, creeping, densely rooting at nodes. Leaves orbicular-peltate, fleshy, 2- 7.5 cm Ø, crenate or crenately lobed, petioles 5-20 (40) cm long. Umbels simple, axillary, peduncles long, erect, ± equaling leaves, flowers white.

Ecology: Coastal, Andean I-II: 0-3000 m. Disturbed areas, riversides.

Distribution: Native to America.

Arequipa: native. ARE, CAR, CAY, UNI.

Arecaceae

Washingtonia filifera (T.Moore & Mast.) H.Wendl. ex de Bary

Description: Solitary palm, up to 20 m tall, stem 80 cm in Ø, covered with a skirt of old leaves. Leaves costapalmate, segments rigid, plicate, split half the length of the blade, segments with or without marginal fibers. Petiole armed with prickles.

Ecology: Coastal, Andean I-II: 0-2500 m. Cultivated in parks and naturalized at low altitude.

Distribution: Native to NW-Mexico, cultivated worldwide in warmer climates.

Arequipa: introduced. ISL.

Use: Ornamental plant.

Note: Villanueva-Almanza et al. (2021) show that W. robusta and W. filifera are the same species. W. robusta is renamed to W. filifera var. robusta.

Asparagaceae

Agave americana L.

Description: Rosulate succulent, 2-4 m in Ø. Leaves lanceolate, bluish-gray, 1-2 m long, apex with terminal spine, margins armed with prickles up to 1 cm long. Inflorescence 2-5 m tall, many-flowered, flowers upright. Flowers 7-10 cm long, yellow. Tepals fused basally. Capsule 4-6 cm long.

Ecology: Andean I-II: 1000-3500 m. Disturbed areas near villages.

Distribution: Native to Mexico, naturalized in warmer areas worldwide.

Arequipa: introduced. ARE, CAY.

Use: Ornamental plant.

Note: In Arequipa, A. americana ssp. americana (= Agave cordillerensis Lodé & Pino).

Asparagus officinalis L.

Description: Erect perennial herb, 1-3 m tall. Branchlets <1 mm wide, leaves reduced to tinny scales. Inflorescence axillary racemes. Flowers nodding, bell-shaped, greenish-white. Fruit red, globular.

Ecology: Coastal, Andean I-II: 500-2000. Cultivated and naturalized near villages and on diches.

Distribution: native to Eurasia.

Arequipa: introduced. CAY.

Use: Valued as a vegetable.

Echeandia eccremorrhiza (Ruiz & Pav.) ined.

Description: Bulbous perennial. Leaves linear, to 2.5 cm broad. Inflorescence many-flowered. Flowers white or yellowish-white, borne in 2-3 at each bract. Tepals oblong, not overlapping, filaments connate, anthers orange-red.

Ecology: Coastal, Andean I-III: 0-4000 m. Cloud forests, lomas, rocky slopes.

Distribution: Ecuador, Peru.

Arequipa: native. ARE, CAM, CAR, CAY, ISL.

Note: Very similar to Trihesperus glaucus.

Furcraea andina Trel.

Description: Similar to Agave americana but with green leaves and pendulous flowers.

Ecology: Andean I-II: 1000-3500 m. Rocky slopes, shrubland, near settlements and in lomas.

Distribution: Venezuela, Colombia, Ecuador, N-Peru.

Arequipa: introduced, adventive. CAR.

Use: Ornamental plant.

Oziroe acaulis (Baker) Speta

Description: Perennial herb. Leaves fleshy, decumbent, 15 x 0.5 cm. Inflorescence sessile, scape subterranean, umbel with 2-7 flowers. Flowers white, filaments subulate, broadest at base.

Ecology: Andean II-III: 3000-4000 m. Grasslands, disturbed areas.

Distribution: Peru, Bolivia, NW-Argentina, N-Chile.

Arequipa: native. ARE, CAY.

Oziroe biflora (Ruiz & Pav.) Speta

Description: Plant 20-95 cm high. Leaves 25-60 cm long, ca. 10 mm wide. Inflorescence an elongate raceme, basally red, green above. Flowers 15-30, usually 2 per node, white. Filaments subulate, broadest at base. Tepals 7- 9 mm long, seeds ca. 3,5 mm long.

Ecology: Coastal: 0-1000 m. Sandy areas, hillsides.

Distribution: Peru, Chile.

Arequipa: native. ARE, CAM, CAR, ISL.

Trihesperus glaucus (Ruiz & Pav.) Herb.

Description: Similar to Echeandia eccremorrhiza, distinguished by the broad elliptic tepals, the club-shaped filaments and sagittate yellow-orange anthers.

Ecology: Coastal, Andean I-II: 500-3500 m. Disturbed areas, shrublands, rocky slopes.

Distribution: Peru, Bolivia, NW-Argentina.

Arequipa: native. CAY. No specimen but several observations. Probably often confused with Echeandia eccremorrhiza.

Asphodelaceae

Aloe vera (L.) Burm.f.

Description: Succulent herb. Leaves erect, pale green, sometimes white spotted, linear-lanceolate, 15-35 × 4-5 cm, margins sparsely prickly. Inflorescences erect, dense, 60-90 cm. Flowers pale yellow.

Ecology: Coastal, Andean I-II: 0-3000 m. Cultivated and naturalized near villages.

Distribution: Native to Arabia and NE-Africa. Widely cultivated and naturalized in dry regions.

Arequipa: introduced. ARE, CAY, ISL, UNI.

Use: Ornamental plant and cultivated for its extracts that are used as an essential ingredient in food, cosmetics, and pharmaceuticals.

Pasithea caerulea (Ruiz & Pav.) D.Don

Description: Bulbous plant, up to 1 m tall. Leaves linear, keeled. Flowers blue, stamens yellow.

Ecology: Coastal: 0-1000 m. Lomas.

Distribution: Peru to C-Chile.

Arequipa: native. CAM, CAR, ISL.

Asteraceae

Achyrocline alata (Kunth) DC.

Description: Herb, 50-70 cm tall, whole plant covered with dense white hairs, stems winged. Leaves sessile, blade linear-lanceolate to oblanceolate, decurrent on stems. Flowers pale yellow.

Ecology: Andean II-III: 2000-4500 m. Disturbed areas, forests, rocky slopes.

Distribution: South America.

Arequipa: native. ARE.

Achyrocline ramosissima Britton

Description: Perennial herb to 0.5 m tall, suffruticose. Stem much-branched, ascending to decumbent, densely lanate, not winged. Leaves sessile, blade linear to linear-lanceolate, 20-30 x 2-9 mm wide, apex acute. Inflorescence glomerate, cymose-paniculate, terminal and axillary.

Ecology: Andean II-III: 2500-4500 m.

Distribution: Peru, Bolivia, NW-Argentina, Uruguay and Venezuela(?).

Arequipa: native. ARE, CAY.

Acmella ciliata (Kunth) Cass.

Description: Herbs, perennial, 30-80 cm tall. Stems decumbent to ascending, aboveground stolons rooting at nodes, green to purple. Blade ovate to broadly ovate, 2.3-7.5 × 1-6 cm, glabrous to sparsely pilose on both surfaces, base truncate or cordate, margins serrate, apex acute. Heads radiate, solitary or 2 or 3, terminal or axillary, peduncles 1- 7.4 cm, moderately pilose. Ray florets 5-10, disc florets >100, tubiform, corollas yellowish-orange. Achenes black, with distinct corklike margin.

Ecology: Coastal, Andean I-II, Amazonian: 0-3000 m. Disturbed areas, riparian zones, seasonally inundated areas.

Distribution: Native to South America, widely naturalized worldwide.

Arequipa: native. CAY.

Use: Acmella species are used in traditional medicine to treat toothache.

Ageratina glechonophylla (Less.) R.M.King & H.Rob.

Description: Shrub, 30-50 cm tall, covered with short, stiff hairs. Stem often purplish-green. Leaves deltoid, 1.5 x 3,5 cm, 3-veined from the base, apex acuminate, margins irregularly crenate-serrulate, petiole up to 3 cm long. Heads 3-9 in a loosely cluster, flowers white.

Ecology: Andean II-III: 2000-4500 m. Rocky slopes, shrublands, grasslands.

Distribution: Colombia, Ecuador, Peru, Bolivia, Argentina, Chile.

Arequipa: native, doubtful. Not specified by province. No voucher, mentioned by Brako & Zarucchi (1993) for Arequipa. Several specimens from Moquegua near the border to Arequipa, labeled as A. azangaroensis (Sch. Bip. ex Wedd.) R.M.King & H.Rob.

Note: A. azangaroensis is synonymized with A. glechnophylla by POWO (2024) and with A. scopulorum by Hassler (2024). A. glechnophylla and A. scopulorum may be conspecific. Here we follow POWO.

Ageratina pentlandiana (DC.) R.M.King & H.Rob.

Description: Shrub. Leaves ovate-lanceolate, entirely glabrous, margins sharply serrate, 3-6 x 1-2.7 cm, glabrous, 3-nerved from slightly above the base, the lateral nerves soon branched, subcuneate at base. Heads ca. 15- flowered, corolla purplish.

Ecology: Andean II-III: 2000-4000 m. Rocky slopes.

Distribution: Peru, Bolivia.

Arequipa: native, doubtful. ARE. No voucher, mentioned by Brako & Zarucchi (1993) for Arequipa. The species is usually found in the wetter areas of the Andes.

Ageratina sternbergiana (DC.) R.M.King & H.Rob.

Description: Perennial herb-subshrub, 1 m tall. Leaves deltoid, 3 x 4 cm, petiole ca. 2 cm, apex acute, base truncate-cordate, margins crenate-serrate. Heads terminal and in leaf-axis, strongly clustered, corolla white.

Ecology: Andean I-III: 500-4500 m. Lomas, riversides.

Distribution: Peru, Bolivia.

Arequipa: native. CAR, UNI.

Aldama dilloniorum (A.J.Moore & H.Rob.) E.E.Schill. & Panero

Description: Shrub to 1 m high, moderately and alternately branched at 30-45° angles. Similar to A. helianthoides but occurring at low altitude near the coast, shrubby habit, adpressed minute hairs on the stem, leaves and bracts. Involucral bracts with a blunt tip.

Ecology: Coastal: 0-500 m. Lomas.

Distribution: Peru.

Arequipa: native, endemic. CAR. Known only from the Lomas de Atiquipa.

Aldama helianthoides (Rich.) E.E.Schill. & Panero

Description: Erect to procumbent subshrub, strongly branched. Leaves ovate, pilose, margins undulate, acute, surface rough. Heads 1.5-2 cm in Ø. Disc and ray florets yellow.

Ecology: Andean I-III: 1000-4000 m.

Distribution: Peru, Bolivia, NW-Argentina, N-Chile.

Arequipa: native. ARE, CAY, CON, UNI.

Note: Flora Argentina (2018) synonymizes Vigueria pazensis with A. helianthoides, based on Robinson & Moore (2004) and molecular data of Magenta et al. (2010).

Aldama lanceolata (Britton) E.E.Schill. & Panero

Description: Smaller than A. helianthoides but erect and branched only from the ground. Leaves lanceolate to oblanceolate. Flowers yellow.

Ecology: Andean II: 2000-3500 m.

Distribution: Peru, Bolivia.

Arequipa: native. ARE, CAY, ISL, UNI.

Ambrosia arborescens Mill.

Description: A shrubby, perennial herb, 1-3 m tall, partly lignified. Leaves petiolate, >10 cm long and wide, pinnatisect, grayish above, whitish beneath, lobes long acuminate. Inflorescence a raceme, yellow female flowers basal and male flowers above, involucral spines ± adpressed.

Ecology: Andean II-III: 1500-4000 m. Disturbed areas, rocky slopes, on nutrient-rich and moist soil.

Distribution: The Andes of Peru, Bolivia, Ecuador and Colombia.

Arequipa: native. ARE, CAY, UNI.

Use: Leaves and fruit used as a medicinal plant to treat diarrhea, headaches, liver problems, etc.

Ambrosia artemisiifolia L.

Description: Annual herb, 20-60 cm tall. Stems erect, hispid-pilose in younger parts, usually branched in upper half. Leaves 1-2-pinnately lobed, 2.5-5.5 cm x 2-3 cm, lobes obtuse, glandular-dotted on both sides. Inflorescence a terminal spike, yellowish-white.

Ecology: Coastal, Andean I-II: 0-3500 m. Disturbed areas.

Distribution: Native to N- and C-America, introduced worldwide.

Arequipa: introduced. ARE, CAS.

Ambrosia artemisioides Meyen & Walp.

Description: Shrub with pinnate linear leaves. Leaf-segments <1.5 mm broad.

Ecology: Coastal, Andean I-III: 500-4000 m. Very dry areas with full sun exposition, rocky slopes.

Distribution: S-Peru and N-Chile.

Arequipa: native. ARE, CAR, CAY, CON, ISL, UNI.

Ambrosia cumanensis Kunth

Description: Perennial herb or subshrub, branched, young stems pilose, older stems strigose with adpressed hairs. Leaves triangular, 2-pinnatifid, upper leaves and bracts entire.

Ecology: Coastal, Andean I, Amazonian: 0-1500 m. Disturbed areas, forests, riversides, inundated areas.

Distribution: Mexico to Brazil and Peru.

Arequipa: native. ARE, CAM, CAR, ISL.

Ambrosia dentata (Cabrera) M.O.Dillon

Description: Upright shrub or subshrub. Leaves sessile, lamina to 2 cm long, unlobed or with few lobe-like marginal teeth. Fruiting involucres 2- to several-flowered, spines uncinate, hooked at the tip, usually slender or gradually tapered.

Ecology: Coastal: 0-500 m. Lomas.

Distribution: Peru.

Arequipa: native. CAR.

Ambrosia pannosa W.W.Payne

Description: Shrub, stem woody, slender, the median internodes of recent growth less than 3 mm thick. Leaves pannose-tomentose, segments >1.5 mm broad at widest point, lamina of cauline leaves <10 cm long. Fruiting involucres 1-5 in small clusters, the spines rigid, emergent.

Ecology: Andean II: 2000-3000 m.

Distribution: Peru.

Arequipa: native. ARE, CAR. Known only from Tingo near Arequipa.

Andicolea graveolens (Sch.Bip.) Mayta & Molinari

Description: Shrubs to 30-50 cm tall, much-branched. Leaves arranged in 2 lines, imbricate on stem, broadly ovate, 5-6 x 34 mm, dorsal surface tomentose, glandular, inner surface lanate. Heads solitary, axillary.

Ecology: Andean III: 3500-4500 m. Shrublands.

Distribution: Peru, Bolivia.

Arequipa: native. ARE, UNI.

Aphyllocladus denticulatus (J.Rémy ex Gay) Cabrera

Description: Shrub up to 1 m high, white tomentose. Leaves oblanceolate, 10-35 mm, margins entire or with 1-3 teeth on each side, trinerved. Apical heads solitary. Involucre 10-12 mm, tomentose bracts. Florets purple, ca. 20, deeply divided into linear segments. Achenes silky.

Ecology: Andean III: 3500-4200 m. Rocky slopes, shrublands.

Distribution: S-Peru, N-Chile.

Arequipa: native. ARE, CAR, UNI.

Aristeguietia ballii (Oliv.) R.M.King & H.Rob.

Description: Shrub with strong odor. Leaves linear, with entire margins, rough to the touch. Heads discoid, many- flowered, florets pink.

Ecology: Andean III: 3500-4000 m. Rocky slopes, shrublands.

Distribution: Peru.

Arequipa: native. ARE, CON, UNI.

Baccharis acaulis (Wedd. ex R.E.Fr.) Cabrera

Description: Rhizomatous herb, aboveground part a small rosette surrounding a solitary head. Leaves linear.

Ecology: Andean II-III: 3200-4600 m. Margins of streams and brackish lagoons.

Distribution: S-Peru, Bolivia, N-Chile, NW-Argentina.

Arequipa: native, expected. No specimen but a not curated observation by the author from RN Salinas y Aguada Blanca in iNaturalist (2024). González et al. (2019) expect the species in Arequipa based on ecological analyses. Inconspicuous species, probably often unnoticed.

Note: Heiden, Antonelli & Pirani (2019) renamed the species from Pseudobaccharis acaulis (Wedd. ex R.E.Fr.) Cabrera to Baccharis acaulis.

Baccharis alnifolia Meyen & Walp.

Description: Shrub, up to 2 m tall. Branching loosely, branches ascending obliquely. Leaf elliptic, with entire margins in lower quarter, serrate in upper leaf, teeth small, narrow, pointing forward. Panicle compact, bracts of lowest panicle branch as long or almost as long as the panicle branch, reaching at least to the first branch.

Ecology: Andean II-III: (2500-)3000-4000 m. Riparian, but also on drier sites.

Distribution: Peru, N-Chile.

Arequipa: native. ARE, CAY.

Note: Brako & Zarucchi (1993) erroneously mention B. alnifolia for the Peruvian coast (0-500 m), probably a confusion with the similar B. scandens. It is likely that the Andean plants with broad leaves and three longitudinal veins were subsumed under B. latifolia by Brako & Zarucchi. See also remarks on B. scandens. B. alnifolia is often confused with Baccharis latifolia Pers., a species of the humid Paramo and the Yungas. The observations of B. latifolia for Arequipa in iNaturalist (2024) seem to be confusions with B. alnifolia / B. scandens, the specimens Galiano 8926 (CUZ) and van der Werfft 20664 (MO) were not seen, but van der Werfft collected a specimen of B. alnifolia near the location of his specimen of B. latifolia.

Baccharis boliviensis (Wedd.) Cabrera

Description: Shrub, up to 1.5 m tall. Leaves linear, needle-like, spatulate. Corolla white.

Ecology: Andean II-III: 3000-4000 m.

Distribution: Peru, Bolivia, NW-Argentina, N-Chile.

Arequipa: native. CAY.

Baccharis caespitosa Pers.

Description: Low creeping cushion-shrub, up to 15 cm tall, stem suffused with pink. Leaves spatulate or linear, 10 x 2 mm, margins entire. Heads solitary or in pairs, 1 cm in Ø, terminal. Corolla cream white.

Ecology: Andean II-III: 3000->4500 m. Rocky slopes, grassland.

Distribution: Peru, Bolivia, NW-Argentina, Chile.

Arequipa: native. ARE, CAY, UNI.

Baccharis genistelloides (Lam.) Pers.

Description: Shrub, 1-2 m tall. Stem leafless, winged, green. Corolla yellowish-white.

Ecology: Coastal, Andean I-III: 500-4500 m.

Distribution: Colombia, Ecuador, Peru, Bolivia.

Arequipa: native. ARE, CON, UNI.

Use: Used in folk medicine.

Baccharis gnidiifolia Kunth

Description: Shrub, 0.4-3 m tall. Tufted indument with curved uniseriate hairs on shoots and leaves. Leaves 1.5-5 x 0.1-0.6 cm, sessile, elliptic, linear or oblanceolate, apex apiculate, margins entire or in the distal half dentate, 3- veined from the base or seemingly 1-veined. Inflorescence corymbose, terminal. Heads pedunculate, corolla white, peduncles 3-40 mm long. Achenes dark, large, papillose.

Ecology: Andean II-III: 1500-4000 m. Dry grasslands and scrub, mostly on slopes.

Distribution: Galapagos to N-Argentina.

Arequipa: native. ARE, CON, ISL, UNI.

Baccharis salicifolia (Ruiz & Pav.) Pers.

Description: Spreading to ascending shrub, up to 4 m tall. Leaves narrowly lanceolate, willowlike, finely serrate. Inflorescence a pyramidal or corymbiform panicle, corolla pinkish-white.

Ecology: Coastal, Andean I-II, Amazonian: 0-3500 m. Riversides, humid places.

Distribution: Mexico and the USA to Argentina and Chile.

Arequipa: native. ARE, CAM, CAR, ISL.

Baccharis scandens Pers.

Description: Glutinous, richly branched shrub, up to 2 m tall. Branches often almost horizontal. Leaf broadly ovate, leaf margins with at least one tooth in lower quarter, teeth coarse, broad, outwardly directed. Panicle spreading, lax, bracts of lowest panicle branch reaching at most to first branch branching.

Ecology: Coastal, Andean I-II: 0-3500 m. Riversides, wet areas.

Distribution: Peru, Bolivia, N-Chile and NW-Argentina.

Arequipa: native. ARE, CAM, CAY, ISL, UNI.

Use: Used as fodder, firewood and in folk medicine.

Note: Often confused with B. alnifolia. Intermediate forms exist between these two species.

Baccharis tola Phil.

Description: Small shrub, up to 80 cm tall, richly branched. Leaves spatulate, oblong or obovate, apex mostly obtuse, margins entire or with up to 3 teeth on each side.

Ecology: Andean II-III: 2500-4000 m. Rocky slopes.

Distribution: Peru, Bolivia, NW-Argentina, N-Chile.

Arequipa: native. ARE, CAY, CON, UNI.

Note: B. tola is often confused with B. tricuneata, a species of northern South America that extends south into C-Peru (Müller, 2013).

Bidens andicola Kunth

Description: Annual herb, up to 1 m tall, erect. Stem quadrangular in Ø. Leaves trifoliate to divided, leaflets serrate. Head several per main axis, yellow, 4 cm in Ø, usually some of the flowers with 8 ray florets.

Ecology: Andean II-III: 3000-4500 m. Disturbed areas, grasslands.

Distribution: Colombia, Ecuador, Peru, Bolivia and NW-Argentina.

Arequipa: native. ARE, CAS, CAY, ISL, UNI.

Bidens exigua Sherff

Description: Annual, erect herbs, 10-40 cm tall, branched, stems glabrous. Leaves opposite, petiolate, ovate, 3-5 cm long, deeply 2-pinnatisect, with 2-3 pairs of primary segments, segments oblong to linear-lanceolate, 1-3 mm wide. Heads solitary, terminal, peduncles 2-6 cm long and glabrous. Involucre cylindrical, 3-4 × 4-5 mm, elongating to 7 mm in fruit. Involucral bracts 2-seriate, glabrous, outer narrowly linear, acute, inner lanceolate, membranous and hyaline-edged. Ray florets 0-5, ca. 2 mm long, yellow. Disc florets few, tubular, yellow. Seeds spindle-shaped, 8- 14 mm long, ± pubescent, dark, with 2-3 short awns.

Ecology: Andean I-II: 500-2500 m. Rocky slopes.

Distribution: Peru, Bolivia, Argentina.

Arequipa: native. UNI. No specimen but several observations in iNaturalist (2024). Specimen Weberbauer 7401 (F) from Moquegua.

Note: B. exigua and B. pseudocosmos are morphologically similar and may be confused.

Bidens pilosa L.

Description: Annual herb, up to 1.8 m tall. Leaves opposite, leaflets pinnate, ovate to lanceolate, margins 3-5 serrate. Petioles slightly winged. Flowerheads on long peduncles, ray florets 0-5, white, disc florets yellow. Fruits slightly curved, stiff, black, tetragonal in cross-section, 1 cm long, with typically 2-3 stiff, heavily barbed distal awns.

Ecology: Coastal, Andean I-III: 0-4000 m. Disturbed areas.

Distribution: Native to the Americas, introduced worldwide.

Arequipa: native. ARE, CAR, CAY, ISL, UNI.

Note: In Peru B. pilosa var. alausensis, var. minor and var. pilosa.

Bidens pseudocosmos Sherff

Description: Annual, erect herbs, 20-40 cm tall, branched, stems ± glabrous to hirsute. Leaves petiolate, 2-8 x 2-3.5 cm, deeply 2-pinnatisect, with 4 pairs of primary segments and linear secondary segments, glabrous or ± pubescent. Segments 1 mm wide. Capitula, solitary, terminal, peduncles 4-10 cm long, glabrous. Involucre campanulate, 5-7 × 8 mm. Involucral bracts 2-seriate, glabrous, outer narrowly linear, acute, inner lanceolate, membranous and hyaline-edged. Ray florets 5 mm long, yellow. Disc florets numerous, tubular, yellow. Seeds spindle-shaped, 10 mm long, ± glabrous, black, with 2 short awns.

Ecology: Andean II: 1500-3500 m. Disturbed areas, riversides.

Distribution: Peru, Bolivia, NW-Argentina, N-Chile.

Arequipa: native, doubtful. ARE. No specimen, one observation from Yura in iNaturalist (2024), probably confused with B. exigua.

Bidens triplinervia Kunth

Description: Perennial herb, up to 70 cm tall, procumbent. Stem round, furcate. Leaves trifoliate to divided, leaflets serrate. One or few long-pedunculate heads on each main branch, head yellow, 5 cm in Ø, 5 ray florets.

Ecology: Andean II-III: 2500-4000 m. Disturbed areas.

Distribution: Peru, Bolivia, NW-Argentina, N-Chile.

Arequipa: native, doubtful. Not specified by province. No voucher, mentioned by Brako & Zarucchi (1993). Species of the humid E- slopes of the Andes.

Burnellia fasciculata (Wedd.) Mesfin & D.J.Crawford

Description: Perennial shrub, up to 0.5 m tall. Leaves opposite, 2-pinnate, fleshy, lobes lineal, ± terete and pointed. Heads usually with 7-9 ray florets, yellow and dark-yellow disc florets.

Ecology: Andean II-III: 2000-4500 m. Rocky slopes.

Distribution: Peru, Bolivia, N-Chile.

Arequipa: native. ARE, CAY, UNI.

Note: C. fasciculata and C. ferreyrae have identical ITS sequences (Crawford et al., 2014) and may be the same species.

Centaurea melitensis L.

Description: Erect plants 10-100 cm tall, branched. Leaves gray-green loosely villous, dotted with minute resinous glands. Basal and proximal cauline leaves petiolate or tapering to the base, cauline leaves long, decurrent. Heads terminal, solitary or in clusters of 2-3, all disc florets. Corollas yellow. Involucres 8-15 mm, loosely cobwebby- tomentose or becoming glabrous. Central spine of principal phyllaries 5-12 mm long, slender, often purple to brown tinged, with lateral spines in 3 to 4 pairs.

Ecology: Coastal, Andean I-II: 0-3500 m. Disturbed areas.

Distribution: Native to the W-Mediterranean region, naturalized worldwide.

Arequipa: introduced. CAY.

Chaetanthera peruviana A.Gray

Description: Annual herb, 5-10 cm, erect, slightly pubescent. Leaves 6-14 x 2-4 mm, upper leaves smaller, opposite then alternate to head, sessile, attenuate, ovate-lanceolate to spatulate, 3-5-dentate, margins ± thickened, revolute and pilose, apex mucronate. Heads solitary, terminal, or axillary, shortly pedunculate. Ray florets 6-16, pistillate, creamy to pinkish-white, pistillate.

Ecology: Andean II-III: 3000-4000 m. Grasslands, rocky slopes.

Distribution: Peru.

Arequipa: native. ARE, CAY, UNI.

Chersodoma antennaria (Wedd.) Cabrera

Description: Dioecious, suffrutescent, perennial herbs, stems stoloniferous, 5-25 cm tall, ascending to decumbent, densely sericeous-tomentose. Leaves petiolate, petiole 1-10 mm long, blades oval to elliptic, 5-25 x 3-15 mm, cinereous, densely tomentose, margins entire to 1-2-dentate. Heads solitary, peduncles 1-5 cm long and 1-3- bracteolate. Involucral bracts 18, purple. Heads with ca. 100 florets. Achenes cylindrical, costate, glabrous, 2 mm long, pappus white.

Ecology: Andean III: 3500->4500 m.

Distribution: Peru to W- & C-Bolivia.

Arequipa: native. ARE. No specimen, a convincing observation from the slope of Chachani in iNaturalist (2024).

Chersodoma arequipensis (Cuatrec.) Cuatrec.

Description: Subshrubs 20-40 cm tall, stems intricately pseudo-dichotomous branched, tortuous, procumbent to ascending, white lanate-tomentose to glabrescent. Leaves with 5-6 mm long petiole, the base expanded into amplexicaul auricles, blades concolorous whitish-green, margins with 2-3 pairs of teeth. Heads solitary or weakly cymose with 2-3 heads, white flowers on a long peduncle. Involucral bracts 8-10, yellowish-green, purple-tinged. Heads with 12-17 flores, corolla yellow. Achenes cylindrical, 2.5 mm long, densely villous with inflated, oblong trichomes. Pappus white.

Ecology: Andean II: 2000-2500 m. Lower and dryer areas than Ch. jodopappa.

Distribution: Peru and N-Chile.

Arequipa: native. ARE, CAY.

Chersodoma jodopappa (Sch.Bip.) Cabrera

Description: Dioecious shrub, much-branched from the base, 10-40 cm tall, erect, the branches densely whitish- tomentose, glabrescent. Leaves with petioles 1-3 mm long, blades lanceolate or elliptic, gray, tomentose, margins revolute. Heads solitary on distal branchlets, peduncles 1-1.5 cm long. Involucral bracts 8. Heads with 10-20 florets, corolla yellow. Achenes cylindrical, 2.5 mm long, costate, glabrous. Pappus rose to yellow.

Ecology: Andean II-III: 3000-4500 m. Shrublands, semiarid thorn and succulent scrub, open rocky slopes.

Distribution: The Andes of Peru, Bolivia, N-Chile and N-Argentina.

Arequipa: native. ARE, CAS, CAY, CON.

Chersodoma juanisernii (Cuatrec.) Cuatrec.

Description: Perennial subshrub, stems much-branched, 10-20 cm long, lanate, densely leafy. Leaves alternate, petiole 6 mm long. Blades ovate, 5-12 x 5-9 mm, margins 4-5-lobed, revolute, white-tomentose. Heads, solitary, terminal, peduncle 3-9 mm long. Involucral bracts 9, heads with 12-16 florets, corolla yellow. Achenes cylindrical, costate, pubescent. Pappus white.

Ecology: Andean I-II: 500-3500 m. Lomas, rocky places, in crevices.

Distribution: Peru.

Arequipa: native. ARE, CAR.

Chionopappus benthamii S.F.Blake

Description: Shrub with reddish stems. Leaves ovoid-rhomboidal, trinerved, vaginate at base, underside tomentose white, margins serrate, petiole short and ± alate. Basal leaves 'bat-like'. Flowers solitary, terminal. Ray florets >80, yellow, 25 x 1 mm. Disc florets reddish-brown.

Ecology: Coastal, Andean I-II: 500-3500 m. Lomas, rocky slopes.

Distribution: Peru.

Arequipa: native. ARE, CAY, CON, UNI.

Chuquiraga spinosa D.Don

Description: Intricately branched, 0.4-1.5 m high with twinned yellow stem spines 7-18 mm long. Leaves orbicular, 5- 20 mm long, mucronate and revolute, glabrous above, white or red-brown silky-hairy beneath. Involucre 3-4.5 by 1.5- 2 cm, campanulate, the bracts variably shaped but all with downy backs and margins, colored chestnut-reddish and contrasting with the yellow florets.

Ecology: Andean III: 3500-4500 m.

Distribution: Peru, N-Chile.

Arequipa: native. ARE, CON, UNI.

Note: In Arequipa, C. spinosa ssp. rotundifolia.

Cichorium intybus L.

Description: A stiffly erect ± robust perennial herb up to 120 cm. tall, from a long stout taproot and a rosette. Inflorescence strongly branched, flowers blue.

Ecology: Andean II: 1500-2500 m. Disturbed areas.

Distribution: Native to Europe and W-Asia, introduced worldwide.

Arequipa: introduced, adventive. ARE. No specimen, one observation from Arequipa.

Use: Edible plant. The roots are used as coffee substitute.

Cotula australis Hook.f.

Description: Decumbent annual or perennial herb, usually <10 cm tall, weakly rooting at nodes, ± hairy. Leaves obovate to oblanceolate, 1-2 cm long, pinnatisect, lobes 3-7-parted. Heads 4-5 mm Ø, peduncles slender, exceeding leaves, involucral bracts oblong to lanceolate, subglabrous. Corolla discoid, florets pale yellow or cream.

Ecology: Coastal, Andean I-II: 0-3000 m.

Distribution: Africa, Australia and New Zealand, naturalized elsewhere.

Arequipa: introduced. ARE, CAR, ISL.

Cotula coronopifolia L.

Description: Annual or perennial herb, glabrous. Leaves bright-green, fleshy, irregularly lobed, 4 cm long, base amplexicaule. Heads flat, discoid, florets bright yellow.

Ecology: Coastal, Andean I-III: 0-4000 m. Muddy, anoxic wetlands and brackish water, salt tolerant.

Distribution: S-Africa, naturalized elsewhere.

Arequipa: introduced. ARE, CAR, ISL.

Cuatrecasasiella argentina (Cabrera) H.Rob.

Description: Cushion plant of high-altitude bogs. Leaves 2-5 mm long. White arachnoid filaments spanned between the leaf tips. Male heads 4 mm long, with 5 florets and 8-10 involucral bracts, pappus white.

Ecology: Andean III: 3500-<4500 m. Wet areas, bogs.

Distribution: S-Peru, Bolivia, N-Chile, NW-Argentina.

Arequipa: native. ARE.

Culcitium humile DC.

Description: Perennial, rhizomatous herb, 5-15 cm tall. Leaves rosulate, 3-8 x 0.5-2 cm, margins entire, densely tomentose on both sides, canescent. Heads discoid, hemispherical, 1.5-2 cm in Ø, solitary, nodding, terminal. Involucral bracts 18-24, whitish-lanose. Disc florets numerous, yellow.

Ecology: Andean III: 3500->4500 m. Rocky slopes, dry areas.

Distribution: Peru, Bolivia, N-Chile, NW-Argentina.

Arequipa: native. ARE, CAS, CAY.

Note: Pruski (2021) synonymized Senecio candollei Wedd. with C. humile. POWO (2024) accepts this synonymy, but keeps the species as C. candollei, although this is not the older name.

Culcitium oligocephalum Cabrera

Description: Erect perennial 30-40 cm tall, densely white-tomentose overall. Main leaves in prominent basal rosettes, spatulate-oblanceolate, entire, obtuse to acute at apex, 12-19 x 1.2-1.6 cm, margins revolute. Stems simple or branched with several sessile and acute smaller cauline leaves. Heads terminal in a cluster of 5-9 heads, peduncle 1-6 cm long. Heads hemispherical, 15 mm in Ø, nodding to semi-erect. Involucral bracts linear, tomentose, florets yellow.

Ecology: Andean I-III: 1000->4500 m. Humid places, between rocks.

Distribution: S-Peru, N-Chile, NW-Argentina.

Arequipa: native. CAY.

Note: The type of Senecio zoellneri, Martic. & Quezada 268 (LP), is from Tarapacá, Arica (N-Chile). The species was recorded for Tacna, but mistakenly identified as C. canescens, a species from N-South America (H. Beltrán, pers. comm.). The same is probably true for the specimen Montesinos 5159 (B) from Chivay and the observation from Caylloma in iNaturalist (2024). Salomon et al. (2018) synonymize C. albifolium Zoellner from N-Chile and Bolivia and C. oligocephalum from NW-Argentina with Senecio keshna Cabrera (orthographic C. keshua). POWO (2024) keeps the species separate, here we follow Salomon et al. (2018).

Culcitium serratifolium Meyen & Walp.

Description: Perennial herb, 8-20 cm tall. Basal leaves rosulate, oblanceolate, margins serrulate-dentate, base attenuate, 5-10 cm long, 5-10 mm wide. Cauline leaves sessile, bract-like. Heads discoid, 2-4 cm in Ø, solitary, terminal, nodding, hemispherical. Involucral bracts 30-40, purplish-green, disc flowers numerous, purplish-yellow.

Ecology: Andean III: 4000->4500 m. Wet places, bofedales.

Distribution: Peru, Bolivia, N-Chile, NW-Argentina.

Arequipa: native. CAY, CON, UNI.

Diplostephium cinereum Cuatrec.

Description: Resinous and aromatic shrub. Leaves linear, 8-13 mm long, margins revolute, dark green, tomentose. Heads terminal, ray florets white, >25, 11-12 mm long, disc florets yellow-orange.

Ecology: Andean III: 3500-4500 m. Polylepis forests, restricted to high elevations.

Distribution: Peru, Bolivia, N-Chile.

Arequipa: native, expected. No specimen, one observation from Huambo in iNaturalist (2024). Expected for Arequipa (O. Vargas, pers. comm.).

Use: Medicinal and ritual plant in N-Chile and Bolivia.

Diplostephium meyenii Wedd.

Description: Shrub, up to 1 m tall. Leaves elliptic, 20-30 mm long, light green, tomentose, margins revolute. Ray florets 5-6 mm long, white, disc florets yellow.

Ecology: Andean II-III: 3000-4500 m. Dry slopes.

Distribution: S-Peru to N-Chile.

Arequipa: native. ARE, CAY, CON.

Note: D. tacorense is an unresolved species and is treated here as a synonym of D. meyenii (pers. comm. O. Vargas).

Eclipta prostrata (L.) L.

Description: Annual, erect or prostate, pubescent herb, often rooting at nodes with opposite, sessile, usually oblong, 2.5-7.5 cm long leaves with white adpressed hairs. Heads 6-8 mm in Ø, solitary, ray florets white, achene compressed and narrowly winged.

Ecology: Coastal, Andean I, Amazonian: 0-1500 m. Riversides, fields, abandoned ponds, roadsides.

Distribution: Native to the Americas, introduced worldwide.

Arequipa: native. CAR, CAS, ISL.

Note: Umemoto & Koyama (2007) divide E. prostrata into 4 species based on minor morphological differences. The American clade is named Eclipta alba (L.) Hassk. and is not yet accepted by POWO (2024).

Encelia canescens Lam.

Description: Small shrub, branched, up to 80 cm tall. Leaves tomentose, silvery. Heads with yellow ray florets and brown disc florets.

Ecology: Coastal, Andean I-II: 0-3000 m. Beach, desert, disturbed areas, lomas.

Distribution: Peru, Bolivia, C- and N-Chile.

Arequipa: native. ARE, CAM, CAR, CAY, ISL.

Use: Used as a medicinal plant to promote lactation and urination.

Note: In Arequipa, E. canescens var. canescens and var. parvifolia.

Erigeron bonariensis L.

Description: Annual herb up to 100 cm tall, central stem often overtopped by distal branches in age, plant ± gray hairy. Secondary venation of cauline leaves not visible, seeming 1-veined, lower cauline leaves lanceolate, coarsely serrate or pinnatilobate, upper leaves linear. Inflorescence a pyramidal raceme-like cluster with numerous heads. Heads slightly glutinous, phyllaries purple, peduncles 1-2 cm long, involucre ca. 5 mm in Ø. Pappus reddish, sordid, or tawny.

Ecology: Coastal, Andean I-III, Amazonian: 0-4000 m. Disturbed areas, riversides, rocky slopes.

Distribution: Native to Central and South America, naturalized worldwide.

Arequipa: native. ARE, CAM, CAR, CAY, CON, ISL, UNI.

Erigeron canadensis L.

Description: Herbs, annual, 0.5-2 m tall. Stems erect, green, sparsely hirsute, branched above, densely leafy, central stem generally exceeding distal branches in age. Basal leaves withered at flowering, lower petiolate, oblanceolate, 6-10 × 1-1.5 cm, apical margins ciliate, base attenuate, margins sparsely serrate or entire, apex acute or acuminate, mid and upper leaves subsessile or sessile, blade linear-lanceolate or linear, smaller, margins usually entire. Heads 3-4 mm in Ø, numerous, in terminal, large paniculiform inflorescences. Involucre subcylindrical, 2.5-4 mm. Ray florets 20-30(-45), white, 0.5-3 mm, disc corolla usually 4-lobed.

Ecology: Andean I-II, Amazonian: 0-3500 m. Disturbed areas, riversides.

Distribution: Native to the Americas, introduced worldwide.

Arequipa: native. ARE, CAY, UNI.

Erigeron deserticolus (Phil.) comb.ined.

Description: Perennial herb, 30 cm tall. Basal leaves petiolate, cauline leaves sessile, ascendent. Blade oblong- lanceolate, margins entire, undulate, apex acute or slightly acuminate. Heads terminal solitary, in loose corymbs of 1-10, Ø 5-7 mm, peduncle 1-2 cm long, phyllaries tomentose, marginal florets white, in 3-4 series and central florets yellow.

Ecology: Andean III: 4000-4500 m.

Distribution: Peru, Bolivia, NW-Argentina, N-Chile.

Arequipa: native. ARE, CAY.

Note: No voucher. E. deserticolus and E. pazensis are morphologically similar, their ecology is about the same, and their ranges overlap. There are intermediate specimens, e.g. K000221799. May be they are conspecific. Conyza deserticola L. (= E. deserticolus) is an unplaced name by POWO.

Erigeron floribundus (Kunth) Sch.Bip.

Description: It shares the ± glabrous, ciliate leaves and small Heads with E. canadensis but has tubular, 5-lobed florets, the laminae of the ray florets are absent or only 0.3 mm long, and it has a dull, greyish-green appearance (plants are more hirsute hairy). In addition, the inflorescence is much more open and never cylindrical.

Ecology: Andean II: 1500-3000 m. Riverbeds, disturbed areas, cultivated land.

Distribution: Native to South America, introduced elsewhere.

Arequipa: native, doubtful. CAY. No specimen but several observations.

Erigeron lanceolatus Wedd.

Description: Herbaceous perennial 5-25 cm tall. Leaves numerous from a basal rosette, lanceolate, 1-10 x 0.1-1 cm, pubescent. Heads solitary, born on a 1-10 cm long pedicel, ray florets white to white rose, disc florets yellow.

Ecology: Andean II-III: 3000->4500 m. Rocky soils.

Distribution: Peru, Bolivia, NW-Argentina.

Arequipa: native. CAY.

Erigeron pazensis Sch.Bip.

Description: Suffrutescent herb, 25-50 cm high, with several basal branches, leafy throughout but without a basal rosette, branches gray-green, striate, hairy when young. Cauline leaves ovate, sessile and ± amplexicaule, strongly ascending, up to 5 cm x 1-4 mm, acuminate, margins entire, surface slightly scabrous, pubescent throughout with a strong central midrib. Heads terminal and axillary, 1-3 solitary heads per peduncle, ray florets white, numerous, arranged in 2-4 series, disc florets tubular, yellow, twice the number of ray florets, corolla 4-5 mm long.

Ecology: Andean II: 3000-3500 m.

Distribution: S-Peru and Bolivia.

Arequipa: native. ARE, CAY. No specimen. Mentioned for RN Salinas y Aguada Blanca by Quipuscoa & Huamantupa (2010) and several observations by the author.

Note: See note on E. deserticola.

Erigeron rosulatus Wedd.

Description: Pulvinate perennial herb, 2-6 cm high. Roots, root-crown and underground stems strong, woody, up to 1 cm in Ø. Leaves crowded in a very dense rosette around short, stubby shoots. Leaves linear-lanceolate, rosulate, 3-12 mm long and 1-3 mm wide, very densely covered with long, multicellular, hairs on both surfaces, petiole 1-5 mm long, expanding gradually into the lamina. Heads borne singly at the apex of the branches, often surrounded by the upper leaves giving an aspect of being depressed. Ray florets white.

Ecology: Andean II-III: 3000->4500 m.

Distribution: Peru, Bolivia, N-Chile, NW-Argentina.

Arequipa: native. ARE, CAY, CON.

Facelis plumosa (Wedd.) Sch.Bip.

Description: Annual herb, 2-7 cm tall. Stems usually branched from the base, slender, lanate, foliaceous to the apex. Leaves sessile, blade narrowly linear, 5-10 x 1 mm, apex attenuate, mucronate, lower surface densely lanate, upper surface laxly lanate, margins revolute. Heads 1-3 in the axils of the upper leaves, sessile. Heads 4-5 x 1.5 mm.

Ecology: Andean II-III: 2500-4500 m.

Distribution: Peru, Bolivia, NW-Argentina, N-Chile.

Arequipa: native. ARE, ISL. No voucher. Mentioned by Brako & Zarucchi (1993) and Quipuscoa & Huamantupa (2010).

Flaveria bidentis (L.) Kuntze

Description: Annual herb, stems erect, to 1 m tall. Leaves opposite, petiolate (proximal, petioles 3-15 mm) or sessile and connate (distal), blades lanceolate-elliptic, 10 × 2 cm, margins serrate or spinulose serrate, 3-veined. Heads 20-100+ in tight subglomeruli in scorpioid cymes, involucres oblong-angular, ca. 5 mm, phyllaries 3, oblong. Ray florets 0 or 1, lamina pale yellow, obliquely ovate, to 1 mm. Disc florets 3-8, corolla tubes ca. 0.8 mm, funnelform.

Ecology: Coastal, Andean I-II: 0-2500 m. Moist places, wastelands or disturbed sites, clay.

Distribution: Native to South America, introduced and naturalized elsewhere.

Arequipa: native. ARE, CAM, CAR, CAY, ISL, UNI.

Galinsoga parviflora Cav.

Description: Annual herb, 40-80 cm tall, branched, glabrous or sparsely pubescent. Blade 2-6 × 1-3.5 cm. Peduncles up to 4 cm, involucres campanulate, Ø 2.5-5 mm. Ray florets 4-5, dull white or pink.

Ecology: Coastal, Andean I-II, Amazonian: 0-3000 m. Disturbed places, crop weed.

Distribution: Native to South America, introduced elsewhere.

Arequipa: native. ARE, CAR, CAY.

Use: In Colombia, it is used as a seasoning in soup ajiaco, also as an ingredient in leaf salads.

Galinsoga quadriradiata Ruiz & Pav.

Description: Annual herb, branched, 30-80 cm tall, covered with coarse hairs. Leaf blades 2-6 × 1.5-4 cm. Peduncles 5-20 mm. Involucres hemispherical to campanulate, Ø 3-6 mm. Ray florets 4-7, usually white, sometimes pink.

Ecology: Coastal, Andean I-II, Amazonian: 0-3500 m. Disturbed places, crop weed.

Distribution: Native to Central and South America, naturalized worldwide.

Arequipa: native. ARE.

Use: Used as culinary herb or cooked as a vegetable.

Galinsoga unxioides Griseb.

Description: Branching annual herb up to 40 cm tall. Stem glabrescent or slightly villous in younger parts. Leaves 1- 3.5 x 0.2-1.5 cm, pubescent on both sides. Heads sessile or subsessile, 3-4 mm in Ø, ray florets white or pink.

Ecology: Andean II-III: 1500-4500 m. Disturbed areas, rocky slopes.

Distribution: Peru, Bolivia, NW-Argentina.

Arequipa: native. ARE, CAY, UNI.

Gamochaeta americana (Mill.) Wedd.

Description: Annual or biennial herb, 15-35 cm tall, stems decumbent-ascending, white pannose. Basal leaves present at flowering. Leaves bicolored, upper side glabrous or glabrate, green, underside white-pannose. Heads initially usually in dense, continuous spiciform arrays 2-20 cm × 10-14 mm (pressed), later branched, interrupted. Involucres (± purplish) 2.5-3 mm, bases glabrous, outer phyllaries elliptic-obovate to broadly ovate-elliptic, apices rounded to obtuse. Bisexual florets 2-3.

Ecology: Coastal, Andean I-III: 0-4000 m. Shrublands, disturbed areas.

Distribution: Tropical and Subtropical America to Falkland Islands.

Arequipa: native. ARE, CAM, CAR, ISL.

Gamochaeta capitata Wedd.

Description: Herb, 10-30 cm tall. Leaves linear-obovate, concolorous. Heads in terminal clusters.

Ecology: Andean II: 2000-3500 m. Disturbed areas, grasslands.

Distribution: Peru, Bolivia, NW-Argentina.

Arequipa: native. ARE, CAY.

Gamochaeta humilis Wedd.

Description: Herb 10 cm tall. Leaves oblanceolate, discolors. Heads interrupted in spikes.

Ecology: Andean II-III: 3000->4500 m.

Distribution: Bolivia, Chile, Peru.

Arequipa: native, expected. No specimen but convincing observations in iNaturalist (2024).

Gamochaeta pensylvanica (Willd.) Cabrera

Description: Erect herbs, up to 40 cm tall, stem simple or branched from the base, with thin grayish to white tomentum. Leaves spatulate, usually narrowed at base, 2-5 x 0.6-1.5 cm, broadly rounded, shortly mucronate at the apex, sparsely lanate on the upper surface, white to grayish tomentose on the lower surface. Heads in short, interrupted spicate clusters, c. 2 mm across, peduncle very short, 0.5 mm long.

Ecology: Coastal, Andean I: 0-1500 m.

Distribution: Native to America, now widespread as weed.

Arequipa: native, doubtful. ISL. Specimens FLSP 444 (CUZ) from Quilca and FLSP 960 (MO) from Lomas de Mejía are marked aff. and cf. respectively. Inconsistent observation from Lomas de Mejía in iNaturalist (2024). Occurrence in Arequipa is probable.

Gamochaeta purpurea (L.) Cabrera

Description: Annual herb, densely pannose-tomentose, stem erect to decumbent-ascending, up to 40 cm tall. Leaves spatulate-obovate, 8 × 2 cm, decreasing in size towards the stem apex, margins entire or subentire, both sides pannose-tomentose. Basal leaves usually withering before flowering. Inflorescence in a terminal spike with leaf-like bracts. Involucral bracts 3-seriate, lanceolate-oblong, 3 mm long. Florets small, numerous, brownish. Bisexual florets 3-4.

Ecology: Coastal, Andean I-III, Amazonian: 0-4500 m.

Distribution: Native to America, introduced to Asia, Europe and Australia.

Arequipa: native. ARE, CAM, CAR, CAY, ISL.

Gochnatia arequipensis Sandwith

Description: Woody, tomentose shrub. Leaves grayish, suborbicular and distinctly 3-veined. Heads solitary on terminal branchlets, with 9-12 tubular, actinomorphic florets. Florets discoid, isomorphic, cream to yellow, corolla to 15 mm long. Involucre 10-15 mm high, 3-layered.

Ecology: Andean II: 2000-3500 m. Dry slopes.

Distribution: S-Peru.

Arequipa: native. ARE, CAY, ISL, UNI.

Grindelia boliviana Rusby

Description: Erect shrub. Similar to G. tarapacana but leaves herbaceous, green, not glaucous. Lower leaves up to 13 cm long. Heads Ø >2.5 cm, ray florets 36-45 per head.

Ecology: Andean II-III: 2500-4000 m.

Distribution: Peru, Bolivia, NW-Argentina.

Arequipa: native. ARE, CAY, UNI.

Use: Leaves used to heal bone fractures (compresses).

Grindelia glutinosa (Cav.) Mart.

Description: Subshrub up to 20 cm tall. Leaves glutinous and broad elliptic to obovate, green. Heads 3.5-5.5 cm in Ø, with 30-38 yellow rays.

Ecology: Coastal, Andean I-II: 0-3500 m. Lomas.

Distribution: Peru, N-Chile.

Arequipa: native. ARE, CAM, CAR, ISL.

Note: It probably occurs only on the coast and in the lomas. The specimens seen from above 1000 m seem to be G. tarapacana (habit, leaf shape, capitula size, number of ligulate flowers).

Grindelia tarapacana Phil.

Description: Branched shrub, up to 1 m tall, resinous, glandular, with leafy branches up to the top. Leaves coriaceous, glaucous, linear to obovate, 1.5-3.8 x 0.5-1.2 cm, margins dentate. Flowers solitary with <25 dark yellow rays, less than 2.5 cm in Ø, outer involucral bracts glabrous.

Ecology: Coastal, Andean I-III: 0-4000 m. Lomas, shrubland.

Distribution: South-Peru and North-Chile.

Arequipa: native. ARE, CAY.

Note: In Peru G. tarapacana var. tarapacana and var. verrucosa.

Gynoxys longifolia Wedd.

Description: Woody shrub, up to 2 m. tall. Branchlets tomentose white. Leaves linear-lanceolate, 10-15 x 2.5-3.5 cm, rugose, green on upper side, greyish-white tomentose beneath. Heads discoid, florets white or yellow, aromatic.

Ecology: Andean III: 3500-4000 m. Protected sites, e.g. canyons.

Distribution: Peru.

Arequipa: native. ARE, CAY.

Gynoxys longistyla (Greenm. & Cuatrec.) Cuatrec.

Description: Shrub, up to 2 m tall. Leaves ovate, grayish-white on both sides, smooth. Heads radiate, florets yellow.

Ecology: Andean III: 3500-4000 m.

Distribution: Peru.

Arequipa: native. ARE.

Heiseria irmscheriana (Bruns) Mesfin

Description: Annual herb 10-30 cm tall, stem herbaceous, hirsute. Leaves ovate, blade hirsute, margins serrate. Heads up to 3.5 cm wide, yellow, with brownish-red center. Phyllaries and disc corolla without glands, paleae und phyllaries black at or near the apex.

Ecology: Coastal: 0-500 m.

Distribution: S-Peru.

Arequipa: native. CAM, ISL.

Note: Known only from the type locality in Mejía.

Helogyne apaloidea Nutt.

Description: Perennial herbs, up to 50 cm tall, much-branched, pubescent, ± glutinous. Leaves often divided in 3 linear lobes. Heads in terminal clusters of more than 1, discoid, involucral bracts obtuse. Florets yellow.

Ecology: Coastal, Andean I-II: 0-3000 m. Lomas, dry hillsides.

Distribution: Peru, N-Chile.

Arequipa: native. ARE, CAR.

Helogyne hutchisonii R.M.King & H.Rob.

Description: Shrublet, up to 40 cm tall, much-branched, covered with whitish pubescence. Leaves elliptical, entire or with some teeth, obtuse. Heads usually solitary, discoid, few flowered, florets yellowish-white.

Ecology: Coastal: 0-500 m. Lomas.

Distribution: Peru.

Arequipa: native. CAR.

Helogyne straminea B.L.Rob.

Description: Branched shrub. Leaves linear, pointed. Heads terminal, clustered, involucral bracts straw-colored, pointed.

Ecology: Andean II: 2000-3500 m.

Distribution: Peru, Bolivia.

Arequipa: native. ARE.

Heterosperma diversifolium Kunth

Description: Annual herbs, 10-30 cm tall, stems ascending, hispid. Leaves pinnatisect in lower part and entire in upper part, petiole 7-10 mm, pilose, 3-5-lobed, lobes broadly lanceolate. Heads terminal and axillary, peduncles 1.5- 4 cm. Phyllaries 2-seriate, outer phyllaries 3 to 5, linear to lanceolate. Ray florets 5-8, female, yellow, ligulate, disc florets 8-18, bisexual, yellow, tubular.

Ecology: Coastal, Andean I-III, Amazonian: 0-4000 m. Disturbed areas, rocky slopes.

Distribution: NW-South America.

Arequipa: native. ARE, UNI.

Heterosperma ferreyrae H.Rob.

Description: Annual herbs, 10-25 cm tall, stem hirsute to pilose. Leaves obscurely petiolate, petiole mildly winged, 8- 12 mm, pilose, pinnatisect, 15-40 mm, 3-lobed, upper lobe obovate, acute 3-dentate, lateral lobe, obovate to lanceolate, 2-5 mm wide. Heads terminal and axillary in leafy, loose cymes, long-pedunculate. Ray florets 6, disc florets ca. 15, corollas yellow.

Ecology: Coastal: 0-1000 m.

Distribution: Peru, N-Chile.

Arequipa: native. CAR.

Note: Unresolved name, could be a synonym.

Heterosperma nanum (Nutt.) Sherff

Description: Annual herb, up to 10 cm tall, branches decumbent, all parts pilose. Leaves petiolate, deltoid, divided, margins dentate. Heads terminal and axillary, ray florets 3-5, yellow, peduncle ± as long as leaves.

Ecology: Andean II: 2500-3000 m. Disturbed areas.

Distribution: Peru, Bolivia, NW-Argentina, N-Chile.

Arequipa: native. CAY. No specimen but a convincing observation from Yanque in iNaturalist (2024).

Heterosperma ovatifolium Cav.

Description: Annual herb, up to 50 cm tall, branches decumbent. Leaves ovate, nearly sessile. Heads terminal and axillary, peduncle 2-3 cm, twice as long than the leaves. Ray florets 4, disc florets >15, corollas yellow.

Ecology: Coastal, Andean I-II, Amazonian: 0-3500 m. Grasslands, rocky slopes.

Distribution: Peru, Bolivia, N-Argentina, N-Chile.

Arequipa: native. ARE, CAY.

Hieracium mandonii (Sch.Bip.) Britton

Description: Perennial, rhizomatous herb, 10-40 cm tall. All parts floccose, hirsute, with long bristles. Basal leaves in a rosette, elliptic-lanceolate or oblanceolate, 6-10 × 1.5-2 cm, base attenuate, apex acute to obtuse, entire or slightly denticulate, cauline leaves (1-2), linear or linear-lanceolate, much smaller. Heads arranged in racemes. Florets yellow, with short ligule, barely protruding from involucre. Achenes black, ca. 3 mm long. Pappus white or yellowish.

Ecology: Andean II-III: 2000-4000 m.

Distribution: Peru, Bolivia, NW-Argentina.

Arequipa: native, doubtful. ARE. No voucher, mentioned for RN Salinas y Aguada Blanca by Quipuscoa & Huamantupa (2010). An occurrence in Arequipa is likely due to the distribution range of the species.

Hypochaeris chillensis (Kunth) Britton

Description: Caulescent herb, 30-70 cm tall, stem erect or ascending, always branched distally. Leaves basal and proximally cauline. Cauline leaves sharply dentate or pinnatifid, if dentate, the teeth bent towards the leaf apex. Heads (2-)4-10, in loose, paniculiform to corymbiform arrays, apex of peduncle usually not distinctly enlarged and glabrous. Budding heads always erect. Phyllaries hirsute, with membranous margins, often reflexed or erect and not adpressed. Corolla yellow.

Ecology: Andean II-III: 2500-4000 m. Disturbed and moist areas, lawns.

Distribution: Native to South America, naturalized in southern North America.

Arequipa: native. ARE, UNI.

Hypochaeris echegarayi Hieron.

Description: Herbs to 5 cm tall. Leaves oblanceolate or oblong, 20-70 x 3-10 mm, apex acute, base attenuated in a broad petiole, pinnatipartite to pinnatisect, margins entire or dentate, hirsute to glabrous. Heads pedunculate (to 2 cm), rarely sessile. Involucre campanulate, corolla white.

Ecology: Andean II-III: 2500->4500 m. Bogs or dry places.

Distribution: Peru, Bolivia, NW-Argentina.

Arequipa: native. ARE.

Hypochaeris elata Griseb.

Description: Similar to H. chillensis and with the same ecological parameters. Basal leaves entire with ± prominently dentate margins, the teeth perpendicular to the leaf axis or slightly bent downwards. Apex of peduncle hirsute and distinctly enlarged, phyllaries always adpressed, without membranous margins. Budding heads typically nodding, at least in early stage.

Ecology: Andean II-III: 2750-4000 m. Shrubland.

Distribution: Native to South America.

Arequipa: native. ARE, CAY.

Note: Brako & Zarucchi (1993) treat it as a synonym of H. chillensis.

Hypochaeris eriolaena (Sch.Bip.) Reiche

Description: Herbs to 5 cm tall. Leaves lanceolate or oblong, entire, often leafless when flowering. Heads sessile, phyllaries lanate, whitish. Corolla white or yellow.

Ecology: Andean III: 4000->4500 m. Dry, stony Andean ridges, with scattered grass.

Distribution: Peru, Bolivia.

Arequipa: native. ARE. No voucher, mentioned for RN Salinas y Aguada Blanca by Quipuscoa & Huamantupa (2010). Occurrence in Arequipa is likely, due to the distribution range of the species (Cuzco, Puno).

Hypochaeris meyeniana (Walp.) Benth. & Hook.f. ex Griseb.

Description: Polymorphic herbs to 8 cm tall. Leaves pinnatipartite to pinnatisect, lobes dentate or entire. Phyllaries green, generally glabrous or sometimes hirsute but not lanate. Heads sessile or rarely with a very short peduncle. Corollas yellow-orange, sometimes white.

Ecology: Andean II-III: 2000->4500 m. Various microhabitats, from bogs to dry places.

Distribution: Peru, Bolivia, NW-Argentina, N-Chile.

Arequipa: native. ARE, CAY, CON, UNI.

Hypochaeris taraxacoides Ball

Description: Herbs to 7 cm tall. Leaves lanceolate or oblong, pinnatifid to lobate. Distinguished by the combination of pedunculate (rarely sessile) cylindrical heads, glabrous phyllaries and leaves, and corollas longer than the involucre, white on the upper surface to dark blue-black at the tips of ligules underneath.

Ecology: Andean II-III: 2500->4500 m. Bogs, grasslands, rocky slopes, shrublands.

Distribution: Colombia to NW-Argentina and N-Chile.

Arequipa: native. ARE, CAY, CON.

Jungia axillaris Spreng.

Description: Perennial subshrub, old branches woody. Leaves orbicular, pedately lobed, pilose. Inflorescence axillary. Heads terminal and axillary, corolla pinkish-white.

Ecology: Andean I-II: 1000-3500 m.

Distribution: Peru.

Arequipa: native. CAY, UNI.

Lactuca serriola L.

Description: Biennial 1-2 m high, stem glabrous or bristly, whitish. Leaves glaucous, margins and midrib with spines. Lower leaves obovate-oblong, usually deeply pinnatifid, rarely margins entire, cauline leaves sessile, auriculate, acute to obtuse auricles, usually held in a vertical plane. Heads yellow, solitary in axils of reduced leaves, pedicellate, involucral bracts 3- or 4-seriate.

Ecology: Andean I-II: 1000-2600 m. Disturbed areas, riversides.

Distribution: Native to Eurasia, introduce worldwide.

Arequipa: introduced, adventive. ARE. Observation in iNaturalist (2024), no other voucher.

Laennecia artemisiifolia (Meyen & Walp.) G.L.Nesom

Description: Herb, up to 30 cm tall, hispid, glandular. Leaves sessile, pinnatisect. Heads in racemes, with purple indument. Radial florets filiform, white, central florets tubiform, yellow.

Ecology: Andean III: 3500-4000 m. Disturbed areas.

Distribution: Peru, Bolivia, N-Chile, NW-Argentina.

Arequipa: native. ARE, CAY, CON, UNI.

Leucheria daucifolia (D.Don) Crisci

Description: Perennial, cespitose herb. Leaves petiolate, blade oblong, 3-5 x 1-1.5 cm, membranous, 2-pinnatisect- pinnatifid. Inflorescence cymose-corymbiform, corolla white, bilobed, the lobes coiled.

Ecology: Andean III: 3500->4500 m.

Distribution: Peru, Bolivia, N-Chile.

Arequipa: native. ARE, CAY, UNI.

Lomanthus abadianus (DC.) B.Nord. & Pelser

Description: Erect suffrutescent herb. Leaves cauline, oblanceolate, margins broadly dentate, incised halfway to midrib, apex acute, blade glabrous above, white-tomentose beneath. Inferior leaves attenuate to a pseudopetiole, upper leaves sessile, auriculate. Heads radiate, terminal in clusters of 1-3, on long peduncles. Involucral bracts 12- 14, ray florets 12-14, disc florets 30-35, both yellow.

Ecology: Coastal, Andean I-II: 0-3500 m. Lomas, shrublands.

Distribution: Peru.

Arequipa: native. ARE, CON, ISL.

Lomanthus arnaldii (Cabrera) B.Nord. & Pelser

Description: Erect, probably annual herb. Leaves cauline, lanceolate, up to 20 cm long, deeply pinnatisect, segments lobed to serrate, sessile, base auriculate, apex acute, blade glabrous above, white-tomentose beneath. Heads radiate, terminal in corymbiform cymes of 30+ heads, on short peduncles. Ray and disc florets yellow.

Ecology: Coastal, Andean I-II: 0-3500 m. Lomas. Shrubland.

Distribution: Peru.

Arequipa: native. ARE, CAY.

Lomanthus icaensis (H.Beltrán & A.Galán) B.Nord.

Description: Perennial herb, 20-30 cm tall, stems prostrate ascending. Basal leaves in rosettes, blade 1-3 cm long, pinnatisect, glabrous above, white-tomentose beneath, petiole 2-4 cm long. Blade segments 2-3-lobed. Cauline leaves few, sessile, smaller than basal leaves. Heads 1-3, radiate, terminal. Ray florets numerous, 10-11 mm long, disc florets, disc florets numerous, both yellow.

Ecology: Coastal: 650-700 m. Desert.

Distribution: S-Peru.

Arequipa: native. CAR.

Lomanthus lomincola (Cabrera) B.Nord. & Pelser

Description: Similar to L. abadianus but differs by shortly lobed and acutely dentate leaves.

Ecology: Andean I: 500-1000 m.

Distribution: Peru.

Arequipa: native. CAR, ISL.

Lomanthus mollendoensis (Cabrera) B.Nord.

Description: Perennial subshrub. Leaves cauline, inferior leaves with attenuate base and short petiole, upper leaves sessile with auriculate base, apex acute, margins coarsely dentate-serrate. Heads radiate, terminal, in corymbiform cymes of 20-40 heads. Ray florets 10-15, disc florets ca. 20-30, both yellow.

Ecology: Coastal: 300-500 m. Lomas.

Distribution: Peru.

Arequipa: native. ARE, CAR, ISL.

Lomanthus okopanus (Cabrera) B.Nord.

Description: Perennial herb, prostrate. Basal leaves rosulate, long petiolate, lamina ovate, 3-5 cm long, deeply lobed to lacerate, margins dentate, glabrous above, white-tomentose beneath. Cauline leaves few, sessile, much smaller than basal leaves. Heads radiate, 2-5 in lax corymbiform cymes. Florets yellow.

Ecology: Coastal: 0-1000 m.

Distribution: Peru.

Arequipa: native. CAR.

Lomanthus subcandidus (A.Gray) B.Nord.

Description: Perennial, erect herb with woody base. Cauline leaves ovate, entire, margins irregularly dentate, base truncate to slightly cordate, apex acute, glabrous above, white-tomentose beneath. Petiole 2 cm long, ± alate. Heads 8-13, radiate, in corymbiform cymes. Involucral bracts ca. 20, ray florets 10-12, whitish-yellow, disc florets ca. 30, yellow.

Ecology: Coastal, Andean I-II: 200-3500 m.

Distribution: Peru.

Arequipa: native. ARE, CAY.

Lomanthus tovarii (Cabrera) B.Nord. & Pelser

Description: Suffrutescent herb to 60 cm tall. Leaves petiolate, oblanceolate, margins lobed to dentate, bluish-green above, white-tomentose beneath. Heads 4-7, radiate, in corymbiform cymes. Ray- and disk-florets bright yellow.

Ecology: Andean II: 1500-3000 m. Rocky slopes.

Distribution: Peru.

Arequipa: native. ARE, CAR, CAY, CON, UNI.

Lophopappus cuneatus R.E.Fr.

Description: Shrubs 30-100 cm tall. Leaves oblong to elliptic, margins planate, entire, acute or obtuse and mucronate or caudate at the apex, attenuate at the base, 5-veined, glabrous or glandular-pubescent, sessile to shortly petiolate, petiole 1-2 mm. Heads solitary, discoid, corolla bilabiate. Florets 6-7, isomorphic, corolla bilabiate, white, sometimes tinged purplish.

Ecology: Andean II-III: 2800-4000 m. Xeric forests, dry valleys and hillsides and in tolares.

Distribution: S-Peru, S-Bolivia, N-Chile and NW-Argentina.

Arequipa: native. ARE, CAY.

Use: Used for stomach aches and cattle wounds. The branches are used to make brooms and for firewood.

Note: Faundez & Macaya (1998) consider the Peruvian clade of L. cuneatus and L. tarapacanus as conspecific, because of the minor morphological differences and the disjunct distribution of L. cuneatus (Peru, N-Chile vs. S- Bolivia, NW- Argentina).

Lophopappus foliosus Rusby

Description: Shrubs 0.1-2 m tall. Leaves linear to linear-lanceolate, margins entire, thickened and ± revolute, entire, acute and spinose at the apex, 3-veined. Heads solitary, discoid, florets 5-8, isomorphic, corollas bilabiate.

Ecology: Andean II-III: 3000-4000 m. Xerophilous or sub-humid shrublands and grasslands.

Distribution: Peru, Bolivia, NW-Argentina.

Arequipa: native. CAY.

Lophopappus tarapacanus (Phil.) Cabrera

Description: Shrub, 0.6-3 m tall, branched, the branches striate, glabrous, lacking spines. Leaves 8-35 x 2-10 mm, lanceolate to elliptic, acute or obtuse and mucronate to shortly spinose at the apex, attenuate at the base, glandular pubescent, 5-veined. Inflorescence of 2-4 heads, sometimes solitary, glomerate, white, florets 6-11.

Ecology: Andean II-III: 2900-4000 m. Open soils in the slopes and deserts.

Distribution: Peru.

Arequipa: native. ARE, CAY, UNI.

Note: Katinas et al.(2013) synonymize L. berberidifo-lius Cabrera with L. tarapacanus, which is followed here.

Lucilia conoidea Wedd.

Description:

Ecology: Andean III: 4000-4500 m. Shrublands.

Distribution: Peru, Bolivia.

Arequipa: native. .

Note: Freire et al. (2015) synonymized Lucilia conoidea with M. kunthiana. These changes are not accepted by POWO (2024) and so we do here.

Lucilia kunthiana (DC.) Zardini

Description: Cespitose, much-branched, decumbent herb. Leaves imbricate, sessile, blade linear to spatu-late, 5-15 x 1-3 mm, marcescent, the base slightly ex-panded and clasping stem, apex obtuse, mucronate, lower surface arachnoid-tomentose, glabrescent, up-per surface densely silvery tomentose. Heads solitary, subsessile.

Ecology: Andean II-III: 3000->4500 m.

Distribution: Venezuela to NW-Argentina.

Arequipa: native. ARE, CON. A species with wide distri-bution. No specimen, a (doubtful?) observation from Condesuyos. The occurrence is likely due to the distri-bution range of the species (Apurimac, Ayacucho, Cuzco, Huancavelica, Puno).

Note: The phylogenetic analysis of Freire et al. (2015) shows that Lucilia is nested within Mniodes. Based on this, Freire et al. merged Lucilia with Mniodes. These changes are not accepted by POWO (2024) and so we do here.

Matricaria chamomilla L.

Description: Branched, erect and smooth stem, 15-60 cm tall. Leaves long and narrow, 2-3-pinnate. Flowers in paniculate flower heads, smelling strongly aromatic. Ray florets white, disc florets yellow. Receptacle swol-len, hollow, lacking scales.

Ecology: Andean II-III: 2000-4000 m. Disturbed areas.

Distribution: Europe and Asia, naturalized worldwide.

Arequipa: introduced. ARE, CAR, ISL, UNI.

Use: Used as a medicinal plant for various purposes.

Misbrookea strigosissima (A.Gray) V.A.Funk

Description: Mat forming and rosetted from a network of creeping woody rhizomes, 2-3 cm tall in bloom. Leaves spatulate, 2-3 cm x 3-5 mm, slightly fleshy. Leaves and involucre covered with 3-5 mm long stri-gose hairs. Heads sessile, solitary, 3-5 cm in Ø, ligules white, disc florets yellow.

Ecology: Andean III: 3500->4500 m. Grasslands, rock outcrops, clearings.

Distribution: Peru to W- & C-Bolivia.

Arequipa: native. CAY. No specimen, a convincing ob-servation from Caylloma.

Mniodes argentea (Wedd.) M.O.Dillon

Description: Minute, cespitose-pulvinulate herb. Stems prostrate, foliaceous, internodes short. Leaves spirally arranged, elliptic, 2-4.5 mm long, subplicate, discolorous, green and sparsely tomentose above, be-neath densely lanate-tomentose. Heads solitary, ter-minal, sessile.

Ecology: Andean III: 3500-4500 m. Disturbed areas.

Distribution: Bolivia, S-Peru.

Arequipa: native. CAS.

Note: No voucher. Observation from Viraco in iNatu-ralist (2024). Herbarium specimen from Moquegua (GBIF).

Mniodes caespititia (Wedd.) Quip. & M.O.Dil-lon

Description: Compact cespitose, perennial herbs to 5 cm tall, cushion plant. Stems much-branched, 2-3 cm long, compacted. Leaves obovate, sessile, 4-5 mm long, both surfaces densely lanate, gray. Inflorescence of solitary heads, terminal, sessile, yellow-whitish.

Ecology: Andean III: 4000-4500 m. Dry valleys, grass-lands.

Distribution: Peru, Bolivia, NW-Argentina, N-Chile.

Arequipa: native. ARE, CAY, CON.

Note: Many collections of M. caespititia were previ-ously classified as Mniodes schultzii (Wedd.) Cabrera, an unresolved name (Quipuscoa & Dillon, 2020), which is here considered as a synonym of M. caespititia.

Mniodes coarctata Cuatrec.

Description: Suffruticose perennial herbs, cespitose, forming large, rounded cushions of stems. Leaves densely imbricate, adpressed on stem, sessile, blade 3-4 mm long, both surfaces incano-villous, ovate to ellip-tic. Inflorescence of solitary heads, terminal.

Ecology: Andean III: 4000->4500 m.

Distribution: Peru, N-Chile.

Arequipa: native. ARE, CAY, CON, UNI.

Mniodes longifolia (Cuatrec. & Aristeg.) S.E.Freire, Chemisquy, Anderb. & Urtubey

Description: Cespitose, acaulescent perennial herbs, rhizomes oblique to horizontal, roots filiform. Leaves rosulate, sessile, blade oblanceolate, 2-5 cm long, 6-10 mm wide, marcescent, base attenuate to a winged pet-iole, apex obtuse, mucronate, the margins entire, lower surface silvery tomentose, upper surface densely lanate. Inflorescence of solitary heads, termi-nal, sessile or pedunculate, the peduncles to 5 mm long.

Ecology: Andean III: 3500->4500 m. Grasslands, shrub-lands.

Distribution: Venezuela to Bolivia.

Arequipa: native, doubtful. ARE. No voucher, mentioned for RN de Salinas y Aguada Blanca (Quipuscoa & Huamantupa, 2010).

Mniodes piptolepis (Wedd.) S.E.Freire, Chemisquy, Anderb. & Urtubey

Description: Compact, subcespitose perennial herbs to 5 cm tall. Stems 5-40 cm long, decumbent, tomentose to glabrate. Leaves rosulate, sessile, blade obovate-spatulate, 6-10 mm long. Inflorescence glomerulate, sessile, terminal, occasionally of solitary heads. Achenes brown-red.

Ecology: Andean II-III: 3000->4500 m. Grasslands, shrublands.

Distribution: Colombia to NW-Argentina and C-Chile.

Arequipa: native. ARE, CAY. No specimen, several observations in iNaturalist (2024), mentioned for RN de Salinas y Aguada Blanca (Quipuscoa & Huamantupa, 2010).

Munnozia lyrata (A.Gray) H.Rob. & Brettell

Description: Erect herb from fleshy root. Stem purplish. Leaves cleft (deeply notched), white below. Flowers yellow.

Ecology: Andean II-III: 2000-4000 m. Shrubland.

Distribution: Peru.

Arequipa: native. CAY. No specimen, an observation from Cabanaconde. Mentioned for Arequipa by Brako & Zarucchi (1993).

Mutisia acuminata Ruiz & Pav.

Description: Clambering shrub, up to 1.5 m tall. Leaves pinnate with elliptic leaflets and a terminal tendril. Heads solitary, terminal, bright orange-red to reddish-yellow, 4-5 cm long.

Ecology: Andean II: 2000-3500 m. Shrubland, agricultural land.

Distribution: S-Peru, Bolivia and N-Chile.

Arequipa: native. ARE, CAS, CAY, CON, UNI.

Use: Medicinal plant. The extracts contain antimicrobial agents.

Note: In Arequipa, M. acuminata var. bicolor and var. hirsuta.

Mutisia arequipensis Cabrera

Description: Suffrutescent vine, climbing or procumbent, branches glabrous, winged, wings dentate. Leaves linear- lanceolate, simple, margins entire with a terminal tendril. Heads terminal, upright, red orange, peduncle 5-10 mm, phyllaries 4-5 seriate.

Ecology: Andean II-III: 2000-4000 m.

Distribution: Peru.

Arequipa: native. ARE, CAY, UNI.

Mutisia comptoniifolia Rusby

Description: Shrubs, stem erect, much-branched, the branches strongly ascending, densely leafy, glabrous. Leaves 5-9 x 1.5-2 cm, narrowly lanceolate, deeply pinnately lobed, glabrous, short petiolate. Heads terminal, flowers exserted to 15 mm, rays few, narrow, scarlet.

Ecology: Andean II: 2000-3500 m. Shrubland, agricultural land.

Distribution: Bolivia, Peru.

Arequipa: native. ARE, CAS, CAY.

Note: Cabrera (1965) suggested that it could be treated as a hybrid, an intermediate between M. orbignyana and M. acuminata var. hirsuta.

Mutisia lanigera Wedd.

Description: Scandent to vining herbs, prostrate, branched, the branches glabrous, winged, the wings conspicuously dentate. Leaves lanceolate, sessile, margins basally dentate, apex long acuminate with tendril. Heads solitary, terminal, peduncle 5-15 mm, phyllaries 5-7 seriate and lanuginose, corollas orange.

Ecology: Andean III: 3500-4500 m.

Distribution: Peru, Bolivia, N-Chile.

Arequipa: native. ARE.

Mutisia orbignyana Wedd.

Description: Erect shrubs, 1-2 m tall. Leaves linear-lanceolate with linear-lanceolate, 3.5-7 x 0.5 cm, acute, glabrous, margins entire. Heads discoid.

Ecology: Andean II-III: 3000-4000 m.

Distribution: Colombia, Ecuador, Peru, Bolivia, NW-Argentina.

Arequipa: native. ARE, CAS, CAY, CON.

Novenia tunariensis (Kuntze) S.E.Freire

Description: Acaulescent herb, small colonies of rosettes from fibrous roots. Leaves 20 x 1 mm, apex long acuminate, mucronate. Heads discoid, sessile, few flowered, florets yellow.

Ecology: Andean II-III: 3000-> 4500 m. Grassland, rocky areas.

Distribution: Peru, Bolivia, NW-Argentina.

Arequipa: native, doubtful. ARE. No voucher, mentioned for RN Salinas y Aguada Blanca. Convincing observation from Apurimac near the border to Arequipa in iNaturalist.

Onoseris humboldtiana Ferreyra

Description: Erect herb, 11-25 cm tall. Stem grayish lanuginose. Upper leaves amplexicaule, margins sinuously dentate. Capitulum terminal or axillary, solitary, peduncle up to >10 cm long, ray florets deep pink.

Ecology: Coastal: 500-1000 m. Lomas.

Distribution: Peru.

Arequipa: native, doubtful. ISL. Known only from the type locality in Ica. Specimen Anderson et al. 7962 (F) from Islay, labeled with vel aff.

Onoseris minima Domke

Description: Small herb, 10-14 cm tall, grayish, sparsely branched. Leaves 6-18, puberulent whitish-gray, sessile, margins sinuate-dentate. Capitulum solitary and terminal with 7-9 ray and 8-10 disc florets.

Ecology: Andean I-II: 1000-2000 m.

Distribution: Peru.

Arequipa: native. ARE.

Note: Rare species, blooming only in wet years.

Onoseris odorata Hook. & Arn.

Description: Annual herb, 6-70 cm tall, gray-lanuginose, branched. Leaves short petiolate, lanceolate, margins unequally sinuate-dentate, attenuate at base, apex acuminate, puberulent when young, then glabrescent. Heads solitary, terminal, peduncle 2.5-23 cm long. Involucral bracts in 5-6 series, attenuated, filiform, recurved. Ray florets pink, ligule 3-lobed, disc florets yellow.

Ecology: Coastal, Andean I-III: 0-4000 m.

Distribution: Peru.

Arequipa: native. ARE, CAM, CAR, ISL.

Ophryosporus bipinnatifidus B.L.Rob.

Description: Shrub up to 1 m tall. Leaves 2-pinnatifid. Heads in long panicles, florets white.

Ecology: Coastal, Andean I: 0-1000 m. Lomas.

Distribution: Peru.

Arequipa: native. ISL.

Ophryosporus heptanthus (Sch.Bip. ex Wedd.) R.M.King & H.Rob.

Description: Loose shrubs, branching erect to spreading, regularly >1 m in height. Leaves triangular-lanceolate. Heads few-flowered, in short panicles. Pappus formed by conspicuous setae of 3-4 mm in length.

Ecology: Andean II-III: 2000-4000 m. Disturbed areas, rocky slopes.

Distribution: Peru, Bolivia, N-Chile.

Arequipa: native. ARE, CAS, CAY.

Ophryosporus hoppii (B.L.Rob.) R.M.King & H.Rob.

Description: Loose shrub, >1 m tall. Leaves linear-lanceolate, green. Heads in dense panicles. Pappus formed by minute 0.1-1 mm long, irregular scales.

Ecology: Coastal, Andean I-II: 0-3000 m. Disturbed areas.

Distribution: Peru, N-Chile.

Arequipa: native. ISL.

Ophryosporus peruvianus (J.F.Gmel.) R.M.King & H.Rob.

Description: Shrub much-branched, >1 m. Leaves lanceolate, margins serrate, yellowish-green. Heads in cylindrical panicles, yellowish-green.

Ecology: Coastal, Andean I-II, Amazonian: 0-3500 m. Desert, disturbed areas, rocky slopes.

Distribution: Ecuador, Peru.

Arequipa: native. ARE, CAY, ISL, UNI.

Ophryosporus pubescens (Sm.) R.M.King & H.Rob.

Description: Shrub, up to 1 m tall. Leaves triangular-ovate, margins broadly crenate-serrate, apex obtuse. Heads in loose panicles.

Ecology: Coastal, Andean I-II: 0-2500 m. Lomas.

Distribution: Peru.

Arequipa: native. CAR.

Oriastrum stuebelii (Hieron.) A.M.R.Davies

Description: Perennial dwarf herb, forming lax cushions. Stems with basal bud cluster, trailing, 4-10 cm, decumbent, glabrescent. Leaves 4-7 x 1-1.5 mm, linear, dilated at bases, sessile, dark green, decussate, nearly connate, indumentum long (1 mm) densely curly lanate hairs, apices with short erect mucro. Heads female or bisexual, sessile, solitary, 8-9 mm, disc Ø 1.9-3 mm, campanulate.

Ecology: Andean III: 3800-5200 m. Shrubby slopes in loose sandy soil or stony ground.

Distribution: Peru, Bolivia, N-Chile.

Arequipa: native. CAY.

Oritrophium limnophilum (Sch.Bip.) Cuatrec.

Description: Leaf blades ovate-spatulate, green on the upper side, glabrous or with some trichomes on the underside over the midrib, with entire margins minutely denticulate. Capitulum terminal, solitary on a long pedicel. Involucre 3-5 seriate. Ligulate florets white, disc florets yellow.

Ecology: Andean II-III: 3000->4500 m. Bogs.

Distribution: Colombia, Ecuador, Peru, Bolivia.

Arequipa: native. CAY. No specimen but a convincing observation in iNaturalist (2024) from Caylloma.

Paquirea lanceolata (H.Beltrán & Ferreyra) Panero & S.E.Freire

Description: Shrubs ca. 1.8 m tall, branched, stems conspicuously leafy. Leaves alternate sessile, coriaceous, blades 4-6 × 0.6-1.3 cm, lanceolate, base attenuate, margins denticulate, apex acute, pinnately veined. Heads many flowered, corolla cream.

Ecology: Andean III: 3600-3700 m. Sandy slopes.

Distribution: Endemic to Arequipa, Peru.

Arequipa: native. ARE, CAS, CAY.

Parastrephia lucida (Meyen) Cabrera

Description: Shrub, up to 1.5 m high. Leaves 4-6 mm long, not adnate to the stem, ascending to spreading-recurving. Disc florets 13-28, ray florets 7-16. Corolla yellow.

Ecology: Andean III: 3500->4500 m: Grasslands, shrublands, humid places in the puna.

Distribution: Peru, N-Chile, Bolivia, NW-Argentina.

Arequipa: native. ARE, CAY, CON, UNI.

Use: Medicinal plant for fever and burns. Also used to dye cords and as firewood.

Parastrephia quadrangularis (Meyen) Cabrera

Description: Shrub, 30-180 cm tall. Leaves 2-5 mm long, straight and tightly adpressed to the stem, the basal portion adnate. Disc florets 3-10, ray florets 3-9. Corolla yellow.

Ecology: Andean II-III: 3000-4900 m. Grasslands, shrublands, specific to the tolares.

Distribution: Peru, Bolivia, NW-Argentina, N-Chile.

Arequipa: native. ARE, CAY, CON.

Pascalia glauca Ortega

Description: Erect perennial herb to 70 cm high. Leaves narrow-lanceolate, 5-10 x, 0.5 cm, sessile, usually entire, rarely with remote teeth near the base, pubescent to scabrous. Heads solitary, 10-15 mm Ø, peduncles 2-6 cm long, involucral bracts lanceolate, 6-10 mm long. Ray florets bright yellow, 10-15 mm long, ligulate oblanceolate, 3- toothed. Disc florets numerous, yellow.

Ecology: Coastal: 0-500 m. Disturbed areas, alluvial plains, sandy areas.

Distribution: Native to E-South America, introduced and naturalized in various parts of the world.

Arequipa: introduced, adventive. ISL. No specimen but a convincing observation from Mollendo in iNaturalist (2024).

Note: Invasive species, acute poisonous for livestock, especially cattle.

Pectis sessiliflora Sch.Bip.

Description: Annual herb, 5-15 cm tall, branched and decumbent. Leaves entire, linear, ciliate at margins. Heads sessile, yellow, 5-7 ray florets and 20 disc florets.

Ecology: Andean II: 2500-3000 m. Fields, disturbed areas.

Distribution: Peru, Bolivia, NW-Argentina.

Arequipa: native. UNI.

Perezia ciliosa Reiche

Description: Small, rosulate herb, up to 5-15 cm tall. Leaves entire, margins distinctly ciliate. Heads on peduncles, solitary or in pairs, corolla blue.

Ecology: Andean III: 3500-4500 m. Grassland and Polylepis-forests.

Distribution: N-Chile and probably also S-Peru, Bolivia.

Arequipa: native, doubtful. ARE, CAY. Species of N-Chile and expected for the extreme south of Peru (Katinas, 2012). Mentioned for RN de Salinas y Aguada Blanca, specimen Delgado et al. 5077 (HUSA) from Chiguata, not seen. All other specimens from Peru are from the E-slopes of the Andes.

Perezia multiflora (Bonpl.) Less.

Description: Erect herb, 5-100 cm tall. Basal leaves rosulate, lanceolate to oblanceolate, with spiny margins, cauline leaves numerous, smaller than basal leaves. Heads 3-30 in dense corymbs, florets 35-50, corolla bluish-white to white.

Ecology: Andean II-III: (1100-)3500->4500 m. Disturbed areas, grasslands, riversides.

Distribution: Ecuador, Colombia, Peru, Bolivia, NW-Argentina, S-Brazil.

Arequipa: native. ARE, CAY, CON, UNI.

Perezia pinnatifida (Bonpl.) Wedd.

Description: Acaulescent herb, with or without rosulate leaves. Leaves variable, oblanceolate, margins entire to deeply lobed, petioles 3-7 cm long, 10-15 pairs of lobes. Flowers solitary, subsessile, 1-5 heads per plant. Corolla variable in color, blue, white, reddish or yellow.

Ecology: Andean II-III: 3000->4500 m. Disturbed areas, in rocks and grassland.

Distribution: Peru, Bolivia, N-Chile and Argentina.

Arequipa: native. ARE, CAY, CON.

Use: Medicinal plant.

Note: Katinas (2012) synonymized P. coerulescens Wedd. and P. pinnatifida, this synonymy is widely accepted, but not by POWO (2024). Here we consider P. coerulescens synonymous with P. pinnatifida.

Perezia pungens (Bonpl.) Less.

Description: Herb, 10-70 cm tall. Basal leaves rosulate, margins ciliate, spiny, cauline leaves similar, sessile to amplexicaule. Heads solitary or 2-8 on a branched stem. Corolla white, pink or blue.

Ecology: Andean III: 3500-4500 m.

Distribution: Peru, Bolivia, Ecuador.

Arequipa: native. ARE, CAY.

Note: Katinas (2012) synonymized P. carduncelloides , P. ciliaris, P. mandonii, P. purpurata and P. sublyrata with P. pungens. This synonymy is accepted by Hassler (2024), but not by POWO (2024). Here they are considered to be conspecific.

Perityle emoryi Torr.

Description: Annual herb, 2 to 60 cm tall, stem small, delicate and simple, or thick, branching, and sprawling.

Usually hairy and glandular. Alternately arranged leaves of various shapes, toothed or divided into lobes, petiolate. Heads hemispherical to bell-shaped, 1 cm wide, ray florets white.

Ecology: Andean I: 500-1000m.

Distribution: S-North America to N-Chile.

Arequipa: native. CAM, CAR, ISL.

Philoglossa peruviana DC.

Description: Small herb. Leaves ovate, margins serrate, 3 prominent longitudinal veins. Heads terminal, ligulate, florets yellow.

Ecology: Coastal, Andean I-II: 0-2500 m.

Distribution: Peru.

Arequipa: native. CAR, ISL.

Picradeniopsis multiflora (Hook. & Arn.) B.G.Baldwin

Description: Plants 3-25 cm tall. Stems ± decumbent to ascending or erect. Leaves 1-3 cm, blades linear or lobed, lobes terete, blade puberulent and gland dotted. Peduncle 5-25 mm. Involucre ± obconic, 5-6 mm. Phyllaries 7-10, green to purple, weakly carinate, oblanceolate to obovate, ± hirsutulous and gland dotted. Ray florets 0. Disc florets 15-30, corolla yellowish, 1-2 mm.

Ecology: Andean II-III: 1000-4000 m. Roadsides, sandy slopes.

Distribution: Disjunct, Madrean and Peru, Boliva, NW-Argentina and N-Chile.

Arequipa: native. ARE, UNI.

Picrosia longifolia D.Don

Description: Herbs erect, green, to 80 cm tall. Stems simple or sparsely branched, glabrous, laxly leafy. Basal leaves oblanceolate, 6-50 × 8-30 cm, base attenuate, apex acute, margins entire or with few teeth, glabrous. Cauline leaves gradually smaller, lanceolate, semi-auriculate, apex attenuate, entire. Heads solitary. Involucre 14-25 × 6-10 mm, with ca. 8 lanceolate, acute, glabrous phyllaries. Flowers white, ligulate, 5-dentate at apex. Pappus yellowish, 10-12 mm long.

Ecology: Coastal, Andean I-II: 0-3000 m. Roadsides and field margins, irrigation channels.

Distribution: Peru, Bolivia, Chile, Argentina.

Arequipa: native. CAY, ISL.

Plazia daphnoides Wedd.

Description: Branched shrubs, glabrous. Leaves oblong or oblong-spatulate, 8-18 x 2-3.5 mm, heads with 5-10 white ray florets and 10-12 purple disc florets. Involucres cylindrical.

Ecology: Andean II-III: 3000-4000 m.

Distribution: S-Peru, Bolivia, Chile, NW-Argentina.

Arequipa: native, expected. No voucher. Expected due to the distribution range (Ayacucho, Puno, Moquegua).

Pluchea chingoyo DC.

Description: Spreading shrubs to 3.5 m tall, stems densely branched, glandular-pubescent. Branches spreading at an angle of >45°. Leaves sessile or petiolate to 1 cm, blade ovate-oval, 1-3 x 0.8-2.5 cm, surfaces glabrous to tomentose, glandular-punctate, base truncate to cordate, apex rounded to obtuse, margins entire to serrate. Heads arranged in densely corymbose-panicles, terminal and axillary, peduncles 1-5 mm long, puberulent.

Ecology: Coastal, Andean I: 0-1000 m. Shrublands.

Distribution: Peru, N-Chile.

Arequipa: native. ARE, CAM, CAR, CAY, ISL.

Polyachyrus annuus I.M.Johnst.

Description: Annual herb. Leaves pinnatifid, green, base amplexicaule. Heads in loose, hemispherical pseudoheads. Florets isomorphic, corolla lavender.

Ecology: Coastal: 0-1000 m. Shrub and cactus vegetation, lomas.

Distribution: S-Peru, N-Chile.

Arequipa: native. CAM, CAR, ISL.

Polyachyrus sphaerocephalus D.Don

Description: Herb 60-70 cm tall. Leaves pinnatisect, lobes dentate, whitish-green, tomentose. Heads in dense, orbicular pseudoheads. Florets isomorphic, fragrant, corolla pinkish-white.

Ecology: Andean II-III: 2500-4000 m. Shrublands.

Distribution: Peru, N-Chile.

Arequipa: native. ARE, CAM, CAY, CON, ISL, UNI.

Porophyllum ruderale (Jacq.) Cass.

Description: Annual herb, slender, 30-50 cm tall. Stems green, hairless, erect, branched. Leaves opposite, glaucous-green, glabrous, petiolate, elliptic to ovate, apex obtuse, blade dotted with embedded oil glands. Heads erect, discoid, florets brown.

Ecology: Coastal, Andean I-II, Amazonian: 0-2000 m. Disturbed areas.

Distribution: Native to South America, introduced to SE-Asia.

Arequipa: native. CAS, UNI.

Use: The leaves are added to salads or cooked to add flavor.

Proustia cuneifolia D.Don

Description: Erect shrubs, up to 2 m tall. Inflorescence axes distally thorny. Leaves oblanceolate, margins shallowly dentate, apex obtuse to cuspidate. Numerous heads in racemes, ray florets pinkish-white, disc florets purple.

Ecology: Coastal, Andean I-III: 0-4000 m. Dry Shrubland, rocky slopes with sparse vegetation.

Distribution: Peru, Bolivia.

Arequipa: native. ARE, CAR.

Pseudelephantopus spiralis (Less.) Cronquist

Description: Perennial, stoloniferous herb, stems pilose to hirsute. Leaves cauline, blades oblanceolate to obovate, acute to obtuse at the apex, attenuate at the base, 3-10 x 1-3 cm, margins crenate, hispid above, punctate and hispid beneath. Inflorescence racemose-spicate, bracteate, clusters of heads subsessile, 5-10 headed. Heads with 4 florets. Corollas 6-7 mm long, whitish to blue-purple.

Ecology: Andean I-II, Amazonian: 0-2500 m. Forests, grasslands.

Distribution: Native to humid tropical America.

Arequipa: introduced. ARE.

Pseudognaphalium cheiranthifolium (Lam.) Hilliard & B.L.Burtt

Description: Perennial herbs, 40-80 cm, stem usually solitary (2 or 3), erect, unbranched or branched at the distal part, uniformly leafy, woolly-glandulous. Stem leaves 30-45(-60) x 3 2-5 mm, lanceolate or linear-lanceolate, margins flat, apex long attenuate, acute, base decurrent 6-13 mm long, basal leaves 40-80 x 3 3-8 mm, oblanceolate- spatulate, apex obtuse, concolorous, adaxial surface glandulous-tomentose. Heads numerous in clusters arranged in corymbs, involucre broadly campanulate, 5 mm in Ø. Corolla yellow.

Ecology: Andean I-III: 1000-4200 m.

Distribution: South America but not in tropical rainforest.

Arequipa: native. ARE, CAY.

Pseudognaphalium dysodes (Spreng.) S.E.Freire, N.Bayón & C.Monti

Description: Annual to perennial herbs, 20-100 cm, stem usually solitary (2 or 3), erect, unbranched or branched at the distal part, uniformly leafy, lanuginose. Leaves 6-9 x3 1-2.5 cm, linear-lanceolate to broadly lanceolate, margins flat, slightly revolute, apex long attenuate, acute, base broadened, subamplexicaule, decurrent on stem, discolorous, above glabrous or glandulous-lanuginose and beneath densely white tomentose. Heads numerous in clusters arranged in corymbs.

Ecology: Andean II-III: 2500-4500 m. Grasslands and paramo.

Distribution: Venezuela to Peru and Bolivia.

Arequipa: native. ARE, CAM, CAR, ISL.

Pseudognaphalium elegans Kartesz

Description: Recognized by its robust habit, strongly discolorous cauline leaves, and large heads with white or rarely yellowish phyllaries. Differs from P. dysodes by its woolly stems and the very discolorous, elliptic to lanceolate, acute leaves.

Ecology: Andean I-II: 1000-3000 m.

Distribution: Mexico to S-Peru.

Arequipa: native. ARE, CAY, CON.

Pseudognaphalium gaudichaudianum (DC.) Anderb.

Description: Similar to P. dysodes, since both have lanuginose stems, discolorous, linear-obong leaves which are long-attenuate at the apex, and involucres 4 mm high. P. gaudichaudianum differs because it has leaves 1-8 mm wide (vs. 10-25 mm wide in P. dysodes).

Ecology: Coastal, Andean I-III: 0-4700 m. Steppes, deserts, shrublands.

Distribution: South America.

Arequipa: native, doubtful. CAR, ISL, UNI. Mentioned by Freire et al. (2018) for Arequipa.

Note: Hassler (2024) and POWO (2024) restrict P. gaudichianum to SE-South America and to Ecuador and Colombia.

Pseudognaphalium lacteum (Meyen & Walp.) Anderb.

Description: Perennial herbs, stems prostrate, branched, densely white- to gray tomentose, leafy to the apex, 5-10(- 20) cm tall. Leaves sessile, blade linear to spatulate, 3-15 x 2-3 mm, rarely decurrent, apex rounded, both surfaces densely white tomentose, the margins entire. Heads corymbose-paniculate in semiglobose clusters of 3-8 heads, terminal and axillary.

Ecology: Andean II-III: 3000-4500 m. Moist areas near high-elevation lakes, bofedales.

Distribution: Peru, Bolivia, NW-Argentina, N-Chile.

Arequipa: native. ARE, CAY, CON.

Pseudognaphalium lanuginosum (Kunth) Anderb.

Description: Perennial herbs, 19-20 cm tall, multi-stemmed, stems erect or ascending, unbranched, more rarely branched at the distal part, whitish-lanuginose. Stem leaves 20-25 x 2-5 mm, linear or linear-oblong, margins flat, apex acute, base clasping and shortly decurrent, basal leaves commonly approximate, linear obovate, apex obtuse, concolorous, whitish-lanuginose on both surfaces. Heads numerous in dense terminal clusters, sometimes arranged in corymbs. Corolla yellow.

Ecology: Andean II-III: 2500-4500 m. Polylepis forests, rocky places.

Distribution: Andes, from Venezuela to NW-Argentina.

Arequipa: native. ARE, CAR.

Pseudognaphalium viravira (Molina) Anderb.

Description: Prostrate to decumbent, perennial herb. Stems thickened lignescent, much-branched at the base, 5-10 cm long. Leaves sessile or slightly decurrent, blade linear to spatulate, 5-15 x 2-2.5 mm, apex rounded to acute, both surfaces densely tomentose. Inflorescence corymbose in globose clusters, terminal. Heads 4 x 4 mm. Involucral bracts 3-5-seriate. Pistillate florets ca. 100, bisexual florets 8-12. Achenes 0.5-0.7 mm long, pappus bristles ca. 3 mm long.

Ecology: Coastal, Andean I-III: 0-4900 m. On dry, sandy soils or in forests in shady places.

Distribution: Andes from Peru to Chile and Argentina.

Arequipa: native. ARE.

Rockhausenia apiculata (Sch.Bip.) D.J.N.Hind

Description: Similar to W. pygmaea, but leaves coriaceous, flat in cross section, shiny, conspicuously veined.

Ecology: Andean II-III: 3000-4500 m.

Distribution: Peru, Bolivia, NW-Argentina.

Arequipa: native. ARE, CAY.

Note: Funk (1997) included R. apiculata as a synonym of R. pygmaea. Hind (2022) disagrees, as the two species have different ecological preferences and can be separated morphologically.

Rockhausenia aretioides (Wedd.) D.J.N.Hind

Description: Rhizomatous herb, rosulate, forming mats, 1-2 cm tall. Leaves spatulate, 3-8 x 2-3 mm, folded in cross section, short fringed at margins (mainly near the base), rather leathery with false petioles and sheathing bases Heads radiate, solitary, sessile to subsessile. Involucral bracts 8-9, ray florets 5-10, white, disc florets 7-22, whitish and purple tipped.

Ecology: Andean III: 4000->4500 m. Dry moorland in the shelter of bunchgrass.

Distribution: Peru, Bolivia, NW-Argentina, N-Chile.

Arequipa: native. ARE, CON. Not mentioned for Arequipa by Calvo et al. (2020), convincing observation from Condesuyos in iNaturalist (2024).

Rockhausenia microphylla (H.Beltrán & S.Leiva) D.J.N.Hind

Description: Small rhizomatous herb, forming lax patches. Leaves slightly fleshy, 2-5 mm long, elliptic to widely oblong, glabrous, apex mucronate. Heads radiate, solitary, on short pedicel. Involucral bracts 8-13, ray florets 8(- 10), white, disc florets 12-23, purplish-white.

Ecology: Andean III: 4000-4900 m. Bogs.

Distribution: Peru to Bolivia.

Arequipa: native. CAY. No specimen but a convincing observation from Caylloma in iNaturalist (2024).

Rockhausenia nubigena (Kunth) D.J.N.Hind

Description: Rhizomatous herb, rhizomes often vertical, sometimes slowly creeping, stout. Plant 3-4 or up to 35 cm tall, smoothly glaucous and glabrous. Leaves overlapping, sessile, 4-30 x 0.3-3 cm, broadly linear, apex obtuse, in rosettes. Heads radiate, sessile or on a peduncle up to 29 cm long, Involucre 0.5-3 cm wide, involucral bracts 12- 27. Ray florets 16-21, white, disc florets 50-400, yellow.

Ecology: Andean II-III: 2500-4500 m. Dwarf moorland, grasslands.

Distribution: Central America to Bolivia.

Arequipa: native. CAY.

Note: The species shows great morphological variability and may be a complex of species, judging by the synonyms. A phylogenetic study is needed (Beltrán, 2017).

Rockhausenia orbignyana (Wedd.) D.J.N.Hind

Description: Rhizomatous herb, rosulate, forming clumps, rarely solitary, 2-10 cm tall. Leaves linear-lanceolate, 2-9 cm long, 1-12 mm wide, folded, light green, margins denticulate, apex minutely tridentate to 3-7 dentate, rarely subentire. Heads radiate, solitary, terminal, sessile to pedunculate up to 7 cm. Involucre 7-13 mm wide, involucral bracts 11-13. Ray florets 11-13, white, usually bluish or purplish beneath. Disc florets 32-61, yellowish to creamy.

Ecology: Andean II-III: 2500-4000 m.

Distribution: Peru, Bolivia.

Arequipa: native. ARE, CAS, CAY, CON.

Rockhausenia pectinata (Lingelsh.) D.J.N.Hind

Description: Unmistakable for having spatulate leaves with long fringed margins (fringes up to 1 mm long), radiate capitula, ray flowers white, of the same size as the phyllaries and purplish to purple disc flowers.

Ecology: Andean III: 4000-4800 m. Grasslands, often with Pycnophyllum molle.

Distribution: Peru, Bolivia.

Arequipa: native. CAS, CAY.

Rockhausenia pinnatifida (J.Rémy) D.J.N.Hind

Description: Rhizomatous herb, rosulate, forming lax clumps or solitary plants, 2-5 cm tall. Rhizome very short, erect. Leaves oblong, 1-7 x 0.5-1.5 cm, 1-2-pinnatisect into 10-15 pairs of very unequal. Heads discoid, solitary, short-pedunculate. Involucre 4-10 mm wide, involucral bracts 13-19. Disc florets 40-60, purplish.

Ecology: Andean III: 4000->4500 m. Wet places and bofedales.

Distribution: Peru, Bolivia, N-Chile and NW-Argentina.

Arequipa: native. ARE, CAS, CAY.

Rockhausenia pygmaea (Gillies ex Hook. & Arn.) D.J.N.Hind

Description: Low rhizomatous herb, rosulate, forming lax clumps or solitary, plant 2-4 cm tall. Leaves narrowly linear, apex acute to obtuse, elliptical to terete in cross section, 1-3 x 0.1 cm, ± woolly only at the base, 0-1-veined. Heads radiate, solitary, sessile to pedunculate. Involucre 6-10 mm wide, involucral bracts 11-13, often contrasting violet tinged. Ray florets 8-15, white, disc florets 22-55, creamy to yellowish.

Ecology: Andean III: 3500->4500 m. Wet places, bofedales, tolerates salinity.

Distribution: Colombia, Ecuador, Peru, Bolivia, Chile and Argentina.

Arequipa: native. ARE, CAS, CAY.

Rockhausenia solivifolia (Sch.Bip.) D.J.N.Hind

Description: Rhizomatous herb, rosulate, forming lax clumps, 2-3 cm tall. Leaves pseudopetiolate, blade oblong, 4.5- 13 x 2.5-9 mm, 1-pinnatisect with 3-4 pairs of lobes. Heads discoid, solitary, terminal, sessile to pedunculate. Involucre 4-5 mm wide, bracts 8-11. Disc florets 16-28, whitish, dark purple in the upper third.

Ecology: Andean III: 4000-4500 m. Bogs.

Distribution: Peru, Bolivia, N-Chile.

Arequipa: native. CAY.

Rockhausenia spathulata (Wedd.) D.J.N.Hind

Description: Rhizomatous herb, rather scapiform, forming lax clumps, 2-5 cm tall. Leaves pseudopetiolate, blade glabrous, spatulate, 5-8 x 2-4 mm. Heads radiate, solitary, on a short peduncle. Involucre 3-6 mm wide, involucral bracts 8-12. Ray florets 8-12, white, usually bluish or purple beneath, disc florets 11-26, whitish, usually purple tipped.

Ecology: Andean III: 4000-4500. Bogs, wetlands.

Distribution: Peru, Bolivia, N-Chile.

Arequipa: native. CAY.

Schkuhria pinnata (Lam.) Kuntze ex Thell.

Description: Annual herb 5-50 cm high, much-branched, the branches glandular and pilose. Leaves alternate, pinnatisect or the upper simple and filiform, 1-6 x 0.1-2 cm, with 3-7 lobes, lobes filiform or divided, pilose or glabrous, glandular. Phyllaries 4-6, glabrous, gland dotted. Ray florets 0-2, disc florets 2-6.

Ecology: Andean I-II: 1000-3500 m. Disturbed areas, grasslands, rocky slopes, shrublands.

Distribution: Tropical and subtropical America.

Arequipa: native, doubtful. ARE. No specimen, unverified but convincing observation in iNaturalist (2024) from Sta. Rita de Siguas.

Senecio acarinus Cabrera

Description: Suffrutescent shrub, up to 25 cm tall. Leaves linear, ± sessile, blade laminar, 10 cm x 2-5 mm, margins with a few solitary teeth, apex acuminate, base attenuate. Inflorescence in terminal clusters of 10-20 heads. Heads discoid, corollas yellow.

Ecology: Coastal: 0-500 m. Lomas.

Distribution: Peru.

Arequipa: native. ISL.

Senecio adenophyllus Meyen & Walp.

Description: Small, much-branched shrublet, old branches covered with dry leaves. Leaves 10-20 x 3-8 mm, densely glandulose-pubescent on both sides, attenuate at base, margins dentate or lobulate. Heads discoid, in cymes of 2- 4, peduncle short. Involucral bracts 10-14. Florets numerous, tubular, yellow.

Ecology: Andean III: 3000->4500 m. Shrublands, Polylepis-forests.

Distribution: Peru, Bolivia, N-Chile, NW-Argentina.

Arequipa: native. ARE, CAS, CAY, UNI.

Senecio algens Wedd.

Description: Perennial, cespitose herb, much-branched, 4-6 cm tall. Leaves spatulate, obtuse, attenuate to pseudopetiole, margins entire, glabrous, 1-3.5 cm long, 2-5 mm wide. Heads discoid, solitary, terminal, peduncle short. Involucral bracts 15, florets numerous, tubular, yellow.

Ecology: Andean III: 4500-5000 m. Among rocks.

Distribution: Peru, Bolivia, NW-Argentina and N-Chile.

Arequipa: native. CAY.

Senecio anastasioi Montesinos

Description: Perennial subshrub, spreading or mat forming, rhizomatous, creeping, low-growing, with dense fibrous roots, forming dense mats 4-10 cm tall and up to 1.8 m in Ø. Cauline leaves oblong-spathulate, asymmetrically along stems with about 10-150 leaves each. Blade thick, lustrous, 7-30 x 1.5-3.0 mm, plain or rarely involute, recurved towards the tip, marginally with scarce trichomes. heads solitary, discoid, peduncle 2-4 mm long. Involucral bracts 12- 14, disc florets 35-40, bright yellow.

Ecology: Andean III: 4700-5000 m.

Distribution: S-Peru.

Arequipa: native. CON. No specimen, convincing observation from Caylloma.

Senecio beltranii P.Gonzáles & Montesinos

Description: Perennial subshrub, cespitose, creeping, rhizomatous, with fibrous roots, forming mats 2-4 cm tall and 5-30 cm Ø. Stems woody, often covered with persistent remains of leaves towards the base. Cauline leaves spathulate-oblanceolate, alternate, 2-3 long, thick, lustrous, involute, pinnatilobate, glabrous, greenish-gray. Peduncle 5-12 mm long, involucral bracts 18-22. Heads solitary, discoid. Florets 44-54, pale yellow.

Ecology: Andean III: 4700-5470 m. Highland and subnival puna.

Distribution: S-Peru.

Arequipa: native. CAY.

Note: Probably a local biotype of S. algens.

Senecio breviscapus DC.

Description: Acaulescent perennial herb. Leaves in basal rosettes, oblanceolate, glabrous, margins dentate to lobed or pinnatisect. Heads sessile, radiate, 1.5-2.5 cm in Ø. Involucral bracts 15-24. Ray florets 12-15, disc florets numerous, both yellow.

Ecology: Andean III: 3500->4500 m. Moist places.

Distribution: Peru, Bolivia, N- and C-Chile, NW-Argentina.

Arequipa: native. ARE. The type of of S. wernerioides, synonym of S. breviscapus, is from the Cordilleras de Arequipa (Cabrera, 1985).

Senecio calcicola Meyen & Walp.

Description: Erect subshrub, up to 0.75 m tall. Stem leafy up to the top. Leaves linear, up to 6 cm long, sessile, ± hastate, midvein prominent, margins entire, revolute. Heads radiate, terminal in clusters of 1-4. Ray florets up to 2 cm long, disc florets numerous, both yellow.

Ecology: Andean I: 500-1000 m. Lomas.

Distribution: Peru.

Arequipa: native. ISL.

Senecio chachaniensis Cuatrec.

Description: Shrub, 1 m tall, glutinous-tomentose to glabrous. Leaves linear, up to 2 cm long, crowded in leaf axis, margins entire or above the middle with 1-2 pairs of teeth, often with blackish point at apex. Heads discoid solitary or in pairs. Involucral bracts ca. 10. Florets numerous, yellow.

Ecology: Andean III: 3500-4500 m. Rocky slopes, open places.

Distribution: S-Peru.

Arequipa: native. ARE.

Senecio clivicola Wedd.

Description: Shrub up to 1 m tall, much-branched, glabrous, densely leafy. Branches brownish, angulate. Leaves linear-lanceolate, sessile, 2-7 cm x 1-10 mm, margins entire, sometimes slightly serrate apically. Inflorescence many-flowered. Heads radiate, 5-6 mm in Ø, pedicel 5-25 mm long. Involucral bracts ca. 13. Ray florets 8-12, 5-6 mm long, disc florets numerous, both yellow.

Ecology: Andean III: 3500-4000 m.

Distribution: S-Peru, Bolivia, NW-Argentina.

Arequipa: native, doubtful. ARE, CAY. Specimens Vargas 8244 (CUZ) from Chala and Vargas 18101 (CUZ) from Mollendo, both in lomas, do not match in ecology. Specimens Montesinos 4553, 5170 (both B) from Chivay and Vargas 18284 (CUZ) from Solitario have few flowers, discoid heads, or deeply dentate leaves (probably confusion).

Senecio comosus Sch.Bip.

Description: Rhizomatous perennial herb, 40-60 cm tall. Stem whitish-tomentose. Basal leaves tufted, linear-lanceolate, 20-35 cm long, 0.5-1.5 cm wide, acute, glossy, green and glabrous above, densely whitish-tomentose beneath. Cauline leaves smaller. Heads radiate or discoid, in terminal many flowered corymbose cymes. Involucral bracts 18-22. Ray florets, if present, 12-15 mm long, disc florets numerous, both yellow.

Ecology: Andean II-III: 2500-4500 m. Rocky places, in stony moraine debris with regular precipitation or run-off.

Distribution: Colombia, Ecuador, Peru, Bolivia.

Arequipa: native. ARE, CAY, UNI.

Note: In Arequipa, S. comosus var. culcitioides. Based on morphological characteristics, this species should be placed in Culcitium.

Senecio crassilodix Cuatrec.

Description: Much-branched shrub, all parts densely whitish-tomentose. Stem densely leafy. Leaves linear, sessile, with prominent midrib, margins entire, revolute. Heads discoid, solitary or in pairs, terminal. Florets many, yellow.

Ecology: Andean III: 4000-4500 m.

Distribution: Peru.

Arequipa: native. ARE, CAS.

Senecio ctenophyllus Phil.

Description: Shrub, up to 80 cm tall, open branched. Leaves oblong-lanceolate, pinnatisect, 1 4 cm long, 1-2 mm wide, glabrous, sessile, margins with 2-5 pairs of segments perpendicular to the rachis. Heads radiate, in lax cymes, peduncle long. Involucral bracts 8-13. Ray florets 5-8, disc florets 12-18, tubular, yellow.

Ecology: Andean II-III: 3000-3600(-4000) m. Shrubland.

Distribution: S-Peru, N-Chile.

Arequipa: native, doubtful. ARE. No specimen but a convincing observation from Chapi. First mention for Peru in Tacna by Morales-Fierro & Beltrán (2021).

Note: Often confused with S. yurensis.

Senecio evacoides Sch.Bip.

Description: Subshrub forming white-tomentose mats or cushions, up to 2 cm tall. Leaves crowded, almost imbricate, entire, 1-2 cm x 3-6 mm, obovate-spatulate, apex obtuse, base attenuate in a long pseudopetiole.

Heads almost sessile, solitary, discoid, 7-8 x 5-6 mm. Involucral bracts 13-20, disc florets numerous, yellow.

Ecology: Andean III: 4000->4500 m.

Distribution: Peru, Bolivia, NW-Argentina.

Arequipa: native. ARE, CAY, CON, UNI.

Senecio gamolepis Cabrera

Description: Subshrub forming sense cespitose mats, measuring a few m2. Leaves succulent, linear-lanceolate, in rosettes, glabrous, apex acute. Heads discoid, pedicel 0.2-2 mm long, florets yellow. Involucral bracts 13.

Ecology: Andean III: 4000-4500 m. Grassland, moist places near glaciers.

Distribution: Peru.

Arequipa: native, expected. No voucher, expected due to the range of the species (Huancavelica, Cuzco, Puno, Moquegua).

Senecio hastatifolius Cabrera

Description: Branched shrub, semi-decumbent, densely covered with glandular multicellular hairs, densely leafy up to the inflorescence. Leaves herbaceous, sessile, oblanceolate, nearly obtuse at apex and attenuate below, the upper ones hastate and semi-auriculate, minutely toothed at margin, glandular-pubescent on both sides, 6-11 x by 1-2.5 cm. Heads discoid, in terminal, lax corymbiform cymes. Pedicels slender, densely hirsute-glandular, 1-4 cm long. florets numerous, yellow, tubular.

Ecology: Andean III: 3500-5000 m.

Distribution: Peru, Bolivia.

Arequipa: native. ARE. No specimen but several convincing observations.

Senecio humillimus Sch.Bip.

Description: Neatly mat-forming perennial. Leaves terete, fleshy, linear-spatulate, apex acuminate to obtuse, 3-10 x 1 mm, green. Heads solitary, discoid, sessile, 5-6 x 3-4 mm. Florets 15-18, yellow, involucral bracts 8.

Ecology: Andean III: 3500->4500 m. Humid areas.

Distribution: Peru, Bolivia, N-Chile, NW-Argentina.

Arequipa: native. ARE, CAY.

Senecio jarae Phil.

Description: Rhizomatous herb, with woody rhizome, 6-20 cm tall. Leaves 4-12 cm long, glabrous, lanceolate, deeply pinnatisect, 5-12 pairs of segments. Heads discoid, solitary, pedunculate, 10-15 x 15-30 mm. Involucral bracts 16- 25, disc florets yellow.

Ecology: Andean III: 4000-4500 m.

Distribution: Peru, Bolivia, N-Chile, NW-Argentina.

Arequipa: native. ARE.

Senecio melanandrus (Wedd.) J.Calvo, A.Granda & V.A.Funk

Description: Cespitose perennial herb, forming tuft mats. Leaves 4-15 mm long, 1.2-2.6 mm wide, linear-oblong to spatulate, apex acute to obtuse, base narrowed, margins entire, somewhat fleshy, greenish or glaucous. Capitulum discoid, solitary, terminal, sessile or subsessile, disc florets whitish.

Ecology: Andean III: 3800-5100 m. Exposed places on bare soils.

Distribution: Peru, Bolivia.

Arequipa: native. CAY.

Senecio moqueguensis Montesinos

Description: Cespitose perennial herb, forming mats, 2-4 cm tall and up to 60 cm in Ø. Leaves 8-12 mm long, 1-2.5 mm wide, oblong-spatulate, apex subpinnatifid, base truncate, ± fleshy, sparsely covered with trichomes, young leaves dark green with yellowish margins, turning light green-greyish with age. Heads solitary, discoid, sessile or subsessile. Involucral bracts 9-12, florets 24-48, yellow.

Ecology: Andean III: >4500 m. Rocky soils.

Distribution: Peru, Bolivia.

Arequipa: native. CAY.

Senecio nutans Sch.Bip.

Description: Aromatic shrub, 30-50 cm tall, resinous, fragrant, glabrous, densely branched. Leaves oblong-lanceolate, succulent, pale green, apex with 1-4 pairs of teeth, margins entire, revolute. Heads discoid, nodding. Involucral bracts 6-12. Florets 12-30, yellow, orange or whitish-pink, tubular.

Ecology: Andean II-III: 3000->4500 m.

Distribution: Peru, Bolivia, N-Chile and N-Argentina.

Arequipa: native. ARE, CAY, CON, UNI.

Use: The infusion of leaves and flowers, in combination with coca leaves, is a remedy for altitude sickness (soroche). The infusion is an analgesic for headaches. One component has been found to be a potent antihypertensive agent.

Senecio pentlandianus DC.

Description: Perennial herb, 50-60 cm tall, stems erect, leavy up to the inflorescence. Leaves linear-oblong, 2.5-6 cm x 2-8 mm, laxly glandulous-pubescent to glabrous, sessile, margins entire and often slightly dentate, revolute (rosemary like). Heads discoid, in cymes of 3-6, peduncle up to 3 cm long, glandular. Involucral bracts 12-14. Florets 20-40, tubular, yellow. Pappus white.

Ecology: Andean III: 3000-4500 m.

Distribution: Peru, Bolivia.

Arequipa: native. Not specified by province. Cabrera (1985) mentions two specimens for Arequipa (not seen).

Senecio pinnatilobatus Sch.Bip.

Description: Perennial herb, forming hemispherical mats 20-40 cm tall, stems densely leavy, glabrous. Leaves spatulate, 2-3 cm long, glabrous, long petiolate, deeply lobbed or dentate with 2-4 teeth on each side. Heads discoid, solitary, terminal. Involucral bracts 12-15. Florets tubular, yellow. Pappus white.

Ecology: Andean III: 4000-4500 m.

Distribution: Peru, Bolivia.

Arequipa: native, doubtful. ARE. No voucher. The species is endemic to the Titicaca region (Cabrera, 1985). Mentioned for RN Salinas y Aguada Blanca by Quipuscoa & Huamantupa (2010). The occurrence in Arequipa is likely, an observation from Moquegua near the border with Arequipa.

Senecio reicheanus Cabrera

Description: Shrublet, 30-150 cm tall, stems densely leavy, glandular. Leaves linear, sessile, 1-5 cm long, 1 mm wide, glandular, margins with 1-4 pairs of teeth on each side or entire. Heads discoid, in few-flowered cymes, pedicel 5-30 mm long. Involucral bracts 13. Florets numerous, tubular, yellow. Pappus white.

Ecology: Andean II-III: 2500->4500 m.

Distribution: Peru, Bolivia, N-Chile.

Arequipa: native. ARE, CAY, CON, ISL, UNI.

Note: Calvo & Saldivia (2022) synonymize S. phylloleptus (S-Peru) with S. reicheanus (N-Chile, Bolivia). This synonymy is not widely accepted, but we accept it here, as does POWO (2024).

Senecio rhizomatus Rusby

Description: Perennial herb, stems simple, ascendent to erect, 20-40 cm tall, glandular-pubescent. Leaves ovate, herbaceous, up to 10 x 1.5 cm, cordate to amplexicaul at base, margins dentate, glandular to glabrous. Heads discoid, long pedunculate, clustered by 3-5 in a terminal inflorescence. Involucre 2-3 cm wide, bracts 18-25. Florets numerous, tubular, yellow.

Ecology: Andean III: 3500-4500 m. At the base of rocks.

Distribution: Peru, Bolivia.

Arequipa: native. ARE, CAY, CON, UNI.

Senecio rudbeckiifolius Meyen & Walp.

Description: Suffrutescent herb or shrub, 50-180 cm tall, stems angular, glabrous. Leaves pinnatisect, 4-13 cm long, 2-5 cm wide. Segments 2-3 pairs, linear-lanceolate, margins entire or finely serrate, glabrous. Heads numerous, radiate, in cymose-corymbose inflorescences. Involucral bracts 16-20, glabrous. Florets bright yellow, ray florets 9- 10 mm long, disc florets tubular.

Ecology: Andean I-II: 500-4500 m. Roadside, agricultural fields.

Distribution: Peru, Bolivia, NW-Argentina.

Arequipa: native. ARE, CAY.

Senecio rufescens DC.

Description: Densely branched small shrub, 40-60 cm tall. Leaves linear-lanceolate, 1.2-3 cm x 1-5 mm, densely glandulous-pubescent, sessile, margins entire, sometimes subdentate, revolute. Heads discoid, solitary or in cymes of 2-4, peduncle 5-20 mm long. Involucral bracts 10-16. Florets numerous, tubular, yellow.

Ecology: Andean III: 3500-5300 m. High puna areas.

Distribution: Peru, Bolivia, NW-Argentina.

Arequipa: native. ARE, CAY, CON.

Senecio scorzonerifolius Meyen & Walp.

Description: Perennial herb, 4-8 cm tall, stems densely leafy at base, leafless above. Leaves linear, 2.5-8 cm long, 0.7-2 mm wide, sessile, margins entire. Heads discoid, solitary terminal. Involucral bracts 13-18. Florets numerous, tubular, white.

Ecology: Andean III: 3500->4500 m.

Distribution: Peru, Bolivia, N-Chile, NW-Argentina.

Arequipa: native. ARE. No specimen but a convincing observation in iNaturalist (2024). Arequipa is part of the species' range.

Senecio smithianus Cabrera

Description: Shrub of the lomas. Leaves linear-lanceolate to ovate, 2-3 cm long, ± succulent, glaucous, amplexicaule. Heads radiate, in cymes of 1-4, long pedunculate. Ray and disc florets yellow.

Ecology: Coastal: 0-1000 m. Lomas.

Distribution: Peru.

Arequipa: native. CAM, CAR.

Senecio spinosus DC.

Description: Spiny shrub, 10-25 cm, intricately branched, spiny, resembling Adesmia. Leaves linear, 5-8 x 1-1.5 mm, glabrous. Heads discoid, sessile, solitary, axillary. Involucral bracts 8-14, florets 15-20 deep yellow to orange-yellow.

Ecology: Andean II-III: 3000-4500 m. Rocky slopes, grasslands.

Distribution: Peru, Bolivia.

Arequipa: native. ARE, CAY, CON.

Senecio trifurcifolius Hieron.

Description: Subshrub, densely branched from the base, glabrous. Leaves very dense, fleshy, trifurcate, linear- cuneiform, 10-12 mm long, apex acute, spiny. Heads discoid, solitary, terminal. Involucral bracts 8, florets 10-12, yellow.

Ecology: Andean III: 4000->4500 m.

Distribution: Peru, N-Chile, expected in Bolivia.

Arequipa: native. ARE. No voucher. Mentioned by Brako & Zarucchi (1993) and for RN Salinas y Aguada Blanca by Quipuscoa & Huamantupa (2010).

Senecio vegetus (Wedd.) Cabrera

Description: Mat-forming dwarf shrub, 1-2 cm tall. Leaves lineal-lanceolate, slightly succulent, margins entire or with 1-3 pairs of teeth, ± lanuginose or glabrous. Heads discoid, solitary, sessile, terminal. Involucral bracts 13, florets numerous, yellow.

Ecology: Andean III: 3500->4500 m.

Distribution: Peru, Bolivia.

Arequipa: native. ARE, UNI. No specimen, several observations.

Senecio violifolius Cabrera

Description: Perennial herb, glabrous. Leaves long petiolate, petiole 5-13 cm long. Blade elliptic to suborbicular, 3-5 x 2.5-3 cm, margins dentate, attenuate at base, obtuse to acute at apex. Stem magenta. Heads discoid, terminal, in cluster of 3-4. Involucral bracts ±20, florets numerous, yellow.

Ecology: Andean III: >4500 m. Stony tracts.

Distribution: Peru, Bolivia.

Arequipa: native. CON, UNI. No specimen, several mentions and observations.

Senecio vulgaris L.

Description: Erect annual herb up to 40 cm tall, growing from a thick taproot. Stems covered in whitish felt-like hairs when young. Leaves alternate, subsessile, pinnately lobed, up to 6 cm long, margins dentate. Heads discoid, in terminal corymbiform cymes. Involucral bracts 20, apex blackish, florets yellow.

Ecology: Andean I-III: 500-4000 m.

Distribution: Native to Europe, introduced worldwide.

Arequipa: introduced. ARE.

Senecio yurensis Rusby

Description: Erect annual herb, much-branched, glaucous, glabrous. Leaves sessile, linear-lanceolate, base hastate. Heads radiate, in terminal cymes of 1 to several heads. Involucral bracts ca. 15, ray florets 4-8, disc florets ca. 20, both yellow.

Ecology: Andean II: 2500-3500 m.

Distribution: Peru, Bolivia.

Arequipa: native. ARE, CAY, UNI.

Sigesbeckia jorullensis Kunth

Description: Erect herbs, stem reddish-green, hirsute and glandular above. Upper leaves lanceolate, sessile, basal leaves deltoid with alate petiole, decurrent. Margins shallowly serrate apex acute at apex. Heads 8 x 4

mm, solitary, terminal, on slender, densely hairy peduncles. Bracts biseriate, outer bracts spatulate, 2 cm long, glandular. Corollas yellow.

Ecology: Andean II-III: 2500-4000 m. Disturbed areas, rocky slopes.

Distribution: Central and South America.

Arequipa: native. ARE, CAR.

Silybum marianum (L.) Gaertn.

Description: Annual or biennial herb to 1.2 m tall. Stems erect, unwinged, grooved, slightly cottony, simple, or branched above. Leaves oblong, glabrous, with strongly spiny margins, pale, green with distinctive white along the veins above. Stem leaves amplexicaul with spiny-ciliate auricles. Heads 4-5 × 1-2 cm, solitary, erect, or nodding. Phyllaries bearing a stout long yellowish spine. Flowers reddish-purple. Achenes blackish, transversely wrinkled with a yellow ring near the apex.

Ecology: Andean II-III: 3000-4000 m. Disturbed areas.

Distribution: Mediterranean to C-Asia and Ethiopia, introduced worldwide.

Arequipa: introduced. CAR.

Soliva mexicana DC.

Description: Small herb, with prostrate branching stems and short rhizomes. Stems up to 5 cm long, rooting at nodes. Leaves pinnatisect, lobes entire. Heads on short peduncle, discoid, yellowish-green, marginal florets pistillate and white, florets in center staminate. Achenes alate, with stylar spine.

Ecology: Andean II-III: 3000->4500 m. Grassland.

Distribution: Native to the Americas.

Arequipa: native. ARE, CAY. No specimen, some observations in iNaturalist (2024) and mentioned for RN Salinas y Aguada Blanca by Quipuscoa & Huamantupa (2010).

Soliva stolonifera (Brot.) Loudon

Description: Annual herbs, unbranched or with prostrate branches rooting at nodes. Leaves 1(-2)-pinnatisect, 1-4 cm long, sparsely hairy. Heads 6-8 mm Ø in fruit. Fruit triangular, 1.5-2 mm long, brown, cottony at the apex. Achenes with spinescent stylar apices 1-1.5 mm long.

Ecology: Andean I-II: 500-3500 m. Disturbed areas.

Distribution: Native to C-Chile and C-Argentina. Introduced to Bolivia and Peru, and North America, Europe and Australia.

Arequipa: introduced, doubtful. ISL. No specimen, mentioned for Lomas de Yuta by Quipuscoa et al. (2016). The seeds cling to fabric and shoes and are transported long distances, therefore, an adventive occurrence in Arequipa is likely.

Sonchus asper (L.) Hill

Description: Annual and sometimes biennial herb, 0.2-2 m tall, containing white latex in all parts. Yellow flowers, <2 cm in Ø. Leaves variable, spiny, base auriculate, auricles round, spirally adpressed (difference to S. arvensis). Achenes compressed, winged, smooth when fully mature.

Ecology: Andean II-III: 1500-3500 m. Field margins, ruderal areas.

Distribution: Native to Europe and the Mediterranean region, introduced worldwide.

Arequipa: introduced, doubtful. ARE. Probably introduced. Several not curated observations in iNaturalist (2024), no specimen.

Use: Edible as a vegetable, used in folk medicine.

Sonchus oleraceus L.

Description: Annual and sometimes biennial herb, 40-150 cm tall, containing white latex. Leaves softly dentate, base auriculate, auricles not spirally adpressed. Yellow flowers, <2.5 cm in Ø. Fruit compressed, not winged, hairless, distinctly rugulose when fully mature, sharply acute, downward-pointing auricles of leaves and bracts.

Ecology: Coastal, Andean I-II: 0-3000 m. Open and disturbed areas, waste places.

Distribution: Native to Europe, N-Africa and W-Asia, introduced worldwide.

Arequipa: introduced. ARE, CAM, CAR, CAY, CON, ISL, UNI.

Use: Used as a famine food and in folk medicine.

Spilanthes leiocarpa DC.

Description: Spreading herbs, rooting at the nodes. Leaves linear-lanceolate, 2-4 x 0.5 cm. Heads terminal, solitary, discoid. Florets pinkish-white.

Ecology: Coastal, Andean I: 0-1000 m. Beach, disturbed areas, riversides.

Distribution: Ecuador, Peru, N-Chile.

Arequipa: native. CAM, ISL.

Stevia cuzcoensis Hieron.

Description: Herbaceous or suffruticose, shortly incurved-puberulent, branched from the base. Leaves rhombic- ovate to lanceolate, petiolate, mostly less than half as wide as long, apex acute, serrate except at cuneate base. Inflorescence rather dense.

Ecology: Andean II-III: 3000-4000 m. Grasslands, riversides, rocky slopes, shrublands.

Distribution: Peru.

Arequipa: native, doubtful. ARE, CAS, CAY, UNI. Specimen Solomon 2862 (MO) from 1977, no modern mention or observation.

Note: Gutierrez et al. (2016) show that Stevia species have a wide morphological variation, and they recognize only one species for Chile (Stevia philippiana Hieron.). It is likely that the number of Peruvian Stevia species is also overestimated.

Stevia herrerae B.L.Rob.

Description: Perennial herb, 0.5-0.8 m tall. Stem and peduncles on upper half with stipitate glands, lower part with simple trichomes. Leaves subsessile to sessile, blades glandular. Inflorescence laxly corymbiform, peduncle 4-8 mm long.

Ecology: Andean II: 1500-3000 m. Rocky slopes, shrublands.

Distribution: Peru.

Arequipa: native. UNI. Not mentioned for Arequipa by Brako & Zarucchi (1993). Bedoya-Cuno et al. (2024) mention it for Cotahuasi.

Stevia hoppii B.L.Rob.

Description: Ascending to erect perennial herb, 0.3 m tall. Stem scabrous, not hairy. Leaves lanceolate-ovate, subsessile, apex acuminate, margins shallowly crenate-serrate. Flowers pinkish-white.

Ecology: Andean II: 2500-3000 m.

Distribution: Peru.

Arequipa: native. ARE, CAR, CAY, CON, UNI.

Stevia melissifolia Sch.Bip.

Description: Erect shrub. Stem hirsute. Leaves obovate-orbiculate, amplexicaul by a broadly petioliform base, margins serrate-crenate. Inflorescence in dense clusters of up to 20 heads.

Ecology: Coastal, Andean I: 200-1500 m. Lomas.

Distribution: Peru.

Arequipa: native. CAR, ISL.

Stevia weberbaueri B.L.Rob.

Description: Procumbent or ascending subshrub, 40-80 cm tall, densely glandular. Inferior leaves opposite, pedicel 0.4-0.6 mm. Upper leaves sessile, lanceolate, acute to obtuse, serrate, base entire, cuneate. Inflorescence corymbiform, dense, heads sessile. Corolla pinkish-purple.

Ecology: Andean II: 3000-3500 m. Rocky slopes.

Distribution: Peru.

Arequipa: native. CAR, CAS, CAY, CON, UNI. Not mentioned for Arequipa by Brako & Zarucchi (1993). Bedoya-Cuno et al. (2024) mention this species based on herbarium analysis and collections.

Symphyotrichum squamatum (Spreng.) G.L.Nesom

Description: Annual or short-lived perennial herb, glabrous or rarely with a few hairs on upper stem, stems erect, terete, sparingly branched below, 0.2-2 m tall.

Lowermost leaves mostly petiolate, elliptic, cuneate, obtuse. Uppermost leaves linear. Inflorescence a many-headed diffuse panicle. Heads 2-5 mm Ø. Involucral bracts unequal, inner bracts narrow-oblong to subulate, acute, green toward apex and along midrib, purplish at apex and margins, 5- 8 mm long, outer bracts not wholly herbaceous, 2 mm long. Ray florets numerous, ligules white to pale purple, 1-2 mm long.

Ecology: Coastal: 0-500 m. Riverbanks, disturbed places.

Distribution: Tropical South America.

Arequipa: native. ISL.

Tagetes dombeyi Schiavinato, D.G.Gut. & Adr.Bartoli

Description: Small annual herb, 10-20 cm tall. Leaves pinnate. Leaf segments spatulate, dentate, 1-2 mm wide. Heads in loose clusters, peduncle usually >10 mm. Flowers yellow, 3-5 ray florets and 6-10 disc florets. Involucre entirely purplish. Pappus scales oblanceolate, widest part in the upper 2/5, margins plumose, lateral projections longer than the scale base.

Ecology: Andean II-III: 1900-3900 m. Hillsides, Rocky soils, roadsides.

Distribution: Peru, Bolivia.

Arequipa: native. CAY.

Tagetes filifolia Lag.

Description: Erect annual, much-branched, 8-50 cm tall, with anise-like odor. Leaves pinnate, pinnae linear, <1 mm wide, margins entire. Heads discoid, florets yellow. Involucres covered with punctiform glands, apex of phyllaries truncated, bearing a small tooth.

Ecology: Andean II: 2000-2500 m. Disturbed areas, fields.

Distribution: Central and South America.

Arequipa: native. ARE, CAY, UNI.

Note:

Tagetes minuta L.

Description: Erect annual herb, 1-2 m tall, strongly aromatic. Leaves glossy green, 3.5-28 cm long, pinnate, 4-8 pairs of pinnae, margins finely serrate, blade with elliptic oil glands. Heads isomorphic, 3-4 fused involucral bracts, 2- 3 white ray florets, 3-4 yellow-orange disc florets. Heads small, 10-15 x 10-20 mm. Inflorescence in a clustered panicle of 20-80 heads. Peduncle 4 mm long.

Ecology: Coastal, Andean I-II, Amazonian: 0-3500 m.

Distribution: Native to South America, from N-Chile and S-Brazil southward, introduced worldwide, an invasive weed in some countries.

Arequipa: introduced, doubtful. Not specified by province. Tagetes minuta is probably not native to Peru (Schiavinato et al., 2017). It is often confused with T. terniflora. Perhaps it is cultivated and has escaped from cultivation here and there.

Use: Used in perfumery, as a spice, as a flavoring in the food and beverage industry, and as a medicinal plant to relieve stomach upset.

Tagetes multiflora Kunth

Description: Small annual herb, 10-20 cm tall. Leaves pinnate. Leaf segments spatulate, dentate, 1-2 mm wide. Heads in loose clusters, peduncle usually >10 mm. Flowers yellow, 3-5 ray florets and 6-10 disc florets. Involucre entirely purplish. Pappus scales oblanceolate, widest part in the upper 2/5, margins plumose, lateral projections longer than the scale base.

Ecology: Andean I-III: 1000-4500 m. Riverside, disturbed areas, fields.

Distribution: Ecuador, Peru, Bolivia, N-Chile, NW-Argentina.

Arequipa: native. ARE, CAY, CON, UNI.

Use: Medicinal herb used to relieve stomach problems.

Tagetes terniflora Kunth

Description: Annual herb, up to 0.8 m tall, erect, branched, aromatic. Leaves 5-15 cm long, pinnate, 3-6 pairs of pinnae, margins dentate, blade with circular oil glands. Heads terminal, 5 fused involucral bracts, often ternate, in dense clusters, dimorphic, most heads many-flowered and some heads uniflorous with one pistillate and lingulate floret. Many-flowered heads with 2-3, usually yellow ray florets and 5-10 tubular disc florets. Peduncle 1-10 mm long.

Ecology: Coastal, Andean I-III, Amazonian: 500-4500 m. Forests, grasslands.

Distribution: The Andes, from Venezuela to Argentina.

Arequipa: native. ARE, UNI.

Tanacetum parthenium (L.) Sch.Bip.

Description: Perennial herb, up to 60 cm tall with pungently scented leaves. Leaves variously pinnatifid, yellowish- green, puberulent, gland-dotted. Inflorescence a compound corymb, heads up to 2 cm in Ø. Ray florets 10-20, white, disc florets yellow.

Ecology: Andean I-III: 500-4000 m. Disturbed areas.

Distribution: Native to SE-Europe, introduced worldwide.

Arequipa: introduced. ARE, UNI.

Use: Medicinal plant.

Taraxacum officinale F.H.Wigg.

Description: A milky, perennial, rosulate herb with a vertical taproot. Leaves sessile, oblanceolate to spatulate, 7 to 20 cm long, entire to pinnatifid, lobes pointed and ± dentate. Heads solitary, yellow, involucre bell-shaped, outer bracts recurved. Peduncle 10-20 cm long, hollow, leafless.

Ecology: Coastal, Andean I-II: 0-3500 m. Disturbed areas.

Distribution: Cosmopolitan, origin Europe to Siberia.

Arequipa: introduced. ARE, CAY, UNI.

Use: Medicinal plant.

Note: More than 2000 polyploid species of Taraxacum have been described. Some are classified as native to Peru, such as Taraxacum cuzcense A.J.Richards. All have similar ecological parameters and are morphologically almost indistinguishable. We limit ourselves to T. officinale, an aggregate that probably includes several microspecies.

Tessaria absinthioides (Hook. & Arn.) DC.

Description: Similar to Tessaria integrifolia but with leaves that are mostly serrated and with a shrubby growth habit.

Ecology: Andean II: 2000-2500 m. Cloud forests, riversides.

Distribution: Peru, Bolivia, S-Brazil, Chile, Argentina.

Arequipa: native. ARE.

Note: Nylinder & Anderberg (2015) show that the genus Tessaria is nested in the Pluchea clade. T. absinthioides is often included in T. integrifolia, both species hybridize.

Tessaria integrifolia Ruiz & Pav.

Description: Shrub or tree, 2-8 m tall, branches spreading at an angle of <45°. Leaves oblanceolate, 5-8 x 0.5-1.2 cm, base attenuate to a pseudo petiole, apex acute to subobtuse, blade grayish, concolorous, margins entire, venation pinnate. Inflorescence corymbose, terminal. Florets purple, with abortive, straw-colored florets.

Ecology: Coastal, Andean I-II, Amazonian: 0-3500 m. Riverbanks, forests.

Distribution: South America.

Arequipa: native. ARE, CAM, CAR, CAY, ISL.

Use: Leaves used as medicine for asthma, kidney and liver diseases. The plant extract has an antiseptic effect. The wood is used for construction and as firewood.

Note: The phylogenetic study of Nylinder & Anderberg (2015) shows that the genus Tessaria is nested in the New World Pluchea clade and should be treated conclusively as Pluchea.

Trixis cacalioides D.Don

Description: Shrub, 0.1-1.5 m tall. Leaves sessile, oblanceolate, base attenuate, margins entire, or slightly dentate. Heads in terminal cymes, 5-7 flowered, corolla yellow (sometimes white), involucral bracts 1-seriate.

Ecology: Coastal, Andean I-II, Amazonian: 0-2500 m.

Distribution: South America.

Arequipa: native. ARE, CAM, CAR, CAY, ISL, UNI.

Viguiera weberbaueri S.F.Blake

Description: Perennial herb, stem glabrous. Leaves subsessile, ovate, margins entire to slightly dentate. Heads solitary, terminal, 8-9 cm in Ø, about 30 rays, yellow.

Ecology: Coastal: 0-500 m. Lomas.

Distribution: Peru.

Arequipa: native. CAM, CAR, CAY, ISL, UNI.

Villanova oppositifolia Lag.

Description: Annual herbs. Leaves pinnate to pinnatisect, lobes linear. Heads terminal in clusters of 2-4. Involucral bracts 1-seriate, cup-shaped, ray florets few, yellow, disc florets yellow.

Ecology: Coastal, Andean I-II: 500-4000 m. Disturbed areas.

Distribution: The Andes from Venezuela to N-Chile.

Arequipa: native. ARE, CAM, CAR, CAY, ISL, UNI.

Werneria ciliolata A.Gray

Description: Pungently aromatic subshrub with creeping rhizomes and thick, fastigiate stems, 4-8 cm tall. Leaves imbricate, especially towards the shoot tips, sessile, triangular, fleshy, glaucous, 3-5 x 3 mm, fringed-toothed. Heads discoid, solitary, sessile, 1-2 cm in Ø, involucres purplish-glaucous. Ray florets yellow, not surpassing the involucre, disc florets numerous, yellow. Reminiscent of a clubmoss when not in bloom.

Ecology: Andean III: 4000->4500 m. Stony tracts.

Distribution: Peru, Bolivia, N-Chile.

Arequipa: native. ARE, CAS, CAY, CON, UNI.

Werneria dactylophylla Sch.Bip.

Description: Suffruticose perennial herb, forming clumps of erect stems or cushions, broccoli-like appearance. Stems 11-24 cm tall, branched, glabrous, usually with leaves restricted to the upper part. Leaves imbricate, finger- like at the apex, at least 9-divided. Heads radiate, erect, sessile. Ray florets 17-22, white, disc florets 37-50, yellowish.

Ecology: Andean III: 4000->4500 m.

Distribution: Peru, Bolivia.

Arequipa: native. ARE, CAY, CON, UNI.

Werneria digitata Wedd.

Description: Suffruticose plant, forming clumps of erect stems. Stems 10-13 cm tall, simple or branched, glabrous, usually with leaves restricted to the upper part. Leaves imbricate, ± flattened just below divisions, the leaf upper portion 3-forked at the apex. Heads radiate, erect, sessile to subsessile. Ray florets 21-36, conspicuously surpassing the involucre, white. Disc florets 49-63, yellowish.

Ecology: Andean III: 3500-4000 m. Rocky slopes, grasslands.

Distribution: Peru, Bolivia.

Arequipa: native. ARE, CAS, CAY, CON.

Werneria esquilachensis Cuatrec.

Description: Suffruticose plant, forming clumps of erect or decumbent stems. Leaves somehow distally arranged, flat, fleshy, apex 3-forked, notched or digitated, margins dentate. Heads discoid, erect, sessile to subsessile. Disc florets ca. 15, purplish-orange.

Ecology: Andean III: 3700-5000 m. Rocky outcrops, steep volcanic slopes, open shrublands.

Distribution: S-Peru, Bolivia, N-Chile.

Arequipa: native. ARE, CAY.

Werneria marcida S.F.Blake

Description: Suffruticose plant, forming dense mats or hummocks, with rhizome-like stems up to 8 cm long covered with old leaves, rather erect, simple or branched from the base. Stems 1-2 cm tall (aerial part), glabrous. Leaves densely imbricate, extending into a glabrous sheath-like base. Heads radiate, erect, sessile to subsessile. Ray florets 14-21, white, disc florets 35-48, yellowish.

Ecology: Andean III: 4500-5350 m. Exposed rocky slopes and cryoturbated soils.

Distribution: C-Peru to N-Bolivia.

Arequipa: native. CAS.

Werneria poposa Phil.

Description: Suffruticose plant, forming clumps of erect stems (rarely dense mats). Stems 2-15 cm tall, simple or branched, arachnoid, usually with functional green leaves restricted to the upper part. Leaves imbricate, extending into a sheath-like base with arachnoid trichomes. Heads radiate, erect, sessile. Ray florets 8-11, white, disc florets 7-18, whitish to creamy.

Ecology: Andean III: 4600-4800 m. Rocky slopes, grassland.

Distribution: Peru, Bolivia, NW-Argentina, N-Chile.

Arequipa: native. ARE, CAY, CON, UNI.

Werneria staffordiae Sandwith

Description: Similar to W. dactylophyllum but stems 30-60 cm tall, with broccoli-like appearance. Leaves imbricate, with a sheath-like base, glabrous or with long silky trichomes, spatulate, with at least 12 finger-like segments at apex. Heads radiate, rather nodding, subsessile to shortly pedunculate. Ray florets 15-28, pale yellow, not exceeding the involucre, disc florets 36-81, yellowish.

Ecology: Andean III: 3900-5200 m. Rocky slopes, exposed grasslands.

Distribution: Peru.

Arequipa: native. CAS, CAY, CON, UNI.

Werneria weddellii Phil.

Description: Suffruticose plant, forming clumps of rather decumbent stems. Stems 3-8 cm tall, simple or branched, glabrous, with leaves rather uniformly arranged. Leaves imbricate, extending into a glabrous sheath-like base. Heads radiate, erect, sessile. Ray florets 9-15, white, rarely pale purplish beneath, disc florets 24-32, whitish to creamy.

Ecology: Andean III: 3800-4700 m. Along the banks of marshes and salt lagoons.

Distribution: S-Peru, Bolivia, N-Chile.

Arequipa: native. CAY.

Xanthium spinosum L.

Description: Annual herb, stem slender up to 1 m tall, with 3 cm long, often trifurcate, yellowish spines in leaf axils. Leaves 10-12 cm long, lobed, the middle lobe much longer than the others, dark green or grayish on top, white underneath.

Ecology: Andean II: 2000-3500 m. Disturbed areas.

Distribution: Native to the Americas, introduced worldwide.

Arequipa: native. ARE, CAY, UNI.

Zinnia peruviana (L.) L.

Description: Annual herb, leaf blade elliptic. Involucres narrowly to broadly campanulate, 9-18 x 10-20 mm. Ray florets red, pink or orange, pappus usually of 1 awn.

Ecology: Coastal, Andean I-II: 0-3500 m. Rocky roadsides, ravines, calcareous soils.

Distribution: Native to South and Central America, cultivated and introduced worldwide.

Arequipa: native. CAY, UNI.

Basellaceae

Anredera diffusa (Moq.) Sperling

Description: Twining vine, not producing tubers. Leaves succulent, elliptic to ovate, base attenuate to subcordate, apex acute. Inflorescences 3-16 cm long, axillary, unbranched to branched above the base. Flowers bisexual, pedicellate. Petals white, becoming reddish. Fruit completely enclosed by persistent perianth.

Ecology: Coastal, Andean I-II: 0-3500 m. Desert, lomas, rocky slopes.

Distribution: C- and S-Peru.

Arequipa: native. ARE, CAS, CAY.

Berberidaceae

Berberis lutea Ruiz & Pav.

Description: Unarmed, evergreen shrub growing up to 3 m tall, wood yellow. Leaves entire, sessile or subsessile, spatulate to obovate, 1-2 cm x 5-9 mm, apex cuspidate. Flowers yellow to orange, fruits black when ripe.

Ecology: Andean II-III: 2500-4500 m. Rocky slopes.

Distribution: The Andes from Colombia to Bolivia.

Arequipa: native. ARE, CON.

Note: Conflicting information about spines and leaf appearance.

Berberis saxicola Lechl.

Description: Glabrous, armed shrub. Leaves in fascicles of 5-12, obovate, attenuate to the short petiole, coarsely spinose-dentate, coriaceous, laxly reticulate-veined, lustrous above, pale waxy-papillose beneath, 5.5 x 2.5 cm. Flowers yellow to orange.

Ecology: Andean II-III: 2500->4500 m. Shrublands.

Distribution: S-Peru and Bolivia.

Arequipa: native, expected. No voucher, expected due to occurrence nearby (Ayacucho, Apurimac, Cuzco, Puno, Moquegua).

Note: May be conspecific with B. lutea, here separated by the presence of spines.

Betulaceae

Alnus acuminata Kunth

Description: Deciduous tree up to 30 m tall, bole cylindrical, up to 50 cm in Ø, bark smooth and gray when young, scaly and ridged when old. Branchlets with lenticels and conspicuous rounded leaf-scars. Leaves alternate, stipules present, petiolate, entire, lanceolate to ovate, thinly leathery, 5-14 × 3-9 cm, cuneate to obtuse or rounded at base, apex acuminate to obtuse, margins serrate, glandular hairy. Inflorescence a many-flowered catkin. Fruit a winged nut in cone-like, ovoid infructescence.

Ecology: Coastal, Andean I-III: 0-4000 m. Disturbed areas, forests, grasslands, riversides, shrublands.

Distribution: Mexico to Bolivia and Chile.

Arequipa: native. CAS, CON, UNI. No specimen but several mentions and convincing observations in iNaturalist (2024).

Bignoniaceae

Argylia radiata (L.) D.Don

Description: Perennial herb, 0.2-1 m tall. Leaves mostly alternate, palmately 5-11-foliate, the primary leaflets ovate or obovate, deeply 2-pinnatifid, segments linear to narrowly oblanceolate. Inflorescence a long pedunculate terminal raceme, with flowers clustered toward apex or along upper third, pedicels 0.4-1 cm long, subtended by a basal linear bracteole. Flowers deeply 5-parted to near base, corolla yellow to red or maroon, when yellow with dark red or purple spots in throat.

Ecology: Coastal: 0-1000 m. Lomas.

Distribution: S-Peru and N-Chile.

Arequipa: native. CAM, CAR, ISL.

Tecoma fulva (Cav.) G.Don

Description: Variable, deciduous shrub 1-4 m high. Leaves pinnate with 3-21 leaflets or simple, rachis winged to unwinged. Inflorescence a terminal raceme, calyx shortly cylindrical, 5-toothed. Corolla bicolorous, inside yellow, outside reddish, tubular, straight to slightly curved. Fruit a linear, glabrous capsule, 5-12 cm long, seeds with white scarious wings.

Ecology: Coastal, Andean I-II: 0-3500 m. Dry valleys.

Distribution: S-Peru, Bolivia.

Arequipa: native. ARE, CAM, CAY, CON.

Note: In Arequipa, T. fulva ssp. arequipensis (composed leaves) and ssp. tanaeciiflora (most leaves simple).

Tecoma stans (L.) Juss. ex Kunth

Description: Shrub or small tree, up to 5 m tall. Leaves 8-25 cm long, leaflets usually 3-7, ovate-lanceolate, 2-10 x 1- 3 cm, apex acuminate, margins toothed, ± glabrous, petiole 1-6 cm long. Inflorescence 5-15 cm long. Calyx campanulate. Corolla tubular, 3.5-5 cm long, bright yellow with reddish lines in throat. Capsule linear-oblong, flattened, 10-30 x 5-20 mm. Winged seeds flat.

Ecology: Andean II: 1500-3500 m. Disturbed areas, dry valleys.

Distribution: Native to the Americas, planted and naturalized in the tropics.

Arequipa: native. ARE.

Use: Ornamental shrub, the wood is used to make tools and is said to relieve toothache.

Note: In Arequipa, T. stans var. sambucifolia.

Boraginaceae

Amsinckia calycina (Moris) Chater

Description: Erect herb, 15-50 cm high, stems strigose and with short fine hairs. Leaves stiffly hairy, apex acute or obtuse, base cuneate to subcordate, margins undulate. Flowers yellow.

Ecology: Andean II-III: 2500-4000 m. Disturbed areas.

Distribution: Native to Argentina and Chile, introduced elsewhere.

Arequipa: native. ARE, CAY.

Cordia lutea Lam.

Description: Small tree with a dark twisted trunk and dark green sandpaper leaves. Flowers in groups, yellow, almost all year long. Fruits round and luminous white, sticky flesh, small black seeds.

Ecology: Coastal, Andean I: 0-1500 m. Disturbed areas, rocky slopes, shrublands.

Distribution: Native to Colombia, Ecuador, Peru and Bolivia.

Arequipa: native. CAM, ISL.

Use: Ornamental shrub.

Cryptantha filaginea (Phil.) Reiche

Description: Annual up to 8 cm tall. Leaves linear, hirsute. Inflorescence bracteate, corolla small, white, limb 0.5-1.5 mm wide. Calyx in fruit 3-4 mm long, hispid along midrib, lobes erect to slightly spreading at tip. Nutlets 4, homomorphic, 1.2-1.5 mm long, base truncate, margins angled, apex acute, surface tuberculate.

Ecology: Coastal, Andean I: 0-1500 m. Lomas.

Distribution: S-Peru and N-Chile.

Arequipa: native. ARE, ISL.

Note: Schwarzer (2007) limits this species to the lomas of N-Chile and S-Peru. Brako & Zarucchi (1993) classify the species as Andean III (probably a confusion with C. peruviana).

Cryptantha granulosa (Ruiz & Pav.) I.M.Johnst.

Description: Small annual, 5-30 cm tall, stem branched, stem and leaves hispid. Leaves linear-oblong, 30 x 5 mm. Inflorescence a scorpioid spike. Corolla showy, to 7 mm wide, white. Style scarcely or not exserted.

Ecology: Coastal: 0-500 m.

Distribution: Peru.

Arequipa: native. CAM, CAR, ISL.

Cryptantha limensis (A.DC.) I.M.Johnst.

Description: Annual herb, up to 15 cm tall. Leaves oblong-lanceolate, up to 10-20 x 5 mm. Inflorescence solitary or in pairs, 4-6 cm long, scorpioid. Inflorescence bracted, corolla 1 mm wide. Calyx in fruit 5-6 mm long, broad lobes spreading at tips.

Ecology: Coastal: 0-500 m.

Distribution: Peru, N-Chile.

Arequipa: native. CAM, CAR, ISL.

Cryptantha peruviana I.M.Johnst.

Description: Annual herb, up to 25 cm tall, all parts short-hispid. Leaves linear, 20 x 2 mm. Inflorescence not bracted. Corolla 0.5-2 mm wide. Style longer than nutlets.

Ecology: Andean II-III: 2000-4000 m.

Distribution: Peru, Bolivia.

Arequipa: native. ARE, CAS, CAY, UNI.

Euploca pilosa (Ruiz & Pav.) Luebert

Description: Subshrub, prostrate or spreading-ascending, older stems ligneous. Leaves elliptic-lanceolate, densely pilose, up to 3-5 cm long. Inflorescence composed of 2-3 spikes. Corolla minute, not at all slightly exserted, white.

Ecology: Andean II: 1500-3000 m. Disturbed areas, lomas, rocky slopes.

Distribution: Peru.

Arequipa: native. CAM, CAR, ISL.

Euploca toratensis (I.M.Johnst.) J.I.M.Melo

Description: Subshrub up to 20 cm tall. Leaves linear to lanceolate, 3-6 x 0.4-0.8 cm. Inflorescence scorpioid, terminal. Flowers white. Similar to E. pilosa but leaves up to 6 cm long and not pilose.

Ecology: Andean II: 2000-2500 m. Shrublands.

Distribution: Peru.

Arequipa: native, doubtful. ARE. No voucher, mentioned for Socabaya by Montesinos (2012). The type Weberbauer 7407 is from near the Arequipa border, so an occurrence is likely.

Heliotropium arborescens L.

Description: Shrub up to 2 m tall, Leaves ovate to oblong-elliptic, 4-8 x 2-4 cm, bullate-rugose in age. Inflorescence 3-10 cm across, flowers fragrant (vanilla), lavender-purple. Fruit dry.

Ecology: Coastal, Andean I-II: 0-3500 m. Forests, lomas.

Distribution: Peru, Bolivia, N-Chile. Introduced in North America, Europe and Asia.

Arequipa: native. ARE, CAR, CAY, CON, ISL, UNI.

Use: Cultivated as an ornamental plant.

Heliotropium curassavicum L.

Description: A low-growing perennial with fleshy gray-blue leaves, stem and branches decumbent. Leaves 15-40 x 4-8 mm, glabrous, oblanceolate or linear-lanceolate, obtuse, nerves faint. Inflorescence terminal, simple or bifurcate, 3-6 cm long, flowers uniseriate, corolla white.

Ecology: Coastal: 0-1000 m. Lomas, beaches, salty habitats.

Distribution: Native to South America, introduced worldwide.

Arequipa: native. ARE, CAM, CAR, ISL.

Heliotropium krauseanum Fedde

Description: Shrub, up to 1 m tall, densely leafy. Leaves sessile, oblong to oblanceolate, 2-5 x 0.5 cm, subcoriaceous. Inflorescence congested, densely flowered, 3-4 cm long. Corolla white.

Ecology: Coastal: 0-500 m. Rocky slopes.

Distribution: Peru, N-Chile.

Arequipa: native. CAM, CAR, CAS, ISL.

Heliotropium lilloi (I.M.Johnst.) Luebert

Description: Shrub or tree-like, up to 3 m tall, branchlet tips and young inflorescences reddish pilose, soon glabrous. Leaves ovate-lanceolate, 8-15 x 3-6 cm, margins entire, at the end of branches getting smaller. Inflorescence in scorpioid spike, densely flowered, flowers white. Berries white.

Ecology: Coastal: 0-500 m. Lomas.

Distribution: Ecuador, Peru, Bolivia, NW-Argentina.

Arequipa: native. CAR, ISL.

Heliotropium microstachyum Ruiz & Pav.

Description: Prostrate perennial herb, much-branched. Stems and leaves strigose, short adpressed pubescent. Leaves oblanceolate, with prominent pinnate veins, gray-green. Inflorescence terminal, clustered, not scorpioid.

Ecology: Andean II: 2000-3000 m. Disturbed areas.

Distribution: Ecuador, Peru, Bolivia and Argentina.

Arequipa: native. ARE, CAS, CAY, UNI.

Johnstonella parviflora (Phil.) Hasenstab & M.G.Simpson

Description: Similar to Cryptantha. Slender, usually with a single stem laxly branched from near base, hispid. Leaves linear, 1-4 x 0.2 cm, hispid. Inflorescence not bracted or only at base. Nutlets thin-margined, dissimilar.

Ecology: Coastal, Andean I-II: 0-3000 m.

Distribution: Peru, N-Chile.

Arequipa: native. ARE, CAM, CAR, CON, ISL.

Nama dichotoma (Ruiz & Pav.) Choisy

Description: A delicate annual, up to 20 cm tall. Stem branching dichotomous. glandular. Leaves linear-elliptic or narrowly spatulate, 1-4 mm wide, glandular viscid. Flowers bell-shaped or tube-shaped, 5 mm long, white to lavender, calyx glandular.

Ecology: Coastal, Andean I-III: 0-4000 m.

Distribution: Western America, from California to C-Chile.

Arequipa: native. ARE, CAM, CAR, ISL.

Note: The genus Nama is also placed in the family Namaceae by some authors.

Pectocarya anomala I.M.Johnst.

Description: Annual, decumbent to ascending herb, densely hirsute. Leaves linear. Flowers clustered axillary, white. Calyx in fruiting star-shaped, in nodding position, above glabrous. Nutlets subulately fringed dorsally as marginally, densely pubescent.

Ecology: Andean II: 2000-2500 m. Grasslands.

Distribution: Peru, N-Chile.

Arequipa: native. ARE, CAY.

Pectocarya linearis (Ruiz & Pav.) DC.

Description: Annual herb, ascending to erect. Leaves linear, 1-4.5 x 0.1 cm, hirsute. Flowers white, axillary. Nutlets smooth dorsally. Calyx above puberulent.

Ecology: Coastal, Andean I-II: 0-3000 m. Rocky slopes.

Distribution: Ecuador, Peru, Argentina and Chile.

Arequipa: native. ARE, CAR, CAY, CON, ISL.

Phacelia pinnatifida Griseb. ex Wedd.

Description: Herb. Leaves lacerate to pinnatifid. Inflorescence secund. Flowers violet.

Ecology: Andean II-III: 1500-4500 m.

Distribution: S-Peru, Bolivia, N-Chile and Argentina.

Arequipa: native. ARE, CAR, CAY.

Phacelia secunda J.F.Gmel.

Description: Prostrate herb, up to 50 cm tall. Leaves imparipinnate. Odd-leaflet 3-4 cm long, 2-3 pairs of smaller leaflets. Inflorescence scorpioid. Flowers bluish-pink.

Ecology: Andean II-III: 3000-4500 m.

Distribution: South America.

Arequipa: native. ARE, CAY.

Plagiobothrys humilis (Ruiz & Pav.) I.M.Johnst.

Description: Small perennial, prostrate to ascending, glabrescent, rooting at nodes. Leaves linear 2-6 x 0.3 mm, greenish, opposite, sessile, widened at base. Flowers very small, sessile, corolla white. Calyx tips reddish.

Ecology: Andean II-III: 2500-4500 m. Grasslands, riversides, shrublands.

Distribution: Peru.

Arequipa: native. ARE, UNI. No specimen, several verified observations in iNaturalist (2024).

Plagiobothrys linifolius (Lehm.) I.M.Johnst.

Description: Annual herb, decumbent to creeping, rooting at nodes. stems up to 15 cm tall. Leaves linear, sessile, 7- 25 x 1-3 mm, ± hairy. Flowers axillary, corolla white, pedicel not elongated in fruit.

Ecology: Andean III: 3500-4500 m. Grasslands, often along lake shores.

Distribution: W-South America.

Arequipa: native. CAY. Specimen Weigend & Schwarzer 691 (HUSA) on the way to Puno, not seen, several specimens from Moquegua near the border to Arequipa.

Plagiobothrys macbridei I.M.Johnst.

Description: Annual herb, branching from base, but not creeping and rooting at nodes. Stem up to 4 cm tall. Lower leaves opposite, even the earliest. Leaves linear, 1.8-2.6 cm long, sparsely hispid-villous. Inflorescence lax, regularly bracted.

Ecology: Andean II-III: 2500-4500 m. Grasslands.

Distribution: Peru.

Arequipa: native. ARE.

Plagiobothrys myosotoides (Lehm.) Brand

Description: Erect herb, rosulate, 10-20 cm tall. Leaves oblong linear, 1-4 x 0.1-0.4 cm, mostly in a basal rosette. Corolla tube at least as long as calyx, corolla white.

Ecology: Andean II-III: 3000-4000 m. Rocky slopes.

Distribution: California to Chile.

Arequipa: native. ARE, ISL.

Tiquilia conspicua (I.M.Johnst.) A.T.Richardson

Description: Procumbent herb, forming mats to 0.5 m across, older stems woody, young stems with spreading or inclined hairs. Leaves green, 7-12 mm long, margins entire. Corolla deciduous, lilac, 7-13 mm long, stamens included or slightly exserted, style attachment gynobasic. Nutlet spheroid.

Ecology: Coastal, Andean I-II: 0-2500 m. Lomas, sandy areas.

Distribution: Peru.

Arequipa: native. CAM, ISL.

Tiquilia dichotoma (Ruiz & Pav.) Pers.

Description: Procumbent herb, forming mats to 2 m across, older stems woody, young stems with spreading hairs. Leaves green, oblong or ovate, margins crenate, blade deeply plicate. Corolla persistent, lilac to milky-white, 5-6 mm long, stamens exserted to 2 mm.

Ecology: Coastal: 0-500 m. Lomas.

Distribution: Peru.

Arequipa: native. CAM, ISL.

Tiquilia elongata (Rusby) A.T.Richardson

Description: Procumbent herb, forming mats to 0.5 m across, young stems with dense spreading or inclined hairs. Leaves narrowly ovate to lanceolate, 10-23 mm long with 3-4 pairs of lateral veins, margins crenate. Corolla deciduous, blue, 5-12 mm long, stamens included. Nutlet spheroid.

Ecology: Andean II: 1500-3000 m. Disturbed areas, shrublands.

Distribution: S-Peru.

Arequipa: native. ARE, CAM, CAS, CAY, ISL.

Tiquilia ferreyrae (I.M.Johnst.) A.T.Richardson

Description: Procumbent, forming mats to 1 m across. Older stems with lurid bark, young stems with dense spreading hairs. Leaves green-gray, ovate, 1 cm long, blade deeply plicate, margins entire. Corolla white, persistent, 7.5-10 mm long, stamens exserted 5 mm. Nutlet hemi-ovoid, ventral surface broadly flattened.

Ecology: Coastal: 0-500 m. Lomas.

Distribution: Peru.

Arequipa: native. CAM, CAR.

Tiquilia grandiflora (Phil.) A.T.Richardson

Description: Procumbent, forming mats to 1 m across, older stems with brown bark, young stems glandular. Leaves green, lanceolate, 1-2 cm long, blade deeply plicate, 3-4 pairs of veins, margins crenate or entire. Corolla blue, purple or salmon, persistent, 8-13 mm long, stamens exserted 8 mm. Nutlet hemi-ovoid, each with a ventral knife-like collar.

Ecology: Coastal, Andean I-II: 0-2500 m. Desert, lomas, open soil.

Distribution: Peru, N-Chile.

Arequipa: native. ARE, CAM, CAS, CAY.

Tiquilia hunteri A.T.Richardson

Description: Semi-erect subshrub, forming mounds to 0.4 m across, older stems with whitish bark, young stems with dense spreading hairs. Leaves green, narrowly ovate, 5-6 mm long with 2-3 pairs of lateral veins, deeply plicate, margins crenate. Corolla deciduous, blue, 5-6 mm long, stamens included. Nutlet spheroid.

Ecology: Coastal: 0-500 m. Disturbed areas.

Distribution: Peru.

Arequipa: native. ARE, CAM.

Tiquilia litoralis (Phil.) A.T.Richardson

Description: Procumbent with elongated internodes to 5 cm long, forming mats to 0.6 m across, older stems woody with whitish bark, young stems with dense spreading hairs. Leaves green to olive-green, ovate, 3-8 mm long, margins entire. Corolla deciduous, sky-blue or milky-white, whit yellow throat, 4-8 mm long, stamens reaching the limb, style attachment gynobasic. Nutlet spheroid.

Ecology: Coastal: 0-1000 m. Disturbed areas, lomas, sandy areas.

Distribution: Peru, N-Chile.

Arequipa: native. ARE, CAM, CAR, CAS, ISL.

Tiquilia paronychioides (Phil.) A.T.Richardson

Description: Procumbent, occasionally semi-erect, forming mats to 2 m across. Leaves obovate to lanceolate, 3-6 mm long, green to gray-green, margins entire. Flowers axillary and solitary. Corolla deciduous, white, stamens included. Nutlet black, granular, unequal in size within a fruit.

Ecology: Coastal, Andean I: 0-1500 m. Desert, disturbed areas, lomas.

Distribution: Ecuador to N-Chile.

Arequipa: native. ARE, CAM, CAR.

Tiquilia simulans (I.M.Johnst.) A.T.Richardson

Description: Procumbent or spreading, forming mats to 1.2 m across, older stems with lurid bark, young stems with spreading hairs. Leaves green, ovate, 7-11 mm long, blade deeply plicate, margins entire. Corolla lilac, blue or white, persistent, 4-6 mm long, stamens exserted 3 mm. Nutlet hemi-ovoid.

Ecology: Coastal: 0-500 m. Lomas.

Distribution: Peru.

Arequipa: native. CAM, CAR.

Brassicaceae

Alshehbazia friesii (O.E.Schulz) Salariato, Zuloaga & Al-Shehbaz

Description: Rhizomatous perennials, rhizome branched terminated by a small rosette. Leaves rosulate, petiolate, lanceolate, 9-16 x 2-3 mm, lobed or rarely coarsely toothed, above sparsely pubescent. Flowers ± sessile, sepals connate, petals creamy white or yellow. Fruit sessile, leaflike, ovate, glabrous.

Ecology: Andean III: 4200-5000 m. Dry puna.

Distribution: Peru, Bolivia, NW-Argentina, N-Chile.

Arequipa: native. CAY.

Alshehbazia werdermannii (O.E.Schulz) Salariato, Zuloaga & Al-Shehbaz

Description: Plants cespitose, rhizomatous, forming dense cushions. Leaves rosulate, ± fleshy, entire, spatulate, 8- 20 x 1-2 mm, above densely pubescent, rarely glabrescent or only ciliate. Flowers on a 3-20 cm long pedicel, sepals free, petals creamy white to yellow. Fruit sessile, suborbicular, 2.5-3 mm in Ø.

Ecology: Andean III: 3800-4800 m. Wetlands.

Distribution: Peru, NW-Argentina, N-Chile.

Arequipa: native. CON.

Aschersoniodoxa cachensis (Speg.) Al-Shehbaz

Description: Perennial herbs, with a thick, woody, branched caudex to 2 cm in Ø, branches ending in rosettes. Basal leaves rosulate, oblanceolate, broadly petiolate, fleshy, entire, tuberculate, purplish-gray. Inflorescence short- pedunculate, flowers white. Fruit 25-38 × 7-10 mm.

Ecology: Andean III: 4000-5100 m. In non-consolidated moraine.

Distribution: Peru, Bolivia, Argentina.

Arequipa: native. CAY.

Brassica rapa L.

Description: Erect, annual to biennial herb up to 1 m tall, with stout taproot, stem simple or branched. Leaves glabrous, glaucous, basal leaves pinnately lobed, lateral lobes 2-4 pairs, terminal lobe obovate, wavy-dentate, cauline leaves sessile, base lobed, generally clasping. Inflorescence an umbel-like raceme, up to 60 cm long. Flowers bisexual, pedicel up to 3 cm long, petals yellow. Fruit linear 4-10 cm long, 2-4 mm wide, beak 0.5-3 cm long.

Ecology: Coastal, Andean I-II: 0-4000 m. Disturbed areas, cultivated fields.

Distribution: Cosmopolitan species, many different cultivars as subspecies.

Arequipa: introduced. ARE, ISL.

Note: In Peru B. rapa ssp. oleifera and ssp. rapa.

Capsella bursa-pastoris (L.) Medik.

Description: Annual or biennial herb, to 40 cm, basal rosette with oblanceolate leaves, entire or deeply pinnatifid. Cauline leaves amplexicaul. Hairs simple or stellate. Flowers 2.5 mm in Ø, white. Fruit 6-9 mm, triangular-obcordate.

Ecology: Andean II-III: 2500->4500 m. Disturbed areas, grasslands, shrublands.

Distribution: Native to SW-Asia, naturalized elsewhere as a weed.

Arequipa: introduced. ARE, CAR, CAY, ISL, UNI.

Use: It is an important wild vegetable in China.

Cremolobus chilensis (Lag. ex DC.) DC.

Description: Annual herb, 3-35 cm tall. Leaves alternate, the lower ones occasionally opposite, subsessile to petiolate, petiole up to 1 cm, blade linear-ovate, serrate to deeply pinnatifid. Pedicel up to 1 cm long when mature, petals white, 2.5-3 mm long, spatulate to distinctly clawed, stigma distinct. Fruit triangular to orbicular, flat or lens shaped in the center, papery, winged to wingless.

Ecology: Coastal, Andean I-III: 0-4500 m.

Distribution: S-Peru, Bolivia, NW-Argentina, N-Chile.

Arequipa: native. ARE, CAY, UNI.

Descurainia athrocarpa (A.Gray) O.E.Schulz

Description: Herb, erect, 20-50 cm tall. Leaves incised, lobed, 1-3 cm long, pale blue-green. Flowers yellow. Infructescence congested capitate, fruit dark purple, dehiscing from apex to base. Sepals and petals persistent after fruit dehiscence.

Ecology: Andean III: 3900-5300 m. Cliff ledges, rock crevices, slopes.

Distribution: Peru, Bolivia.

Arequipa: native. CAS, CAY.

Descurainia depressa (Phil.) Reiche

Description: Biennial subshrub, 2-30 cm tall, stems prostrate to decumbent from a caudex. Leaves 1-2-pinnate, pubescent. Raceme bracteate at least basally.

Flowers yellow-orange. Fruit linear, adpressed to rachis, dehiscing from apex to base, 4-8 mm. Sepals caducous before or during fruit development. Seeds biseriate.

Ecology: Andean III: 3500-5000 m. Moist grassy plain, steep rocky slopes, disturbed areas, roadsides, riverbanks.

Distribution: C- and S-Peru, Bolivia, Chile, NW-Argentina.

Arequipa: native. ARE. No specimen but mentioned for Arequipa by Al-Shehbaz (2012).

Descurainia myriophylla (Kunth ex DC.) R.E.Fr.

Description: Annual herbs, eglandular or sparsely to densely glandular along racemes, densely pubescent. Stems 30-120 cm tall, erect, simple at base, often branched at top. Basal leaves withered early, cauline leaves (1-)2- pinnatisect, pubescent, 3-12 cm long, petioles 0.5-2 cm long, lateral lobes 4-10 on each side, smaller and often with narrower terminal lobes. Racemes ebracteate, 20-150-flowered. Sepals purplish, rarely yellowish, petals yellowish to creamy-white. Fruits linear, 1-1.8 cm × 1.5-2 mm, erect, adpressed to rachis, terete, not torulose.

Ecology: Andean II-III: 2500-4000 m.

Distribution: Colombia, Ecuador, Peru, Bolivia, NW-Argentina, N-Chile.

Arequipa: native. ARE, CAY, CON, UNI.

Descurainia stricta (Phil.) Reiche

Description: Annual herbs, eglandular or sparsely to densely glandular in racemes, canescent or pubescent. Stems 10-50 cm tall, erect, much-branched from base. Basal leaves withered early, lower cauline leaves pinnatifid to 1- or 2-pinnatisect, 1.5-6 cm long, petioles 0.3-2 cm long, last segment entire, upper cauline leaves sessile, smaller. Racemes ebracteate, dense, 25-110-flowered. Sepals purplish or yellowish, petals yellow. Fruits linear, 1-1.6 cm × 1.7-2.2 mm, erect, adpressed to rachis.

Ecology: Andean II-III: 3000->4500 m.

Distribution: Bolivia, N-Chile, Peru.

Arequipa: native. ARE, CAY.

Dictyophragmus englerianus (Muschl.) O.E.Schulz

Description: Annual herb, stems erect, simple or branched, 4-25 cm tall. Lower leaves subsessile, oblong, cauline leaves few, auriculate to amplexicaule. Flowers white, in dense racemes. Fruits linear-lanceolate, straight, 1.5-3 x 0.3-0.4 cm. Seeds broadly winged all around, suborbicular.

Ecology: Coastal: 0-500 m.

Distribution: Peru.

Arequipa: native. CAM, CAR, ISL.

Dictyophragmus lactucoides (Förther & Weigend) Al-Shehbaz

Description: Annual herb, stems erect, 12-20 cm tall, main stem far overtopped by the 3-5 lateral branches. Cauline leaves sessile, widely lanceolate. Flowers white, in dense racemes. Fruits narrowly linear, straight, 2-3 x 0.15 cm. Seeds wingless, oblong.

Ecology: Coastal: 300-1000 m. Lomas.

Distribution: S-Peru (Arequipa).

Arequipa: native. CAR.

Diplotaxis muralis (L.) DC.

Description: Usually scapose, hairy. Stem 20-50 cm. Leaves in basal rosette pinnately lobed, cauline leaves short-petiolate to sessile if present. Flowers yellow, sepals 3-5.5 mm, petals 5-8(10) mm. Fruit (1.5)2-4 cm long, 1.5- 2.5 mm wide, pedicel 8-20 mm. Seed ovoid, 20-36, 0.9-1.3 mm.

Ecology: Andean I-III: 1000-4000 m. Disturbed areas, dry streambeds, cultivated fields.

Distribution: Native to Europe and N-Africa, introduced worldwide.

Arequipa: introduced. ARE.

Draba lapaziana Al-Shehbaz

Description: Herbs scapose, cespitose, forming cushions, 1-2 cm tall. Caudex with several to numerous fine branches each terminating in a rosette and covered with some leaf remains of previous years. Leaves rosulate, thin, oblanceolate to spatulate, 5-8 x 1-2 mm, subsessile, attenuate at base, entire, subacute to obtuse at apex, ± tomentose on both surfaces. Inflorescences 2-4-bracteate umbels, bracts unequal, early deciduous or rarely persisting till fruit maturity. Sepals oblong, 1.7-2.2 x 0.7 mm, persistent. Petals white, oblanceolate, 1.1-1.4 x 0.3-0.4 mm, narrowed to a clawlike base. Fruits oblong, 2.4-3 x 1.1-1.4 mm, not compressed. Seeds 4-7 per locule, ovate, brown.

Ecology: Andean III: >4500 m.

Distribution: Peru, Bolivia, NW-Argentina.

Arequipa: native. CAY.

Draba soratensis Wedd.

Description: Annual herb, scapose, pubescent. Leaves oblong to lanceolate-spatulate, obtuse, margins entire or 1- serrate. Inflorescence long pedunculate. Flowers white. Fruit oblong-lanceolate.

Ecology: Andean III: 4000-5000 m.

Distribution: Peru to Bolivia.

Arequipa: native. CAR, UNI. No specimen but a convincing observation from La Unión in iNaturalist (2024).

Draba werffii Al-Shehbaz

Description: Herbaceous perennials. Caudex simple or few- to several-branched, not stoloniferous. Basal leaves rosulate, margins entire or subapically 1-dentate on each side, not ciliate, cauline leaves absent. Raceme ebracteate, 4-14-flowered, corymbose. Petals blue to bluish-purple. Fruit lanceolate to oblong, 8-11 x 2-3.2 mm, strongly compressed, not twisted, ± sessile.

Ecology: Andean III: 3600-4250 m. Scrublands, forests.

Distribution: Peru.

Arequipa: native. ARE.

Eudema arequipa (Al-Shehbaz, A.Cano, M.A.Cueva & Salariato) Al-Shehbaz, Salariato, A.Cano & Zuloaga

Description: Perennial herbs, rosulate, minutely puberulent on leaves, not pulvinate. Basal leaves with slender, glabrous petioles, blade orbicular, 4-6 x 3-5 mm, margins entire. Fruiting pedicels solitary, erect to ascending, 0.5-1 cm long. Sepals oblong, deciduous shortly after flowering, petals purple. Fruits linear, straight, 5-7 x 2 mm, ovules 20-30 per ovary.

Ecology: Andean III: 4700-4900 m. Between rocks.

Distribution: Peru, Bolivia, NW-Argentina.

Arequipa: native. CON.

Eudema calycinum (Desv.) Al-Shehbaz, Salariato, A.Cano & Zuloaga

Description: Perennial herbs, rosulate, glabrous to sparsely pubescent. Basal leaves with thick, ciliate petioles, 5-10 mm long, blade linear to lanceolate, 5-15 x 1-4 mm, margins entire or rarely obscurely 3-dentate. Fruiting pedicels straight to arcuate, 1-4 cm long. Sepals oblong, erect to spreading, deciduous shortly after flowering, petals creamy- white. Fruits linear-oblong, smooth, straight to slightly curved, 7-15 x 2 mm.

Ecology: Andean III: 4000->4500 m. Grasslands.

Distribution: Peru, Bolivia, NW-Argentina.

Arequipa: native. ARE, UNI.

Eudema monimocalyx (Gilg & Muschl. ex Hosseus) Al-Shehbaz, Salariato, A.Cano & Zuloaga

Description: Perennial herbs, rosulate, densely pubescent, with simple, straight trichomes. Basal leaves with thick, ciliate petioles, 4-7 x 1.5-2 mm, blade ovate, 4-6 x 2-3.5 mm, densely pubescent on adaxial surface, beneath glabrous to sparsely pubescent, margins entire. Fruiting pedicels straight to arcuate, pubescent, 1-3 cm long. Sepals oblong, erect, persistent even after fruit dehiscence, petals white. Fruits linear-oblong, smooth, straight to slightly curved, 5-12 x 2 mm.

Ecology: Andean III: 4000-4500 m.

Distribution: S-Peru, Bolivia, NW-Argentina.

Arequipa: native. ARE, UNI.

Exhalimolobos weddellii (E.Fourn.) Al-Shehbaz & C.D.Bailey

Description: Annual to biennial herbs, densely pubescent throughout. Stems erect, often solitary from base, branched at top, 40-90 cm tall, with 3-7-radiate trichomes. Basal leaves early deciduous, cauline leaves sessile, auriculate to amplexicaule, oblong to lanceolate, 2-8 × 0.5-3 cm. Racemes elongated during fruiting, fruiting pedicels horizontal to divaricate-ascending. Sepals oblong, green to purplish, petals white. Fruits narrowly oblong-oblong to linear, 0.8-1.5 cm × 1.5-2 mm. Seeds oblong, 2-seriate or multiseriate.

Ecology: Andean II-III: 2500-4000 m.

Distribution: South America, from Peru to S-Argentina.

Arequipa: native. ARE.

Lepidium bipinnatifidum Desv.

Description: Annual herb, stems leafy, much-branched from the base, decumbent or prostrate, stems 10-30 cm long, hirsutulous. Leaves pinnatifid or the lowest 2-pinnatifid, cauline leaves auriculate. Racemes numerous, dense short and narrow. Petals 0.3-1.5 mm, sepals longer. Fruits winged, longer than wide, 2-5 mm, glabrous, faintly reticulated.

Ecology: Andean II-III: 2500-4500 m. Grasslands, shrublands.

Distribution: NW-South America, naturalized in W-Mexico and California.

Arequipa: native. ARE, CAY, UNI.

Lepidium chichicara Desv.

Description: Similar to L. 2-pinnatifidum but basal leaves oblanceolate, long petiolate, cauline leaves sessile, linear, margins serrate. Flowers white, stamens 2.

Ecology: Andean I-II: 500-3000 m. Rocky slopes.

Distribution: Ecuador, Peru, Chile and Bolivia.

Arequipa: native. ARE, CAY, CON, UNI.

Lepidium crassius (C.L.Hitchc.) Al-Shehbaz

Description: Densely pubescent perennial with hirsute stems 30-50 cm tall. M middle and upper cauline leaves auriculate. Fruiting pedicels narrowly winged, pubescent all around. Petals obovate, 2-3 × 1-1.8 mm, 2 median and 2 lateral stamens. Fruits without apical notch.

Ecology: Coastal, Andean I: 500-1000 m. Rocky slopes.

Distribution: Peru.

Arequipa: native. CAM, CAR, UNI.

Lepidium didymum L.

Description: Annual or biennial herb with decumbent or ascending, glabrous green stems, up to 40 cm long. Leaves 1-2-pinnate and alternate, up to 5 cm long. Flowers inconspicuous, the 4 yellowish-white petals very short or absent, with 2 (rarely 4) stamens. Fruit notched in the middle, looks like eyeglasses.

Ecology: Coastal, Andean I-II: 0-2500 m. Open and disturbed areas, waste places.

Distribution: Native to South America, now cosmopolitan.

Arequipa: native. ARE.

Lepidium latifolium L.

Description: Perennial herb, 0.3-1.5 m tall. Long creeping rhizome. Rosette of leaves 10-30 x 2.5-5 cm, margins dentate, long petiolate. Cauline leaves alternate, 2.5-7.6 cm long, oblong. Inflorescence a terminal, dense cluster of white flowers. Capsule round to elliptic, hairy, 1.5 mm in Ø, single-seeded.

Ecology: Andean II: 2200 -2500 m. Open and disturbed areas, waste places.

Distribution: Native to Eurasia, introduced to Australia and the Americas.

Arequipa: introduced. ARE.

Lepidium strictum (S.Watson) Rattan ex B.L.Rob.

Description: Annual to biennial decumbent to ascending, up to 20 cm long. Basal leaves 2-pinnately lobed, cauline leaves short petiolate, pinnately lobed. Inflorescence crowded, petals 0.2-0.5 mm, white, stamens 2. Fruits 2.5-3.3 mm wide, ovate to round, tip winged.

Ecology: Andean II: 2500-3500 m.

Distribution: Disjunct, California/Oregon and S-Peru/Chile.

Arequipa: native. ARE.

Lepidium virginicum L.

Description: Annual or biennial weed 30-50 cm high, erect, hairy. Basal leaves lyrate-pinnatisect to pinnate, to 9 cm long, hirsute with shortly curved hairs. Cauline leaves reducing to oblanceolate, toothed or entire, margins ciliate. Inflorescence a dense, elongating raceme. Sepals 0.7-1 mm long. Petals 1-2 mm long, white. Stamens usually 2, rarely 3 or 4. Fruit ± circular, flat, 3-4 mm long, 3 mm wide, wings only in upper half, notch shallow, pedicel spreading, 4-5 mm long, shortly hairy above.

Ecology: Coastal, Andean I-II: 0-3500 m. Disturbed areas.

Distribution: Native to North America, introduced elsewhere.

Arequipa: introduced. ARE.

Lepidium werffii Al-Shehbaz

Description: Perennial herbs, puberulent, caudex simple, 0.5-2 cm in Ø, not fleshy. Stems 4-25 cm, decumbent to ascending, branched above. Basal leaves up to 11 cm long, oblanceolate, entire or pinnatifid, margins

dentate to incised, cauline leaves not auriculate. Racemes terminal. Petals white, obovate, longer than sepals, stamens 4. Fruiting pedicels narrowly winged, fruit broadly rhombic, up to 5.5 mm long, distinctly wider than long.

Ecology: Andean II-III: 2500-4500 m. Scrub and puna vegetation.

Distribution: S-Peru.

Arequipa: native. ARE.

Note: Often confused with Lepidium meyenii Walp. from C-Peru.

Lobularia maritima (L.) Desv.

Description: Suffruticose herb. Stem 5-25 cm. Leaves linear or lanceolate-oblanceolate, 1.6 -2.5 cm × 2-3 mm, base attenuate, tip acute. Racemes dense, elongated in fruit, 4-8 cm long. Fruiting pedicels 4.5-6 mm. Sepals often tinged purplish, 1.5-1.7 mm, petals broadly obovate, white, 2.5 × 1.6-2 mm, abruptly clawed. Fruit 2.3-2.7 × 1.6-2 mm.

Ecology: Andean II: 2000-3000 m. Pathways, disturbed areas near villages.

Distribution: Native to the Mediterranean.

Arequipa: introduced. ARE.

Machaerophorus arequipa Al-Shehbaz, A.Cano, M.A.Cueva & Salariato

Description: Herbs suffrutescent perennials or subshrubs, not glaucous, slightly canescent. Basal leaves not developed, cauline leaves petiolate, not auriculate at base, finely pinnatisect into narrowly linear lobes, usually grooved beneath and densely pubescent along groves, fleshy, entire. Racemes 10-40-flowered, petals white. Fruit terete, cylindrical, torulose, seeds uniseriate.

Ecology: Andean II: 2000-2400 m. Shrublands.

Distribution: S-Peru (Arequipa).

Arequipa: native. ARE.

Machaerophorus laticarpus Al-Shehbaz, A.Cano, M.A.Cueva & Salariato

Description: Subshrub, not glaucous, canescent when young, puberulent with age, up to 1 m tall. Basal leaves not developed, cauline leaves petiolate, 3-6 cm long, upper side floccose, underside puberulent, 3-5 lobes on each side. Racemes 25-56-flowered, petals yellow. Fruit lanceolate to oblong-lanceolate, not torulose, seeds biseriate.

Ecology: Andean II: 3200-3400 m. Shrublands.

Distribution: S-Peru (Arequipa).

Arequipa: native. UNI.

Mancoa hispida Wedd.

Description: Herbs short-lived perennials or biennials, pubescent throughout with 2- or 3-rayed trichomes.

Stems 2- 8 from rosette, 2-11 cm long, prostrate to decumbent. Basal leaves rosulate, later withered, oblanceolate to spatulate, 1-3 cm × 2-7 mm, 3-6 lateral lobes on each side, surfaces densely pubescent. Cauline leaves sessile. Racemes 5- 12-flowered. Petals white, spatulate 1.5-2.5 × 0.4-0.8 mm. Fruits oblong or ovate, rarely suborbicular, 5-10 × 2.5-4 mm, valves densely pubescent.

Ecology: Andean III: 3500 ->4500 m. Grassy slopes, sandy open ground, field margin, gravelly slopes.

Distribution: Peru, Bolivia, N-Chile and NW-Argentina.

Arequipa: native. ARE, CAY.

Mancoa laevis Wedd.

Description: Herbs annual, glabrous throughout except leaf margin. Stems 2-6 from rosette, 1.5-5 cm long, prostrate to decumbent, simple, glabrous. Basal leaves rosulate, later withered, petiole 1-5 mm long, oblanceolate to spatulate, lyrate, 0.8-2.5 cm × 2-3 mm. Cauline leaves sessile, not auriculate, glabrous. Racemes ebracteate, 5-18- flowered, subcorymbose, only slightly elongated in fruit, fruiting pedicels 1-4 mm long, glabrous. Sepals ovate, 1.6 × 0.7 mm, persistent, glabrous, petals white, oblanceolate, 1.2-2.5 × 0.5-0.7 mm. Fruits oblong, 6 × 3 mm, valves glabrous, not keeled.

Ecology: Andean III: 3500 ->4500 m. Grassy slopes, sandy open ground, field margin, gravelly slopes.

Distribution: Peru, Bolivia, N-Chile and NW-Argentina.

Arequipa: native. ARE.

Mathewsia densifolia Rollins

Description: Plants 10-20 cm tall, densely covered throughout with a whitish to tawny indument of highly branched trichomes. Leaves densely clustered above the naked woody foot, sessile or short-petiolate, non-auriculate, oblong, pinnately lobed, 2-4 x 1 cm, lobes linear-oblong to broader. Inflorescence pedunculate, petals yellow, 12-15 x 2.5-3 mm, pedicels widely spreading, 3-5 mm long. Fruit sessile, divaricate, flattened, oblong, 15 x 3-5 mm.

Ecology: Coastal: 0-1000 m. Rocky slopes.

Distribution: Peru.

Arequipa: native. ISL.

Mathewsia peruviana O.E.Schulz

Description: Plants suffruticose at base, 50-60 cm tall, whitish pubescent throughout, leafy below the inflorescences. Leaves sinuate dentate to irregularly lobed, narrowly obovate to oblong, sessile, upper auriculate and amplexicaule, lower auriculate, 4-12 x 1-4 cm. Inflorescences 10-30 cm long, pedunculate, peduncle 5 cm long, petals yellowish- white, 14-18 x 4 mm, pedicels ascending, 6-10 mm long. Fruits, flattened, narrowly ovate, rounded below, tapered above, 20-30 x 4-5 mm.

Ecology: Coastal: 0-500 m. Lomas.

Distribution: Peru.

Arequipa: native. CAR, ISL.

Matthiola incana (L.) W.T.Aiton

Description: Perennial or biennial herbs, densely tomentose with white stellate trichomes, about 20-75 cm tall. Stem erect, woody near the base, ascending, leafy, simple or sparsely branched from the base. Basal leaves simple, in dense rosettes, oblanceolate to linear, 3-9 x 0.5-2 cm, base cuneate to attenuate, margins entire, rarely sinuate, apex acute, petiole 0.5-2 cm long. Cauline leaves gradually smaller, base not auriculate. Flowers white to pink, petals much longer than sepals. Fruit dehiscent, linear, sessile, terete, erect or slightly curved, torulose, with 2-3 hornlike appendages.

Ecology: Coastal, Andean I-III: 0-4000 m. Disturbed areas.

Distribution: Native to SE-Europe, cultivated and naturalized elsewhere.

Arequipa: introduced, adventive. ARE, CAR.

Mostacillastrum dianthoides (Phil.) Al-Shehbaz

Description: Perennial herbs, with unbranched caudices, often with remnants of leaves from previous years, glabrous. Stems erect, 6-40(-90) cm tall. Basal leaves rosulate, linear to lanceolate, 3-14 cm × 2-5 mm, ± entire, cauline leaves sessile, auriculate, entire, becoming smaller to the top. Racemes corymbose, elongated during fruiting, fruiting pedicels 4-10 mm long, ascending to divaricate, straight. Petals white. Fruits teretes, 1.5-4 cm × 1.5- 2 mm, erect to ascending. Seeds 1-seriate.

Ecology: Andean III: 4000-4500 m. Rocky slopes.

Distribution: S-Peru, N-Chile, NW-Argentina.

Arequipa: native. ARE, CAS.

Mostacillastrum ferreyrae (Förther & Weigend) Al-Shehbaz

Description: Annual herb 20-30 cm tall. Stems branched 1-3 cm above ground. Leaves petiolate, 20-30 x 7-10 mm. Raceme terminal, ebracteate, 10-30-flowered, flowers white. Fruit subsessile, terete, narrowly linear, 40-50 x 1.5 mm. Seeds 0.8-1.1 mm long, coarsely reticulate.

Ecology: Coastal: 350-960 m. Lomas, sandy and rocky slopes.

Distribution: S-Peru, N-Chile.

Arequipa: native. CAR.

Mostacillastrum gracile (Wedd.) Al-Shehbaz

Description: Biennial herb, branching above, ± hispid below with bifurcate hairs. Leaves pinnatifid to pinnatisect, flat, with simple trichomes. Racemes lax, pedicels 3.5-9 mm long. Sepals 3-4 mm long, petals white, 4-5.5 mm long, with broad claws. Fruits 1.2-2.8 cm x 1.5 mm, valves 1-nerved.

Ecology: Andean II-III: 2500-4000 m. Shrublands.

Distribution: S-Peru, N- and C-Chile.

Arequipa: native. ARE, CON, UNI.

Mostacillastrum morrisonii (Al-Shehbaz) Al-Shehbaz

Description: Perennial herb, 20 cm tall. Plants densely pubescent. Leaves irregularly crenate. Flowers white. Fruits 5-6.5 cm long.

Ecology: Coastal: 0-500 m. Lomas.

Distribution: S-Peru.

Arequipa: native. ISL. Known only from the type locality in Mollendo.

Mostacillastrum oleraceum (O.E.Schulz) Al-Shehbaz

Description: Glabrous, woody at the base, erect, to 55 cm high, branched from the axils of the leaves. Leaves petiolate, ovate, acute, coarsely dentate, base cuneate. Racemes lax, 20-30-flowered, pedicels 4-5 mm long, to 15 mm in fruit. Sepals 4 mm long, obtuse, petals yellowish-white, 6.5 mm long. Fruit 4-4.5 cm x 1.5 mm, valves 3-nerved. Seeds 1-seriate.

Ecology: Andean II: 3000-3500 m. Disturbed areas.

Distribution: S-Peru.

Arequipa: native. CAY.

Mostacillastrum pectinifolium (Al-Shehbaz) Al-Shehbaz

Description: Plant densely branched. Leaves pectinate, above sulcate. Sepals 1.8-2.5 mm long, petals 2.5-3.5 mm long, flowers white.

Ecology: Andean II: 2500-3500 m.

Distribution: Peru, N-Chile.

Arequipa: native. ARE, CAY.

Nasturtium officinale W.T.Aiton

Description: Glabrous, (semi-)submersed aquatic herb. Stem branched. Leaves pinnate, 1-9 leaflets, terminal leaflet suborbicular to ovate, larger than lateral leaflets. Raceme terminal, Petals <5 mm, white, anthers yellow. Fruit protruding, 1-2 cm long, arching ascending.

Ecology: Coastal, Andean I-III: 500-4000 m. Aquatic herb.

Distribution: Native to Eurasia.

Arequipa: introduced. ARE, CAR, CAY.

Use: Rich in vitamins and minerals, valued as a food and medicinal plant.

Neuontobotrys amplexicaulis (Kuntze) Al-Shehbaz

Description: Divaricately branched herb, up to 60-70 cm tall. Lower leaves obovate-elliptic, sinuate-dentate, 6 cm long, middle and upper cauline leaves entire, glabrous, glaucous. Raceme corymbose-congested, petals pale lilac. Fruiting pedicels often strongly recurved or reflexed, fruits often strongly curved, not adpressed to rachis.

Ecology: Andean I: 1000-1500 m. Rocky slopes.

Distribution: Peru, N-Chile.

Arequipa: native. CAR.

Neuontobotrys camanaensis Al-Shehbaz & A.Cano

Description: Herbaceous perennial, branched. Stems 15-35 cm long, simply pubescent. Leaves all cauline, broadly spatulate to broadly oblanceolate, base auriculate, margins repand to dentate, apex obtuse. Raceme ebracteate, flowers yellow. Fruits linear, 1.7-3.2 cm x 1.5 mm, torulose.

Ecology: Coastal: 500-850 m. Rocky slopes.

Distribution: Peru (Arequipa).

Arequipa: native. CAM.

Neuontobotrys lanatus (Walp.) Al-Shehbaz

Description: Suffruticose, with ascending or diffuse, much-branched, ash-gray pubescent stems. Leaves lanceolate, acuminate, deeply clasping, dentate, pubescent like the stems. Racemes dense, 50-120-flowered. Petals pinkish- white.

Ecology: Andean II: 2500-3000 m. Rocky slopes.

Distribution: S-Peru, C- & N-Chile.

Arequipa: native. ARE, CAM, ISL, UNI.

Neuontobotrys schulzii (Al-Shehbaz) Al-Shehbaz

Description: Stems 10-60 cm tall. Basal leaves not rosulate, petiolate, pinnatisect, 4-7 cm long, lateral lobes oblong to linear, upper cauline leaves narrowly linear, 2-3.5 cm x 0.5-1.5 mm, base strongly sagittate-amplexicaul, margins entire to dentate or rarly pinnatisect. Raceme rachis straight. Fruit straight, fruiting pedicels divaricate, straight, 6-8 mm long.

Ecology: Andean II: 2500-3100 m. Sandy hills.

Distribution: Peru.

Arequipa: native. ARE.

Raphanus raphanistrum L.

Description: Sparsely to densely hairy annual or biennial herb, up to 1 m tall, reflexed-hairy. Leaves 3-15 x 1-5 cm, oblong, obovate, or oblanceolate, lyre-shaped to pinnately lobed, margins dentate. Flowers white, pink or sometimes yellow, petals 15-25 x, 4-7 mm, claw 8-14 mm. Fruits cylindric to narrowly lanceolate, 5-10 mm in Ø, segmented, pedicel ascending to spreading, 0.7-2.5 cm. Seed 3 mm, ovoid to oblong.

Ecology: Coastal, Andean I-II: 0-3500 m. Cultivated and naturalized in disturbed areas.

Distribution: Cosmopolitan.

Arequipa: introduced. ARE, CAR, ISL, UNI.

Use: Raphanus raphanistrum ssp. sativus (L.) Domin is grown mainly for its fleshy root (radish).

Note: Crop of ancient cultivation in the Mediterranean (before 2000 BC), from where it spread worldwide.

Rorippa nana (Schltdl.) J.F.Macbr.

Description: Rosette-forming plant, developing ascending stems 10 cm long when older. Leaves pinnate, leaflets entire or 1-2 dentate. Flowers at first ± sessile in the rosette, later in ± sessile racemes, yellow. Fruit 5-7 x 2.5-3 mm.

Ecology: Andean III: 3500-4000 m.

Distribution: The Andes of South America.

Arequipa: native. ARE, CAY.

Sisymbrium irio L.

Description: Annual herb, branched from the ground. Basal leaves not rosulate, petiolate, up to 15 cm long. Cauline leaves variously shaped, sometimes lobed, never filiform or narrowly linear. Flowers yellow. Fruit erect or ascending, 2-4(-5) cm long.

Ecology: Coastal, Andean I-II: 0-3500 m. Abandoned fields, waste places, roadsides and orchards.

Distribution: Native to Eurasia, introduced worldwide.

Arequipa: introduced. ARE, ISL.

Use: Edible leaves and flowers. The seeds are used in folk medicine as a stimulant.

Sisymbrium officinale (L.) Scop.

Description: Annual herb, up to 1 m tall. Stems erect, branched above, hirsute. Basal leaves usually rosulate, 2-7 cm long. Flowers yellow. Fruit subulate-linear, 10-14 x 1-1.5 mm, adpressed to rachis.

Ecology: Andean II: 2000-3000 m. Disturbed areas, roadsides, fields, pastures.

Distribution: Native to Europe and N-Africa, introduced worldwide.

Arequipa: introduced. ARE, CAY.

Use: Edible leaves and seeds. Used in folk medicine to soothe sore throat.

Sisymbrium orientale L.

Description: Annual herb, stem hairy, branching, to about 30 cm tall. Basal leaves divided into deep lobes or dentate leaflets. Cauline leaves lanceolate with small separate lobes near the base. Raceme terminal, flowers light yellow. Fruit up to 10 cm long.

Ecology: Coastal, Andean I-II: 0-2500 m. Disturbed areas.

Distribution: Native to Eurasia, introduced in temperate regions.

Arequipa: introduced. ARE, CAR, ISL.

Thlaspi arvense L.

Description: Herb 10-60 cm high, erect, simple or branched, glabrous, slightly fetid. Basal leaves petiolate, obovate or oblanceolate, 5 cm long, ± dentate. Stem leaves reducing, auriculate. Sepals 2 mm long. Petals 3-4 mm long, white. Fruit circular, 10 mm Ø, valves strongly compressed, winged, wings widening towards apex forming a deep notch.

Ecology: Andean II-III: 3000-4000 m. Fields, disturbed areas.

Distribution: Native to Eurasia.

Arequipa: introduced. CAS. No specimen but a convincing observation from Andagua in iNaturalist (2024).

Weberbauera arequipa Al-Shehbaz & Montesinos

Description: Annual herb, ca. 6 cm tall, stems decumbent, glabrous with few simple trichomes restricted to the petiolar bases, several branched above base. Cauline leaves oblanceolate, petiolate, 8-12 x 3- 5 mm, petioles 2-4 mm, blade glabrous, base attenuate, margins dentate. Racemes ebracteate, elongated slightly in fruit. Sepals green, persistent, petals white, not clawed. Fruiting pedicels mostly divaricate. Fruit oblong to lanceolate, 5-7 x 1.3-1.6 mm, not torulose, straight, 4-seeded.

Ecology: Andean III: 4000-4500 m.

Distribution: S-Peru.

Arequipa: native. ARE.

Weberbauera ayacuchoensis Al-Shehbaz, A.Cano & Trinidad

Description: Perennial herb, caudex ± fleshy, surculose, naked, slender, to 15 cm, often swollen at rosette attachments of previous years. Trichomes simple, 0.4-1 mm. Stems decumbent, 2-5 from caudex apex, 1.5-4 cm, glabrous. Basal leaves rosulate, petiolate, broadly orbicular to spatulate, 6-10 × 5-8 mm, ciliate, hirsute, dentate, cauline leaves few, similar to basal ones. Racemes basally bracteate, dentate, fruiting pedicels 1-2 mm. Sepals persistent, petals pale yellow. Fruits broadly oblong to ovoid, 4-5 mm × 1.7-2 mm, torulose, glabrous.

Ecology: Andean III: 3800-4450 m.

Distribution: Peru (Arequipa and Ayacucho).

Arequipa: native. CAY.

Weberbauera minutipila Al-Shehbaz

Description: Perennial herb, stems decumbent, 2-7 cm long, trichomatous. Basal leaves filiform to linear, 0.5-1.5 mm wide, coarsely dentate-serrate, cauline leaves ovate to lanceolate. Inflorescence ebracteate, sepals caducous, petals white. Fruits narrowly oblong, sub-torulose, 6-9 x 1 mm.

Ecology: Andean III: 3500-4000 m. Rocky slopes.

Distribution: S-Peru, Bolivia.

Arequipa: native. ARE.

Weberbauera peruviana (DC.) Al-Shehbaz

Description: Annual herb, much-branched, his-pidulous. Stem 10-40 cm long. Leaves elliptic, petio-late, margins dentate. Racemes 12-60-flowered, sepals obtuse, petals white or pinkish. Fruit recurved, 1-2.5 cm x 1.5 mm, pubescent.

Ecology: Andean II-III: 2500-4500 m. Rocky slopes.

Distribution: S-Peru, Bolivia and NW-Argentina.

Arequipa: native. CAY.

Bromeliaceae

Puya cahuachensis A.Galán, J.Montoya, Vicente Orell. & E.Linares

Description: Plants flowering up to 1.7 m tall. Leaves with brownish spines of 2-7 mm. Peduncle 1-1.5 m long, 1.5-3.5 cm in Ø, straight, grayish. Peduncle bracts foliaceous, lower ones up to 4-5 x longer than the internodes, upper ones shorter than the internodes. Inflorescence dense, branches 4-8 cm long, 1-2 cm apart, the soft white woolly pubescence does not fully mask the grayish color of the rachis. Flowers distinctly pedicellate, sepals oblong, acute, greenish, petals green to bluish.

Ecology: Andean II: 3200-3300 m. Between eruptive rocks.

Distribution: S-Peru (Arequipa).

Arequipa: native. CAR.

Puya colca-canyonese sp. nov. ined.

Description: Stout perennial herb, up to 1.5 m tall in vegetative growth. Similar to Puya webebaueri Mez., but flowers more spreading and leaves lepidote beneath.

Ecology: Andean II-III: 2500-4000 m. Rocky slopes.

Distribution: S-Peru.

Arequipa: native. CAY. Specimen from Cabanaconde and Cruz del Condor, several observations from Colca in iNaturalist (2024).

Note: The species will be described in Phytotaxa (E.J. Gouda, pers. comm.).

Puya colcaensis Treviño, Quip. & Gouda

Description: Plant perennial, flowering 1.5-2 m high. Leaves narrowly triangular, 50-80 x 2.8-4.2 cm, sparsely lepidote to glabrous on the upper side surface and densely lepidote beneath, coriaceous, with sharp apex. Inflorescence paniculate, cream-whitish, fertile part 69-105 cm long, peduncle 88-105 cm long, 1.4-1.6 cm Ø. Flowers with a pedicel of 5-12 mm, 1.6-2.6 cm

apart. Petals 7.7-8 × 1.6-1.8 cm, narrowly oblong, bluish-green, the margins entire, glabrous, apex acute or obtuse.

Ecology: Andean III: 3700 m. Steep slopes.

Distribution: S-Peru (Arequipa).

Arequipa: native. CAY.

Puya cylindrica Mez

Description: Herb, up to 1 m tall. Leaves green, typically bromeliad, with serrate margin. Inflorescence cylindrical, dense, many-flowered, branches, bracts and sepals pinkish-green, petals green. Floral bracts about equaling the sepals. Indument ferruginous.

Ecology: Andean II: 2500-3000 m. Dry valleys.

Distribution: Endemic to S-Peru.

Arequipa: native. ARE, CAY, CON, UNI.

Puya ferruginea (Ruiz & Pav.) L.B.Sm.

Description: Leaves long, very thin silvery-green, twisting towards the tips giving the plant a wind-blown asymmetrical appearance. Inflorescence lax, few-flowered, branches and bracts ferruginous. Petals white to greenish-yellow.

Ecology: Coastal, Andean I-III: 500-4000 m. Lomas, rocky slopes.

Distribution: Ecuador, Peru, Bolivia.

Arequipa: native. ARE, CAY, UNI.

Puya hoxeyi Janeba

Description: Plants in flower up to 4 m high, clumps of rosettes to 2 m in Ø. Leaves numerous, 100 x 6 cm, lanceolate, acute, pointed, glabrous. Above green, a little shiny, below green with the white vertical lines of the internal fibers clearly visible. Margins serrate, marginal spines brown at tip, paler towards the base, 5 mm long at the base, the first few basal spines retrorse, otherwise antrorse. Inflorescence 1.5 m to 2.5 m tall, 8 cm in Ø, with 20- 100 racemose branches holding the flowers, peduncle bracteate. Flower tubular, 45 x 15-20 mm, not opening widely. Petals 3, yellow/greenish, elongated with mucronulate apex, 40 x 15 mm.

Ecology: Coastal: 500-1000 m. Lomas.

Distribution: Peru.

Arequipa: native, expected. Type locality in Ilo (Moquegua), near the border to Arequipa.

Puya raimondii Harms

Description: Giant herb, up to 3 m tall in vegetative growth. Leaves borne on stout stem. Inflorescence 9-10 m tall, with more than 3000 flowers. Flowers bluish-white.

Ecology: Andean III: 3500-4500 m. Rocky slopes, shrublands.

Distribution: Peru, Bolivia.

Arequipa: native. CAY, UNI.

Note: P. raimondii is the largest Bromeliaceae. The reproductive cycle of this endangered species is approx. 40 years.

Tillandsia capillaris Ruiz & Pav.

Description: Caulescent lithophyte or epiphyte, very variable, up to 16 cm long. Roots present. Stems many from a single point, massed, simple or branched. Leaves 1-4 cm long, pruinose-lepidote, sheaths elliptic, thin, blades erect to spreading, linear, <2m thick. Scape often pseudo-axillary, 0-8 cm long, slender, glabrous, developed almost wholly after flowering. Inflorescence 1-2-flowered. Floral bracts ovate, with 3+ nerves, equaling or exceeding the sepals. Sepals lanceolate, 8 mm long, connate posteriorly, petals linear, stamens deeply included, exceeding the pistil.

Ecology: Coastal, Andean I-III: 500-4000 m. Lomas, rocky slopes, shrubland.

Distribution: Ecuador to C-Chile and E-Argentina.

Arequipa: native. CAM, CAR, CAY, ISL, UNI.

Tillandsia hirta W.Till & L.Hrom.

Description: Epiphyte on cacti and shrubs. Leaves distinctly imbricate, 2-3 cm long, bent downwards, hirtulous. Inflorescence axillary, peduncle ca. 4-5 cm long, 1-2 flowered.

Ecology: Andean II: 2000-2500 m. Shrublands.

Distribution: S-Peru, Bolivia, NW-Argentina.

Arequipa: native. ARE.

Tillandsia latifolia Meyen

Description: Plant usually caulescent, very variable, flowering to 60 cm high, mostly rootless. Stems prostrate, often branched. Leaves to 20 cm long, covered with adpressed ash-gray scales. Inflorescence branched, spikelets densely 6-12-flowered, peduncle and bracts red, petals pink.

Ecology: Coastal, Andean I-II: 0-3000 m. Beach, lomas, rocky slopes.

Distribution: Ecuador, Peru.

Arequipa: native. ARE, CAR, CAY, ISL.

Tillandsia paleacea C.Presl

Description: Terrestrial plant, caulescent with a thick appearance, quickly branching and forming clumps. Branches decumbent to ascending, up to 60 cm long and 20 cm high. Leaves thin and flexible. Inflorescence few flowered, unbranched, peduncle 4-8 cm long. Flowers densely packed, almost sessile, 1 cm wide, petals deep purplish-blue.

Ecology: Coastal, Andean I-II: 0-3000 m. Lomas, rocky slopes.

Distribution: Colombia, Peru, Bolivia, N-Chile.

Arequipa: native, doubtful. Not specified by province. No voucher, mentioned by Brako & Zarucchi (1993), confusion with T. purpurea?

Tillandsia purpurea Ruiz & Pav.

Description: Plant rooting, up to 50 cm tall, forming clumps up to 80 cm in Ø. Leaves rosulate or along the stem, arranged in several rows, 10-20 cm long, covered with fine cinereous or fuscous spreading scales. Inflorescence branched, pyramidal, to 9 cm long, spikelets laxly 7-flowered. Peduncle green, bracts pink, petals white with purplish- blue at apex.

Ecology: Coastal, Andean I-II: 0-3000 m. Lomas, rocky slopes, desert dry shrublands.

Distribution: Ecuador, Peru.

Arequipa: native. ARE, CAM, CAR, ISL.

Tillandsia recurvata (L.) L.

Description: Epiphytic or terrestrial herb, 5-15 cm long, often growing spheroid. Similar to T. capillaris but peduncle and leaves longer and inflorescence 1-5-flowered.

Ecology: Coastal, Andean I-III, Coastal: 500-4500 m. Dry valleys, lomas.

Distribution: Southern North America to Chile and Argentina.

Arequipa: native. CAM, CAR, ISL.

Tillandsia usneoides (L.) L.

Description: Plant hanging from trees in branching strands up to 8 m long. Roots lacking from the first. Stem <1 mm thick, internodes 3-6 cm long with only the base covered by the leaf, the pseudo-axillary branches very short, concealed by the basal leaf, bearing 2-3 leaves. Leaves 25-50 mm long, densely lepidote, sheaths elliptic, 8 mm long, blades filiform, <1 mm thick. Inflorescence a single flower, pseudo-terminal. Floral bract ovate, shorter than the sepals. Petals linear, 11 mm long, pale green or blue, stamens deeply included.

Ecology: Coastal, Andean I-III: 0-4000 m. Disturbed areas, forests, lomas.

Distribution: Southern North America to Chile and Argentina.

Arequipa: native. ARE, CAY, UNI.

Use: Medicinal plant for the relief of rheumatism, lung and liver diseases. The plant is also said to repel fleas and other insects. Traded as an ornamental plant and for decoration.

Note: It is typically spread by fragments, blown by the wind, or spread by birds (nesting).

Tillandsia virescens Ruiz & Pav.

Description: Plant short-stemmed, few or not branched, 8-10 cm long. Leaves distichous, lax on the stem and spaced 4-7 mm apart, bent, densely gray scaled. Inflorescence 1-flowered, very seldom 2-flowered. Flowers light yellow to dark brown, often flecked.

Ecology: Andean I-III: 1000-4000 m. Dry valleys, rocky slopes.

Distribution: Peru, Bolivia, N-Argentina, Chile.

Arequipa: native. ARE, CAY.

Cactaceae

Airampoa soehrensii (Britton & Rose) Lodé

Description: Prostrate cactus, forming colonies up to 1 m in Ø. Segments flattened, very spiny, orbicular, 4-6 cm in Ø, often purplish. Spines slender and erect, up to 5 cm long, yellow to brown. Flowers yellow, style green. Fruit spineless, purple.

Ecology: Andean III: 3500-4000 m. Disturbed areas, rocky slopes.

Distribution: S-Peru, N-Chile and W-Bolivia.

Arequipa: native. ARE, CAY, UNI.

Use: The fruits are used as a food colorant (chicha morada). Used in folk medicine.

Armatocereus matucanensis Backeb.

Description: Columnar cactus, up to 2-3 m long, much-branched. Stems articulate, with 5-9 ribs, radial spines 10-15, <1 cm long, 1-4 central spines, up to 12 cm long. Flower white. Fruit up to 15 cm long, gray-violet spines.

Ecology: Andean I-II: 500-3000 m. Rocky slopes, shrublands.

Distribution: S-Peru.

Arequipa: native. ARE, CAR, CAY, UNI.

Note: This species is often split. For Arequipa, the most common species is A. riomajensis, which is synonymized with A. matucanensis by POWO (2024).

Armatocereus procerus Rauh & Backeb.

Description: Strongly articulate columnar cactus, up to 7 m tall. Stem usually unbranched and solitary, 8-10 ribs, radial spines 15-20, up to 2 cm long, up to 4 central spines, 12 cm long. Flowers 5 cm in Ø, white. Fruit 7 cm long, white spines.

Ecology: Coastal: 500-1000 m. Desert, rocky slopes.

Distribution: C-Peru.

Arequipa: native. CAR. Occurrence in the northern coastal desert of Caravelí at the border to Ica.

Austrocylindropuntia floccosa (Salm-Dyck) F.Ritter

Description: Clump-forming cactus forming loose to dense cushions, with abundant long silvery hairs, projecting well above the cushion. Flowers yellow to deep orange.

Ecology: Andean II-III: 3000->4500 m. Dry valleys, rocky slopes, shrublands.

Distribution: Peru, Bolivia.

Arequipa: native. ARE, CAY, CON, UNI.

Austrocylindropuntia subulata (Muehlenpf.) Backeb.

Description: Erect, tree-like branching cactus up to 4 m tall with round cylindrical joints. Leaves nearly cylindrical, up to 13 cm long, short-lived, spines light yellow, 1-2 per areole and 8 cm long. Flowers red, followed by reddish fruits up to 10 cm long.

Ecology: Coastal, Andean I-II: 0-3500 m. Disturbed areas, rocky slopes.

Distribution: Peru, Bolivia, Argentina.

Arequipa: native. ARE, CAS, CAY, UNI.

Use: Planted as boundary hedges and on top of walls.

Note: In Arequipa, A. subulata ssp. exaltata.

Browningia candelaris (Meyen) Britton & Rose

Description: Erect, branched, or columnar cactus, up to 6 m tall, with a straight spiny trunk topped by ± spineless thinner branches, spreading like a candelabra. Flowers white, opening at night. Fruits fleshy, yellow when ripe.

Ecology: Andean II: 2000-3000 m. Rocky slopes, shrublands, desert. Receives rainfall probably only during El Niño years.

Distribution: S-Peru, N-Chile.

Arequipa: native. ARE, CAR, CAS, CAY.

Corryocactus aureus (Meyen) Hutchison

Description: Forms large colonies by underground growing shoots. The upright, non-segmented, cylindrical stems up to 20 cm long and 3-5 cm in Ø, the decumbent stems up to 2 m long. 5-8 ribs, up to 1 cm high, ± notched at the base. Spines brown to blackish, the 1-2 central spines up to 6 cm long. The 9-11 radial spines unequal long. Flowers orange to orange-red, up to 4 cm long. Fruits green to red, 2 cm in Ø.

Ecology: Andean II: 1500-3500 m. Inter-Andean valleys and matorral.

Distribution: S-Peru.

Arequipa: native. ARE, CAR, CAY, UNI.

Corryocactus brachypetalus (Vaupel) Britton & Rose

Description: Densely branched from the base, stems up to 4 m tall, 6-10 cm in Ø, in large colonies. 7-8 ribs. 20 spines, main spine 10-16 cm long, other spines usually shorten than 1 cm, occasionally twisted, spines first black, then becoming pale. Deep orange flowers, 4-6 cm in Ø. Fruits greenish-yellow, 6-7 cm in Ø, initially spiny.

Ecology: Coastal: 0-1000 m. Lomas.

Distribution: S-Peru.

Arequipa: native. CAM, CAR, ISL.

Corryocactus brevistylus (K.Schum. ex Vaupel) Britton & Rose

Description: Erect, branched cactus, up to 5 m tall. Branches with 7-8 ridges, spines up to 24 cm long, longest spine often pointing down. Flowers yellow, fruit pear-like 12 cm in Ø, with plenty of small spines.

Ecology: Andean II: 2000-3500 m. Rocky slopes, shrubland.

Distribution: S-Peru, N-Chile.

Arequipa: native. ARE, CAR, CAS, CAY, CON, UNI.

Use: Edible fruit. The juice is said to regulate blood pressure and cure liver disease.

Corryocactus dillonii A.Pauca & Quip.

Description: Similar to C. aureus, can be distinguished by the presence of small areoles, yellowish young spines, shorter and less rigid central spines, yellowish flowers, larger tepals (0.7-1.0 x 2.3-2.6 cm), slightly rough seeds with shallow pores.

Ecology: Coastal: 500-1000 m. Lomas.

Distribution: S-Peru.

Arequipa: native. ISL.

Corryocactus quadrangularis (Rauh & Backeb.) F.Ritter ex D.R.Hunt, N.P.Taylor & G.J.Charles

Description: Erect to decumbent cactus, 1-1.5 m tall. Stem 4-5 cm in Ø, 4-5 alate ribs. 3 central spines, 4.5-6 cm long. Flowers yellow to brilliant red, 4-5 cm long. Fruits reddish-green.

Ecology: Andean II: 3000-3500 m.

Distribution: S-Peru.

Arequipa: native. CAR.

Cumulopuntia corotilla (K.Schum. ex Vaupel) E.F.Anderson

Description: Globular cactus forming ± dense clusters up to 25 cm tall, root napiform. Stem segments with 23-30(- 32) areoles. 0-11 spines per areole, longest spine 5.8 cm. Flowers at flowering pale yellow to orange.

Ecology: Andean II-III: 3000-3800 m. Grasslands, rocky slopes.

Distribution: S-Peru.

Arequipa: native. ARE, CAY.

Note: See note on C. sphaerica.

Cumulopuntia dimorpha (C.F.Först.) A.Pauca & Quip.

Description: Globular cactus, erect shrub, solitary. Similar to C. sphaerica but stem segments with 16-25 areoles. Flowers yellow, becoming orange in age.

Ecology: Andean II-III: 1900-3700 m. Shrubland.

Distribution: S-Peru, N-Chile.

Arequipa: native. ARE, CAY, CON.

Note: See note on C. sphaerica.

Cumulopuntia ignescens (Vaupel) F.Ritter

Description: Globular cactus forming cushions up to 50 cm tall, root napiform. Stem segments with <17-23 areoles, spines 0-13 per areole, 1-7 cm long. Flowers red.

Ecology: Andean III: 3300-4700 m. Grasslands, rocky slopes.

Distribution: Bolivia, S-Peru, N-Chile.

Arequipa: native. ARE, CAY.

Cumulopuntia ignota (Britton & Rose) F.Ritter ex D.R.Hunt

Description: Globular cactus forming cushions, root napiform. Stem segments with 14-22 areoles, spines 0-9 per areole, 0.5-5.6 cm long. Flowers yellow or orange. Fruit globose or subglobose.

Ecology: Andean III: 3500-4700 m. Grasslands, rocky slopes.

Distribution: S-Peru.

Arequipa: native. ARE.

Note: Several authors consider it to be a synonym of C. corotilla or C. sphaerica.

Cumulopuntia leucophaea (Phil.) Hoxey

Description: Globular cactus, solitary or in lax cushions. Stem segments with 48-74 areoles, spines 0-12 per areole, 0.5-2.3 cm long. Flowers yellow. Fruit subglobose.

Ecology: Coastal: 0-1000 m. Sandy soil.

Distribution: C-Peru to N-Chile.

Arequipa: native. ARE, CAY, ISL.

Note: Several authors consider it to be a synonym of C. sphaerica.

Cumulopuntia sphaerica (C.F.Först.) E.F.Anderson

Description: Sprawling cactus, prostrate or semi-erect, moderately branched, fragile, root fibrous. Stem segments spherical, with 28-60 prominent areoles. 0-22(-29) spines per areole, longest spines 1-3 cm long, sometimes bent. Flowers yellow at flowering.

Ecology: Coastal, Andean I-II: 0-3000(-3700) m. Grasslands, rocky slopes, dry valleys, lomas.

Distribution: S-Peru to C-Chile.

Arequipa: native. ARE, CAR, CAS, CAY, ISL, UNI.

Note: Following Hunt (2006), this complex includes the microspecies C. corotilla, C. crassicylindrica, C. ignota, C. mistiensis, C. tumida, C. unguispina and C. zehnderi. POWO (2024) synonymizes C. mistiensis with C. corotilla and C. crassicylindrica, C. ignota, C. tumida and C. unguispina with C. sphaerica. The Caryophyllales Network (2015ff.) accepts C. ignota but synonymizes C.

dimorpha with C. sphaerica. The two disputed species C. dimorpha and C. ignota are treated here separately. The phylogenetic studies of Ritz et al. (2008) show that C. sphaerica and C. leucophaea are very close genetically.

Pauca-Tanco & Quipuscoa (2020) divide C. sphaerica into C. sphaerica s.str. (above 2000 m), C. unguispina (coastal, S-Arequipa), C. tumida (N-Arequipa) and C. dimorpha based on differences in areoles, spines and segments.

Cumulopuntia zehnderi (Rauh & Backeb.) F.Ritter

Description: Sprawling cactus, cushion-forming, up to 30 cm tall, root fibrous. Stem segments with 12-28 woolly areoles. 0-13 spines per areole, longest spines 8 cm long, usually bent. Flowers yellow at flowering, turning orange with age.

Ecology: Andean II: 3000-3500 m. Dry valleys, rocky slopes.

Distribution: S-Peru.

Arequipa: native. CAR.

Cylindropuntia imbricata (Haw.) F.M.Knuth

Description: Erect cactus, 1-2.5 m tall, very thorny. Stems cylindrical, 10-40 cm long and 2-3 cm thick, segments firmly attached. Spines up to 3 cm long, 10-30 on areole, covered with sheaths. Flowers reddish-pink, 5-7 cm in Ø, with numerous tepals. Fruit yellow, fleshy, ovoid, 2.5-5 cm long.

Ecology: Coastal: 0-1000 m. Lomas, disturbed places.

Distribution: Native to SW-USA and Mexico, introduced worldwide.

Arequipa: introduced, doubtful. CAR. No voucher, mentioned for Lomas de Atiquipa by Pauca & Quipuscoa (2015). Occurrence is likely.

Cylindropuntia tunicata (Lehm.) F.M.Knuth

Description: Densely branched shrub, up to 0.6 m tall. Stem segments break off easily. Spines 5-12, up to 6 cm long, straw-colored, covered with a papyraceous sheath. Flowers yellow-green. Fruit 2-5 cm long, single.

Ecology: Coastal, Andean I-II: 0-3000 m. Disturbed areas, dry valleys, rocky slopes.

Distribution: Native to North America, introduced to subtropical areas.

Arequipa: introduced. CAY, ISL, UNI.

Eriosyce islayensis (C.F.Först.) Katt.

Description: Small variable cactus with very woolly areoles and dark spreading spines. Globose to short-cylindrical, gray-green, 5 to 15(-70) cm tall, 5-10 cm in Ø. Ribs 19-21 low and obtuse, ± tuberculate.

Ecology: Coastal, Andean I: 0-1500 m. Dry valleys, lomas, rocky slopes.

Distribution: S-Peru and N-Chile.

Arequipa: native. ARE, CAM, CAR, CAS, ISL.

Eulychnia ritteri Cullmann

Description: Treelike cactus, 2-3 m high, branched laterally, when old quite spineless below but very spiny toward the top. Growth irregularly, parts of the plants are usually dead or dying. Side branches strong and erect, 12-15 ribs and large areoles of 5-10 mm in Ø with 12-20 stout spines. Flower-buds globular, covered with long, white, silky hairs, flowers pink.

Ecology: Coastal: 0-1100 m.

Distribution: S-Peru (northern lomas of Arequipa).

Arequipa: native. CAR.

Haageocereus bylesianus (Andreae & Backeb.) Lodé

Description: A small columnar cactus, stems up to 8 cm long, 2 cm in Ø, 12-14 ribs. Spines numerous, 4-5 mm long, rust-colored, changing to gray with age. Flowers nocturnal, white, to 6 cm long.

Ecology: Coastal: 0-1000 m. Rocky slopes, desert, lomas.

Distribution: S-Peru (Arequipa).

Arequipa: native. CAM, CAR.

Haageocereus decumbens (Vaupel) Backeb.

Description: Decumbent cactus, the whole stem lies on the ground, stems 5 cm in Ø, with 20 ribs. Central spines 1- 2.5 cm long, brown. Flowers white, 8 cm long and 4.5 cm in Ø, fruits reddish-brown.

Ecology: Coastal: 0-500 m. Dry valleys, lomas.

Distribution: S-Peru, N-Chile.

Arequipa: native. CAM, CAR, ISL.

Note: Haageocereus australis Backeb. is mentioned by Ostolaza (2019) for Atico and Mollendo. An unresolved species, synonymized with H. decumbens by POWO (2024), as we do here.

Haageocereus platinospinus (Werderm. & Backeb.) Backeb.

Description: Decumbent cactus with erect apex, 13-15 ribs. Central spines 1-4, up to 7 cm long, initially brown, then silvery. Flowers white, 7 cm long, fruit orange.

Ecology: Andean I-II: 1000-2000 m. Dry valleys, rocky slopes, shrublands.

Distribution: S-Peru.

Arequipa: native. ARE, CAY.

Note: Haageocereus pluriflorus Rauh & Backeb. is mentioned by Ostolaza (2019) for Majes. An unresolved species, synonymized with H. platinospinus by POWO (2024), as we do here.

Lobivia maximiliana (Heyder ex A.Dietr.) Backeb. ex Rausch

Description: Globular to short cylindrical cactus, very variable, often forming colonies, 5 cm in Ø and up to 20 cm tall, 12-18 ribs, spines 7-12 per areole, 3-5 cm long, ± erect. Flowers red, sometimes with yellow spots, 3-4 cm long.

Ecology: Andean II-III: 2500-4000 m. Rocky slopes, shrubland.

Distribution: Peru and Bolivia.

Arequipa: native. CAY. No specimen but convincing observations from Caylloma in iNaturalist (2024).

Lobivia pampana Britton & Rose

Description: Clustered, globular cactus, 5-11 cm in Ø, 17-21 ribs, spines 5-20 per areole, 5 cm long, curved. Flowers red, sometimes with yellow spots, 6 cm long. Involucral bracts with long white hairs.

Ecology: Andean III: 3500-4000 m. Grasslands, rocky slopes.

Distribution: S-Peru.

Arequipa: native. ARE, CAY, UNI.

Loxanthocereus jajoanus (Backeb.) Backeb.

Description: Columnar cactus, erect to slightly decumbent, branched from base, stems 0.5-1.5 m tall, Ø 3-6 cm, 12- 17 ribs, central spines 1-4.6 cm long, 20+ radial spines 0.6-1 cm long. Flowers orange to red, up to 6 cm long. Fruit globose.

Ecology: Coastal, Andean I-II: 500-3000 m. Dry valleys, rocky slopes, shrublands.

Distribution: Peru.

Arequipa: native. ARE. Type from Arequipa is lost. Several mentions.

Loxanthocereus sextonianus (Backeb.) Backeb.

Description: Columnar cactus, generally prostrate, stems up to 1.5 m, up to 10 cm in Ø. 12-14 ribs, 1 to 4 brownish central spines, 4 cm long. 8 to 30 radial spines. Red flowers, up to 6 cm long, followed by green fruits.

Ecology: Coastal, Andean I: 0-1500 m. Desert, dry valleys.

Distribution: S-Peru.

Arequipa: native. ARE, CAR, ISL.

Note: Loxanthocereus gracilis (Akers & Buining) Backeb. is mentioned by Ostolaza (2019) for the northern coast of Arequipa. An unresolved species, synonymized with H. sextonianus by POWO (2024), as we do here.

Matucana haynii (Otto ex Salm-Dyck) Britton & Rose

Description: Globular cactus, unbranched, 14-30 ribs. Spines numerous, varying in size and color. Flower apical, 4-6 cm long, deep red or salmon pink.

Ecology: Andean II: 2000-3000 m. Dry valleys, rocky slopes.

Distribution: Peru.

Arequipa: native. CAR. Mentioned for prov. Caravelí by Pauca & Quipuscoa (2017) and dept. Arequipa by Ostolaza (2014), no specimen.

Note: In Arequipa, M. haynii ssp. hystrix

Melocactus peruvianus Vaupel

Description: Solitary globose cactus, glaucous green. Ribs 10-15. Spines extremely variable yellowish to dark brown or black turning gray as they age. Cephalium with white wool and reddish bristles. Flowers pink to red.

Ecology: Coastal, Andean I: 0-1500 m. Dry valleys, rocky slopes.

Distribution: S-Ecuador, Peru.

Arequipa: native. CAR.

Neoraimondia arequipensis (Meyen) Backeb.

Description: Erect, branched cactus up to 10 m tall and 40 cm in Ø. Branches with 5-8 ridges, pale green. Areoles growing continuously, long and conical. Flowers white or pink, fruits purple, globular.

Ecology: Coastal, Andean I: 500-2000 m. Desert, rocky slopes.

Distribution: C- and S-Peru.

Arequipa: native. ARE, CAM, CAR, CAS, CAY, CON, ISL, UNI.

Use: Harvested for fuel.

Opuntia ficus-indica (L.) Mill.

Description: Perennial shrub, erect, up to 5 m tall, with thick, succulent, oblong to spatulate, spiny stems, segments up to 50 cm long, epidermis waxy. Flowers yellow or red. Fruits pale green to red.

Ecology: Coastal, Andean I-II: 500-3000 m. Rocky slopes, shrublands, disturbed areas.

Distribution: Native to Mexico, naturalized in the subtopics worldwide.

Arequipa: introduced. ARE, CAY, UNI.

Use: Widely grown for its fleshy fruits and as a host for cochineal insects, which produce red and purple dyes.

Note: It has been domesticated in Mexico at least 9000 years before present.

Opuntia pubescens H.L.Wendl. ex Pfeiff.

Description: Plant decumbent to ascending, <50 cm tall. Segments unequal, terete or flattened, <15 cm long. Flowers yellow to orange.

Ecology: Andean I-II: 1000-3000 m. Dry valleys, rocky slopes, shrublands.

Distribution: Mexico to Bolivia.

Arequipa: native, doubtful. CAR, CAY, ISL.

Note: Ostolaza (2019) lists O. infesta and O. pestifer as independent species and doubts the occurrence of O. pubescens in Peru.

Opuntia stricta (Haw.) Haw.

Description: Perennial shrub, up to 2 m tall. Stems green to bluish-green, flattened, 10-25 x 7.5-15 cm. Spines yellowish. Flowers bright yellow. Fruits pear-shaped, nearly spineless, purple at maturity, pulp intense purple, sour tasting.

Ecology: Coastal: 500-1000 m. Cultivated and naturalized in disturbed areas near villages.

Distribution: Native to Central America, introduced worldwide.

Arequipa: introduced, doubtful. CAR. No voucher, mentioned for the Lomas de Atiquipa by Pauca et al. (2018). Probably introduced.

Oreocereus hempelianus (Gürke) D.R.Hunt

Description: Short cylindrical cactus, basally branched, 12-24 ribs, very spiny. 1-6 central spines 2-5 cm long, 10-15 radial spines like divergent needles. Flowers purple-red, terminal.

Ecology: Andean II-III: 2500-4000 m. Rocky slopes, shrublands.

Distribution: S-Peru and N-Chile.

Arequipa: native. ARE, CAY.

Oreocereus leucotrichus (Phil.) Wagenkn.

Description: A small shrubby columnar cactus, forming dense broad clumps covered in spines, with abundant white to black silky hairs 5-10 cm long, thicker on the more recent vegetation. Flowers diurnal, terminal in small groups, 5- 10 x 3-5 cm, corolla with dark red or purplish-red tepals and several dark purple stamina.

Ecology: Andean II-III: 2000-4000 m. Rocky slopes, shrublands.

Distribution: S-Peru and N-Chile.

Arequipa: native. ARE, CON.

Note: O. hendrcksenianus (synonym of O. leucotrichus) is treated as a separate species by some authors.

Trichocereus chalaensis Rauh & Backeb.

Description: Columnar cactus, with erect stems, up to 4 m tall. Stems cylindrical, 15 cm in Ø. Ribs 8, flat. Central spines 2-3 cm long, 6-10 radial spines, 1 cm long. Flowers white, to 17 cm long. Fruits covered with black hairs.

Ecology: Coastal: 500-1000 m. Lomas.

Distribution: Peru.

Arequipa: native. ARE, CAR.

Trichocereus cuzcoensis Britton & Rose

Description: Erect, branched cactus, up to 6 m tall, 7-8 rounded ribs. Multiple sturdy spines, up to 12 cm long. Flowers white, 15 cm long, open day and night. Fruit greenish.

Ecology: Andean II-III: 2500-4000 m.

Distribution: Peru.

Arequipa: native. CAS, CAY, UNI.

Note: Schlumpberger & Renner (2012) showed that Echinopsis is not monophyletic, neither are the synonymized genera Lobivia and Trichocereus. Unfortunately, they did not include E. cuzcoensis and E. peruviana in the analysis. Both species have an overlapping distribution, and the descriptions show minor differences.

Trichocereus macrogonus (Salm-Dyck) Riccob.

Description: Columnar cactus up to 5-6 m tall, 6-15 cm in Ø. Stems light to dark green, often bluish, 4-8 rounded ribs. Areoles with 6-8 yellow to brown, unequal spines, 1-4 cm long. Flowers white, fragrant, around 25 cm long, at the end of the stem, open at night. Fruits green, 5 cm long.

Ecology: Andean II: 2000-3500 m. Shrubland, rocky slopes.

Distribution: Peru.

Arequipa: native. ARE, CAR, CAY, ISL, UNI.

Use: It contains the psychoactive alkaloid mescaline and is used in folk medicine and as a drug. The name San Pedro is attributed to Saint Peter, the holder of the key to heaven.

Note: Whether T. macrogonus should be split into a northern species (T. pachanoi Britton & Rose, Ecuador to C- Peru) and a southern species (T. schoenii Rauh & Backeb., S-Peru) is under discussion.

Weberbauerocereus cephalomacrostibas (Werderm. & Backeb.) F.Ritter

Description: Erect, branched, or columnar cactus up to 2 m tall, basally with several stems. Branches with 8 ribs, 1-4 main spines up to 12 cm long. Flowers white, fruit globular, orange or red.

Ecology: Coastal, Andean I: 500-1000 m. Desert, rocky slopes, lomas.

Distribution: S-Peru.

Arequipa: native. ARE, ISL.

Weberbauerocereus rauhii Backeb.

Description: Erect columnar cactus, older plants with one single basal stem, branched laterally, 4-6 m tall, stems with 23 ribs. Spines variable, 6-8 sturdy central spines, 4-7 cm long, radial spines 20. Flowers reddish, 10 cm long. Fruits 3 cm in Ø, woolly hairy.

Ecology: Coastal, Andean I-II: 500-2500 m. Rocky slopes, shrublands.

Distribution: S-Peru.

Arequipa: native. CAR, CAY, UNI.

Weberbauerocereus weberbaueri (K.Schum. ex Vaupel) Backeb.

Description: Erect, columnar cactus, up to 4 m tall, usually several basal stems. Stems with 14-22 ribs, hairy and spiny, spines maroon, up to 6 cm long. Flowers slightly asymmetric, white or rose, up to 10 cm long. Fruits hairy, yellow to orange.

Ecology: Andean II: 1500-3000 m. Rocky slopes.

Distribution: S-Peru.

Arequipa: native. ARE, CAR, CAS, CAY.

Use: Edible fruits.

Calceolariaceae

Calceolaria ajugoides Kraenzl.

Description: Erect or ascending subshrub, 25-80 cm tall. Inflorescence and distal parts of stems densely glandular- puberulous. Leaves ternate, herbaceous or subcoriaceous, lanceolate, 1-4.5 x 0.5-3 cm, margins deeply serrate or dentate. Corolla bright yellow, the neck speckled with red or entirely red, bag unspotted or with some red spots.

Ecology: Coastal, Andean I-II: 0-3500 m. Rocky slopes.

Distribution: S-Peru (Ayacucho and Arequipa).

Arequipa: native. CAR, CAS.

Calceolaria angustiflora Ruiz & Pav.

Description: Subshrub 0.2-1 m tall. Leaves ternate, 2-5(-8) x 1-3.5(-5) cm, ovate, margins sharply serrate to biserrate, lower surface glabrous or sparsely pilose. Inflorescence Corolla bright yellow, with red dots in the throat.

Ecology: Andean I-II: 1000-3500 m. Shrublands.

Distribution: Peru.

Arequipa: native. ARE, CAR.

Calceolaria chelidonioides Kunth

Description: Erect or ascending herb, 0.1-2 m tall, stems rather succulent. Leaves 5-25 x 4-19 cm long, 2-4 pairs of broad lobes. Petioles up to 12 cm long, conspicuously connate across the nodes with 2-11 mm wide wings. Corolla pale to bright yellow.

Ecology: Andean I-II: 1000-3500 m. Grasslands, seasonally inundated areas.

Distribution: Central and South America.

Arequipa: native. CAR, CON, ISL.

Calceolaria colquepatana Pennell

Description: Shrubby plant, 10-50 cm tall, stems purplish, puberulous or tomentose with white hairs.

Leaves ovate, obtuse, basally truncate to cordate, 7-14 x 5-10 mm, margins crenate-dentate, with 6-7 teeth on each side, upper surface puberulous or pilose, stalked glands on the lower leaf surface. Corolla deep yellow or golden yellow, internally with a crossband of red spots in the throat.

Ecology: Andean II-III: 3000-4000 m. Grasslands, shrublands.

Distribution: Peru.

Arequipa: native. ARE, CAY.

Calceolaria cuneiformis Ruiz & Pav.

Description: Densely branched shrub, 0.2-1.2 m tall. Stem dark purple to black, distally glandular-villose. Leaves lanceolate, 10-18 x 3-7 mm, obtuse or subacute at apex, cuneate at base, margins serrate or deeply crenate, upper surface hirsute, lower surface puberulous. Flowers pale yellow.

Ecology: Andean II-III: 2500-4500 m. Grasslands, shrublands.

Distribution: Peru, Bolivia.

Arequipa: native. CON, UNI.

Note: In Arequipa, C. cuneiformis ssp. xerophila.

Calceolaria dichotoma Lam.

Description: Herb, 5-50 cm tall, pilose. Leaves petiolate, 1-4.5 x 0.5-3 cm, blade ovate, margins crenate or serrate, densely pilose on both sides. Corolla yellow, throat with reddish crossband or several spots.

Ecology: Coastal, Andean I-II, Amazonian: 0-2000 m. Lomas.

Distribution: Colombia, Ecuador, Peru, Bolivia.

Arequipa: native. CAR, ISL.

Note: It is probably sometimes confused with C. utricularioides, a species that Molau (1988) restricts to N- and C- Peru.

Calceolaria engleriana Kraenzl.

Description: Shrub up to 1.5 m tall. Leaves lanceolate, rarely ovate, upper surface olive green, glandular hirsute, lower surface grayish, reticulate, margins finely serrulate, revolute, seeming entire. Inflorescence composed of 1-2(- 4) pairs of 2-12-flowered cymes. Sepals yellowish-green to light greenish-yellow, ovate. Corolla bright yellow, in Peru usually with faint red dots in the sides of the throat.

Ecology: Andean II-III: 2000-4500 m. Forests, grasslands, shrublands.

Distribution: Peru, Bolivia.

Arequipa: native. ARE, CAY, CON, UNI.

Calceolaria inamoena Kraenzl.

Description: Small-leaved shrub 0.1-1 m tall. Stems purplish and glandular-hairy with whitish hairs. Leaves fasciculate, ovate to ± orbicular, 3-15 x 2-12 mm, subcoriaceous, ± aceriform, truncate to cordate at base, 3-

9- lobate or deeply crenate, margins deflexed or revolute. Flowers yellow, unspotted.
Ecology: Andean II-III: 2000-4500 m. Shrublands and grasslands.
Distribution: S-Peru, N-Chile.
Arequipa: native. ARE, CAS, CAY, UNI.
Note: In Arequipa C. inamoena ssp. inamoena (shrublands) and C. inamoena ssp. millefolia (graslands).

Calceolaria lobata Cav.

Description: Ascending to erect herb, 0.2-1 m tall. Leaves suborbicular to deltoid, 2-9 cm in Ø, palmately lobed (1/3 of radius), lobes irregularly crenate-dentate. Flowers yellow, with dark spots inside the bag.
Ecology: Andean II-III: 1500-4500 m. Forests, shrublands.
Distribution: Peru, Bolivia.
Arequipa: native. ARE, CON, UNI.

Calceolaria parvifolia Wedd.

Description: Erect to ascending, low shrub, distally parts tomentose or villous with white hairs. Leaves ovate, 0.5-1.8 x 0.3-1.2 cm, upper surface olive to grayish-green, lower surface with sessile glands. Flowers pale yellow.
Ecology: Andean III: 3500-4000 m. Grasslands.
Distribution: Peru, Bolivia, NW-Argentina.
Arequipa: native, doubtful. CAY. Specimens van der Werff 20728, 20886 and 20872 (all MO) from Chivay, not seen. The occurrence so far north could indicate a confusion with C. colquepatana.

Calceolaria pinnata L.

Description: Erect or ascending herb, 5-50 cm tall, stem succulent, pale green to purplish. Leaves 4-15 x 3-12 cm, pinnatisect to pinnate with 4+ pairs of pinnae, pinnae pinnatifid to pinnatisect. Flowers yellow, rarely white, stamens orange.
Ecology: Coastal: 0-1000 m. Lomas.
Distribution: Peru to N-Chile.
Arequipa: native. CAR, ISL.

Calceolaria pisacomensis Meyen ex Walp.

Description: Erect or ascending shrub or subshrub. Leaves sessile, decussate, 2.5-4.5 x 1-3 cm, blade ovate to triangular, margins serrate to crenate. Inflorescence brownish glandular-hirsute, corolla lemon-yellow at base, otherwise orange to brownish-red, unspotted.
Ecology: Andean II-III: 2000-4500 m. Shrublands, rocky slopes.
Distribution: Peru.
Arequipa: native. ARE.

Calceolaria plectranthifolia Walp.

Description: Erect or ascending herb, pilose to hirsute. Leaves sessile or with winged petioles, margins serrate. Flowers yellow, bag bent upwards and pressed against the head.
Ecology: Andean II-III: 1500->4500 m. Shrublands.
Distribution: Peru, Bolivia, NW-Argentina.
Arequipa: native. ARE.

Calceolaria rhacodes Kraenzl.

Description: Ascending herb or subshrub, 0.5-1 m tall, whole plant pubescent grayish. Leaves ovate, 3-7 x 3-5 cm, margins broadly dentate to lobed, the margins of lobes serrate. Flowers yellow, throat with a large purple blotch inside.
Ecology: Andean II: 2500-3500 m. Rocky slopes.
Distribution: Peru.
Arequipa: native. CON, UNI.

Calceolaria rugulosa Edwin

Description: Erect or ascending subshrub. Inflorescence and distal parts of stem villous or hirsute with coarse whitish hairs. Leaves ternate or less often decussate, petiolate, blades herbaceous, ovate, 3-7 x 2-5 cm, sharply serrate or biserrate, upper surface olive green. Inflorescence of 2-4 pairs or triplets of 4-12 flowered cymes. Corolla light yellow or greenish-yellow, with a large purple spot or 1-2 crossbands in the throat.
Ecology: Coastal, Andean I-III, Amazonian: 0-4000 m. Lomas, rocky slopes.
Distribution: Peru.
Arequipa: native, doubtful. CAM, CAR, ISL. Several specimens from 1840 to 1950, no modern voucher. Mentioned for the Lomas de Yuta by Quipuscoa et al. (2016), but listed as only recorded from Lambayeque in the checklist or the Peruvian Lomas by Dillon et al. (2011).

Calceolaria tenuis Benth.

Description: Erect to decumbent herb, 5-50 cm tall. Leaves 3-16 x 1-16 cm, 3 pairs of lobes. Petioles up to 5 cm long, connate across the nodes with 0.3-5 mm wide wings. Inflorescence 3.5-15 cm long, flowers pale yellow, usually unspotted.
Ecology: Andean II-III: 2500-4000 m. Wet soil, seepages and stream banks, usually in disturbed areas.
Distribution: Ecuador, Peru, Bolivia.
Arequipa: native. ARE, CAY.

Calyceraceae

Calycera pulvinata J.Rémy

Description: Sessile or almost so, with prostrate basal rosettes. Leaves spatulate, crenate, the blades 5-25 x

4-12 mm on petioles 1.5-4.5 cm long. Inflorescences solitary or clustered. Involucral bracts seven, fused at the base to form a shallow bowl 2-3 cm across in which the hemispherical flowerhead of slightly smaller dimensions sits. Seed spines up to 1 cm long.

Ecology: Andean III: 3500-4000 m. Riversides.

Distribution: Peru, Bolivia, N-Chile, NW-Argentina.

Arequipa: native, doubtful. ARE. No voucher, mentioned for RN Salinas y Aguada Blanca by Quipuscoa & Huamantupa (2010). An occurrence in Arequipa is likely due to the distribution range of the species in Peru (Cuzco, Puno, Moquegua).

Campanulaceae

Lobelia decurrens Cav.

Description: Stout erect herb, 0.8-1.2 m tall. Stem densely leafy with decurrent, irregularly toothed, linear leaves. Inflorescence a leafy raceme. Corolla 3 cm long, purplish-lavender, pubescent, the lower lip 3-lobed, the upper lobes linear.

Ecology: Andean I-II: 1000-2500 m.

Distribution: Peru.

Arequipa: native. ARE, CON, UNI.

Lobelia oligophylla (Wedd.) Lammers

Description: A dwarf herb, rooting at nodes, glabrous, densely cespitose. Leaves orbicular to oval, 2-7 mm in Ø, petiole long. Flowers solitary, axillary, calyx lobes triangular, corolla white, barely tinted with a pinkish-lavender, 8-10 mm long, tube slightly funnel-shaped, lobes lanceolate, 3-5 mm long.

Ecology: Andean III: 3500-4500 m. Marshes, wet places.

Distribution: Ecuador, Peru, Bolivia, Argentina, Chile.

Arequipa: native. ARE, CAY, CON, UNI.

Lysipomia mitsyae Sylvester & D.Quandt

Description: Minute glabrous short-lived monocarpic herb, 2-5 mm tall. Stem 0.3-2 mm long. Cotyledons persistent, subsessile, elliptic, 1-3 x 0.5 mm. Leaves usually absent. Flowers emerging from cotyledon axil, usually 1 per plant, 1.5-2.5 mm long, corolla white.

Ecology: Andean III: 4000->4500 m. Ephemeral in heavily grazed grasslands, only present in March-May.

Distribution: Peru.

Arequipa: native. CAY.

Siphocampylus candollei E.Wimm.

Description: Herb, >50 cm tall. Leaves ovate to oblong, 7-10 x 2-4 cm, irregularly dentate. Inflorescence a terminal raceme, corolla orange to red, 4 cm long.

Ecology: Andean II: 2000-3500 m. Disturbed areas, grasslands, rocky slopes.

Distribution: Peru.

Arequipa: native. CAY, UNI.

Triodanis perfoliata (L.) Nieuwl.

Description: Annual herb 0.1-1 m tall. Stems erect, slender, unbranched or with a few long branches, soft-hairy or coarsely stiff-hairy below, usually rough, bristly-hairy above. Leaves alternate, sessile, amplexicaul, 0.5-3 x 0.5-2.5 cm, rounded, palmately veined, margins dentate, stiff hairy. Flowers sessile, 5-merous, actinomorphic, lilac, corolla divided to the middle.

Ecology: Coastal: 0-500 m.

Distribution: America, from Canada to Argentina.

Arequipa: native. ISL.

Note: In Peru T. perfoliate ssp. biflora.

Wahlenbergia peruviana A.Gray

Description: Small prostrate, perennial subshrub. Stems mostly sparsely pubescent. Leaves elliptic, inconspicuously denticulate to the base, ciliate. Corolla campanulate, white to light blue, 5 mm long.

Ecology: Andean II-III: 3000-4500 m. Grasslands, rocky slopes, shrublands.

Distribution: Peru, Bolivia, NW-Argentina.

Arequipa: native. ARE, CAY.

Cannaceae

Canna indica L.

Description: Tuberous, perennial herb, up to 1.5 m tall. Stem glabrous. Leaves large, petiole sheathing. Inflorescence branching, flowers irregular, red, lip orange and spotted with red. Fruit a capsule, 3-5 x 2-3 cm, bluntly 3-angled, softly spiny.

Ecology: Coastal, Andean I: 0-2000 m. Disturbed areas.

Distribution: Native to tropical America, introduced and naturalized in the tropics worldwide.

Arequipa: introduced, adventive. ARE, CAR, UNI.

Use: Ornamental, edible tubers.

Caprifoliaceae

Dipsacus sativus (L.) Honck.

Description: Erect biennial, grooved, with prickles on ridges especially up the stem, up to about 2 m tall. Basal leaves elliptic-oblong, crenate but not lobed, narrowed to winged petiole, up to 40 cm long, prickles mostly on midvein of lower surface. Stem leaves smaller, sessile, lanceolate-triangular, connate at base. Heads terminal, ovoid- cylindrical. Corolla pale purple to pinkish-purple. Receptacle bract ending in a recurved, stiff spine, involucre bract generally ± spreading or reflexed.

Ecology: Andean I-II: 1500-2500 m. Disturbed areas.

Distribution: Native to N-Africa, Europe and W-Asia, introduced to North and South America, Australia and New Zealand.

Arequipa: introduced. ARE.

Use: Ornamental plant.

Lonicera japonica Thunb.

Description: Twining vine climbing up to 10 m high or more in trees. Leaves opposite, simple ovate, 3-8 x 2-3 cm. Flowers very fragrant, turn from white to yellow as they age. Fruit a globose dark blue berry 5-8 mm Ø, numerous seeds.

Ecology: Andean II: 2500-3000 m. Shrublands.

Distribution: Native to China, Korea and Japan, introduced as ornamental shrub all over the world, naturalized and invasive in many countries.

Arequipa: introduced. CAY.

Use: Ornamental plant.

Valeriana chaerophylloides Sm.

Description: Lax, diffuse annual herb, looks like Apiaceae. Leaves imparipinnate, leaflets 1-5 jugate, coarsely dentate or pinnatifid. Inflorescence axillary on petiole as long as the leaves. Flowers white.

Ecology: Coastal, Andean I-II, Amazonian: 0-2500 m. Lomas, rocky slopes.

Distribution: Panama to Peru.

Arequipa: native. CAR, ISL.

Valeriana coarctata Ruiz & Pav.

Description: Rosulate herb. Leaves linear, acute, ciliolate. Inflorescence up to 12 cm long, densely clustered (mint- like), interrupted on lower part of stem, flowers pinkish-white.

Ecology: Andean III: 3500->4500 m. Grasslands, rocky slopes.

Distribution: Peru, Bolivia.

Arequipa: native, expected. Occurrence in Moquegua near the border to Arequipa.

Valeriana globularis A.Gray

Description: Low cespitose herb. Leaves rosulate, linear or linear-spatulate, 2-4 cm long. Inflorescence a single globose head, flowers white-lilac.

Ecology: Andean III: 4000->4500 m. Grasslands, rocky slopes.

Distribution: Peru.

Arequipa: native. ARE, CAY, UNI.

Valeriana interrupta Ruiz & Pav.

Description: Herb, branched from the base, stem appears interrupted by the nodes, reddish. Leaves deeply incised to pinnate, leaflets narrow. Inflorescence a cyme, flowers white.

Ecology: Andean II-III: 2000-4500 m. Forests, grasslands, rocky slopes.

Distribution: Ecuador, Peru, N-Chile.

Arequipa: native. ARE, CAY, ISL.

Valeriana nivalis Wedd.

Description: Cespitose herb with a thick root. Stems several, erect. Leaves spatulate, 1.5-3 cm long, subabruptly tapering to the petiole. Inflorescence solitary, terminal, globose, flowers pale purple.

Ecology: Andean III: 3500->4500 m. Grasslands, rocky slopes.

Distribution: Peru, Bolivia, N-Chile.

Arequipa: native. ARE, CAS, CAY.

Valeriana pinnatifida Ruiz & Pav.

Description: Erect herb, >30 cm tall. Stem straight, glabrous. Basal leaves 2-pinnate, leaflets broad. Cauline leaves petiolate, ovate, entire, irregularly crenate-dentate or pinnatifid.

Ecology: Coastal, Andean I-II, Amazonian: 0-4000 m. Lomas, rocky slopes.

Distribution: Peru.

Arequipa: native. CAY, ISL. No specimen, convincing observations from Caylloma and Islay in iNaturalist (2024).

Valeriana radicata Graebn.

Description: Erect herb, glabrous throughout, 20-50 cm tall, with a thick, elongate root. Leaves imparipinnate, 2-5- jugate, the terminal one much larger than the terminal one, leaflets broadly dentate-serrate. Panicle rather contracted, the lower branches elongate, flowers in dense clusters.

Ecology: Andean III: 3500-4500 m.

Distribution: Peru.

Arequipa: native. CAY. No specimen. Observed by the author in Uyo Uyo (iNaturalist, 2024).

Valeriana rhizantha A.Gray

Description: Rosette-forming perennial with a long taproot, ± stemless, no stolons. Leaves subrotund, lying flat on the ground, glaucous, verrucose. Inflorescence a dense, acaulescent head. Flowers whitish-yellow to pinkish.

Ecology: Andean III: 4000->4500 m.

Distribution: Peru.

Arequipa: native. ARE. No specimen but a convincing observation from Uyo Uyo in iNaturalist (2024). The occurrence in Arequipa is likely due to the distribution in Moquegua.

Valeriana rubra L.

Description: Perennial herb to 70 cm high, glabrous. Younger leaves glaucous. Lower cauline leaves oblanceolate, 3.5-12 x 1-5 cm, apex acute, margins entire, upper cauline leaves ovate-acuminate with toothed margins, petiolate to sessile and auriculate. Inflorescence terminal and axillary, thyrses with numerous sessile flowers, bracts linear. Calyx dentate. Corolla pink to rose-red, rarely white. Tube slender, 7-11 mm long, basal spur 4-8 mm long. Fruit with feathery pappus.

Ecology: Andean II: 2100-2500. Cultivated land, disturbed areas.

Distribution: Native to the W-Mediterranean, cultivated as ornamental and naturalized here and there.

Arequipa: introduced, adventive. Not specified by province. No voucher, mentioned as adventive by Montesinos & Zegarra (2019).

Use: Ornamental plant.

Valeriana wandae (Graebn. & Tessend.) Christenh. & Byng

Description: Cespitose perennial with stolons, ± acaule. Leaves acuminate, lying flat on the ground, verrucose. Inflorescence a dense, acaulescent head. Flowers whitish-yellow.

Ecology: Andean III: 4000->4500 m. Rocky slopes.

Distribution: Peru.

Arequipa: native. CAY. No specimen but convincing observations from Caylloma in iNaturalist (2024).

Caricaceae

Carica augusti Harms

Description: Single-stemmed glabrous tree. Leaves long-petiolate, oblong-elliptic or lanceolate, 20 x 8 cm white- glaucous beneath, margins irregularly dentate, apex acuminate, base rounded. Fruit long ovoid, narrowly winged.

Ecology: Andean II: 2500-3000 m. Forests.

Distribution: Peru (Ayacucho).

Arequipa: native, doubtful. CAY, UNI. No voucher, mentioned by Galán de Mera & Linares (2012) for La Unión. Species of the E-slopes of the Andes. Probably cultivated here and there.

Use: Edible fruits. Medicinal plant.

Vasconcellea candicans (A.Gray) A.DC.

Description: Semi-lignose tree, up to 8 m tall. Leaves broad ovoid to orbicular, base cordate, hairy below. Flowers greenish to purplish, plants dioecious. Fruit ellipsoidal, yellow green at maturity, 10-18 x 4-6 cm, many seeds.

Ecology: Coastal, Andean I-II: 0-3000 m. Forests, lomas, riversides.

Distribution: Peru.

Arequipa: native. CAR, CAS, CAY, ISL, UNI.

Use: Edible fruits.

Caryophyllaceae

Arenaria acaulis Montesinos & Kool

Description: Cushion forming herb with a short stem, ciliate short branches. Leaves ovate always covered with dense, long trichomes, never glabrous. Pedicels shorter than or equal to the size of the calyx, sepals broadly ovate, glabrous. Corolla tubular in early anthesis, becoming narrowly campanulate, pale white-greenish, with an inconspicuous midrib, petals glabrous.

Ecology: Andean III: > 4500 m. Rocky slopes.

Distribution: S- Peru.

Arequipa: native. CAY. No specimen but a convincing observation in iNaturalist (2024).

Arenaria digyna Willd. ex D.F.K.Schltdl.

Description: Perennial, cespitose herbs, procumbent to suberect. Leaves 5 x 1 mm, oblanceolate, not succulent, keel absent, not acuminate, margins ciliate. Pedicels more than a quarter of the length of the calyx. Flowers solitary, 2-3 mm long, calyx a bit shorter than petals, petals 5, white, stamens 10. Capsule 2 mm long, 6-valved.

Ecology: Andean III: 3500->4500 m. Altiplano and bofedales.

Distribution: Colombia to Bolivia and N-Chile.

Arequipa: native. CAY.

Note: A. serpylloides (= A. serpens) was erroneously synonymized with A. digyna by Macbride in Flora of Peru (1936ff.) and Brako & Zarucchi (1993). POWO (2024) considers A. digyna to be a Mexican species, Montesinos & Teillier (2022) mention it for N-Chile and Hassler (2024) for W-South America, from Ecuador to Bolivia. The type Humboldt s.n. (HAL 136848!) is from Peru.

Arenaria orbignyana Wedd.

Description: Perennial, cespitose herbs, procumbent. Leaves 3-4 x 1 mm, spathulate-lanceolate, ± fleshy, margins serrulate-ciliate. Flowers solitary, petals shorter than sepals, white, stamens 5.

Ecology: Andean III: 3500-4000 m.

Distribution: Peru, Bolivia.

Arequipa: native. ARE.

Arenaria rivularis Phil.

Description: Perennial herb, cespitose. Stems 2.5-3.5 cm long, prostrate, glabrous. Leaves 2-3 x 1 mm, glabrous, succulent, elliptic-triangular. Flowers solitary, terminal, petals white, longer than calyx, stamens 10.

Ecology: Andean III: 3500-4500 m. Plains and high Andean salt marshes.

Distribution: Peru, W-Bolivia, NW-Argentina, N-Chile.

Arequipa: native, doubtful. ARE. No specimen, a convincing observation from RN Salinas y Aguada Blanca (iNaturalist, 2024).

Cardionema ramosissima (Weinm.) A.Nelson & J.F.Macbr.

Description: Prostrate matted perennial, stems up to 30 cm long, pubescent, usually covered at the base with old, dead leaves and stipules. Leaves opposite, linear, spine-tipped, crowded, 5-10 mm long. Flowers sessile, inconspicuous, sepals densely woolly, spine-tipped.

Ecology: Andean II-III: 2000-4500 m. Grasslands, rocky slopes.

Distribution: W-North America to C-Chile and Argentina.

Arequipa: native. ARE, CAY, ISL, UNI.

Cerastium behmianum Muschl.

Description: Stems ascending, 3-5 cm long, sparsely if at all branched. Leaves 0.5-2 cm long, 1-3 mm wide, pubescent when young, glabrate in age, oblong to ovate, apex acute, margins ciliate. Flowers few, solitary, on pubescent pedicels, petals equal to sepals, white.

Ecology: Andean III: 4000-4500 m. Grasslands.

Distribution: Peru, Bolivia.

Arequipa: native. ARE, CAY.

Cerastium peruvianum Muschl.

Description: Low cespitose herb, glandular. Stems 3-8 cm long, erect to ascending. Basal leaves oblanceolate to spatulate, base attenuate apex acute, cauline leaves sessile, linear-lanceolate. Inflorescence several flowered, flowers solitary, pedicellate. Sepals lanceolate, petals notched to 1/5 of their length, white.

Ecology: Andean III: 4000->4500 m. Rocky slopes.

Distribution: Ecuador to Bolivia.

Arequipa: native, doubtful. CAY. No specimen. Observations from Callyoma and prov. Arequipa in iNaturalist (2024). Likely to occur due to the distribution range of the species.

Colobanthus quitensis (Kunth) Bartl.

Description: A little, tufted herb with grass-like, linear, rigid or soft leaves 1-1.5 cm long. Flowers solitary on peduncles as long or longer than leaves. Petals absent, capsule exserted.

Ecology: Andean III: 3500-4000 m. Grasslands.

Distribution: Andes from Colombia to Patagonia and Antarctica.

Arequipa: native. ARE, UNI.

Drymaria divaricata Kunth

Description: Erect or ascending annual or perennial herb, up to 30 cm tall, minutely but obviously glandular- pubescent. Leaves opposite, blades ovate to reniform, 5-14 x 5-20 mm, margins entire or undulate, apex acute, base truncate to cordate, stipules incised, 1.5-2 mm long, petiole often alate, 2-6 mm long, stipules incised, 1.5-2 mm long. Inflorescences terminal, several-flowered lax cymes, peduncle 1-6 cm long, pedicels 1-10 mm long. Petals 4-6 mm long, white, as long as sepals, bifid about half their length.

Ecology: Coastal, Andean I-II: 0-3000 m.

Distribution: Peru, Bolivia.

Arequipa: native. ARE, CAR, ISL.

Note: In Arequipa, D. divaricata var. stricta.

Drymaria fasciculata A.Gray

Description: Erect annual herb, up to 25 cm tall, poorly foliate. Leaves opposite, blades glabrous, ovate, 8-18 x 4-10 mm, sessile, margins ± undulate, apex acute, base truncate, stipules entire, 1 mm long. Inflorescences terminal, dichotomously branched, 3-many-flowered, peduncle 2-6 cm long, glabrous to densely glandular. Petals 2.5-7 mm long, white, as long as sepals, bifid about half their length.

Ecology: Andean II-III: 1500-4000 m.

Distribution: Peru.

Arequipa: native. ARE, CAY, CON.

Drymaria paposana Phil.

Description: Erect or ascending annual herb, up to 30 cm tall. Leaves opposite, blades glabrous to glandular-puberulent, ovate, 5-20 x 4-15 mm, subsessile, apex acute, base truncate to subcordate, lacerate stipules mostly longer than petioles. Inflorescences terminal, subcapitate, 3-many-flowered, peduncle 1-4 cm long, pedicels 0.5-3 mm long. Petals 5-7 mm long, white, exceeding the sepals, bifid little more than half their length.

Ecology: Coastal: 0-1000 m. Lomas.

Distribution: S-Peru, Bolivia.

Arequipa: native. ARE, CAM, CAR, ISL.

Note: In Arequipa, D. paposana var. serrulata (southern lomas) and var. weberbaueri (Caravelí lomas).

Herniaria sp.

Description: Plants annual, Portulaca-like habit. Stems prostrate, short. Leaves green, opposite, alternate distally, blade elliptic to obovate, hirsute to glabrescent. Inflorescence axillary, leaf-opposed on short branches. Flowers with densely pubescent calyx, in shortly pedunculate glomeruli.

Ecology: Andean II-III: 3000-4200 m. Disturbed areas.

Distribution: S-Peru.

Arequipa: native. ARE, CAY.

Note: This species was previously addressed as Herniaria hirsuta, then as Herniaria austroamericana Chaudhri & Rutish., but it is probably an undescribed species (D. Montesinos, pers. comm.).

Paronychia andina A.Gray

Description: Small perennial, prostrate shrub. Leaves sessile, elliptic-oblong to ovate, 2.5-4 x 1.25-2 mm, closely imbricate, coriaceous, often concave with revolute margins, margins ciliate. Stipules larger than the leaves. Flowers densely clustered into subglobose terminal or subterminal glomeruli, 3-6 mm in Ø, 3-7-flowered.

Ecology: Andean II-III: 2500-4500 m. Grasslands, riversides.

Distribution: Peru, Bolivia.

Arequipa: native. ARE, CAY.

Note: Iamonico & Montesinos (2023) rename this species to P. compacta, due to the ambiguity of the types. Here, I do not follow this change, because the binomial P. andina is widely accepted.

Paronychia arequipensis (Chaudhri) Montesinos & Iamonico

Description: Cespitose perennial shrub, up to 1.2 m tall. Stems pubescent, with long internodes. Leaves narrowly ovate, 8 x 2-2.5 mm. Flowers green-yellow. Sepals awned.

Ecology: Andean II: 2000-3500 m. Rocky slopes, shrublands.

Distribution: S-Peru.

Arequipa: native. ARE.

Note: P. microphylla var. arequepensis was raised to species level by Iamonico & Montesinos (2023), due to its isolated occurrence and morphological parameters. This change has not yet been accepted by POWO (2024) but is followed here.

Paronychia muschleri Chaudhri

Description: Small perennial herb with woody tap root. Stems prostrate, congested, basally woody and much-branched, 5-15 cm long, usually with short internodes. Leaves 5-9 x 1.5-2 mm, obovate to oblanceolate, apex acute, shortly mucronate, glabrous, margins very shortly ciliate. Stipules 4-5 mm long, mostly shorter than leaves, finely ciliate. Flowers in 3-7-flowered lateral clusters, 1.5-2 mm long, glabrous.

Ecology: Andean II-III: 2000-4500 m. Cloud forests, disturbed areas, grasslands, shrublands.

Distribution: Peru, Bolivia, NW-Argentina.

Arequipa: native. ARE, CAY, UNI.

Paronychia ubinensis Montesinos

Description: Perennial herb with woody stem, branched at base. Leaves 1.8-2.4 x 0.6-0.9 mm, brown-red to lilac- red, glabrous, linear to narrowly elliptic, apex acute, with few minute bristle hairs at the tip, stipules bristly, mostly equaling the leaves. Flowers axillary, petals dark red to purple.

Ecology: Andean III: 4000-4500 m. Rocky slopes.

Distribution: Peru.

Arequipa: native, doubtful. ARE, CAY. No specimens, several observations in iNaturalist (2024) from Montesinos. Iamonico & Montesinos (2023) do not mention the species for Arequipa.

Polycarpon tetraphyllum (L.) L.

Description: Small annual, glabrous, green and generally not turning red. Four leaves per node, (pseudo-)whorled, leaves either petiolate and obovate or sessile and spatulate, 4-15 mm long, 2-7 mm wide. Pedicels 2-4 mm long. Sepals 2-3 mm long. Petals minute, white. Stamens 3. Capsule ovoid, seeds 15.

Ecology: Andean II: 2500-3000 m. Disturbed areas, riversides, shrublands.

Distribution: Native from Europe to India, introduced to the Americas and Australia.

Arequipa: introduced. ARE. No specimen but a convincing observation from the city of Arequipa.

Pycnophyllum bryoides (Phil.) Rohrb.

Description: Small shrub, forming cushions of 15-25 cm in Ø, 8-10 cm high. Leaves spirally arranged, adpressed, 1.6-2 mm long, light brown to yellowish brown, oblong to ovate, margins minutely membranous. Flowers single, terminal, 5 small petals, apically divided.

Ecology: Andean III: 4000-4500 m. Grasslands, rocky slopes, shrubland.

Distribution: Peru, W-Bolivia and N-Chile.

Arequipa: native. ARE.

Pycnophyllum molle J.Rémy

Description: Large shrub, forming cushions up to 3 m in Ø, 50 cm high. Leaves imbricate, adpressed, 1.8-3 mm long, yellowish to light brown, wide ovate to obovate, margins membranous, apex obtuse to rounded, mucronate. Calyx wide obovoid, strongly enlarged. Flowers single, terminal, bracts 4, petals 5, 1-1.2 mm long, oblong to ovate, translucent, apex obtuse to truncate.

Ecology: Andean III: 4000->4500 m. Grasslands, rocky slopes, shrublands.

Distribution: Peru, NW-Argentina, Bolivia.

Arequipa: native. ARE, CAS, CAY, CON, UNI.

Pycnophyllum tetrastichum J.Rémy

Description: Large shrub, forming cushions of 1 m in Ø, 30 cm high. Leaves opposite and decussate to imbricate, adpressed, 1.2-2.5 mm long, sheath light brown, blade yellowish green to yellowish, ovate to lanceolate, margins membranous, apex acute to obtuse, sometimes mucronate. Calyx cylindrical, bracts 4-6. Flowers single, terminal, sepals yellowish, 2-3.3 mm long, margins partially involute, apex acute, petals absent.

Ecology: Andean III: 4000->4500 m. Grasslands, rocky slopes.

Distribution: Peru, Bolivia, NW-Argentina, N-Chile.

Arequipa: native. ARE, CAY, CON, UNI.

Note: Timana (2017) synonymizes P. glomerulatum Mattf. with P. tetrastichum, Hassler (2024) and KEW (2024) do not. The facts presented by Timana are convincing, so we follow his proposal here.

Silene gallica L.

Description: Annual herb, 15-45 cm tall, stem hairy, viscid-glandular distally. Leaves oblanceolate to spatulate- petiolate, 0.5-5 cm × 3-15 mm. Inflorescences open, with racemose branches, internodes and bracts usually ca. equaling fruiting calyx, 1-5 mm, longer proximally. Flowers 5-8 mm Ø, calyx prominently 10-veined, hispid. Petals white or pink, often with dark spot or dark pink throughout, clawed.

Ecology: Coastal, Andean I-III: 0-4000 m. Cultivated areas, grassland.

Distribution: Native to the Mediterranean. Widely naturalized in grasslands all over the world.

Arequipa: introduced, adventive. ARE. No specimen, convincing observation from Cayma in iNaturalist (2024).

Silene genovevae Bocquet

Description: Annual herb, branched from the base. Leaves in basal rosettes, linear-lanceolate. Flowers in few flowered cymes, white.

Ecology: Andean II-III: 2500->4500 m. Grasslands, rocky slopes.

Distribution: Peru, Bolivia.

Arequipa: native. CAY.

Silene mandonii (Rohrb.) Bocquet

Description: Perennial, cespitose herb. Leaves in basal rosettes, linear lanceolate, densely ciliate. Flowers solitary on short, stout pedicel. Pedicel elongating up to 10 cm in fruit, often bitten off by herbivores.

Ecology: Andean III: >4500 m. Grasslands, rocky slopes.

Distribution: S-Peru, Bolivia, NW-Argentina and N-Chile.

Arequipa: native. CAY, UNI.

Spergularia andina Rohrb.

Description: Stems prostrate or ascending, 2-5 cm long. Leaves short-mucronulate, glabrescent, 10-15 mm long, 0.5 mm broad, stipules ovate-lanceolate, long-acuminate, 3-4 times longer than broad, 5-6.5 mm long, connate below. Flowers few, terminal and solitary, petals white, as long as sepals.

Ecology: Andean II-III: 2500-4500 m. Disturbed areas.

Distribution: Ecuador, Peru, Bolivia.

Arequipa: native. ARE. No specimen, a convincing observation from prov. Arequipa (iNaturalist, 2024).

Spergularia collina I.M.Johnst.

Description: A loosely, dichotomously branched annual, erect with ascending branches, 5-10 cm tall, abundantly but shortly glandular-villous. Leaves spreading, linear, 1-2 cm long. Stipules ovate, ± lacerate, 1 mm long. Flowers axillary, pedicel 1-2 cm long. Petals as long as sepals, white.

Ecology: Coastal: 0-500 m. Lomas.

Distribution: S-Peru.

Arequipa: native. CAM, CAR, ISL.

Spergularia congestifolia I.M.Johnst.

Description: Prostrate herb, much-branched, stems up 3-8 cm long, glandular-villous. Leaves 3-9 mm long, strongly congested, linear, mucronulate, falcate, hispidulous and ± glandular, longer than the internodes. Stipules black, equaling the leaves, deeply laciniate. Flowers 3-12 in lax cymes, pedicels 5-12 mm long, sepals glandular-villous, petals white, slightly longer than sepals.

Ecology: Coastal: 0-1000 m. Lomas.

Distribution: S-Peru.

Arequipa: native. CAM, CAR, CAY, ISL.

Spergularia fasciculata Phil.

Description: Perennial herb up to 20 cm tall. Stems numerous, erect, glandular-puberulent. Leaves to 1 cm long, mucronate, lower internodes much longer than the leaves. Stipules ovate, acuminate, finally mostly lacerate at the tip. Flowers few, pedicellate, sepals broadly hyaline-margined, glandular, petals white, ovate, acute to obtuse, 5 mm long.

Ecology: Coastal, Andean I-II: 0-3500 m. Lomas, open rocky slopes, waysides.

Distribution: Peru, Bolivia, N-Chile.

Arequipa: native. ARE, CAR, CAS, CAY, CON, ISL, UNI.

Stellaria cuspidata Willd. ex D.F.K.Schltdl.

Description: Prostrate to ascending herb. Leaves ovate-oblong, acuminate, ± cordate at base, glabrous, 4 cm long, lower leaves with petiole longer than blade. Sepals narrowly lanceolate, enlarging to 7-8 mm in fruit, petals 5-8 mm, 1.5-2 times as long as sepals, white.

Ecology: Coastal, Andean I-III, Amazonian: 0-4000 m. Disturbed areas, lomas, rocky slopes, shrublands.

Distribution: North and South America.

Arequipa: native. ARE, CAR, CAY, UNI.

Note: For Arequipa S. cuspidata ssp. cuspidate and ssp. prostrata (Brako & Zarucchi, 1993, Quipuscoa & Huamantupa, 2010, Hassler, 2024). Both subspecies differ in chromosome number but not morphologically (Flora of North America, 1993ff.).

Stellaria media (L.) Vill.

Description: Annual herb. Stem round, hairy on a longitudinal line. Leaves acute, ovate. Tiny white flowers, Ø 5 mm, in the leaf axils or in terminal clusters, five deeply notched petals that look like ten, five green sepals longer than the petals. The flowers close at night and open in the morning.

Ecology: Coastal, Andean I-III: 0-4500 m. Grasslands, damp places.

Distribution: Native to Eurasia, introduced worldwide.

Arequipa: introduced. ARE, CAY. No specimen but several convincing observations.

Stellaria weberbaueri (Muschl.) Montesinos & Borsch

Description: Perennial, densely cespitose, suffrutescent shrub, 10-15 cm in Ø, dioecious. Leaves imbricate, 3.5-4.5 mm long, light yellowish-brown to green on top, blade slightly incurved, lanceolate to triangular, margins ciliate. Inflorescence terminal, petals light cream, translucent.

Ecology: Andean III: 4000-4500 m.

Distribution: Peru, Bolivia.

Arequipa: native. ARE.

Note: Pycnophyllopsis weberbaueri (Muschl.) Timaná was placed in the genus Stellaria by Montesinos & Borsch (2023) based on phylogenetic analyses.

Stellaria weddellii Pedersen

Description: Similar to Stellaria media, distinguishable by its undulate leaf margin. Petals white.

Ecology: Andean III: 3600-4600 m.

Distribution: Peru, Bolivia, N-Chile, NW-Argentina.

Arequipa: native, expected. No voucher, expected due to distribution range (D. Montesinos, pers. comm.).

Celastraceae

Maytenus verticillata (Ruiz & Pav.) DC.

Description: Tree or treelet, up to 8 m tall. Stem terete, blackish, branches aggregate-verticillate. Leaves crowded, often glaucous, up to 10 cm long, oblong-lanceolate or oblong-obovate, margins serrulate, lustrous on both sides. Inflorescence 1-5-flowered, style obsolete, stigmas 3. Flowers greenish-white. Fruit 3-valved. Seeds orange-yellow.

Ecology: Andean I-II: 500-3500 m. Lomas, rocky slopes.

Distribution: Colombia to Bolivia.

Arequipa: native, doubtful. CAR. Specimen Dillon 3834 (F) from Lomas de Chaparra (Caravelí), labeled as Maytenus octogona (L'Hert.) DC. (obviously a mistake) and redetermined as Maytenus verticillata (probable).

Ceratophyllaceae

Ceratophyllum demersum L.

Description: Perennial, submersed aquatic plant, occasionally branching but with a single branch produced per node. Leaves mid-dark green, rigidly sessile, in whorls of 6-8, dichotomously (1-3-forked) divided into linear segments with 4 or 5 prominent marginal teeth. Roots absent but leafy branches sometimes modified as rhizoids, stems break easily, and the pieces continue to grow separately.

Ecology: Coastal, Amazonian: 0-500 m. Submersed.

Distribution: Cosmopolitan.

Arequipa: introduced, doubtful. CAR. Mentioned for Acarí (Caravelí) by Montesinos & Mondragón (2016).

Cleomaceae

Cleome chilensis DC.

Description: Annual herb, unarmed. Leaves palmately compound, 3-5 leaflets. Flowers white, 1 cm long.

Ecology: Coastal, Andean I-II, Amazonian: 0-3000 m. Disturbed areas, rocky slopes.

Distribution: Peru, N-Chile.

Arequipa: native. CAM, CAR, ISL.

Commelinaceae

Commelina tuberosa L.

Description: Prostrate herb up to 40 cm tall. Tuberous roots. Leaf-sheath hispid. Bracts well developed. Petals clearly overlapping and only shortly nailed, blue.

Ecology: Coastal, Andean I-II: 0-3500 m. Lomas, rocky slopes, swamps, terra firme forests.

Distribution: Peru, Bolivia.

Arequipa: native. CAM, ISL, UNI.

Convolvulaceae

Convolvulus arvensis L.

Description: Trailing herbaceous perennial. Leaves sagittate, flowers pink, white or striped pink and white, 1.5-2.5 cm long, with 2 small linear bracts. Calyx with truncate sepals.

Ecology: Coastal, Andean I-II: 0-3500 m. Disturbed areas.

Distribution: Native to Eurasia, introduced worldwide.

Arequipa: introduced, adventive. CAY. Specimen Lilljekvist 2060894 (LD) from Caylloma (not seen), no observation and no mention. The occurrence in Arequipa is likely.

Convolvulus hermanniae L'Hér.

Description: Trailing velvety villous perennial herb, up to 1.5 m tall. Leaves sagittately cordate-lanceolate or

oblong- ovate, 5 x 1.5-2 cm. Flowers solitary or in pairs, petals 1 cm long, white. Calyx with acute sepals. Fruits brown.
Ecology: Coastal, Andean I-II: 0-3000 m. Rocky slopes, shrublands.
Distribution: Ecuador, Peru, Bolivia, N-Argentina.
Arequipa: native. ISL.

Cressa truxillensis Kunth

Description: A branching, ash-gray sericeous subshrub. Leaves sessile, oblong, 4-6 x 2-3 mm. Flowers axillary, solitary, calyx deeply parted, corolla white.
Ecology: Coastal, Andean I-II, Amazonian: 0-2500 m.
Distribution: SW-USA to Patagonia.
Arequipa: native. CAM, ISL.

Cuscuta acutiloba Engelm.

Description: Stems medium to slender. Flowers 2.5-3 mm long, purple or reddish, pedicellate. Calyx as long as the corolla tube.
Ecology: Coastal, Andean I-II: 0-500(-2800) m.
Distribution: Peru, Bolivia.
Arequipa: native. ARE.
Note: Brako & Zarucchi (1993) classified the species as coastal, but the specimen Pennell 13242 (MO) from prov. Arequipa is from an altitude of 2800 m.

Cuscuta cockerellii Yunck.

Description: Stems slender. Flowers 4-6 mm long, white, ± glandular, in dense globose clusters. Calyx about half as long as corolla, imbricate at base.
Ecology: Andean II: 2000-2500 m. Shrubland, a parasitic climber growing on a different host.
Distribution: Type Cockerell 1231001.21 from Yura, distribution unknown.
Arequipa: native. ARE, CAY.

Cuscuta foetida Kunth

Description: Stem medium to coarse. Leaves (scales) developed. Flowers sessile, 4-7 mm long in dense clusters, these sometimes about 2.5 cm in Ø.
Ecology: Andean II-III: 2500-4000 m.
Distribution: Ecuador, Peru.
Arequipa: native. CAM, CAS.

Cuscuta grandiflora Kunth

Description: Stems medium in size, sometimes reddish. Leaves (scales) not developed. Flowers in loose cymose clusters, large, 4-6 mm long, yellowish to white.
Ecology: Andean II-III: 2000-4000 m. Damp places.
Distribution: Andes from Venezuela to Bolivia and NW-Argentina.
Arequipa: native. CAM, CAR.

Cuscuta hitchcockii Yunck.

Description: Stems nearly filiform. Flowers red, glandular, 2-4 mm long, shortly pedicellate, in paniculate cymes. Calyx lobes free to base.
Ecology: Coastal: 0-500 m.
Distribution: Peru.
Arequipa: native, endemic. CAR, ISL.

Dichondra argentea Willd.

Description: Perennial herb, stems silvery-canescent. Leaves herbaceous, blades reniform, 5-13 x 12-20 mm, pubescent on both sides, apex obtuse to slightly emarginate, base truncate-cordate, blade decurrent to petiole, margins entire, petioles 1-2 cm long. Flower cream, on recurved peduncles, calyx campanulate, sepals linear- oblong, 2-2.6 mm long, pubescent.
Ecology: Andean II: 2000-3000 m. Disturbed areas.
Distribution: Mexico to Argentina.
Arequipa: native. CAY, UNI.

Dichondra microcalyx (Hallier f.) Fabris

Description: Creeping herb. Leaves round, ear-like, cordate, pubescent on the underside but not sericeous. Flowers white, solitary, 5 mm in Ø.
Ecology: Coastal, Andean I-III, Amazonian: 0-4000 m.
Distribution: Native to South America.
Arequipa: native. ARE, CAR, ISL.

Dichondra sericea Sw.

Description: Similar to D. argentea but leaves ± orbicular, not decurrent to petiole, bicolorous, variable in pubescence.
Ecology: Coastal, Andean I-II: 150-2750 m. Disturbed areas.
Distribution: Mexico to Argentina.
Arequipa: native, doubtful. CAR, ISL. Several specimens that look like D. microcalyx.
Note: D. argentea and D. sericea can be confused, it is questionable if both species occur in Arequipa.

Evolvulus argyreus Choisy

Description: Prostrate herb, 10-35 cm long, branched from the base, densely sericeous with adpressed shining hairs. Leaves subsessile, lanceolate to linear. Flowers bright blue.
Ecology: Coastal, Andean I-II: 0-3500 m.
Distribution: South America.
Arequipa: native, doubtful. CAR, ISL. Species of N-Peru and the E-slopes of the Andes, 3 specimens from the lomas of Islay and Caravelí (not seen), confusion with E. lanatus?

Evolvulus lanatus Helwig

Description: Perennial from a thick woody root, prostrate or ascending, up to 15 cm long. Whole plant densely lanate. Leaves subsessile, longer than the internodes, obovate to elliptic. Corolla blue.

Ecology: Coastal: 0-500 m.

Distribution: Peru.

Arequipa: native. CAM, CAR, ISL.

Evolvulus villosus Ruiz & Pav.

Description: Perennial from a thick woody root, prostrate or ascending, up to 40 cm long. Stem and leaves often glabrescent in age. Leaves petiolate, petiole 1.5 mm long. Pedicels generally longer than the leaves, sepals long villous. Corolla blue.

Ecology: Coastal, Andean I-II: 0-2000 m.

Distribution: Peru.

Arequipa: native. CAR.

Ipomoea alba L.

Description: Vigorous perennial vine. Leaves 5-15 x 4-14 cm, sometimes lobed to about one third, cordate at base. Corolla tubiform, white with greenish midrib, 5-12 cm long, 4-5 cm in Ø.

Ecology: Coastal, Andean I, Amazonian: 0-1500 m. Disturbed areas.

Distribution: Native to tropical America, introduced pantropically.

Arequipa: introduced. ARE, CAM, CAS, CAY.

Use: Ornamental plant.

Ipomoea cairica (L.) Sweet

Description: Perennial herb, growing from a tuberous rootstock. A prostrate creeper or twining into other vegetation. Leaves palmate with 5 to 7 lobes. Flowers pink or purple.

Ecology: Coastal: 0-1000 m. Disturbed areas.

Distribution: Tropical and subtropical Africa, Asia and Australia. Naturalized in Europe and America.

Arequipa: introduced. ARE, ISL.

Use: Edible tuber, stems and leaves, cooked. Caution should be exercised due to the presence of hydrogen cyanide. The whole plant is used to treat external infections. Ornamental plant.

Ipomoea carnea Jacq.

Description: Shrub growing 1-3 m tall, with an erect or ascending habit when growing in open positions, in shady positions it adopts a climbing habit with twining stems up to 5 m long. Flowers lilac or pink, darker inside, funnelform, 7-9 cm, tube and mid petaline bands mealy outside.

Ecology: Coastal, Andean I-II: 0-2000 m. Along rivers and canals, sometimes on beaches.

Distribution: Native to South America, introduced in the tropics worldwide.

Arequipa: native. ARE, CAR, ISL.

Ipomoea dumetorum Willd.

Description: Slender, twining, perennial herb, stems glabrous. Leaves petiolate 4.5-11 x 1.5-6 cm, ovate, cordate, the auricles almost touching, surfaces glabrous. Flowers solitary, axillary, sepals with dark spots on underside, corolla 2 cm in Ø, bluish-purple.

Ecology: Coastal, Andean I-II, Amazonian: 0-3500 m. Bogs, disturbed areas, riversides.

Distribution: S-North America and South America.

Arequipa: native. ARE, CAR, ISL, UNI.

Note: Listed as a synonym of I. tricolor by Flora of Peru (at least the Peruvian species).

Ipomoea nationis (Hook.) G.Nicholson

Description: Perennial climbing herb with root tubers, stems glabrous or pilose. Leaves petiolate, 2-8.5 × 1.3-8 cm, ovate-deltoid to suborbicular, acute, base truncate to cordate, the auricles rounded to subacute, margins entire or undulate, glabrous except at apex of petiole and on main veins beneath, petioles 1-7 cm. Inflorescence of long pedunculate, 1-3-flowered axillary cymes, peduncles 6-20 cm, corolla scarlet, glabrous.

Ecology: Coastal, Andean I-II: 0-2700 m. Rocky slopes.

Distribution: Peru.

Arequipa: native. ARE, CAY, UNI. No specimen for Arequipa but several observations.

Ipomoea plummerae A.Gray

Description: Glabrous perennial herb, stems usually several, branched near the base, decumbent or ascending, up to 30 cm long. Leaves petiolate, small, digitately divided into 5-7 linear leaflets. Leaflets 3-30 x 1-3 mm. Flowers solitary, corolla funnel-shaped, tube white, limb dark pink, 2 cm in Ø.

Ecology: Andean II-III: 2500-4000 m. Shrublands.

Distribution: Mexico, W-South America to N-Argentina.

Arequipa: native. ARE, CAY. No specimen, several observations in iNaturalist (2024) and mentioned by Brako & Zarucchi (1993).

Ipomoea purpurea (L.) Roth

Description: Herbs annual, twining, axial parts short pubescent and long retrorse hirsute. Stems 2-3 m. Leaves orbicular or broadly ovate, 4-18 x 3-16 cm, ± strigose, base cordate, margins entire or ± 3-lobed, apex acute. Inflorescence 1-5-flowered, corolla 4 cm in Ø, red, reddish-purple, or bluish-purple, with a fading to white center, funnelform, 4-6 cm, glabrous.

Ecology: Andean II: 1500-2500 m. Disturbed areas, riversides.

Distribution: Probably native to South America, naturalized elsewhere.

Arequipa: native. ARE, CAR, CAS, ISL.

Use: Ornamental plant. Seeds used medicinally with an effect similar to LSD.

Jacquemontia unilateralis (Roem. & Schult.) O'Donell

Description: Suffrutescent perennial, non-climbing. Leaves ovate to narrowly elliptic, base truncate, rounded or slightly cordate. Flowers deep blue throughout, corolla funnel-shaped.

Ecology: Coastal, Andean I: 0-2000 m. Lomas, rocky slopes.

Distribution: Tropical America.

Arequipa: native. CAR, ISL.

Crassulaceae

Crassula connata (Ruiz & Pav.) A.Berger

Description: Erect annual, 2-6 cm tall, red in age. Leaves opposite, ovate to oblong, succulent, connate at base. Flowers 4-merous, reddish-white.

Ecology: Andean I-III, Amazonian: 0-4500 m.

Distribution: North and South America.

Arequipa: native. ARE, CAR, ISL.

Echeveria peruviana Meyen

Description: Rosettes on rhizome, no stem. Leaves in rosettes, widely obovate when young, narrowly obovate in older plants, sessile, almost horizontal, 2-6 cm long, 0.7-1.7 cm wide at the base. Flowers red.

Ecology: Andean II: 2000-3500 m.

Distribution: Peru, Bolivia, NW-Argentina.

Arequipa: native, expected. Mentioned for Tacna and Moquegua, an occurrence in Arequipa is not reported but quite possible.

Echeveria vulcanicola Pino, Montesinos & Matusz.

Description: A succulent glabrous herb with conspicuous stems up to 30 cm long and 2-2.5 cm in Ø, swollen at base up to 4 cm in Ø, branched. Leaves on top of the stem, oblong to narrowly obovate, sessile, strongly attached, slightly incurved, 3-8 (-10) cm long.

Ecology: Andean II: 3500 m. On the cracks of black lava stones.

Distribution: S-Peru.

Arequipa: native. CAS.

Kalanchoe delagoensis Eckl. & Zeyh.

Description: Perennial succulent plant, 0.2-2 m tall. Stems erect, glabrous. Leaves verticillate or opposite, terete, 1- 13 x 0.5 cm, reddish-green to gray-green, margins with propagules.

Ecology: Andean II: 1500-2500 m. Disturbed areas near villages.

Distribution: Native to Madagascar, cultivated and naturalized in the tropics and subtopics.

Arequipa: introduced. ARE.

Sedum ignescens Pino & Montesinos

Description: Perennial, succulent, cespitose plant, glabrous, forming mats to 3-15 cm in Ø. Leaves imbricate, reddish, 2-3.5 x 1.5-2.5 mm. Flowers yellowish-green with some reddish dots.

Ecology: Andean II-III: 2500-4000 m. Rock crevices, mossy rocks.

Distribution: S-Peru.

Arequipa: native. ARE, CAY, UNI.

Cucurbitaceae

Apodanthera mandonii Cogn.

Description: Vine that develops from a perennial rootstock, tendrils 1-2-fid. Leaves deeply lobed to palmate, whitish underside. Flowers yellow, pedicels and hypanthium of staminate flowers densely hirsute. Fruit fleshy, indehiscent, white-spotted green, smooth.

Ecology: Andean II: 2000-3500 m. Grasslands, rocky slopes.

Distribution: Ecuador, Peru, Bolivia.

Arequipa: native. CAY, UNI.

Cyclanthera mathewsii Arn. ex A.Gray

Description: Vine. Leaves obtusely lobed in adult leaves, sometimes deeply lobed in young leaves, the underside of blade green. Flowers greenish-white. Fruit pale green, spiny.

Ecology: Coastal, Andean I-II: 0-3500 m.

Distribution: Ecuador, Peru.

Arequipa: native. CAR, ISL.

Sicyos baderoa Hook. & Arn.

Description: Annual creeper with fistulous, weak and hairless stems, tendrils forked 2-4 times. Leaves deeply cordate, up to 5 cm long, 5-angled or slightly 3-5-lobed. Flowers capitately crowded, pedicels 1-3 mm, corolla white. Fruits ovoid, tuberculate or ± villous.

Ecology: Coastal, Andean I-III, Amazonian: 0-4000 m. Disturbed areas, lomas.

Distribution: The Andes, from Venezuela to C-Chile.

Arequipa: native. ARE, CAM, CAR, CAS, ISL, UNI.

Sicyos kunthii Cogn.

Description: Similar to S. baderoa but leaves short-hirsutulous beneath, staminate racemes about as long as the leaves.

Ecology: Andean I-II: 750-2700 m. Riverbanks, disturbed places.

Distribution: Central and South America.

Arequipa: native, doubtful. CAM. Specimen Anonymous 1091 (MO, HUSA) from Camaná with swapped coordinates, no mention and no observation. Mentioned by Dillon et al. (2011) for the lomas of Peru.

Cyperaceae

Bolboschoenus maritimus (L.) Palla

Description: Variable perennial herb, 15-180 cm tall, erect. Stem sharply triangular, with horizontally creeping rhizome forming hard, round tubers at the nodes. Leaves stiff, flat, long, (1-)2-12 mm wide, the uppermost overtopping the flower clusters. Flowers in terminal clusters or umbels, rust-colored to brownish, spikelets solitary or in groups of 3-10, ovoid to oblong-ovoid, round 1-2(-4) cm long.

Ecology: Coastal: 0-500 m. Beaches, disturbed areas, brackish habitats.

Distribution: Cosmopolitan.

Arequipa: native. CAM, ISL.

Carex collumanthus (Steyerm.) L.E.Mora

Description: Low-growing plant. Leaf and bract blades less than 2 cm long, occasionally with prickle-hairs on the upper surface (especially on or near the midvein). Inflorescence near the base of the plant, largely buried under the leaves, elongated culms lacking. Spikes 3-4. Fruits oblong-elliptical, 4.6-6.2 mm long, 2-3 mm wide, several-veined, yellowish-brown to yellow at maturity.

Ecology: Andean III: 4400-4700. Moist to wet sites in high andean moors, bofedales.

Distribution: Venezuela to Bolivia.

Arequipa: native, doubtful. ARE. No voucher. Specimen Ruthsatz 10132 (MIN) from S-Peru, mentioned by Wheeler (2002). An inconspicuous species that is probably often overlooked.

Carex ecuadorica Kük.

Description: Similar to C. melanocystis but with long runners and subtending bract longer than glume.

Ecology: Andean III: 3500-4500 m. Humid puna.

Distribution: Ecuador, Peru, Bolivia, NW-Argentina.

Arequipa: native, doubtful. ARE. No voucher, mentioned for RN Salinas y Aguada Blanca by Quipuscoa & Huamantupa (2010), probably confusion with C. melanocystis.

Carex humahuacaensis G.A.Wheeler

Description: Plants perennial, low growing, forming ± dense mats. Rhizomes slender, creeping. Fertile culms 2-2.5 cm tall, brownish-tinged and somewhat fibrillose at the base. Leaves 9-13, basal, exceeding the culms, blades 2-4 x 0.1- 0.4 cm, stiff, midvein prominent on underside. Inflorescences inconspicuous, 1.5-2 cm tall, all spikes buried among leaves at base of plant. Spikes

3-4, the terminal one staminate, the lateral ones pistillate. Fruits 3.9-4.6 mm long, yellowish-brown to brown, abruptly contracted into a 0.4-0.6 mm long beak.

Ecology: Andean III: 3500-4500 m. Wetlands, bofedales. Tolerates salty conditions.

Distribution: C-Peru to NW-Argentina.

Arequipa: native. ARE.

Carex melanocystis É.Desv.

Description: Plants colonial, runners 1-2 cm. Culms usually curved, bluntly trigonous, 1-15 cm long. Leaves basal, sheaths brown to dark brown, blades involute, ± equal to culms, 0.5-2 mm wide. Inflorescences 0.5-1.6 cm, spikes 3- 7, in dense ovoid to hemispherical head. Subtending bract as long as glume. Anthers 1.1-2 mm. Fruits pale yellowish-brown proximally, darker brown on beak and distally, ± inflated, ovate to broadly ovate, 3.2-5.1 × 1.6-2.3 mm, papery or leathery, beak 0.5-1 mm.

Ecology: Andean III: 4000->4500 m. Shores, rocky places.

Distribution: Peru to Argentina.

Arequipa: native. ARE.

Carex vallis-pulchrae Phil.

Description: Perennial small herb, in loose to dense clumps, 3.5-4 cm tall. Rhizomes creeping, elongate, often branched, ferruginous. Leaves (3)5-9, mostly basal, blades 1-4 cm long, 0.3-0.8 mm wide, filiform, erect or somewhat spreading. Leaf margins smooth. Inflorescence a single, terminal spike, staminate part 2-3-flowered, pistillate part 2-4-flowered. Fruits 3.4-5 mm long.

Ecology: Andean II-III: 2000-4000 m. Bogs, margins of steams and small lakes.

Distribution: C-Andes in Peru and Bolivia to S-Patagonia.

Arequipa: native. CAY.

Cyperus alternifolius L.

Description: Erect, semi-aquatic herb, up to 2 m tall, roots fibrous. Culm trigonous to subterete, solid, glabrous, apically scabrid. Leaves at the base, short, almost reduced to sheaths. Inflorescence a compound umbel with 8-18 rays, each with 1-4 raylets, spikes globose. Involucral bracts 11-18, leaf-like. Spikelets 3-9 per raylet, flattened, greenish-white.

Ecology: Coastal, Andean I: 0-1500 m.

Distribution: Native to Africa, introduced in the tropics and subtropics worldwide.

Arequipa: introduced, adventive. CAM.

Use: Ornamental plant. Discussed as a nitrogen remover in wastewater.

Cyperus articulatus L.

Description: Robust perennial, 40-140 cm tall. Culms to 25 cm apart, 1(-3) together, terete, glabrous, pith-filled with conspicuous transverse septa 5-50 mm apart. Leaves reduced to sheaths. Inflorescence simple to compound, rays 5-8, 1-10 cm long, spikes generally 1, broadly ovoid to ± umbellate. Involucral bracts usually 2, 0.2-5 cm long, longest appearing to be continuation of culm Spikelets in loosely digitate clusters, scales deciduous.

Ecology: Coastal, Andean I-II: 0-2000 m.

Distribution: Pan(sub)tropical.

Arequipa: native. CAS.

Cyperus difformis L.

Description: Annual herbs, cespitose. Culms trigonous, smooth, 10-40 cm tall, 2-3 mm Ø. Leaves 2-7, flat, 7-22 x 0.2-0.4 cm. Inflorescence simple, rays 1-5, 5 cm long, spikes dense, globose, 12-23 mm in Ø. Involucral bracts 2-4, 1-22 cm long, leaf-like, longest ± erect. Spikelets 30-120, greenish-brown to purplish-brown, flattened.

Ecology: Coastal, Andean I, Amazonian: 0-1500 m.

Distribution: Native to the tropics and subtropics of the Old World, introduced to the Americas.

Arequipa: introduced. ARE. No specimen, a single observation in iNaturalist (2024). Mentioned by Brako & Zarucchi (1993).

Cyperus eragrostis Lam.

Description: Herbs, perennial, cespitose, rhizomatous. Culms trigonous to terete, 40-60 cm tall, glabrous. Leaves flat or V-shaped. Inflorescence simple, spikes globose, rays 3-10. Involucral bracts 4-8, 3-50 cm long, leaf-like. Spikelets 20-70, green, becoming brown, flattened.

Ecology: Coastal, Andean I-II, Amazonian: 0-3000 m. Disturbed areas.

Distribution: Native to South America, introduced elsewhere.

Arequipa: native. ISL.

Cyperus hermaphroditus (Jacq.) Standl.

Description: Herbs, perennial, rhizomatous. Culms trigonous to terete, 30-100 cm tall, glabrous. Leaves V- or W- shaped, 15-90 cm long. Inflorescence simple, rays 5-10, spikes ovoid to oblong cylindric. Involucral bracts 4-8, 8-40 cm long, leaf-like. Spikelets 50-150, green becoming brown, oblong to linear, quadrangular.

Ecology: Coastal, Andean I-III, Amazonian: 0-4000 m.

Distribution: Central and South America.

Arequipa: native. ARE, CAR, ISL.

Cyperus laevigatus L.

Description: Herbs, perennial, rhizomatous. Culms trigonous to terete, 8-40 cm tall, glabrous. Leaves present or reduced to sheaths, base of culm covered with reddish sheath 5-25 mm. Inflorescence appearing lateral, spikes digitate. Involucral bracts 2, 1-8 cm long, longer bract erect, appearing as continuation of culm. Spikelets 2-7, whitish-green to reddish, compressed-turgid, ellipsoid to oblong-lanceolate.

Ecology: Coastal: 0-500 m. Saline plains, beaches, disturbed areas.

Distribution: Tropics and subtropics of the New and Old World.

Arequipa: native. CAR.

Cyperus ochraceus Vahl

Description: Herbs, perennial, cespitose. Culms trigonous, 20-70 cm tall, glabrous. Leaves 3-11, V-shaped, 12-40 cm long. Inflorescence simple to compound, rays 5-10, spike loosely to densely hemispherical, 15-30 mm in Ø. Involucral bracts 4-8, 10-36 cm long, leaf-like. Spikelets 10-30, greenish, grayish, or yellowish, oblong to linear- oblong, compressed.

Ecology: Coastal, Andean I, Amazonian: 0-1000 m. Damp, disturbed soils, ditches.

Distribution: Southern North America to N-Argentina and Peru.

Arequipa: native. CAR, ISL.

Cyperus rotundus L.

Description: Herbs, perennial, stoloniferous, stolons bearing tubers. Culms trigonous, 10-35 cm tall, glabrous. Leaves V-shaped. Inflorescence compound, spikes broadly ellipsoid, rays 4-6. Involucral bracts 3-5, 1-10 cm long, leaf-like. Spikelets 3-7 per raylets, reddish-brown, slightly flattened.

Ecology: Coastal: 0-500 m. Disturbed areas, lawns. Grows best in moist, fertile soils.

Distribution: Native to Africa, introduced all over the world, pantropical weed.

Arequipa: introduced. ISL. No voucher. Mentioned for Mejía by Arce Condori (2010) and for the lomas of Peru by Dillon et al. (2011).

Use: Consumed as a famine food and used in folk medicine to prevent tooth decay in Egypt.

Cyperus seslerioides Kunth

Description: Herbs, perennial, cespitose, rhizomes knotty. Culms roundly trigonous, 4-30 cm tall, glabrous. Leaves 1-5, flat. Inflorescence simple, spikes sessile, hemispheric to spheric, rays absent. Involucral bracts 2-5, 1-3 cm long, horizontal to reflexed, parallel to culm. Spikelets 20-60, whitish to greenish or brownish, ovoid, flattened.

Ecology: Andean II-III: 2000-4500 m. Seeps, damp meadows in forests.

Distribution: Texas to Peru and N-Argentina.

Arequipa: native. ARE.

Cyperus tacnensis Nees & Meyen

Description: Herbs, perennial, cespitose. Culms compressed-trigonous, up to 30 cm tall, ± glaucous. Leaves V- shaped. Inflorescence compound, spikes hemispherical, rays 6-8, raylets 4-5. Involucral bracts 2-5, 1-10 cm long, spreading to erect. Spikelets 20-30, reddish-brown, flattened. Anthers 1.5 mm long.

Ecology: Coastal, Andean I-II: 0-2500 m.

Distribution: Peru, Bolivia.

Arequipa: native. ARE. No specimen but a not curated observation by the author from Sta. Rita de Siguas in iNaturalist (2024). Mentioned for Arequipa by Bracko & Zarucchi (1993).

Note: Some authors synonymize C. tacnensis with C. rigens, others separate them into C. tacnensis for Peru and Bolivia and C. rigens for SE-South America. POWO (2024) and Hassler (2024) mention both species for Peru. According to Pedersen (1972) only C. tacnensis is listed here, because there are no vouchers of C. rigens for   Arequipa.
C. tacnensis is often confused with C. laetus.

Eleocharis albibracteata Nees & Meyen ex Kunth

Description: Extensively creeping, ligneous rootstock. Culms 2-15 cm, slender, rigid and sulcate, sometimes recurved. Sheats deep-brown, rigid, with scarious edge. Spikelets 2-5 mm long, broadly ovate, 3-6 flowered, almost black, scales obtuse, black mingled with reddish-brown, sometimes with a whitish region near the margin. Achenes yellow or pale.

Ecology: Andean II-III: 2000-4500 m.

Distribution: Mexico to Argentina and Chile.

Arequipa: native. ARE.

Eleocharis geniculata (L.) Roem. & Schult.

Description: Tufted herb, up to 45 cm tall, rhizomes and stolons absent. Stem 0.5-1 mm Ø, green, bended. Open sheath base brown, becoming straw-colored above, tip oblique, 1-toothed. Spikelet 2-8 mm, much wider than stem, ovate to ± round, 10-many-flowered. Flower bracts ovate-obtuse, greenish with dark brown sides, lower deciduous. Style 2-branched.

Ecology: Coastal, Andean I, Amazonian: 0-1500 m. Wet soil, marshes, streambanks, brackish water.

Distribution: Worldwide in the tropics and subtropics.

Arequipa: native. CAR, ISL, UNI.

Isolepis cernua (Vahl) Roem. & Schult.

Description: A tufted annual with fibrous roots and very slender culms. Leafless except for one short filiform blade from the upper sheath. Spikelet solitary, ovoid, 2-20 mm long, brownish.

Ecology: Andean II-III: 3000-4000 m. Riversides.

Distribution: Native to the Americas and was introduced to Europe, Asia and Australia.

Arequipa: native. ARE.

Phylloscirpus deserticola (Phil.) Dhooge & Goetgh.

Description: Cespitose, small herbs, 2-12 cm tall, rhizomes creeping, branching. Culms terete, glabrous, covered with old brown leaf sheaths. Leaves ca. 0.5 x 2 cm, scabrous toward the tip, apex acuminate, often orange. Inflorescence a single spikelet or a terminal, dense head of 2-6 spikelets.

Ecology: Andean III: 4000-5000 m. Bofedales.

Distribution: Ecuador to NW-Argentina.

Arequipa: native. ARE, CAY.

Rhodoscirpus asper (J.Presl & C.Presl) Lév.-Bourret, Donadío & J.R.Starr

Description: Perennial herb, 30-160 cm tall, forming tufts or mats. Rhizomes short creeping. Aerial vegetative parts light yellowish-green. Culms solitary or loosely clumped, erect, obtusely trigonous. Leaves basal numerous and (0)1- 3 cauline. Ligule wider than long, rounded to cordate, margins ciliate. Leaf-sheaths loose, dark to pale reddish-brown basally. Inflorescence terminal, compound, congested to open at maturity, 12 × 10 cm, spikelets (20+) in dense glomeruli.

Ecology: Coastal, Andean I-II: 200-3500 m. Dry valleys, beaches, pastures, cultivated areas, often in or near flowing water.

Distribution: Peru, Bolivia, Argentina, Chile.

Arequipa: native. ARE.

Schoenoplectus americanus (Pers.) Volkart

Description: Rigid and rather stout, 0.5-2.5 m tall, growing from long rootstocks. Culms 3-10 mm in Ø, sharp trigonous. Leaves 1-3, keeled, ± as long as sheath. Inflorescence head-like, inflorescence bract erect, 1-6 cm long. Spikelets 2-20, oblong-ovoid, acute, about 1 cm long.

Ecology: Coastal, Andean I: 0-1500 m. Disturbed areas, riversides, tolerant to brackish water.

Distribution: Throughout tropical and temperate America.

Arequipa: native. ARE, CAR, ISL.

Use: Used for braiding and construction.

Schoenoplectus californicus (C.A.Mey.) Soják

Description: Knotty rhizomes. Stems 1-4 m tall, dark green, smooth, slightly 3-angled. Leaf blades absent, just sheaths. Inflorescence terminal, panicle-like, inflorescence bract, erect, 1-8 cm long. Spikelets 25-150+, narrowly ovoid, numerous, deep brown.

Ecology: Coastal, Andean I-III: 0-4500 m. Brackish to fresh marshes, shores, often emergent in water.

Distribution: California to Patagonia.

Arequipa: native. CAY. No specimen but a convincing observation from Caylloma in iNaturalist (2024).

Use: Used to make boats (caballito de totora), basketry, food, fodder, and fertilizer.

Schoenoplectus pungens (Vahl) Palla

Description: Perennial herb, culms 1-2 m tall, rhizome long. Culms 1-6 mm in Ø, sharp trigonous to concave. Leaves 2-6, 3-sided to flat, 2-5x as long as sheath. Inflorescence head-like, inflorescence bracts 1-2, erect, 1-20 cm long. Spikelets 1-5, oblong-ovoid, acute, about 1 cm long.

Ecology: Coastal, Andean I-II: 0-2500 m. Marshy areas, ponds and lakes.

Distribution: The Americas, Europe and Australia.

Arequipa: native. ARE, ISL.

Use: Used for braiding and construction.

Note: In Arequipa, S. pungens var. badius.

Zameioscirpus muticus Dhooge & Goetgh.

Description: Small, perennial herbs, 1.5-5 cm tall. Culms 0.8-4 cm long, erect, terete, unbranched, 0.2-0.45 mm wide below the inflorescence, maximum width 0.6-1.15 mm, inferior part conspicuously inflated. Leaves shorter than to as long as the culms, ligulate. Blades 3.5-10.1 cm x 0.3-0.7 mm, margins smooth, tips not awned. Old leaf sheaths brown, persistent at the base. Inflorescence a single terminal spikelet, 3.5-6 x 1.2-3 mm, yellow or brown.

Ecology: Andean III: 4000-4800 m. Bofedales.

Distribution: W-Peru to NW-Argentina.

Arequipa: native, expected. No voucher, said to be common in S-Peru, probably often unnoticed (Ruhsatz, 2012).

Ephedraceae

Ephedra americana Humb. & Bonpl. ex Willd.

Description: Shrub or creeping sub-shrub, up to 3.5 m tall, densely branched, olive green. Leaves scale-like and arranged in rings spaced along the stem, up to 3 mm long, pink to red. Seeds covered by 6-9 fleshy, pink to red bracts.

Ecology: Andean I-III: 500-4500 m. Shrub, rocky areas.

Distribution: Ecuador to Argentina.

Arequipa: native. ARE, CAR, CAY, ISL, UNI.

Use: The stems are depurative and diuretic. The whole plant contains ephedrine, an alkaloid that is used for the treatment of asthma. The cones are eaten sometimes, and the plant is used as fodder for cattle in hard times.

Note: E. breana Phil. is often considered as a synonym of E. americana, as suggested by Martínez Carretero (2018). POWO (2024) does not accept this synonymy.

The two species do not differ clearly in morphology and show little difference in the phylogenetic study of Ickert-Bond & Wojciechowski (2004). Here the 2 species are treated as synonyms.

Ephedra rupestris Benth.

Description: Dwarf shrub up to 60 cm tall, prostrate, forming dense cushions. Female cone with 3-6 fleshy, red or white bracts.

Ecology: Andean II-III: 2500-4500 m. Grasslands, rocky areas.

Distribution: Ecuador, Peru, Bolivia and NW-Argentina.

Arequipa: native. ARE, CAY, CON.

Note: Martínez Carretero (2018) treats E. rupestris as synonymous with E. chilensis, a species from Patagonia. The two species differ in morphology and phylogenetic studies, so we treat them as separate species, although Hassler (2024) considers them synonymous.

Escalloniaceae

Escallonia angustifolia C.Presl

Description: Shrub, up to 4 m tall. Leaves linear, >2 cm long, apex obtuse, base attenuate, whitish-green. Inflorescence terminal, a panicle, sepals reddish-green, lobes cuspidate, petals white.

Ecology: Andean II: 2000-2500 m. Disturbed sites, riparian.

Distribution: S-Peru, N-Chile, NW-Argentina.

Arequipa: native. ARE.

Escallonia myrtilloides L.f.

Description: Spreading shrub with elongate, straight branches. Leaves coriaceous, 1 x 0.5 cm, margins finely serrate. Flowers axillary, 5-merous, petals white.

Ecology: Andean II-III: 2000-4500 m.

Distribution: The Andes of South America.

Arequipa: native. ARE, UNI.

Euphorbiaceae

Croton alnifolius Lam.

Description: Shrub up to 1.5 m tall. Leaves elliptic, 3-6 x 2.5-4.5 cm, base and apex obtuse, margins entire to slightly crenate, stellate above, tomentose beneath, soon glabrate, petiole 1-2 cm long. Inflorescence an axillary raceme, 10-20 cm long, dense, flowers white, male flowers with 12-20 stamens. Ovary stellate tomentose, capsule 7.5 mm long, glabrescent.

Ecology: Coastal, Andean I: 0-1500 m.

Distribution: Ecuador, Peru.

Arequipa: native. ARE, CAM, CAR, ISL.

Croton ruizianus Müll.Arg.

Description: Shrub. Leaves ovate to orbicular, 5-6 cm long and nearly as wide, whitish-gray, densely villous,

basal glands 2. Raceme short, few-flowered, male flowers with 30-40 stamens. Ovary densely stellate-to-mentose, capsule subglobose, 6 mm in Ø.
Ecology: Coastal, Andean I-II: 500-3500 m. Lomas, shrubland.
Distribution: Ecuador, Peru, Bolivia.
Arequipa: native. ARE, CAM, CAR, CAY, ISL, UNI.

Euphorbia apurimacensis Croizat

Description: Shrub or small tree, 3-4 m tall, bark pale brown, all parts with milky latex. Leaves elliptic-obovate, shortly mucronate acuminate, base attenuate, pinnately veined. Flowers single, 3 x 5 mm, glabrous, erect.
Ecology: Andean II: 1500-2000 m.
Distribution: Peru.
Arequipa: native. ARE, CAY.

Euphorbia hinkleyorum I.M.Johnst.

Description: Subshrub, erect to decumbent, densely branched, stems reddish. Leaves sessile, orbicular, irregularly dentate-serrate.
Ecology: Andean II: 3000-3500 m.
Distribution: Peru, Bolivia, NW-Argentina.
Arequipa: native. ARE.

Euphorbia hirta L.

Description: Prostrate or ascending annual herb, all parts densely hairy. Branches up to 50 cm. Leaves ovate, 1-4 cm, often purple, margins finely toothed. Inflorescences terminal and axillary, spherical, 10-15 mm in Ø, ± sessile. Cyathia unisexual, whitish, tinged with purple.
Ecology: Coastal, Andean I, Amazonian: 0-1500 m. Disturbed areas.
Distribution: Probably of American origin, introduced worldwide.
Arequipa: native. CAR, ISL.

Euphorbia hypericifolia L.

Description: Erect to ascending annual, stems glabrous, reddish. Leaves elliptic, 1.3 cm long, usually shorter, blade glabrous, distally serrulate-denticulate, usually reddish-purplish, palmately veined at base, pinnate distally. Cyathia in dense, axillary and terminal glomeruli, bracts white. Capsules glabrous, 1-1.5 mm in Ø, with forked styles.
Ecology: Coastal, Andean I, Amazonian: 0-1500 m. Dry disturbed habitats.
Distribution: Native to the Americas, introduced elsewhere.
Arequipa: native, doubtful. ISL. Specimen FLSP 1210 (HUSA) from Islay (not seen), no mention and no observation, but mentioned for the lomas of Peru by Dillon et al. (2011).

Euphorbia lasiocarpa Klotzsch

Description: Erect to ascending herb, branches 0.3-1 m long. Stem pilose to tomentose. Leaves opposite, ovate to oblong, sometimes falcate, to 4 cm long, apex obtuse to acute, base oblique-rounded, margins serrulate, blade often with red spot in center, pilose to sericeous, palmately veined from the base. Cyathia in dense, usually terminal, capitate glomeruli, appendage white or pink, glands, yellow or pink, circular or oblong. Capsules subglobose, pilose to sericeous with yellow hairs.
Ecology: Coastal, Andean I-II, Amazonian: 0-2000 m. Disturbed areas.
Distribution: Tropical America.
Arequipa: native. ARE, CAS.

Euphorbia meyeniana Klotzsch

Description: Small, erect herb 7-12 cm high, ash-gray pubescent. Leaves suborbicular, base obliquely cordate, margins serrate, 2-3 lines long. Cyathia several at apex of branches, sparsely pilose, appendage unequal, white. Capsule pilose, nodding
Ecology: Andean I-II: 1000-2000 m. Desert, dry shrubland.
Distribution: Peru, Bolivia, Paraguay, N-Chile.
Arequipa: native. CAR, ISL.

Euphorbia peplus L.

Description: Erect, annual herb, growing up to 30 cm tall. Leaves quite variable, pear-shaped to triangular or rhombic, margins entire.
Ecology: Andean II: 2500-3500 m. Gardens, wasteland.
Distribution: Eurasia, introduced to the Americas.
Arequipa: introduced. ARE.

Euphorbia peruviana L.C.Wheeler

Description: Prostrate annuals, stems and leaves pubescent. Leaves orbicular. Cyanthia solitary, >5 mm in Ø, glands yellow, entire, appendage conspicuously dentate, white.
Ecology: Andean I-II: 500-2500 m. Open soils, disturbed areas.
Distribution: S-Peru.
Arequipa: native. CAR.

Euphorbia serpens Kunth

Description: Fragile, glabrous, loosely prostrate herb, often rooting at the nodes. Stipules whitish. Leaves glabrous, 5-10 × 4-6 mm, ovate to lanceolate, base obtuse to cordate, apex emarginate, margins entire. Cyathia solitary at distal nodes, gland appendage conspicuous, reniform, entire.
Ecology: Andean I-III: 1000-4500 m. Grasslands, disturbed areas.

Distribution: North and South America, introduced worldwide.

Arequipa: native. ARE, CAR, CAY, ISL.

Jatropha macrantha Müll.Arg.

Description: Succulent shrub, much-branched, stems gray. Leaves palmately 3-lobed, clustered at end of stems, base deeply cordate to slightly peltate. Flowers star-like, bright red, capsule brown, not armed.

Ecology: Andean II: 1500-2500 m. Shrubland.

Distribution: Ecuador, Peru.

Arequipa: native. ARE, CAR, CAY, ISL, UNI.

Use: Latex is used in folk medicine.

Ricinus communis L.

Description: Herbs erect, often single-stemmed but sometimes branched, 2-5 m tall, younger parts glaucous, the whole plant often reddish or purplish. Leaf-blade palmately 7-11-lobed, 30-50 × 30-50 cm, margins serrate. Inflorescence to 30 cm long, flowers apetalous. Capsule ellipsoid, 1.5-2.5 cm, echinate, spines to 5 mm, sometimes smooth. Seed 7-12 mm, grayish-brown with darker markings.

Ecology: Coastal, Andean I: 0-1500 m. Wasteland.

Distribution: Native to tropical Africa, introduced worldwide.

Arequipa: introduced. ARE, CAY, ISL.

Use: Several medicinal uses, the most important is the laxative effect of castor oil. It has been suggested as a treatment for cancer. Castor oil from the seeds is a valuable industrial oil and is used as a non-toxic lubricant and hydraulic oil.

Fabaceae

Adesmia augusti J.F.Macbr.

Description: Shrub to 1.5 m tall, very spiny, not glutinous. Leaves well-petiolate, 3-4 pairs of 5-11 mm long leaflets. Flowers 7-8 mm long, inner petals yellow with brown stripes, outer petals pubescent dull yellow with brown stripes.

Ecology: Andean II-III: 2500-3500 m. Shrublands.

Distribution: Peru.

Arequipa: native. ARE.

Adesmia melanthes Phil.

Description: Strongly armed shrub. Leaves very small. Similar to A. spinosissima, from which it differs in its petiolate leaves, its 1-5 mm long leaflets, generally arranged in 3 pairs, its non-white-sericeous pubescence and its 8-9 mm flowers.

Ecology: Andean II: 2000-3000 m. Rocky slopes, shrublands.

Distribution: Disjunct, S-Peru to N-Chile and S-Chile.

Arequipa: native. ARE.

Note: Unresolved species, may be a hybrid of A. spinosissima and A. augusti. Flora of Peru (1936ff.) include the Peruvian specimens in A. spinosissima.

Adesmia miraflorensis J.Rémy

Description: Much-branched, thorny shrub, up to 2 m tall, not glutinous. Leaves well-petioled, with 6-10 pairs of leaflets. Flowers racemose, yellow, glabrous, 12-15 mm long.

Ecology: Andean II-III: 3000-4000 m. Riversides, rocky slopes.

Distribution: S-Peru, W-Bolivia and NW-Argentina.

Arequipa: native. ARE, CAY.

Adesmia muricata (Jacq.) DC.

Description: Herbs 10-60 cm tall, ascending, erect, pubescent, setulose, not spiny. Leaves 1.5-6 cm long, leaflets 4- 8 pairs, often dentate. Segments of pod muricate.

Ecology: Andean II: 2000-3000 m. Rocky slopes, shrublands.

Distribution: Peru, Bolivia, Argentina, Uruguay and Brazil.

Arequipa: native. CAM, CAR.

Adesmia spinosissima Meyen ex Vogel

Description: Similar to A. miraflorensis but distinctly smaller (up to 60 cm tall), with subsessile leaves and minute leaflets. Flowers 6 mm long.

Ecology: Andean II-III: 3000-4500 m. Rocky slopes, shrublands. Tolerating saliferous ground.

Distribution: S-Peru, Bolivia, NW-Argentina and N-Chile.

Arequipa: native. ARE, CAY, CON.

Adesmia verrucosa Meyen

Description: Tall shrub, densely glandular glutinous except the flowers and pods, spineless. Leaves with 9-12 pairs of broadly obovate or subrotund, crenate leaflets, leaflets 4 mm long. Flowers in terminal racemes, yellow. Pods to 2 cm long with 8+ segments, shaggy plumose.

Ecology: Andean III: 3500-4000 m. Rocky slopes, shrublands.

Distribution: S-Peru, N-Chile.

Arequipa: native. ARE.

Astragalus arequipensis Vogel

Description: Herb, forming close tufts, up to 10 cm tall. Leaves long petiolate, gray, 7-14 pairs of leaflets. Inflorescence pedunculate, flowers 6-7 mm long, bluish.

Ecology: Andean II-III: 2500-4500 m. Grasslands.

Distribution: Peru, Bolivia, NW-Argentina and N-Chile.

Arequipa: native. ARE, CAY, UNI.

Astragalus confinis I.M.Johnst.

Description: Cespitose perennial. Leaflets linear-oblong, 2-5 mm long, 2-4 pairs, widely spaced. Inflorescence pedunculate, 2-5 flowers. Ovary and fruit glabrous.

Ecology: Andean II: 2000-2500 m.

Distribution: S-Peru, N-Chile.

Arequipa: native. ARE. Known in Peru only from the type locality.

Astragalus diminutivus (Phil.) Gómez-Sosa

Description: Cespitose herb, 2-8 cm Ø, caudex 1-2(-3) cm, entire plant silvery-strigose. Leaves to 1-2 cm long, 5-8 pairs of elliptic to subcircular leaflets with obtuse to slightly notched apex. Flowers 1(-2), sessile, 8 mm long, blue to light blue, calyx 3-4 mm long. Legume unilocular.

Ecology: Andean II-III: 2750-4300 m. Grasslands.

Distribution: Peru, Bolivia, NW-Argentina, N-Chile.

Arequipa: native. ARE.

Note: In Arequipa, A. diminutivus var. dielsii.

Astragalus garbancillo Cav.

Description: Tufted perennial or subshrub, decumbent to ascending, up to 80 cm tall. Leaflets 10-15 pairs, green and ashy-pilose. Flowers 1cm long, in many flowered, pedunculate inflorescences, white to purple. Pods ovoid- oblong, usually black pilose.

Ecology: Andean II-III: 2000->4500 m. Grasslands, shrubland, waysides.

Distribution: Peru, Bolivia, Chile and W-Argentina.

Arequipa: native. ARE, CAS, CAY, CON, UNI.

Use: The resin of the plant is used to treat toothache and bone fractures. An infusion of the plant is used to relieve abdominal pain. The smoke of the dried plant is used as an insecticide to treat livestock and wool.

Astragalus micranthellus Wedd.

Description: Prostrate herb to 6 cm across. Leaves with 6-8 pairs of leaflets, leaflets truncate. Inflorescence 0.5-1 cm long pedunculate, shorter than the leaves. Flowers 3-4 mm long, purple.

Ecology: Andean III: 3500-4500 m. Grasslands, rocky slopes.

Distribution: Peru, Bolivia, NW-Argentina, N-Chile.

Arequipa: native. ARE.

Astragalus minimus Vogel

Description: Cushion forming plant. Leaves 1-2 cm long ± ashy-strigillose or glabrate above, 5-6 pairs of 1 mm long leaflets. Flowers 5 mm long.

Ecology: Andean III: 4000->4500 m. Grasslands, shrublands.

Distribution: Peru, Bolivia, N-Chile and Argentina.

Arequipa: native. ARE. The type is from "Alto de Toledo" at 5000 m (Puno or Arequipa?). No other vouchers.

Astragalus neobarnebyanus Gómez-Sosa

Description: Stout prostrate perennial. Leaves with 7-8 pairs of linear-obovate leaflets. Inflorescence 1-5 flowered, peduncles 3.5-5 cm, longer than the leaves. Flowers purple, banner pink to white.

Ecology: Coastal: 0-1000 m. Lomas.

Distribution: Peru.

Arequipa: native. CAR, ISL.

Astragalus peruvianus Vogel

Description: Densely leafy, small perennial with prostrate, 2-3 mm thick stems, forming small mats from a central rootstock. Stipules straw-colored. Leaves up to 4 cm long. Leaflets ovate to lanceolate, flat, folded, 1-2 mm long, 4- 10 pairs, green with whitish hairs. Flowers solitary rarely twinned, sessile from the axillary leaf stipules, 6-10 mm long, lilac with a white basal blotch bearing radiating veins on the rounded standard.

Ecology: Andean III: 4000->4500 m. Grasslands.

Distribution: Peru, Bolivia and NW-Argentina.

Arequipa: native. ARE, CAY.

Astragalus pusillus Vogel

Description: Prostrate perennial. Leaves petiolate, 6-8 pairs of well-spaced, 3-5 mm long elliptic to obcordate leaflets, apex emarginate, green, white-hairy when young. Flowerheads acaulescent, intermixed with broad, white papery sheaths. Flowers 4-8 sessile, 6-7 mm long, pale lilac-blue, standard dark-veined. Pods sparsely hairy, 5-6 mm long, ovate-oblong, 4-seeded.

Ecology: Andean III: 3500->4500 m.

Distribution: Peru, Bolivia, N-Chile and Argentina.

Arequipa: native. ARE.

Astragalus richii A.Gray

Description: Plants diffuse. Stem slender, elongated to >10 cm. Leaflets ashy-sericeous. Leaflets 7-12 pairs, elliptic. Flowers lilac, 10 mm long.

Ecology: Andean II-III: 2000-4000 m. Rocky slopes.

Distribution: Peru.

Arequipa: native. ARE.

Astragalus triflorus (DC.) A.Gray

Description: Delicate annual, suberect to erect. Leaves with 4-6 pairs of leaflets. Inflorescence 1-3 flowered, flowers 5 mm long.

Ecology: Coastal: 0-1000 m. Disturbed areas, lomas, rocky slopes.

Distribution: Peru, Bolivia, N-Chile.

Arequipa: native. CAM, CAR, ISL.

Astragalus uniflorus DC.

Description: Depressed herb, stem 1-3 cm long. Leaves with 3-12 pairs of leaflets, glabrous. Flowers usually solitary, 18 mm long.

Ecology: Andean III: 3500-4500 m. Grasslands.

Distribution: C-Peru to NW-Argentina.

Arequipa: native. CAY, UNI.

Astragalus weddellianus (Kuntze) I.M.Johnst.

Description: Modest perennial plant, branched from base. Leaves 1-1.5 cm long, glabrous, 3-6 pairs of leaflets. Leaflets 1-2 mm long, glabrous. Flowers, sessile 7 mm long.

Ecology: Andean II-III: 2000-4500 m. Rocky slopes.

Distribution: S-Peru, Bolivia, NW-Argentina.

Arequipa: native. ARE.

Calliandra taxifolia (Kunth) Benth.

Description: Low shrub, much-branched. Leaves small, crowded, the 2-4 pairs of pinnae 1-3 cm long. Leaflets 2-4 mm long. Flowers crowded at foliate nodes, yellow to red. Pods erect, thin, with large margins.

Ecology: Coastal, Andean I-II: 500-3500 m. Forests, lomas, grasslands, rocky slopes.

Distribution: Colombia, Ecuador and Peru.

Arequipa: native. CAR, ISL.

Crotalaria incana L.

Description: Annual, much-branched, herbaceous to shrubby, pubescent, 0.3-1.2 m tall. Leaves trifoliolate, leaflets obovate or orbicular, 1.5-4 x 1-2.5 cm, stipules small, setaceous, deciduous. Racemes 5-many-flowered, lateral and terminal. Flowers greenish to brownish yellow, standard often slightly reddish, calyx pubescent and deeply lobed. Pods sessile, pendulous, pubescent with spreading brownish hairs, 2.5-3.5 cm long, slightly inflated.

Ecology: Coastal, Andean I-II: 500-2000 m.

Distribution: Tropical and subtropical America, Africa and India.

Arequipa: native. CAR.

Dalea boliviana Britton

Description: Prostrate to ascending herb, stems 10-45 cm long, from a perennial root-crown or shortly forked woody caudex. Leaves pinnate, with 5-8 pairs of leaflets, leaflets ± glandular-punctate beneath. Flowers blue or purple, keel white, calyx villous.

Ecology: Andean II-III: 2500-4000 m.

Distribution: NW-Argentina, Bolivia, Peru.

Arequipa: native. ARE, UNI.

Dalea coerulea (L.f.) Schinz & Thell.

Description: Openly branched shrub, 1.5-3(-5) m tall, growing at humid places. Leaves pinnate, with 6-8 pairs of leaflets, 10-15 mm long, half as wide. Peduncles shorter than the subtending leaves, spike elongating to 20 cm. Flowers blue to violet.

Ecology: Andean II: 2000-3000 m. Humid places.

Distribution: Colombia, Ecuador, Peru.

Arequipa: native. ARE.

Dalea cylindrica Hook.

Description: Simple branched perennial herb, subdecumbent, stems >50 cm. Leaves with 4-7 pairs of leaflets, leaflets 6-15 x 2-5 mm. Inflorescence 2-8 cm long, flowers lilac to purple.

Ecology: Andean I-III: 500-4000 m.

Distribution: Ecuador, Peru.

Arequipa: native. ARE, CAR, CAY, ISL, UNI.

Dalea exilis DC.

Description: Sprawling perennial herb, remotely branched. Leaves pinnate, 1.5-4.5 cm long with 5-8 pairs of leaflets, leaflets 4.5 x 1.5 mm, glabrous. Flowers 8 mm long, blue-purple, calyx densely villous.

Ecology: Andean II-III: 2500-4000 m. Grasslands, rocky slopes.

Distribution: Bolivia and Peru.

Arequipa: native. ARE, ISL, UNI.

Dalea onobrychis DC.

Description: Erect herb. Leaves with 4-6 pairs of obovate, oblong-obovate, or -elliptic, obtuse leaflets. Peduncles of the inflorescence 4 x longer than leaf, narrowly margined rachis. Flowers violet.

Ecology: Coastal, Andean I-II: 0-3500 m.

Distribution: Peru.

Arequipa: native. UNI. Specimen Weberbauer 6854, not seen. Marked with (?) in Flora of Peru) and a doubtful observation by the author from Cotahuasi. Mentioned by Brako & Zarucchi (1993) for Arequipa. Specimen Orellana et al. AO-870 (K) from Ica near the border to Arequipa.

Dalea pennellii (J.F.Macbr.) J.F.Macbr.

Description: Similar to D. boliviana, sprawling herbs, prostrate to ascending. 4-5 pairs of leaflets, leaflet 6-10 x 3 mm, ± adpressed strigillose, obscurely punctate beneath. Flowers mulberry purple, banner yellowish-white.

Ecology: Andean II-III: 2500-4000 m.

Distribution: N-Chile, Peru.

Arequipa: native. ARE.

Desmodium scorpiurus (Sw.) Poir.

Description: Straggling, climbing, or procumbent herb with small blue flowers. Terminal leaflet 2 x 1 cm, rounded at each end, thinly pubescent. Racemes up to 10 cm long, few-flowered. Flowers 4 mm long. Pod 5-

8-seeded, segments 2x as long as broad, not deeply incised.

Ecology: Coastal, Amazonian: 0-500 m. Open dry places.

Distribution: Native to tropical America, now pantropical.

Arequipa: native. CAM, CAR. No specimen, a convincing observation from Majes in iNaturalist (2024). Pantropical weed recorded for Ica.

Desmodium uncinatum (Jacq.) DC.

Description: Prostrate or sprawling perennial, Stems, leaves covered in dense hooked hairs readily adhering to clothing. Leaves 3-foliolate, leaflets ovate, acute-acuminate at apex, with a broad whitish marking centered on the main vein on the upper surface, petiole 1-5 cm, rachis 0.5-1.5 cm. Inflorescence terminal, 10-40 cm. Standard purple, fading to bluish-green. Fruits clearly constricted between the 3-8 articles.

Ecology: Andean I-II: 500-2500 m. Disturbed areas, riverine vegetation.

Distribution: Native to tropical South America, introduced and naturalized worldwide in the highland tropics and humid subtopics.

Arequipa: native. CAS.

Erythrostemon gilliesii (Hook.) Klotzsch

Description: Unarmed, evergreen shrub, <4 m. Stem ± glandular-hairy. Leaves 2-pinnate, 10-20 cm, glabrous, stipules small, persistent, leaflets 7-11 pairs, <8 mm. Inflorescence <10 cm, many-flowered, glandular-hairy. Sepals 1.5-2 cm, oblong-elliptic, glandular, petals yellow with orange marks, stamens 8-10 cm, ovary densely glandular- hairy. Fruit dehiscent, 6-12 x 1.9-2 cm, oblong, flat, twisted when mature, gland dotted. Seeds 6-10, ovate, brown.

Ecology: Coastal, Andean I-II: 0-2500 m. Disturbed areas.

Distribution: Native from C-Chile to Uruguay, cultivated and naturalized worldwide.

Arequipa: introduced. ARE, ISL.

Use: Ornamental plant.

Genista monspessulana (L.) L.A.S.Johnson

Description: Erect shrub, 1-5 m tall, unarmed, all parts but the main stem ± hirsute. Leaves trifoliolate, petiolate, leaflets acute to obtuse, 0.5-2 x 0.2-1.5 cm. Flowers yellow, standard ± glabrous. Pod silvery hirsute.

Ecology: Andean II: 3000-3500 m.

Distribution: Native to the Mediterranean, cultivated in temperate and montane tropical regions.

Arequipa: introduced. ARE, CAY. Often cultivated, adventive.

Geoffroea decorticans (Gillies ex Hook. & Arn.) Burkart

Description: Deciduous tree, up to 8 m tall. Trunks mottled green. Leaves imparipinnate, leaflets 3 pairs, obcordate. Flowers yellow, papery. Fruits fleshy, oval pods, single-seeded. Fruit initially green, deep orange when ripe.

Ecology: Coastal: 0-500 m.

Distribution: Southern South America.

Arequipa: native. ARE, CAR.

Hoffmannseggia arequipensis Ulibarri

Description: Distinguished from H. miranda by stipules with small teeth, comparatively small non-coriaceous leaves with prominent veins, (4)6-9(10) pair of leaflets per pinna and the yellow long trichomes on the petal claw (purple in H. miranda).

Ecology: Coastal, Andean I-II: 0-3000 m. Lomas, rocky slopes.

Distribution: Peru.

Arequipa: native. ARE, CAR.

Note: Simpson et al. (2004) suggest that H. miranda and H. arequipensis are one variable species. The same authors accepted H. arequipensis as a separate species in 2006.

Hoffmannseggia miranda Sandwith

Description: Subshrub, stem bases woody with stems often branching basally. Flowers orange, trichomes on petals purple. Fruits dehiscent with each valve twisting around itself.

Ecology: Coastal, Amazonian: 0-500 m. Lomas, rocky slopes.

Distribution: Peru.

Arequipa: native. CAM, CAR, ISL.

Hoffmannseggia prostrata Lag. ex DC.

Description: Completely herbaceous plant. Flowers yellow to orange. Flowers remain upright after flowering. Mature fruits straight with a pointed tip, tardily dehiscent with each valve twisting around itself.

Ecology: Coastal, Andean I-II, Amazonian: 0-3000 m. Lomas, riversides.

Distribution: Peru, Chile.

Arequipa: native. ARE, CAM, CAR, CAS, CAY, ISL.

Hoffmannseggia viscosa (Ruiz & Pav.) Hook. & Arn.

Description: Plants spreading, often decumbent with branches woody at the base, stems and foliage often sticky. Flowers borne on leafy branches. Fruits with parallel margins and curved into an arc with a rounded tip.

Ecology: Coastal, Andean I-II, Amazonian: 0-2500 m. Rocky slopes, lomas.

Distribution: Colombia, Ecuador, Peru, Bolivia, N-Chile.
Arequipa: native. CAR, CAY, ISL, UNI.

Indigofera humilis Kunth

Description: Prostrate herbaceous perennial. Leaves imparipinnate, petiole and leaflets whitish strigose, leaflets 4-5 pairs, usually alternate. Inflorescence axillary, few-flowered. Flowers orange-red.

Ecology: Coastal, Andean I-II, Amazonian: 0-3500 m. Lomas.

Distribution: Peru.

Arequipa: native. CAR. No specimen but a convincing observation from Atiquipa in iNaturalist (2024).

Indigofera suffruticosa Mill.

Description: Shrubs or shrublets, erect, 0.8-1.5 m tall. Stems grayish-green, with adpressed trichomes. Leaves 5-10 cm, leaflets usually alternate, 4-7 pairs. Corolla red, keel as long as wings, hairy. Pods deflexed and upwardly falcate, 1-1.5 cm, hairy.

Ecology: Coastal, Andean I, Amazonian: 0-1500 m. Disturbed areas, forests, riversides, rocky slopes.

Distribution: Native to tropical America, widely cultivated.

Arequipa: native. CAR.

Use: A major source of blue dye indigo.

Inga feuillei DC.

Description: Large tree with a spreading crown, up to 18 m tall. Leaves pinnate, rachis alate. Flowers white. Fruit a long pod with thickened edges, green. A white, sugar-rich pulp surrounding the seeds.

Ecology: Coastal, Andean I-II, Amazonian: 0-3000 m. Disturbed areas, dry valleys, forests, often cultivated.

Distribution: Native to Ecuador, introduced to W-South America.

Arequipa: introduced, adventive. ARE, CAR, CAY, UNI.

Use: Edible pulp.

Leucaena leucocephala (Lam.) de Wit

Description: Shrubs or small trees, 2-6 m tall. Branchlets pubescent, with brown lenticels. Pinnae 4-8 pairs, rachis pubescent with black glands at the location of lowest pinnae, leaflets 5-15 pairs, linear-oblong, main vein close to upper margin, base cuneate, margins ciliate, apex acute. Heads axillary, white. Legume straight, flat, leathery, pubescent, beak acute. Seeds 6-25, brown, glossy, ovoid, flat.

Ecology: Coastal, Andean I, Amazonian: 0-1500 m. Disturbed areas.

Distribution: Native to Central America, cultivated and naturalized widely in (sub-)tropical regions.

Arequipa: introduced. ARE, ISL.

Use: Young leaves, pods, and green seeds are eaten raw or steamed. The dried seeds are fermented or roasted and used as a coffee substitute. The decoction of the root and bark is used as an abortifacient. Fodder plant, green manure. Industrial use of gum.

Lupinus ananeanus Ulbr.

Description: Densely congested perennial 5-8 cm tall, all over very villous-hirsute. Petiole 2-6 cm, 7-12 pairs of leaflets. Peduncles almost not developed, racemes 2 cm long, concealed among the leaves, densely flowered, pyramidal with 9-10 mm long, ovate bracts. Flowers white to pale purple, 1.2 cm long.

Ecology: Andean III: 4000->4500 m. Grasslands.

Distribution: Chile, Peru.

Arequipa: native. ARE, CON. No specimen, several observations and mentioned for Arequipa by Brako & Zarucchi (1993).

Lupinus arequipensis C.P.Sm.

Description: Subshrub, 60-90 cm tall. Stems and petioles densely lanate with whitish-brown hairs, glabrescent when old. Leaves with 7-9 pairs of leaflets, petiole 5-12 cm long, leaflets 70 x 16 mm, lanate, villous, glabrate. Peduncle 6- 8 cm long, raceme 10-15 cm long, dense. Flowers deep purple, banner with a yellow spot in the center.

Ecology: Coastal: 0-500 m. Lomas, shrublands.

Distribution: Peru.

Arequipa: native. CAM, ISL.

Lupinus ballianus C.P.Sm.

Description: Low shrub, 20-40 cm tall, procumbent to ascending, branchlets and petiole loosely pubescent. Stipules 12-18 mm long. Leaves with 7-8 pairs of leaflets, petiole 3-4 cm long, leaflets 40 x 7 mm, pubescent, grayish-green. Peduncle 3-5 cm long, raceme 10 cm long, lax. Flowers red-purple, keel dark-tipped, banner with deep yellow central spot.

Ecology: Andean II-III: 3000-4000 m. Rocky slopes.

Distribution: Peru.

Arequipa: native. ARE, CAY.

Lupinus cuzcensis C.P.Sm.

Description: Tufted perennial herb. Petiole up to 16 cm long, 7-9 pairs of leaflets, leaflets 60 x 10 mm. Racemes dense, exceeding the leaves. Flowers yellow, 13 mm long, showy.

Ecology: Andean III: 4000-4500 m. Rocky slopes, riversides.

Distribution: Peru.

Arequipa: native. CAY, UNI. Specimen Weberbauer 6874 from La Unión probably destroyed, mentioned in Flora of Peru (1936ff.).

Lupinus eriocladus Ulbr.

Description: Shrub, 60-80 cm tall. Leaves with usually 7 pairs of leaflets, petiole 1-2.5 cm long, leaflets 20 x 9 mm, densely tomentose-villous on both sides. Peduncle short, raceme elongate, lax.

Ecology: Andean II-III: 3000-4500 m. Grasslands, rocky slopes.

Distribution: Peru.

Arequipa: native. ARE.

Lupinus hinkleyorum C.P.Sm.

Description: Shrub, 30-60 cm tall. Stem distinctly branched. Stem and petioles densely villous with spreading hairs up to 8 mm long. Leaves with 5-8 pairs of leaflets, petiole 3-4 cm long, leaflets 22 x 4 mm, densely sericeous to villous. Peduncle 5 cm long, raceme much longer. Inflorescence with 10-12 verticils, flowers blue with white.

Ecology: Andean II: 2000-2500 m. Shrublands.

Distribution: Peru.

Arequipa: native. ARE. Type from prov. Arequipa. No modern vouchers.

Note: May be synonymous with L. munzianus.

Lupinus misticola Ulbr.

Description: Herb, 10-30 cm tall. Leaves with 8-10 pairs of leaflets, petiole 10-20 cm long, leaflets 50 x 7 mm, folded, sparsely adpressed sericeous villous. Peduncle 20 cm long, raceme 10-15 cm long, dense. Flowers reddish- purple.

Ecology: Andean III: 3500-4000 m. Grasslands, shrublands.

Distribution: Peru.

Arequipa: native. ARE.

Lupinus mollendoensis Ulbr.

Description: Ephemeral annual herb of the lomas, <10 cm tall. Young stem and petioles hairy. Raceme with few flowers, lax. Flowers purple.

Ecology: Coastal: 0-1000 m. Lomas.

Distribution: Peru.

Arequipa: native. CAM, CAR, ISL.

Note: The type of Weberbauer seems to be a very young plant. Markedly similar to L. arequipensis. Both could be the same species.

Lupinus munzianus C.P.Sm.

Description: Shrub, 20-40 cm tall. Stem and petioles densely villous with spreading hairs up to 8 mm long. Leaves with 7-9 pairs of acuminate leaflets, petiole 2-4 cm long, leaflets 20-50 x 4 mm, densely sericeous to villous. Peduncle 15 cm long, raceme 15 cm long, many-flowered. Flowers rich purple with a deep yellow spot on the banner.

Ecology: Andean II: 2500-3000 m. Shrublands, riversides.

Distribution: Peru.

Arequipa: native. ARE. Type from prov. Arequipa. No modern vouchers.

Note: May be synonymous with L. hinkleyorum.

Lupinus proculaustrinus C.P.Sm.

Description: Treelike shrub, up to 3 m tall. Prominent stipules along 1/3 of the petiole. Banner deep blue with yellow and brown markings. Wings blue, keel bluish-white. Fruit green, with silvery hairs.

Ecology: Andean II: 2500-3500 m. Rocky slopes.

Distribution: Peru.

Arequipa: native. ARE, UNI. No specimen but convincing observations from prov. Arequipa.

Lupinus saxatilis Ulbr.

Description: Subshrub forming low clumps, 20-30 cm tall, branches minutely adpressed sericeous. Stipules 7-10 mm long. Leaves with 5-7 pairs of leaflets, petiole 2.5-7 cm long, leaflets 30 x 6 mm, glabrate above, subsericeous beneath. Peduncle 2-5 cm long, raceme 3-6 cm long, 2-6 verticils. Flowers blue, pedicel up to 1.5 cm long.

Ecology: Andean III: 3500-4200 m. Rocky slopes.

Distribution: Peru.

Arequipa: native. ARE.

Lupinus tarapacensis C.P.Sm.

Description: Shrub, 20-40 cm tall. Stems and petioles densely silky tomentose. Leaves with 6-7 pairs of leaflets, petiole 1.5-3 cm long, leaflets 25 x 6 mm, sericeous with adpressed whitish hairs, mucronate. Peduncle 3 cm long, raceme several times longer, 10 verticils. Flowers pedicelled < 1 cm, ultra-marine with a yellow-orange boss to standard petal.

Ecology: Andean II-III: 3000-4000 m. Grasslands.

Distribution: S-Peru, N-Chile.

Arequipa: native. CAY. Specimen Weberbauer 6894 from Chivay. Mentioned for Arequipa by Brako & Zarucchi (1993).

Note: May be synonymous with L. tomentosus.

Lupinus tomentosus DC.

Description: Shrub, 30-60 cm tall, much-branched. Branches erect, thickly clothed with a short silky tomentum. Leaves with 5-7(-10) pairs of leaflets, petiole 4-6 cm long, leaflets 45 x 10 mm, densely adpressed sericeous. Stipules narrowly linear, acute, bearded at the points, attached to the base of the petioles. Peduncle at least 3 cm long, usually much longer, raceme long, many verticils. Flowers at first straw-colored, tinged and striped with light blue, afterward changing to a bluish-purple, becoming darker by age.

Ecology: Andean II-III: 3000-4000 m. Grasslands, rocky slopes, shrublands.

Distribution: Peru, Bolivia, NW-Argentina.

Arequipa: native, doubtful. Not specified by province. Mentioned for Arequipa by Brako & Zarucchi (1993). No other vouchers.

Note: May be synonymous with L. tarapacanus.

Medicago lupulina L.

Description: Inflorescence with 10-50 yellow flowers. Pods sickle-shaped, spiraling less than 180°.

Ecology: Coastal, Andean I-III, Amazonian: 0-4000 m. Disturbed areas, rocky slopes.

Distribution: Native to Eurasia, introduced worldwide.

Arequipa: introduced. ARE, CAY.

Medicago polymorpha L.

Description: Prostrate plant. Inflorescence axillary, 2-10 yellow flowers. Pods discoid, coiled in 1.5-2.5 spirals, 15 spines or tubercles in each row.

Ecology: Coastal, Andean I-III: 0-4000 m. Disturbed areas, grasslands.

Distribution: Native to Eurasia, introduced worldwide.

Arequipa: introduced. ARE, CAY, ISL.

Medicago sativa L.

Description: Herbaceous perennial, up to 1m tall. Inflorescences in dense racemes with 10-35 purple to whitish-blue flowers. Pods coiled in 2-3.5 spirals, not spined.

Ecology: Coastal, Andean I-III: 0-4500 m. Disturbed areas, grasslands.

Distribution: Native to the Near East and C-Asia.

Arequipa: introduced. ARE, CAR, CAY, ISL.

Melilotus albus Medik.

Description: Annual or biennial herb up to 2.5 m tall. Stems round, glabrous, branched. Leaves alternate, trifoliate, sparsely distributed along the stem. Inflorescence a raceme of white flowers. Pods ovoid, blackish.

Ecology: Coastal, Andean I-II: 500-3000 m.

Distribution: Native to Eurasia, introduced worldwide.

Arequipa: introduced. ARE, CAR, CAY, ISL.

Melilotus indicus (L.) All.

Description: Annual or biennial herb up to 50 cm tall. Stems angled, glabrous. Leaves alternate, trifoliate, leaflets serrate. Inflorescence a dense raceme of yellow flowers. Pods ovoid, brown.

Ecology: Coastal, Andean I-III, Amazonian: 0-4000 m. Cloud forests, disturbed areas, lomas.

Distribution: Native to Eurasia, introduced worldwide.

Arequipa: introduced. ARE, CAM, CAR, CAY, ISL.

Mimosa albida Humb. & Bonpl. ex Willd.

Description: Shrub, erect, climbing or decumbent, up to 4 m tall. Stems hispid, armed with prickles. Leaves pedate, leaflets strongly oblique. Flowers pink in axillary heads. Pod oblong, 1.5-3.5 x 0.5 cm, 1-6 seeded.

Ecology: Coastal, Andean I-II: 0-2500 m. Shrublands.

Distribution: Central and South America.

Arequipa: native. CAM, CAR.

Note: In Arequipa, M. albinda var. willdenowii.

Neltuma alba (Griseb.) C.E.Hughes & G.P.Lewis

Description: A thorny, semideciduous tree with a rounded crown, 5-15 m tall. Short bole and densely branched. Leaves 1-3-jugate, glabrous, pinnae 6-14 cm long, with 25-50 pairs of leaflets, these linear, 5-17 x 1-2 mm.

Ecology: Andean I-II: 1000-2000 m. Rocky slopes.

Distribution: South America, introduced to Sudan.

Arequipa: native, doubtful. ISL. Mentioned by Brako & Zarucchi with specimen Gallardo s.n., observation in iNaturalist (2024) from Punta de Bombon (cultivated or confused with N. pallida?). Arequipa is at the northern limit of the species' distribution.

Neltuma calderensis (A.Galán, E.Linares, J.Montoya & Vicente Orell.) C.E.Hughes & G.P.Lewis

Description: A thorny, semi deciduous tree, 1-4 m tall. Short bole and densely branched, branches spiny, often with sugary excretions. Leaves 2-jugate, glaucous. Inflorescence 4-7 x 1.4-2 cm. Flowers yellowish-green, Pods falcate, frequently twisted at maturity, in groups up to 7.

Ecology: Andean II: 1500-2500 m. Shrubland, often cultivated.

Distribution: S-Peru.

Arequipa: native. ARE.

Note: New binomial by Galán de Mera et al. (2019) for P. andicola in Arequipa. Galán de Mera et al. (2019) separated P. andicola from P. laevigata var. andicola based on the geographic range and restricted P. andicola to Cuzco. The species from Arequipa was renamed P. calderensis.

Neltuma pallida (Humb. & Bonpl. ex Willd.) C.E.Hughes & G.P.Lewis

Description: Shrub or tree, 10 m tall. Leaves 2-pinnate with 3 pairs of secondary leaf axis. Inflorescence cylindrical, >10 cm. Flowers white. Most common Neltuma in coastal Peru.

Ecology: Coastal, Andean I: 0-1500 m. Desert, rocky slopes, semi-deciduous forests.

Distribution: Native to W-South America, from Colombia to Bolivia, introduced to Papua New Guinea and Australia.

Arequipa: native. ARE, CAR.

Otholobium pubescens (Poir.) J.W.Grimes

Description: Shrub or tree, 0.5-15 m tall, ± pubescent. Leaves pinnately 3-foliolate, petiole 2.5-8 cm long, leaflets 3-5 x 1-1.5 cm. Inflorescence short-elliptic to elongate, with 7-35 nodes and three flowers per node, the lowest node sometimes with a branch. Petals blue to purple, rarely white.

Ecology: Andean II-III: 2500-4000 m. Rocky slopes.

Distribution: Ecuador, Peru and Bolivia.

Arequipa: native. ARE, CAY, UNI.

Use: Medicinal plant used in the treatment of diabetes. Bakuchiol, a substance isolated from O. pubescens, shows insulin properties.

Note: The genus Otholobium is doubtful and is listed as unplaced.

Parasenegalia visco (Lorentz ex Griseb.) Seigler & Ebinger

Description: Small tree, not armed. Leaves 2-pinnate 15-20 × 7-15 cm, 5-16 pairs of pinnae, leaflets 14-48 pairs per pinna. Heads 15-20 mm in Ø, 1 to many per shoot, yellow. Pods 1-4 per head, chartaceous, 10-17 × 1.5-2.5 cm, apex apiculate, valves light brown, reticulated.

Ecology: Andean II: 1500-2000 m. Disturbed areas.

Distribution: Native to Peru, Bolivia, Argentina and Chile, introduced to Africa and Europe.

Arequipa: native. ARE, CON.

Paraserianthes lophantha (Vent.) I.C.Nielsen

Description: Straggly evergreen shrub or small tree to 8m high but usually smaller. Twigs slightly ribbed, with narrow raised ribs running down the stem from the base of each leaf. Leaves compound, of numerous small leaflets. Flowers cream carried in bottlebrush clusters. Pods broad and flat, 8-12 cm long, containing 6-12 black shiny seeds.

Ecology: Andean II: 2000-3500 m. Disturbed areas, cultivated.

Distribution: Native to Western Australia, naturalized in New Zealand and the Americas.

Arequipa: introduced. ARE.

Use: Used as an ornamental plant and planted as a shade tree or for soil improvement in agriculture and forestry.

Note: Very fast-growing tree, considered as a weed in Chile.

Parkinsonia aculeata L.

Description: Tree <12 m, with spines at nodes. Leaves ± sessile, pinnea 2-4, main axis <30 cm, conspicuous, persistent as ribbon, leaflets 30-60, scattered, 3-5 x 1-1.5 mm, elliptic, ephemeral. Flowers 2 cm wide, sepals reflexed, petals round, banner red-spotted at base, all red in age. Fruit 4-12 cm, thickened, leathery.

Ecology: Coastal, Andean I: 0-1000 m. Desert, forests.

Distribution: Native to North and South America, introduced worldwide.

Arequipa: native. CAR, ISL.

Note: Hybridizes with P. praecox. In Caravelí the hybrid P. aculeata × P. praecox may be present.

Poissonia weberbaueri (Harms) Lavin

Description: Low stocky much-branched shrub, younger branches and leaves ashy silky pubescent. Leaves imparipinnate, 4 pairs of suborbicular leaflets, 5-9 x 3-7 mm. Inflorescence a lax raceme. Flowers pedicellate, 1.5 cm long, petals brick-red.

Ecology: Coastal: 0-500 m.

Distribution: S-Peru.

Arequipa: native. CAM, CAR, ISL.

Senna bicapsularis (L.) Roxb.

Description: Shrubs, erect to 3 m tall, much-branched. 3-4 pairs of leaflets, rachis of leaves with a gland between the lowest pair of leaflets. Corolla yellow, fertile stamens 7.

Ecology: Coastal, Andean I, Amazonian: 0-1500 m. Disturbed areas, rocky slopes.

Distribution: Native to tropical America, widely cultivated in the tropics.

Arequipa: native. CAR, ISL.

Use: Ornamental plant.

Note: In Arequipa, S. bicapsularis var. augusti.

Senna birostris (Dombey ex Vogel) H.S.Irwin & Barneby

Description: Shrub up to 2 m tall. Leaves pilose-puberulent with 6-14 pairs of leaflets. Flowers golden yellow, showy.

Ecology: Coastal, Andean I-III: 0-4000 m. Lomas, riversides, shrublands, forests.

Distribution: Ecuador, Peru, Chile, Bolivia and Argentina.

Arequipa: native. ARE, CAR, CAY, UNI.

Note: In Arequipa, S. birostris var. arequipensis and var. birostris.

Senna brongniartii (Gaudich.) H.S.Irwin & Barneby

Description: Shrub. Leaves with 1-2 pairs of leaflets, often trifoliate. Flowers yellow.

Ecology: Coastal: 0-1500 m. Desert, lomas.

Distribution: Peru, N-Chile.

Arequipa: native. ARE, CAR, ISL.

Senna didymobotrya (Fresen.) H.S.Irwin & Barneby

Description: Hairy shrub, up to about 5 m tall. Leaves up to 50 cm long. Leaflets in 8-18 pairs, oblong-elliptic,

2-6.5 cm long. Inflorescence an erect, axillary raceme, 20-30 flowered, up to 50 cm long. Flowers yellow, ± sessile. Pods oblong, flattened, 8-12 cm long, dehiscent and transversally septate.

Ecology: Coastal, Andean I-II: 0-2500 m. Disturbed places.

Distribution: Native to tropical Africa, introduced in subtropical and tropical areas of Australia, America and Asia.

Arequipa: introduced, adventive. ARE.

Use: A garden ornamental, a medicinal plant and a nitrogen-fixing cover crop.

Spartium junceum L.

Description: Perennial shrub, 2-4 m tall. Stems cylindrical, rush-like, green when young, maturing into woody branches. Leaves caducous, oblong-linear to lanceolate, 2-2.5 cm long. Inflorescence an open terminal raceme with several flowers. Corolla yellow, 20-25 mm long. Fruit linear, dehiscent, 5-10 x 5 mm, 10 to 18 black seeds.

Ecology: Andean II-III: 1500-4000 m. Disturbed areas, riversides, rocky slopes.

Distribution: Native to the Mediterranean basin, introduced.

Arequipa: introduced. ARE, CAR, ISL, UNI.

Tara spinosa (Molina) Britton & Rose

Description: Stocky shrub or tree with spreading spinose gray-barked branchlets. Leaves up to 10cm long, with 8 pairs of subsessile, elliptic leaflets. Flowers reddish-yellow. Pods often red or reddish.

Ecology: Coastal, Andean I-II: 0-3000 m. Lomas, shrublands, often cultivated in settlements.

Distribution: Endemic to Peru, cultivated elsewhere.

Arequipa: native. ARE, CAM, CAR, CAY, ISL, UNI.

Use: Medicinal plant. The pods are an excellent source of tannins used in the manufacture of automotive and furniture leather. The seeds are processed into tara gum, a thickening agent used in the food industry.

Trifolium amabile Kunth

Description: Perennial, pubescent herbs, cespitose or cushion-like. Inflorescence axillary, racemose, subsessile, peduncle 1,5-4 mm long, with 14-20 flowers. Flower subtended by a linear-filiform bract, with a pedicel up to 4 mm long. Calyx 5 mm long. Corolla white, pinkish or purplish.

Ecology: Coastal, Andean I-III: 0-4500 m. Cloud forests, disturbed areas, elfin forests, grasslands, lomas.

Distribution: Native to South America, introduced to W-Asia.

Arequipa: native. ARE.

Note: Molinari-Novoa (2016) proposes to split the T. amabilis complex of Peru into 3 species, T. amabilis, T. peruvianum and T. absconditum. The Arequipa strain would belong to T. peruvianum.

Trifolium polymorphum Poir.

Description: Creeping, stoloniferous, perennial herb, pubescent. Leaflets emarginate. Inflorescence of 2 types, aerial and hypogeal. Aerial flower heads small, pink, or rose, on long peduncle. Most of the seed is set by the small white, subterranean flowers in axillary, ovoid heads at the nodes.

Ecology: Coastal, Andean I: 0-1000 m. Disturbed areas, grasslands.

Distribution: Southern South America, introduced to Texas.

Arequipa: native. ISL.

Trifolium pratense L.

Description: Erect or decumbent perennial, moderately hairy to glabrescent, branches to 50 cm long. Leaves 3- foliolate, leaflets elliptic to obovate or ± circular, 15-50 x 7-15 mm, 1-2 times as long as broad, entire. Inflorescence globose or ovoid, 1-27 mm Ø, >100-flowered, terminal, usually sessile. Flowers sessile. Calyx tube 10-veined, teeth erect in pod. Corolla 15-18 mm long, longer than calyx, pink to purple, persistent. Pod ± ovoid, 2-4 mm long, 1- seeded, included in calyx.

Ecology: Andean II: 3200-3300 m. Cultivated as pasture.

Distribution: Eurasia, introduced worldwide.

Arequipa: introduced, adventive. Not specified by province. No voucher, mentioned as adventive herb by Montesinos & Zegarra (2019).

Trifolium repens L.

Description: Perennial herb, stoloniferous, glabrous. Stems 10-30 cm, prostrate, rooting at nodes. Leaves long petiolate, palmately 3-foliolate, stipules ovate-lanceolate, membranous, apex subulate, leaflets obovate to ovate, 6- 20 × 8-16 mm, lateral veins 13 pairs, prominent on both surfaces, base cuneate, apex emarginate to broadly rounded. Flowers 20-50, in a terminal, globose head, 1.5-4 cm. Corolla white, rarely pink-tinged, 5-12 mm, fragrant. Legume linear-oblong.

Ecology: Coastal, Andean I-II: 0-3500 m. Disturbed areas, grasslands, lomas.

Distribution: Cosmopolitan.

Arequipa: introduced. ARE, CAR, ISL.

Vachellia karroo (Hayne) Banfi & Galasso

Description: Shrub or tree 2-20 m tall. Conspicuous white thorns, up to 10(-25) cm long. Leaves with <8 pairs of pinnae. Flowers yellow. Pod dehiscent, green to brown when mature, flat, constricted between the seeds, 16 x 1 cm, split open at maturity.

Ecology: Coastal: 0-1000 m. Disturbed places, hedges.

Distribution: Native to South Africa, naturalized in dry regions of America, Asia, Africa and Australia.

Arequipa: introduced, adventive. ISL. Frequent in Nazca. No specimen but several convincing observations from coastal areas (iNaturalist, 2024).

Vachellia macracantha (Humb. & Bonpl. ex Willd.) Seigler & Ebinger

Description: A spiny acacia. Feathery pinnate leaves protected with big straight spines, >11 pairs of pinnae. Small yellow flowers. Pods like flat beans.

Ecology: Coastal, Andean I-II, Amazonian: 0-3000 m. Beach, disturbed areas, rocky slopes.

Distribution: Central and South America.

Arequipa: native. ARE, CAR, CAY, CON, ISL, UNI.

Vicia andicola Kunth

Description: Annual herb, climber with tendrils. Leaves pinnate, 4-6 pairs of leaflets, leaflets 15-22 x 3-6 (-10) mm, tendril furcate, stipules unequal. Raceme 5-6 flowered, flowers 1 cm long, purple, keel white. Pod 5-6-seeded.

Ecology: Andean II-III: 2000-4000 m. Riversides, shrublands.

Distribution: The Andes of South America, from Venezuela to Chile.

Arequipa: native. ISL.

Vicia graminea Sm.

Description: Annual herb, climbing with tendrils. Leaves pinnate, 1-3 pairs of leaflets, leaflets linear, 25 x 2 mm, tendril simple, stipules equal. Racemes 1-5 flowered, flowers 6-10 mm long, whitish-violet. Pod 4-8-seeded.

Ecology: Coastal, Andean I-III: 0-4000 m. Disturbed areas.

Distribution: South America, from Colombia to Uruguay.

Arequipa: introduced. ARE, CAY.

Vicia lomensis J.F.Macbr.

Description: Annual herb, climbing with tendrils. Leaves pinnate, leaflets linear-oblong, 15 x 4 mm, tendril simple, stipules broad and dentate. Flowers whitish- violet, solitary, subsessile. Pod up to 10-seeded.

Ecology: Coastal, Andean I-II: 0-3000 m. Cloud forests, disturbed areas, lomas, rocky slopes.

Distribution: Peru.

Arequipa: native, doubtful. CAM, CAR, ISL.

Vigna luteola (Jacq.) Benth.

Description: Glabrous to densely hairy perennial climber with adventitious roots below the stems. Leaflets 3, ovate- elliptic to ovate-lanceolate, acute or acuminate. Inflorescence an axillary, 3-8-flowered raceme. Flowers yellow or greenish. Fruit a linear pod, slightly constricted between the seeds and densely adpressed pubescent.

Ecology: Coastal, Andean I-II, Amazonian: 0-2500 m. Disturbed areas, forests, lomas, riversides.

Distribution: Pantropical.

Arequipa: native. CAM, CAR.

Weberbauerella brongniartioides Ulbr.

Description: Erect ephemeral herb or subshrub, 25-35 cm tall, deep tubers. Stems, leaves and flowers densely covered with pustular glands. Leaves imparipinnate, with 8-18 pairs of leaflets, leaflets ovate, 6-20 x 3-14 mm. Flowers pedicellate, corolla orange, standard petal with yellow margins and center. Ovary pubescent to villous.

Ecology: Coastal: 0-1000 m. Forests, lomas.

Distribution: Peru.

Arequipa: native. CAM, ISL. Endemic to prov. Camaná and Islay.

Weberbauerella raimondiana Ferreyra

Description: Subshrub, erect, 20-35 cm tall, deep tubers. Stems, leaves and flowers densely covered with pustular glands. Imparipinnate leaves with 14-34 pairs of leaflets, leaflets ovate-elliptic, 2-7.5x 1.6-4 mm. Flowers yellow with oranges stripes. Ovary glabrous.

Ecology: Coastal, Andean I: 0-1800 m. Lomas.

Distribution: Peru (S-Ica and N-Arequipa).

Arequipa: native. CAM, CAR.

Francoaceae

Balbisia meyeniana Klotzsch

Description: Branched shrub, up to 1 m tall, densely tomentose. Leaflets narrowly obovate, margins flat, lamina subglabrous to finely pubescent, glaucous in the living state, fresh green when dried. Flowers terminal, yellow to orange, calyx finely puberulent, beneath flushed with red.

Ecology: Andean II-III: 1500-4000 m. Lomas, rocky slopes, shrublands.

Distribution: S-Peru, Bolivia, NW-Argentina.

Arequipa: native, doubtful. ARE, CAY.

Note: Weigend (2011) states that "in Peru the species appears to be restricted to the two southwestern departments Moquegua and Tacna and has not so far been found further north, in spite of copious collections now available from Arequipa". Nevertheless, some specimens from Arequipa match the description of B. meyeniana.

Balbisia verticillata Cav.

Description: Branched shrub to 1 m tall. Whole plant ash-colored and densely tomentose. Leaves mostly 3-parted to base with linear divisions, leaflets narrowly elliptic-acuminate (often linear in dried material) with recurved margins and dense, sericeous to villous pubescence. Flowers terminal, yellow-orange, calyx densely pubescent to sericeous, usually not flushed with red.

Ecology: Andean II: 2500-3000 m. Rocky slopes.

Distribution: Peru, Bolivia.

Arequipa: native. ARE, CAY, UNI.

Use: Used in folk medicine to treat coughs.

Note: Formerly Balbisia was placed in Geraniaceae, then in Vivianaceae. APG IV (2016) circumscribes the family Francoaceae anew and includes Vivianaceae and Melianthaceae.

Frankeniaceae

Frankenia chilensis C.Presl ex Schult. & Schult.f.

Description: Low shrub, 15-20 cm tall, stem nodose, ashy puberulent. Leaves 5 mm long, ovate to orbicular, blunt, margins revolute. Corolla white, petals 7-10 mm long, crenulate at apex. Calyx 5-7 mm long, setose or sparsely hirsute.

Ecology: Coastal: 0-500 m.

Distribution: S-Peru, N and C-Chile.

Arequipa: native. ARE, CAM, CAR, CAS, ISL.

Gentianaceae

Centaurium erythraea Rafn

Description: Herb, 20-60 cm tall, Branched only at the top. Lower leaves in a persistent rosette, cauline leaves opposite, sessile. Flowers in terminal, umbellate inflorescence. Pedicels 0-2 mm, inflorescence ± flat-topped, corolla pink, lobes 4.5-8 mm.

Ecology: Coastal: 0-500 m. Disturbed areas, lomas.

Distribution: Native to Eurasia, introduced worldwide.

Arequipa: introduced, doubtful. ISL. Specimen Dillon 3732 (F) from Islay (sheet blank) and Angulo 2588 (F) from Mollendo (not seen), no observation and no mention. Occurrence in Arequipa is likely.

Cicendia quadrangularis (Lam.) Griseb.

Description: Annual herb. Leaves ovate to lanceolate. Flowers 4-merous, corolla yellow. Calyx ± amplified, deeply furrowed, square.

Ecology: Coastal: 0-1000 m. Lomas, rocky slopes.

Distribution: W-North America and South America.

Arequipa: native. ISL.

Gentiana sedifolia Kunth

Description: Small annual herb. Leaves linear to narrowly elliptic. Flower solitary, 1 cm wide. Corolla 5-merous, lobes triangular, connate to 2/3 of the length, white to blue, ornamental spots on tube and base of lobes.

Ecology: Andean II-III: 1500->4500 m.

Distribution: Mexico to Bolivia.

Arequipa: native. ARE, CAY, CON.

Note: In Arequipa, an intermediate between G. sedifolia and G. gayi grows with tetramerous flowers but with ornamental spots in the corolla tube (Pfanzelt & von Hagen, 2016).

Gentianella potamophila (Gilg) Zarucchi

Description: Small herb, up to 10 cm tall. Leaves oblanceolate, basal. Flower long petiolate, whitish-violet. Corolla 5-merous, petals overlapping.

Ecology: Andean III: 3500-4000 m. Grasslands, riversides.

Distribution: Peru.

Arequipa: native. ARE, CAY.

Gentianella sandiensis (Gilg) J.S.Pringle

Description: Dense rosette of lanceolate acute leaves dilated at base, fleshy, 11-13 x 1.5-3 mm, sterile cauline much smaller. Flowers solitary on erect-spreading peduncle-like leafless stems of 1.5-4 cm. Calyx campanulate, acute lobes shorter than the tube. Corolla 13-14 mm long, obovate acute lobes twice as long as tube. Flowers white, violet or green veins or exterior band, tube often yellow.

Ecology: Andean II-III: 3000-4000 m. Grasslands.

Distribution: Peru, Bolivia.

Arequipa: native. ARE. No specimen but several observations in iNaturalist (2024) and mentioned for RN Salinas y Aguada Blanca by Quipuscoa & Huamantupa (2010).

Gentianella scarlatina (Gilg) J.S.Pringle

Description: Small annual herb, 5-10 cm tall. Leaves 2-3 cm long, acute, without pseudopetiole. Flowers solitary, erect, 17-19 mm long, lobes acute, lanceolate. Corolla scarlet and yellow.

Ecology: Andean III: 3500-4000 m. Grasslands.

Distribution: Peru, Bolivia.

Arequipa: native. CAY. No specimen but a convincing observation in iNaturalist (2024).

Geraniaceae

Erodium cicutarium (L.) L'Hér.

Description: Hairy, sticky annual. Leaves pinnate to pinnate-pinnatifid, on long petioles, leaflets sessile, incised up to the main vein. Flowers pink, often with dark spots on the bases, petals 5-8 mm long. Seedpod long, shaped like the bill of a stork, bursts open spirally when ripe.

Ecology: Coastal, Andean I-III: 500-4500 m. Grasslands, arable land, waste areas, roadsides.

Distribution: Native to the Mediterranean, introduced worldwide.

Arequipa: introduced. ARE, CAM, CAR, CAS, CAY, ISL, UNI.

Erodium geoides A.St.-Hil.

Description: Annual herb. Leaves pinnatifid, with <5 pairs of lobes, glandular and densely pilose-hirsute even from the base of the many branching stems. Flowers pink, petals with distinctly purple veins.

Ecology: Coastal, Andean I-II: 0-3500 m. Disturbed areas.

Distribution: S-Peru, N-Chile, N-Argentina, Uruguay, S-Brazil.

Arequipa: native. ARE, ISL, UNI.

Erodium malacoides (L.) L'Hér.

Description: Annual or biennial herb. Stem decumbent. Leaves entire, ovate to oblong, cordate, margins crenate or shallowly lobed. Bracts at least 3 below each umbel. Flowers pink to purple. Fruit <3 cm, glandular.

Ecology: Andean II-III: 1500-3500 m. Disturbed areas, lomas.

Distribution: Native to S-Europe and W-Asia, introduced all over the world.

Arequipa: introduced. CAM, CAR, ISL.

Erodium moschatum (L.) L'Hér.

Description: Hairy annual herb, stem glandular. Leaves pinnate, on long petioles, leaflets petiolate, margins dentate but not deeply incised. Flowers pink, petals 13-15 mm long. Seedpod long, storkbill-like.

Ecology: Coastal, Andean I-III, Amazonian: 0-4000 m. Open and disturbed areas, waste places.

Distribution: Native to Europe.

Arequipa: introduced. ARE, CAR, ISL.

Geranium core-core Steud.

Description: Herbs 15-75 cm tall. Cymules 2-flowered, solitary, peduncles 1-3 cm long, with retrorse, adpressed, eglandular. Petals purplish, 4-6 x 1.7-3.5 mm, entire or slightly notched, without claw, glabrous on both sides, ciliate on the basal margin. Readily distinguishable by its peduncles and pedicels with retrorse-adpressed, eglandular hairs, by its leaves not deeply divided and short petals.

Ecology: Andean II: 2500-3500 m. Roadsides, cultivated grounds, beaches, grassy hillsides, riversides, shrublands.

Distribution: Ecuador to Chile and Argentina, introduced to Great Britain and California.

Arequipa: native. ARE.

Geranium diffusum Kunth

Description: Ascending or suberect perennial herb, 25 cm tall. Leaves adpressed hispidulous, round-reniform, palmately 5-7 parted, the lobes apically trilobulate. Peduncles axillary, solitary, 1-flowered. Petals pinkish-white, with pink veins.

Ecology: Andean II: 2000-3500 m.

Distribution: The Andes from Venezuela to Peru.

Arequipa: native. ISL.

Geranium fallax Steud.

Description: Herbs 20-100 cm tall, always with glandular hairs, sometimes restricted to pedicels, sepals and fruits but usually spread over all the inflorescence and stem. Inflorescence terminal and axillary, 2-flowered cymules, peduncles 3-5 cm long, with glandular hairs. Petals purplish, 5-8 x 2-6 mm, entire or slightly notched, without claw, glabrous on both sides, ciliate basally.

Ecology: Andean II-III: 2500-4000 m.

Distribution: S-Peru, Bolivia and NW-Argentina.

Arequipa: native. ARE, CAR, CAY, ISL.

Geranium limae R.Knuth

Description: Herbs 10-60 cm tall, stem erect to ascending, with patent to retrorse, not adpressed, eglandular hairs. Cymules 2-flowered, solitary, peduncles 2-4 cm long, with patent to retrorse, eglandular hairs. Petals purplish, 5.5- 7.5 x 2-5 mm, entire, without claw, glabrous on both sides, ciliate on the basal margin.

Ecology: Coastal, Andean I-II: 0-3500 m. Coastal desert, on hillsides in sandy or rocky areas.

Distribution: Peru.

Arequipa: native. CAM, CAR, ISL.

Geranium sessiliflorum Cav.

Description: Herbs 2-9 cm tall, usually stemless with all cymules borne on the rootstock apex. Petals purplish. Basal leaves orbicular to polygonal, palmatifid with 5-7 segments, cordate with basal leaf segments downward. Cymules 1- flowered, solitary, peduncles 0-0.5 cm long, with retrorse, adpressed, eglandular hairs. Petals purplish or white, 6-13 x 1.5-5 mm, entire, rarely emarginate, notch 1 mm deep, without claw, glabrous.

Ecology: Andean III: 3000->4500 m. River shores, turfy bogs, grasslands, clearing of Polylepis forest.

Distribution: C- and S-Peru, Bolivia, Chile and Argentina.

Arequipa: native. ARE, CAS, CAY, CON.

Hypseocharis pimpinellifolia J.Rémy

Description: Acaulescent rosette herb, 5-6 cm tall, glabrous. Leaves 10-20 cm long, 1-2-pinnate with up to 13 pairs of ternate leaflets, leaflets with 3-4 incised blunt lobules on each side, the terminal leaflet at most marginally larger than the lateral ones. Flowers usually solitary, usually white, sometimes red, orange or yellow, 15 anthers. Fruit capsular.

Ecology: Andean III: 2000-4000 m. Shrublands, grassland.

Distribution: S-Peru, Bolivia. NW-

Arequipa: native. ARE, CAY.

Note: Hypseocharis bilobata Killip, Hypseocharis malpasensis R.Knuth, H. pedicularifolia R.Knuth and H. pilgeri R.Knuth are synonymized with H. pimpinellifolia by Lozada et al. (2020), as we do here.

Grossulariaceae

Ribes brachybotrys (Wedd.) Jancz.

Description: Shrub with reddish-gray branchlets. Leaves crowded, 1.5-3 cm wide, rotund, 3-lobed, finely serrate, pilose. Raceme 1 cm long, little glandular but densely pilose. Flowers few, small. Similar to R. ovalifolium but leaves wider and lobes longer.

Ecology: Andean II-III: 2500->4500 m. Forests, rocky slopes.

Distribution: S-Peru, Bolivia.

Arequipa: native, doubtful. ARE. No voucher, mentioned for RN Salinas y Aguada Blanca by Quipuscoa & Huamantupa (2010). The occurrence in Arequipa is likely (occurring in Cuzco, Puno, Moquegua).

Ribes ovalifolium Jancz.

Description: Branched shrub, 0.9-2 m tall. Leaves deciduous, ovate to subelliptic, 7-20 x 5-15 mm, 3-veined, obscurely to distinctly 3-lobed, margins serrate. Inflorescence a terminal racemes, pendulous, 8 mm long in flower, 2-3 flowered, peduncle 2.5-3 mm long. Flowers subsessile, rotate, 5.5-6 x 4-5 mm, reddish-brown.

Ecology: Andean II-III: 2500-4000 m. Rocky slopes, rock surfaces.

Distribution: Peru.

Arequipa: native. ARE, UNI.

Haloragaceae

Myriophyllum aquaticum (Vell.) Verdc.

Description: Dioecious, much-branched aquatic herb. Leaves opposite, verticillate, 1.5-3.5 cm long, blade finely pinnately-dissected, segments 5-8 mm long. Submersed and emergent leaves similar. Flowers solitary in axils, female flowers 4-merous, very shortly pedicellate, white.

Ecology: Andean I-II, Amazonian: 0-3500 m. Floodplain lagoons, lakes, ponds, marshes, fens and ditches.

Distribution: Native to South America, introduced all over the world.

Arequipa: native. ARE.

Myriophyllum quitense Kunth

Description: Aquatic, monoecious herb, the emerging flowering branches to 1 m long. Leaves 1-2.5 cm long, in whorls of 4-5, pinnately-dissected, segments >10 mm. Inflorescence an emergent spike, bracts triangular, entire or finely serrate.

Ecology: Andean II-III: 1500-4500 m. Lakes.

Distribution: Native from Canada to Patagonia.

Arequipa: native. CAS.

Hydrocharitaceae

Elodea potamogeton (Bertero) Espinosa

Description: Submersed aquatic plant with ebony stem. Leaves crowded in whorls of 3-4, 15 x 2-4 mm, obtuse, minutely serrulate. Flowers solitary, female flowers pedicellate from a spathe, male flowers sessile.

Ecology: Andean I-III, Amazonian: 0-4500 m. Lakes and slow-moving water.

Distribution: Ecuador, Peru, Bolivia and Chile.

Arequipa: native. ARE, CAY.

Vallisneria sp.

Description: Submersed aquatic plant, stoloniferous, stems short. Leaves linear from base. Flowers long, pedunculate, white, with long spathe, peduncle screwed.

Ecology: Andean II: 1800 m. Lakes, in shallow water.

Distribution: Not known.

Arequipa: native. CAS. Observed in Mamacocha, Ayo by the author (iNaturalist, 2024).

Hypericaceae

Hypericum silenoides Juss.

Description: Annual or perennial herb, 10-60 cm tall, erect, stems reddish-green. Leaves sessile, linear-lanceolate, 0.6-3 x 0.1-0.6 cm, decurrent at base, apex acute to rounded. Inflorescences corymbose. Flowers 1-1.5 cm in Ø, 5- merous, sepals linear-lanceolate, unequal, petals oblong-lanceolate, 4-8 x 1-2 mm, yellow to reddish, with scattered linear glands. Capsules ovate-cylindric, ca. 5 x 2 mm.

Ecology: Coastal, Andean I-II: 500-3000 m. Rocky slopes.

Distribution: Central and South America.

Arequipa: native. ARE, CAR, ISL.

Iridaceae

Hesperoxiphion pardalis (Ravenna) Ravenna

Description: Bulbous herb, up to 50 cm tall. Cauline leaves 20 x 1-2 cm. Spathe 4 cm long, green. Corolla yellow- orange, inner petals with red dots.

Ecology: Andean II: Shrublands.

Distribution: Peru.

Arequipa: native. UNI. No specimen, a convincing observation from Alca in iNatralist (2024).

Mastigostyla cyrtophylla I.M.Johnst.

Description: Bulbous perennial. Basal leaves reduced to brownish sheaths, or if produced, linear, 30 cm long, 1 mm wide. Cauline leaves 1-2, 30 cm long, 2-4 mm wide. Spathes not subtended by cauline leaves, the plants obviously caulescent. Flowers pale blue or violet, sometimes with darker spots.

Ecology: Andean II: 2000-3500 m. Disturbed areas.

Distribution: Peru, N-Chile.

Arequipa: native. ARE.

Mastigostyla hoppii R.C.Foster

Description: Bulbous plant. Spathe subtended by cauline leaves, the plants appearing acaulescent. Flowers solitary, terminal, with 2 unequal bracts, one very long, the other as long as the corolla. Petals violet.

Ecology: Andean II: 2500-3000 m. Rocky slopes.

Distribution: Peru.

Arequipa: native. ARE.

Olsynium junceum (E.Mey. ex C.Presl) Goldblatt

Description: Stems 30-60 cm long, terete, ending in a long whip-like bract subtending one sessile or several pedunculate flower clusters. Basal leaves 2-3, glaucous. Spathes narrow, often <2 cm long, few- to many-flowered. Flowers 8-12 mm long, pink or white, ± nodding, filaments elongate, connate. Capsule globose.

Ecology: Andean II-III: 3000->4500 m. Beach, grasslands, rocky slopes.

Distribution: Peru, Bolivia, Chile, Argentina.

Arequipa: native. ARE, CAY.

Sisyrinchium chilense Hook.

Description: Tufted, evergreen perennial. Leaves erect, flattened, linear, glaucous. Flowers in clusters on long pedicel, veined, lilac. Fruits globose, erect when ripe.

Ecology: Andean II-III: 2000-4000 m. Grasslands.

Distribution: Ecuador, Peru, Bolivia, Chile, Argentina.

Arequipa: native, doubtful. ARE. No specimen, mentioned for RN Salinas y Aguada Blanca by Quipuscoa & Huamantupa (2010), a doubtful observation of a fruiting plant from Lomas de Atiquipa (iNaturalist, 2024), the altitude of the site and the pendulous fruits suggest a confusion with S. micranthus.

Sisyrinchium jamesonii Baker

Description: Densely tufted, with narrow stems winged above, up to 20 cm tall, rather longer than the linear rigid leaves. Spathes 2-4-flowered, the outer over 2.5 cm long. Flowers 8 mm long, tepals yellow. Fruits oblong.

Ecology: Andean II-III: 3000-4000 m. Rocky slopes.

Distribution: NW-Venezuela to NW-Argentina.

Arequipa: native. UNI. No specimen, a convincing observation from La Unión (iNaturalist, 2024). Occurrence likely, due to distribution range.

Sisyrinchium micranthum Cav.

Description: Grass-like annual or perennial 10-60 cm high with a short rhizome. Stems flattened to winged. Leaves linear, sword-shaped, 5-20 cm x 1.5-5 mm. Spathe bracts narrow-lanceolate, subequal, mostly to 5 cm long, acuminate. Pedicels 2.5-4 cm long. Perianth 10-15 mm long, 10-20 mm Ø, with tepals spreading widely or reflexed at apex, white to cream with purple or blue center, externally purple-striped, pubescent below. Capsule globose, 3-8 mm Ø, glabrous, brown, pendulous.

Ecology: Andean I-II: 500-2500 m. Disturbed areas, lomas.

Distribution: Native to South America, introduced worldwide as ornamental.

Arequipa: native. CAR, ISL.

Use: Ornamental plant.

Tigridia arequipensis Montesinos, A.Pauca & Revilla

Description: Basal leaves 12-70 mm long and 1.5-9 mm wide. White to pale white (or pale lilac) flowers, outer tepals with purplish-maroon and dark yellow spots and stripes, and inner tepals with pale purplish and bluish spots and stripes. Differs from T. raimondii by having longer basal leaves, narrower and larger bracts and outer tepals ovate, longer fruits and grows at higher altitude.

Ecology: Andean II: 2500-3150 m. Rocky slopes, xerophytic shrubland.

Distribution: S-Peru (vicinity of the city of Arequipa).

Arequipa: native. ARE.

Tigridia raimondii Ravenna

Description: It differs from T. arequipensis in that the basal leaves are 15-30 cm long and 14-18 mm wide, and the flowers are white with a purple basal section. Plant of the lomas at low altitude.

Ecology: Coastal: 0-1000 m. Lomas.

Distribution: S-Peru (Lomas de Atiquipa).

Arequipa: native. CAR. No voucher. Mentioned for Arequipa by Brako & Zarucchi (1993) and for the Lomas de Atiquipa by Talavera et al. (2017).

Juncaceae

Distichia muscoides Nees & Meyen

Description: Perennial forming dense, hard cushions up to several meters in Ø. Leaves 1-2 cm long, arranged in two rows, strongly imbricate, cylindrical with an acute point. Male flowers subterminal, pedicel 1-2 cm long, female flowers hidden in the sheath of the supporting leaf, with only stigmas projecting above the shoot apex.

Ecology: Andean III: 4000->4500 m. Bofedales.

Distribution: Colombia, Ecuador, Peru, Bolivia.

Arequipa: native. ARE, CAY, CON.

Use: The dry blocks of the plant are an important fuel for people living at high altitudes.

Juncus balticus Willd.

Description: Plants 50-170 cm tall, robust in dense stands. Rhizome with 1-4 cm long internodes. Stem 3-10 mm thick, terete, not compressed or twisted, often interrupted. Inflorescence much-branched, 2-15 cm, dense, with 20 to several hundred flowers, partly congested. Bracteoles ovate, 1.5-2 mm long, acute, reddish-brown. Tepals subequal, 4.5-6 mm long, narrowly lanceolate, red-brown, margins narrow. Capsule ellipsoid-ovoid, trigonous, shorter or equaling perianth, acuminate.

Ecology: Andean II-III: 2500-4000 m. Grasslands, riversides.

Distribution: Europe and the Americas.

Arequipa: native. ARE, ISL.

Note: In Arequipa, J. balticus ssp. andicola.

Juncus bufonius L.

Description: Annual, variable in appearance. Generally, a green clumping grass-like rush, with many thin stems wrapped with few thread-like leaves. Flowers borne in inflorescences and in the joint where the inflorescence branches off of the stem. Flower folded within tough bracts and sepals.

Ecology: Coastal, Andean I-III, Amazonian: 0-4000 m. Disturbed areas, shrublands, bogs.

Distribution: Cosmopolitan.

Arequipa: native. ARE, ISL.

Juncus ebracteatus E.Mey.

Description: Easily recognized by its long creeping, stolon-forming rhizome, the globular clusters of flowers in the inflorescence, the protruding styles that remain as long beaks on the mature capsules and the light yellow or white anthers, several times longer than the filaments.

Ecology: Andean II-III: 2000-4000 m. Rocky slopes, shrublands.

Distribution: Mexico to Bolivia.

Arequipa: native. ARE, CAY. No specimen but a convincing observation from Yanque in iNaturalist (2024).

Juncus imbricatus Laharpe

Description: Stiffly clumped tufts to 70 cm tall. Leaves 1-2 per stem, very slender, wiry, channeled, shorter than the stem, growing from the base. Inflorescence 2.5-6 cm long, few branched, overtopped by a rigid leaf-like extension. Flowers solitary along branchlets. Capsules 6-6.5 mm long.

Ecology: Andean III: 3500-4500 m. Grasslands, disturbed areas.

Distribution: Native to South America, introduced to Mexico, South Africa, Australia and Europe.

Arequipa: native. CAR, ISL.

Juncus stipulatus Nees & Meyen

Description: Stems 1-5 cm high. Inflorescence simple or compound, 1 to several heads. Capsule a little longer than the brownish perianth, style elongate.

Ecology: Andean II-III: 3000->4500 m. Riversides, bofedales.

Distribution: Native to South America.

Arequipa: native. CAY. No specimen, an not curated observation from Sallialli by the author (iNaturalist, 2024). A common species in bofedales, probably often remained unnoticed.

Luzula racemosa Desv.

Description: Perennial, cespitose herbs, 5-45 cm high, without stolons. Culms erect, slightly curved, 0.5-2 mm Ø. Basal leaves 5 to several to each culm, in a loose or sometimes squarrose rosette, sheaths short, blades apically acute, 5-25 x 0.1-0.4 cm, flat. Cauline leaves 0-4 to each culm, to 15 cm long, margins pilose, especially in young and cauline leaves. Inflorescence a terminal, erect or nodding, spike-like raceme of 3 to several flower clusters, 1-6 x 0.5-2 cm.

Ecology: Andean III: 3500->4500 m. Grasslands, rocky slopes.

Distribution: Central and South America.

Arequipa: native. ARE, CAY.

Oxychloe andina Phil.

Description: Dioecious perennial, 8-15 cm tall forming large cushions. Leaves linear, spreading vertically, deep green to brownish-olive, up to 4 cm long, sharply pointed, membranous basal sheaths. Male flowers petiolate 3-4 cm, tepals straw-colored with transparent margins, females hidden amongst the foliage, slightly larger with broader tepals. Capsules ovoid, 8-10 by 4-5 mm, lustrous reddish-chestnut.

Ecology: Andean III: 3500-4800 m. Bofedales.

Distribution: Colombia to Argentina.

Arequipa: native. ARE, CAY, CON.

Patosia clandestina (Phil.) Buchenau

Description: Dioecious perennial, forming cushions, stems densely covered with old leaves. Leaves linear, 6-10 x 1 mm, broadly vaginated at base, tip mucronate, pungent. Male flowers subterminal, peduncle 3 cm long, tepals 5 mm long, female flowers hidden amongst the foliage. Capsule ovate, long acuminate.

Ecology: Andean III: 3500-4500 m. Bofedales.

Distribution: S-Peru, Bolivia, NW-Argentina, N-Chile.

Arequipa: native. CAY. No specimen, observation from Saillalli by the author. Ruhsatz (2012) notes that the species is common in the bofedales of S-Peru and probably often remained unnoticed or mistaken for Oxychloe andina.

Juncaginaceae

Triglochin striata Ruiz & Pav.

Description: Slender, erect rhizomatous perennial to 30 cm high. Leaves 30 cm x 1-3 mm, arising in distinct tufts separated by several nodes along an extensive rhizome. Inflorescence a spike-like raceme usually 2-15 cm long and <1 cm wide. Flowers 2 mm long, on short pedicels.

Ecology: Coastal, Amazonian: 0-500 m. Emergent.

Distribution: Pansubtropical.

Arequipa: native. ARE.

Krameriaceae

Krameria lappacea (Dombey) Burdet & B.B.Simpson

Description: Hemiparasitic shrub, up to 90 cm tall. Rootstock with wrinkled reddish-brown smooth bark and brown woody core, parasitized with haustoria. Branches silky when young, becoming woody and knotty with age. Leaves opposite, ovate, 1-2 cm long, silky. Flowers pink to red, 4-merous, unequal petals.

Ecology: Coastal, Andean I-III: 500-4000 m. Disturbed areas, lomas, rocky slopes, shrublands.

Distribution: Ecuador, Peru, Bolivia, N-Chile, NW-Argentina.

Arequipa: native. ARE, CAR, CAS, CAY, UNI.

Use: Medicinal plant. Extracts of the root have been used to improve the color, body and astringency of red wine.

Lamiaceae

Clinopodium bolivianum (Benth.) Kuntze

Description: Much-branched aromatic undershrub with erect or ascending branches, 0.5-2 m tall. Leaves oblong, 1- 2 cm long, more than 3x as long as broad, shortly petiolate or subsessile, margins entire. Flowers usually solitary in leaf axils. Calyx tube cylindrical, 2-3 x as long as the teeth. Corolla white, corolla tube 5-8 mm long, clearly exceeding the calyx.

Ecology: Andean II-III: 3000-4500 m. Rocky slopes, shrublands.

Distribution: Peru, Bolivia, NW-Argentina.

Arequipa: native. ARE, UNI.

Use: Medicinal plant.

Clinopodium gilliesii (Benth.) Kuntze

Description: Much-branched aromatic undershrub with stiff, erect branches, 0.6-2 m tall, shortly but uniformly pubescent. Leaves oblong, often narrower than 4 mm, more than 3x as long as broad. Corolla white, tube 1-2 mm long, about equaling the calyx.

Ecology: Andean III: 4000->4500 m. Grasslands, rocky slopes, shrublands.

Distribution: Peru, Bolivia, N-Chile, N-Argentina.

Arequipa: native. CAY.

Hedeoma mandoniana Wedd.

Description: Perennial subshrub, to 10-20 cm tall. Leaves sessile to short petiolate, appearing imbricate, ovate to obovate, 5-10 x 4-7 mm, margins crenate. Inflorescence 1-5 flowered, flowers 20-25 mm long, white tinged with pink.

Ecology: Andean II-III: 2500-4000 m. Rocky slopes.

Distribution: Peru, Bolivia, NW-Argentina.

Arequipa: native. ARE, CAY, UNI.

Use: Medicinal plant.

Leonotis nepetifolia (L.) R.Br.

Description: An erect, loosely branched annual with strongly angled stems, to 3 m tall. Leaves smooth, triangular. Flowers in rounded, spiny clusters encircling the stems. Tubular flowers orange and furry, 2.5 cm long and curve downward.

Ecology: Coastal, Andean I: 0-2500 m. Roadsides, overgrazed pastures, disturbed sites, waste areas, waterways and floodplains.

Distribution: Native to tropical and subtropical E-Africa, widely naturalized in the tropics worldwide.

Arequipa: introduced. ARE, CAR.

Use: Ornamental and medicinal plant.

Lepechinia lamiifolia (Benth.) Epling

Description: Erect shrub, up to 2 m tall. Leaves ovate, base cordate, apex obtuse to acute, margins crenate. Inflorescence an axillary lax panicles. Flowers red to purple, 1.5-3 cm long.

Ecology: Coastal, Andean I-II: 500-3000 m. Disturbed areas, forests, lomas, rocky slopes, shrublands.

Distribution: Ecuador, Peru.

Arequipa: native. CAR.

Lepechinia meyenii (Walp.) Epling

Description: Procumbent-ascending perennial herb from ligneous caudex, 20-40 cm tall. Leaves ovate, margins crenate, apex and base obtuse, petiole 3-6 mm. Flowers in foliose, 3-6-flowered glomeruli. Corolla tube 3.5-4.5 mm long, white.

Ecology: Andean II-III: 2500-4000 m. Disturbed areas, rocky slopes, shrublands.

Distribution: N-Argentina, Bolivia, Peru.

Arequipa: native. ARE, CAY, UNI.

Marrubium vulgare L.

Description: Herbaceous perennial, up to 25-45 cm tall. Leaves 2-5 cm long, densely rugose surface, gray tomentose. Flowers white, borne in axillary whorls on the upper part of the main stem.

Ecology: Coastal, Andean I-III: 500-4000 m. Disturbed areas, grasslands, lomas, rocky slopes, shrublands.

Distribution: Native to the Mediterranean, introduced in warmer climates.

Arequipa: introduced. ARE, CAM, CAR, CAY, ISL, UNI.

Use: Medicinal plant used to treat infections and ectoparasites and as a natural insect repellent in agriculture.

Mentha aquatica L.

Description: Perennial, stoloniferous herb. Leaves glabrous to hispidulous, green, margins serrate, strongly aromatic. Flowers axillary and terminal in upper part of stem, clustered in pseudoheads. Corolla pink or purple.

Ecology: Andean II: 3000-3500 m. Wet places, grassland.

Distribution: Native to Europe, introduced worldwide.

Arequipa: introduced. ARE, CAY.

Mesosphaerum sidifolium (L'Hér.) Harley & J.F.B.Pastore

Description: Subshrub, 1 m tall, very variable. Leaves 3-5 cm long, ovate, apex obtuse, base slightly cordate, margins finely serrate, underside white pulverulent. Inflorescence an axillary 2-6-flowered head on a 10-30 mm long filiform peduncle. Flowers 4 mm long, purplish-white.

Ecology: Coastal, Andean I-II: 0-3500 m. Disturbed areas, lomas, rocky slopes.

Distribution: Tropical America.

Arequipa: native. CAR, ISL.

Minthostachys mollis (Benth.) Griseb.

Description: Pungent, ± villous-tomentose perennial, 0.4-1 m tall. Leave ovate, 20-35 x 10-13 mm, margins serrulate. Inflorescence axillary, in sessile or short pedunculate heads of 8-10 flowers. Flowers white, purple or lavender, 4-4.5 mm long.

Ecology: Coastal, Andean I-II: 500-3500 m. Disturbed areas, rocky slopes, shrublands.

Distribution: Venezuela, Colombia, Ecuador, Peru and Bolivia.

Arequipa: native. CAY, UNI.

Use: Used as a medicinal plant with effects similar to peppermint (upset stomach, colds).

Minthostachys spicata (Benth.) Epling

Description: Subshrub, decumbent, branched perpendicularly. Stem thin, quadrangular in Ø, woody. Leaves ovate, 40 x 25 mm, apex acute to acuminate, base obtuse, margins serrate. Inflorescence axillary and terminal in dense clusters. Flowers white, underlip violet spotted.

Ecology: Andean II-III: 2500-4000 m. Rocky slopes, shrublands.

Distribution: Ecuador, Peru.

Arequipa: native. CAR, ISL.

Salvia haenkei Benth.

Description: Slender shrub, 1-2.5 m tall. Leaves oblong-elliptic, 4-14 x 1-4 cm, margins regularly crenulate, bullate- rugulose above. Trichomes of leaves and branches branched. Inflorescence with whorls of mostly 3 flowers. Flowers 2.5-3 cm long, bright red, the outer surface of the corolla tube hairy.

Ecology: Andean II: 2500-3000 m. Shrublands.

Distribution: Peru, Bolivia.

Arequipa: native. ARE, UNI.

Salvia hispanica L.

Description: Annual herb, up to 1 m tall, the whole plant gray-puberulent. Leaves ovate, 5-8 x 3-6 cm, cuneate at base, gray-green on both surfaces. Inflorescence of terminal racemes 5-17 cm long, rahter dense, only the lowest whorls not confluent, whorls 6-20-flowered. Bracts persistent, calyx gray-pubescent, 8 mm long, elongating to 11 mm in fruit, not sticky but rather bulbous at base. Corolla blue, 8 mm long, the upper lip shorter than the lower.

Ecology: Andean I-II: 1000-3000 m. Disturbed areas, cultivated land.

Distribution: Native to Mexico, cultivated and introduced worldwide.

Arequipa: introduced. ARE, UNI.

Use: The sprouted seeds are eaten in salads, soups, stews, etc. The seeds are rich in protein and oil.

Salvia oppositiflora Ruiz & Pav.

Description: Low sprawling to erect half-shrub, up to 50 cm tall, glandular. Leaves deltoid-ovate, 2.5 x 1-3 cm, ± glabrous, margins serrate. Inflorescence short, whorls 1-2-flowered. Flowers 2.5 cm long, red to yellow, with reddish stamens protruding from the lower lip.

Ecology: Andean II-III: 2000-4000 m. Cloud forests, disturbed areas, riversides, rocky slopes, shrublands.

Distribution: Peru.

Arequipa: native. CAR, CAY.

Note: S. oppositiflora, S. striata and S. tubiflora are very similar in morphology and ecology.

Salvia paposana Phil.

Description: Erect annual herb. Leaves ovate, sessile, wrinkled, margins dentate. Inflorescence a terminal raceme, often with some short axillary racemes, whorls 1-4-flowered. Flowers deep blue, calyx eglandular.

Ecology: Coastal: 0-1500 m. Lomas, rocky slopes.

Distribution: Peru, N-Chile.

Arequipa: native. CAR.

Note: Similar to S. rhombifolia but restricted to low altitudes.

Salvia rhombifolia Ruiz & Pav.

Description: Herbaceous plant up to 50 cm tall, glandular-hispid. Leaves ovate, up to 4 cm long, blade green above, paler below, petiole 10-15 mm long. Inflorescence terminal, 3-6(-12) flowers per whorl, corolla blue, calyx glandular.

Ecology: Coastal, Andean I-III: 0-4000 m. Disturbed areas, lomas, riversides, rocky slopes, shrublands.

Distribution: Peru, N-Chile.

Arequipa: native. CAR, ISL.

Salvia striata Benth.

Description: Shrub. Leaves obovate, underside whitish, base obtuse, mostly <25 mm wide, apex obtuse to acute, margins crenate. Inflorescence of terminal racemes 10-30 cm long. Whorls 1-2-flowered. Corolla red, pubescent, upper lip slightly longer than the lower. Calyx blackish.

Ecology: Coastal, Andean I-II: 0-3000 m. Disturbed areas, shrublands.

Distribution: Peru.

Arequipa: native. CAM.

Note: See note on S. oppositiflora.

Salvia tubiflora Sm.

Description: Shrub. Leaves obovate, beneath white villous-lanate, base cordate, 4.5-8 x 4-6 cm, apex obtuse to acute, margins crenate, petiole up to 3 cm long. Inflorescence of terminal racemes 10-30 cm long. Whorls 2- flowered. Corolla red, 3-4.5 cm long, pubescent, upper lip slightly longer than the lower. Calyx hairy.

Ecology: Coastal, Andean I-II: 0-3000 m. Disturbed areas, lomas, rocky slopes, shrublands.

Distribution: Peru, N-Chile.

Arequipa: native. CAR, ISL.

Note: See note on S. oppositiflora.

Stachys aperta Epling

Description: Annual herb. Leaves ovate, 2-3 cm long, margins crenate-serrate, cordate at base, acute at apex, petiole as long as the blade. Inflorescence spike-like, leavy. Calyx turbinate in flower, 7-8 mm long, even in fruit. Corolla pink, tube 6-8 mm long, annulate above the base.

Ecology: Andean II: 2000-3000 m. Disturbed areas, riversides, shrublands.

Distribution: Peru, Bolivia.

Arequipa: native. CAM.

Stachys arvensis (L.) L.

Description: Herb with stems decumbent to ascending, much-branched, 20-60 cm long, hirsute. Leaves ovate to narrowly ovate, 1-3.5) x 0.6-3 cm, hirsute, margins crenate, apex obtuse to acute, base truncate to subcordate, petioles 0-2 cm long. Flowers usually 3-6 in verticillasters, in terminal, leafy, spike-like inflorescences. Calyx tinged purple, campanulate, 3-5 mm long in flower, twice as long in fruit, hirsute. Corolla pink, rose, or blue, 3.5-5 mm long, inserted in calyx, upper lip erect.

Ecology: Coastal: 0-500 m. Disturbed areas, rocky slopes, lomas.

Distribution: Native to Eurasia, introduced worldwide.

Arequipa: introduced. CAR.

Stachys herrerae Epling

Description: Perennial herb. Leaves ovate, base subcordate, apex obtuse, margins regularly crenate. Flowers axillary, pale purple, calyx 3-6.5 mm long, corolla tube 5-7 mm long, longer than lower lip.

Ecology: Andean I-III: 500-4000 m. Disturbed areas.

Distribution: Peru.

Arequipa: native. CAR, ISL. Mention for the Lomas de Atiquipa and specimen from Mejía. The picture in Talavera et al. (2017) clearly shows the elongated calyx in fruit.

Linaceae

Linum prostratum Dombey ex Lam.

Description: Branched perennial subshrub (ssp. parvum annual). Leaves linear, 1 cm long, lower subopposite, upper alternate, gradually reduced. Flowers typically opposite, subsessile to petiolate, petals yellow.

Ecology: Coastal, Andean I-III: 0-4000 m. Rocky slopes.

Distribution: Peru and N-Chile.

Arequipa: native. CAM, CAS, ISL.

Note: In Arequipa, L. prostratum var. parvum (lomas) and var. prostratum (Andes).

Loasaceae

Caiophora andina Urb. & Gilg

Description: Annual erect herb, stems 10-60 cm long, all parts with stinging hairs. Few flowered inflorescence (1-2 flowers), flowers balloon-shaped, salmon-pink. Fruit not to slightly twisted. Morphologically intermediate between C. cirsiifolia and C. rosulata.

Ecology: Andean III: 3500-5000 m. Dry puna.

Distribution: Peru, Bolivia.

Arequipa: native. ARE.

Note: Ackermann (2011) considers it to be an unresolved species that may belong to Caiophora chuiquitensis (Meyen) Urb. & Gilg.

Caiophora carduifolia C.Presl

Description: Twining herb up to 7 m long. Habit Clematis-like. Leaves lobed, not pinnatisect. Flowers campanulate, 4-5-merous, 40 mm long, petals yellow to pale orange, nectar scales green, sometimes yellow. Fruit twisted.

Ecology: Andean II: 3000-3500 m. Rocky slopes.

Distribution: Peru.

Arequipa: native. ARE, CAS, CAY.

Caiophora cirsiifolia C.Presl

Description: Perennial, twining herb up to 5 m tall, without basal rosette. The whole plant with scattered stinging hairs. Leaves pinnate-pinnatifid to 2-pinnatifid, with 6-7 lobes on each side, margins serrate. Flowers orange, nectar scales orange. Fruit twisted.

Ecology: Andean II-III: 2400-3700 m. Shrubland, hedges, road banks and dry-stone walls.

Distribution: Peru and N-Chile.

Arequipa: native. ARE, CAS, CAY.

Use: Used as a medicinal plant for headaches, fever, and upset stomach. The plant is also used in traditional ceremonies for exorcism.

Caiophora deserticola Weigend & Mark.Ackermann

Description: Suffrutescent, sparsely setose perennial herb. Leaves pinnate-pinnatifid to 2-pinnatifid with (4-) 7-8 lobes on each side and only one pair of free leaflets, ovate to lanceolate in outline, scarcely covered with stinging hairs, whitish beneath. Flowers 5-7-merous, corolla pink, pedicel 2-3 cm long, petals usually connate.

Ecology: Andean II-III: 2500-4000 m. Arid habitats, roadsides, crevices, between rocks.

Distribution: S-Peru to N-Chile.

Arequipa: native. ARE. Presumably, the voucher of C. chuiquitensis from Arequipa by Brako & Zarucchi (1993) should be placed here.

Caiophora pentlandii (Paxton) Paxton & Lindl.

Description: Decumbent herb, 0.5 m tall, not climbing, internodes over 5 cm long. Leaves pinnate, narrowly ovate in outline, densely covered with stinging hairs, green beneath. Corolla bright red-orange, flowers decumbent, nectar scales white, petals free. Fruit not twisted.

Ecology: Andean II-III: 2500-3000 m. Arid habitats, shrubland.

Distribution: S-Peru.

Arequipa: native. CAY.

Caiophora rosulata (Wedd.) Urb. & Gilg

Description: Perennial, acaulescent rosulate plant, 5-20 cm tall. Flowers red, nectar scales white. Capsule up to 16 mm long, not twisted.

Ecology: Andean III: 3500->4500 m. Grasslands, rocky slopes, at the base of rocks and grass tussocks.

Distribution: S-Peru, Bolivia, N-Chile.

Arequipa: native. ARE, CAY.

Loasa nitida Desr.

Description: Procumbent annual herb. Leaves opposite, ovate to orbicular, deeply palmately lobed, venation ± palmate. Flowers yellow to orange, with blood-red nectaries in the center.

Ecology: Coastal: 0-500 m.

Distribution: Peru, N-Chile.

Arequipa: native. CAR, CAS.

Mentzelia scabra Kunth

Description: Perennial herbs, 1-3 m tall. Leaves simple, hastate, margins serrate, petiole 2-5 cm long. Flowers yellow-orange, petals overlapping, stamens not in fascicles.

Ecology: Coastal, Andean I-II, Amazonian: 0-3000 m. Rocky slopes, 0-3500 m.

Distribution: The Andes from Venezuela to Chile.

Arequipa: native. ARE, CAR, CAY, ISL.

Note: In Arequipa, M. scabra ssp. chilensis.

Nasa chenopodiifolia (Desr.) Weigend

Description: Annual herb 10-110 cm tall. Leaves only shallowly lobulate, margins irregularly crenate-dentate. Petals white (8-)10-15 mm long, capsule 4x as long as wide, plants dwarfed and unbranched, or branched from base, densely covered with short, yellowish setae.

Ecology: Andean I-II: 1000-3500 m. Disturbed habitats.

Distribution: Peru.

Arequipa: native. UNI.

Nasa urens (Jacq.) Weigend

Description: Annual herb of the lomas. Leaves 2-pinnate. Corolla bright yellow, nectar scales white and grayish- green.

Ecology: Coastal: 0-500 m. Lomas.

Distribution: Peru, N-Chile.

Arequipa: native. CAM, CAR, ISL.

Presliophytum arequipense Weigend

Description: Shrub, 50-100 cm tall. Leaves 2-6 x 1.5-4.5 cm, ovate, margins lobate. Uniseriate glandular hairs rare or absent. Sepals 5, broadly ovate, 5-12 x 5-10 mm. Petals subequal in size to sepals, cream to greenish-white, tinged darker greenish beneath.

Ecology: Coastal: 0-100 m. Cliffs, dry riverbeds, lomas.

Distribution: S-Peru (restricted to coastal Arequipa).

Arequipa: native. ARE, CAR, ISL, UNI.

Presliophytum incanum (Graham) Weigend

Description: Shrub, 50-150 cm tall. Leaves 4-10 x 2.5-8 cm, ovate with (0-)3-6 lobes on each side, margins crenate or toothed. Uniseriate glandular hairs rare or absent. Sepals 5, ca. 1.5-2 × as long as wide. Petals at least 1.5 × as long as sepals, white.

Ecology: Coastal, Andean I-II: 0-3000 m. Lomas, shrubland, disturbed areas.

Distribution: Peru.

Arequipa: native. ARE, CAM, CAR, CAY.

Loranthaceae

Ligaria cuneifolia (Ruiz & Pav.) Tiegh.

Description: Densely branched hemiparasitic shrub growing on trees, cacti and shrubs. Leaves 1 cm long, apically rounded, mucronate, faintly 3-nerved. Flowers 3 cm long, yellowish-red.

Ecology: Andean II-III: 1000-4000 m. Disturbed areas, dry valleys, forests, rocky slopes.

Distribution: Peru, Bolivia, N-Chile and NW-Argentina.

Arequipa: native. ARE, CAS, CAY, UNI.

Use: The stems are depurative and diuretic. The whole plant contains ephedrine, an alkaloid used to treat asthma. The cones are sometimes eaten, and the plant is used as fodder for cattle in hard times.

Lythraceae

Ammannia coccinea Rottb.

Description: Erect, sometimes branched annual herb, up to 60 cm tall. Leaves opposite, sessile, lanceolate-linear, >6 cm long. Flowers 4-merous, sessile, axillary, in small clusters. Petals dark pink.

Ecology: Coastal, Amazonian: 0-500 m. Riversides.

Distribution: N-America to Ecuador, Transcaucasus to Iran, N-Australia.

Arequipa: introduced. CAM. One convincing observation from Camaná in iNaturalist (2024).

Malvaceae

Abutilon arequipense Ulbr.

Description: Shrub, up to 2 m tall. Younger branches and leaves densely yellowish-ash-gray lanate-tomentose, bark gray. Leaves cordate-oblong, obtuse or subacute, margins crenate. Corolla violet, soon reflexed, darker toward base within.

Ecology: Andean II: 2500-3000 m.

Distribution: Peru, known only from Arequipa province Cotahuasi (La Union).

Arequipa: native. UNI.

Abutilon bivalve (Cav.) Dorr

Description: Shrub, 1-2 m tall. Leaves cordate-ovate, apex long acuminate, base cordate, margins crenate-serrate, both sides densely villous. Calyx deeply parted. Petals 10-12 mm long, dark yellow, becoming orange after flowering.

Ecology: Coastal, Andean I-II: 0-3000 m. Forests, rocky slopes.

Distribution: Mexico to Argentina.

Arequipa: native. CAR.

Acaulimalva engleriana (Ulbr.) Krapov.

Description: Acaulescent herb. Leaves up to 3 cm long, obovate, slightly 3-5-lobed. Flowers white, sometimes with pale pink tip, petals 15-35 mm long.

Ecology: Andean II-III: 2500-4500 m. Disturbed areas, grasslands, rocky slopes.

Distribution: Peru.

Arequipa: native. UNI.

Acaulimalva nubigena (Walp.) Krapov.

Description: Acaulescent, rosulate herb. Flowers long pedicellate, corolla purple to white with purple marks in center. Highly variable in leaf shape. Calyx completely glabrous calyx. Mericarps indehiscent, with the margins of the dorsum irregularly toothed.

Ecology: Andean III: 3500-4500 m. Wet meadows.

Distribution: S-Peru, Bolivia, NW-Argentina.

Arequipa: native. CAY.

Acaulimalva rhizantha (A.Gray) Krapov.

Description: Leaves orbicular, 1.5 cm in Ø, margins apically crenate. Peduncle 2 cm long, shorter than the petiole. Flowers white, petals <10 mm long.

Ecology: Andean II-III: 2500->4500 m. Grasslands.

Distribution: C-Peru to W-Bolivia.

Arequipa: native. ARE, UNI. Specimen Rauh & Hirsch P-631 from Tincopalca at 4200 m is indicated for Arequipa by Krapovickas (1974). At this altitude the locality must be in Puno. Mentioned for Arequipa, no other specimens.

Cristaria multifida (Dombey ex Cav.) Cav.

Description: Erect herb, up to 2 m tall. Leaves ovate to elliptic, often trifid, the lobes ± deeply pinnate. Flowers whitish or lilac, distinctly pending on an expanding distal end of pedicel (= caprocrater).

Ecology: Coastal, Andean I-II, Amazonian: 0-2500 m. Disturbed areas, roadsides.

Distribution: Peru, Bolivia and Chile.

Arequipa: native. ARE, CAM, CAR, CAY, ISL.

Note: Only C. multifida is recognized for Peru by Schneider (2013).

Fuertesimalva chilensis (A.Braun & C.D.Bouché) Fryxell

Description: Annual herb. Stem and petiole hispid. Stipules narrowly lanceolate, 13 x 2-4 mm. Leaves orbicular, divided or very deeply lobbed, margins serrate, often glabrous with a bristle terminating each tooth, petiole 1-2.5 cm long. Inflorescence scorpioid, peduncle 2-6 cm long. Flowers white with purplish base or pale violet. Pistils 15, mericarps laterally reticulate.

Ecology: Coastal, Andean I-II: 0-3000 m. Lomas.

Distribution: Peru, Chile.

Arequipa: native. ARE, CAR, ISL, UNI.

Fuertesimalva corniculata (Krapov.) Fryxell

Description: Annual herb with stellate hairs. Leaves 3 x 3 cm , 3-5-lobed, margin crenate, both sides stellate. Inflorescence axillary with 2-3 flowers. Petals 4 x 3 mm, glabrous. Mericarp 4.5 x 2.8 mm, gibbous, indehiscent, sides reticulate.

Ecology: Coastal, Andean I-II: 500-3000 m.

Distribution: Peru.

Arequipa: native. ARE, CAY, UNI.

Fuertesimalva echinata (C.Presl) Fryxell

Description: Annual herb, hirsute. Leaves orbicular, lobed to one third. Inflorescence 2-6-flowered, subequal to the subtending leaf. Pistils 7-10. Frutescence echinate and mericarps 2 x 2 mm.

Ecology: Coastal, Andean I-II: 0-4500 m. Disturbed areas, grasslands, rocky slopes.

Distribution: Peru, Bolivia, NW-Argentina.

Arequipa: native, doubtful. ARE. No voucher. Mentioned for RN Salinas y Aguada Blanca by Quipuscoa & Huamantupa (2010). Expected due to the distribution range in S-Peru (Cuzco, Puno, Moquegua).

Note: Easily confused with F. corniculata.

Fuertesimalva limensis (L.) Fryxell

Description: Annual herb at medium elevations. Plant stellate pubescent. Stipules ovate-lanceolate, 6 mm long. Leaves ovate to subrotund, ± clearly 3(-5) lobed, often glabrate, especially above, petiole 1-2.5 cm long. Peduncle 2-4 cm long. Flowers pink, sometimes white with purplish base. Pistils 12, mericarps rugulose, lacking an appendix (= endoglossum).

Ecology: Andean II: 1500-3000 m. Disturbed areas, rocky slopes.

Distribution: Mexico to Bolivia.

Arequipa: native. ARE, UNI.

Fuertesimalva pennellii (Ulbr.) Fryxell

Description: Annual herb, ± stellate-pilose. Leaves trilobed, 5 x 5-6 cm, the middle lobe 1-1.5 cm wide at base, dilated to 2-2.5 cm at tip, blade bright green, margins serrate-crenate. Inflorescence axillary. Flowers pink or purple, petals 7-8 mm long, ca. 10 pistils. Mericarps 3 x 3 mm.

Ecology: Andean I-II: 2500-3500 m. Rocky slopes, disturbed areas.

Distribution: Peru.

Arequipa: native. ARE, CAY.

Note: Hill (1982) treats it as an unresolved species, the type from Arequipa is lost.

Fuertesimalva peruviana (L.) Fryxell

Description: Similar to F. limensis but growing at low altitude and mericarps with a small internal appendix.

Ecology: Coastal: 0-1000 m. Desert, disturbed areas, grasslands, lomas.

Distribution: Peru, Chile.

Arequipa: native. CAM, CAR, ISL.

Gaya atiquipana Krapov.

Description: Small shrub, all parts velutinous, without long hairs (difference to G. mollendoense). Leaves ovoid, margins dentate-serrate. Flowers yellow to orange, petals 15-20 mm long. Mericarps 12-14, all isodiametric (difference to G. mollendoense).

Ecology: Coastal: 250-600 m. Lomas.

Distribution: S-Peru (Arequipa).

Arequipa: native. CAR.

Gaya mollendoensis Krapov.

Description: Subshrub ca. 30-40 cm tall. Leaves obtuse, wide, both sides velutinous and with longer hairs on the veins. Flower all but calyx orange-yellow, petals 15 mm long. Mericarps 10-11, longer than wide.

Ecology: Coastal: 400-600 m. Lomas.

Distribution: S-Peru (Arequipa).

Arequipa: native. CAR, ISL.

Gaya weberbaueri Ulbr.

Description: Shrub 1 m high. Plant yellowish velutinous-pubescent. Inflorescence 3-flowered, petals 10 mm long, orange.

Ecology: Andean I-II: 1000-2500 m.

Distribution: S-Peru (Arequipa).

Arequipa: native. ARE, CAR.

Note: The specimens Dillon & Dillon 3805 (F), Anderson 7996 (F, incl. photo), and Dillon et al. 8872 (F) probably represent G. atiquipana. G. weberbaueri is restricted to N- and C-Peru (Brako & Zarucchi, 1993; Krapovickas, 1996, León et al., 2006). The taxonomy of Gaya species at mid-altitudes in S-Peru remains unclear. However, since G. weberbaueri is frequently used for the species in Arequipa, we will adopt this name.

Herissantia crispa (L.) Brizicky

Description: Prostrate herbs, with spreading hairs. Leaves 5 x 3.5 cm, ovate-cordate, acuminate, base deeply cordate, crenate. Lower leaves petiolate, upper leaves immediately below inflorescence sessile and amplexicaule. Flowers axillary, solitary, white or yellow. Fruit subglobose, fragile-walled, lantern-like.

Ecology: Coastal, Andean I-II: 0-2000 m. Riversides, shrublands.

Distribution: Native to tropical Americas, introduced all over the tropics.

Arequipa: native. CAY.

Malva assurgentiflora (Kellogg) M.F.Ray

Description: Perennial shrub, stem generally 1-4 m tall. Leaves long petiolate, blade 5-15 cm in Ø, 5-7 lobed. Flowers showy, petals 2.5-4.5 cm, rose to deep purple, calyx >10 mm.

Ecology: Andean II-III: 2000-3800 m. Disturbed areas.

Distribution: Native to USA, naturalized here and there.

Arequipa: introduced, adventive. ARE.

Use: Cultivated as an ornamental plant.

Malva parviflora L.

Description: Annual herb, trailing or ascending. Leaves orbicular or reniform, 2-7 cm long, crenate, undulate, or 5-7- lobed Flowers 1-4 in the leaf axils, short-pedicellate, calyx 3-4 mm long, accrescent to 7-8 mm in fruit, petals lavender or white, 4-5 mm long Fruit nearly glabrous, mericarps ca. 10, rugose or wrinkled dorsally and winged at the angle between the dorsal and lateral walls.

Ecology: Coastal, Andean I-III, Amazonian: 0-4000 m. Rocky slopes.

Distribution: Native to S-Europe, introduced worldwide.

Arequipa: introduced. ARE, CAR, ISL, UNI.

Malvastrum coromandelianum (L.) Garcke

Description: Erect or decumbent subshrub to 1 m tall, most parts pilose and adpressed stellate pubescent. Leaves ovate to ovate-lanceolate 3-7 x 1-4 cm, sparsely hairy above, beneath pilose, margins coarsely dentate, apex and base acute or obtuse. Inflorescences axillar, solitary flowers, flowers sometimes congested toward branch tips, petals yellow, epicalyx lobes filiform, early caducous. Schizocarp 6 mm in Ø, mericarps 8-12(-14), reniform, sparsely pubescent, 2.5 mm in Ø.

Ecology: Coastal, Andean I-II, Amazonian: 0-2500 m. Disturbed areas, riversides, shrublands.

Distribution: Probably originating in America, now pan-tropical.

Arequipa: native. CAY. No specimen, a convincing observation from Sangalle in iNaturalist (2023).

Use: The whole plant is used medicinally.

Modiola caroliniana (L.) G.Don

Description: Annual to short-lived perennial herb rooting at the nodes, stems ± prostrate, <25 cm. Leaves ovate, 3-4 x 2-3 cm, 3-7-lobed, the lobes toothed, petiole 2-7 cm. Peduncles shorter than leaves. Epicalyx segments 3, linear- lanceolate, 5 mm long. Sepals 5-6 mm long, hairy. Petals red or orange-red, slightly longer than the calyx. Mericarps 15-25, reniform, black, bristly on top.

Ecology: Coastal, Andean I-II: 200-2500 m. Gardens and waste places.

Distribution: Native to southern South America, introduced worldwide.

Arequipa: introduced. ARE.

Use: Ornamental plant.

Nototriche anthemidifolia (Remy) A.W.Hill

Description: Small rosette-forming perennial herb, Ø 5 cm. Leaves all basal, 2-pinnatifid, pinnately divided, grayish- tomentose on the upper side, glabrous on the underside. Stipules distinctly hairy, whitish-brown. Flowers rose-pink, 10-12 mm long, emerging from below the middle of the petioles, calyx tube glabrous.

Ecology: Andean III: 3500->4500 m. Grasslands.

Distribution: Peru, Bolivia, N-Chile, N-Argentina.

Arequipa: native. ARE, CAY.

Nototriche argentea A.W.Hill

Description: Perennial, cespitose herb with sericeous tomentose indument. Petiole 2-3.5 cm long, leaves palmately incised. Flowers lavender, calyx hirsute.

Ecology: Andean III: 3500->4500 m. Grasslands.

Distribution: Peru, N-Chile.

Arequipa: native. ARE, CAY.

Nototriche argyllioides A.W.Hill

Description: Rosette-forming perennial with a stout rhizome. Leaves rosulate, multilobed, palmately divided at base, lobes ± 2-pinnatifid, incised-crenate, ashy puberulent above, glabrous beneath. Flowers pink-violet, 18-20 mm long, calyx tube glabrous.

Ecology: Andean III: 4000->4500 m. Disturbed areas.

Distribution: Peru, Bolivia.

Arequipa: native. ARE, CAY. Specimen Hill 80 (K) from Vincocaya in 1903. No contemporary vouchers.

Nototriche azorella A.W.Hill

Description: Dense cushion plant, comprising tightly packed cylindrical rosettes, scarcely 1 cm across. Leaves imbricated, orbicular, 4 mm wide, divided into 5-7 lobes. Flowers white tinged pink, petals truncate, 7 mm long.

Ecology: Andean III: >4500 m. Rocky slopes.

Distribution: S-Peru and N-Chile.

Arequipa: native. ARE, CAY, CON.

Nototriche digitulifolia A.W.Hill

Description: Similar to N. orbignyana but primary leaf-lobes digitiform or 5-7.

Ecology: Andean III: 4000-4500 m.

Distribution: Peru.

Arequipa: native, doubtful. ARE. Specimen Müller 2007 (1972), not seen. No contemporary observation.

Nototriche longirostris (Wedd.) A.W.Hill

Description: Perennial herb in lax rosettes. Leaves long petiolate, pinnatifid, 1-1.5 cm long, oblong-lanceolate, blade above glabrous, segments entire. Corolla 15 mm long, petals obovate, retuse.

Ecology: Andean III: 4000->4500 m. Bogs, disturbed areas, grasslands, rocky slopes.

Distribution: Peru, Bolivia.

Arequipa: native. ARE.

Nototriche mandoniana (Wedd.) A.W.Hill

Description: Leaves in lax rosettes, green, long petiolate, lanceolate, 1-2-pinnatifid, above stellate-tomentose, beneath glabrous, ciliate, segments linear. Flowers 20 mm long, pinkish-violet, petals retuse, calyx tube finely stellate.

Ecology: Andean III: 4000-4500 m.

Distribution: S-Peru, Bolivia.

Arequipa: native. ARE, CON.

Nototriche meyenii Ulbr.

Description: Loosely clustered rosettes, 5-10 cm in Ø, the rosulate leaves congested hemispherical. Leaves silvery tomentose, reniform, 14 x 7 mm, flabellate-5-7-fid, segments 3-5-lobate. Flowers up to 8-15 mm long, white, sometimes with dark violet stripes.

Ecology: Andean III: 3500->4500 m. Rocky slopes.

Distribution: S-Peru, Bolivia, N-Chile.

Arequipa: native. ARE.

Note: Rodriguez et al. (2018) synonymizes N. borussica with N. meyenii, although N. borussica is the older name. Here we follow him and prefer the more commonly used name.

Nototriche nigrescens A.W.Hill

Description: Perennial herb, forming cushions. Leaves crowded, white ash-gray, beneath blackish, petiolate, flabellate, segments lobate, crenate, obtuse. Flowers subsessile, corolla 20 mm long, white to pale purple, anthers dark purple. Calyx 12 mm long.

Ecology: Andean III: 4000->4500 m.

Distribution: Peru.

Arequipa: native. ARE. No specimen, two observations.

Nototriche obcuneata (Baker f.) A.W.Hill

Description: Small herb. Leaves reniform, with palmately proliferations on the apex, succulent, glaucous. Flowers white, with red marks on the apex of petals.

Ecology: Andean III: 4000->4500 m. Rocky slopes.

Distribution: Peru, Bolivia, N-Chile.

Arequipa: native. ARE, CAY, CON, UNI.

Nototriche orbignyana (Wedd.) A.W.Hill

Description: Rosettes depressed. Leaves silvery tomentose, hemispherically congested. Corolla dark blue to deep purple. Calyx lobes entire.

Ecology: Andean III: 4000-4500 m.

Distribution: S-Peru, Bolivia, N-Chile.

Arequipa: native. CAY. Flora of Peru (1936ff.) mentions specimen Copeland #s.n. (type of M. copelandii = synonym of N. orbignyana).

Note: Unresolved species. The images of specimen d'Orbigny, #s.n. (F) and specimen Meyen, #s.n. (B) are the same. Therefore, N. orbignyana is a doubtful species and could be a synonym of N. meyenii.

Nototriche pedatiloba A.W.Hill

Description: Similar to N. obcuneata but leaves incised up to half and flowers pinkish-white.

Ecology: Andean III: 4000->4500 m. Grasslands.

Distribution: S-Peru.

Arequipa: native, doubtful. ARE, CAY, UNI. Specimens Hill 78 (K) from Vincocaya and HUSA 7006 from El Simbral (second not seen), no contemporary observation. Confusion with N. obcuneata?.

Nototriche pediculariifolia (Meyen) A.W.Hill

Description: Rosette forming herb. Leaves gray-tomentose, long petiolate, palmatisect, segments 5-7-fid, incised- crenate. Calyx densely hairy, flowers white, marked purple and maroon, petals linear, 15 mm long.

Ecology: Andean III: 4000->4500 m. Grasslands, rocky slopes.

Distribution: S-Peru, Bolivia, N-Chile.

Arequipa: native. ARE, CAY, UNI.

Nototriche pellicea A.W.Hill

Description: Depressed, cespitose, pulvinate, hirsute. Leaves oblong-orbiculate, 4.5 mm long, 6 mm wide, deeply 5- 7-parted, the segments digitately lobulate with 30-50 narrowly linear-oblanceolate lobules. Calyx cylindric- campanulate, 7-7.5 mm long, tube glabrous, lobes stellate-hirsute. Corolla white, 1 cm long, petals overlapping, with purple lines.

Ecology: Andean III: 4000-> 4500 m. Grasslands, rocky slopes.

Distribution: Peru, Bolivia.

Arequipa: native, doubtful. ARE. Specimen van der Werff H. 20630 (MO) from Pampa de Arrieros (seen a doubtful image of the plant) and several doubtful observations.

Nototriche salina B.L.Burtt & A.W.Hill

Description: Similar to N. longirostris but in borax habitats (halophile).
Ecology: Andean III: 3500-4500 m. Saline habitats.
Distribution: S-Peru.
Arequipa: native. ARE. Known only from the type locality in Las Salinas (Arequipa).

Nototriche sepaliloba Hochr.

Description: Cushion plant. Scare information available.
Ecology: Andean III: >4500 m. Stony tracts.
Distribution: S-Peru, type from Huancavelica.
Arequipa: native, doubtful. ARE. No specimen but mentioned by Brako & Zarucchi (1993) and for RN Salinas y Aguada Blanca by Quipuscoa & Huamantupa (2010).

Nototriche staffordiae B.L.Burtt & A.W.Hill

Description: Globosely rosulate plant. Leaves 4-5 mm long, 10 mm wide, 9-dissected more than medially, the divisions 5-lobed. Corolla reddish-brown, greenish striped, petals not overlapping. Calyx lobes trifid.
Ecology: Andean III: >4500 m. Rocky slopes.
Distribution: S-Peru.
Arequipa: native. CAY.

Nototriche sulphurea A.W.Hill

Description: Cespitose perennial. Leaves laxly rosulate, ovate, 20-25 x 20-25 mm, pinnatifid, composed of oblong segments, base cordate, apex acuminate, margins crenate or entire, stellate sulphury-tomentose, petiole 40 mm long. Flowers solitary and axillary, calyx with 5 triangular oval sepals, 11 mm long, corolla with 5 obovate and bluish petals, 20-35 mm.
Ecology: Andean III: 3500->4500 m. Grasslands
Distribution: Peru, Bolivia.
Arequipa: native. CAY. No herbarium species, an observation in iNaturalist (2024) and distribution range Apurimac, Cuzco, Puno.

Nototriche turritella A.W.Hill

Description: Rosettes turret-like. Leaves cuneate or cuneate-reniform, trifid, deeply incised and minutely lobed, though only the dissection makes this apparent, for they are bound by their woolly covering into taught columns, the mature leaf tips glabrous and slightly projecting. Flowers pale lilac with deep purple centers, petals obovate, 13-15 mm long.
Ecology: Andean III: 3500->4500 m. Grasslands.
Distribution: S-Peru, Bolivia, N-Chile.
Arequipa: native. ARE.

Palaua camanensis Ferreyra & Chanco

Description: Low subshrubs, erect, ca. 15 cm high. Leaves lobed to pinnate-pinnatifid, canescent. Flowers up to 4 cm in Ø, solitary, pink to lilac. Most like P. mollendoensis but calyx with several parallel striae below the indument (inconspicuous in P. mollendoensis), calyx 12 x 18 mm (7-11 x 8-11 mm in P. mollendoensis), the apex of calyx lobes acute to acuminate (obtuse to acute in P. mollendoensis), the base of calyx has only a small reddish-brown part (calyx red in P. mollendoensis).
Ecology: Coastal: 0-500 m. Lomas.
Distribution: Endemic to the Lomas de Camaná.
Arequipa: native. CAM.

Palaua dissecta Benth.

Description: Annual or perennial herb, prostrate, decumbent, ascending or erect, to 30 cm high. Leaves (2-)pinnatifid, lobes 5-7(-9), basal lobes forming a sagittate base. Flowers solitary, pink, petals 7-22 mm. Calyx 5-7 mm.
Ecology: Coastal, Andean I: 0-1200 m. Lomas, rocky slopes.
Distribution: C- and S-Peru, N-Chile.
Arequipa: native. CAM, CAR, ISL.
Note: The P. dissecta complex consists of P. camanensis, P. mollendoensis, P. dissecta and P. weberbaueri (Huertas, 2010).

Palaua guentheri Bruns

Description: Herb, 20 cm tall. Leaves lobed nearly to the rachis, lobes linear. Flowers 5-8 mm long, in dense corymbiform inflorescences.
Ecology: Coastal: 0-500 m. Lomas.
Distribution: S-Peru.
Arequipa: native. ISL.

Palaua inconspicua I.M.Johnst.

Description: Erect annual herb, 20 cm tall. Branches green. Leaves orbicular-deltoid, 2 cm in Ø, slightly lobed- dentate. Stipules 1-2 veined. Flowers solitary or in racemes of 2-4 flowers, white, up to 1 cm in Ø.
Ecology: Coastal: 0-1000 m. Lomas.
Distribution: S-Peru, N-Chile.
Arequipa: native. ISL.
Note: P. inconspicua and P. modesta are closely related based on to morphological and molecular data, and both taxa can be considered conspecific if a broad species concept is applied (Huertas, 2010).

Palaua modesta (Phil.) Reiche & Johow

Description: Prostrate annual herb, branches reddish-brown. Leaves orbicular-deltoid, 2 cm in Ø, ± lobed-dentate. Stipules 3-5 veined. Flowers solitary or in racemes of 2-4 flowers, white, 1 cm in Ø.

Ecology: Coastal: 0-1000 m. Lomas, deserts.

Distribution: S-Peru, N-Chile.

Arequipa: native. CAR, ISL.

Palaua mollendoensis Ulbr.

Description: Robust annuals or perennial herbs or subshrubs, decumbent to erect, to 1 m tall. Leaves deltoid, biggest >10 cm in Ø, trilobed to the base, lobes pinnatifid to 2-pinnatifid. Flowers clustered at the end of branches, petals 10-30 mm, lilac with a white center. Calyx 7-11 mm.

Ecology: Coastal, Andean I: 500-1000 m. Lomas.

Distribution: S-Peru.

Arequipa: native. CAM, ISL.

Palaua moschata Cav.

Description: Small shrub up to 1 m tall, perennial, entire plant hirsute. Leaves ovate to deltoid, biggest >5 cm in Ø, in outlines trilobed, base cordate, apex obtuse, margins crenate, both sides tomentose, underside pale. Calyx tomentose, >10 mm long.

Ecology: Coastal: 0-1000 m. Lomas.

Distribution: Peru, N-Chile.

Arequipa: native. CAM, CAR.

Palaua sandemanii (Sandwith) Fryxell

Description: Suffrutescent herb, not branched, to 50 cm tall. Basal leaves in rosettes, slightly trilobate, petiole as long as the blade, Cauline leaves gray-green, 0.5 cm long, nearly sessile, trilobed, margins entire. Flowers pinkish- lilac, 15 mm in Ø, pedicel 2-3 cm long, thin.

Ecology: Coastal: 0-500 m. Lomas.

Distribution: S-Peru.

Arequipa: native. CAM, CAR.

Palaua tomentosa Hochr.

Description: Sprawling, decumbent herb, 50 cm tall. Leaf blades orbicular, rhombiform or deltoid, chartaceous, grayish-green. Flowers light pink. Calyx <10 mm long, calyx base laxly tomentose or hispid, shallowly winged, venation of calyx lobes inconspicuous or with one dark midvein.

Ecology: Coastal: 0-1000 m. Lomas, rocky slopes.

Distribution: S-Peru.

Arequipa: native. CAM, CAR, ISL.

Palaua trisepala Hochr.

Description: Perennial herb or subshrub, decumbent or erect, 1 m tall. Leaves ovate, both sides of blade tomentose, gray-green, base cordate, apex acute, margins crenate, 20 x 35 mm. Flowers 3 cm in Ø, solitary, on pedicels longer than petiolate leaf, bluish-lilac, calyx trilobed.

Ecology: Coastal: 0-1000 m. Lomas.

Distribution: S-Peru.

Arequipa: native. CAM, CAR, ISL.

Palaua velutina Ulbr. & A.W.Hill

Description: Herbs or perennial subshrubs, decumbent, rarely prostrate or erect, to 30 cm long. Leaves ovate- elliptic, lobate, both sides of blade tomentose, gray, base and apex obtuse, margins crenate, 2 x 3 cm. Flowers >3 cm in Ø, pink, calyx >10 mm.

Ecology: Coastal: 0-500 m. Lomas.

Distribution: S-Peru.

Arequipa: native. CAM, ISL.

Palaua weberbaueri Ulbr.

Description: Spreading annual herb to 30 cm tall. Leaves palmately trilobed to trisect, basal lobes forming a truncate base. Flowers pink to purple.

Ecology: Coastal, Andean I: 500-1000 m. Lomas.

Distribution: S-Peru.

Arequipa: native. CAM, CAR, ISL.

Sida jatrophoides L'Hér.

Description: Annual herbaceous or shrubby plant. From 15-150 cm high. Stem, petiole and pedicels densely pubescent. Leaves palmately lobed with 7-9 well-defined lobes, underside of blade purple, basal leaves different from the cauline, petioles 1.5x longer than leaves at maturity. Flowers pinkish-violet, petal edges with some yellowish shade. Mericarps 7-9.

Ecology: Coastal, Andean I-II: 0-2000 m. Lomas, rocky slopes.

Distribution: Galapagos, Peru.

Arequipa: native, expected. No voucher. Mentioned for the lomas of Peru by Dillon et al. (2011) and the lomas of Piura to Tacna by Lleellish et al. (2015).

Sida oligandra K.Schum.

Description: Annual herb, glandular-hirsute, 20-400 cm tall. Leaves deeply 3-5-lobed, some of them incised 4/5 of blade, lobes oblanceolate, margins crenate-serrate. Inflorescence axillary, 1-many-flowered. Flowers dark pink. Mericarps 5, appendage long.

Ecology: Coastal, Andean I-II: 0-3000 m. Disturbed areas, lomas, rocky slopes.

Distribution: Peru, Bolivia.

Arequipa: native. ARE, CAR, CON, ISL, UNI.

Sida rhombifolia L.

Description: Leaves arranged spirally, leaf blade rhombic to oblong-lanceolate or obovate, rarely linear-lanceolate, base broadly cuneate, margin dentate, apex obtuse to acute. Flowers solitary, axillary. Pedicel 1-2.5 cm, densely stellate tomentose. Corolla 1 cm in Ø, petals yellow, obovate, 8 mm, base attenuate, apex rounded.

Ecology: Coastal, Andean I-III, Amazonian: 0-4500 m. Disturbed areas, dry valleys, forests, grasslands, riversides, rocky slopes.

Distribution: Native to the subtropics and tropics of America, introduced worldwide.

Arequipa: native, doubtful. ARE, CAR. No voucher, mentioned for Arequipa by Brako & Zarucchi (1003), for RN Salinas y Aguada Blanca by Quipuscoa & Huamantupa (2010) and for Acarí by Montesinos & Modragón (2013).

Sida spinosa L.

Description: Erect undershrub, stellate pubescent. Leaves elliptic-obovate, 1-2 x 0.5-1 cm, stipule filiform, c. 5 mm long, petiole 5-10 mm long, often with 1-2 tubercles or small spines just below the base. Flowers usually solitary, sometimes paired or more, pedicel 0.5 cm, in fruit 1-2 cm long. Calyx 5 lobed, 4-6 mm long, lobes 2-3 x 1-2 mm, triangular, acute-acuminate. Corolla yellow, slightly longer than calyx. Fruit depressed globose, pubescent above.

Ecology: Coastal, Andean I-II, Amazonian: 0-2500 m. Disturbed areas, lomas, rocky slopes.

Distribution: Native to tropical America, introduced in Africa, Asia and Australia.

Arequipa: native. CAY. No voucher for Arequipa. The occurrence in Arequipa is likely, due to the distribution range of the species in S-Peru (Apurimac, Cuzco, Moquegua).

Sidastrum paniculatum (L.) Fryxell

Description: Plants 1-3 m. Stems roughly stellate-hairy, hairs somewhat ferruginous. Leaves slightly discolorous, 3- 13 × 1-7 cm (reduced distally), apex acute, petiolate proximal, distal subsessile. Inflorescence forming in age a many-flowered (>30), terminal panicle. Pedicels capillary, 1-3.5 cm, often subtended by 3 stipuliform bracts. Flowers reddish-purple, calyx 2-3 mm, stellate-hairy. Petals 3 mm, reflexed, glabrous, staminal column purple, subequal to corolla, hairy, style exserted, pallid. Schizocarps 4-5 mm in Ø, puberulent, mericarps 2.5 mm.

Ecology: Coastal, Andean I-II: 0-2000 m. Disturbed areas, dry valleys, forests, rocky slopes.

Distribution: Tropical and subtropical America.

Arequipa: native. CAR. No specimen, several convincing observations.

Tarasa capitata (Cav.) D.M.Bates

Description: Erect shrub, up to 2 m tall. Leaves 3-5-lobate, lobes entire, acute, folded like a fan. Inflorescence axillary, much longer than the leaves, many-flowered, often distinctly scorpioid. Flowers pink, petals 1-2 cm long. Mericarps 13, dehiscent, with 2 apical awns.

Ecology: Andean II-III: 1500-4000 m. Disturbed areas, rocky slopes.

Distribution: Peru, Bolivia, NW-Argentina.

Arequipa: native. ARE.

Note: Tarasa hornschuchiana (Walp.) Krapov. has broader sepals and the same distribution and ecology as T. capitata. T. hornschuchiana is here considered synonymous with T. capitata (S. Knapp, pers. comm.).

Tarasa congestiflora (I.M.Johnst.) Krapov.

Description: Herb, 25-30 cm tall. Leaves rhombic, entire, margins dentate in the upper half. Long and dark stalked stellate hairs on the calyx (as all Tarasa sp.). Flowers usually white.

Ecology: Andean II: 2000-3000 m. Disturbed areas.

Distribution: Peru, N-Chile.

Arequipa: native. ARE.

Tarasa nototrichoides (Hochr.) Krapov.

Description: Decumbent annual herb. Leaves deeply incised, <4 cm long, hirsute. Flowers white, <5 mm long, as long as the calyx. Peduncle fused to petiole.

Ecology: Andean III: 4000-4500 m. Highly manured areas in pasture.

Distribution: Peru and Bolivia.

Arequipa: native. ARE, CAY. Holotype of Tinopalca.

Tarasa operculata (Cav.) Krapov.

Description: Perennial herb or subshrub, entire plant canescent stellate. Leaves ovate to truncate, 3(-5)-lobed but not deeply incised, the middle lobe clearly longer than the lateral ones. Petals white with purple base, protruding at right angles to the style.

Ecology: Coastal, Andean I-III: 0-4000 m. Disturbed areas, dry valleys, lomas, rocky slopes.

Distribution: Peru, N-Chile.

Arequipa: native. ARE, CAS, CAY, ISL, UNI.

Tarasa tenella (Cav.) Krapov.

Description: Erect annual herb, up to 60 cm tall. Leaves deeply incised, hirsute, >5 cm long. Flowers white, <5 mm long, as long as the calyx. Peduncle free.

Ecology: Andean III: 3500-4000 m. Rocky slopes.

Distribution: Peru, Bolivia, N-Chile, NW-Argentina.

Arequipa: native. ARE.

Tarasa tenuis Krapov.

Description: Annual herb, decumbent to ascending, 20-30 cm tall. Leaves trilobed. Flowers white to pale blue or violet, >1cm in Ø, petals 7 mm, longer than the calyx.

Ecology: Andean II: 2000-3000 m.

Distribution: Peru, Bolivia.

Arequipa: native. ARE, CAS, CAY, CON.

Tarasa thyrsoidea Krapov.

Description: Perennial herb or subshrub. Leaves deltoid, slightly 5-lobed, margins crenate and undulate, petiole shorter than blade. Inflorescence a terminal thyrse, composed of terminal and axillary peduncles. Flowers lavender.

Ecology: Coastal: 0-500 m. Lomas, rocky slopes.

Distribution: Peru.

Arequipa: native. CAR.

Urocarpidium albiflorum Ulbr.

Description: Sparsely and minutely stellate pubescent but becoming glabrous, often somewhat branched, 10-50 cm tall, stipules membranous, persisting. Petioles 1-1.5 cm long. Leaves ovate-rhombic, obtuse or subtruncate at the broadly cuneate base, subacute, often indistinctly 3-5-lobed, 3.5 x 2 cm, irregularly serrate. Peduncles axillary, to 10 cm long, the white to pale pink, subsessile flowers congested toward the tips, petals glabrous, obovate, 4 mm long equaling the calyx. Mericarps reticulate with 6-7 mm long awn.

Ecology: Coastal, Andean I-II: 0-2500 m. Lomas, rocky slopes, shrublands.

Distribution: Peru, N-Chile.

Arequipa: native. ARE, CAS.

Note: Phylogenetic analyses show an unresolved position of Urocarpidium, monotypic genus, sister to Fuertesimalva and Palaua (Tate, 2011).

Waltheria indica L.

Description: Similar to W. ovata but leaves ovoid-oblong and less tomentose.

Ecology: Coastal: 0-500 m. Forests.

Distribution: Native to tropical America, now pantropical.

Arequipa: native, doubtful. Not specified by province. No voucher, mentioned for Arequipa by Brako & Zarucchi (1993).

Waltheria ovata Cav.

Description: Shrub 0.5-2 m tall, much-branched, younger stems covered with minute stellate hairs. Leaves alternate, simple, blade ovate or slightly 3-lobed, 1-6 cm long, both surfaces covered with grayish hairs, margins serrate. Flowers in axillary, ± sessile clusters, corolla yellow, 5 petals 6-7 mm long. Fruit a capsule, roundish, 2-3 mm long.

Ecology: Coastal, Andean I-II: 0-3000 m. Disturbed areas.

Distribution: Ecuador, Peru, N-Chile.

Arequipa: native. CAR, CAS, ISL, UNI.

Molluginaceae

Hypertelis cerviana (L.) Thulin

Description: Diffuse ephemeral or annual herb mostly less than 10 cm high. Leaves verticillate, linear, mostly <10 mm long and 1-5 mm wide. Perianth segments <2.5 mm long, obtuse, greenish with a membranous margin. Stamens 5. Capsule globose. Seeds numerous, angular, with a single low ridge on the keel.

Ecology: Coastal, Amazonian: 0-500 m. Disturbed areas.

Distribution: Native to Eurasia and Australia, introduced to America.

Arequipa: introduced, adventive. CAR. No specimen, a convincing observation from Caravelí in iNaturalist (2024).

Mollugo verticillata L.

Description: Annual, mat-forming, <50 cm Ø, glabrous. Stem at base often with pinkish swelling, forked unequally. Leaves in whorls of 3-8, 5-40 mm. Inflorescence axillary, 2-6 flowered, sepals 1.5-2.5 mm, oblong-elliptic, petals greenish-white. Fruit ovoid-ellipsoid.

Ecology: Coastal, Andean I: 0-1000 m. Beach, lomas, rocky slopes.

Distribution: Tropical America, introduced to southern North America.

Arequipa: native. ARE, CAM, CAR.

Montiaceae

Calandrinia acaulis Kunth

Description: Neatly tufted and rosette-forming, 2-5 cm tall, acaulescent. Leaves linear, often slightly spatulate, acute, 2-8 x 0.1-0.2 cm, often reddish at base. Flowers 2-3 cm Ø, white (later pinkish), 5 petals, stamens 5-10, pedicels 0.2-2 cm long.

Ecology: Andean II-III: 2500-4500 m. Rocky slopes.

Distribution: Mexico to N-Chile and NW-Argentina.

Arequipa: native. ARE, CAY, UNI.

Calandrinia alba (Ruiz & Pav.) DC.

Description: Glabrous, with slender, often elongated stems and remote, lanceolate-spatulate leaves to 3.5 x 0.8 cm, strongly narrowed to a petiole-like base 1 cm long, rounded to acute at apex. Flowers white. Pedicels in fruit 8 mm long, capsule exserted about 1 mm, seeds round, turgid, scarcely lustrous, distinctly punctulate, 1 mm wide.

Ecology: Coastal, Andean I-III: 0-4000 m. Disturbed areas, grasslands, rocky slopes (all herbarium specimens except 1 from the coast).

Distribution: Peru.

Arequipa: native. CAR, ISL.

Calandrinia carolinii Hershk. & D.I.Ford

Description: Like C. acaulis but with 12-20 stamens.

Ecology: Andean III: >4500 m. Rocky slopes.

Distribution: Peru, Boliva, N-Chile, NW-Argentina.

Arequipa: native. CAY.

Note: Personal observations show that plants with 10 and 12-20 stamens grow almost together in the same place. The species status of C. carolinii is questionable.

Calandrinia ciliata (Ruiz & Pav.) DC.

Description: Annual herb, glabrous. Stems 20-30 cm tall, ascending. Leaves 5-7 cm long, alternate, linear-lanceolate to slightly spatulate, entire, acute, lower leaves long attenuate, upper leaves shorter and less attenuate, membranous, shortly ciliate at margin. Inflorescence of solitary flowers, axillary along the stem, peduncles up to 3 cm long, sepals ca. 6 mm long, ovate, acute, shortly ciliate at margin, petals equal in length to sepals, purplish or white, stamens 5. Fruit an ovoid, 3-valved capsule, at maturity surpassing the calyx.

Ecology: Coastal, Andean I-III: 0-4500 m. Disturbed areas, grasslands, rocky slopes.

Distribution: North and South America, introduced to South Africa and Australia.

Arequipa: native. ARE, CON, ISL.

Cistanthe calycina (Phil.) Carolin ex Hershk.

Description: Succulent, prostrate herb, reddish or green. Leaves oblanceolate to spatulate, succulent. Inflorescence terminal, with conspicuous imbricate bracts. Flowers in clusters of <10 flowers, petals <2 mm, bright purple.

Ecology: Coastal, Andean I-II: 0-3000 m. Disturbed areas, lomas.

Distribution: S-Peru, N-Chile.

Arequipa: native. ARE, CAR.

Cistanthe celosioides (Phil.) Carolin ex Hershk.

Description: A much-branched, prostrate to ascending herb with dwarf-shrub growth. Leaves spatulate, succulent, grayish-green. Flowers in dense glomeruli with >30 flowers, involucral bracts persistent, petals ± 1 mm long, yellow to bright purple.

Ecology: Coastal, Andean I-II: 0-2000 (-3000?) m. Dry, arid areas, roadsides.

Distribution: S-Peru, N-Chile.

Arequipa: native. ARE, CAY, ISL.

Cistanthe lingulata (Ruiz & Pav.) Hershk.

Description: An herb ± 20 cm high, branching from the base, with glabrous, succulent and lingulate leaves 3-4 mm wide. Inflorescence on a 2-4 cm long peduncle, crowded. Flowers magenta, Ø ca. 10 mm with showy petals, stamens 10-15, pedicels ascending and 5-10

mm long, sepals 3-4 mm long, bracts 2 mm long. Fruit a capsule, finally about 5 mm long.

Ecology: Coastal: 0-1000 m. Disturbed areas, lomas.

Distribution: Peru.

Arequipa: native. CAR, ISL.

Note: Hershkovitz (1991) considers C. lingulata and C. weberbaueri to be intergrade. They may be local forms of the same species.

Cistanthe paniculata (DC.) Carolin ex Hershk.

Description: A stout, glabrous, glaucous perennial, sometimes over 40 cm high. Leaves obovate-oblong, acuminate, fleshy, the largest 10 x 6 cm. Flowers racemose, the pedicels 3-7 x longer than the conspicuous bracts, sepals suborbicular, 8-10 mm long, petals bright pink or yellow, up to 1 cm long. Capsule longer than the calyx, seeds black, opaque, densely hispidulous or glandular-echinate.

Ecology: Coastal, Andean I-II: 0-2500 m. Lomas, rocky slopes.

Distribution: Native to Peru, introduced in Hawaii, Sri Lanka and Cape Verde Isl.

Arequipa: native. ARE, CAM, CAR, CON, ISL.

Cistanthe weberbaueri (Diels) Carolin ex Hershk.

Description: Annual or biennial, much-branched at the base, the ascending or erect stems 5-30 cm high. Leaves spatulate, >10 cm wide. Inflorescence long pedunculate, leafless. Flowers purple, pedicels slender, 5-8 mm long, sepals 3 mm long. Capsule 4 mm long.

Ecology: Andean I-II: 1000-2000 m. Disturbed areas, rocky slopes.

Distribution: Peru.

Arequipa: native, endemic. ARE, CAR, ISL.

Note: See note on C. lingulata.

Montia fontana L.

Description: Tufted annual herb with ascending to erect stems to 20 cm long, sometimes rooting at the nodes. Leaves sessile and stem-clasping, linear to lanceolate, oblanceolate or spatulate, 1-2 cm long, 2-4 mm wide, acute, base dilated, membranous, shortly sheathing the stem. Pedicels 2-10 mm long. Sepals 1-1.5 mm long. Petals 1.5-2 mm long, white, unequal. Stamens 3, or rarely 5. Stigmas 3. Capsule globose to obovoid, 1-2 mm long, as long as sepals, seeds black.

Ecology: Andean II-III: 2000-4500 m. Bogs, grasslands, riversides, seasonally inundated areas.

Distribution: Cosmopolitan.

Arequipa: native. ARE.

Montiopsis cumingii (Hook. & Arn.) D.I.Ford

Description: Annual, prostrate herb 10-20 cm in Ø, with numerous basal branches. Basal leaves up to ca. 3 x 0.2 cm, linear-lanceolate, entire, cauline leaves smaller and spreading along branches, with scattered trichomes, glandular. Flowers with pedicels 2-5 mm long, petals 3-5, white or lilac, stamens 3-5. Fruit a capsule equal to the calyx, 10-15 seeds.

Ecology: Andean II-III: 2000-4500 m.

Distribution: Peru, Bolivia, Chile, Argentina.

Arequipa: native. ARE.

Moraceae

Ficus carica L.

Description: Shrubs, 3-10 m tall, much-branched. Bark grayish-brown. Leaves alternate, petiole strong, 2-5 cm, blade broadly ovate, usually with 3-5 ovate lobes, 10-20 × 10-20 cm, thickly papery, beneath densely gray pubescence, above scabrous. Fruits axillary, solitary, purplish-red to yellow when mature.

Ecology: Andean II: 1500-2000 m. Disturbed areas.

Distribution: Native to the Mediterranean, cultivated worldwide.

Arequipa: introduced, adventive. ARE, CAM, ISL.

Use: Cultivated for its edible fruits.

Myricaceae

Myrica pavonis C.DC.

Description: Dioecious shrub or tree, crown globose, up to 12 m tall, trunk grayish-brown, up to 80 cm in Ø, rough bark with transverse fissures and suberose protuberances. Leaves aromatic, linear-lanceolate to spatulate, margins entire or serrated, base attenuated, apex acute or obtuse. Flowers apetalous, male flowers gathered in catkins. Fruit a globose drupe covered with waxy and verrucous papillae.

Ecology: Coastal, Andean I-II: 0-3000 m. Lomas, riparian forests.

Distribution: Peru, N-Chile.

Arequipa: native. ARE, ISL.

Myrtaceae

Myrcianthes ferreyrae (McVaugh) McVaugh

Description: Shrub to small tree. Leaves opposite, elliptic, entire, apex acute to slightly acuminate, base acute, coriaceous, lateral veins pinnate perpendicularly to the main vein. Flowers white, axillary, in groups of 2-3.

Ecology: Coastal: 0-1000 m. Lomas.

Distribution: Peru.

Arequipa: native. CAR.

Note: It is likely that the specimens of Myrcianthes minimifolia (McVaugh) McVaugh from Caravelí, Ferreyra 13423 and 13481 (MICH), can be attributed to this species.

Nyctaginaceae

Allionia incarnata L.

Description: Perennial herb, sometimes annual, stem and leaves sparingly glandular puberulent to spreading viscid- villous. Stems often reddish, 0.1-1.5 m. Leaves progressively reduced distally. Flowers in axillary clusters, peduncle 3-25 mm, involucre 3-parted, ovoid when mature. Corolla deep pink to magenta, 5-15 mm. Fruit deeply convex.

Ecology: Coastal, Andean I-II: 0-2500 m.

Distribution: Native to the Americas.

Arequipa: native. ARE, CAM, CAR.

Boerhavia coccinea Mill.

Description: Perennial herb, often woody at base. Stems prostrate to decumbent, much-branched, 30-150 cm long, minutely pubescent. Leaves broadly lanceolate to ovate, glandular-hirsute to glabrate. Inflorescence axillary or terminal, forked unequally 3-6 times, open, branches divergent, terminating in compact subumbellate or capitate 5- flowered clusters. Flowers pink, rarely white or yellow, corolla bell-shaped, 1-3 mm in Ø. Fruit 3-5-ribbed with trichomes.

Ecology: Coastal, Andean I-II: 0-2000 m. Disturbed areas, lomas.

Distribution: Tropical and subtropical America, Africa and Australia.

Arequipa: native. CAR, UNI.

Bougainvillea spinosa (Cav.) Heimerl

Description: Thorny, much-branched shrub, 2-4 m tall. Leaves linear-spathulate, 10 x 1 mm, ± sessile. Bracts whitish-red, smaller than the flowers.

Ecology: Andean II: 1500-2500 m. Dry shrublands with cacti.

Distribution: S-Peru, Bolivia, NW-Argentina, Paraguay.

Arequipa: native, expected. No voucher, mentioned for Caylloma by Galán de Mera et al. (2009). Occurrence in southern Arequipa likely, due to the distribution range in S-Peru (Moquegua and Tacna).

Colignonia parviflora (Kunth) Choisy

Description: Shrub 1-2 m tall. Leaves deltoid-ovate, base obtuse-truncate, apex acute, margins entire, uppermost cauline leaves sometimes white. Inflorescence axillary, a many-flowered, dense umbel, not bracteate. Flowers white, 3 mm long, tube minute, so that the petals appear free.

Ecology: Andean I-III: 1000-4000 m. Humid places.

Distribution: Colombia, Ecuador, Peru.

Arequipa: native. ARE, CAY. No specimen but several convincing observations.

Commicarpus tuberosus (Lam.) Standl.

Description: Perennial herb, much-branched, suberect to scandent, the stems up to 2 m long. Stems and leaves glabrous, conspicuously cuspidate. Corolla lilac, pink, or purplish-red, funnel-shaped, puberulent at tip in bud, otherwise glabrous, stamens 3. Fruit terete, 10-ribbed with large viscid gland.

Ecology: Coastal, Andean I-II, Amazonian: 0-2500 m.

Distribution: Colombia, Ecuador, Peru, W-Bolivia.

Arequipa: native. CAY, UNI. No specimen but several convincing observations in iNaturalist (2024).

Mirabilis elegans (Choisy) Heimerl

Description: Annual herb, apparently erect, up to 1 m tall. Stem thick, ± hollow, densely branched from the base. Leaves on very stout and short petiole, the uppermost sessile. Blade deltoid-ovate, 4-9 cm long, glabrous, margins entire, truncate or rounded at base, rounded to acute at apex. Inflorescence large and open. Involucres narrow, 6.5- 10 mm long, densely villous. Corolla 5-merous, pinkish-red, 15 mm long, stamens usually 3, exserted. Fruit shorter than the involucre.

Ecology: Coastal, Andean I-II: 0-3000 m. Desert, rocky slopes.

Distribution: Peru, N- and C-Chile.

Arequipa: native. ARE, CAM, CAR.

Mirabilis expansa (Ruiz & Pav.) Standl.

Description: Ephemeral herb, suberect to subscandent. Stems up to 2.5 m long, not hollow, densely branched from the base. Leaves rather thick on stout and usually short petiole. Blade ovate-orbicular, 2.5-5 cm long, pilosulous to glabrate, margins entire, truncate to acute at base and often decurrent, obtuse at apex. Inflorescence rather small. Involucres in small, dense clusters, 4-5 mm long, villous. Corolla 5-merous, pinkish-red, 6 mm long, stamens usually 3, not exserted. Fruit shorter than the involucre.

Ecology: Andean II-III: 3000-4000 m. Rocky slopes.

Distribution: S-Peru, Bolivia, N-Chile.

Arequipa: native. ARE, CAR, ISL.

Mirabilis intercedens Heimerl

Description: Ascending herb, up to 1 m tall. Branches stout, puberulent below or glabrous. Leaves short petiolate, the upper ± sessile, rhombic-ovate to deltoid-ovate, obtuse at apex, obtuse or truncate and ± oblique at base, sometimes decurrent, rather fleshy, glabrous, 4-6 cm long. Inflorescence dichotomous, densely leafy. Involucres pedunculate, villous, 4-5 mm long, lobes equaling the tube. Corolla purplish-red, 12-16 mm long, stamens 5. Fruit subglobose, much exceeding the involucre.

Ecology: Andean I-II: 500-3000 m. Rocky slopes, shrublands.

Distribution: Peru.

Arequipa: native. ARE, CAR, ISL.

Mirabilis jalapa L.

Description: Long-lived perennial herb growing up to 2 m high, with tuberous root. Leaves ovate, oblong to triangular, to 9 cm long, acute, petiole 4 cm long. Flowers in groups of 3-7, sessile, opening in the afternoon, tubular, fragrant, 3-6 cm long, white, pink or red.

Ecology: Coastal, Andean I-II, Amazonian: 0-2500 m. Disturbed areas, gardens, backyards.

Distribution: Native to Central America, introduced all over the world.

Arequipa: introduced. ARE.

Use: Ornamental plant. An edible crimson dye is made from the flowers to color foods, cakes, and jellies.

Mirabilis ovata (Ruiz & Pav.) F.Meigen

Description: Coarse and stout herbs, erect, up to 1 m tall, branches densely viscid-villous. Leaves thick and fleshy, ovate or deltoid ovate, 2-3 cm long, rounded or subtruncate at base, obtuse to acute at apex. Inflorescence lax and open, involucres short-pedunculate, densely viscid-villous, in fruit 2 cm broad, lobes obtuse. corolla purplish-red, 1 cm long, stamens 3, short exserted. Fruit 5 mm, conspicuously 5-angled.

Ecology: Andean I-II: 500-2500 m. Lomas, disturbed areas.

Distribution: Peru, Chile, Argentina.

Arequipa: native. CAR.

Mirabilis prostrata (Ruiz & Pav.) Heimerl

Description: Much-branched herb, erect or procumbent, stems up to 2 m long. Leaves thin, ovate or ovate-deltoid, often elongate, 3-7 cm long, truncate or subcordate at base, acute to long acuminate at apex. Inflorescence open, often leavy-bracted, involucres 4-5 mm long, narrow, densely villous. Corolla purplish-red or pink, 7-10 mm long, stamens 3. Fruit 3 mm long, glabrous, nearly smooth.

Ecology: Coastal, Andean I-III: 0-4000 m.

Distribution: The Andes from Colombia to C-Chile.

Arequipa: native. ARE, CAR, CAY, ISL, UNI.

Onagraceae

Camissonia dentata (Cav.) Reiche

Description: Plants ascending, decumbent or prostrate. Leaves linear to broadly elliptic, 0.3-2.8 cm. long, 0.07-0.45 cm. wide, usually denticulate but more rarely subentire. Petals 4, <6 mm. long.

Ecology: Andean II-III: 2000-4000 m.

Distribution: Peru, Argentina and Chile.

Arequipa: native. ARE.

Epilobium denticulatum Ruiz & Pav.

Description: Ascending herb, 20-70 cm tall. Leaves mostly opposite, alternate above and in the inflorescence, margins weakly dentate. Flowers pink, nodding.

Ecology: Andean II-III: 2000-4500 m. Grasslands, along rivers.

Distribution: Central and South America.

Arequipa: native. ARE, CAY, UNI.

Epilobium fragile Sam.

Description: Low, densely cespitose, stems many, 10 cm tall. Leaves all opposite or alternate in the inflorescence, subentire with 0-3 obscure teeth on each side. Flowers erect at flowering.

Ecology: Andean II-III: 2500->4500 m.

Distribution: Peru, Bolivia, N-Chile.

Arequipa: native. ARE.

Epilobium pedicellare C.Presl

Description: Prostrate herb, to 1.2 m tall, with leafy runners. Leaves mostly alternate, opposite only near the base, margins distinctly dentate. Flowers erect, pink, inflorescence congested.

Ecology: Andean II: 2000-3500 m. Moist places.

Distribution: Peru, Bolivia.

Arequipa: native. ARE, CAR.

Ludwigia hexapetala (Hook. & Arn.) Zardini, H.Y.Gu & P.H.Raven

Description: Polymorphic branched perennial aquatic herb. Leaves alternate, lanceolate with distinct veins, 4-12 cm long. Flowers yellow, 5-6 petals, (12-)15-23 mm long, capsule hairy.

Ecology: Coastal, Andean I-II, Amazonian: 0-2000 m. Seasonally inundated areas.

Distribution: Native to Central and South America. Introduced worldwide.

Arequipa: native. CAM.

Note: Formerly L. grandiflora var. hexapetala. Separated from L. grandiflora by ploidy. 2n = 80.

Ludwigia octovalvis (Jacq.) P.H.Raven

Description: Upright woody herb or small shrub (1-4 m) with rounded stems and leaves, usually pubescent. Leaves quite variable, ovate to linear-lanceolate, margins entire. Flowers mostly 4-merous, petals yellow, often emarginate, sepals 4. Fruit 20-45 mm long, hairy and rounded in cross-section.

Ecology: Coastal, Andean I-II, Amazonian: 0-2500 m. Along streams, ponds, or lakes, disturbed sites.

Distribution: Cosmopolitan.

Arequipa: native. ARE, CAM, ISL.

Ludwigia peploides (Kunth) P.H.Raven

Description: A creeping or floating herb with rounded stems and leaves mostly glabrous. Leaves of the flowering stems obovate-oblong to elliptical-oblong, shiny, glossy, 3-6 cm long. Petals usually 5, yellow, 10-18 mm long. Fruit 10- 30 mm long, mostly hairless and rounded in cross-section.

Ecology: Coastal, Amazonian: 0-500 m. Water, swamps.

Distribution: Native to South America, introduced worldwide in subtropical and tropical climates.

Arequipa: native, doubtful. CAR. No specimen, a convincing observation from Lomas de Atiquipa in iNaturalist (2024).

Note: In Arequipa, L. peploides ssp. montevidensis.

Oenothera arequipensis Munz & I.M.Johnst.

Description: Prostrate ascending herb, not forming a rosette, usually branched near the base. Leaves lanceolate to oblanceolate, gradually narrowed to the short petiole. Flowers yellow at first, later orange-red, floral tube 1-3 cm long. Capsule cylindrical, only indistinctly fused with the subtending leaf if at all. Seeds 1-1.3 mm long, brown. Plant of the lomas (0-1000 m).

Ecology: Coastal: 0-1000 m. Lomas.

Distribution: Peru, N-Chile.

Arequipa: native. ARE, CAM, CAR, ISL.

Oenothera nana Griseb.

Description: Perennial from a thick, often branching rootstock, up to 7 cm tall in bloom. Leaves linear to lanceolate, 3-7 cm long, short-petiolate, entire or with wavy-toothed or crisped margins, finely bristly-hairy. Flowers sessile, 1 cm long, yellow aging orange or orange-brown.

Ecology: Andean II-III: 3000->4500 m. Disturbed areas, grasslands.

Distribution: Peru, Bolivia, N-Chile and Argentina.

Arequipa: native. ARE.

Oenothera peruana W.Dietr.

Description: Erect annual or biennial, forming a rosette, 0.5-1 m tall, unbranched or somewhat obliquely branched. Entire plant densely strigose. Rosette leaves very narrowly elliptic, acute, gradually narrowed to the petiole, 15-20 cm long, 2-5 cm wide. Inflorescence unbranched. Flowers bright yellow, petals 1.5-2.5 cm long, floral tube 1.2-2 cm long. Capsule 6-9 mm thick.

Ecology: Andean II: 2500-3500 m. Disturbed areas.

Distribution: Peru, N-Chile.

Arequipa: native. ARE, CAY.

Oenothera punae Kuntze

Description: Annual or perennial herb, forming a rosette, with a prostrate main stem and prostrate branches arising from the rosette, the branches 5-25 cm m tall. Rosette leaves linear to very narrowly elliptic, acute, narrowly cuneate at the base, 2-5 cm long, 1-3 cm wide. Inflorescence unbranched. Petals very broadly obovate, yellow to orange, often flushed with red, 0.4-1 cm long.

Ecology: Andean III: 3500-4500 m. Grasslands.

Distribution: Peru, Bolivia, NW-Argentina.

Arequipa: native. ARE, CAY.

Oenothera rosea L'Hér. ex Aiton

Description: The erect or ascending stem up to 65 cm tall. Leaves lanceolate, subentire, margins corrugate. Flowers solitary, generally pink, seldom dark yellow, petals 5-12 mm long.

Ecology: Andean II-III: 1500-4000 m. Disturbed areas.

Distribution: Native to South America, introduced elsewhere.

Arequipa: native. ARE, CAR, CAY, ISL, UNI.

Use: Used to heal lacerations (plaster of leaves mixed with salt).

Oenothera sandiana Hassk.

Description: Erect annual herb, not forming a rosette, much-branched, 50-100 cm tall. Leaves 5-10 x 1-3 cm. Petals very broadly obovate, yellow, sometimes reddish at the base and along the veins, 1.5-2.5 cm long, floral tube 2.5- 5.5 cm long. Capsule 4-7 mm thick.

Ecology: Andean II-III: 1500-4000 m. Rocky slopes.

Distribution: Ecuador, Peru, N-Chile and Bolivia.

Arequipa: native. ARE.

Oenothera verrucosa I.M.Johnst.

Description: Erect annual herb, not forming a rosette, simple or branched from the base upward, 15-40 cm tall. Leaves elliptic to lanceolate, acute, only the lowermost shortly petiolate, the remainder sessile, 4-8 x 0.5-1.5 cm, longer than the capsule the subtend. Inflorescence simple or branched. Petals very broadly obovate, 0.3-1 cm long, yellow. An important characteristic is the pedicellate appearance of its erect capsules, due to their abrupt narrowing towards the base.

Ecology: Andean II: 2000-3000 m. Rocky slopes.

Distribution: Peru.

Arequipa: native. ARE.

Orchidaceae

Aa weddelliana (Rchb.f.) Schltr.

Description: Small terrestrial orchid. Leaves present before flowering, narrowly oblong, 11 x 1.7 cm, acute to acuminate. Inflorescence a slender spike up to 50 cm long, enclosed by 10-13 translucent sheaths. Flowers white with pink-brown tones.

Ecology: Coastal, Andean I-III: 300-3800 m. Lomas, stony and rocky soils.

Distribution: Peru, Bolivia, NW-Argentina.

Arequipa: native. CAR.

Myrosmodes gymnandra (Rchb.f.) C.A.Vargas

Description: Plant small, 3.0-6.8 cm tall. Inflorescence 4.3-8.0 cm long (10.0-17.0 cm long during fruit maturation), scape covered by broad hyaline sheaths, obtuse, somewhat mucronate, spike conic to cylindrical-conic, 1.0-4.0 cm long, 14-27 flowers. Floral bracts suborbicular-ovate, obtuse to rounded at the apex, with entire margins, shorter than the flowers, 10.7-18.0 x 10-18 mm. Easily distinguished from all other species of the genus by the long column, the straight dorsal sepal (never revolute) and lip with a narrow opening.

Ecology: Andean III: 4000-4800 m. Wetland, bogs.

Distribution: Peru, Bolivia, NW-Argentina.

Arequipa: native. CAS.

Myrosmodes nervosa (Kraenzl.) Novoa, C.A.Vargas & Cisternas

Description: Like M. gymnandra but smaller, inflorescence only 3 cm long at flowering (M. gymnandra 4.3-8.0 cm long) and lip simple without a short mid lobe.

Ecology: Andean II-III: 3100-4900 m. Wetland.

Distribution: Peru, N-Chile.

Arequipa: native. ARE. Observation from prov. Arequipa, mentioned in Trujillo et al. (2016).

Myrosmodes nubigena Rchb.f.

Description: Plant small, 1.7–3.5 cm tall. Inflorescence 3.0–8.0 cm long, elongating to 13.0 cm in fruit. Ovary not rostrate at apex, lip with a cuneate base, perianth insertion on ovary not oblique. Floral bract with the upper margin crenulate, petals with the upper margin lacerate-fimbriate, lip shallowly 3-lobed in front (or with a conspicuous mid lobe), clinandrium margin dentate to lacerate.

Ecology: Andean III: 3700-4700 m. Grasslands and wetland.

Distribution: NW-Venezuela to W-South America

Arequipa: native, doubtful. ARE. No voucher, mentioned for RN Salinas y Aguada Blanca by Quipuscoa & Huamantupa (2010). The occurrence is likely, there are specimens in Cuzco near the border to Arequipa.

Orobanchaceae

Castilleja arvensis Schltdl. & Cham.

Description: Hemiparasitic herb, stems erect, densely hairy, usually unbranched. Leaves green, sessile, alternate, 3-veined, linear-elliptic, lanceolate, or oblanceolate. Inflorescence a spike, with red-tipped, petal-like bracts. Flowers green, corolla bilabiate, tubular, hirsute.

Ecology: Coastal, Andean I-II: 0-3500 m. Grasslands, lomas.

Distribution: Mexico to N-Argentina.

Arequipa: native. CAY, ISL.

Castilleja profunda T.I.Chuang & Heckard

Description: Annual plant, hirsute or hirsute-hispid, not glandular. Bracts margined with lavender. Corolla yellow with lavender helmet. Calyx subequal cleft to one-half of its length.

Ecology: Coastal, Andean I-II: 0-3500 m. Desert, lomas.

Distribution: Peru.

Arequipa: native. CON, ISL.

Castilleja pumila (Benth.) Wedd.

Description: Cespitose perennial herb, sometimes minute. Leaves sessile, 1-3.5 cm long, margins entire or lobed. Inflorescence a short spike or the flowers axillary and solitary on dwarfed plants. Bracts wider than the leaves, red tipped, flowers green.

Ecology: Andean II-III: 3000->4500 m. Grasslands, boggy spots.

Distribution: The Andes from Colombia to N-Argentina.

Arequipa: native. ARE, CAY, CON, UNI.

Neobartsia bartsioides (Hook.) Uribe-Convers & Tank

Description: Perennial herb, stems erect and branched. Leaves linear, pilose to hispid, margins serrate. Inflorescence dense, subspicate, upper bracts prominent, lanceolate to subulate, acute or acuminate, entire or laciniate-dentate, strigose with mostly eglandular hairs. Bracts purple, corolla red-purple, stamens yellow.

Ecology: Andean II-III: 2000-4500 m.

Distribution: Peru.

Arequipa: native. ARE.

Neobartsia camporum (Diels) Uribe-Convers & Tank

Description: Branched, erect subshrub. Leaves linear-oblong, margins serrate-crenate. Inflorescence loose, the bracts smaller than leaves, crenate, dentate or entire. Corolla with orange-red helmet and bright yellow lip, the helmet 1.4-1.7 × the length of the lip.

Ecology: Andean II-III: 2500-4500 m. Dry valleys.

Distribution: Peru.

Arequipa: native, doubtful. CAY. Specimen Aedo & Galán de Mera 11053 (MA) from Chivay. No other voucher and no mention. The presence in Arequipa is likely, due to the distribution range of the species in S-Peru (Apurimac, Ayacucho, Cuzco, Puno).

Neobartsia diffusa (Benth.) Uribe-Convers & Tank

Description: Small herb, branched from the base, up to 15 cm tall. Leaves linear-oblong, 4-7 mm long, serrate. Flowers deep yellow, helmet pubescent.

Ecology: Andean II-III: 3000->4500 m. Grasslands.

Distribution: Peru, Bolivia, N-Argentina.

Arequipa: native, doubtful. ARE. No voucher, mentioned for RN Salinas y Aguada Blanca (Quipuscoa & Huamantupa, 2010). The occurrence in Arequipa is likely, due to the distribution range of the species in S-Peru (Cuzco, Puno).

Neobartsia elongata (Wedd.) Uribe-Convers & Tank

Description: Perennial herb, densely branched from the base, up to 15-40 cm tall. Leaves oblong-ovate, serrate. Corolla 8 mm long, yellow, externally glabrous or sparsely hispid, helmet equal to or longer than the lower lip, helmet brown. Calyx 5-6 mm long.

Ecology: Andean II-III: 3000->4500 m. Rocky slopes.

Distribution: Peru, Bolivia, NW-Argentina.

Arequipa: native. ARE, CAY.

Neobartsia pedicularoides (Benth.) Uribe-Convers & Tank

Description: Erect herb up to 30 cm tall, much-branched. Stems violaceous, leaves green or violaceous, the upper ones or bracts and the corolla lightly violaceous.

Ecology: Andean II-III: 2500->4500 m. Bogs, grasslands.

Distribution: Andes from Venezuela to Bolivia.

Arequipa: native. CAY. No specimen, two observations from Caylloma in iNaturalist (2024), mentioned by Brako & Zarucchi (1993). The occurrence in the high Andean bofedales of Arequipa is likely.

Neobartsia peruviana (Walp.) Uribe-Convers & Tank

Description: Erect herb, up to 30 cm tall, branched, glandular-tomentose. Leaves linear, obtuse, margins crenate- dentate. Flowers axillary, subsessile, yellow, margins of helmet brownish. Calyx 18-20 mm long.

Ecology: Andean II-III: 3000->4500 m. Rocky slopes.

Distribution: Peru, Bolivia, N-Chile and Argentina.

Arequipa: native. ARE, CAY.

Neobartsia serrata (Molau) Uribe-Convers & Tank

Description: Suffrutescent. Leaves deeply serrate, apex acuminate. Corolla pale yellow, margins of helmet purplish- brown.

Ecology: Andean II-III: 2000-4000 m. Grasslands, shrublands.

Distribution: Peru, N-Chile.

Arequipa: native. ARE, CON.

Neobartsia weberbaueri (Diels) Uribe-Convers & Tank

Description: Vine, the stem twining among shrubs, up to 2 m long. Leaves lanceolate-oblong, 15-20 x 4-5 mm. Calyx and corolla dark purple, lip sometimes yellow-green, anthers protruding.

Ecology: Andean II-III: 2500-4500 m.

Distribution: Ecuador, Peru.

Arequipa: native. ARE.

Orobanche tacnaensis Mattf.

Description: Parasitic herb. Stems stout and sometimes branched from the base, ± squamate, to 25 cm tall, glabrous below, glandular-pilose above. Flowers 2 cm long, sessile, strongly constricted above stamen insertion.

Ecology: Andean II: 2500-3000 m. Shrublands.

Distribution: Peru.

Arequipa: native. ARE.

Orobanche weberbaueri Mattf.

Description: Parasitic herb. Stems stout and branched above or slenderly from the little incrassate base, sparsely squamate, to 15 cm tall, glabrate toward base, early densely ferruginous glandular pubescent toward apex, including the upper part of the corollas. Flowers 2.5 cm long, the upper subsessile, the lower on pedicels to 1 cm.

Ecology: Coastal: 0-500 m. Lomas.

Distribution: Peru.

Arequipa: native. CAR, ISL.

Oxalidaceae

Oxalis corniculata L.

Description: Stem usually prostrate, rooting at nodes. Leaves alternate, green or reddish. Stipules triangular. Pedicel folded back after flowering, fruits erect and densely hairy, 12-25 mm long. Flowers bright yellow with orange markings. Plant never with whitish runners.

Ecology: Andean I-II, Amazonian: 0-3500 m. Disturbed areas, rocky slopes.

Distribution: Origin probably Mexico to Peru, cosmopolitan in present time.

Arequipa: native. ARE, CAR, ISL.

Use: Used medicinally.

Oxalis latifolia Kunth

Description: Perennial herb with numerous ovoid bulbs. Leaflets up to 7.5 × 5 cm, triangular in outline, apex emarginate with a broad shallow sinus. Flowers in pseudo-umbels. Petals 10-16 mm, purple. Capsule broadly cylindrical, slightly longer than the sepals.

Ecology: Coastal, Andean I-III: 0-4000 m. Disturbed areas, lomas, rocky slopes.

Distribution: Native to Tropical America, a widespread almost cosmopolitan weed.

Arequipa: native. CAR.

Oxalis laxa Hook. & Arn.

Description: Small usually ± tufted annual ordinarily only a few cm tall, in varying degree setulose-villous and ± glandular, especially the nearly filiform divaricate pedicels. Petioles to 8 cm long. Sepals often rubescent below and marginally, nearly linear, acute, 3-4 mm long, equaled by the spatulate yellow petals.

Ecology: Coastal, Andean I-II: 500-3500 m. Disturbed areas, rocky slopes.

Distribution: Peru, Chile and Argentina.

Arequipa: native. ARE.

Note: Hassler (2024) synonymize O. micrantha Bertero ex Colla and O. micrantha Bojer ex Progel. POWO (2024) synonymizes O. micrantha Bertero ex Savi with O. laxa Hook. & Arn. and O. micrantha Bojer ex Progel with O. corniculata L. POWO's opinion is followed here.

Oxalis lomana Diels

Description: Perennial herb 5 cm and less, entire plant tomentose. Rhizome small, bulbiform. Flowers yellow, clearly smaller than the trifoliate leaves.

Ecology: Coastal: 0-1000 m. Disturbed areas, lomas, rocky slopes.

Distribution: Peru.

Arequipa: native. CAR, ISL.

Oxalis megalorrhiza Jacq.

Description: Perennial subshrub. Rhizome or caudex becoming stout, often producing a scarred and succulent stem. Leaves fleshy, petiole to 8 cm long, glabrous Leaflets 11-18 x 10-20 mm long, above glaucous green, beneath appearing crystalline. Inflorescence 2-5-flowered, peduncle to 10 cm long. Flowers yellow, 18 mm in Ø, nearly as large than the trifoliate leaves, sepals triangular, tip obtuse, not apparently sagittate, margins not undulate.

Ecology: Coastal, Andean I-II: 0-3000 m. Disturbed areas, dry valleys, elfin forests, grasslands, lomas, shrublands, rocky slopes.

Distribution: Peru.

Arequipa: native. ARE, CAM, CAR, CAS, CAY, ISL, UNI.

Note: O. megalorrhiza of the lomas seems to be polyphyletic (L. Gaspar, pers. comm.).

Oxalis mollendoensis J.M.H.Shaw

Description: Perennial herb with globose to napiform tuber. Stem short, to 5 cm, densely covered with imbricate bases of stipules and petioles. Leaves and petiole succulent, petiole up to 17 cm long, tomentose. Leaflets obovate- cuneate, 7-25 x 7-30 mm, underside pubescent to tomentose. Inflorescence longer than the leaves, umbelliform, 1- 5-flowered, peduncle to 22 cm long. Flowers 35-40 mm in Ø, petals yellow to orange, to 25 mm long,

Ecology: Coastal: 500-900 m.

Distribution: S-Peru.

Arequipa: native. CAM, ISL.

Oxalis nubigena Walp.

Description: Annual acaulescent herb forming small cushions. Leaves varying from light green to dark red. Flowers white.

Ecology: Andean II-III: 2500-4500 m. Disturbed areas, grasslands, rocky slopes.

Distribution: Peru.

Arequipa: native. CAY. No herbarium species. Observed in Pinchollo by the author (iNaturalist, 20024).

Oxalis pachyrrhiza Wedd.

Description: Perennial subshrub with slightly succulent stems, up to 10 cm tall, covered by sclerified, pubescent remains of stipules and leaf bases. Leaves apically crowded together, leaflets 5-12 mm long. Flowers smaller than leaves, petals yellow. Sepals triangular, 4-5 mm long, base hastate, apex obtuse, glabrous, margins undulate.

Ecology: Andean II-III: 2000-4000 m. Shrublands.

Distribution: Ecuador, Peru, Chile, Bolivia and NW-Argentina.

Arequipa: native. ARE, CAY, UNI.

Note: Shaw (2023) restricts O. pachyrrhiza to high altitudes.

Oxalis peruviana Norlind

Description: Procumbent or ascending shrub, very irregularly branched stems, 30-60 cm long. Branches densely leafy apically, often naked below. Leaflets 3-4 mm x 2-3 mm, glaucous. Inflorescence 1-3-flowered, peduncle up to 3.5 cm long. Flowers up to 7 mm long, petals deep yellow to orange.

Ecology: Coastal, Andean I: 500-1500 m. Lomas, rocky slopes.

Distribution: Peru.

Arequipa: native. CAR. Type locality in Ayacucho. Specimen Weigend & Förther 97/721 (F) from Caravelí is convincing but labeled as O. dombeyi. No other observation or mention.

Note: Often confused with O. peruviana Standley ex Kunth (= O. dombeyi).

Oxalis sepalosa Diels

Description: Small erect subshrubs to 40 cm. Stems thick, terete, 2 cm in Ø, basally swollen, arising from subterranean swollen tuber. Leaves crowded at the branch tips, numerous, petiole to 20 cm long, leaflets green, succulent, obcordate 30-35 x 35-50 mm, adpressed-sericeous beneath. Flowers 2-3 cm in Ø, petals narrowly ovate, yellow, to 20 cm long.

Ecology: Coastal: 0-1000 m. Rocks and sandy coastal lomas.

Distribution: S-Peru.

Arequipa: native. CAM, ISL. Type Weberbauer 1519 (B) from Arequipa.

Note: Formerly included in O. pachyrrhiza. Shaw (2023) resurrects the species and includes O. buchtienii as a synonym.

Papaveraceae

Argemone mexicana L.

Description: Annual herb, up to 1.5 m tall, with a slightly branched tap root. Stems branched and usually extremely prickly, exuding a yellow juice when cut. Leaves thistle-like, alternate, sessile, serrate, margins spiny, gray-white veined. Stem oblong in cross-section. Flowers terminal, solitary, bright yellow, 2.5-5 cm Ø. Fruit a prickly ovoid capsule, 2.5-5 x 1.5-3 cm. Spines of capsule 4-7 mm long, the sepal horns 5-10 mm long.

Ecology: Coastal, Andean I-II: 0-3000 m. Disturbed areas, lomas.

Distribution: Native to Mexico, naturalized worldwide.

Arequipa: introduced. CAR, ISL, UNI.

Argemone subfusiformis Ownbey

Description: Like A. mexicana but petals pale yellow and capsule elongate subfusiform, 3-5.5 x 1-1.5 cm. Spines of capsule 5-15 mm long, the sepal horns 10-14 mm long.

Ecology: Coastal, Andean I-III: 0-4000 m. Disturbed areas, lomas, rocky slopes, shrublands.

Distribution: South America.

Arequipa: native. ARE, CAR, ISL.

Eschscholzia californica Cham.

Description: Herbaceous perennial, caulescent, erect or spreading, 5-60 cm, glabrous, sometimes glaucous. Leaves basal and cauline, 1-4-ternate-dissected, lobes obtuse or acute. Inflorescences cymose or 1-flowered, buds erect. Flowers yellow to orange, usually with orange spot at base, 2-6 cm. Capsules oblong, dehiscent from base, 3- 9 cm long.

Ecology: Andean II: 3000-3500 m. Disturbed areas.

Distribution: Native to the SW-USA and N-Mexico, naturalized worldwide.

Arequipa: introduced. ARE, CAY.

Fumaria capreolata L.

Description: Annual herb, 10-80 cm tall. Leaves lobed, the lobes also lobed. Flowers 9-14 mm long, almost white with purple or black-red tips. Fruits smooth, pedicels curved.

Ecology: Coastal, Andean I-II: 0-3000 m. Disturbed areas, roadsides.

Distribution: Native to Europe, naturalized worldwide.

Arequipa: introduced. CAR.

Passifloraceae

Malesherbia angustisecta Harms

Description: Shrub. Leaves pinnatisect, grayish. Flowers white-yellowish, receptacle tubular, stamens inserted.

Ecology: Andean II: 1500-2500 m. Dry valleys, rocky slopes.

Distribution: S-Peru.

Arequipa: native. ARE, CAM.

Note: Malesherbia are placed in a separate family, Malesherbiaceae, by some authors.

Malesherbia arequipensis Ricardi

Description: Shrub. Leaves pinnatisect, grayish. Flowers white-yellowish, receptacle obconical, stamens exserted.

Ecology: Coastal, Andean I-II: 0-1500 (-3000) m. Dry valleys, lomas, rocky slopes.

Distribution: S-Peru.

Arequipa: native. ARE, CAR.

Note: Similar to M. angustisecta and probably often confused with this species. It is likely that M. arequipensis grows only in the lomas and M. angustisecta above the abiotic zone (>1500 m).

Malesherbia fatimae Weigend & H.Beltrán

Description: Erect shrub up to 1 m tall, sparsely branched, branches erect, densely tomentose. Leaves elliptical to obovate, 10-55 × 5-15 mm, apex rounded, base rounded to cuneate, margins irregularly crenate to lobulate, planar to slightly incurved, densely white and tomentose. Inflorescence a loose racemes 5-6 cm long. Flowers axillary, pedicels 3-4 mm long, densely pubescent. Tube cylindrical, 30-35 long, red, basally green. Petals 5, scarlet red. Stamens 5, anthers long exserted. Fruit a capsule, 25-30 mm long, glabrous.

Ecology: Andean I: 1500 m.

Distribution: S-Peru (Arequipa).

Arequipa: native. CON.

Malesherbia haemantha Harms

Description: Subshrub. Leaves pinnatisect, with thick, pinkish glandular hairs on the margins.

Ecology: Andean II: 2100-3500 m. Dry valleys, rocky slopes, shrublands.

Distribution: S-Peru.

Arequipa: native, doubtful. CAR. Known only from 2 localities in Caravelí, it has not been collected since 1957. Unresolved species, may be of hybrid origin.

Malesherbia tenuifolia D.Don

Description: Shrub with lax branching. Leaves pinnatisect, densely tomentose. Inflorescence lax, tube yellowish-red, petals yellow, stamens exserted.

Ecology: Andean II: 1500-2400 m.

Distribution: S-Peru, N-Chile.

Arequipa: native. CON.

Passiflora foetida L.

Description: Ill-scented, climbing herbaceous plant, stem 1.5-6 m long. Leaves entire or 3-lobed, not beyond half to mid-rib, margins ± entire. Flowers 2-5 cm in Ø. Sepals and petals white, filaments purplish, at least at base. Fruit <2 cm in Ø. Bracts linear-lanceolate, entire, not surrounding the fruit.

Ecology: Coastal, Andean I-II, Amazonian: 0-2500 m. Disturbed areas, crops.

Distribution: Native to South America, naturalized in the tropics.

Arequipa: native. CAR, ISL.

Use: Edible fruit when fully ripe, the unripe fruit contains cyanide and can be poisonous. Cooked young leaves are used as an ingredient in soups. Medicinal plant. Substances in the leaves are said to deter feeding insects.

Note: A very variable plant with 38 varieties.

Passiflora peduncularis Cav.

Description: Vine. Stipules semi-ovate, coarsely serrate. Leaves 3-lobed to below the middle, 5-10 x 6-13 cm, cordate, margins serrate. Sepals and petals pure white, corona with several series, filaments 1 cm long, peduncle 10-15 cm. Column with anthers and style symmetrical in the center of the flower. Fruit 3-4 cm in Ø.

Ecology: Andean I-II: 500-3500 m. Lomas, cloud forests.

Distribution: Peru.

Arequipa: native. ARE, CAR.

Passiflora pinnatistipula Cav.

Description: Vigorous vine. Stem with pinnate stipules. Leaves 3-lobed, up to 12 cm long. Flowers pink, spreading. Fruit yellow.

Ecology: Andean II-III: 2500-4000 m. Rocky slopes, shrublands.

Distribution: Native from Colombia to Peru.

Arequipa: introduced, adventive. CAY. Often cultivated and probably adventive.

Passiflora suberosa L.

Description: Perennial tendrilous vine, glabrous, stem purplish. Leaves variable, entire or 3-lobed not beyond half to mid-rib, 4-8 cm long, acute. Bracts linear-lanceolate, entire, not surrounding the fruit. Flowers solitary in axils, 18-25 mm in Ø, sepals yellow-green, petals lacking, filaments white. Pedicel 1.5-2.5 cm. Fruit subglobose, purple, Ø 1 cm.

Ecology: Coastal, Andean I, Amazonian: 0-1500 m. Disturbed areas, shrublands, rocky slopes, seasonally inundated areas.

Distribution: Tropical America.

Arequipa: native. CAR.

Note: For Arequipa P. suberosa ssp. litoralis.

Passiflora tripartita (Juss.) Poir.

Description: Vigorous vine. Climbing vine, stems terete, entire plant softly pubescent. Leaves 3-lobed, grayish or yellowish-tomentose beneath. Flowers pendulous, dark pink, calyx not inflated. Fruits oblong-ovoid, 6-7 cm long, yellowish.

Ecology: Andean II: 2500-3500 m. Forests, riversides, rocky slopes, shrublands.

Distribution: Native from Panama to C-Chile.

Arequipa: introduced, adventive. ARE, CAY, UNI. Often cultivated and probably adventive.

Note: In Arequipa, P. tripartita var. mollissima.

Passiflora trisecta Mast.

Description: Vine with ferruginous-villous indument. Stipules ovate-lanceolate, 1-2.5 cm long. Leaves 4-9 x 1.5-3 cm, trifoliate with petiolate leaflets, margins serrate. Bracts 2.5-3.5 cm long, free to base, deeply fimbriate-laciniate. Sepals and petals pure white, flowers 4-6 cm in Ø. Column with anthers and style asymmetrical, on one side of the flower. Fruit globose, 5 cm in Ø.

Ecology: Andean II-III: 2500->4500 m. Forests, grasslands, shrublands.

Distribution: Peru, Bolivia.

Arequipa: native. CAY. Specimen Raimondi 11479 (B) from Huasca is the only evidence.

Phrymaceae

Erythranthe glabrata (Kunth) G.L.Nesom

Description: Prostrate herb. Leaves deltoid to orbicular, dentate, palmately veined. Flowers yellow with orange blotches, calyx connate, conspicuously compressed at the sides.

Ecology: Andean I-III: 1000-4500 m. Riversides.

Distribution: Central and South America.

Arequipa: native. ARE, CAR, CAY, ISL, UNI.

Note: Barker et al. (2012) revived the binomial E. andicola for populations of Mimulus glabratus (= Erythranthe glabrata) from Colombia to southern South America. POWO (2024) considers E. andicola to be a synonym of E. glabrata, as we do here.

Phyllanthaceae

Andrachne microphylla (Lam.) Baill.

Description: Annual herbs up to 30 cm tall. Leaves orbicular, glandular, long petiolate. Flowers green. Fruits 3-lobed, tuberculate.

Ecology: Coastal, Andean I: 0-1500 m. Lomas, desert. Grows after heavy rains.

Distribution: Disjunct, Mexico and Peru.

Arequipa: native. CAR. No specimen, a convincing observation.

Phytolaccaceae

Phytolacca bogotensis Kunth

Description: Herb up to 2 m tall. Stems violet. Leaves broadly lanceolate to elliptic, 18 × 7 cm, base obtuse to attenuate, apex acute to acuminate. Inflorescence a raceme, dense, to 20 cm long. Peduncle to 5(-8) cm, pedicel 2- 4 mm. Sepals 5, white to red, oblong to ovate, subequal. Berries greenish-brown becoming dark brown, 5-8 mm Ø.

Ecology: Coastal, Andean I-III: 100-4000 m. Disturbed areas, forests, riversides, rocky slopes.

Distribution: Tropical America.

Arequipa: native. CAR. No specimen, mentioned for the Lomas de Atiquipa by Talavera et al. (2017).

Piperaceae

Peperomia galioides Kunth

Description: An erect to ascending, branched subshrub, moderately to densely puberulent. Leaves verticillate, 3-9 at a node, succulent, oblong to ovate. Inflorescence a terminal spike or in whorls of 3-6 spikes, 4-7 cm long. Fruits globose, 1 mm long.

Ecology: Andean I-III: 500-4000 m. Rocky slopes.

Distribution: Tropical America.

Arequipa: native. ARE, CAM, CAR, CAY.

Use: Used in folk medicine to treat many ailments (earaches, burns, hemorrhoids, hair loss).

Peperomia peruviana Dahlst.

Description: Small acaulescent herb with an underground tuber. Leaves ovate to orbicular, peltate. Inflorescence a linear spike.

Ecology: Andean II: 2500-3000 m. Rocky slopes.

Distribution: Peru.

Arequipa: native. ARE, CAR, CAY.

Note: P. scutellifolia is mentioned for Arequipa by Brako & Zarucchi (1993). Probably a confusion with P. peruviana.

Plantaginaceae

Bacopa monnieri (L.) Wettst.

Description: Stems creeping, rooting at nodes, succulent, glabrous. Leaves sessile, succulent, oblong-oblanceolate, 0.8-2 cm x 3-6 mm, margins entire or rarely dentate, apex rounded. Flowers axillary. Corolla pinkish-white with purple center, 8-10 mm, obscurely 2-lipped. Capsule narrowly ovoid, enveloped in persistent calyx, apex acute.

Ecology: Coastal: 0-1000 m. Lomas.

Distribution: Pantropical.

Arequipa: native. ARE, CAR, ISL.

Callitriche heteropoda Engelm. ex Hegelm.

Description: Submersed aquatic plant with some floating terminal leaves. Stem and leaf scales present. Leaf bases connate. Lingulate leaves 0.6 × 5.0-5.5 mm, expanded submersed or floating leaves broadly ovate, 2.3-4.9 × 1.3-3.4 mm, petiole 1.6-4.7, 3-veined. Fruit not strumose, subsessile, 1.1 mm in Ø, grayish-brown to maroon when mature, unwinged.

Ecology: Andean II-III: 3000-4500 m. Floating, submersed in ponds.

Distribution: Peru, Bolivia, NW-Argentina.

Arequipa: native. ARE.

Cymbalaria muralis G.Gaertn., B.Mey. & Scherb.

Description: Procumbent herb, glabrous, branches, petioles and pedicels often reddish. Leaf blade reniform to ovate, 5-30 x 5-35 mm, base cordate, margins with 5-11 teeth, upper surface dark green, lower surface gray-green, petiole to 5 cm long. Pedicels 1-6 cm long in flower, elongating and decurved in fruit. Sepals 2-2.5 mm long. Corolla 6-8.5 mm long, purple, glabrous outside, pubescent within tube on lower side, spur decurved, 2-3 mm long. Anthers white. Capsule oblong, glabrous.

Ecology: Andean II: 2200-2400 m. Wet gravel, walls, epiphytic on palm trees.

Distribution: Native to SC-Europe, introduced all over the world.

Arequipa: introduced. CAY.

Galvezia elisensii M.O.Dillon & Quip.

Description: Much-branched shrubs, branches erect to arching, 0.8–1.2 m long. Leaves opposite in whorls of 3, elliptic to elliptic-lanceolate, 15-25 x 3-5 mm, pedicles 8-12 mm long, stout, straight to slightly curved. Flowers red, corolla tubular, 8-9 mm long, deep red.

Ecology: Coastal, Andean I-II: 0-1500 m. Cloud forests, disturbed areas, lomas, rocky slopes.

Distribution: S-Peru.

Arequipa: native. CAR.

Nuttallanthus canadensis (L.) D.A.Sutton

Description: Annual or biennial herb, erect. Leaves alternate on flowering stems, opposite on the short non-flowering stems, blade linear, 2.5-4 x 0.3 cm. Inflorescence a raceme, terminal. Flowers with long spurs, blue to lilac with white throat.

Ecology: Coastal, Andean I-II: 0-3500 m. Grasslands, lomas.

Distribution: North and South America, introduced to Eurasia.

Arequipa: native. CAR, ISL.

Ourisia muscosa Benth.

Description: Mat-forming, prostrate perennial herbs 0.6-1.3 cm tall. Leaves orbicular, long petiolate. Floral bracts placed directly below each flower and surround it, thus covering the calyx. Corollas white, yellow inside the corolla tube, petals 5 mm long, regular with completely reflexed lobes.

Ecology: Andean III: 3800-5000 m. Bogs, moist meadows, edges of small lakes and rivulets.

Distribution: The high Andes of Colombia, Ecuador, Peru, Bolivia, and N-Chile.

Arequipa: native. ARE, CAY, CON.

Plantago australis Lam.

Description: Perennial herb, acaulescent, primary root usually disappearing ultimately. Leaves rosulate, narrowly elliptic, 8-30 x 3.5-6 cm, margins subentire to denticulate and ± ciliate, blades subglabrous to villose. Spikes 12-25 cm long, peduncle ± equal in length. Bracts narrowly ovate-lanceolate, 2.5 mm long, sparsely pubescent.

Ecology: Andean I-III: 500-4000 m. Disturbed areas, grasslands, shrublands.

Distribution: Tropical America, introduced in Australia, New Zealand and Hawaii.

Arequipa: native. ARE.

Use: Medicinal plant.

Note: In Arequipa, P. australis ssp. hirtella.

Plantago lanceolata L.

Description: Perennial herb, rosulate, tap rooted, hairs short. Leaves 10-30 x 1-3 cm, lanceolate to oblong, tapered to petiole, margins finely dentate. Inflorescence 20-80 cm long, spike 2-8 cm long, ovoid, peduncle grooved.

Ecology: Andean I-II: 500-4000 m. Disturbed areas, grasslands.

Distribution: Native to Eurasia, introduced worldwide.

Arequipa: introduced. ARE, CAR, CAY.

Use: Medicinal plant. From the leaves is obtained a good fiber, suitable for textiles.

Plantago limensis Pers.

Description: Acaulescent, annual herb with slender root. Leaves rosulate, linear-lanceolate or lanceolate, 3-11 x max. 0.5 cm, margins slightly undulate or remotely and obtusely denticulate, ashy, indument sericeous to villous- lanuginose, younger parts even floccose. Peduncles 4-19 cm long, spike cylindric, 0.5-6 cm long. Bracts two-thirds as long as the calyx. Fruit 2-seeded.

Ecology: Coastal, Andean I-II: 0-3500 m. Cloud forests, lomas, rocky slopes.

Distribution: Peru, N-Chile.

Arequipa: native. ARE, CAM, CAR, ISL.

Plantago linearis Kunth

Description: Stemless herb. Leaves narrow-linear, 10-20 x 0.2 cm, erect, glabrous or sparsely pilose. Inflorescence 0.8-1.2 cm long, pedicel longer than leaves. Corolla lobes shortly narrowed from a broad base, shortly acute ot apiculate-acute.

Ecology: Andean I-III: 1500-4000 m. Grasslands, rocky slopes.

Distribution: Peru, Ecuador, Colombia, Venezuela.

Arequipa: native, doubtful. ARE. No voucher. Mentioned for RN Salinas y Aguada Blanca.

Plantago major L.

Description: Perennial herb, acaulescent. Leaves rosulate, 5-18 cm long, widely elliptic to ± cordate, narrowed abruptly to petiole, entire or finely dentate. Inflorescence 3-7, 5-60 cm long. Spike 3-20 cm long, linear cylindric.

Ecology: Coastal, Andean I, Amazonian: 0-1500 m. Disturbed areas, lomas. Tolerant to trampling and soil compaction.

Distribution: Native to Europe and Asia, widely naturalized worldwide.

Arequipa: introduced. ARE, CAR, ISL.

Use: Used in folk medicine around the world. Scientific studies have shown that plantain extract has a wide range of biological effects.

Plantago myosuros Lam.

Description: Small annual herb with slender, elongated root. Leaves rosulate, numerous, lanceolate, 4-10 x 0.5-2 cm, ± sparsely villous, margins subentire or dentate. Inflorescence a dense spike, 4-15 cm long, peduncle equaling length of spike. Bracts narrowly lanceolate, pubescent.

Ecology: Coastal, Andean I-III: 0->4500 m. Disturbed areas, grasslands, rocky slopes, seasonally inundated areas, shrublands.

Distribution: South America, introduced to Madeira and Australia.

Arequipa: native. CAY, ISL.

Plantago nubicola (Decne.) Rahn

Description: Perennial(?) herb with a stout root, stems short, thick, densely yellowish-lanate above in the axils of persistent leaves. Leaves densely rosulate, linear, 1-6 cm long, the younger densely yellowish-silky-villous, becoming glabrescent and finally glabrate. Inflorescence globular. Peduncles 1-2.5 cm long, the spikes 0.5-1 cm long. Bracts 5-5.5 mm long, sepals 2 mm long. Corolla lobes obtuse. Fruit an indehiscent nutlet, 1-seeded.

Ecology: Andean II-III: 2000-4500 m. Grasslands.

Distribution: S-Peru, Bolivia, N-Argentina.

Arequipa: native, doubtful. ARE. Specimen Guenther & Buchtien 1734 (not seen) doubtful because of the elevation. Mentioned for Arequipa by Brako & Zarucchi (1993). Herbarium specimens from Puno and Moquegua, near the border of Arequipa.

Plantago sericea Ruiz & Pav.

Description: Shrubs, forming dense clumps, with a branching, decumbent to ascending stem. Leaves densely crowded at the branchlet tips, persisting, ± curving, narrowly linear, 2-5 x 0.1 cm, silky-pilose. Spikes dense, 1 cm long, peduncles 10-15 cm long. Fruit 2-seeded.

Ecology: Andean II-III: 1500-4500 m. Grasslands, rocky slopes, shrublands.

Distribution: The Andes of Venezuela to Bolivia.

Arequipa: native. ARE.

Note: In Arequipa P. sericea ssp. sericans, P. sericea ssp. sericea and P. sericea var. lanuginosa.

Plantago tubulosa Decne.

Description: A high mountain shrub, cespitose, stems subterranean, rosette appearing acaulescent, often forming dense mats, often found in bofedales. Leaves densely rosulate, rigid, lanceolate, 1-4 cm long, coarsely dentate or ± entire. Inflorescence 1-2 flowered, ± sessile. Flowers unisexual, calyx 7-10 mm long, corolla tube 10-12 mm long, anthers or stigmas long exserted. Capsule 1-3 cm long.

Ecology: Andean II-III: 3000->4500 m. Bogs, grasslands, rocky slopes.

Distribution: Ecuador, Peru, Bolivia, NW-Argentina, N-Chile.

Arequipa: native. ARE.

Veronica anagallis-aquatica L.

Description: Perennial herb, glabrous, stem generally decumbent and rooting at nodes, 10-60 cm tall. Leaves sessile, 2-8 cm, elliptic to ovate, margins entire to serrate. Axillary racemes with >30 flowers. Flowers lavender to blue, violet lined.

Ecology: Andean II-III: 2000-4000 m. Wet and disturbed areas as banks of streams and drainage channels.

Distribution: Native to Eurasia, naturalized worldwide.

Arequipa: introduced. ARE, CAY.

Veronica peregrina L.

Description: Annual herb with one or several erect stems to 20 cm high, pilose with glandular hairs. Leaves narrow- elliptic to spathulate, 0.3-1.1 cm long, 0.5-5 mm wide, subsessile, apex obtuse, base attenuate, margins entire or finely toothed. Flowers solitary in the axils of leaf-like bracts, pedicels 2-3 mm long in fruit. Calyx lobes 2.5-6.5 mm long in fruit, glabrous. Corolla 1.5 mm long, pale lavender or white. Capsule oblate, 2-3 x 2.5-3.5 mm, apex truncate, glabrous.

Ecology: Andean II: 2500-3500 m. Disturbed areas.

Distribution: Native to North America, introduced to South America and Europe.

Arequipa: introduced. Not specified by province. No specimen, several convincing observations of Arequipa and S-Peru.

Veronica persica Poir.

Description: Annual herb, roots fibrous. Stems hairy, often prostrate rooting at nodes. Lower leaves opposite, petiolate, coarsely dentate, broad-ovate, up to 2 cm long, upper leaves similar but alternate and sessile. Flowers solitary, calyx 4-lobed, green, 6-8 mm long, corolla blue, 8-12 mm. Fruit on an elongate pedicel.

Ecology: Coastal, Andean I-III: 0-4000 m. Disturbed areas, lomas.

Distribution: Native to Eurasia, introduced worldwide.

Arequipa: introduced. ARE, CAR, CAS, UNI.

Veronica serpyllifolia L.

Description: Herbaceous perennial to 30 cm high. Creeping, ascendent or with prostrate branches. Leaves opposite, 1-1.5 cm long, glabrous, shiny, margins entire to slightly dentate. Flowers on short pedicel, petals gray-blue to blue and irregular in shape, 0.6 cm wide.

Ecology: Andean II-III: 2500-4000 m. Shrublands.

Distribution: Native to Eurasia, introduced worldwide.

Arequipa: introduced. ARE. No specimen but a convincing observation from Uchumayo in iNaturalist (2024).

Plumbaginaceae

Limonium hyblaeum Brullo

Description: Perennial herb up to 25 cm tall with many small rosettes, often forming a hemispherical cushion or extensive mat. Leaves densely clustered in basal rosettes, broad-spatulate, tapered at the base and rounded or obtuse at the apex, glabrous, 2-3.5 x 0.7-1.6 cm broad. Flowering stems not winged. Calyx 4.5-5.5 mm long, pink, tube hairy at the base, lobes at least half as long as the tube, spreading to recurved in fruit. Corolla funnel-shaped with recurved lobes ± truncate, lavender.

Ecology: Coastal, Andean I-II: 0-2500 m. Saline and dry soils.

Distribution: Native to Sicily.

Arequipa: introduced, adventive. ARE. Restricted to Miraflores and Socabaya at 2300 m, introduced in the 1930s (Montesinos & Zegarra, 2019). Specimens Ferreyra 12040 and 11950 (MO) from the same area near Arequipa, labeled as L. bellidifolium, are likely L. hyblaeum.

Plumbago caerulea Kunth

Description: Herb. Cauline leaves elliptic-spatulate, base auriculate and perfoliate, apex acute, margins entire. Inflorescence terminal. Flowers blue, 5-merous. Calyx conspicuously glandular.

Ecology: Coastal, Andean I-II: 0-3500 m. Disturbed areas, forests, grasslands, lomas.

Distribution: The Andes from Venezuela to C-Chile and NW-Argentina.

Arequipa: native. CAM, CAR, ISL, UNI.

Poaceae

Aciachne pulvinata Benth.

Description: Perennial, cushion forming. Culms 1-3 cm long. Sheaths longer than adjacent culm internode. Ligule an eciliate membrane, 0.7 mm long. Blades aciculate, conduplicate, 0.4-1 cm long, 0.3-0.4 mm wide, stiff, venation with 3 vascular bundles, surface pubescent, hairy above, apex acute, pungent. Inflorescence with 1 fertile spikelets, shorter than basal leaves, with 1 fertile florets, lanceolate, subterete, 3 mm long.

Ecology: Andean III: 3500-4500 m. Bogs, grasslands, rocky slopes.

Distribution: The Andes from Venezuela to Bolivia.

Arequipa: native. ARE, UNI. No specimens. Mentioned for RN de Salinas y Aguada Blanca by Quipuscoa & Huamantupa (2010). Observation from La Unión in iNaturalist (2024).

Agrostis breviculmis Hitchc.

Description: Perennial, cespitose. Culms erect, 4-10 cm long. Culm-internodes distally glabrous. Ligule an eciliate membrane. Blades involute, 1-3 cm long, 1-1.3 mm wide. Blade surface glabrous. Panicle spiciform, linear, 1-3 cm long. Spikelets solitary. Fertile spikelets pedicellate.

Ecology: Andean II-III: 3000->4500 m. Disturbed areas, grasslands.

Distribution: The Andes, from Venezuela to Chile.

Arequipa: native, doubtful. ARE. No voucher, mentioned for RN Salinas y Aguada Blanca (Quipuscoa & Huamantupa, 2010). The occurrence is likely.

Agrostis stolonifera L.

Description: Sprawling stoloniferous, rooting at the nodes, mat or turf forming grass, culms 30-60 cm long. Can be submersed in clear fast-flowering streams. Blade 1-20 cm long and 1-8 mm wide with an acute tip. Ligule 2-6 mm long, membranous. Panicles usually upright or bending upwards in a many-flowered open or contracted panicle 3-28 × 0.5-2.5 cm.

Ecology: Coastal, Andean I-II: 0-2500 m. Disturbed areas, wet places, riversides.

Distribution: Native to Europe and N-Africa, now cosmopolitan.

Arequipa: introduced, doubtful. ARE. No voucher, mentioned for RN Salinas y Aguada Blanca (Quipuscoa & Huamantupa, 2010). The introduction to Arequipa is probable.

Agrostis tolucensis Kunth

Description: Perennial, cespitose, clumped densely. Culms erect, slender, 8-32 cm long, 0.5 mm Ø, 1-3-nodded. Leaves mostly basal. Leaf-sheaths smooth, glabrous on the surface. Ligule 2.8-4.5 mm long. Blades erect, filiform, 5-10 cm x 0.7-1 mm. Blade surface ribbed, scaberulous. Inflorescence a panicle. Peduncle 4-8 cm long.

Ecology: Andean II-III: 2500-4500 m. Disturbed areas, grasslands.

Distribution: Central and South America, from Mexico to NW Argentina.

Arequipa: native. ARE, CAY.

Aristida adscensionis L.

Description: Annual, or perennial. Culms erect, geniculately ascending, or decumbent, 10-100 cm long. Culm-internodes distally glabrous. Culm-nodes glabrous. Ligule a fringe of hairs. Blades flat, or conduplicate, 5-20 cm long, 1-3 mm wide. Recognized by its long, parallel-sided, flattened lemma often exserted from the glumes.

Ecology: Coastal, Andean I-III: 0-4000 m. Disturbed areas, lomas.

Distribution: Worldwide distribution, origin uncertain.

Arequipa: native. ARE, CAM, CAR, CAY, CON, ISL.

Use: Used as fodder.

Arundo donax L.

Description: Perennial herb, up to 10 m tall, with hollow stems 2-3 cm in Ø. Leaves alternate, 30-60 cm long and 2-6 cm wide with a tapered tip, gray-green, hairy at the base. All parts of this reed are distinctly bigger than Phragmites australis.

Ecology: Coastal, Andean I-II: 0-3000 m. Disturbed areas, riversides, wetlands.

Distribution: Native to the Mediterranean Basin and E-Asia. Introduced in subtropical and tropical regions worldwide.

Arequipa: introduced. ARE, CAM, CAR, ISL.

Use: Used for construction, tool making, and ornamental purposes such as flutes. Discussed as a potential biofuel and energy crop.

Avena sterilis L.

Description: Annual grass, up to 1.8 m tall. Leaf-blades to 45 cm long and 4-18 mm wide, with ligules up to 6 mm long. Panicle loose up to 30 cm long and 20 cm wide. Spikelets 35-50 mm long. Dorsal awns 5-8 cm long.

Ecology: Coastal, Andean I-II: 500-3500 m. Disturbed areas, lomas.

Distribution: Cosmopolitan.

Arequipa: introduced. ARE.

Bouteloua simplex Lag.

Description: Plants annual. Culms 3-35 cm, usually decumbent and unbranched. Ligules 0.1-0.2 mm of short hairs. Blades 2-8 cm long, 0.5-1.5 mm wide, flat to involute, above mostly glabrous, beneath often pilose. Panicles usually with only 1 branch terminating the culm, branches 1-2.5 cm, persistent, straight, arcuate, with 30-80 spikelets. Spikelets pectinate, with 1 bisexual floret and 1-2 rudimentary florets. Glumes glabrous. Lowest lemmas 2.5-3.5 mm, 3-awned, awns stout and flattened, lateral awns shorter than the central awns.

Ecology: Andean II-III: 1500-4000 m. Disturbed areas, grasslands, shrublands.

Distribution: The Americas.

Arequipa: native. ARE, CAS, CAY, CON.

Bromus berteroanus Colla

Description: Short lived annual, 30-60 cm tall, often tufted. Ligule 1-3 mm. Leaf sparsely to densely hairy, blade 2-9 mm wide. Inflorescence 8-30 cm, ± open, branches generally ascending. Spikelet 15-20 mm not strongly flattened. Glumes glabrous, lower 8-16 mm, 1(3)-veined, upper 10-18 mm, 3(5)-veined. Florets 3-9. Lemma 6-15 mm, back rounded, hairy, teeth 1-3 mm, awn-like to acuminate, awn (7)14-22 mm, bent, twisted below middle.

Ecology: Andean II-III: 2500-4000 m. Disturbed areas, forests, shrublands.

Distribution: North and South America.

Arequipa: native. ARE, CAY, CON, UNI.

Bromus catharticus Vahl

Description: Long-lived annuals, biennials, or perennials. Leaf glabrous or sparingly hairy, ligule 2-5 mm, blade 2-9 mm wide. Inflorescence 10-30 cm, upper branches erect to ascending, lower branches ascending to spreading. Spikelet 15-30 mm, glumes glabrous, scabrous, or occasionally hairy, lower 6-12 mm, upper 8-14 mm, 5-7(9)- veined. Lemma 10-17 mm, glabrous or slightly hairy, veins (9)11-13-veined, prominent for most of their length, awn 0-3.5 mm, anthers 0.5-1.3 mm.

Ecology: Coastal, Andean I-III: 500-4500 m. Disturbed areas, grasslands, lomas, rocky slopes.

Distribution: Native to South America, introduced elsewhere.

Arequipa: native. ARE, CAM, CAR, ISL, UNI.

Bromus lanatus Kunth

Description: Plants perennial, erect, loosely tufted, the culms glabrous or pubescent, 60 cm high or less. Leaf-blades flat, 2-12 mm wide, villous on the upper surface and sometimes also beneath. Panicles loose and open, nodding, rather few-flowered, the axis and the slender flexuous branches velvety-pubescent to hairy. Lemmas villous.

Ecology: Andean II-III: 3000-4500 m. Disturbed areas, grasslands, rocky slopes, shrublands.

Distribution: The Andes from Venezuela to N-Chile.

Arequipa: native. ARE, CAY.

Bromus striatus Hitchc.

Description: Annual. Culms erect, 50-80 cm long. Culm-internodes distally glabrous, or pubescent. Leaf-sheaths striately veined, pilose, with reflexed hairs. Blades 10-15 cm long, 4-8 mm wide, flat, surface pubescent, sparsely hairy. Panicle open, nodding, 10-15 cm long. Spikelets 2.5-3 cm long. Lemma awns 12-18(-25) cm long.

Ecology: Coastal: 0-500 m. Desert, disturbed areas, lomas.

Distribution: Peru (endemic to Arequipa).

Arequipa: native. CAM, ISL.

Bromus villosissimus Hitchc.

Description: Perennial, cespitose, densely clumped, dwarf grass. Culms decumbent, 5-10 cm long. Leaf-sheaths pubescent. Ligule an eciliate membrane. Leaf-blades involute, 20-50 x 1-2 mm, surface hirsute, hairy on both sides, apex obtuse. Inflorescence a simple panicle, condensed, 1-2 cm long. Fertile spikelets 3-8, woolly.

Ecology: Andean III: >4500 m. Grasslands, rocky slopes.

Distribution: Peru, Bolivia.

Arequipa: native. CAY.

Cenchrus brownii Roem. & Schult.

Description: Erect to straggling annual grass. Its stems initially horizontal, sometimes rooting at the nodes, and then growing upright to 25-110 cm high. Blades 4-13 mm wide. Inflorescence dense, green, 3-12 cm long, with many densely crowded burrs. Similar to C. echinatus, but the majority of the outer bristles equal or slightly exceeding the inner, flattened bristles.

Ecology: Coastal, Amazonian: 0-500 m. Disturbed areas.

Distribution: Native to tropical America, now pantropical.

Arequipa: introduced, doubtful. CAR, ISL. Species of the humid tropics. Specimens FLSP 2312 (HUSA) from Islay and Anonymous 2443 (MO) from Caravelí (both not seen). Mentioned for the lomas of Peru by Dillon et al. (2011) and for Arequipa by Galán de Mera et al. (2009). Occurrence unlikely, confusion with C. echinatus?

Cenchrus clandestinus (Hochst. ex Chiov.) Morrone

Description: Stoloniferous and rhizomatous perennial, 30-40 cm tall, stolons much-branched and adpressed to the ground, forming a turf. Vegetative shoots lower-growing than sterile shoots and with shorter leaves. Sheath pale yellowish-green, with dense hairs. Ligule a dense rim of hairs. Blade linear, 15-20 cm long and 2-5 mm wide, folded when young, flat when mature, glabrous, margins finely serrate, apex acute to attenuate. Inflorescence a terminal cluster of 2-4 ± sessile spikelets, almost enclosed in the uppermost leaf sheath. Spikelets linear-lanceolate, 10-20 mm long, 2-flowered, the lower sterile, the upper bisexual or functionally male or female. Stamens 3, exserted on filaments to 5 cm long, white feathery.

Ecology: Andean II-III: 2500-4000 m. Disturbed areas, riversides.

Distribution: Native to Eritrea and Abyssinia, cultivated and naturalized worldwide in the tropics and subtropics.

Arequipa: introduced. ARE, ISL.

Use: Planted for ornamental purposes and as pasture for cattle.

Note: Similar to Stenotaphrum secundatum (Walter) Kuntze, a valued and widely cultivated lawn grass. Introduced from Kenya to Cuzco Valley in 1928.

Cenchrus echinatus L.

Description: Plants annual. Culms 20-100 cm, ascending from a geniculate base. Sheaths shorter to equaling the internodes, compressed, densely pilose. Ligules 0.7-1.7 mm. Blades 4-20+ cm long, 2-10 mm wide, adaxial surfaces sparsely pilose. Panicles 2.5-12 cm, fascicles 5-10 x 3.5-6 mm, imbricate, outer bristles 10-

20, terete, usually half as long as the inner bristles, inner bristles in 1 whorl, flattened, not grooved, mostly erect, fused for at least 1/2 their length into a globose cupule, shortly pubescent, often purple at maturity. Spikelets 2-3 per fascicle, 4.8-7 mm. Anthers 0.8-2.4 mm.

Ecology: Coastal, Andean I-II: 0-3000 m. Disturbed areas, dry valleys.

Distribution: Native to North and South America, naturalized in the tropics and subtropics worldwide.

Arequipa: native. ARE, CAR, ISL.

Cenchrus spinifex Cav.

Description: Plants annual or short-lived perennial, tufted. Culms 30-100 cm, geniculate. Sheaths compressed, glabrous or sparsely pilose. Ligules 0.5-1.4 mm. Blades 3-28 cm long, 3-7.2 mm wide, glabrous or sparsely long- pilose above. Panicles 3-5 cm, fascicles 5.5-10.2 x 2.5-5 mm, imbricate, ovoid to globose, outer bristles, when present, mostly flattened, inner bristles in more than 1 whorl, 8-40, fused at least 1/2 their length, forming a distinct cupule, yellowish to mauve or purple, flattened. Spikelets 2-4 per fascicle, 3.5-5.9 mm, glabrous. Anthers 0.5-1.2 mm.

Ecology: Coastal: 0-1000 m. Lomas.

Distribution: Native to America, introduced to Europe, Africa and Asia.

Arequipa: native. CAM, CAR, ISL.

Chloris radiata (L.) Sw.

Description: Plants annual, with dense fibrous root growth, not stoloniferous. Culms 30-60 cm, erect or decumbent. Sheaths usually glabrous. Ligules membranous, shortly ciliate. Blades 10-30 cm long, to 10 mm wide, usually pilose, occasionally glabrous or scabrous. Panicles with 5-15, evidently distinct branches in 1-2 whorls. Branches 4.5-8 cm, ascending to spreading, open after flowering, green. Spikelets with 1 bisexual and 1 sterile floret. Lower glumes 0.7- 1.6 mm, lemma awned. Second florets 0.4-0.7 long, about 0.1 mm wide, borne on an equally long or longer rachilla segment, not or inconspicuously bilobed, awned, awns 3-5 mm.

Ecology: Coastal, Amazonian: 0-1000 m. Disturbed areas, forests on white sand, lomas, riversides.

Distribution: North and South America, introduced in E-Europe.

Arequipa: native. CAR.

Chloris virgata Sw.

Description: Plants annual, usually tufted, occasionally stoloniferous. Culms 10-100+ cm. Sheaths usually glabrous. Ligules to 4 mm, erose or ciliate. Blades to 30 cm long, to 15 mm wide, basal hairs to 4 mm, otherwise usually glabrous, occasionally pilose. Panicles digitate, with 4-20, silvery due to silky hairs on upper

lemma margins. Branches 5-10 cm, erect and not opening. Spikelets strongly imbricate, with 1 bisexual and 1-2 sterile floret(s). Lower glumes 1.5-2.5 mm, lemma awned. Second florets 1.4-2.9 mm long, 0.4-0.8 mm wide, awns 3-9.5 mm. Third florets greatly reduced, awnless.

Ecology: Coastal, Andean I-II: 0-2500 m. Lomas, rocky slopes.

Distribution: Worldwide in the tropics and subtropics.

Arequipa: native. CAR.

Cinnagrostis breviaristata (Wedd.) P.M.Peterson, Soreng, Romasch. & Barberá

Description: Perennial, cespitose. Culms erect, 6-25 cm long. Sheaths glabrous on surface. Ligule an eciliate membrane, 2-4.5 mm long, pubescent on abaxial surface. Blades conduplicate or involute, 2-6 cm x 2-3 mm, stiff, surface smooth or scaberulous, rough beneath, pubescent, hairy above. Blade margins ciliate. Blade apex obtuse. Panicle open, oblong, 2-4 cm long. Spikelets solitary, pedicellate, lanceolate, laterally compressed, 3.5-4.7 mm long. Lemma lanceolate, 3.5-4 mm long, membranous, without keel, apex dentate, 3-4-fid, 1-awned. Awn dorsal, arising 0.66 way up back of lemma, straight, 1-2 mm long. Rhachilla extension 1.5-2 mm long, pilose, with 2-2.5 mm long hairs.

Ecology: Andean III: 3500-4500 m. Grasslands, rocky slopes.

Distribution: Peru, Bolivia, N-Chile, NW-Argentina.

Arequipa: native. ARE.

Cinnagrostis brevifolia (J.Presl) P.M.Peterson, Soreng, Romasch. & Barberá

Description: Perennial, cushion forming. Culms 15-45 cm long. Leaf-sheaths glabrous, ligule an eciliate membrane, 1.5-2 mm long, truncate. Blades curved or flexuous, filiform, involute, 2-8 cm x 1 mm, surface scaberulous, rough on the upper side, margins ciliate, apex pungent. Panicle subtended by an inflated leaf-sheath, contracted, oblong, 3.5-7 cm x 1.2-1.8 cm. Spikelets with 1 fertile floret, pedicels glabrous.

Ecology: Andean III: 4000-4500 m. Disturbed areas, grasslands, rocky slopes.

Distribution: Peru, Bolivia, N-Chile, NW-Argentina.

Arequipa: native, doubtful. ARE. No voucher, mentioned for RN Salinas y Aguada Blanca (Quipuscoa & Huamantupa, 2010). The presence is probable due to the distribution of the species.

Cinnagrostis curvula (Wedd.) P.M.Peterson, Soreng, Romasch. & Barberá

Description: Perennial, cushion forming. Rhizomes short. Culms erect, 5-30 cm long. Sheaths glabrous on surface. Ligule an eciliate membrane, 1.2-2 mm long. Blades curved, involute, 1-4 cm x 1-2 mm, stiff, surface scaberulous, rough on both sides. Blade margins pubescent. Blade apex pungent. Panicle spiciform, linear, 1-3 x 0.7-1.6 cm. Spikelets solitary, pedicellate, lanceolate, laterally compressed, 4.6-5 mm long, breaking up at maturity. Lemma lanceolate, 3-3.7 mm long, membranous, without keel, apex dentate, 3-fid, 1-awned. Awn dorsal, arising 0.4-0.5 way up back of lemma, straight or flexuous, 1.5 mm long. Rhachilla extension 1.8-2.3 mm long, pilose.

Ecology: Andean III: 4000-4500 m. Grasslands.

Distribution: Peru, Bolivia, N-Chile, NW-Argentina.

Arequipa: native. ARE, CAY.

Cinnagrostis filifolia (Wedd.) P.M.Peterson, Soreng, Romasch. & Barberá

Description: Perennial, cespitose. Culms erect, 20-50 cm long, ligule membranous. Blades flexuous, filiform, involute, 6-15 cm x 1 mm, coriaceous, scabrous, apex acuminate. Panicle open, elliptic, 5-12 cm x 1-2.5 cm, branches spreading, branching divaricately. Spikelets 1-flowered, pedicels pubescent.

Ecology: Andean III: 3500-4000 m. Rocky slopes, grasslands.

Distribution: NW-Argentina, Bolivia, Peru.

Arequipa: native, doubtful. ARE. No voucher, mentioned for RN Salinas y Aguada Blanca (Quipuscoa & Huamantupa, 2010). The presence is probable due to the distribution of the species.

Cinnagrostis heterophylla (Wedd.) P.M.Peterson, Soreng, Romasch. & Barberá

Description: Perennial. Culms erect or geniculately ascending, 8-60 cm long. Sheaths glabrous or puberulous. Ligule an eciliate membrane, 1.5-4 mm long, truncate. Blades 3-15 cm x 2-5 mm, hairy on both sides. Panicle contracted or spiciform, linear, 2-20 x 0.6-2 cm, branches adpressed or ascending. Spikelets solitary, pedicellate, with 1 fertile floret, lanceolate, laterally compressed, 3.5-4.5 mm long. Glumes persistent, similar, exceeding apex of florets, firmer than fertile lemma. Lemma lanceolate, 3-3.7 mm long, membranous, without keel. Lemma scabrous, glabrous or puberulous, apex dentate, 3-4-fid, 5-awned. Principal awn dorsal, geniculate, 4.3-5.8 mm long, column twisted. Lateral awns apical, 0.5 mm long. Rhachilla extension 0.6-1 mm long, pubescent, with 0.5-1 mm long hairs.

Ecology: Andean II-III: 3000-4500 m. Grasslands, riversides.

Distribution: Venezuela, Colombia, Ecuador, Peru, Bolivia, NW-Argentina.

Arequipa: native. ARE, CAY, CON, UNI.

Cinnagrostis jamesonii (Steud.) P.M.Peterson, Soreng, Romasch. & Barberá

Description: Perennial. Rhizomes elongated, culms erect, 6-24 cm long. Culm-internodes purple. Ligule membranous, 1-2 mm long, lacerate. Blades curved, involute, 1-4.5 cm x 1-2 mm, surface pubescent, hairy above, margins pubescent, apex obtuse, or abruptly acute. Panicle subtended by inflated leaf-sheaths, spiciform, ovate, 1.8-4 cm x 1-1.5 cm. Spikelets 1-flowered.

Ecology: Andean III: 3500->4500 m. Rocky slopes.

Distribution: The Andes from Venezuela to Peru.

Arequipa: native, doubtful. ARE. No voucher, mentioned for RN Salinas y Aguada Blanca (Quipuscoa & Huamantupa, 2010). The presence is probable due to the distribution of the species.

Cinnagrostis lagurus (Wedd.) P.M.Peterson, Soreng, Romasch. & Barberá

Description: Perennial, cespitose. Culms 2.5-4 cm long. Sheaths glabrous on surface. Ligule an eciliate membrane, 0.5-1.2 mm long. Blades involute, 1-2 cm x 1 mm, surface pubescent, hairy on both sides. Panicle subtended by an inflated leaf-sheath, contracted, ovate, 0.7-1.7 x 0.5-0.7 cm. Spikelets solitary, pedicellate, lanceolate, laterally compressed, 4.7-5.2 mm long. Lemma lanceolate, 3.5-4 mm long, membranous, without keel, apex dentate, 3-4-fid, 1-awned. Awn dorsal, arising 0.33 way up back of lemma, geniculate, 4.5-5.5 mm long, column twisted. Rhachilla extension 2.3-2.8 mm long, pilose, with 2.5-3 mm long hairs.

Ecology: Andean III: >4500 m. Grasslands, rocky slopes.

Distribution: NW-Argentina, Bolivia, Peru.

Arequipa: native. ARE, CAY.

Cinnagrostis minima (Pilg.) P.M.Peterson, Soreng, Romasch. & Barberá

Description: Perennial, cushion forming. Culms erect, 2-5 cm long. Sheaths glabrous on surface. Ligule a ciliolate membrane, 0.3 mm long, truncate. Blades curved, conduplicate or involute, 0.5-1 cm x 1 mm. Blade margins ciliate. Blade apex obtuse. Panicle spiciform, linear, 1-1.5 x 0.4-0.6 cm, with few spikelets. Spikelets solitary, pedicellate, lanceolate, laterally compressed, 5-5.5 mm long. Lemma lanceolate, 3.7-4 mm long, membranous, without keel. Lemma surface smooth or scaberulous, apex erose, 1-awned. Awn dorsal, arising 0.1-0.2 way up back of lemma, geniculate, 4.5-4.8 mm long, column twisted. Rhachilla extension 0.6-0.8 mm long, pubescent, with 0.6-1 mm long hairs.

Ecology: Andean III: 4000->4500 m. Grasslands.

Distribution: Peru, Bolivia, NW-Argentina.

Arequipa: native. ARE, CAY, CON.

Cinnagrostis preslii (Kunth) P.M.Peterson, Soreng, Romasch. & Barberá

Description: Perennial, cespitose. Culms erect, 4-10 cm long. Sheaths glabrous on surface. Ligule an eciliate membrane, 0.4-0.8 mm long. Blades conduplicate, 1-3 cm 1 mm, herbaceous or coriaceous, surface glabrous. Panicle spiciform, linear, 1.5-2.5 x 0.6-0.8 cm. Spikelets solitary, pedicellate, lanceolate, laterally compressed, 4.7- 5.4 mm long. Lemma lanceolate, 3-4 mm long, membranous, without keel, apex dentate, 2-3-fid, 1-awned. Awn dorsal, arising 0.1 way up back of lemma, geniculate, 4-4.5 mm long, column twisted. Rhachilla extension 0.4-0.5 mm long, glabrous.

Ecology: Andean II-III: 2000-4500 m. Grasslands, rocky slopes.

Distribution: Peru.

Arequipa: native, doubtful. CAY. Specimen Aedo & Galán de Mera 11010 (MA) from Caylloma, no mention and no observation. Confusion with C. vicunarum?

Cinnagrostis rigescens (J.Presl) P.M.Peterson, Soreng, Romasch. & Barberá

Description: Perennial, cespitose. Rhizomes short. Culms 10-30 cm long. Sheaths glabrous on surface. Ligule an eciliate membrane, 0.5-1 mm long. Blades flat or involute, 4-10 cm 1-3 mm, surface smooth or scaberulous. Panicle spiciform, linear, 4-10 x 0.5-0.8 cm. Peduncle smooth or scaberulous above. Spikelets solitary, pedicellate, lanceolate, laterally compressed, 4-5 mm long. Lemma lanceolate, 3.5-4.3 mm long, membranous, without keel. Lemma surface smooth or scaberulous, rough above, apex entire or dentate, 2-fid, obtuse, 1-awned. Awn dorsal, arising 0.5 way up back of lemma, straight, 2-3.5 mm long. Rhachilla extension 0.5-0.8 mm long, pilose.

Ecology: Andean III: 3500->4500 m. Grasslands, riversides, swamps.

Distribution: The Andes from Colombia to N-Chile and NW-Argentina.

Arequipa: native. ARE, CAY.

Cinnagrostis rigida (Kunth) P.M.Peterson, Soreng, Romasch. & Barberá

Description: Perennial, cespitose. Rhizomes short. Culms erect, 40-70 cm long. Leaf-sheaths smooth, or scaberulous. Ligule an eciliate membrane, 4-7 mm long. Blades involute, 13-30 cm x 2-3 mm, coriaceous, stiff, surface scabrous. Inflorescence a panicle. Panicle contracted, linear, 12-20 cm x 1.5-2.3 cm, branches adpressed. Spikelets 1-flowered.

Ecology: Andean II-III: 2500->4500 m. Grasslands, rocky slopes.

Distribution: The Andes from Ecuador to Patagonia.

Arequipa: native, doubtful. ARE. No voucher, mentioned for RN Salinas y Aguada Blanca (Quipuscoa & Huamantupa, 2010). The presence is probable due to the distribution of the species.

Cinnagrostis setiflora (Wedd.) P.M.Peterson, Soreng, Romasch. & Barberá

Description: Perennial, cespitose. Culms 9-12 cm long. Ligule an eciliate membrane, 3-4.5 mm long. Blades conduplicate, or involute, 1-2 mm wide. Blade surface glabrous. Panicle spiciform, oblong, 3-4 cm x 0.8-1 cm.

Ecology: Andean III: 4000->4500 m. Grasslands.

Distribution: Ecuador, Peru, Bolivia, NW-Argentina.

Arequipa: native, doubtful. ARE. No voucher, mentioned for RN Salinas y Aguada Blanca (Quipuscoa & Huamantupa, 2010). The presence is probable due to the distribution of the species.

Cinnagrostis spicigera (J.Presl) P.M.Peterson, Soreng, Romasch. & Barberá

Description: Perennial, stoloniferous. Culms 12-50 cm long. Sheaths glabrous on surface. Ligule an eciliate membrane, 1.5-2 mm long. Blades curved or flexuous, involute, 3-9 cm 2-3 mm, surface puberulous, hairy above. Blade margins pubescent. Panicle subtended by an inflated leaf-sheath. Panicle spiciform, linear, 2-6 x 0.8-1.5 cm. Panicle branches pubescent. Spikelets solitary, pedicellate, lanceolate, laterally compressed, 6-6.6 mm long. Lemma lanceolate, 5-5.2 mm long, membranous, without keel, apex truncate, 1-awned. Awn dorsal, arising one third way up back of lemma, geniculate, 4-5 mm long, column twisted. Rhachilla extension 2-2.3 mm long, villous, with 3.2-4.3 mm long hairs.

Ecology: Andean III: 4000-4500 m. Grasslands.

Distribution: Andes from Colombia to NW-Argentina.

Arequipa: native. CAY, UNI.

Cinnagrostis vicunarum (Wedd.) P.M.Peterson, Soreng, Romasch. & Barberá

Description: Perennial, cespitose. Culms erect, 5-25 cm long. Culm-internodes smooth or scaberulous. Sheaths glabrous on surface. Ligule an eciliate membrane, 0.5-2 mm long. Blades curved or flexuous, filiform, involute, 2-4.5 cm 1 mm, surface scaberulous. Blade margins scabrous. Blade apex acute. Panicle spiciform, linear or oblong, 2-6 x 0.7-1.1 cm. Spikelets solitary, pedicellate, lanceolate, laterally compressed, 5.5-6.5 mm long. Lemma lanceolate, 3.5-4 mm long, membranous, without keel, apex dentate, 3-4-fid, 1-awned. Awn dorsal, arising 0.2-0.25 way up back of lemma, geniculate, 5.5-6 mm long, column twisted. Rhachilla extension 0.5 mm long, pubescent, with 0.8-1.3 mm long hairs.

Ecology: Andean II-III: 3000->4500 m. Grasslands, rocky slopes.

Distribution: Peru, Bolivia, N-Chile, NW-Argentina.

Arequipa: native. ARE, CAY, UNI.

Note: It is said to be one of the main sources of food for vicuñas.

Cinnagrostis violacea (Wedd.) P.M.Peterson, Soreng, Romasch. & Barberá

Description: Perennial, mat-forming. Culms erect, 6-25 cm long. Culm-internodes smooth. Leaf-sheaths smooth. Ligule membranous, ciliolate, 1.6-3.2 mm long, obtuse to acute. Blades conduplicate, 1.5-13 cm x 2-3 mm, venation with subepidermal sclerenchyma attached to veins below, surface scabrous, rough on both sides, glabrous or puberulous, margins scabrous, apex acute to acuminate. Panicle contracted, linear, 2-5.5 cm x 0.5-0.8 cm, branches 0.8-1 cm long. Spikelets 1-flowered.

Ecology: Andean II: 2000-2500 m. Rocky slopes.

Distribution: NW-Argentina, Bolivia, Peru.

Arequipa: native, doubtful. Not specified by province. No voucher, mentioned for Arequipa by Brako & Zarucchi (1993). The presence is probable due to the distribution of the species.

Cortaderia echinata H.P.Linder

Description: Cushion-forming grass, 20-30 cm tall. Leaves pungent (vegetable hedgehogs).

Ecology: Andean III: 3500-4500 m. Rock ledges.

Distribution: Peru.

Arequipa: native. ARE, CAY.

Note: Testoni & Linder (2017) separated the Peruvian clade from C. pungens as a separate species, C. echinate, which differs in its shattered leaves and longer spikelets.

Cortaderia jubata (Lemoine) Stapf

Description: Perennial bunchgrass, apomictic (Barker et al., 2003), or dioecious (GrassBase, 2002ff.). Leaves long, razor-edged, sheaths longer than adjacent culm internode, ligule a fringe of hairs. Inflorescence a purplish to white panicle, several meters tall.

Ecology: Andean II: 1500-3500 m. Disturbed areas, riversides, moist places.

Distribution: The Andes from Ecuador to Chile. Introduced and naturalized elsewhere.

Arequipa: native. ARE, CAY, UNI.

Use: A decoction is used in folk medicine to treat menstrual disorders. Pampas grass is often grown as an ornamental shrub. The plant is also used in construction and as cattle fodder.

Note: Barker et al. (2003) postulate that all South American species of Cortaderia sect. Cortaderia (pampas grass) belong to C. selloana (dioecious, eastern South America) and C. jubata (apomictic, western South America).

Cottea pappophoroides Kunth

Description: Perennial, cespitose. Culms erect or geniculately ascending, 30-70 cm long. Culm-internodes distally pilose with capitate hairs. Leaf-sheaths without keel, with capitate hairs, ligule a fringe of hairs, 0.5 mm long. Blades 8-20 cm x 3-6 m, surface pilose, hairy on both sides, apex attenuate. Panicle elliptic, 10-20 cm x 4-6 cm, branches pilose. Spikelets 5-10 mm (including the awns). Glumes 4-5 mm. Lemmas with both awns and awned teeth, not forming a pappus-like crown.

Ecology: Coastal, Andean I-II: 0-2500 m.

Distribution: Southern North America to central South America.

Arequipa: native. CAS.

Cynodon dactylon (L.) Pers.

Description: Stoloniferous, usually also rhizomatous herb. Culms 5-40(50) cm. Sheaths glabrous or with scattered hairs, collars usually with long hairs. Ligules 0.5 mm of hairs. Blades 1-6 cm long, 2-4 mm wide, flat at maturity. Panicles with 4-6 branches in a terminal whorl, branches 2-6 cm, axes triquetrous. Spikelets 2-3.2 mm. Lower glume 1.5-2 mm, upper glume 1.4-2.3 mm. Lemma 1.9-3.1 mm, awnless.

Ecology: Coastal: 0-1000 m. Open and disturbed areas, waste places.

Distribution: Native to Africa, introduced worldwide.

Arequipa: introduced. ARE, CAM, CAR, ISL.

Deschampsia chrysantha (J.Presl) Saarela

Description: Perennial herb up to 50 cm tall, growing in bogs, grasslands and rocky places at high altitude. Ligule an eciliate membrane, 7–18 mm long. Blades stiff, 5-15 cm long, 2-3 mm wide, surface scabrous, with a pungent apex. Panicle oblong, spiciform, 2.5-6 x 1.3-2 cm. Lemma 1-awned, awn 2-3 mm long.

Ecology: Andean III: 4000->4500 m. Bogs, grasslands, rocky slopes.

Distribution: Peru, Bolivia, N- and C-Chile, NW-Argentina.

Arequipa: native. ARE, CAY.

Deschampsia eminens (J.Presl) Saarela

Description: Perennial, cespitose, clumped densely. Rhizomes short. Culms erect or geniculately ascending, 35-100 cm long. Leaf-sheaths scaberulous. Ligule membranous, 10-18 mm long, acuminate. Blades involute, 10-30 cm x 2- 3 mm, stiff. Inflorescence an open panicle, ovate, 10-20 cm long, with spikelets clustered towards branch tips. Primary panicle branches drooping. Spikelets solitary.

Ecology: Andean III: 3500-4500 m. Bogs, riversides.

Distribution: The Andes from Colombia to C-Chile.

Arequipa: native, doubtful. ARE. No voucher, mentioned for RN Salinas y Aguada Blanca (Quipuscoa & Huamantupa, 2010). The occurrence in Arequipa is likely.

Deschampsia ovata (J.Presl) Saarela

Description: Perennial, cespitose. Culms erect, 12-40 cm long, internodes distally glabrous. Sheaths glabrous on surface, ligule membranous, 6-15 mm long, acuminate. Blades flat or involute, 3-10 cm x 1.3-4.5 m, surface glabrous. Inflorescence a panicle, subtended by an inflated leaf-sheath. Panicle capitate, ovate, 0.2-0.6 cm x 0.2-0.3 cm. Spikelets solitary. Pedicels pubescent.

Ecology: Andean III: 4000->4500 m. Disturbed areas, grasslands, rocky slopes.

Distribution: Ecuador, Peru, Bolivia.

Arequipa: native. ARE, CAY, CON, UNI.

Digitaria californica (Benth.) Henrard

Description: Perennial, cespitose. Culms erect, 50-100 cm long. Sheaths densely pilose. Ligule an eciliate membrane. Blade 2-12 cm long, 2-5 mm wide, surface glabrous or puberulent. Inflorescence with 5-8 unilateral racemes, 3-5 cm long. Spikelets lanceolate, dorsally compressed, 3-4 mm long, with hairs extending 2-4 mm beyond apex.

Ecology: Coastal, Andean I: 500-1000 m.

Distribution: USA to subtropical America.

Arequipa: native. CAR.

Diplachne fusca (L.) P.Beauv. ex Roem. & Schult.

Description: Ascending to erect grass, 30-150 cm tall, forms large tussocks. Stems smooth, hollow. Leaf-sheaths 10-25 cm long. Ligules 2-8 mm, membranous, attenuate, becoming lacerate at maturity. Blades 3-50 cm x 2-7 mm, inrolled along the edges and wiry. Panicles 10-105 cm long, with 3-35 racemose branches, bases of the panicles sometimes remaining enclosed in the upper leaf sheaths at maturity, branches 1.5-20 cm, ascending to reflexed. Spikelets 5-12 mm, with 6-20 florets, with at least 2 fertile florets. Lemmas 2-6 mm, sometimes with a dark spot near the base, apices acute to truncate, usually awnless, mucronate, sometimes 1-awned.

Ecology: Coastal, Amazonian: 0-1000 m. Disturbed areas, riversides, saline habitats.

Distribution: Native to the Americas, introduced worldwide.

Arequipa: native. ARE, CAM, CAR, ISL.

Note: In Arequipa, D. fusca ssp. uninervia.

Distichlis humilis Phil.

Description: Perennial, cespitose grass, forming dense stands. Rhizomes elongated, culms 2-5 cm long. Leaves distichous, sheaths longer than adjacent culm internode. Ligule membranous. Blades conduplicate, 7-15 cm x 1 mm wide, coriaceous, stiff, surface ribbed, apex pungent. Inflorescence dioecious, a raceme or only 1-3 fertile spikelets.

Ecology: Andean III: 3500-4000 m.

Distribution: Peru, Bolivia, N-Chile, NW-Argentina.

Arequipa: native. ARE, CAR.

Distichlis spicata (L.) Greene

Description: Perennial, rhizomatous and sometimes stoloniferous grass forming dense monotypic stands. Culms 10- 60 cm tall. Leaves distinctly distichous, with salt crystals, upper leaf sheaths strongly overlapping, upper blades stiff, glabrous, ascending to horizontally spreading, equaling or exceeding the panicles. Panicles dioecious, 1-7 cm long, with 2-20 spikelets. Spikelets 5-20 mm long, with 5-20 florets.

Ecology: Coastal: 0-500 m. Seashores, lomas, tidal marshes. Tolerant of salty soils and trampling.

Distribution: Native from Canada to Patagonia. Naturalized elsewhere.

Arequipa: native. ARE, CAM, CAR, ISL.

Use: Used as forage. Studies have shown that saltgrass is a good choice for salt-water irrigated pastures. Total dry matter yield was 9 t/ha with high protein content.

Echinochloa colonum (L.) Link

Description: Plants annual, erect or decumbent, cespitose or spreading, rooting from the lower cauline nodes. Culms 10-70 cm, lower nodes glabrous or hispid, hairs adpressed, upper nodes glabrous. Sheaths glabrous. Ligules absent. Blades 8-22 cm long, 3-6 mm wide, mostly glabrous, sometimes hispid. Panicles 2-12 cm, erect, primary branches 5-10, 0.7-2(4) cm, erect to ascending, spikelike, somewhat distant, without secondary branches, axes glabrous or sparsely hispid. Spikelets 2-3 mm, pubescent to hispid, tips acute to cuspidate, florets awnless.

Ecology: Coastal, Amazonian: 0-500 m.

Distribution: Native to Africa and S-Asia, introduced worldwide.

Arequipa: introduced. CAR.

Echinochloa crus-pavonis (Kunth) Schult.

Description: Plants annual or short-lived perennials. Culms 30-150 cm, nodes glabrous. Sheaths glabrous, often purplish. Ligules absent. Blades 12-60 cm long, 10-25 mm wide, glabrous. Panicles 10-30 cm, erect or drooping, nodes sparsely hispid, internodes glabrous, primary branches to 14 cm, nodes sometimes sparsely hispid, internodes usually glabrous, secondary branches to 3 cm. Spikelets 2.5-3.4 x 1.2-1.4 mm. Lower lemmas awnless or awned, awns 3-10(15) mm, curved.

Ecology: Coastal, Andean I, Amazonian: 0-1000 m. Rice paddies, inundated areas.

Distribution: Native to Eurasia and Africa, introduced worldwide.

Arequipa: introduced. CAR.

Eleusine indica (L.) Gaertn.

Description: Plants annual. Culms 30-90 cm, erect or ascending, somewhat compressed. Sheaths conspicuously keeled, margins often with long, papillose-based hairs, particularly near the throat. Ligules 0.2-1 mm, truncate, erose. Blades 15-40 cm long, 3-7 mm wide, with prominent, white midveins, margins and/or above often with hairs. Panicles with 4-10 branches, often with 1 branch attached as much as 3 cm below the terminal whorl. Spikelets 4-7 x 2-3 mm, with 5-7 florets, florets awnless.

Ecology: Coastal, Andean I-II, Amazonian: 0-2500 m. Disturbed areas.

Distribution: Native to tropical and subtropical Africa and Asia, introduced worldwide.

Arequipa: introduced. CAM.

Enneapogon desvauxii P.Beauv.

Description: Plants perennial. Culms 20-45 cm, 1 mm thick, ascending to erect from a hard knotty base, often branching, nodes pubescent. Sheaths usually shorter than the internodes, more or less pubescent. Ligules 0.5 mm. Blades 2-12 cm long, 1-2 mm wide, ± hairy, soon involute. Panicles 2-10 cm, spikelike, grayish-green or lead-colored. Spikelets mostly 5-7 mm, 3-6-flowered, usually only the lowest floret bisexual. Glumes 3-5 mm, subequal, thin, puberulent. Lemmas awned but without awned teeth, the awns forming a pappuslike crown.

Ecology: Andean II: 2000-2500 m.

Distribution: Subtropical America, Africa and Asia.

Arequipa: native. ARE, CAS, UNI.

Eragrostis attenuata Hitchc.

Description: Perennial, cespitose. Culms geniculately ascending, or decumbent, 20-80 cm long. Sheaths longer than adjacent culm internode, glabrous on surface, or pilose, oral hairs ciliate. Ligule a fringe of hairs, 1 mm long. Blades 2-4(-7) cm long, 1(-3) mm wide, surface pilose, hairy on both sides, with tubercle-based hairs. Panicle rachis glabrous. Panicles 10-25 x 0.2-0.5 cm, closed, densely flowered, branches adpressed, flowered to the base. Spikelets 3-4 flowered, dark reddish-purple.

Ecology: Coastal: 0-1000 m.

Distribution: S-Peru, N-Chile.

Arequipa: native. ARE, CAM, CAR, ISL.

Eragrostis cilianensis (All.) Vignolo ex Janch.

Description: Annual, cespitose. Warty glands on the keels of lemma and usually on the keels of glumes and blade-margins. Culms erect, or geniculately ascending, 10-90 cm long. Ligule a fringe of hairs. Leaf-blades 5-15 cm long, 2-8 mm wide. Inflorescence a panicle, ovate, 4-30 cm long. Panicle branches stiff, glandular. Spikelets solitary, oblong, or ovate, laterally compressed, 3-20 x 2-4 mm, pedicellate with 5-60 fertile florets. Glumes deciduous, similar, shorter than spikelet. Fertile lemma ovate, or orbicular, (1.3-)2-2.8 mm long, chartaceous, yellow to gray.

Ecology: Coastal, Andean I: 0-1500 m. Disturbed areas, cultivated fields.

Distribution: Native to Europe, Asia and Africa, introduced worldwide.

Arequipa: introduced. CAS.

Eragrostis ciliaris (L.) R.Br.

Description: Annual, cespitose. Culms erect, 5-60 cm long. Ligule a fringe of hairs. Leaf-blades 2-12 cm long, 1-5 mm wide. Inflorescence a spiciform panicle, linear or oblong, 1-20 cm long. Spikelets solitary, pedicellate, with 6-12 fertile florets, ovate, laterally compressed, 1.8-3.2 mm long. Glumes deciduous, similar, shorter than spikelet. Fertile lemma elliptic to oblong, 0.9-1.5 mm long, membranous, keeled, 3-veined, midvein ciliolate (at least a few hairs on upper lemmas). Palea keels tuberculate, ciliate. Stamens 2.

Ecology: Coastal, Andean I, Amazonian: 0-1500 m.

Distribution: Pantropical but not in Australia.

Arequipa: native. ISL.

Eragrostis lurida J.Presl

Description: Perennial. Culms geniculately ascending or decumbent, 15-60 cm long. Ligule a fringe of hairs. Leaf- blades 2-3 mm wide. Panicle open, lanceolate, dense, 5-20 x 0.5-5 cm, with spikelets clumped along branches, branches stiff, ascending and spreading to divaricate, 1-5 cm long. Spikelets solitary, pedicellate, with 6-10 fertile florets, oblong, laterally compressed, 4-5 mm long. Glumes deciduous, similar, shorter than spikelet. Lemma ovate, 1.8-2 mm long, membranous, keeled, 3 -veined, scabrous, rough above.

Ecology: Andean I-II: (1000-)2000-3800 m. Dry rocky hillsides, sandy roadsides, rocky alluvial fans.

Distribution: Colombia, Ecuador, Peru, Bolivia.

Arequipa: native. CAR, CAY.

Eragrostis mexicana (Hornem.) Link

Description: Annual, cespitose. Culms geniculately ascending, 15-120 cm long, with ring of glandular depressions below nodes. Sheath glabrous or papillate-soft-hairy on upper margins, often with glandular depressions on veins. Ligule a fringe of hairs. Blade 5-25 x 0.4-0.7 cm, flat, occasionally hairy below, midvein rarely glandular. Inflorescence (5)10-35 cm, open, axis below nodes, branches, spikelet stalks generally sparsely glandular. Spikelet ovate to oblong in outline, 1.5-2.4 mm wide, gray-green to red.

Ecology: Coastal, Andean I-II: 0-2500 m. Roadsides, cultivated fields, disturbed areas, lomas.

Distribution: Native to the Americas, introduced to Europe and Australia.

Arequipa: native. ARE, CON, ISL, UNI.

Eragrostis nigricans (Kunth) Steud.

Description: Cespitose annual. Culms decumbent, 15-40 cm long. Ligule a fringe of hairs. Leaf-blades 5-13 cm long, 2-3 mm wide, surface glabrous. Panicle open, lanceolate, 5-16 cm long, primary branches ascending. Spikelets arranged in glomeruli, pedicellate, with 2-4(-6) fertile florets, oblong, laterally compressed, 3-3.8 mm long. Glumes deciduous, similar, shorter than spikelet. Lemma ovate, 1.9-2 mm long, membranous, purple or black, keeled, 3-veined. Seeds with a shallow or deep ventral groove.

Ecology: Andean I-II: 1000-3800 m. Rocky slopes, near cultivated fields, roadsides.

Distribution: Colombia, Ecuador, Peru, Bolivia, NW-Argentina, N-Chile.

Arequipa: native. ARE, CAM, CAR, CAS, CAY, CON, ISL.

Eragrostis pectinacea (Michx.) Nees

Description: Cespitose annuals. Culms 10-80 cm, erect to geniculate or decumbent below, glabrous. Leaf-sheaths overlapping below, hirsute at the apex. Ligule a fringe of hairs, 0.5-1 mm long. Leaf-blades 5-18 cm long, 1-3 mm wide, surface scaberulous, rough above. Panicle open, ovate, 7-20 cm long, primary branches simple, branches bearded in axils. Spikelets adpressed, solitary, pedicellate, 5-18 flowered, linear, laterally compressed, 5-8 x 1-1.5 mm. Glumes deciduous, similar, shorter than spikelet. Fertile lemma ovate, 1.5-2 mm long, membranous, keeled, 3 - veined. Caryopses without a ventral groove.

Ecology: Coastal, Andean I-II: 0-2600 m. Disturbed areas, roadsides, sidewalks, cultivated fields.

Distribution: Native to the Americas, introduced to Europe.

Arequipa: native. CAR.

Eragrostis peruviana (Jacq.) Trin.

Description: Annual. Culms decumbent, 10-35 cm long. Ligule a fringe of hairs. Leaf-blades 3-6 cm long, 2-3 mm wide, surface pubescent, hairy on both sides. Panicle spiciform, oblong, or ovate, continuous, 2-5 cm long. Spikelets solitary, pedicellate, 5-10 flowered, oblong, laterally compressed, 2.8-3 mm long. Glumes deciduous, similar, shorter than spikelet. Fertile lemma ovate, 1.8-2 mm long, membranous, keeled, 3 -veined. Palea keels soft and silky, not tuberculate at base. Stamens 3.

Ecology: Coastal, Andean I-II: 0-3000 m.

Distribution: Peru, N-Chile.

Arequipa: native. ARE, CAM, CAR, CAY, ISL.

Eragrostis virescens J.Presl

Description: A cespitose annual. Culms ascending to erect, 50-70 cm long, slender. Leaf-blades 3-6 mm wide. Panicle open, reaching 1/3 of the culm length.

Culms 20-70 cm long. Spikelets linear to linear-lanceolate, 0.7-1.4 mm wide. Seed without shallow depression on the ventral side.

Ecology: Coastal, Andean I-II: 0-2500 m.

Distribution: Native to the Americas, Introduced to Europe and Africa.

Arequipa: native. ARE, ISL.

Note: Sometimes treated as a subspecies of E. mexicana.

Eragrostis weberbaueri Pilg.

Description: Perennial, cespitose. Culms erect, slender, 20-30 cm long, nodes silky pilose. Ligule a fringe of hairs. Leaf-blades flat or involute, 3-8 cm long, 2-3 mm wide, apex attenuate, silky pilose. Panicle spiciform, linear to oblong, interrupted, 3-6 cm long, rachis densely pilose. Spikelets solitary, pedicellate, 4-6 flowered, lanceolate, laterally compressed, 3-4 mm long. Glumes deciduous, similar, shorter than spikelet. L lemma ovate, 1.8-2 mm long, membranous, keeled, 3 -veined. Lemma surface scaberulous, apex acute. Apical sterile florets resembling fertile though underdeveloped.

Ecology: Andean I-II: 1000-3500 m.

Distribution: Peru, N-Chile.

Arequipa: native. ARE, CAM, CAR, CAY.

Eriochloa procera (Retz.) C.E.Hubb.

Description: Cespitose perennial to 0.8 m high. Leaves with sheath smooth, glabrous. Ligule 0.75 mm long. Blade 1- 5 mm wide. Inflorescence 6-15 cm long, with 5-25 racemes 1.5-4 cm long. Spikelets neatly and closely arranged, solitary or paired, the longer one up to 2 mm long, silky.

Ecology: Coastal: 0-500 m. Wetlands, roadsides, paddy fields.

Distribution: Native to tropical Asia and Australia, introduced to Africa and South America.

Arequipa: introduced. Not specified by province. Specimen Anderson 826 (MO) from 1948, mentioned by Brako & Zarucchi (1993), no other vouchers.

Eriochloa punctata (L.) Ham.

Description: Plants perennial, rhizomatous, often flowering the first year and resembling an annual. Culms 30-150 cm, erect or decumbent, not rooting at the lower nodes, internodes glabrous, nodes 3-10, glabrate. Sheaths occasionally inflated, glabrous, often purplish at maturity. Ligules 0.4-1 mm. Blades 10-50 cm long, 4-10 mm wide, linear, flat, straight, spreading, glabrous on both surfaces. Panicles 9-22 x 1-10 cm, contracted, rarely open, rachises scabrous to densely pubescent, branches 8-20, 1-6 cm long, adpressed or divergent, glabrous, not winged, with 28-60 spikelets. Spikelets mostly in unequally pedicellate pairs, 4.5-5.7

x 0.9-1.4 mm. Lower lemmas 4.3-5.5 mm long, lanceolate, setose, 5-7-veined, acuminate, awnless or mucronate.

Ecology: Coastal, Andean I, Amazonian: 0-1500 m.

Distribution: Native to the Americas, introduced in Java.

Arequipa: native. CAM, CAR.

Festuca chrysophylla Phil.

Description: Perennial, cespitose, tussok rhomboidal, often forming circular colonies with bare soil in the center. Culms erect, 30-65 cm long, 1-noded. Sheaths glabrous on surface. Ligule a ciliolate membrane, 0.3-0.5 mm long, bilobed. Blades filiform, conduplicate, 7-40 cm x 0.7-1 mm. Panicle spiciform, linear, 6-10 cm long. Spikelets with 3-5 fertile florets, with diminished florets at the apex, oblong, laterally compressed, 7-10 mm long. Rhachilla internodes 0.8-1 mm long, scaberulous. Lemma elliptic, 5.3-6.5 mm long, without keel, 5-veined, surface glabrous or puberulous, hairy below, apex acuminate. Awn 0-0.5 mm.

Ecology: Andean III: 3500->4500 m. Grasslands. Characteristic grass of the dry Puna.

Distribution: Peru, Chile, Bolivia, Argentina.

Arequipa: native. ARE, CAY.

Note: F. orthophylla represents more than 90% of the biomass in many dry altiplano ecosystems (Fernandez Monteiro, 2012).

Festuca dolichophylla J.Presl

Description: Perennial, cespitose. Culms erect, 65-110 cm long, 2-4 mm Ø. Culm-nodes constricted, brown, glabrous. Leaves mostly basal. Sheaths glabrous on surface. Ligule an eciliate membrane, 1-2.5 mm long. Blades erect, filiform, conduplicate, 25-50 cm long, 2-5 mm wide. Panicle open, elliptic, nodding, 10-25 cm long, 3-4 cm wide, branches ascending, 2-nate, 6-12 cm long. Spikelets with 5-7 fertile florets, with a barren rhachilla extension or with diminished florets at the apex, oblong, laterally compressed, 10-17 mm long, disarticulating below each fertile floret. Lemma lanceolate, 6.5-8 mm long, chartaceous, purple, without keel, 5-veined. Lemma lateral veins distinct, surface scaberulous, apex acuminate, muticous or awned. Awn 0-1 mm long. Rhachilla extension 3-4 mm long. Apical sterile florets rudimentary.

Ecology: Andean III: 3500-4500 m.

Distribution: Costa Rica to Peru.

Arequipa: native. ARE, CON.

Festuca floribunda (Pilg.) P.M.Peterson, Soreng & Romasch.

Description: Perennial, cespitose. Basal innovations intravaginal. Culms 6-30 cm long. Ligule an eciliate membrane. Blades filiform, convolute, 3-8 cm long, 0.5-1.5 mm wide, surface pubescent, hairy above. Panicle

shorter than basal leaves, embraced at base by subtending leaf, contracted, oblong, 3-5 cm long. Spikelets with 2-3 fertile florets, with diminished florets at the apex, oblong, laterally compressed, 10-20 mm long, disarticulating above glumes but not between florets. Rhachilla internodes zig-zag, 1.3-1.8 mm long. Glumes persistent, similar, shorter than spikelet. Lemma lanceolate, 8-11 mm long, membranous, keeled, 5-veined, surface scaberulous, apex lobed, 2-fid, with linear lobes, incised 0.2 of lemma length, 1-awned. Awn dorsal, arising 0.66 way up back of lemma, straight, 6-10 mm long overall.

Ecology: Andean III: 4000->4500 m. Grasslands.

Distribution: Peru, Bolivia, NW-Argentina, N-Chile.

Arequipa: native. ARE, CAY, UNI.

Festuca myuros L.

Description: Annual, culms solitary or cespitose, erect or decumbent, 10-70 cm long, 2-3-noded. Leaf-sheaths without keel, smooth, glabrous on surface. Ligule an eciliate membrane, 0.3-1 mm long. Blades, 2-15 cm long, 0.5-3 mm wide, surface pubescent, hairy above, margins scaberulous, apex attenuate. Panicle contracted, linear, nodding, 5-30 cm long, branches adpressed. Spikelet-pedicels oblong, 1-3 mm long. Spikelets with 3-10 fertile florets, with diminished florets at the apex, oblong or cuneate, laterally compressed, 7-10 mm long, disarticulating below each fertile floret. Rhachilla internodes eventually visible between lemmas, scaberulous. Glumes persistent, similar, shorter than spikelet, thinner than fertile lemma. Lemma lanceolate, 5-7 mm long, chartaceous, without keel, 5- veined, surface scaberulous, apex acuminate, 1-awned. Awn 5-15 mm long.

Ecology: Andean II-III: 2000->4500 m. Disturbed areas, grasslands, lomas.

Distribution: Native to N-Africa, Europe and SW-Asia, introduced worldwide.

Arequipa: introduced. ARE, CAS, CAY, ISL.

Festuca peruviana Infantes

Description: Perennial, cespitose, clumped densely. Culms erect, 5-12 cm long, 2-noded. Sheaths smooth, glabrous on surface. Ligule a ciliolate membrane, 0.5 mm long. Blades curved, conduplicate, 5 cm x 1-2 mm, surface scaberulous, rough above, margins scaberulous, apex obtuse or abruptly acute. Panicle contracted, linear, 2-4 cm long, branches adpressed. Spikelets with 3-4 fertile florets, with diminished florets at the apex, oblong, laterally compressed, 6.5-8 mm long. Glumes persistent, similar, shorter than spikelet. Lemma oblong, 5.5-6 mm long, chartaceous, without keel, 5-veined, surface scaberulous, rough in lines, glabrous or puberulous, apex acuminate. Awn 1 mm long.

Ecology: Andean III: 4000-4500 m.

Distribution: Bolivia, Peru.

Arequipa: native. CAY.

Festuca procera Kunth

Description: Perennial, cespitose, clumped densely. Culms geniculately ascending, straight, 80-120 cm long, 2-3- nodded. Sheaths tight, not thickened at base, without keel, striately veined, smooth or scaberulous, glabrous on surface. Ligule membranous, 1 mm long, white, entire, truncate. Blades straight, conduplicate, linear in section or angular in section, 20-45 cm x 2-3 mm, coriaceous, stiff or firm, glaucous to gray-green. Panicle open, oblong- ovate, continuous, loose, straight, 18-25 cm long, branches ascending, moderately to profusely divided, 7-14 cm long. Spikelets with 4-7 fertile florets, elliptic or ovate, laterally compressed, 9-15 mm long. Lemma oblong, symmetrical, 6-7 mm long, dark green or purple, without keel, 5-veined. Awn stiff, 0.5-2 mm long. Apical sterile florets resembling fertile though underdeveloped.

Ecology: Andean II-III: 3000-4000 m.

Distribution: Colombia, Ecuador, Peru.

Arequipa: native. ARE.

Festuca rigescens (J.Presl) Kunth

Description: Perennial, cespitose. Culms erect, 15-25 cm long, 1-noded, with 0.25 of their length below uppermost node. Leaf-sheaths glabrous on the surface. Ligule a membrane, 0.5-1 mm long, truncate. Blades involute, 4-10 cm x 1-2 mm, pubescent, hairy above, margins scaberulous, apex obtuse. Panicle contracted, linear, 4-7 cm long, branched adpressed. spikelets with 2-3 fertile florets, with a barren rhachilla extension, oblong, laterally compressed, 6-7 mm long. Lemma 1-awned, awn 1 mm long.

Ecology: Andean III: 3500->4500 m.

Distribution: Peru, Bolivia, N- and C-Chile, NW-Argentina.

Arequipa: native, doubtful. ARE. No voucher, mentioned for RN Salinas y Aguada Blanca by Quipuscoa & Huamantupa (2010). Occurrence in Arequipa is probable due to the distribution range of the species.

Gynerium sagittatum (Aubl.) P.Beauv.

Description: Erect grass, the straight and vertical stems up to 4-5 m tall. Leaves up to 2 m long, 4-6 cm wide, forming a large fan. Inflorescence rise up to 9 m, panicle pale, plume-like, 1 m long.

Ecology: Coastal, Amazonian: 0-1000 m. Riversides, wet areas.

Distribution: Tropical and subtropical America.

Arequipa: native. CAR, ISL.

Hordeum muticum J.Presl

Description: Perennial, cespitose. Butt sheaths pubescent. Culms erect or geniculately ascending, 15-50 cm long, 2- 6-nodded. Leaf-sheath auricles absent. Ligule an eciliate membrane, 0.5-2 mm long. Blades 3-12 cm x 1.5-3 mm, surface scabrous, glabrous, or pilose. Inflorescence composed of single racemes, bilateral, 20-90 cm long. Rachis fragile at the nodes, flattened.

Ecology: Andean II-III: 3000-4500 m. Disturbed areas, riversides.

Distribution: Peru, Bolivia, N- Chile, N-Argentina.

Arequipa: native. ARE.

Note:

Jarava ichu Ruiz & Pav.

Description: Perennial, cespitose, clumped densely. Culms 80-100 cm long, internodes smooth to scaberulous. Sheaths wider than blade at the collar, smooth, or scaberulous, glabrous on surface, oral hairs bearded. Ligule an eciliate membrane, 0.7 mm long, truncate. Blades erect to ascending, filiform, conduplicate, 15-45 cm long, 0.5-0.9 mm wide, stiff, surface scaberulous. Panicle contracted, lanceolate to elliptic, 18-40 cm long, branches with spikelets almost to the base. Spikelets solitary, pedicellate, with 1 fertile floret, 7-9 mm long. Lemmas hairy throughout, hairs on the lower portion about 0.15 mm, sparse, adpressed, pappus hairs 3-4 mm, silky. Principal lemma awn bigeniculate, 11-13 mm long, with 5-8 mm long limb and twisted column, not plumose.

Ecology: Andean I-III: 1000-4500 m. Grasslands, rocky slopes, shrubland, in wetter places than Festuca chrysophylla.

Distribution: Mexico to S-South America.

Arequipa: native. ARE, CAY, CON, UNI.

Use: Important fodder plant for livestock. Also used in roofing and construction (component in adobe bricks).

Note: The genus delimitation in the tribe Stipeae is still under discussion. Here we follow POWO (2024), who accepts Jarava as a homogeneous genus with nine species. The other species of Jarava s.l. are included in the genus Stipa.

Koeleria spicata (L.) Barberá, Quintanar, Soreng & P.M.Peterson

Description: Plants perennial, with both fertile and sterile shoots, cespitose. Culms 10-50 cm, 2-5-noded, densely hairy below the inflorescence. Ligule an eciliate membrane, 1 mm long. Leaves 4-10 cm long, 1-4 mm wide, often tufted, gradually acuminate, glabrous. Panicle spike-like, oblong-ovate, 1.5-3 cm long. Spikelets variegated, mostly 2- flowered. Spikelet axis only sparsely hairy. Glumes short 2-pointed, with a long awn inserted in the upper third.

Ecology: Andean II-III: 2000->4500 m. Grasslands, rocky slopes.

Distribution: Eurasia and the Americas.

Arequipa: native. ARE, CAM, CAR, CAY, ISL, UNI.

Lolium perenne L.

Description: Perennial, turf-forming grass. Culms tufted, erect or spreading, sometimes prostrate and rooting from lower nodes, 30-90 cm tall, 3-4-noded. Leaf blades soft, 5-20 cm × 3-6 mm, glabrous, young blades folded, auricles to 3 mm, ligule 2-2.5 mm. Raceme stiffly erect, 10-30 cm, rachis glabrous, smooth, spikelets 0.8-2 cm, alternate and edgewise along 2 sides of rachis, less than their own length apart, awnless.

Ecology: Coastal, Andean I-II: 500-3000 m. Disturbed areas, meadows.

Distribution: Native to N-Africa and Eurasia, introduced worldwide.

Arequipa: introduced. ARE. Observation in iNaturalist (2024), no other voucher.

Use: Excellent forage and lawn grass for walk-on areas.

Lorenzochloa bomanii (Hauman) Romasch.

Description: Perennial, cespitose, clumped densely. Culms 20-40 cm long. Leaf-sheaths 7-10 cm long. Ligule a membrane, 4-6 mm long. Blades filiform, convolute, 10-17 cm x 0.5 mm, surface ribbed. Inflorescence a panicle, open, linear, 15-20 cm x 2-3 mm, bearing few spikelets. Primary panicle branches 2-4-nate, 4-6 cm long. Spikelets with 1 fertile floret, without rhachilla extension, lanceolate, subterete, 4.5 mm long. Fertile lemma lanceolate, subterete, 3 mm long, membranous, without keel, 5-veined. Lemma surface pubescent. Lemma 1-awned, awn flexuous, 45-60 mm long.

Ecology: Andean II: 3000-3500 m.

Distribution: Peru, Bolivia, NW-Argentina, N-Chile.

Arequipa: native. ARE, CAY.

Lorenzochloa mucronata (Griseb.) Romasch.

Description: Perennial. Culms erect, or geniculately ascending, 60-100 cm long. Culm-nodes pubescent. Sheaths glabrous on surface, outer margins hairy. Ligule an eciliate membrane, 0.5-1 mm long. Blades ± flat, 10-20 cm x 1-3 mm, surface sparsely hairy on both sides. Panicle open, linear, dense, nodding, 10-30 cm long, branches adpressed, or ascending, 1-2-nate. Spikelets pedicellate, with 1 fertile floret, without rhachilla extension, lanceolate, subterete, 8-10 mm long, breaking up at maturity. Floret callus elongated, 2 mm long, pubescent, acute. Glumes persistent, similar, exceeding apex of florets, thinner than fertile lemma. Lemma apex with a short cylindrical neck and surmounted by a ring of hairs, 0.5 mm long, 1-awned. Awn bigeniculate, 35-50 mm long, column twisted. Fruit pubescent to the summit, 5-6 mm long.

Ecology: Andean I-II: 1000-3500 m. Disturbed areas, lomas.

Distribution: Mexico to NW-Argentina.

Arequipa: native. ARE, CAY, ISL.

Lorenzochloa obtusa (Nees & Meyen) Romasch.

Description: Perennial, cespitose, densely clumped grass. Culms 30-60 cm long. Blades filiform, 10-25 cm long, 0.5- 1 mm wide, stiff. Blade surface ribbed, scabrous. Similar to Jarava ichu but smaller, spikes up to 15 cm long. Spikelets with 1 fertile florets, without rhachilla extension, lanceolate, subterete, 2.5–3 mm long. Lemma 1 -awned, awn straight, or flexuous, 8-10 mm long overall. Palea 1 length of lemma, 2 -veined, without keels.

Ecology: Andean II-III: 3000-4500 m. Grasslands, often growing together with Jarava ichu at high altitude.

Distribution: Peru, Bolivia, NW-Argentina.

Arequipa: native. ARE, CAY, CON.

Lorenzochloa rigidiseta (Pilg.) Romasch.

Description: Perennial, cespitose. Culms erect, 30-50 cm long. Ligule an eciliate membrane, 4 mm long. Blades filiform, convolute, 25-45 cm long, 0.5 mm wide, surface ribbed, scabrous, apex pungent. Panicle narrow, branches adpressed, axis and pedicels puberulous or scabrous-puberulous. Glumes ± unequal, 4 to 5 mm, long, purplish, rather firm, broad, acute at apex, obscurely nerved, glabrous. Lemma 4 mm long, awn not twisted, falcate, scabrous or scabrous-pubescent, 1 to 2 cm long, the base gradually widened and as wide as the summit of the lemma.

Ecology: Andean III: 3500-4000 m. Grasslands, disturbed areas.

Distribution: Peru, Bolivia, N-Chile.

Arequipa: native, doubtful. ARE. No voucher, type from Puno. Mentioned for RN Salinas y Aguada Blanca by Quipuscoa & Huamantupa (2010), occurrence likely.

Melica scabra Kunth

Description: Perennial. Culms decumbent, 30-80 cm long. Sheaths tubular for much of their length, closed. Ligule an eciliate membrane, 2.5-4 mm long, erose. Blades 5-10 cm long, 2.5-7 mm wide, surface and margins scabrous. Panicle open, ovate, 5-18 cm long, branches ascending, or spreading. Spikelets pedicellate, with 1-2 fertile florets, diminished florets at apex. Spikelets cuneate, laterally compressed, 8-12 mm long, falling entire. Glumes dissimilar, lower wider than upper, reaching apex of florets, thinner than fertile lemma. Fertile lemma lanceolate, 6.5-8 mm long, chartaceous, without keel, 6-8-veined, surface scabrous, rough below. Apical sterile florets barren, in a clump, cuneate. Anthers 3. Fruit with adherent pericarp.

Ecology: Andean II-III: 2000->4500 m. Disturbed areas, grasslands, riversides.

Distribution: Colombia, Ecuador, Peru, Bolivia.

Arequipa: native. ARE.

Microchloa kunthii Desv.

Description: Perennial, rarely annual, erect and cespitose, or decumbent and mat-forming. Culms 5-60 cm, nodes glabrous. Leaves mostly basal, blades stiff, narrowly linear, 1-6 cm, ca. 1 mm wide. Panicles with a terminal, solitary, slender spikelike branch, rachis curved or falcate. Spikelets solitary, with 1-2 florets, terete to dorsally compressed, florets bisexual. Glumes subequal, exceeding the florets, 1-veined. Lemma 1.5-3 mm, mucronate.

Ecology: Andean II: 2000-3000 m. Disturbed areas.

Distribution: Tropical and subtropical America, Africa and Asia.

Arequipa: native. ARE, CAY.

Muhlenbergia coerulea (Griseb.) Mez

Description: Densely cespitose perennials. Culms 50-100 cm tall. Ligules 8-15 mm long, firm below, strongly decurrent, often lacerate. Leaf blades 15-40 cm long, 2-5 mm wide, tightly involute, glabrous to scaberulous below mostly scaberulous near above. Panicles 10-25 x 0.6-1.5 cm, narrow, spike-like, usually plumbeous to reddish- purple. Glumes not awned.

Ecology: Andean II-III: 2500-4000 m. Grasslands, riversides.

Distribution: The Andes, from Venezuela to N-Argentina.

Arequipa: native. ARE, CAR.

Muhlenbergia fastigiata (J.Presl) Henrard

Description: Rhizomatous herb. Culms 2-4 cm tall. Ligules 0.6-1.4 mm long, membranous, decurrent. Leaf blades tightly involute, 2-8.5 mm long, arcuate, apex often pungent. Panicles 1-2 cm long, 1-5 mm wide, contracted, narrow, usually exserted with 3-9 nodes, primary branches 0.4-1 cm long, ascending, solitary at a node. Glumes acute, florets mucronate.

Ecology: Andean I-III: 1000-4500 m. Disturbed areas, grasslands.

Distribution: Colombia to N-Chile and NW-Argentina.

Arequipa: native. ARE, CAY.

Muhlenbergia ligularis (Hack.) Hitchc.

Description: Loosely tufted herb. Culms 2-12 cm tall. Ligules 0.6-2.5 mm long, membranous to hyaline. Leaf blades 3-25 mm long, flat or folded, prominently veined, thick, firm, usually with whitish-thickened midvein and margins, conspicuously crystalline or spiculate on both surfaces. Panicles 1-3 cm long, 3-14 mm wide, long exserted or included in the uppermost sheath, loosely contracted. Glumes acute, florets mucronate.

Ecology: Andean II-III: 3000-4500 m. Disturbed areas, grasslands.

Distribution: Central and W-South America.

Arequipa: native. ARE, CAY.

Muhlenbergia peruviana (P.Beauv.) Steud.

Description: Tufted annuals. Culms 3-27 cm tall, erect, glabrous. Leaf-sheaths usually longer than the internodes, smooth or scabridulous. Ligules 1.5-3 mm long, membranous, acute. Leaf blades 1-5 cm long, 0.6-1.5 mm wide, flat to involute, smooth or scabridulous beneath. Panicles 2-8 x, 0.3-3.4 cm, contracted or open, primary branches 1–5 cm long. Upper glumes 3-veined, the apex broad and truncate, usually 2- or 3-toothed. Florets awned.

Ecology: Andean II-III: 2500->4500 m. Forests, grasslands.

Distribution: S-North America to W-South America.

Arequipa: native. ARE, CAY.

Munroa decumbens Phil.

Description: Annual, mat forming, stoloniferous. Culms decumbent, 6-12 cm long, rooting at nodes, internodes alternately elongated and bunched. Lateral branches ample. Ligule a fringe of hairs. Blades 1.5-2.2 cm x 1.5-2 mm, stiff, apex acuminate. Inflorescence of only a few spikelets, shorter than basal leaves, subtended by 2 inflated leaf- sheath, embraced at base by subtending leaf.

Ecology: Andean II: 2000-2500 m. Rocky slopes.

Distribution: Peru, Bolivia, N-Chile, NW-Argentina.

Arequipa: native. ARE.

Nassella asplundii Hitchc.

Description: Perennial, cespitose. Culms erect, 30-60 cm long, 2-noded, lateral branches lacking. Ligule an eciliate membrane, 1-2 mm long. Blades filiform, involute, 1-6 cm long, 1 mm wide, surface glabrous, apex attenuate. Panicle open, oblong, 8-15 cm long, branches spreading, 2-5 cm long. Panicle branches capillary, smooth. Spikelets pedicelled, pedicels scabrous, with 1 fertile floret, without rhachilla extension, ovate, subterete, 4 mm long. Glumes persistent, similar, exceeding apex of florets, thinner than fertile lemma. Lemma oblong, subterete, 2.5 mm long, coriaceous, lightly keeled, surface pilose, 1-awned. Awn eccentric, bigeniculate, 15 mm long overall, with twisted column, deciduous, column ciliate.

Ecology: Andean III: 3500-4000 m. Riversides.

Distribution: Peru, Bolivia, N-Chile.

Arequipa: native, doubtful. ARE. No voucher, mentioned for Arequipa by Brako & Zarucchi (1993) and for RN Salinas y Aguada Blanca by Quipuscoa & Huamantupa (2010). Occurrence is likely.

Nassella brachyphylla (Hitchc.) Barkworth

Description: Perennial, cespitose. Culms erect, 20-40 cm long, internodes distally glabrous. Sheaths glabrous on surface. Ligule an eciliate membrane, 2-5 mm long. Blades involute, 3-5 cm long, 1 mm wide, stiff, surface ribbed, scabrous, rough above, apex pungent. Panicle open, linear, 3-10 cm long, branches ascending, 2-4 cm long, with 1- 3 fertile spikelets on each lower branch. Spikelets pedicellate, with 1 fertile floret, without rhachilla extension, lanceolate, subterete, 10 mm long. Floret callus evident, 1.5 mm long, pubescent, pungent. Glumes persistent, similar, exceeding apex of florets, thinner than fertile lemma. Lemma surmounted by a ring of hairs, 3 mm long, 1- awned. Awn bigeniculate, 20 mm long overall, with twisted column. Middle segment of lemma awn 5 mm long, puberulous, column 5 mm long, pubescent.

Ecology: Andean II-III: 3000-4500 m. Grasslands.

Distribution: The Andes, from Colombia to NW-Argentina.

Arequipa: native, doubtful. ARE. No voucher, mentioned for RN Salinas y Aguada Blanca by Quipuscoa & Huamantupa (2010). Occurrence is likely.

Nassella depauperata (Pilg.) Barkworth

Description: Perennial, cespitose. Basal innovations intravaginal. Culms 20-40 cm long, 2-noded. Culm-nodes glabrous. Sheaths 3-4 cm long, glabrous on surface, or puberulous. Leaf-sheath oral hairs ciliate. Ligule an eciliate membrane, truncate. Blades 2-7(-9) cm long, surface pubescent, hairy above, margins pubescent. Panicle contracted, linear, 7-15 cm long, branches 4-5 cm long, bearing spikelets almost to the base. Spikelets pedicellate, with 1 fertile floret, without rhachilla extension, lanceolate, subterete, 6 mm long. Floret callus evident, pubescent, acute. Glumes similar, exceeding apex of florets, thinner than fertile lemma. Lemma surmounted by a ring of hairs, 1-awned. Awn bigeniculate, 13 mm long overall, with twisted column, column pubescent. Fruit 2.5-3 mm long.

Ecology: Andean II-III: 2500-4000 m. Riversides.

Distribution: The Andes, from Colombia to NW-Argentina.

Arequipa: native. ARE, CAY.

Nassella inconspicua (J.Presl) Barkworth

Description: Perennial, cespitose. Culms 20-50 cm long. Ligule an eciliate membrane, 0.5-1 mm long, erose. Blades 4-10 cm long, 2.5-3.5 mm wide, surface ribbed, glabrous, apex abruptly acute. Panicle contracted, linear, or lanceolate, interrupted, 7-10 cm long, branches adpressed, or ascending, bearing spikelets almost to the base. Spikelets pedicelled, with 1 fertile floret, without rhachilla extension, lanceolate, subterete, 3.5-4 mm long. Floret callus evident, pubescent, acute. Glumes persistent, similar, exceeding apex of florets, thinner than fertile lemma. Lemma surmounted by a ring of hairs, 1-awned. Awn bigeniculate, 5-8 mm long overall, with twisted column.

Ecology: Andean II-III: 3000-4000 m. Grasslands.

Distribution: The Andes, from Colombia to NW-Argentina.

Arequipa: native, doubtful. ARE. No voucher, mentioned for RN Salinas y Aguada Blanca by Quipuscoa & Huamantupa (2010). The occurrence in Arequipa is likely, due to the distribution range of the species in S-Peru (Ayacucho, Cuzco, Puno, Moquegua).

Nassella mexicana (Hitchc.) R.W.Pohl

Description: Perennial, cespitose. Culms geniculately ascending, or decumbent, 20-30 cm long. Sheaths glabrous on surface. Leaf-sheath oral hairs lacking. Ligule an eciliate membrane. Blades filiform, involute, 5-20 cm long, 0.2- 0.3 mm wide, surface scabrous, rough above, apex pungent. Panicle open, linear, 5-10 cm long, branches adpressed, or ascending. Spikelets pedicellate, with 1 fertile floret, without rhachilla extension, lanceolate, subterete, 10 mm long. Floret callus evident, white pubescent, obtuse. Glumes persistent, similar, exceeding apex of florets, thinner than fertile lemma. Lemma surmounted by a ring of hairs, 0.5 mm long, 1-awned. Awn bigeniculate, 10 mm long overall, with a straight or slightly twisted column, column puberulous.

Ecology: Andean II-III: 2500-4000 m. Disturbed areas, grasslands, riversides.

Distribution: Mexico to NW-Argentina.

Arequipa: native, doubtful. ARE. No voucher, mentioned for RN Salinas y Aguada Blanca by Quipuscoa & Huamantupa (2010). Occurrence is likely.

Nassella nardoides (Phil.) Barkworth

Description: Perennial, cespitose. Culms erect, 8-10 cm long. Ligule a fringe of hairs, 0.2-1 mm long. Blades filiform, convolute, 2-4 cm x 0.5 mm, stiff, surface ribbed, pubescent, hairy above. Panicle open, lanceolate, 2-3 cm long, bearing few spikelets. Spikelets pedicellate, pedicels 10-20 mm long, glabrous, with 1 fertile floret, without rhachilla extension, lanceolate, subterete, 7-11 mm long. Floret callus evident, 0.8 mm long, pubescent, hairy in a ring above, acute. Glumes persistent, similar, exceeding apex of florets, thinner than fertile lemma. Lemma with conspicuous apical hairs, apex awned, 1-awned. Awn bigeniculate, 8-14 mm long, with twisted column, column 3.5 mm long, puberulous. Fruit 3.5 mm long.

Ecology: Coastal, Andean I-II: 0-2000 m. Lomas.

Distribution: Peru, Bolivia, NW-Argentina, N-Chile.

Arequipa: native. ARE, CAY, ISL.

Nassella neesiana (Trin. & Rupr.) Barkworth

Description: Perennial, cespitose. Culms erect, 30-100 cm long, 3-4-noded, internodes distally glabrous. Lateral branches lacking. Sheaths glabrous on surface to pubescent. Ligule an eciliate membrane. Blades flat, or convolute, 2-5.5 mm wide, surface ribbed, glabrous to pubescent. Panicle open, lanceolate, 5-30 cm long, branches ascending. Spikelets pedicellate, with 1 fertile floret, without rhachilla extension, lanceolate, sub-terete, 10-22 mm long. Floret callus elongated, curved, 4 mm long, pubescent, pungent. Glumes persistent, similar, exceeding apex of florets, thinner than fertile lemma. Lemma with a membranous corona and surmounted by a ring of hairs, 1 mm long, 1-awned. Awn bigeniculate, 40-50 mm long overall, with twisted column. Fruit 8 mm long.

Ecology: Andean II: 2000-3000 m. Rocky slopes.

Distribution: South America, introduced to S-Europe, South Africa and Australia.

Arequipa: native. ARE, CAR.

Nassella pubiflora (Trin. & Rupr.) É.Desv.

Description: Perennial, cespitose. Culms erect, or geniculately ascending, 50-90 cm long, 2-4-noded. Lateral branches lacking. Sheaths smooth, or scaberulous, glabrous on surface. Ligule an eciliate membrane. Blades flat, or convolute, 10-20 cm long, 2-3 mm wide, surface glabrous, apex attenuate. Panicle open, or contracted, linear, or lanceolate, 10-20 cm long, branches adpressed, or ascending. Spikelets pedicellate, with 1 fertile floret, without rhachilla extension, ovate, subterete, 3.5-5 mm long. Floret callus pubescent, obtuse. Glumes persistent, similar, exceeding apex of florets, thinner than fertile lemma. Lemma surface pilose, margins convolute, covering most of palea. Lemma obtuse, 1-awned. Awn eccentric, flexuous, 15 mm long overall, deciduous. Fruit 1.5-2 mm long.

Ecology: Andean II-III: 2500-4500 m. Disturbed areas, grasslands, rocky slopes.

Distribution: Ecuador, Peru, Bolivia, NW-Argentina, N-Chile.

Arequipa: native. ARE, CON, UNI.

Nassella smithii (Hitchc.) Barkworth

Description: Perennial, cespitose. Culms 25-60 cm long, 3-noded. Ligule an eciliate membrane, 1-1.5 mm long. Blades involute, 8-15 cm x 1-2 mm, surface ribbed, apex pungent. Inflorescence a contracted panicle, linear, 8-15 cm long. Primary panicle branches appressed. Spikelets solitary.

Ecology: Andean II-III: 3000-4500 m. Grasslands.

Distribution: Peru.

Arequipa: native, doubtful. Not specified by province. No vouchers. Mentioned for RN Salinas y Aguada Blanca by Quipuscoa & Huamantupa (2010). Presence in Arequipa is likely, specimens from Huancavelica, Puno, Tacna.

Pappostipa chrysophylla (É.Desv.) Romasch.

Description: Perennial, cespitose, stolons present. Butt sheaths yellow, glabrous, or pubescent. Culms erect, or geniculately ascending, 15-40 cm long, 2-4-noded, internodes distally glabrous, or pubescent, nodes glabrous. Ligule an eciliate membrane, 5 mm long. Blades

filiform, conduplicate, 10-30 x 0.5 mm, surface scaberulous, glabrous. Inflorescence a panicle, 10-20 fertile spikelets, open, linear, dense, 6-9 x 0.5-1 cm. Spikelets solitary. Fertile spikelets pedicellate. Pedicels 5-10 mm long, scabrous, ciliate, hairy at tip. Lemma apex bifid. Awn geniculate, 25-60 mm long, plumose.

Ecology: Andean II-III: 3000-4500 m. Rocky slopes.

Distribution: Peru, Argentina, Chile.

Arequipa: native. ARE. Specimen Quipuscoa et al. 3804 (HUSA) from prov. Arequipa (not seen).

Paspalum candidum (Humb. & Bonpl. ex Flüggé) Kunth

Description: Annual. Culms decumbent, or rambling, 50-100 cm long, 1-2 mm Ø, rooting from lower nodes. Culm- internodes hollow, smooth, distally glabrous, nodes brown, glabrous, or pubescent. Lateral branches lacking. Leaf- sheaths smooth, glabrous on surface, outer margins glabrous, or hairy. Ligule a ciliolate membrane, 1-2.8 mm long. Blade base broadly rounded. Blades lanceolate, 4-10 cm long, 8-21 mm wide, flaccid, surface glabrous or pubescent, apex abruptly acute. Panicle open, 4-8 cm long, few branched, primary panicle branches single, bearing few spikelets. Spikelets solitary, 1 basal sterile floret, 1 fertile floret, without rhachilla extension. Spikelets ovate, dorsally compressed, 3.3-4.5 mm long. Glumes dissimilar, reaching apex of florets, thinner than fertile lemma.

Ecology: Andean I-II: 500-3500 m. Lomas, rocky slopes.

Distribution: Tropical and subtropical Central and South America.

Arequipa: native. CAR, ISL.

Paspalum flavum J.Presl

Description: Annual. Culms 20-60 cm long. Lateral branches ample, arising from lower culm. Ligule an eciliate membrane. Blades lanceolate, 5-10 cm long, 5-10 mm wide, surface glabrous. Inflorescence composed of numerous, 1-2 cm long racemes.

Ecology: Coastal, Andean I-II: 0-3000 m. Disturbed areas, lomas.

Distribution: Peru.

Arequipa: native. CAR, ISL.

Paspalum vaginatum Sw.

Description: A widely creeping perennial with rhizomes and/or stolons. Leaves flat, elongate, with scabrous margins. Inflorescence a panicle with usually a pair of branches.

Ecology: Coastal: 0-500 m. Disturbed areas, salt marshes.

Distribution: Pantropical and subtropical.

Arequipa: native. ISL.

Use: Fodder grass, although not of great quality. Used to stabilize frequently flooded, saline soils. Can take up heavy metals, so its use as a filter or buffer plant for phytoremediation is being evaluated.

Phalaris canariensis L.

Description: Annual, cespitose. Culms erect, or geniculately ascending, 20-120 cm long. Leaves cauline. Ligule an eciliate membrane, 3-8 mm long. Blades 5-25 cm long, 4-12 mm wide. Panicle spiciform, ovate, 1.5-6 x 1.2-2.2 cm. Spikelets sessile, with 2 basal sterile florets, 1 fertile floret, no rhachilla extension. Spikelets obovate, laterally compressed, 6-10 mm long, disarticulating below each fertile floret. Glumes persistent, similar, exceeding apex of florets, thinner than lemma. Fertile lemma elliptic, laterally compressed, 5-6 mm long, shiny, keeled, 5-veined, surface pubescent. Palea surface pubescent, hairy on back.

Ecology: Coastal: 0-500 m. Disturbed areas.

Distribution: Native to Mediterranean Europe, introduced to the Americas.

Arequipa: introduced. ISL.

Phragmites australis (Cav.) Trin. ex Steud.

Description: Stout plants, commonly 2-4 m high, with stolons and rhizomes, often forming large colonies. Leaves long, 1-3 cm wide, distichous. Panicles plumelike, usually purplish, 20-40 cm long.

Ecology: Coastal, Andean I: 0-1500 m. Watersides, ponds.

Distribution: Cosmopolitan.

Arequipa: native. ISL.

Note: In Peru Phragmites australis ssp. berlandieri.

Piptochaetium bicolor (Vahl) É.Desv.

Description: Perennial, cespitose. Culms erect or geniculately ascending, 50-80 cm long, 3-noded. Sheaths glabrous. Ligule an eciliate membrane, 1-2 mm long, truncate. Blades flat or convolute, 12-25 cm x 1-2 mm, with 7- 11 secondary veins, surface ribbed, glabrous. Panicle open, ovate, nodding, 10-20 cm long. Primary panicle branches 3-5-nate, whorled at lower nodes, 5-10 cm long. Spikelets with 1 fertile floret, oblong, laterally compressed, 10-12 mm long. Glumes persistent, similar, exceeding apex of florets, thinner than fertile lemma, 5-veined. Upper glume apex 1-awned, awn 2 mm long. Lemma obovate, laterally compressed, gibbous, 5-6.5 mm long, dark brown, apex with a ring of hairs, awned. Awn eccentric, bigeniculate, 40-50 mm long, with twisted column. Palea 1 length of lemma, 2-veined.

Ecology: Coastal: 0-1000 m. Lomas.

Distribution: N-Argentina, Uruguay, C- and S-Chile.

Arequipa: native. CAR. Mentioned as a new species for the Lomas de Atiquipa by Muñuico et al. (2024b).

Piptochaetium montevidense (Spreng.) Parodi

Description: Perennial, cespitose. Culms 30-65 cm long. Sheaths glabrous. Ligule an eciliate membrane, 2 mm long, 0.5 mm long on basal shoots. Blades filiform, conduplicate, 5-15 cm x 0.5 mm, surface glabrous, or pubescent. Panicle contracted, exserted, or embraced at base by subtending leaf., linear, 2-10 cm long, 1-2 cm wide. Spikelets with 1 fertile floret, oblong, laterally compressed, 3-3.5 mm long. Glumes persistent, similar, exceeding apex of florets, thinner than fertile lemma. Fertile lemma orbicular, laterally compressed, gibbous, 1.5-2 mm long, indurate, dark brown, without keel. Lemma awn eccentric, flexuous, 7-8 mm long, deciduous. Palea 1 length of lemma, 2- veined.

Ecology: Coastal, Andean I-III: 0-4000 m. Grasslands. Invasive in coastal areas.

Distribution: Native to E-South America (pampas), introduced to SE-Australia.

Arequipa: native. CAR, ISL.

Poa adusta J.Presl

Description: Perennial, cespitose. Culms geniculately ascending, 10-30 cm long. Sheaths smooth, ligule membranous, 0.5-2 mm long, truncate. Blades involute, 5-16 cm x 2-3 mm, glabrous to scaberulous beneath with prickles or hooks usually restricted to the leaf margin. Panicle open, ovate, 4-8 cm long. Spikelets with 2-3 fertile florets. Lemma pubescent and hairy below.

Ecology: Andean II-III: 3000->4500 m. Grasslands.

Distribution: Colombia, Peru, Bolivia.

Arequipa: native. ARE, CAY.

Poa aequigluma Tovar

Description: Perennial, cespitose, forming dense cushions. Culms erect to falcate, 4-6 cm long. Leaves mostly basal. Leaf-sheaths glabrous, ligule membranous, 0.5-1 mm long, erose, truncate. Blades conduplicate, 1-2.5 cm x 1-2 mm, apex abruptly acute. Panicle linear, 1.5-2 cm long and 0.3-0.4 cm wide. Spikelets 3.5-4 mm long. Lower glume 3.6-4 mm long.

Ecology: Andean III: 4000->4500 m. Grasslands, rocky slopes.

Distribution: Peru, Bolivia, NW-Argentina.

Arequipa: native. ARE, CAY.

Poa annua L.

Description: Annual, rarely persisting for 2 years, cespitose. Butt sheaths herbaceous. Culms erect or geniculately ascending, 3-30 cm long, 2-4-nodded. Leaves mostly basal. Leaf-sheaths keeled, smooth. Ligule membranous, 2-5 mm long, obtuse. Blades flat or conduplicate, 1-14 cm x 1-5 mm, surface smooth, glabrous, margins scaberulous, apex obtuse to abruptly acute, hooded. Panicle open, pyramidal or ovate, 1–12 cm long, primary branches spreading. Spikelets loosely arranged.

Ecology: Coastal, Andean I-III, Amazonian: 0-4000 m. Disturbed areas, grasslands, lomas, riversides. Tolerant to trampling.

Distribution: Cosmopolitan.

Arequipa: native. ARE, CAR.

Note: One of the most widespread plant species in the world.

Poa brevis Hitchc.

Description: Perennial, cespitose. Culms erect, 4-6 cm long. Leaf-sheaths glabrous on surface, ligule membranous, 0.5-1 mm long, erose, truncate. Blades conduplicate, 1-2.5 cm x 1-2 mm, surface glabrous, margins smooth, apex abruptly acute. Panicle linear, 2-4 x 0.3 cm, with 2 fertile florets. spikelets 3-3.3 mm long. Apex of upper glume and lemma acute. Lemma 2.5-2.7 mm long.

Ecology: Andean III: 3500-4500 m. Rocky slopes.

Distribution: Peru.

Arequipa: native. ARE, CON.

Poa gilgiana Pilg.

Description: Perennial, cespitose. Rhizomes short. Culms erect, 25-35 cm long, 3-4-nodded. Lateral branches lacking. Leaf-sheaths keeled. Ligule membranous, 2-3.5 mm long, erose. Blades conduplicate, 8-25 cm x 3-4 mm wide, stiff. Panicle open, ovate, 8-13 cm long, spikelets clustered towards branch tips, branches scabrous. Spikelets with 3-4 fertile florets. Lemmas glabrous, smooth or scabrous. Callus glabrous. Apical sterile florets resembling fertile though underdeveloped.

Ecology: Andean III: 4000-4500 m. Disturbed areas, grasslands.

Distribution: Bolivia, Peru.

Arequipa: native. ARE, CAR.

Poa gymnantha Pilg.

Description: Perennial, cespitose. Culms erect, 13-25 cm long. Sheaths glabrous on surface, ligule membranous, 4-8 mm long. Blades involute, 5-12 cm x 2-3 mm wide, stiff, surface puberulous, hairy above, apex acute, margins scabrous. Panicle contracted, lanceolate 3-5.5 x 0.8-1.3 cm. Spikelets with 2-3 fertile florets. Lower glume lanceolate, 3.5-3.8 mm long. Lemma ovate, 4.6-5 mm long.

Ecology: Andean III: 3500->4500 m. Disturbed areas, grasslands.

Distribution: The Andes from Peru to Argentina, Mexico.

Arequipa: native. ARE, CAY.

Poa horridula Pilg.

Description: Perennial, with extravaginally shoots. Culms erect, geniculate, ascending, robust, 30-90 cm long, 3-4- nodded. Sheaths glabrous. Ligule membranous, 3-6 mm long, truncate. Blades 10-40 cm x 5-10 mm, flat, surface rough on both sides, margins scabrous. Panicle open, 16-30 x 7-10 cm, primary branches 7-15 cm long, branches capillary, flexuous, scaberulous. Spikelets with 3-5 fertile florets, with diminished florets at the apex. Lemmas, at least of the upper florets, pubescent or villous in their lower half. Callus glabrous.

Ecology: Andean II-III: 3000-4500 m. Grasslands, rocky slopes, shrublands.

Distribution: Colombia, Ecuador, Peru, Bolivia.

Arequipa: native. CAY.

Poa humillima Pilg.

Description: Perennial, cespitose. Culms erect, 1-4 cm long. Leaf-sheaths glabrous on surface, ligule membranous, 0.5-0.8 mm long, truncate. Blades conduplicate, 0.5-3 cm x 1-2 mm, stiff, surface glabrous, margins scabrous, apex obtuse. Panicle shorter than basal leaves, embraced at base by subtending leaf, spiciform, oblong or ovate, 0.5-1.5 cm long. Spikelets with 3-4 fertile florets. Lemma glabrous, apex obtuse.

Ecology: Andean III: 4000->4500 m.

Distribution: Peru, Bolivia, NW-Argentina, N-Chile.

Arequipa: native. ARE, CAY.

Poa kurtzii R.E.Fr.

Description: Perennial, cespitose. Culms erect, 30-60 cm long. Sheaths smooth, ligule membranous, 5-8 mm long, acute. Blades convolute, 10-25 cm x 1-2 mm, scabrous beneath, apex pungent. Inflorescence a dioecious, open panicle, ovate, 8-20 cm long. Primary panicle branches 4-8 cm long. Spikelets with 3-4 fertile florets. Lemmas glabrous to scabrous-pubescent.

Ecology: Andean II-III: 2500-4500 m. Grasslands, rocky slopes.

Distribution: Peru, Bolivia, NW-Argentina, N-Chile.

Arequipa: native. ARE, CAR, CAY.

Poa laetevirens R.E.Fr.

Description: Perennial, cespitose, clumped loosely. Rhizomes elongated. Culms decumbent, 5-15 cm long, 3-6-noded. Sheaths keeled, glabrous on surface. Ligule an eciliate membrane, 1-2 mm long, erose, obtuse. Blades flat or conduplicate, 3-8 cm x 1-2 mm, glabrous, margins smooth, apex abruptly acute. Panicle shorter than basal leaves, embraced at base by subtending leaf, spiciform, elliptic, 3-5 x 0.5-1 cm. Spikelet with 3-4 fertile florets. Lemma 2.5 mm long.

Ecology: Andean III: >4500 m. Grasslands.

Distribution: S-Peru, Bolivia, NW-Argentina.

Arequipa: native. ARE.

Poa lepidula (Nees & Meyen) Soreng & L.J.Gillespie

Description: Perennial, cushion forming. Culms 3-12 cm long. Ligule membranous, lacerate. Blades 1-5 cm x 0.5-2 mm. Panicle shorter than basal leaves, embraced at base by subtending leaf, spiciform, oblong to ovate, 0.7-3 cm long. Spikelets with 3-8 fertile florets. Fertile lemma flabellate, translucent, apex emarginate. Glumes translucent.

Ecology: Andean II-III: 2000->4500 m. Grasslands, rocky slopes.

Distribution: Peru, Bolivia, NW-Argentina, N-Chile.

Arequipa: native. ARE, CAY, CON.

Poa macusaniensis (E.H.L.Krause) Refulio

Description: Annual. Culms slender, 3-7 cm long. Leaf-sheaths glabrous, ligule membranous, 0.7-1.7 mm long. Blades ascending, involute, 1-4 cm x 1-2 mm, surface smooth, glabrous, apex acute. Panicle shorter than basal leaves, spiciform, oblong, 1-2 cm long. Spikelets with 2 fertile florets. Fertile lemma ovate, papery, apex dentate. Glumes membranous, with translucent margins.

Ecology: Andean II-III: 2500-4500 m.

Distribution: Peru, Bolivia, NE-Argentina.

Arequipa: native. ARE, CAY.

Poa marshallii Tovar

Description: Perennial, cespitose. Culms 30-40 cm long, 1-noded. Leaves mostly basal. Ligule an eciliate membrane, 2-3 mm long, truncate. Leaf-blades filiform, conduplicate or involute, 15-20 cm x 1 mm, stiff. Leaf-blade apex acute. Panicle contracted, linear to oblong, 5.5-8 cm long, 0.7-1 cm wide. Primary panicle branches adpressed, scabrous. Spikelets with 2-3 fertile florets, pedicellate. Glumes and lemma coriaceous.

Ecology: Andean III: Grasslands.

Distribution: Peru, Bolivia.

Arequipa: native. CAY.

Poa pearsonii Reeder

Description: Perennial, cespitose, clumped densely. Culms geniculately ascending, 18-40 cm long, 2-3-nodded. Sheaths smooth, glabrous on surface. Ligule membranous, 8-15 mm long, acuminate. Blades involute, 8-20 cm x 2- 4 mm, stiff, surface scabrous, rough beneath, pubescent, densely hairy on both sides, apex acuminate. Panicle open, elliptic, 6-12 cm long, peduncle smooth, branches ascending, scabrous. Spikelets with 2-3 florets. Lemmas scabrous. Callus glabrous.

Ecology: Andean III: >4500 m. Grasslands.

Distribution: Peru, Bolivia, N-Chile.

Arequipa: native. ARE, CAY.

Poa perligulata Pilg.

Description: Perennial, cespitose. Culms geniculately ascending, 3.5-7 cm long. Leaf-sheaths glabrous, ligule membranous, 1-3.5 mm long, truncate. Blades conduplicate, 1-3 cm x 1-2 mm. Blade surface smooth, or scaberulous, rough above. Blade margins scaberulous, apex abruptly acute or acute. Inflorescence a spiciform panicle, ovate, 2-2.5 cm x 0.5-0.7 cm. Panicle ovate, 2-2.5 cm long and 0.5-0.7 cm wide. Spikelets 4-4.5 mm long. Lower glume 2.5-2.8 mm long.

Ecology: Andean III: 4000-4500 m. Grasslands.

Distribution: Peru, Bolivia, NW-Argentina, N-Chile.

Arequipa: native. ARE, CAY.

Poa spicigera Tovar

Description: Perennial, cespitose, clumped loosely. Culms erect, 10-16 cm long, 2-nodded. Sheaths glabrous, ligule membranous, 2.5-3.5 mm long, erose, truncate. Blades 3-5 cm x 1-2.5 mm, surface glabrous, margins smooth. Panicle spiciform, linear-oblong, 3.5-5.5 cm x 0.7-0.8 cm. Spikelet with 2 fertile florets. Lemma 4-4.5 mm long.

Ecology: Andean II-III: 2000-4500 m. Bogs, shrublands.

Distribution: Bolivia, Peru.

Arequipa: native. ARE, CAY, CON.

Polypogon elongatus Kunth

Description: Perennial, cespitose. Culms erect to geniculately ascending, 60-100 cm long, nodes constricted, brown. Ligule an eciliate membrane, 4-8 mm long, lacerate. Blades 15-30 cm long, 4-15 mm wide. Peduncle 10-30 cm long, glabrous. Panicle contracted, lanceolate, or elliptic, 10-30 x 1-7 cm, branches scabrous. Spikelets pedicellate, pedicels filiform, 3-5 mm long, scabrous. Spikelets with 1 fertile floret, cuneate, laterally compressed, 2-3 mm long, deciduous with the pedicel. Glumes similar, exceeding apex of florets, firmer than fertile lemma, gaping. Upper glume 1-awned, awn 1-2 mm long. Lemma ovate, 1-1.5 mm long, membranous, 1-awned, awn 1.2-2 mm long.

Ecology: Coastal, Andean I-II, Amazonian: 0-2000 m. Cloud forests, disturbed areas, riversides.

Distribution: Central and South America.

Arequipa: native, doubtful. CAR, ISL. Species of the E-slopes of the Andes. Specimen Anonymous 1891 (MO) from Caravelí, not seen. Mentioned for the lomas of Peru by Dillon et al. (2011) and for Arequipa by Brako & Zarucchi (1993).

Polypogon interruptus Kunth

Description: Perennial, cespitose. Culms decumbent, 30-80 cm long. Ligule an eciliate membrane, 2-5 mm long. Blades 4-6 mm wide. Panicle contracted or spiciform, oblong, ± interrupted, 5-15 cm long. Spikelets pedicellate, pedicels cuneate, 0.5-1 mm long, with 1 fertile florets, cuneate, laterally compressed, 2.5-3 mm long, deciduous with the pedicel. Glumes similar, exceeding apex of florets, firmer than fertile lemma, gaping. Upper glume 1-awned, awn 3-5 mm long. Lemma oblong, 1 mm long, cartilaginous, shiny, 5-veined, truncate, 1-awned, awn 1.5-2 mm long.

Ecology: Coastal, Andean I-II, Amazonian: 0-3500 m. Disturbed areas, lomas, riversides.

Distribution: North and South America.

Arequipa: native. ARE, CAY, ISL.

Polypogon monspeliensis (L.) Desf.

Description: Annual, culms solitary or cespitose. Culms erect to decumbent, 6-80 cm long. Ligule an eciliate membrane, 3-15 mm long. Blades 5-20 cm long, 2-8 mm wide, surface ± rough. Panicle spiciform, oblong to ovate, 1.5-16 x 1-3.5 cm, branches scabrous. Spikelets pedicellate, pedicels linear, 0.5 mm long, scabrous, with 1 fertile florets, no rhachilla extension, oblong, laterally compressed, 2-3 mm long, falling entire. Glumes similar, exceeding apex of florets, firmer than fertile lemma. Lemma oblong, 1-1.5 mm long, hyaline, 5-veined, muticous or awned, awn 0-2 mm long.

Ecology: Coastal, Andean I-II: 0-3000 m. Disturbed areas, humid places.

Distribution: Native to Eurasia and N-Africa, naturalized worldwide.

Arequipa: introduced. ARE, ISL.

Polypogon viridis (Gouan) Breistr.

Description: Perennial, cespitose. Culms geniculately ascending to decumbent, 30-100 cm long. Ligule an eciliate membrane, 2-4 mm long. Blades 2-17 cm long, 2-7 mm wide, surface scaberulous. Panicle contracted, oblong to ovate, interrupted, 2-8 x 0.5-4 cm, branches whorled, spikelets almost to the base. Spikelets pedicellate, with 1 fertile florets, no rhachilla extension, elliptic, laterally compressed, 1.75-2 mm long, deciduous with the pedicel. Glumes similar, exceeding apex of florets, firmer than fertile lemma. Lemma oblong, 1-1.5 mm long, hyaline, not keeled and awnless.

Ecology: Coastal, Amazonian: 0-500 m. Riversides.

Distribution: Native to Eurasia, introduced worldwide.

Arequipa: introduced, doubtful. Not specified by province. Mentioned for lomas in Arequipa by Lleellish et al. (2015). No voucher. An occurrence in Arequipa is likely.

Rostraria trachyantha (Phil.) Soreng

Description: Annual, cespitose. Culms erect, 20-70 cm long. Sheaths glabrous. Ligule an eciliate membrane, 2-3 mm long. Blades 3-6 mm wide, glabrous. Panicle spiciform, linear to oblong, interrupted, loose, 6-7 cm long, rachis pilose. Spikelets pedicellate, with 3 fertile florets, obovate, laterally compressed, 3-4 mm long, disarticulating below each fertile floret. Rhachilla internodes pubescent. Glumes persistent, similar, shorter

than spikelet, shiny. Lemma oblong, 3-4 mm long, membranous, keeled, 5-veined, midvein ciliate, surface pubescent with tubercle-based hairs, apex truncate, 1-awned. Awn subapical, 1.5-2 mm long.

Ecology: Coastal: 0-1000 m. Lomas, the dominant grass on the interior desert hills.

Distribution: Peru, Chile.

Arequipa: native. CAM, ISL.

Setaria parviflora (Poir.) Kerguélen

Description: Tufted perennials, erect, spreading, or ascending. Culms to 1.2 m tall, smooth except below inflorescence, glabrous. Sheaths compressed, sometimes scabrous upwards, glabrous, ligule densely ciliate, 1 mm long. Blade flat, straight, acuminate, 25 x 0.8 cm, scabrous, often glaucous, lower surface glabrous. Inflorescence long-exserted, erect, cylindrical, dense, 1-10 cm long, branches very short, bristles 1-6 x longer than spikelets.

Ecology: Coastal, Andean I-II, Amazonian: 0-3500 m. Disturbed areas, lomas.

Distribution: Native to America, introduced worldwide.

Arequipa: native. CAM, CAR, ISL.

Sorghum halepense (L.) Pers.

Description: Rhizomatous perennial to 2 m high. Leaves with ligule membranous, 1-3 mm long, with hairs to 2 mm long. Blade flat, to 2 cm wide. Inflorescence open, 25-45 x 3-15 cm, axils with hairs to 1 mm long. Spikelets paired, 4-6 mm long, florets 2, lower spikelet sessile, dorsally compressed, upper spikelet pedicellate, narrower than lower spikelet, terminal spikelets in triplets. Sterile lemma 2–5 mm long, acute or 2-lobed, sometimes with an awn to 7 mm long.

Ecology: Coastal, Andean I, Amazonian: 0-1200 m. Disturbed areas, fields, humid places.

Distribution: Native to N-Africa and SW-Asia, introduced worldwide.

Arequipa: introduced. CAR. No specimen, several observations in iNaturalist (2024) and mentioned by Montesinos & Zegarra (2019).

Sporobolus indicus (L.) R.Br.

Description: Perennial, cespitose herb. Culms erect, 30-100 cm long. Ligule a fringe of hairs, 0.2-0.5 mm long. Leaves basal and cauline. Blades ± flat, 10-30 cm long, 1-5 mm wide, glabrous. Panicle narrowly contracted or spiciform, linear, 10-30 x 0.6-1 cm. Primary branches 0.5-2 cm long, bearing solitary spikelets almost to the base. Spikelets pedicellate, with 1 fertile floret.

Ecology: Coastal, Andean I-II, Amazonian: 0-3500 m. Disturbed areas.

Distribution: Native to the Americas, introduced worldwide in warmer areas.

Arequipa: native. ARE.

Sporobolus virginicus (L.) Kunth

Description: Perennial herb, rhizomatous, stoloniferous. Flowering stems erect to decumbent, round, hollow, hard and smooth, 10-60 cm tall. Alternately one long and 2 short internodes. Sheats hairy near the blade. Ligule a short rim, hairy. Blades conspicuously distichous, pale green, linear, apex pointed, glabrous, rigid, 4-16 cm x 1.5-4 mm. Inflorescence 3-10 cm long, contracted, dense, primary branches 0.5-2 cm long, adpressed, spikelet-bearing to base. Spikelets 2-3 mm long, yellowish-white, 1-flowered.

Ecology: Coastal: 0-500 m. Beach.

Distribution: Tropics and subtropics, worldwide.

Arequipa: native. CAM, ISL.

Use: The species is a good sand binder.

Stipa annua Mez

Description: Annual, cespitose. Culms erect, 10-20 cm long, 2-noded. Ligule an eciliate membrane, 0.2 mm long. Blades ± flat, 2-5 cm long, 0.5-1.5 mm wide, surface ribbed, puberulous, hairy beneath. Inflorescence a panicle, embraced at base by subtending leaf, contracted, linear, 4-8 cm long. Spikelets solitary, pedicellate, with 1 fertile florets, lanceolate, subterete, 7 mm long. Lemma apex surmounted by a ring of hairs, 0.5-1.5 mm long, 1-awned. Principal lemma awn geniculate, 20-30 mm long overall, with twisted column. Column of lemma awn 10-15 mm long, hirtellous.

Ecology: Coastal, Andean I-II: 500-2500 m. Lomas, rocky slopes.

Distribution: Peru, N-Chile.

Arequipa: native. ARE. Specimen Pennell 13139 (F) from Tingo.

Stipa pachypus Pilg.

Description: Perennial grass, 20-150 cm long. Leaves 2-6 cm long, glaucous, narrowly linear. Ligule short, membranous. Panicle contracted, 8-13 cm long. Lemma hairy, pappus-like, 1- awned, awn glabrous.

Ecology: Coastal: 0-1000 m. Lomas.

Distribution: S-Peru.

Arequipa: native. CAM, ISL.

Note: Based on morphological differences, Muñuico et al. (2024) split Stipa pachypus into Jarava disticha, distributed in the lomas of Ica and Arequipa, and J. pachypus, endemic to the lomas of S-Arequipa.

Stipa plumosa Trin.

Description: Perennial, cespitose. Culms erect to ascending, 30-300 cm long, nodes glabrous. Lateral branches ample. Ligule an eciliate membrane, 0.2-2 mm long. Blades convolute, 10-20 cm long, 0.5-4 mm wide, surface ribbed. Inflorescence a panicle, embraced at base by subtending leaf, linear, 10-25 cm long. Spikelets solitary, with 1 fertile floret, without rhachilla extension, lanceolate, subterete, 8-10 mm long. Fertile lemma lanceolate, subterete, 5- 7 mm long, surface pubescent, 1-awned. Awn curved, 30-40 mm long, with a straight or slightly twisted column, limb plumose, with 2-2.5 mm long hairs.

Ecology: Andean II: 1500-3500 m. Rocky slopes.

Distribution: Peru, Bolivia, Argentina, Chile.

Arequipa: native. ARE, CAY.

Tragus berteronianus Schult.

Description: Loosely tufted annual, culms 5-60 cm high, erect or ascending, proximal internodes of the primary branches not longer than the second internode. Blades 1-6 cm long, 2-5 mm wide. Inflorescence 2-15 cm long, compactly cylindrical, congested. Lower spikelet 2-3(33) mm long, upper glume 5-veined, with 5 longitudinal rows of spinelike projections.

Ecology: Coastal, Andean I-II, Amazonian: 0-2500 m. Desert, disturbed areas, lomas.

Distribution: Pansubtropical.

Arequipa: native. ARE, CAM, CAR, CAS, ISL.

Tragus racemosus (L.) All.

Description: Plants annual. Culms 5-40 cm. Ligules 0.5-1.3 mm, hairy. Panicles 2-11 cm long, 7-13 mm wide, rachis pubescent, branches with 3-5 (rarely 2) spikelets, proximal spikelets on the branches 3.8-6.6 mm long. Upper glume 7-veined, with 6-7 longitudinal rows of spinelike projections.

Ecology: Coastal: 0-1000 m. Lomas.

Distribution: Native to the Mediterranean, introduced worldwide.

Arequipa: introduced. CAM, CAR, ISL.

Tripogonella spicata (Nees) P.M.Peterson & Romasch.

Description: Perennial, cespitose. Culms erect, 10-30 cm long. Leaves mostly basal. Sheaths without keel, glabrous on surface. Ligule a ciliolate membrane, 0.1-0.2 mm long. Blades filiform, involute, 3-10 cm long, 1 mm wide, surface glabrous, margins sparsely pubescent. Inflorescence an unilateral raceme, 4-10 cm long, with 10-22 spikelets. Spikelet with 5-14 florets, pedicels 0.5 mm long, packing broadside to rachis, regular, 2-rowed. Glumes unequal, exceeded by the basal florets, 1-veined. Lemmas 3-veined, with a short awn (0.2-1 mm).

Ecology: Andean II: 1500-2500 m. Rocky slopes.

Distribution: Central and South America.

Arequipa: native. CAR.

Polemoniaceae

Cantua buxifolia Juss. ex Lam.

Description: Erect, branching shrub or small tree up to 5 m tall, ± pubescent throughout. Leaves simple, 0.6-4 cm long, elliptic, margins entire to ± dentate toward apex. Inflorescence of few to many flowers in lax terminal corymbs. Flowers 5.5-8 cm long, calyx irregularly dentate. Corolla usually pink to purple, sometimes white or yellow, lobes bilobate, retuse-apiculate or rounded and irregularly dentate, spread 90°.

Ecology: Andean II-III: 2500-4000 m. Shrubland, often cultivated in gardens and parks.

Distribution: Peru and Bolivia.

Arequipa: native. ARE, CAS, CAY.

Use: Cultivated in pre-Columbian times. The Incas dedicated it to the god Sun and the flower was used in ceremonies. The leaves and branches are used as a yellow dye and the wood is used for carving, becoming bright yellow when polished.

Cantua candelilla Brand

Description: Erect, branching shrub or small tree, 1.5-4 m tall, glandular villous. Leaves alternate and fascicled on short axillary shoots, quite small but occasionally up to 3 cm long, soft herbaceous, lanceolate to oblanceolate- ovate, sessile, entire to irregularly dentate, glandular-pubescent on both sides. Inflorescence composed of few to many flowers. Corolla narrowly funnelform, 5-8 cm long, reddish-orange, lobes truncate, spread ± 45°.

Ecology: Andean II: 2500-3500 m. Volcanic slopes, alluvial soils along washes and ravines.

Distribution: S-Peru.

Arequipa: native. ARE, CAY.

Cantua volcanica J.M.Porter & Prather

Description: Erect, branching shrub, 0.2-1.0 m tall, glandular throughout. Leaves alternate, usually fasciculate, linear to linear-lanceolate, 5-35 x 1-6 mm, pinnatisect. Corolla funnelform, 2-3 cm long, the tube 1-1.8 cm long, 5-6 mm in Ø, blue or bluish-violet.

Ecology: Andean II: 2000-3500 m. Shrubland, rocky slopes.

Distribution: S-Peru.

Arequipa: native. ARE.

Dayia glutinosa (Phil.) J.M.Porter

Description: Annual herb, glandular- pubescent, 5-20 cm tall. Leaves narrowly linear, 5-20 x 0.5-1 mm, once dissected. Flowers axillary, solitary or in pairs, pedicel 1-3 cm long. Corolla 6-12 mm long, white, blue or violet.

Ecology: Andean I-II: 500-3500 m. Desert, dry shrubland.

Distribution: Peru, N-Chile.

Arequipa: native. ARE, CAR, CAY.

Gilia laciniata Ruiz & Pav.

Description: Plants erect, to 20 cm tall, with a central leader, ± glandular-pubescent. Leaves 1-2-pinnately dissected, segments 1-3 pairs, rachis and segments 2 mm wide. Flowers axillary. Corolla pink to purple, throat without spots. Corolla tube short, included in calyx. Calyx in fruit 7-8 mm long.

Ecology: Andean I-III: 1000-4000 m.

Distribution: Andes from Ecuador to Tierra del Fuego.

Arequipa: native. ARE, CAR, CAY, ISL, UNI.

Gilia lomensis V.E.Grant

Description: Differs from C. laciniata in the low decumbent habit and the pinnately cleft leaves with broad segments.

Ecology: Coastal: 0-500 m. Sandy hills and lomas.

Distribution: Peru.

Arequipa: native. CAR.

Microsteris gracilis (Hook.) Greene

Description: Stem spreading to erect, 1-12 cm tall, branching, ± glandular-pubescent. Leaves lanceolate, opposite on lower stems, alternate near the top, margins entire. Flowers 5-12 mm long, axillary in 1-3. Corolla white to purple, tube yellowish-white, lobes bilobate.

Ecology: Andean III: 3500-4500 m.

Distribution: The Americas.

Arequipa: native. ARE, CAY, UNI.

Polygalaceae

Monnina macbridei Chodat

Description: Completely herbaceous annual herb, simple or sparsely branched distally. Leaves linear, lowermost spathulate, margins not revolute, acute, 15-45 x 2-5 mm, petiole 1 mm long. Inflorescence many-flowered, 2-25 cm long. Outer sepals unequal in shape and size, pale blue, glabrous, wings pale blue, keel yellow anterior yellow. Fruit glabrescent.

Ecology: Andean II: 2000-3500 m. Sandy sites on dry slopes.

Distribution: S-Peru.

Arequipa: native. ARE, UNI.

Monnina macrostachya Ruiz & Pav.

Description: Herbaceous annual, 3-45 cm tall. Stem erect, terete, slightly furrowed by decurrent leaves, branched all along the axis, stipules absent. Leaves changing along the axis, basally obovate, then elliptic-spathulate and distally lanceolate, blade acuminate, 1-5 cm long, 0.5- 2.5 cm wide, petiole 1-2.5 mm long. Inflorescence many-flowered, 1.5-30 cm long. Flowers 4-

6 mm long, blue, sometimes reddish. Fruit 2 x 2 mm, irregularly winged.

Ecology: Coastal, Andean I-II: 0-3500 m. Lomas, rocky slopes.

Distribution: Peru.

Arequipa: native. ARE, CAM, CAR, ISL, UNI.

Monnina pterocarpa Ruiz & Pav.

Description: Annual (or perennial?) herb, ca. 0.5-2.5 m tall. Stems erect, terete, simple below, loosely branched above. Leaves stipulate, petiole 1-3 mm long, blade lanceolate, 2-6.5 x 0.2-3.2 cm. Inflorescence 15-40 cm long. Flowers pink to violet, keel yellow, at least the tip. Fruit 4-7 mm in Ø, always with a broad, entire wing.

Ecology: Coastal, Andean I-II: 0-2200 m. Rocky slopes, in sand and gravel in dry areas.

Distribution: Peru.

Arequipa: native. ARE, CAR.

Monnina ramosa I.M.Johnst.

Description: Perennial herb, 15-25 cm tall, much-branched from the base. Leaves linear, 8-30 x 1-2 mm, margins revolute. Inflorescence 5-20 cm long. Flowers 3.5-5 mm long, wings bluish-white, keel yellow, at least apically. Fruit puberulous.

Ecology: Andean II: 2300-3500 m. Desert.

Distribution: S-Peru.

Arequipa: native. ARE, CAY.

Monnina salicifolia Ruiz & Pav.

Description: Shrub, up to 2.5 m tall. Leaves usually elliptic, apex obtuse, sometimes acute, 7 x 2.5 cm, glabrate in age, with 4-5 lateral veins. Inflorescence conical, acute, simple. Flowers purple, keel yellow.

Ecology: Andean II-III: 1500->4500 m. Disturbed areas, grasslands, riversides, rocky slopes, semi-deciduous forests, shrublands.

Distribution: Colombia, Ecuador, Peru, Bolivia.

Arequipa: native. CAR.

Polygonaceae

Chorizanthe commissuralis J.Rémy

Description: Erect herb, annual, branched, 5-25 cm, with distinctive nodes, fragile, pilose. Basal leaves lanceolate, cauline leaves lineal, apex uncinate. Inflorescence much-branched. Flowers white, petals as long as sepals.

Ecology: Andean II: 2000-2500 m. Rocky slopes, shrubland.

Distribution: S-Peru, Chile, Argentina.

Arequipa: native. ARE. No specimen, some convincing observations.

Muehlenbeckia fruticulosa (Walp.) Standl.

Description: Erect, deciduous shrub, up to 1 m tall, densely branched, glabrous. Leaves mostly conspicuously longer than broad, mostly cuneate at the base, never cordate or hastate-lobate. Flowers in leaf axils, short pedicellate, white.

Ecology: Andean II: 2500-3000 m. Riversides, rocky slopes.

Distribution: Peru, Bolivia, N-Chile.

Arequipa: native. ARE, UNI.

Note: Could be conspecific to M. volcanica.

Muehlenbeckia hastulata (Sm.) I.M.Johnst.

Description: Scandent shrub. Leaves ± hastate-lobate, not cordate at base. Racemes simple. Flowers in simple or paniculate racemes.

Ecology: Andean II-III: 2500-4000 m. Disturbed areas, dry valleys, riversides, rocky slopes.

Distribution: Peru, Chile, Argentina.

Arequipa: native. ARE, CAY, UNI.

Muehlenbeckia tamnifolia (Kunth) Meisn.

Description: Large scandent shrub, vine-like. Leaves elliptic, short petiolate and cordate at base, margins entire, apex cuspidate. Inflorescence an axillary panicle. Flowers pinkish-white.

Ecology: Andean II-III: 1500-4000 m. Disturbed areas, elfin forests, rocky slopes, shrublands.

Distribution: Central and western South America.

Arequipa: native. ARE.

Persicaria hydropiperoides (Michx.) Small

Description: Perennial amphibious or terrestrial herb, 0.2-1 m tall. Leaves oblong and smooth on the edges. Stems distinctly jointed. rooting at the joints. Leaves alternate. Flowers small and tightly clustered, white or pink. Terrestrial to completely submersed in water with only flowers visible above the surface.

Ecology: Coastal, Andean I-III, Amazonian: 0-4000 m. Disturbed areas, emergent, forests, riversides.

Distribution: North and South America.

Arequipa: native. ARE, CAR.

Polygonum aviculare L.

Description: Prostrate and much-branched annual herb, very resistant to trampling. Leaves elliptical to lanceolate- linear, 15-40 x 2-12 mm, green to grayish-green, sometimes glaucous, entire, obscurely nerved, top acute to rounded, base gradually narrowing into a short petiole or subsessile. Flowers small and white.

Ecology: Coastal, Andean I-III: 2000-4000 m. Disturbed areas, forests.

Distribution: Native to Eurasia, introduced worldwide.

Arequipa: introduced. ARE, CAM.

Rumex acetosella L.

Description: Perennial herb, dioecious. Stems usually numerous from rhizome, erect or ascending, 15-40 cm tall, slender, finely grooved, branched above middle. Basal leaves hastate, 2-4 cm × 3-6(-10) mm, glabrous, cauline leaves smaller upward. Petiole short or in upper cauline leaves nearly absent. Flowers unisexual. Pedicel 2-2.5 mm, articulate near base of tepals. Achenes brown, shiny, broadly ovoid, trigonous, 1-1.5 mm.

Ecology: Andean II-III: 2000->4500 m. Disturbed areas, rocky slopes, shrublands.

Distribution: Native to Europa, introduced worldwide.

Arequipa: introduced, adventive. Not specified by province. No voucher, mentioned as adventive herb by Montesinos & Zegarra (2019).

Rumex conglomeratus Murray

Description: Perennial herb, 30-80 cm high, mostly erect, often branching from below the middle. Basal and lowermost stem leaves petiolate, petiole usually as long as lamina, lamina oblong to ovate, to 6 x 3 cm, flat, apex rounded to obtuse, base slightly cordate or truncate. Flower whorls distant and each whorl mostly subtended by a leaf, whorls many-flowered. Perigon lobes entire.

Ecology: Andean II: 2000-3000 m. Disturbed areas, riversides.

Distribution: Native to Eurasia, introduced worldwide.

Arequipa: introduced. ARE, ISL.

Rumex crispus L.

Description: Robust perennial herb, 50-150 cm high, erect. Basal and lower stem leaves petiolate, petiole shorter than lamina, lamina lanceolate, to 30 x 6 cm, undulate, apex acute, base cuneate. Inflorescence branches usually erect, whorls close together, rarely lowermost subtended by a leaf.

Ecology: Coastal, Andean I-II: 1000-3500 m. Disturbed areas.

Distribution: Native to Eurasia, introduced worldwide.

Arequipa: introduced. ARE, CAY, ISL.

Rumex cuneifolius Campd.

Description: Perennial, stoloniferous herb, <25 cm tall. Leaves obovate or obovate-elliptic, margins entire or crisped. Inflorescence paniculate, pedicels articulated and swollen.

Ecology: Andean I-III: 1000-4000 m. Riversides.

Distribution: Native to South America, introduced to North America.

Arequipa: native. ARE, CAY, UNI.

Rumex obtusifolius L.

Description: Perennial herb, 0.6-1 m high, branching from the middle. Basal leaves with petioles usually longer than lamina, stem leaves with shorter petioles. Lamina oblong-ovate, 30 x 10 cm, flat, apex obtuse or broad-acuminate, base cordate. Inflorescence branched, branches forming angle of <60° with 1. order stem. Bracts only in lower part of the inflorescence. Flower whorls ± close together, without subtending leaf, 15-40-flowered. Fruiting valves 3.5-6 mm long, with 2-3 teeth in the lower half.

Ecology: Andean II-III: 2000-4000 m. Disturbed areas, riversides.

Distribution: Africa, Europe and Asia, introduced worldwide.

Arequipa: introduced. CAR.

Rumex pulcher L.

Description: Perennial herb, stems erect, often flexuous in distal part, branched in distal 2/3, 20-60 cm. Leaf blade oblong to ovate-oblong, contracted near middle, 4-15 × 3-5 cm, less than 4x as long as wide, base normally truncate or weakly cordate, occasionally rounded, margins entire, flat or undulate, rarely slightly crisped, apex obtuse or subacute. Inflorescences terminal, occupying distal 2/3 of stem or more, usually lax and interrupted, broadly paniculate, branches forming angle of 60-90° with 1st-order stem. Flowers 10-20 in rather dense whorls. Perigone lobes distinctly dentate.

Ecology: Coastal, Andean I: 0-1500 m. Waste places, roadsides, shores, fields, meadows.

Distribution: Native to Europe, introduced worldwide.

Arequipa: introduced. ARE, ISL.

Pontederiaceae

Pontederia crassipes Mart.

Description: Herb, usually free-floating but occasionally rooted in mud, 4-40 cm. Leaves 4-8, clustered on short rhizomes, erect, very variable, in free-floating plants petiole 5-10 cm, markedly swollen below, lamina 4-9 cm wide, ovate to circular, in older plants rooted in mud, the petiole is longer and not swollen. Inflorescence an 8-flowered showy spike, raised above leaves. Perianth tube 20 mm, curved, limb 5-7 cm in Ø, slightly 2-lipped, pale mauve with the central lobe broadest with large blue area and central yellow spot.

Ecology: Coastal, Amazonian: 0-1000 m. Ponds, still water.

Distribution: Native to tropical South America, naturalized worldwide.

Arequipa: native. ARE, ISL.

Portulacaceae

Portulaca nivea Poelln.

Description: Decumbent perennial herb, much-branched, soft white-pilose throughout. The minute, crowded leaves completely hidden by stem hairs. Flowers magenta.

Ecology: Coastal: 0-1000 m. Lomas.

Distribution: S-Peru, probably also in N-Chile.

Arequipa: native. CAM, CAR, ISL.

Portulaca oleracea L.

Description: Prostrate annual herb, much-branched. Stems reddish or purple, ± glabrous. Leaves sessile, flat, linear-oblong to spathulate, glabrous. Flowers sessile, 3 mm across, terminal, in groups of 1 to several, surrounded by a cluster of crowded leaves, petals yellow.

Ecology: Coastal, Andean I-II: 0-2800 m. Disturbed places, roadsides.

Distribution: Pantropical.

Arequipa: native. ARE, CAR, ISL, UNI.

Use: Edible leaves.

Portulaca perennis R.E.Fr.

Description: Perennial, fleshy, much-branched at the base, with often numerous stems. Stems thick, short pubescence to glabrous. Leaves alternate, terete, 5-10 mm long, linear-spatulate, many of them persistent, the shorter involucral ones scarcely exceeding the purple flowers or the capsules. Capsule dehiscent at the base, seeds black, radiantly stellate-tuberculate.

Ecology: Andean II: 1500-2500 m. Disturbed areas, rocky slopes, shrublands.

Distribution: S-Peru, Bolivia, NW-Argentina.

Arequipa: native. ARE, CAM, CAR.

Portulaca pilosa L.

Description: Ascending-erect annual with long soft hairs in leaf-axils, stem reddish. Leaves 6-20 x 1-4 mm, narrowly linear, apex acute. Flowers terminal, solitary or 2-4, sessile. Sepals 2-4 x 1-2.5 mm. Petals 4-6, bright yellow or pink, obovate. Capsules 2-3 mm across, globose.

Ecology: Coastal, Andean I-III: 0-4000 m. Disturbed areas, forests, shrublands, riversides.

Distribution: Native of tropical America, now pantropical.

Arequipa: native. ARE, CAR, CAS, ISL.

Use: Medicinal plant.

Note: Doubtful if P. lanuginosa Kunth, mentioned for Arequipa, is a separate species or a synonym of P. macbridei or P. pilosa.

Portulaca tingoensis J.F.Macbr.

Description: Small herb, simple or branching near the base. Branches strongly enlarged toward the tips, 2-8 cm long. Leaves alternate, caducous, flat, oblong-linear, 3-8 x 1-2 mm, stipular hairs not many, ± the same length as the leaf. Flowers solitary, pinkish-purple,

densely involved in hairs 5-7 mm long. Capsules sometimes axillary, solitary. Seeds iridescent, bluntly tuberculate, the scales radiate at the base, 0.6 mm broad.

Ecology: Andean II: 2000-2500 m. Rocky slopes.

Distribution: S-Peru, W-Argentina, probably also in Bolivia and N-Chile.

Arequipa: native. ARE.

Potamogetonaceae

Stuckenia striata (Ruiz & Pav.) Holub

Description: Submersed aquatic plant, abundantly branched, especially on upper portions. Leaves acute, apiculate, cuspidate, rarely rounded, 3-5-veined, 5-21 cm x 0.4-5.1(-8.5) mm, basal sheaths not swollen. Spikes cylindrical, 1.3-4.5 cm long, pedunculate. Fruit distinctly beaked.

Ecology: Coastal, Andean I-III: 0-4000 m. Waters of alkaline rivers, canals and ponds.

Distribution: NW-USA to S-Argentina.

Arequipa: native. CAM, CAS, ISL.

Zannichellia palustris L.

Description: Submersed rooting aquatic plant. Stem up to 50 cm long, filiform. Leaves 1-8 cm long and up to 1.5 mm wide, acuminate, opposite or whorled. Flowers axillary, without calyx and corolla, 1 male and 1-6 female flowers surrounded by a common spathe. Fruit 2-4 mm long, flat, sickle-shaped, style about half as long as the fruit.

Ecology: Andean II: 2000-2500 m. Submersed.

Distribution: Cosmopolitan.

Arequipa: native. ARE.

Primulaceae

Lysimachia arvensis (L.) U.Manns & Anderb.

Description: Prostrate, annual herb, up to 50 cm long. Stem 4-angled. Leaves soft green, 4-25 mm x 3-15 mm, opposite. Flowers 5-merous, reddish-orange or purplish-blue. Capsule rounded, 3-6 mm x and 3-5 mm, pale brown when ripe.

Ecology: Coastal, Andean I-III: 0-4000 m. Disturbed areas.

Distribution: Native to Eurasia, introduced worldwide.

Arequipa: introduced. ARE, CAR, ISL.

Samolus parviflorus Raf.

Description: Plants light green, not stoloniferous, 10-80 cm tall. Stems erect or ascending. Leaves basal and cauline, sessile or petiolate. Blade obovate to spatulate or elliptic, 1.5-15 cm, base decurrent, cuneate, apex rounded to obtuse. Inflorescences terminal and axillary, racemose to paniculate. Pedicels spreading or ascending, 1.5-14 mm, glabrous, leafy bract at mid-length. Corolla white, 1.2-3 mm, lobes oblong, longer than tube, apex rounded or emarginate. Capsules 2-3 mm.

Ecology: Coastal: 0-500 m. Disturbed areas, humid places.

Distribution: The Americas and Japan.

Arequipa: native. ISL.

Note: Some authors treat it as a synonym of Samolus valerandi L.

Ranunculaceae

Clematis millefoliolata Eichler

Description: Vine. Gray tomentose, or the 6-angled, striate stem finally glabrescent. leaves 10-12 cm long and broad, the ultimate segments only 2-4 mm long, linear-lanceolate. Panicles scarcely equaling the leaves, 3-7- flowered. Corolla yellow.

Ecology: Andean II-III: 3000-4000 m.

Distribution: Peru, Bolivia.

Arequipa: native. ARE.

Clematis peruviana DC.

Description: Stems terete, ± purplish, pubescent. Leaves 2-pinnatifid, segments sometimes ternate, sometimes 3- parted, ovate at base, acutely incised, dentate above, villous beneath. Peduncles axillary, equaling the leaves, sparsely branched, few-flowered. Flowers whitish-yellow.

Ecology: Andean II-III: 2500-4000 m.

Distribution: Peru.

Arequipa: native. ARE, UNI.

Halerpestes uniflora (Phil. ex Reiche) Emadzade, Lehnebach, P.J.Lockh. & Hörandl

Description: Perennial, stoloniferous aquatic plant. Leaves ovate, 3-8 x 3-6 mm, margins entire, almost veinless. Flowers solitary, axillary, pedicel 2.5-4 cm long, sepals 3, petals 7 and ca. 100 carpels (ca. 100), corolla yellow.

Ecology: Andean III: 4000->4500 m. Bofedales, lakes.

Distribution: Peru, Bolivia, Argentina and Chile.

Arequipa: native. ARE, CAY. No specimen, several observations.

Ranunculus trichophyllus Chaix

Description: Perennial submersed aquatic herb, without laminate leaves. Capillary leaves rarely more than 4 cm long. Flowers emergent, white. Pedicel in fruit usually less than 40 mm.

Ecology: Andean III: 3500-4500 m.

Distribution: Native to the Northern Hemisphere, introduced elsewhere.

Arequipa: introduced. CAY.

Rhamnaceae

Colletia spinosissima J.F.Gmel.

Description: Much-branched shrub, to 2 m tall, nearly leafless, branches and leaves decussately opposite. Leaves minute, margins serrate. Flowers whitish-red, pendulous. Fruit capsular.

Ecology: Andean II-III: 1500-4000 m. Disturbed areas, rocky slopes.

Distribution: Ecuador, Peru, Bolivia, Chile, Uruguay, Argentina.

Arequipa: native. ARE, CAY, UNI.

Use: The bark and branches are used as soap or shampoo. The wood burns even when fresh and is used in clay bread ovens.

Scutia spicata (Humb. & Bonpl. ex Schult.) Weberb.

Description: Shrub, stem green with perpendicular, stout spines, the spines 2.5-6 cm long. Leaves few, ovate, 2.5 cm long, deciduous. Flowers sessile, minute, yellowish. Fruit globose, drupiform.

Ecology: Coastal, Andean I-II: 0-3000 m. Desert, lomas.

Distribution: Ecuador, Peru.

Arequipa: native. CAM, CAR.

Rosaceae

Alchemilla diplophylla Diels

Description: Rhizomatous amphibious herb, floating on water or growing in damp places, 1-2 cm high. Leaves 3-8 x 3-5 mm, folded like an accordion, petiole 2 cm long, margins of the young leaves sparsely pilose. Flowers almost glabrous.

Ecology: Andean III: 4000->4500 m. Bogs, moors.

Distribution: Peru and Bolivia.

Arequipa: native. ARE, CAY, CON.

Alchemilla pinnata Ruiz & Pav.

Description: Rosulate herb, producing long stolons, usually softly gray-villous, stem reddish. Leaves 2-pinnatisect, 1.5-6 cm long, pale green. Flowers solitary or in few-flowered glomeruli, axillary and terminal. Petals greenish-yellow, stigma and stamens yellow.

Ecology: Andean II-III: 2000->4500 m. Cloud forests, disturbed areas, grasslands, shrublands.

Distribution: Central and western South America.

Arequipa: native. ARE, CAY.

Kageneckia lanceolata Ruiz & Pav.

Description: Shrub or small tree. Leaves oblanceolate, margins serrate, coriaceous. Flowers 5-meorus, pinkish- white. Fruits woody, star shaped.

Ecology: Andean II-III: 2000-4000 m. Disturbed areas, dry valleys, forests, rocky slopes, shrublands.

Distribution: Peru, Bolivia, NW-Argentina.

Arequipa: native. ARE, UNI.

Polylepis fjeldsaoi T.Boza & M.Kessler

Description: Trees 2-10 m tall. Leaves having just one lateral leaflet pair, margins crenate. Lower leaflet surfaces densely pilose, villous or tomentose without pannose hairs.

Ecology: Andean III: 3300-4800 m. Arid mountains of Caravelí.

Distribution: S-Peru.

Arequipa: native. CAR.

Polylepis microphylla (Wedd.) Bitter

Description: Shrubs and trees 1.5-8 m tall. Leaves strongly congested at the branch tips, imparipinnate with 3-6 pairs of lateral leaflets, obtrullate in outline.

Ecology: Andean II-III: 3150-4550 m. Disturbed areas, shrublands.

Distribution: Ecuador, Peru.

Arequipa: native. Not specified by province. No voucher. Mentioned for Ica and Arequipa by Boza Espinoza & Kessler (2022).

Polylepis rugulosa Bitter

Description: Shrubby tree up to 8 m tall, growing at high altitude. Reddish-brown bark. Leaves imparipinnate with 1 pair of lateral leaflets. Upper leaflet surfaces strongly rugose, lower surfaces densely tomentose with a dense underlying layer of very short, white pannose hairs. Leaves and inflorescences virtually lack yellow glandular hairs.

Ecology: Andean III: 3400-4500 m. Dry mountains, forests.

Distribution: S-Peru, N-Chile.

Arequipa: native. ARE, CAY, CON, UNI.

Use: Excellent ornamental value, often grown in parks and gardens. Used for timber.

Note: Molecular analyses suggest that hybridization probably occurs between all species of the genus as long as they grow in geographic proximity (Kessler & Schmidt-Lebuhn, 2006). Kerr (2004) also found evidence of hybridization between Polylepis and its sister genus Acaena. Using the genus concept consistently, most Polylepis species should be considered as a single species.

Rubus ulmifolius Schott

Description: Spreading semi-deciduous shrub forming dense thickets. Main stems strongly angled, prickles present, glaucous when young. Leaves palmately compound, with 3-5 leaflets, leaflet similar to leaves of Ulmus glabra. Flowers usually pale pink. Fruits ripening red, maturing black.

Ecology: Coastal, Andean I-II: 0-3000 m. Disturbed areas, hedges, shrubland.

Distribution: Cosmopolitan.

Arequipa: introduced. ARE, CAY.

Rubus urticifolius Poir.

Description: Branchlets, petioles, and inflorescence with prickles and densely hirsute, little if at all glandular. Plants ± densely reddish-setose. Flowers white, berries purple to black.

Ecology: Coastal, Andean I-III, Amazonian: 0-4000 m. Cloud forests, disturbed areas, lomas, riversides.

Distribution: Central and South America.

Arequipa: native. CAR.

Sanguisorba minor Scop.

Description: Herbaceous, perennial, 25-60 cm tall. Stems simple or branched above and sparsely pilose. Lower leaves cauline and pinnate with the stipules adherent to the petiole, 10-15 cm long. Leaflets crenate-toothed, alternate, pinnately compound and reduced upward. Inflorescence a dense spike, 10-25 mm long. Each flower subtended by a papery bract, greenish or white to red-purple, petals absent.

Ecology: Andean II-III: 2500-4000 m. Disturbed areas.

Distribution: Native to Europe, introduced worldwide.

Arequipa: introduced, adventive. CAY. No specimen, one observation and some mentions. Cultivated in the Andes.

Use: Medicinal plant. Edible leaves.

Tetraglochin cristata (Britton) Rothm.

Description: Branched shrub with spiniform leaves, sometimes with 1 or 2 leaflets. Fruit winged, cristate.

Ecology: Andean II-III: 3000-4500 m. Disturbed areas, grasslands, rocky slopes.

Distribution: S-Peru, N-Bolivia.

Arequipa: native. ARE, CAY, UNI.

Rubiaceae

Arcytophyllum thymifolium (Ruiz & Pav.) Standl.

Description: Small shrub, 20-30 cm tall, branched from the base. Leaves ericaceous, opposite and decussate, linear-elliptic, sessile, 8-10 x 2-2.5 mm, margins entire, apex acute, blade coriaceous, glabrous. Inflorescence terminal, clusters of several sessile flowers. Flowers 4-merous, 7 mm in Ø, corolla white.

Ecology: Coastal, Andean I-III: 0-4000 m. Disturbed areas, dry valleys, grasslands, lomas, rocky slopes, shrublands.

Distribution: Colombia, Ecuador, Peru.

Arequipa: native. ARE, CAM, CAR.

Galium aparine L.

Description: Slender and branched, annual herb, stems up to 1 m long. Stem square. Leaves sessile in whorls of 4-8 at the nodes, 1-3 cm long, simple, narrowly elliptic to lanceolate, mucronated, margins and stem retrosely scabrous. Flowers axillary, in cymes of 2-5, white. Fruit green, dry, uncinate-hispid, 0.8 mm long.

Ecology: Coastal, Andean I-III, Amazonian: 0-4000 m. Shrublands, hedges, disturbed places.

Distribution: Native to Eurasia, introduced worldwide.

Arequipa: introduced. ARE, CAS, CAY, UNI.

Note: Some specimens observed by the author in Arequipa have the characteristics of Galium spurium L., similar to G. aparine but stems max. 40 cm long, inflorescence 3-9-flowered, small barbs on stems, also native to Eurasia and Africa. This species is not mentioned for Peru but it is likely to have been introduced.

Galium arequipicum Dempster

Description: Perennial herb, decumbent, stem up to 35 cm long, stem and leaves with long scattered hairs. Leaves 4, ovate, to 8 mm long, variously 1-3-nerved. Inflorescence axillary, peduncle as long as leaves. Flowers not always involucrate, yellow or greenish.

Ecology: Coastal: 0-500 m.

Distribution: S-Peru (Arequipa).

Arequipa: native. CAM, CAR, ISL.

Note: An insolated and morphologically uniform clade, similar to Galium corymbosum.

Galium corymbosum Ruiz & Pav.

Description: Densely branched, perennial herb, prostrate, stems 25 cm long. Stem and leaves glabrate to hispid. Leaves 2-10 mm long, linear-lanceolate, acuminate, margins entire. Flowers yellowish. Fruits green or white.

Ecology: Andean II-III: 2000-4500 m.

Distribution: Venezuela to N-Chile and NW-Argentina.

Arequipa: native. ARE, CAR, CAY, UNI.

Galium hypocarpium (L.) Endl. ex Griseb.

Description: Weak perennial herb with stems up to 1 m, the branches retrose hispidulous. Leaves in whorls of 4, oblong elliptic, 5-15 mm long. flowers pedicellate, corolla white. Fruit fleshy, orange-red.

Ecology: Andean I-III: 500-4500 m.

Distribution: Central and South America.

Arequipa: native. ARE, CAM, CAR, UNI.

Randia rotundifolia Ruiz & Pav.

Description: Woody tree or shrub up to 3 m tall. Apex of lateral branches armed with thorns. Leaves obovate, 1-2 cm long, apex obtuse, base obtuse to cuneate. Flowers terminal, solitary, sessile, corolla white. Fruits orbicular, yellowish, pulp black.

Ecology: Coastal, Andean I-II: 0-3000 m. Rocky slopes. lomas.

Distribution: Peru.

Arequipa: native. CAR, ISL.

Richardia lomensis (K.Krause) Standl.

Description: Low annual herb, hispid throughout. Leaves ovate to ovate-oblong. Inflorescence in terminal heads few- flowered. Flowers usually 6-merous, corolla white, tubular, lobes acute.

Ecology: Coastal: 0-500 m. Desert, lomas.

Distribution: S-Peru.

Arequipa: native. ISL.

Ruppiaceae

Ruppia maritima L.

Description: Aquatic plant with floating leaves just below the water surface. Stems elongated, densely branched. Leaves 2-10 cm × ca. 0.5 mm, with conspicuous midvein, apex acuminate or acute, sheaths 2-10 × ca. 0.4 mm. Spikes 2-flowered, 2-4 cm, peduncles filiform, shortly accrescent after flowering. Anthers elliptic. Carpels 4-6. Fruitlets obliquely ovoid, 2 × 1.5 mm, beak 0.2 mm, pedicel 0.5-1.7 cm.

Ecology: Coastal, Andean I: 0-2000 m. Submersed.

Distribution: Cosmopolitan.

Arequipa: native, doubtful. Not specified by province. No voucher. Mentioned for Ica and Arequipa, an occurrence in Arequipa is likely.

Salicaceae

Salix humboldtiana Willd.

Description: Tree up to 25 m tall. Leaves deciduous, linear, acuminate, serrulate. Catkins appearing with the leaves. Fruit a dark ovoid capsule, 4-5 mm long, with numerous very small and long hairy seeds.

Ecology: Coastal, Andean I-III, Amazonian: 0-4000 m. Riversides, disturbed areas, forests.

Distribution: Mexico to Argentina.

Arequipa: native. ARE, CAM, CAR, ISL.

Use: The bark contains salicin and is used as a flu remedy and antirheumatic like aspirin. Charcoal is used in folk medicine for flatulence, nausea, and indigestion. The light wood is used for wickerwork and for making doors, windows and fruit crates.

Santalaceae

Dendrophthora mesembryanthemifolia Griseb.

Description: Parasitic shrub. Stems succulent, reddish-brown to green, articulate. Leaves opposite, reddish-green, oblanceolate, succulent. Flowers in axillary spikes, 1 mm long. Berries white.

Ecology: Andean II-III: 2000-4000 m.

Distribution: Ecuador, Peru, Bolivia.

Arequipa: native. CAY.

Sapindaceae

Cardiospermum corindum L.

Description: Annual herbaceous climber, stem villous. Leaves pinnate to 2-ternate, leaflets up to c. 4 × 2 cm, terminal leaflets usually distinctly petiolate, lowest pair ternate, bright green above, gray-green below, finely velvety, margins scalloped to lobed. Flowers white to creamy yellow, 4-6 mm long, in axillary branched, many-flowered heads, peduncle 6 cm long, with paired coiled tendrils below the flowers. Fruit a membranous, inflated capsule, 2-3.5 cm long, 3-angled. Seed Ø 5 mm, with a small white reniform scar (hilum).

Ecology: Andean II: 1500-3000 m. Riversides, rocky slopes, shrublands.

Distribution: Tropical and subtropical America, Africa and Asia.

Arequipa: native. CAM, CAR, CAY, ISL.

Note: Similar to C. halicacabum and some authors synonymize it with C. halicacabum.

Cardiospermum halicacabum L.

Description: Annual climber. Stems ± woody, with tendrils, glabrous or puberulent. Leaves 2-ternate, puberulent, margins coarsely serrate, leaflets subsessile, terminal leaflets usually decurrently petiolate if at all. Flowers in axillary heads, usually 3-flowered, white with a yellowish center. Fruit a membranous, inflated capsule, green, drying to brown, more than 2 cm long. Seeds round and black with a broad kidney-shaped scar (hilum).

Ecology: Coastal, Andean I-II, Amazonian: 0-3500 m. Disturbed areas, riversides, rocky slopes, shrublands.

Distribution: Native to South America and Africa, introduced worldwide. Species of humid tropical and subtropical areas.

Arequipa: native. CAS.

Use: Leaves and young shoots are cooked and eaten like spinach.

Sapindus saponaria L.

Description: Small tree, up to 9-10 m tall, deciduous, rounded, symmetrical crown. Leaves pinnately compound, 30 cm long, 6-13 leaflets, each 10 cm long, leaf rachis winged. Inflorescence a triangular-shaped panicle, 30 cm long, at the tip of a current year's shoot, flowers small, creamy-white. Fruit weird, orange-brown, globular, leathery, persistent, 2 cm across, contains a single black seed.

Ecology: Coastal, Andean I-II, Amazonian: 0-200 m. Disturbed areas, semi-deciduous forests, lomas.

Distribution: Native to Central and South America, introduced to tropical Africa and SE-Asia.

Arequipa: native. CAR.

Serjania sp.

Description: Vine, climbing with tendrils. Leaves trifoliate, leaflets conspicuously veined, margins serrate. Fruits schizocarps with broadly winged mericarps.

Ecology: Coastal: 0-500 m. Lomas.

Distribution: Genus native to Central and South America.

Arequipa: introduced. ISL. No specimen but a convincing observation from the Lomas de Matarani in iNaturalist (2024).

Schoepfiaceae

Quinchamalium chilense Molina

Description: Herbaceous hemiparasite, perennial, branched from the base. Roots with small adventive roots and haustoria (organs to connect to other plants). Leaves linear, fleshy, green to reddish. Inflorescence terminal, clusters of more than 20 dark orange flowers, 5-merous.

Ecology: Coastal, Andean I-III: 0-4000 m. Disturbed areas, lomas, shrublands.

Distribution: Peru, Bolivia, Argentina and Chile.

Arequipa: native. ARE, CAM, CAR, CAY.

Note: Lopez et al. (2015) synonymized all Quinchamalium under Q. chilense.

Scrophulariaceae

Alonsoa meridionalis (L.f.) Kuntze

Description: Herbs or subshrubs, usually much-branched, stems and branches usually 4-angled and striate-ribbed, glabrous to subglabrous. Leaves opposite, usually petiolate, ovate-lanceolate, at least some of the leaves >2 cm, margins densely serrate to dentate. Inflorescence a terminal raceme, bracteate, bracts alternate. Corolla orange to dark red, lips very unequal.

Ecology: Coastal, Andean I-III: 0-4000 m. Forests, lomas, rocky slopes.

Distribution: C-Mexico to Bolivia, C-Chile.

Arequipa: native. ARE, CAR, CAY, ISL, UNI.

Buddleja coriacea J.Rémy

Description: Tree or shrub with a dense crown and irregular trunk, branching almost at ground level. Leaves entire, 1-4 × 0.5-1.5 cm, coriaceous, glabrous above, canescent beneath. Inflorescence with 3-8 pairs of head-like cymes, in axils of bracts, peduncles 0.5-3 cm long, cymes with 8-12 sessile flowers. Corolla deep yellow turning orange, fragrant.

Ecology: Andean II-III: 3000-4500 m. Disturbed areas, shrubland.

Distribution: Peru and Bolivia.

Arequipa: native. ARE, CAY.

Use: The young shoots are used for fodder and the wood is used for firewood and construction.

Buddleja incana Ruiz & Pav.

Description: Dioecious tree or shrub, 4-12 m tall, trunk <50 cm at base, bark brownish and furrowed, rounded crown. Branches subquadrangular, tomentose. Leaves lanceolate-oblong, 7-15 x 1-3 cm, subcoriaceous to coriaceous, above glabrescent and often rugose, below white or yellowish tomentose, petiole 1-2 cm long. Inflorescence a 15- 40-flowered panicle with 2-3 orders of branches. Corolla yellow turning orange.

Ecology: Andean II-III: 3000-4500 m. Disturbed areas, grasslands, riversides, shrublands.

Distribution: Colombia, Ecuador, Peru, Bolivia.

Arequipa: native. CAY.

Use: Used in folk medicine to treat warts and as firewood.

Limosella aquatica L.

Description: Stoloniferous annual, 3-5 cm tall, fascicled, glabrous, rooting from nodes. Leaves basal, petiole 1-4(-9) cm, blade broadly linear to narrowly spatulate, 3-15 mm, fleshy, margins entire, apex obtuse. Flowers 3-10, white, pale lilac, or reddish, 2-3.5 mm.

Ecology: Andean I-III: 1700->4500 m. Along streams, wet grassland, sometimes submerged or floating.

Distribution: Cosmopolitan.

Arequipa: native. ARE. No specimen, a not curated observation from San Juan de Tarucani (iNaturalist, 2024). Observations from Puno and Moquegua near the border to Arequipa.

Note: Perhaps future studies will show that L. australis and L. aquatica can be considered as single species.

Limosella australis R.Br.

Description: Semi-aquatic small perennial herb, often stoloniferous. Leaves erect to ascending, linear, 1-6 cm long, cylindrical, obtuse, lamina not differentiated. Pedicels slender, erect in flower, sometimes recurved and usually 8-25 mm long in fruit. Corolla 2 mm long, white, lobes often purple externally, stamens attached at 1 level.

Ecology: Andean III: 3500-4500 m. Seasonally inundate areas, muddy soils.

Distribution: Cosmopolitan.

Arequipa: native, doubtful. ARE, CAY.

Myoporum laetum G.Forst.

Description: Evergreen shrub or small tree to 15 m. Leaves ovate, heavily spotted with translucent dots. Flowers borne in lateral 2-6-flowered clusters, white with purple dots. Fruit pink.

Ecology: Coastal: 0-500 m. Planted for shelter or hedging in sandy areas (beaches, dunes).

Distribution: Native to New Zealand, planted and naturalized in coastal areas worldwide.

Arequipa: introduced. ARE, ISL. Cultivated and probably naturalized in coastal areas.

Solanaceae

Cestrum auriculatum L'Hér.

Description: Branched shrub, glabrous except for the often finely puberulent peduncles. Leaves paired, main leaf oblong-ovate, 10 cm long, acuminate, narrowed at base, secondary leaf small lunate, auriculate. Inflorescence a terminal or axillary corymbose panicle. Flowers nearly sessile, 2.5 cm long, yellowish-green with blackish or reddish- brown overtones.

Ecology: Coastal, Andean I-II: 0-2000 m. Disturbed areas, lomas.

Distribution: Native to Ecuador, Peru and N-Chile.

Arequipa: native. ARE, CAR, UNI.

Use: Ornamental shrub.

Cestrum tomentosum L.f.

Description: Shrub with whitish tomentose branches. Leaves elliptic, underside whitish tomentose. Flowers green. Fruits green.

Ecology: Andean I-II: 700-3500 m. Lomas, shrublands.

Distribution: Central and South America, from Mexico to Peru.

Arequipa: native. CAR.

Datura stramonium L.

Description: Bushy annual glabrous weed with ovate lobed leaves, lobes dentate. Flowers 7-10 cm long, tubular, usually white, sometimes purplish-tinged. Fruit an erect capsule with more than 100 slender spines of various lengths, randomly distributed, or almost no spines. Seeds black or gray.

Ecology: Coastal, Andean I-II, Amazonian: 0-3000 m. Disturbed areas.

Distribution: Native to Central America introduced and naturalized worldwide.

Arequipa: introduced. ARE, CAR, CAY, ISL, UNI.

Use: The root is smoked with tobacco as a remedy for asthma. The whole plant is toxic due to tropane alkaloids, a powerful hallucinogen and deliriant.

Dunalia spinosa (Meyen) Dammer

Description: Much-branched, spiny shrub. Leaves oblanceolate, margins entire, apex obtuse, base attenuate. Flowers solitary, axillary, tubular, 2-3 cm long, corolla purple, stamens not exserted. Fruit green fruit, red-orange when ripe.

Ecology: Andean II-III: 2500-4500 m. Shrublands.

Distribution: S-Peru, N-Chile.

Arequipa: native. ARE, CAS, CAY, CON, UNI.

Use: Medicinal plant.

Exodeconus flavus (I.M.Johnst.) Axelius & D'Arcy

Description: Prostrate herb, stem hispid. Leaves broadly cordate to subrotate or pseudopeltate with ± entire margins. Flowers 2.5-3.5 cm long, yellow, axillary, shortly petiolate.

Ecology: Andean II: 1500-2500 m.

Distribution: S-Peru, N-Chile.

Arequipa: native. ARE.

Exodeconus integrifolius (Phil.) Axelius

Description: Prostrate herb, hairy. Leaves broadly ovate to suborbicular, margins ± entire. Flowers 1.2-1.4 cm long, white. Calyx in fruit ribbed by protruding transverse veins.

Ecology: Andean I-II: 1600-2700 m.

Distribution: S-Peru, NW-Argentina, N-Chile.

Arequipa: native. ARE.

Exodeconus pusillus (Bitter) Axelius

Description: Leaves lanceolate to rhomboid. Flowers 1.5-2.0 cm long, white to pale violet. Calyx in fruit thin, smooth.

Ecology: Coastal, Andean I-II: 100-2500 m. Desert, xeric shrublands.

Distribution: S-Peru, N-Chile.

Arequipa: native. ARE.

Fabiana stephanii Hunz. & Barboza

Description: Shrub up to 1.80 m tall with ascending branches and glandular pubescence, sometimes glabrescent, very resinous. Leaves 1-2.5 mm, alternate, sessile, linear to cylindrical, succulent. Flowers solitary or 2-3 flowers, with 1-2 mm pedicels. Calyx 3-4.5 × 2.3-2.5 mm, with cylindrical tube at maturity and finely dentate. Corolla 9-13 × 1.5-3 mm, funnel-shaped, yellowish with purplish nerves. Capsule 4-6 × 2-3 mm, ovoid, 4-valved. Seeds 0.6-0.8 × 0.2-0.4 mm, brownish to almost black.

Ecology: Andean II-III: 3000-4000 m. Volcanic ash, rocky places.

Distribution: Peru, Bolivia, N-Chile and NW-Argentina.

Arequipa: native. CAR.

Jaborosa squarrosa (Miers) Hunz. & Barboza

Description: Acaulescent herb. Leaves in basal rosettes, lanceolate, deeply divided, coriaceous. Flowers white to lilac. Fruits yellow.

Ecology: Andean III: 3500-4500 m. Grassland.

Distribution: S-Peru, Bolivia, N-Chile and NW-Argentina.

Arequipa: native. ARE, CAY.

Jaltomata atiquipa Mione & S.Leiva

Description: Shrub, up to 1.5 m tall. Corolla white, with a ring of green maculae at the base. Inflorescence pedunculate and flowers pedicellate.

Ecology: Coastal: 0-1000 m. Lomas.

Distribution: S-Peru.

Arequipa: native. CAR.

Jaltomata diversa (J.F.Macbr.) Mione

Description: Shrub. Shrub, up to 1.5 m tall. Corolla white, with a ring of green maculae at the base. Inflorescence sessile, flowers pedicellate.

Ecology: Andean II: 3000-3500 m.

Distribution: Peru.

Arequipa: native. ARE, CAY, CON, UNI.

Jaltomata quipuscoae Mione & S.Leiva

Description: Small herbaceous plant, 0.5-0.7 m tall. Flowers solitary, corolla purple, broadly crateriform and 5-lobed, nectar pools on the corona, lacking radial staminal-corolla thickenings, stigma punctiform, mature fruits whitish.

Ecology: Coastal, Andean I-II: 300-2600 m. Roadsides.

Distribution: S-Peru.

Arequipa: native. ARE, CAR.

Leptoglossis acutiloba (I.M.Johnst.) Hunz. & Subils

Description: Annual herb, 10-30 cm tall. Stems glabrous. Basal leaves rosulate, cauline leaves linear-lanceolate, sessile. Flowers with long tube, usually purplish, calyx 5-lobed, with few capitate hairs.

Ecology: Andean II: 1500-2500 m.

Distribution: S-Peru.

Arequipa: native. ARE.

Note: L. acutiloba is reestablished as proposed by Fernandez & Quipuscoa (2021), POWO (2024) lists this species as a synonym of L. lomana.

Leptoglossis albiflora (I.M.Johnst.) Hunz. & Subils

Description: Annual herb, 10-30 cm tall. Basal leaves rosulate but early deciduous, cauline leaves linear-lanceolate, sessile. Flowers with long tube, white, calyx distinctly 7-11-lobed.

Ecology: Andean II: 1500-2500 m.

Distribution: Peru.

Arequipa: native. ARE. No specimen but a convincing observation from Chapi in iNaturalist (2024).

Leptoglossis lomana (Diels) Hunz.

Description: An ephemeral herb, 5-30 cm tall, rosulate, sparsely branched, ± woody at base. Basal leaves ovate, up to 2 cm long, cauline leaves linear. Inflorescence a compressed cyme, few to many flowered, depending on water availability. Flowers cream or yellow, calyx 5-lobed.

Ecology: Coastal: 0-1000 m. Lomas.

Distribution: Peru.

Arequipa: native. CAM, CAR, ISL.

Note: Leptoglossis acutiloba (I.M.Johnst.) Hunz. & Subils, a synonym of L. lomana by POWO (2024), is reestablished by Fernandez & Quipuscoa (2021) and treated here as a separate species. L. ferreyraei Hunz. & Subils is treated here as synonym of L. lomana, as proposed by Fernandez & Quipuscoa (2021).

Lycianthes lycioides (L.) Hassl.

Description: Shrub, up to 2 m tall, often intricately branched. Leaves glabrous to tomentose, broadly obovate. Flowers bluish-violet with yellow center.

Ecology: Coastal, Andean I-III: 1000-4000 m. Disturbed areas, forests, rocky slopes.

Distribution: The Andes, from Venezuela to Argentina and Chile.

Arequipa: native. ARE, CAY, CON.

Use: Edible fruits without commercial relevance.

Lycium americanum Jacq.

Description: Erect shrub, 1-4 m tall, thorny. Leaves succulent, glabrous, 0.5-6 x 0.5-3 cm. Flowers tubular, funnel- shaped, corolla white, whitish-green to violet with dark purple venation, tube 5-8 mm long, lobes 2.5-4.5 mm long. Fruit a globose berry, 5-13 mm long, red when ripe, 5-13 seeds per locule.

Ecology: Coastal, Andean I: 0-1500 m. Forests, lomas.

Distribution: S-Peru, N-Chile.

Arequipa: native. CAR. No voucher for Arequipa, mentioned for the Lomas de Atiquipa by Talavera et al. (2017).

Note: Lycium is a poorly studied genus, confusion between the species is possible.

Lycium distichum Meyen

Description: Erect shrub, 1.5-2 m tall, thorny. Leaves pubescent, 0.3-1.5 x 0.2-0.7 cm. Flowers spreading, corolla white to light lilac, tube 10-16 mm long, lobes 1.5-2.4 mm long, corolla slightly pubescent toward base. Fruit an ovoid berry, 5 mm long, orange to red when ripe, 3-7 seeds per locule.

Ecology: Coastal, Andean I-II: 0-3500 m. Forests on white sand.

Distribution: S-Peru, N-Chile.

Arequipa: native. ARE.

Lycium stenophyllum J.Rémy

Description: Erect shrub, 0.8-2 m tall, thorny. Leaves pubescent, 0.2-2 x 0.1 cm. Flowers spreading, corolla white to whitish-blue, tube 12-16 mm long, lobes 1-2 mm long, corolla glabrate but lobes usually ciliate. Fruit an ovoid berry, color unknown, 3-4.5 mm long, 1-5 seeds per locule.

Ecology: Coastal, Andean I: 0-2000 m. Lomas, deserts.

Distribution: S-Peru, N-Chile.

Arequipa: native. ARE, ISL.

Nicandra physalodes (L.) Gaertn.

Description: An annual shrub growing to 1 m tall, spreading branches. Leaves ovate, mid-green, toothed and waved. Flowers upward-facing, bell-shaped and 5 cm or more across, corolla pale blue and white with purple spots at the base, bell-shaped and 3.5-5 cm in Ø. Cherry-like, green-brown berries, within green or black-mottled calyces, resembling a lantern.

Ecology: Andean I-II: 0-3000 m. Disturbed areas near villages.

Distribution: Native to South America, introduced worldwide.

Arequipa: native. ARE, CAR.

Nicandra yacheriana S.Leiva

Description: Suffruticose shrub, 0.6-0.7 m tall. Leaves regularly dentate with 6-7 blunt teeth on each side. Similar to N. physalodes but flowers smaller, 1.5-2 cm in Ø, base of calyx rounded, inner sides of corolla without purple spots, mucronate white anthers, greenish ovary with yellow nectary disc and a dark purple ring on distal area, white style.

Ecology: Coastal, Andean I-II: 500-2400 m. Rocky slopes, shrublands.

Distribution: Peru.

Arequipa: native. ARE, CAR, UNI.

Nicotiana benavidesii Goodsp.

Description: Subligneous herb, 2-3 m tall, sparsely branched. Stem to 3 cm in Ø. Leaves ovate, base cordate, sinus often closed. Corolla 4-5 cm long, glabrous, light greenish-yellow, stamens exserted.

Ecology: Andean II: 2000-2500 m.

Distribution: Peru.

Arequipa: native. ARE.

Nicotiana glauca Graham

Description: Shrub, 1-6 m tall. Stems and leaves glabrous, glaucous. Bell-like flowers in a racemiform inflorescence, yellow to slightly greenish, corolla 2-5 cm long, at least 3x as long as the calyx. Fruit a broadly ovoid capsule.

Ecology: Coastal, Andean I-II: 0-3000 m. Disturbed areas, roadsides.

Distribution: Native to South America. Probably introduced to Peru from Argentina, now widely distributed in warm temperate climates worldwide.

Arequipa: introduced. ARE, CAM, CAR, ISL.

Use: A strong narcotic containing nicotine. The leaves are used to make acaricides.

Nicotiana knightiana Goodsp.

Description: Similar to N. paniculata, distinguishable by the dark green margins of the corolla and the perennial habit.

Ecology: Coastal: 0-1000 m. Lomas.

Distribution: Peru.

Arequipa: native. CAM, CAR, ISL.

Nicotiana paniculata L.

Description: A coarsely foliose bushy annual, rarely 3 m tall, with many vigorous erect or spreading branches. Stem pulverulent-puberulent, viscidly pubescent above, especially in inflorescence. Flowers uniformly yellowish-green, corolla 2-5 cm long, at least 3x as long as the calyx, stamens inserted in corolla.

Ecology: Andean I-II: 500-3000 m. Riversides, disturbed areas.

Distribution: Peru.

Arequipa: native. ARE, CAM, CAR, CAS, CAY, ISL, UNI.

Nicotiana rustica L.

Description: Shrub, 50-120 cm tall. Leaves broad ovate, all petiolate. Corolla yellow, less than twice as long as calyx.

Ecology: Coastal, Andean I-III: 0-4000 m. Disturbed areas.

Distribution: Native to South America, cultivated and introduced worldwide.

Arequipa: native. ARE, CAY.

Nicotiana tomentosa Ruiz & Pav.

Description: Shrub 3-5 m tall, whole plant tomentose. Leaves 30 cm long, petiole alate. Inflorescence terminal, flowers white or pink, stamens well-exserted, ascending.

Ecology: Coastal, Andean I-II: 1000-3500 m. Disturbed areas, forests.

Distribution: Native to Peru, Ecuador, cultivated in warm climates.

Arequipa: native. ARE, CAR.

Nicotiana undulata Ruiz & Pav.

Description: Undulate leaves with prominent lateral veins. Flowers whitish-yellow, corolla <2 cm, less than 2x as long as calyx.

Ecology: Andean I-III: 1000-4500 m. Disturbed areas.

Distribution: Peru, Bolivia, N-Chile, NW-Argentina.

Arequipa: native. CAY, UNI.

Nolana adansonii (Schult.) I.M.Johnst.

Description: A perennial shrub, 0.5 m tall and 1.5 m wide, brownish stems. Leaves triangular-reinform, ± cordate at base, petiole as long as the blade. Flowers bluish-lilac, 2 cm long, calyx 3-5 mm in Ø.

Ecology: Coastal: 0-500 m. Lomas.

Distribution: S-Peru, N-Chile.

Arequipa: native. ARE, CAM, CAR, ISL.

Nolana arenicola I.M.Johnst.

Description: Similar to N. humifusa ssp. spathulata. An annual spreading subshrub, 0.2-0.4 m tall. Leaves long petiolate, ovate, 10-20 x 6-10 mm, apex acute. Flowers pale blue, calyx 4-6 mm, pedicel <12 mm.

Ecology: Coastal: 500-1000 m. Lomas.

Distribution: S-Peru, N-Chile.

Arequipa: native. CAM, CAR, ISL.

Nolana arequipensis M.O.Dillon & Quip.

Description: Similar to N. tarapacana ssp. thinophila. Perennial herb. Leaves succulent, 10 x 2-4 mm, terete. Corolla white, internally light purple, 3-6 mm wide, calyx 2-3 mm wide.

Ecology: Coastal: 100-300 m. Desert, inland habitats.

Distribution: S-Peru.

Arequipa: native. CAR.

Nolana bombonensis Quip. & M.O.Dillon

Description: Similar to N. lycioides. A subshrub, 25-50 cm tall, 1.5 m in Ø, densely lanuginose to arachnoid. Leaves linear oblong, 5-8 x 1-2 mm. Flowers pale lilac or light lavender, no purple nectar guides in the inner throat. Calyx 3-4 mm wide, lobes unequal, obtuse or blunt. Corolla 15-20 mm long and wide.

Ecology: Coastal: 0-100 m. Sandy places near the sea.

Distribution: S-Peru.

Arequipa: native. ISL.

Note: May be a synonym of Nolana sedifolia ssp. confinis.

Nolana callae Quip. & M.O.Dillon

Description: Prostrate, forming mats to 1.2 m in Ø. Elliptic-lanceolate leaves 15-30 x 5-12 mm, the base of petioles with conspicuous sheath, abaxial and adaxial surfaces lanuginose. Flowers lavender, pedicles to 35 mm long. Calyx with narrowly deltoid to linear lobes and 5 mericarps, 4-7 mm wide.

Ecology: Coastal: 200-500 m. Dry, rocky slopes at the lower part of the Lomas de Jesus (Islay).

Distribution: S-Peru.

Arequipa: native. ISL. Known only from the type locality Punta Bonbon.

Nolana chancoana M.O.Dillon & Quip.

Description: A much-branched annual herb, succulent, prostrate, glabrous, 25-30 cm tall. Leaves narrowly spatulate, succulent, long petiolate, 10-40 x 2-6 mm. Flowers lavender to blue. Calyx globose, inflated, 6-10 mm wide.

Ecology: Andean I: 600-1200 m. Lomas.

Distribution: S-Peru (Arequipa).

Arequipa: native. CAM, CAR.

Nolana chapiensis M.O.Dillon & Quip.

Description: A perennial succulent herb, glutinous, 50 cm tall, 1.2 m in Ø. Leaves elliptic, 8-15 x 2-3.5 mm. Flowers purple. Calyx 3-3.5 mm wide. Corolla tube 16-20 mm long, 5-8.5 mm wide.

Ecology: Andean II: 2300 m. Shrublands.

Distribution: S-Peru.

Arequipa: native. ARE.

Nolana crassulifolia Kunze ex Walp.

Description: Leaves fleshy, with revolute margins and therefore appearing terete, 8-16 mm long, 1-2 mm wide, laxly pubescent. Corolla white to bluish, <1 cm long. Calyx 3-4 mm wide.

Ecology: Coastal: 0-1000 m. Lomas.

Distribution: Peru, Chile.

Arequipa: native. CAM, CAR, ISL.

Note: In Peru N. crassulifolia ssp. revoluta.

Nolana gracillima (I.M.Johnst.) I.M.Johnst.

Description: Annual herb. Leaves 8-25 x 1-2 mm, strictly glabrous. Flowers lavender, dark purple center, pedicles 5- 13 mm long, mericarps 5. Calyx 2-4 mm wide. Corolla tube 2.5-3.5 mm long.

Ecology: Andean I: 1000-2000 m. Desert.

Distribution: S-Peru, N-Chile.

Arequipa: native. CAR, CON, ISL.

Nolana humifusa (Gouan) I.M.Johnst.

Description: Annual to perennial herb. Leaves entire, ovate-spatulate, often ± succulent, 20 x 10 mm. Flowers lavender with dark purple center. Calyx 8-10 mm.

Ecology: Coastal, Andean I-II: 0-3000 m. Lomas, rocky slopes.

Distribution: Peru.

Arequipa: native. ARE, CAM, CAR, CON, ISL.

Note: In Arequipa, N. humifusa ssp. humifusa, ssp. spathulata and var. plicata.

Nolana inflata Ruiz & Pav.

Description: Plants with a persistent basal rosette of long petiolate leaves, up to 100 x 40 mm, cauline leaves smaller, margins dentate to crenate. Flowering shoots with petiolate, lanceolate bracts subtending flowers, the inflorescence open, flowers not congested, densely hirsute-glandular to arachnoid or lanuginose, calyx globose or bladder-inflated, 10 mm wide, lobes equal, triangular, acute. Corollas deep purple to lavender.

Ecology: Coastal: 0-1000 m. Lomas.

Distribution: Peru.

Arequipa: native. CAM, CAR, ISL.

Nolana laxa (Miers) I.M.Johnst.

Description: Semi succulent, decumbent, annual herb, much-branched. Leaves lanceolate to oblanceolate or spatulate, 15-30 x 10-20 mm, strigose. Corolla bluish-violet, 10-15 mm in Ø, calyx 3-4 mm wide.

Ecology: Coastal, Andean I-II: 0-3000 m. Lomas, rocky slopes.

Distribution: S-Peru.

Arequipa: native. Not specified by province. Quipuscoa & Dillon (2018) mention N. johnstonii, a synonym of N. laxa. Distribution range from Lima to Tacna.

Nolana lycioides I.M.Johnst.

Description: Annual or perennial sparsely branched subshrub, up to 0.35 m tall. Leaves linear to oblong-elliptic, 5-9 x <1 mm, sericeous or glabrous. Flowers white to pale blue, sessile or short pedicellate. Calyx 2-3 mm wide. Corolla 2 mm long and 3 mm wide.

Ecology: Coastal: 0-1000 m. Lomas, rocky slopes.

Distribution: S-Peru, N-Chile.

Arequipa: native. ARE, CAM, CAR, ISL.

Note: Synonymized with Nolana sedifolia ssp. confinis by Hassler (2024).

Nolana quicachaensis Quip. & M.O.Dillon

Description: A perennial shrub, 0.3-1 m tall, stem of dense wood, 3-4 cm in Ø. Differs from N. lycioides by fasciculate, sessile leaves, terete linear to narrowly spatulate, 5-9 x 0.5-1.5 mm, pubescent with short glandular trichomes. The solitary axillary flowers form a weak, terminal raceme, calyx lobes narrowly deltoid, long attenuate, 4- 6 mm long, 3-4 mm wide. Corollas 10-12 mm wide, 18-20 mm long, with yellow tube and white lobes, 15-18 mericarps.

Ecology: Andean I: 1500 m. Shrublands.

Distribution: S-Peru.

Arequipa: native. CAR.

Nolana scaposa Ferreyra

Description: Plants with a persistent basal rosette of long petiolate leaves, margins strictly entire. Flowering shoots with sessile ovate bracts subtending flowers, the flowers congested in a dense inflorescence, densely villous pubescence, calyx campanulate, deeply unequally lobed, corolla white to pale blue. Calyx 5 mm wide.

Ecology: Coastal: 0-1000 m.

Distribution: S-Peru.

Arequipa: native. CAM, CAR.

Nolana tarapacana (Phil.) I.M.Johnst.

Description: An annual or perennial herb growing near the ocean. Leaves glabrous, succulent. 10-15 x 1-2 mm, Ø round. Flowers lavender to purple, dark purple center, 10-40 mm wide. Calyx 1-2 mm wide. Corolla tube 10 mm long.

Ecology: Coastal: 0-1000 m. Lomas, beach.

Distribution: S-Peru, N-Chile.

Arequipa: native. CAM, CAR, ISL.

Note: In Arequipa, N. tarapacana ssp. thinophila.

Nolana tricotiflora Quip. & M.O.Dillon

Description: Perennial shurb. Differs from all other members of the genus with a unique combination of characters. Its erect crooked, woody stems up to 50 cm tall, 1-2 cm in Ø, numerous spirally arranged leaves, elliptic to linear- lanceolate, 10-14 x 2-3 mm and the inflorescence in a terminal, 3-branched, scorpioid cyme. Corolla purple, calyx 8- 12 mm wide.

Ecology: Coastal: 500-800 m. Lomas.

Distribution: S-Peru.

Arequipa: native. CAM.

Physalis peruviana L.

Description: Weak annual erect herbs, to 60 cm high. Leaves alternate, broadly ovate, obliquely cordate at base, entire or shortly lobed, 3-nerved from the base, petiole to 3 cm long. Flowers axillary, solitary, pedicels 1 cm long, slender, calyx campanulate, accrescent in fruit, membranous, corolla campanulate, yellowish with deep brown center.

Ecology: Coastal, Andean I-III: 0-4500 m. Disturbed areas.

Distribution: Native to South America, from Chile to Colombia and Venezuela.

Arequipa: introduced. ARE, CON. A species of humid climate (E-Andes), often cultivated for its edible fruit, naturalized near villages.

Use: Medicinal plant, an infusion is used for abdominal disorders. The edible fruits are a good source of carotene and ascorbic acid. The unripe fruits contain enough solanine to cause gastroenteritis and diarrhea if ingested.

Salpichroa glandulosa (Hook.) Miers

Description: A scrambling, woody shrub, up to 1 m tall, all parts densely glandular-hairy. Leaves 1-2 x 0.8-1 cm,

ovate, apex acute, base truncate to cordate. Flowers 5 cm long, greenish-yellow, tube slightly funnel-shaped, broadest at top, ± glabrous outside, anthers not exserted, calyx lobes sublinear, 1 cm long.

Ecology: Andean II-III: 2500->4500 m. Forests, shrublands.

Distribution: Peru, Bolivia.

Arequipa: native. CAY.

Salpichroa hirsuta (Meyen) Miers

Description: Shrub. Flowers 3-4 cm long, greenish-yellow, tube funnel-shaped, broadest at top, hirsute outside in upper half, lobes mucronate, anthers exserted, calyx lobes sublinear, short.

Ecology: Andean II-III: 3000->4500 m.

Distribution: Peru to C-Bolivia.

Arequipa: native, expected. Recorded from Moquegua near the border to Arequipa (Gonzáles, 2021).

Salpichroa ramosissima Miers

Description: Shrub often with square nodose branches, eglandular. Leaves elliptic-oblong, decurrent into petiole, half as wide as long. Flowers 1-1.5 cm long, whitish-yellow, tube broadest at base, glabrous outside, lobes reflexed at anthesis.

Ecology: Andean II-III: 2500->4500 m. Disturbed areas, grasslands.

Distribution: Ecuador, Peru, Bolivia, NW-Argentina.

Arequipa: native, doubtful. ARE, ISL. No voucher, mentioned for RN Salinas y Aguada Blanca (Quipuscoa & Huamantupa, 2010) and Lomas de Yuta (Quipuscoa et al., 2016).

Salpichroa tristis Walp.

Description: Depressed shrub with granular-glandular indument. Leaves very variable. Flowers 1-1.5 cm long, greenish-yellow to deep yellow, tube with glandular trichomes on the outside, lobes up to 7 mm long.

Ecology: Andean II-III: 2000-4500 m. Disturbed areas.

Distribution: The Andes, from Venezuela to NW-Argentina and N-Chile.

Arequipa: native. ARE, CAY, UNI.

Note: In Arequipa, S. tristis var. lehmannii (= S. lehmannii).

Solanum acaule Bitter

Description: Tuber bearing herb, rosulate, acaulescent. Leaves pinnate, with 4-6 pairs of leaflets, subequal except for the most proximal 1-2 pairs. Inflorescence with 1-4 flowers, unbranched. Corolla 1.4-2.5 cm in Ø, pale blue to more rarely white.

Ecology: Andean III: 3500-4500 m. Disturbed areas, forests, grasslands.

Distribution: Ecuador, Peru, Bolivia and NW-Argentina.

Arequipa: native. ARE, CAR, CAY, CON, UNI.

Solanum acroscopicum Ochoa

Description: Tuber bearing herb, 0.3-2 m tall, erect. Stems 5-20 mm in Ø at base, unwinged or narrowly winged, glabrous, light green or with purple dots in the basal part. Leaves pinnate, with 4-5 pairs of leaflets, subequal except for the mot proximal pair. Inflorescence with 10-40 flowers. Corolla 2.5-4 cm in Ø, lilac to purple.

Ecology: Andean II-III: 2500-3900 m. On rocky cracks and humid places and in the shade, in rich soils, among herbs or shrubs.

Distribution: Peru.

Arequipa: native. ARE, CON, UNI.

Solanum americanum Mill.

Description: Erect or weakly scrambling annual or perennial, up to 1.5 m tall, subwoody and branching at base. Older stems nearly glabrous. Leaves simple, the blades 3.5-10.5 x 1-4.5 cm, ovate to elliptic, concolorous, margins ± entire, sparsely pubescent on both sides. Inflorescence unbranched, 4-6-flowered. Flowers white, calyx reflexed. Fruits deep purple, pedicels remain on the mature fruit and drop off.

Ecology: Coastal, Andean I-II: 0-2500 m. Disturbed habitats, in dry areas close to water sources.

Distribution: Probably native to the Americas, now a pantropical and -subtropical weed.

Arequipa: native. ARE, CAM, CAR, ISL.

Use: Leaves used like spinach and ripe berries eaten, either raw or cooked.

Solanum arequipense Bitter

Description: Annual or perennial subwoody herb, up to 1 m tall, branching at base. Stems green to purple tinged, spreading to decumbent, eglandular. Leaves entire, ovate to rhomboidal, margins sinuate-dentate or entire, ± pubescent, apex acute. Inflorescence furcate, 6-14 flowers. Corolla 1.2-2 cm in Ø. Flowers white with a green central portion, stellate. Fruit a globose berry, 5-6 mm, grayish-green at maturity.

Ecology: Coastal, Andean I-III: 0-4500 m. Disturbed areas, forests.

Distribution: Peru.

Arequipa: native. ARE, CAR, ISL.

Note: Knapp et al. (2023) separate S. arequipense from S. furcatum due to the distribution range and morphological characteristics. S. furcatum is restricted to temperate Chile and Argentina.

Solanum basendopogon Bitter

Description: Upright shrub up to 3 m tall. Stems slender and strong, ± densely strigose with appressed hairs. Leaves simple, ovate, margins entire, pseudostipules present. Inflorescence highly branched, with 10-50 flowers. Corolla 1.5-2.5 cm in Ø, white to pale

purple. Fruit deep green at base, light green and often spotted distally.

Ecology: Andean II: 2000-3800 m. Rocky slopes, disturbed areas.

Distribution: Peru.

Arequipa: native, doubtful. CAR. The type Martinet 324 (P), dated 1878, is from Cerro Prieto, Chala Alta, no modern vouchers or observations. The type locality could also be between Cerro Prieto in Ica and Chala Alta in Arequipa.

Solanum boliviense Dunal

Description: Tuber bearing herb, rosulate, acaulescent. Leaves pinnate, with 0-4 pairs of leaflets and a terminal leaflet much larger than the laterals. Inflorescence with 2-4 flowers, usually forked. Corolla 2.5-3.8 cm in Ø, violet.

Ecology: Andean III: 3500-4500 m. Disturbed areas, shrublands, grasslands.

Distribution: Ecuador, Peru, Bolivia and NW-Argentina.

Arequipa: native. CAY.

Solanum candolleanum Berthault

Description: Tuber bearing herb, 0.3-1.2 m tall, ascending to erect. Stems 1.5-12 mm in Ø at base, unwinged or with wings up to 4 mm, green to purple or green and purple mottled, subglabrous to moderately to densely pubescent with short non-glandular hairs, sometimes glaucous. Leaves pinnate, with 2-6 pairs of leaflets, subequal except for the most distal pair, which is larger. Inflorescence with 3-20 flowers. Corolla 2.5-4 cm in Ø, purple violet to light blue.

Ecology: Andean III: 2000-4000 m. Rocky slopes, grassland.

Distribution: Peru.

Arequipa: native. ARE, CAR, CAS, CAY, CON.

Solanum chilense (Dunal) Reiche

Description: Densely grayish pubescent herb, up to 1 m tall. Leaves pinnate, 5-7 pairs of leaflets, leaflets not markedly oblique. Inflorescence 1-branched and regularly bifurcate, 20-50-flowered. Flowers yellow, the anther tubes straight. Fruits green with purple stripes.

Ecology: Coastal, Andean I-II: 0-3000 m. Lomas, hyperarid rocky plains, coastal deserts.

Distribution: S-Peru, N-Chile.

Arequipa: native. ARE, CAM, CAR, CON, ISL.

Solanum cochabambense Bitter

Description: Shrub, scandent, vinelike, eglandular. Stems 1.5-5 m long. Leaves simple, elliptic to narrowly elliptic, 3.5-16 x 1.5-8 cm, discolorous, slightly pubescent above, densely pubescent beneath, margins entire or occasionally shallowly toothed. Inflorescence several times branched, with 10-80+ flowers. Flowers, corolla 2-3 cm in Ø, pale violet to whitish-violet, with a pale greenish-yellow eye, stellate. Fruits globose, 1-1.2 cm in Ø, green and maturing purplish-black, translucent when ripe.

Ecology: Andean I-III: (500-)1000-4000 m. Roadsides, landslips, treefalls.

Distribution: Peru to NW-Argentina.

Arequipa: native. CAY. Specimen Särkinen et al. 4108 (NHMUK) from Chivay, observation from Ayo in iNaturalist (2024).

Use: Medicinal and fodder plant.

Note: The population in the lomas of Atiquipa and Camaná occurs at unusually low altitude. It was probably introduced as a fodder and medicinal plant when the lomas were used as seasonal pastures.

Solanum corneliomulleri J.F.Macbr.

Description: Spreading, erect to decumbent perennial herb, woody a base. up to 1 m tall. Leaves pinnate, 3-5 pairs of leaflets, basal leaflets half the size of the rest. Inflorescence simple or 1-branched, 8-16-flowered. Flowers yellow, the anther tubes strongly curved. Fruits green with purple stripes.

Ecology: Coastal, Andean I-II: 0-3500 m. Landslides and rocky slopes.

Distribution: C- to S-Peru.

Arequipa: native. ARE, CAY, ISL.

Solanum corymbosum Jacq.

Description: Annual to short-lived perennial subwoody herb, up to 50 cm tall, branching at base. Stems green, decumbent, older stems glabrous. Leaves entire, ovate-lanceolate, margins entire, base and apex acute. Inflorescence 4-7-branched, 20-50 flowers. Corolla 0.5-1 cm in Ø. Flowers white or purple, rotate-stellate. Fruit globose berry, 0.5 cm, orange to red when ripe.

Ecology: Andean I-II: 500-2500 m. Disturbed areas.

Distribution: Native to Peru, introduced to Mexico.

Arequipa: native. CAR, UNI.

Solanum edmondstonii Hook.f.

Description: Perennial, erect shrub 60-100 cm tall, sometimes arising from a thickened woody rootstock up to 12 mm thick. Stems pale brown, woody, glabrous to sparsely pubescent with simple, uniseriate, multicellular hairs up to 0.6 mm long, glabrescent. Leaves entire, deeply lobed, sometimes lobes lobbed, ± fleshy. Inflorescence unbranched or 1-branched, 1-11 flowers. Corolla 1.5-2.5 cm in Ø. Flowers white to whitish-pink, not lobed.

Ecology: Coastal: 300-1100 m. Sandy and rocky lomas.

Distribution: Peru, restricted to coastal Ica and N-Arequipa (prov. Caravelí).

Arequipa: native. CAR.

Solanum filiforme Ruiz & Pav.

Description: Shrub, scandent, 1.5-3+ m tall. Leaves entire, ovate, pubescent on both sides, margins entire, base truncate to slightly cordate. Inflorescence branched, with 10-20(?) flowers. Flowers white, corolla rotate. Fruits globose, 2 cm in Ø, whitish-green with brown stripes.

Ecology: Coastal, Andean I-II: 500-2500 m. Lomas, shrublands.

Distribution: Peru, Bolivia.

Arequipa: native, doubtful. CAR. Mentioned by Talavera et al. (2017) for Atiquipa, anonymous specimen. A species of the humid puna and the E-slopes of the Andes.

Note: One of the wild forms of the pepino fruit.

Solanum fragile Wedd.

Description: Herb or shrublet from a woody base, to 40 cm tall, easily breaking from the woodstock. Stems and leaves glandular pubescent, sticky to touch. Leaves simple and shallowly toothed, the blades 1.2-7 x 0.7-4.5 cm, ovate to rhomboid, widest in the lower half, discolorous. Inflorescences internodal, forked or sometimes branched, 3-5 cm long, with 9-12 flowers. Corolla 1.5-1.6 cm in Ø. Flowers white or violet, with a green eye along midvein of the lobe, stellate. Fruit globose, 0.5-0.8 cm, green when ripe.

Ecology: Andean II-III: 2200-4500 m. Grasslands, rocky slopes.

Distribution: Peru, Bolivia, N-Chile, NW Argentina.

Arequipa: native. ARE, CAR, CAS, CAY, ISL, UNI.

Note: S. grandidentatum is almost identical to S. fragile and is probably a polyploid variant of this species.

Solanum grandidentatum Phil.

Description: Similar to S. fragile but lacking a woody rootstock and calyx lobes 1-1.5 mm long, tip acute.

Ecology: Coastal, Andean I-III: 200-4500 m. Disturbed areas, along streams.

Distribution: Ecuador, Peru, Bolivia, N-Chile, NW Argentina.

Arequipa: native. ARE, CAY. Knapp et al. (2023) mention the species for Arequipa.

Solanum interandinum Bitter

Description: Small shrub or woody herbs to 1 m tall, the branches erect, always woody at base. Leaves narrowly elliptic, 1.8-8 cm long and 0.8-4 cm wide, discolorous, pubescent. Inflorescence forked or several times branched, 10-20-flowered, the branches persisting on the plant after fruiting. Corolla 0.8-1.4 cm in Ø, pale violet or white and violet striped. Fruit whitish green when immature, ripening to green and sometimes blotched black-purple.

Ecology: Andean II-III: 2800-4000 m.

Distribution: Andes from Venezuela to Bolivia.

Arequipa: native, doubtful. Not specified by province. S. interandinum is said to be endemic to Ecuador by POWO (2022) and Hassler (2024). Solanaceae Source (2022) and Tropicos.org (2023) indicate a broader distribution.

Solanum marginatum L.f.

Description: Sparingly shrub to 1.5 m, dull green to silvery-white, pubescent with minute stellate hairs, the hairs dense on lower leaf-surface, sparse on upper surface except along margins. Prickles to 1 cm, scattered on most parts. Leaves elliptic, 10-12 x 8-10 cm, discolors, lobed. Inflorescence short, 2-10-flowered. Corolla rotate, 30-40 mm in Ø, white or very pale mauve, male flower slightly smaller. Berry globular, 30-40 mm Ø, pale or deep yellow.

Ecology: Andean II: 2000-2900 m. Grasslands or open woodland, wasteland and roadsides.

Distribution: Native to Ethiopia and Eritrea, introduced to Australia and the Americas.

Arequipa: introduced, adventive. ARE.

Solanum medians Bitter

Description: Tuber bearing herbs, 0.2-0.6 m tall, erect. Leaves odd-pinnate, the blades 8-15 x 5-15 cm, medium to dark green and sometimes tinged with purple beneath, lateral leaflet pairs 1-4, subequal and rapidly decreasing in size to the base. Inflorescence terminal, flowers nodding. Corolla 2.8-3.5 cm in Ø, dark blue to violet and typically with a green central star on both sides.

Ecology: Andean I-III: 500-4000 m. Lomas, rocky slopes.

Distribution: Peru, N-Chile.

Arequipa: native. ARE, CAM, CAR, CAY, ISL, UNI.

Solanum montanum L.

Description: Perennial herb, often scandent or prostrate, arising from a subterranean spheroidal swollen caudex. Leaves simple, ovate to elliptic, ± succulent, deeply lobed and pubescent. Inflorescence unbranched or 1-branched, 4-8-flowered. Flowers white to purple, corolla 1.5-2.6 cm in Ø, pentagonal. Fruits globose, green to whitish.

Ecology: Coastal, Andean I-II: 0-3500 m. Lomas, sandy or rocky ground

Distribution: Peru, Chile.

Arequipa: native. ARE, CAM, CAR, ISL.

Solanum multifidum Lam.

Description: Annual, erect or prostrate herb up to 50 cm tall. Stems fleshy, becoming woody, up to 12 mm in Ø at the base, glabrous or rarely sparsely pubescent with simple, uniseriate, multicellular hairs up to 0.6 mm long. Leaves 1- 2-pinnate or rarely 3-pinnate, fleshy, usually with a narrow wing 1-6 mm wide along the midrib, 3-6 pairs of pinnae. Inflorescence much-branched, with 8-23 flowers. Corolla 11-16 mm in Ø, pentagonal to rotate. white, sometimes lilac.

Ecology: Coastal, Andean I: 0-1000 m. Lomas, riversides.

Distribution: Peru.

Arequipa: native. CAM, CAR, ISL.

Solanum nitidibaccatum Bitter

Description: Sprawling annual herb to 50 cm, pubescent with glandular hairs. Leaves ovate, the lamina to 6 x 2-4 cm, slightly discolors, entire or shallowly lobed, base cuneate, petiole 5-15 mm long. Inflorescence short, unbranched, internodal, 2-7-flowered, peduncle 1 cm long. Calyx 1.5-2.5 mm long, enlarged in fruit. Corolla stellate, 12-14 mm Ø, white. Berry globular, 5-8 mm Ø, green to purplish-green when ripe.

Ecology: Andean II: 1500-2500 m. Agriculture land, sandy soil, seasonal washes.

Distribution: Native to temperate South America and temperate western North America, introduced elsewhere.

Arequipa: introduced. ARE. No specimen but convincing observations from Arequipa in iNaturalist (2024).

Use: Medicinal plant.

Solanum nitidum Ruiz & Pav.

Description: Shrub up to 1.5 m tall. Leaves lustrous, oblong-lanceolate, acute, 10-15 cm long with prominent veins. Inflorescence branched, 10-20-flowered. Calyx grayish pubescent. Corolla lilac-purple, stellate. Berries dark red when mature.

Ecology: Andean II-III: 2500-4000 m. Grasslands, dry riverbeds, rocky slopes.

Distribution: Ecuador, Peru, Bolivia.

Arequipa: native. ARE, CAS, CAY, UNI.

Use: Medicinal plant used to treat wounds and colds.

Solanum paposanum Phil.

Description: Perennial, erect or procumbent shrub up to 60 cm tall, stems woody, sparsely pubescent, internodes winged. Leaves decurrent on the petiole and the stem, simple, margins deeply incised with 4-5 pairs of regularly spaced lobes, lobes serrate, up to 15 mm long. Inflorescence branched 3-times, 6-8-flowered. Calyx lobes 1-2 mm wide. Corolla 2-3.2 cm in Ø, pentagonal, blue, lilac, or purple. Fruits green with black markings.

Ecology: Andean I-II: (200-) 1500-3500 m.

Distribution: S-Peru to C-Chile.

Arequipa: native. ARE, ISL.

Solanum pennellii Correll

Description: Spreading perennial herb, woody at the base, to 1 m tall. Leaflets orbicular, terminal leaflet wider than long, but not larger than the laterals. Flowers yellow. Fruit green, sparsely to moderately pubescent.

Ecology: Coastal, Andean I-II: 0-2300 m. Dry rocky hillsides and sandy areas.

Distribution: Peru and N-Chile, Ecuador(?).

Arequipa: native, doubtful. CAR, CON.

Solanum pentlandii Dunal

Description: Annual or perennial subwoody herb, up to 1 m tall, subwoody branching at base. Stems strongly angled with wings to 1.5 mm wide, ± glabrous. Leaves entire, ovate to broadly elliptic, margins ± regularly dentate, glabrous and shiny, apex acute. Inflorescence several times branched, 10-20 flowers. Corolla 0.9-1.2 cm in Ø. Flowers violet- blue, often edged white, with a green eye, stellate. Fruit a globose berry, 8-10 mm, dark green, white stripped at maturity.

Ecology: Andean II-III: 2400-5200 m. Disturbed areas, high nitrogen environments.

Distribution: C- & S-Peru, Bolivia.

Arequipa: native. CAY.

Solanum peruvianum L.

Description: Spreading, erect to decumbent perennial herb, uniform, dense, velvety pubescence with only scattered short glandular hairs. Leaves pinnate, 2-4 pairs of leaflets, terminal leaflet markedly larger than the laterals. Inflorescence 1-branched and bifurcate, 8-20-flowered. Flowers yellow, the anther tubes strongly curved. Fruits green with purple stripes.

Ecology: Coastal: 0-800 m. Lomas, coastal deserts, field edges, river valleys.

Distribution: Peru and Chile.

Arequipa: native. ARE, CAM, CAR, CAS, CAY, CON, ISL.

Note: S. chilense and S. peruvianus are promising gene pools for the improvement of the tomato genome.

Solanum pimpinellifolium L.

Description: Annual to biennial herb. Leaves pinnate, 2-4 pairs of leaflets, size of leaflets slightly increasing to tip of leaf. florescence simple or 1-branched, 7-30-flowered. Flowers yellow, the anther tubes ± straight, corolla stellate, divided >¾. Fruits bright red, calyx lobes in fruit strongly reflexed and parallel to fruiting pedicel.

Ecology: Coastal, Andean I-II: 0-1500 m. Disturbed areas, dry places.

Distribution: Ecuador, Peru, N-Chile.

Arequipa: native. ARE, CAM, CAR, CON, UNI.

Solanum pseudocapsicum L.

Description: Shrub to 2 m high, glabrous or sparsely hairy on new growth. Leaves oblong to lanceolate or narrow- elliptic, 4-10 x 1-3 cm, margins entire and slightly undulate, concolorous, green, glabrous, thin,

petiole 5-15 cm long. Inflorescence unbranched, 1-3-flowered, peduncle 0-10 mm long, pedicels 10 mm long. Calyx 4-5 mm long, lobes 2- 3 mm long, glabrescent. Corolla deeply stellate, 10-15 mm Ø, white. Berry 10-15 mm Ø, bright orange-red.

Ecology: Coastal, Andean I-II: 0-2600 m. Disturbed areas, dry places, near villages.

Distribution: Native to Central and South America, cultivated worldwide, often escaped in tropical and subtropical areas.

Arequipa: introduced. ARE. Cultivated and adventive.

Use: Ornamental plant with poisonous berries.

Solanum radicans L.f.

Description: Creeping herb to sprawling subshrub, 0.2-0.75 m high, branches occasionally rooting at the lower nodes. Stems strongly angled to winged from the decurrent leaf bases, with occasional spinescent process along the angles, eglandular. Leaves simple and deeply 5-lobed, blades 2.5-14 x 2.5-6 cm, elliptic to ovate, ± glabrous. Inflorescence usually unbranched, occasionally forked, 2-7 cm long, 10-20-flowered. Corolla 1-1.2 cm in Ø, rotate- stellate, pale violet to purple, with a greenish-yellow central star. Fruit globose, 0.5-1 cm, orange-yellow or slightly greenish-yellow when mature.

Ecology: Andean I-III: 500-4000 m. Riversides, disturbed areas, near villages.

Distribution: Ecuador, Peru, Bolivia, N- & C-Chile.

Arequipa: native. ARE, CAR, CAY, CON, ISL, UNI.

Use: Medicinal plant

Solanum weddellii Phil.

Description: Tiny annual herbs to 0.2 m high, usually appearing as a prostrate rosette. Stems terete, eglandular. Leaves simple and 5-lobed, lobes reaching ± halfway to midrib, blades 1-3 x 0.5-1.5 cm, narrowly elliptic, ± fleshy, sparsely pubescent, eglandular. Inflorescence unbranched, 0.3-1 cm long, with 3-5 flowers. Corolla 0.6 cm in Ø, rotate, purple or white, with a large greenish-yellow, purple-edged central star. Fruiting calyx partly to completely enclosing the berry. Fruit globose, 0.5-0.7 cm, pale green to whitish-green when mature.

Ecology: Andean II-III: 2300-4500 m. Sandy soil among gravel.

Distribution: S-Peru, Bolivia, N-Chile, NW-Argentina.

Arequipa: native. CAY.

Talinaceae

Talinum spathulatum A.Gray

Description: Perennial from a thick, rather fleshy root. Stems stout, up to 1 m high, reddish, little branched, or simple. Leaves numerous, thick and fleshy, obovate to oblanceolate, up to 10 cm long, rounded at the apex, attenuate to the base. Inflorescence an elongated and much-branched panicle. Flowers on pedicels 15 mm long, petals bright yellow, sepals very unequal, almost orbicular, capsule obtusely triquetrous, reddish. Seeds black and shining, minutely papillose.

Ecology: Coastal, Andean I-II, Amazonian: 100-2500 m. Disturbed areas, riparian vegetation.

Distribution: Tropical America.

Arequipa: native. ARE, UNI.

Note: T. spathulatum is synonymized with Talinum paniculatum by POWO (2024). Here, T. spathulatum is listed as a separate species because of the obvious morphological differences (D. Ferguson, pers. comm.). T. paniculatum s.str. does not occur in Arequipa.

Tropaeolaceae

Tropaeolum ferreyrae Sparre

Description: Annual herb. Leaves orbicular, peltate, regularly sinuate, red margined. Leaves, spurs and sepals densely pubescent. Petals up to 2 cm long, reddishorange, spur long and straight, flowers fragrant.

Ecology: Andean I-II: 500-2500 m.

Distribution: Peru.

Arequipa: native. CAR.

Tropaeolum majus L.

Description: Annual herb with trailing stems up to 1 m long, glabrous. Leaves ± orbicular, 3-15 cm in Ø, green to blue-green above, paler beneath, distinctly peltate. Petiole 5-30 cm long, attached in the center of the lamina. Flowers yellow to red, often with darker markings at the base of the petals, 2.5-6 cm in Ø, with a basal spur 2.5-3 cm long. Fruit 2 cm in Ø, 3 one-seeded segments.

Ecology: Coastal, Andean I-II: 0-3000 m. Disturbed areas.

Distribution: Only known in cultivation, probably originating in the Andes.

Arequipa: introduced, adventive. ARE, ISL. Cultivated and adventive.

Use: Ornamental, widely grown as a garden plant and often used in salads. Medicinal plant.

Note: Of hybrid origin, probably derived from T. ferreyrae × T. minus.

Tropaeolum minus L.

Description: Similar to T. majus but all parts distinctly smaller. Central vein of peltate leaf ending in a mucronate tooth. Petals distinctly ciliate on proximal half, spur almost straight and dark orange to red.

Ecology: Coastal, Andean I-II: 0-3500 m.

Distribution: Ecuador, Peru.

Arequipa: native, doubtful. CON, UNI.

Tropaeolum peregrinum L.

Description: Perennial climber, up to 4 m long, glabrous. Leaves orbicular, 2-5 cm across, slightly peltate, pedately lobed into 5 lobes. Flowers bright yellow, lower petals small, upper petals much larger, fringed, spur curled.

Ecology: Andean II-III: 2500-4000 m. Disturbed areas.

Distribution: Native to Ecuador and Peru.

Arequipa: native. ARE, UNI. No specimen but a convincing observation.

Use: Ornamental plant.

Tropaeolum seemannii Buchenau

Description: Similar to T. peregrinum, but sparsely pilose. Leaves deeply lobed, lobes acute. Corolla yellow to orange, upper petals incised at least halfway or more, spur not curled.

Ecology: Andean II-III: 2000-4000 m. Disturbed areas.

Distribution: Peru, Bolivia.

Arequipa: native. ARE.

Note: Included in the T. peregrinum group by Andersson & Andersson (2000).

Tropaeolum tuberosum Ruiz & Pav.

Description: Tuber forming climber. Leaves lobed, peltate. Flowers orange-red, spur straight, linear, petals entire and not spread.

Ecology: Andean II-III: 2500-4000 m. Disturbed areas, rocky slopes.

Distribution: The Andes, from Colombia to Bolivia.

Arequipa: native. ARE, CAY, UNI.

Typhaceae

Typha domingensis Pers.

Description: Robust, erect, perennial herb, 1.5-3 m tall, round stem. Leaves linear, pointed, 8-22 x 0.5-1.5 cm, convex beneath, divided into a great number of compartments. Flowers on a brown, cigar-like spike, the males on the top 15-30 cm, and the females on a slightly shorter section below, separated by 0.5-12 cm without flowers. Flowers interspersed with long, fluffy, woolly hairs.

Ecology: Coastal, Andean I: 0-1500 m. In shallow water of lakes, ponds, rivers, swamps, channels.

Distribution: Worldwide in tropical and subtropical areas.

Arequipa: native. CAR, ISL.

Use: Edible roots.

Note: The phylogenetic analysis by Zhou et al. (2018) shows that T. angustifolia, T. capensis, and T. domingensis form a paraphyletic clade in which the three species are not clearly separable. T. angustifolia is mentioned by different authors for Peru and in some cases also for Arequipa (Brako & Zarucchi, 1993, Arce

Condori, 2010, Montesinos & Zegarra, 2019, and others). Here, we adopt the view of POWO (2024) and Jepson eFlora (2024) that T. angustifolia is restricted to Eurasia and that the South American clade belongs to the tropical and subtropical T. domingensis.

Urticaceae

Parietaria debilis G.Forst.

Description: Polygamous annual herb with tender stems. Leaves alternate, ovate, 1-3 x 0.5-2 cm, both surfaces sparsely pubescent to ± glabrous, lateral veins in 3 pairs, petiole as long as lamina. The lower axils mainly female with upper axils mixed male, female and bisexual flowers.

Ecology: Coastal, Andean I-II, Amazonian: 0-3000 m. Disturbed areas, lomas.

Distribution: South America, S-Africa and Australia.

Arequipa: native. ARE, CAM, CAR, CAY, ISL, UNI.

Soleirolia soleirolii (Req.) Dandy

Description: Slender creeping herb with stems ± succulent, rooting at the nodes, forming dense flat mats, finely pubescent. Leaves alternate, ± circular to reniform, mostly 2-4 x 3-5 mm, sparsely hairy, 3-veined from base, petiole 1 mm long. Flowers 1 mm long. Achene ellipsoid, 1 mm long.

Ecology: Andean II: 1500-3000 m. Damp, shady places, riverbanks, irrigation canals.

Distribution: Native to the W-Mediterranean.

Arequipa: introduced. ARE.

Urtica echinata Benth.

Description: Dioecious perennial (annual?) herb, erect, up to 40 cm tall, densely covered throughout with stinging hairs. Leaves opposite, ovate, petiolate, thick, rugose-bullate, margins coarsely dentate. Inflorescence cymose.

Ecology: Andean III: 3500->4500 m. Disturbed areas, grasslands, rocky slopes.

Distribution: Colombia to N-Chile and NW-Argentina.

Arequipa: native. ARE, CAY, CON.

Urtica flabellata Kunth

Description: Small annual plant with stinging hairs. Leaves opposite, flabellate-incised, the segments lobed. Inflorescence few-flowered, ± sessile.

Ecology: Andean II-III: 1000->4500 m. Disturbed areas, elfin forests, grasslands, riversides, rocky slopes.

Distribution: Ecuador to NW-Argentina and N-Chile.

Arequipa: native. ARE, CAY, CON, UNI.

Urtica leptophylla Kunth

Description: Erect perennial herb, 0.6-1.5 m tall. Stems with numerous stinging hairs ca. 1-1.5 mm long and very short, adpressed trichomes. Leaves opposite, thin, ± flat, on 4-6 cm long petiole, lamina widely ovate, 6-10 x 3-7 cm, base truncate to subcordate, margins serrate with 15-20 teeth on each side. Inflorescence usually an elongate spike, irregularly interrupted, with female and male flowers.

Ecology: Andean I-III: 1000->4500 m. Disturbed areas, elfin forests, grasslands, riversides, rocky slopes.

Distribution: Venezuela to Peru and Bolivia.

Arequipa: native. CAY.

Note: Weigend et al. (2005) points out that the name Urtica magellanica is often erroneously employed for introduced, weedy U. urens from Europe, but also for at least three different native species, most commonly for U. leptophylla, the most widespread and abundant Urtica species in Peru.

Urtica trichantha (Wedd.) Acevedo & L.E.Navas

Description: Slender herb, decumbent, with stinging hairs. Leaves opposite, ovate-orbicular, long petiolate, margins crenate-dentate. Inflorescence globular, shorter than petiole.

Ecology: Andean III: 4000-4500 m.

Distribution: Peru, Bolivia, N-Chile.

Arequipa: native. ARE, CON.

Urtica urens L.

Description: Annual herb with stinging hairs, to 75 cm tall, branching at the base. Leaves opposite, incised-dentate. Flowers in glomeruli.

Ecology: Coastal, Andean I-III, Amazonian: 0->4500 m. Disturbed areas, forests, lomas.

Distribution: Native to Africa, Asia, Europe, widely introduced elsewhere.

Arequipa: introduced. ARE, CAM, CAR, CAY, ISL, UNI.

Use: Medicinal plant.

Note: In South America, the name U. magellanica, restricted to Patagonia, is often erroneously employed for U. urens (Weigend et al., 2005).

Verbenaceae

Aloysia arequipensis Siedo

Description: Shrubs 0.5-1.5 m in height. Leaves opposite, rarely 3-whorled, 1-2 x 0.5-1.2 cm, elliptic, margins finely serrate apically, basally entire, upper side strigose, underside tomentose, with an understory of sub-sessile, glandular trichomes. Inflorescence ± loosely spicate, mostly terminal. Corolla weakly zygomorphic, lavender to pink, often with a whitish limb, tube 2-3 mm long. Fruit obovoid, 1.0-1.5 mm long and wide, glabrous, apically bilobed.

Ecology: Andean II-III: 2000-3000 m. Rocky slopes.

Distribution: S-Peru.

Arequipa: native. ARE, UNI.

Note: Aloysia scorodonioides var. hypoleuca (Briq.) Moldenke and Aloysia spathulata (Hayek) Moldenke have been recorded for Arequipa. The specimens from Arequipa all correspond to A. arequipensis, described by Siedo (2012) as an endemic species to Arequipa.

Citharexylum flexuosum (Ruiz & Pav.) D.Don

Description: Spiny shrub, 3-5 m tall, branches terete, spines 1-3 cm in some leaf axils. Leaves opposite, elliptic, 5 x 2 cm, chartaceous or membranous , margins entire, apex obtuse, base acute, glabrous above, hirsute on veins beneath. Inflorescence an axillary or terminal raceme with 12-20 white flowers.

Ecology: Coastal, Andean I-II: 0-3500 m. Lomas, disturbed areas.

Distribution: Bolivia, Peru.

Arequipa: native. CAR, ISL.

Citharexylum weberbaueri Hayek

Description: Shrub 1-3 m tall. Stems and branches initially tetragonal, terete at maturity, with spines 0.5-4 cm long in leaf axils. Leaves opposite, coriaceous, petioles 0.2-1.5 cm, 1-3(-6) × 1-2 cm, elliptic, obovate, or obtuse, apex acute or mucronate, base cuneate, margins entire, sometimes slightly serrate near apex, blade coriaceous, upper side glabrous, underside densely hirsute when young, glabrous at maturity. Inflorescence in axillary racemes, few- flowered, with 1-6 white flowers.

Ecology: Andean I-II: (1000-)2000-3000 m. Rocky places or dry gravelly slopes and bushy areas.

Distribution: Peru.

Arequipa: native. ISL. The type of C. megacanthum, Williams 2544 (NY) is from Mollendo.

Note: O'Leary et al. (2021) synonymize C. megacanthum with C. weberbaueri. C. weberbaueri and C. flexuosum are morphologically similar, occur in similar habitats and are probably often confused.

Glandularia laciniata (L.) Schnack & Covas

Description: Distinguished from the rest of the Glandularia taxa from Peru by its divided leaf blades, 3 or 5-parted with lobbed or parted segments.

Ecology: Andean II-III: 3000-4000 m.

Distribution: Peru, Bolivia and Chile.

Arequipa: native. ARE, CAY.

Glandularia microphylla (Kunth) Cabrera

Description: Creeping, suffruticose shrub forming mats up to 40 cm in Ø, barely over 10 cm tall, rooting on branches. Leaves coriaceous, sessile. Blades pinnately triparted to doubly pinnately lobed, 4-12 x 2-9 mm, narrowly ovate to broadly elliptic, acuminate, base decurrent, margins entire. Inflorescence unbranched, up to 3 cm when fruiting. Flowers white to violet.

Ecology: Andean II-III: 3000-4000 m.

Distribution: Ecuador to C-Chile.

Arequipa: native. CAY, CON.

Junellia clavata (Ruiz & Pav.) O'Leary & Múlgura

Description: Prostrate shrub, distinguished by its 3-sected leaf blades and the presence of glandular connective appendix.

Ecology: Coastal, Andean I-II: 300-3500 m. Disturbed areas, forests, lomas, rocky slopes.

Distribution: C- and S-Peru.

Arequipa: native. ARE, CAM, CAR, ISL.

Junellia fasciculata (Benth.) N.O'Leary & P.Peralta

Description: Suffrutescent shrub, 50-70 cm tall, much-branched, densely hirsute with glandular hairs. Leaves 3-5- sected or 3-5-parted, never from the base, occasionally 2-pinnatisect or parted. Inflorescence many-flowered, in dense spikes. Corolla white to purplish-pink.

Ecology: Coastal, Andean I-III: 0-4500 m. Disturbed areas, lomas, forests, grasslands.

Distribution: Peru.

Arequipa: native. CAM, ISL.

Note: Flora of Peru (1936ff.) mentions Hierobotana inflata for Peru, however, the specimen is J. fasciculata (O'Leary & Moroni, 2014).

Junellia juniperina (Lag.) Moldenke

Description: Much-branched pubescent shrub, 1 m tall, the opposite leaves deeply 3-parted, the segments acerose or spinose in age. Flowers white to purple, with yellow spots. Distinguished by its prickly leaves

Ecology: Andean II: 2500-3000 m. Rocky slopes.

Distribution: S-Peru and Bolivia, N-Chile, NW-Argentina.

Arequipa: native. ARE, CAY, CON.

Junellia minima (Meyen) Moldenke

Description: Flat cushion habit. Leaves 4-5 mm x 1-2 mm, imbricate. Inflorescence of only 2-3 flowers, calyx puberulent.

Ecology: Andean III: 4000-4500 m. Grasslands.

Distribution: Peru, Bolivia, N-Chile, NW-Argentina.

Arequipa: native. ARE, CAY, UNI.

Lantana reptans Hayek

Description: Dwarf shrub, up to 20 cm tall, decumbent stems up to 1 m long, often rooting at nodes. Stem terete hirsute, eglandular. Leaves ovate, margins crenate, base attenuate. Inflorescence on long peduncle, heads not elongated after anthesis. Flowers deep yellow.

Ecology: Andean I-II: 1000-3000 m. Shrublands.

Distribution: Ecuador to Peru

Arequipa: native. CAY. No specimen, observation from Caylloma in iNaturalist, mentioned for Arequipa by Binder (2002).

Lantana scabiosiflora Kunth

Description: Shrub 0.8 m tall, erect to procumbent (due to browsing). Leaves subrhombic-ovate, margins coarsely crenate-serrate, apex acute, upper side strigose, underside softly pilose, petiole 1-2 cm long. Inflorescence with peduncle to 3x longer than leaf. Heads elongated after anthesis. Flowers yellow or white with yellow center.

Ecology: Coastal, Andean I-II: 0-3500 m. Rocky slopes.

Distribution: Ecuador, Peru.

Arequipa: native. ARE, CAM, CAR, CON, ISL.

Lantana sprucei Hayek

Description: Shrub, up to 1.5 m tall. Young branches often reddish, hirsute. Leaves ovate, apex acuminate, margins finely crenate, underside whitish, upper side rugose, grayish-green. Heads often enveloped by the long bracts when youn, in fruit elongated up to 13 cm. Flowers whitish-violet, sometimes with yellow thorat. Fruits fleshy, whitish when ripe.

Ecology: Coastal, Andean I-II: 0-2500 m. Rocky slopes, shrublands, disturbed areas.

Distribution: Ecuador to Bolivia

Arequipa: native. ARE, CAR, UNI.

Note: The species is easily confused with Aloysia arequipensis. Binder (2002) synonymizes L. zahlbruckneri and L. svensonii with L. sprucei based on morphological parameters. This synonymy is not accepted by POWO (2024). Here we follow Binder, because her proposal combines 3 unresolved species, all with adjacent distributions.

Mulguraea arequipensis (Botta) N.O'Leary & P.Peralta

Description: Distinguished by its long inflorescence and yellow to reddish corollas, with long tubes.

Ecology: Andean II-III: 2200-4300 m. Rocky slopes.

Distribution: Peru, N-Chile.

Arequipa: native. ARE, CAY, UNI.

Phyla nodiflora (L.) Greene

Description: Stems prostrate, intricately branched. Leaves entire, serrate toward the apex. Flowering heads oblong- ovoid, cylindrical in age, corolla white.

Ecology: Coastal, Andean I: 0-1000 m. Grasslands, lomas.

Distribution: Native to South America and southern North America. Naturalized in warmer climates all over the world.

Arequipa: native. ARE, CAM, CAR, CAY, ISL, UNI.

Use: Used in folk medicine to relieve stomach pain and clean wounds.

Note: In Arequipa P. nodiflora var. minor (synonym P. canescens Kunth) and var. nodiflora.

Pitraea cuneato-ovata (Cav.) Caro

Description: Perennial herb, 50-80 cm tall, branched. Rhizome tuberous. Leaves elliptic, serrate. Inflorescence a terminal spike, 10-30 cm long. Flowers white to purple, fragrant, tube 8-14 mm long, calyx 6-10 mm long.

Ecology: Coastal, Andean I: 0-1500 m. disturbed areas, fields.

Distribution: Peru, Bolivia, Chile, N-Argentina, S-Brazil.

Arequipa: native. ISL.

Verbena glabrata Kunth

Description: Erect herb or suffruticose plant, 0.3-1 m, stems subglabrous to scarcely strigose, internodes 3-5 cm long. Leaves sessile, entire, ovate-elliptic, apex acute base amplexicaule, margins serrate. Recognized by the sub- conical inflorescences. Corolla white, lilac colored, tube 3.5-5 mm long, outside densely puberulous.

Ecology: Andean I-III: 1000-4000 m. Disturbed areas.

Distribution: Colombia, Ecuador, Peru.

Arequipa: native. ARE.

Verbena hispida Ruiz & Pav.

Description: Erect shrub with semi-amplexicaul, oblong leaves. Inflorescence a terminal spike, flowers pink to lavender. Distinguished by its densely hispid glandular pubescence.

Ecology: Andean II-III: 2500-4000 m. Disturbed areas, grasslands.

Distribution: Ecuador to Chile and Paraguay.

Arequipa: native. ARE, CAY.

Verbena litoralis Kunth

Description: Herb or suffruticose plant, 0.8-3 m, erect, subglabrous to slightly strigose. Leaves sessile to subpetiolate, 5-12 x 1-3 cm, entire, ovate-elliptic or obovate, apex acute or obtuse, base cuneate, margins irregularly serrate, subglabrous to strigose on both surfaces. Inflorescences cylindrical, up to 18 cm long in fructification. Floral bracts 1.5-4 mm long, ovate, subglabrous. Calyx 2-3.5 mm long, brief triangular teeth, strigose. Corolla white, purple or lilac, tube 3-5 mm long, densely puberulous.

Ecology: Coastal, Andean I-III, Amazonian: 0-4000 m. Disturbed areas, forests, grasslands, rocky slopes, shrublands.

Distribution: Mexico to Chile and Argentina, introduced to Australia and Africa.

Arequipa: native. ARE, CAM, CAR, CAY, ISL, UNI.

Viburnaceae

Sambucus peruviana Kunth

Description: Medium-sized, spreading tree or shrub. Leaves, opposite, pinnate, petiole to 6 cm. Flowers in terminal compound cymes, typically dome-shaped to flat to 30 cm wide, cream-white. Fruits dark-purple, depressed globular, 4 mm in Ø, calyx persistent at the apex.

Ecology: Andean II-III: 2000-4000 m. Disturbed areas, forests, rocky slopes, shrublands.

Distribution: Central America to Bolivia and Peru.

Arequipa: native. ARE, CAY. Often cultivated.

Use: Edible berries, used in traditional medicine as a diaphoretic and to treat sore throat.

Note: Galán de Mera et al. (2020) postulate that S. peruviana and S. nigra are heterotypic synonyms and that the species was introduced from Europe.

Violaceae

Viola granulosa Wedd.

Description: Inconspicuous, acaulescent and well camouflaged herb. Leaves whitish-brown, spatulate, obtuse, margins crenate and ciliate, surface warty, uneven due to protruding veins. Pedicel as long as the leaves. Flowers brownish, sepals linear, superior and lateral petals lanceolate, inferior obovate, emarginate, without spur.

Ecology: Andean III: >4000 m. Rocky places.

Distribution: Peru, Bolivia, N-Chile.

Arequipa: native. ARE.

Viola longibracteolata P.Gonzáles, Ed.Navarro & J.M.Watson

Description: Perennial, acaulous, rosulate herb, 5-9 cm long. Leaves densely imbricate in the rosette, 10-13 mm long, brown, thick succulent. Blade scarcely wider than pseudopetiole, margin 5 crenate along entire outer margin. Flowers solitary, corolla white above, dark purple beneath.

Ecology: Andean III: >5000 m. Rocky soils, cryoturbated soils.

Distribution: S-Peru.

Arequipa: native. ARE. Known only from the type locality (Gonzáles et al., 2022).

Viola micranthella Wedd.

Description: Acaulescent annual with a thin rhizome. Leaves in a rosette, lemon-green, sessile or attenuate into a short petiole, linear or lanceolate-spatulate, subobtuse, glabrous, margins entire, 9-20 x 2-4 mm. Stipules linear, 2-4 mm long, membranous. Flowers very small, white, pedicels shorter than the leaves, the superior petals obovate, the inferior emarginate, spur

at least as long as a third of the inferior petal, the 2 foremost stamens with long appendages engaged into the spur.

Ecology: Andean II-III: 1500-4500 m. Grasslands, shrublands.

Distribution: Peru, Bolivia, NW-Argentina.

Arequipa: native. ARE, CAY.

Viola ornata P.Gonzáles, Montesinos & J.M.Watson

Description: Perennial acaulous rosulate herb, 2 x 2.5 cm. Leaves densely imbricate in the rosette, 10-13 mm long, brown, thick-succulent. Blade widely flabellate, 2x wider than pseudopetiole, margin 11 crenate along entire outer margin. Flowers solitary, corolla white.

Ecology: Andean III: >4500 m. Sandy volcanic soils.

Distribution: S-Peru.

Arequipa: native, expected. Mentioned for Moquegua on the border of Arequipa.

Viola weberbaueri W.Becker

Description: Annual herb with thin root, rosette up to 6 cm. Leaves glabrous, linear-spatulate, subobtuse, margins entire, 15-30 (45) x 1.5-3 mm. Stipules small, very narrow or wanting. Pedicels shorter than leaves, sepals linear- oblong, subobtuse, glabrous, trinerved, membranous at the margin. Flowers pale yellow with brown strips, inferior petal carinate, emarginate, spur short.

Ecology: Coastal: 0-2000 m. Disturbed areas, dunes.

Distribution: Peru.

Arequipa: native. ARE, ISL.

Zygophyllaceae

Bulnesia retama (Gillies ex Hook. & Arn.) Griseb.

Description: Perennial shrub, up to 2 m tall, branched from base. Branches terete, grayish-green, leafless after flowering. Flowers yellow. Fruit alate, lantern-like, glabrous.

Ecology: Coastal, Andean I: 0-1500 m. Desert, rocky slopes.

Distribution: S-Peru, N-Argentina.

Arequipa: native. CAR. Restricted to Caravelí near the border with Ica.

Kallstroemia parviflora Norton

Description: Stem prostrate to decumbent, <1 m, strigose to glabrous. Stipules 5-7 mm. Leaflets 6-10. Flowers pedicellate, pedicel larger than subtending leaf, sepals persistent, taller than fruit body, adpressed, petals 6-12 mm, orange to yellow. Fruit 4-6 mm wide, ovoid, style to 3x as long as fruit.

Ecology: Andean I-II: 1000-3500 m. Disturbed areas, rocky slopes, shrublands.

Distribution: Native to SW North America. Introduced to Colombia, Ecuador and Peru.

Arequipa: introduced, adventive. UNI. No specimen, an observation from Cotahuasi (iNaturalist, 2024).

Larrea divaricata Cav.

Description: Xerophytic shrub up to 3 m tall, sparsely branched. Leaves 4-18 mm long, bifid, segments lanceolate to falcate, leathery, evergreen, sticky-resinous. Flowers 25 mm wide, yellow.

Ecology: Coastal: 500-1000 m. Desert.

Distribution: Peru, Chile and Argentina.

Arequipa: native. CAM, CON, UNI.

Tribulus terrestris L.

Description: Mostly annual. Stem ± silky or adpressed hairy, also sharply bristly. Stipules 1-5 mm, leaflet-like, leaflets 6-12. Flower <5 mm wide, pedicel generally <subtending leaf. Fruit 5 mm, <1 cm wide, ± flat, hairy, gray or ± yellow, prickles 4-7 mm, spreading, hairy to glabrous.

Ecology: Coastal, Andean I-II: 0-2000 m. Disturbed areas.

Distribution: Native to Europe, S-Asia, Africa and Australia, introduced in the Americas.

Arequipa: introduced. ARE, CAR.

Use: Used in folk medicine.

Zygophyllum chilense (Hook. & Arn.) Christenh. & Byng

Description: Perennial subshrub, up to 30 cm tall, entire plant hirsute, branched rectangularly. Leaves trifoliate, leaflets ovoid, acute, apex spiny. Stipules spiny. Flowers pink, 5 petals.

Ecology: Coastal, Andean I-II: 500-2500 m. Disturbed areas.

Distribution: Peru, N-Chile.

Arequipa: native. ARE, CAM.

Cited Literature

Ackermann M (2011). Studies on systematics, morphology and taxonomy of Caiophora and reproductive biology of Loasaceae and Mimulus (Phrymaceae). PhD Thesis, Freien Universität Berlin.

Acuña R & Weigend M (2017). A taxonomic revision of the western South American genus Presliophytum (Loasaceae). Phytotaxa, 329(1): 051-068.

Aedo C, Navarro C, Alarcon ML (2005). Taxonomic revision of Geranium sections Andina and Chilensia (Geraniaceae). Botanical Journal of the Linnean Society, 149: 1-68.

Agudelo CA (2008). Amaranthaceae. Flora de Colombia No. 23. Instituto de Ciencias Naturales, Universidad Nacional de Colombia. Bogotá.

Alonso MA & Crespo MB (2008). Taxonomic and nomenclatural notes on South American taxa of Sarcocornia (Chenopodiaceae). Ann. Bot. Fennici, 45: 241-254.

Alonso MA, Crespo MB & Freitag H (2017). Salicornia cuscoensis (Amaranthaceae / Chenopodiaceae), a new species from Peru (South America). Phytotaxa, 319(3): 254-262.

Al-Shehbaz IA & Montesinos-Tubée D (2009). Weberbauera arequipa (Brassicaceae), a New Species from Peru. Novon, 19(3): 281-283.

Al-Shehbaz IA (1990). A revision of Weberbauera (Brassicaceae). Journal of Arnold Arboretum, 71.

Al-Shehbaz IA (2004). A Synopsis of the South American Neuontobotrys (Brassicaceae). Novon, 14(3): 253-257.

Al-Shehbaz IA (2006a). The genus Sisymbirum in South America, with synopses of the genera Chilocadamum, Mostacillastrum, Neuontobotrys and Polypsecadium (Brassicaceae). Darwiniana, 44(2): 341-358.

Al-Shehbaz IA (2006b). Sisymbrium lactucoides belongs to Dictyophragmus (Brassicaceae). Harvard Papers in Botany, 11(1): 89-90.

Al-Shehbaz IA (2010). A synopsis of the South American Lepidium (Brassicaceae). Darwiniana, 48(2): 141-167.

Al-Shehbaz IA (2012). A synopsis of the South American Descurainia and description of the Colombian new speceis D. cleefii (Brassicaceae). Harvard Papers in Botany, 17(2): 221-229.

Al-Shehbaz IA (2018). A monograph of the South American species of Draba (Brassicaceae). Annals of the Missouri Botanical Garden, 103: 463-590.

Al-Shehbaz IA, Salariato DL, Cano A & Zuloaga FO (2023). A Revised Generic Delimitation of the South American–Endemic Tribe Eudemeae (Brassicaceae). Annals of the Missouri Botanical Garden, 108(1): 250-287.

Andersson L, & Andersson S (2000). A Molecular Phylogeny of Tropaeolaceae and Its Systematic Implications. Taxon, 49(4): 721-736.

APG IV - The Angiosperm Phylogeny Group (2016). An update of the Angiosperm Phylogeny Group classification for the orders and families of flowering plants: APG IV. Botanical Journal of the Linnean Society, 181(1): 1-20.

Arce Condori K (2010). Guia para la Florula del Santuario Nacional Lagunas de Mejia. Ministerio del Ambiente Peru.

Axelius B (1994). The genus Exodeconus and some comments on its relation with Nicandra (Solanaceae). Plant Systematics and Evolution, 193(1/4): 153-172.

Bailey DC, Al-Shehbaz IS, Govindarajulu R (2007). Generic Limits in Tribe Halimolobeae and Description of the New Genus Exhalimolobos (Brassicaceae). Systematic Botany, 32: 140-156.

Balslev H (1983). New Taxa and Combinations in Neotropical Juncus (Juncaceae). Brittonia, 35(3): 302-308.

Barker NP, Linder HP, Morton CM & Lyle M (2003). The Paraphyly of Cortaderia (Danthonioideae; Poaceae): Evidence from Morphology and Chloroplast and Nuclear DNA Sequence Data. Annals of the Missouri Botanical Garden, 90(1): 1-24.

Bartoli A, Tortosa RD (1999). Revisión de las especies sudamericanas de Grindelia. Kurtziana, 27(2): 327-359.

Bedoya-Cuno M, Dillon MO & Quipuscoa-Silvestre V (2024). Análisis taxonómico del género Stevia (Asteraceae, Eupatorieae) en Arequipa, Perú. Acta Botanica Mexicana, 131.

Beltrán H & Galán de Mera A (1996). Senecio [sect. Senecio] ser. Lomincola nova y notas corológicas y taxonómica sobre Senecio sect. Senecio (Asteraceae) para los Andes centrals del Perú. Bot. Complut., 21: 99-111.

Beltrán H (2017). Sinopsis del género Werneria (Asteraceae:Senecioneae) del Perú. Arnaldoa, 24(1): 45-62.

Beltrán H, Roque J, Caceres C (2018). A synopsis of the genus Malesherbia (Passifloraceae) in Peru. Revista peruana de Biología, 25(3): 229 240.

Binder M (2002). Verbena L., Glandularia Gmel., Junellia Moldenke, Lantana L. und Lippia L. (Verbenaceae) in Peru. PhD Thesis, Ludwig-Maximilians-Universität München.

Boza Espinoza TE, Kessler M (2022). A monograph of the genus Polylepis (Rosaceae). PhytoKeys, 203: 1-274.

Brako L and Zarucchi JL (1993). Catalogue of the Flowering Plants and Gymnosperms of Peru. Monogr. Syst. Bot. Missouri Bot. Gard., 45. Missouri Botanical Garden.

Brignone NF, Denham SS & Pozner R (2016). Synopsis of the genus Atriplex (Amaranthaceae, Chenopodioideae) for South America. Australian Systematic Botany, 29: 324-357.

Cabrera AL (1965). Revisión del género Mutisia (Compositae). Opera Lilloana, 13: 1-227.

Cabrera AL (1985). El genero Senecio (Compositae) en Bolivia. Darwiniana, 26(1/4): 79-217.

Calvo J & Moreira-Muñoz A (2020). Taxonomic revision of the Andean genus Xenophyllum (Compositae, Senecioneae). PhytoKeys, 158: 1-106.

Calvo J & Saldivia P (2022). Taxonomic and nomenclatural notes on Chilean Senecio (Asteraceae, Senecioneae), including a new particular species. Phytotaxa, 556(1): 76-86.

Calvo J, Moreira-Muñoz A & Funk VA (2020). Taxonomic Revision of the Neotropical Genus Werneria (Compositae, Senecioneae). Smithsonian Institution Scholarly Press. Book.

Cano Bellido YA (2006). Diversidad Taxonómica y Distribución de las Pteridophyta en las Lomas de Atiquipa-Arequipa, 2005. PhD Thesis, Universidad Nacional de San Augustin de Arequipa.

Caryophyllales Network (2024ff.). A global synthesis of species diversity in the angiosperm order Caryophyllales.

Chaudhri MN (1968). A revision of the Paronychiinae. PhD Thesis, Utecht-University.

Christenhusz MJM, Bangiolo L, Chase MW, Fay MF, Husby CE, Witkus M & Viruel J (2019). Phylogenetics, classification and typification of extant horsetails (Equisetum, Equisetaceae). Botanical Journal of the Linnean Society, 189(4): 311-352.

Crawford DJ, Tadesse M, Kimball RT, Carrillo-reyes P, Sanchez-Vega I & Mor ME (2014). Coreopsis sect. Pseudoagarista (Asteraceae: Coreopsideae): Molecular phylogeny, chromosome numbers, and comments on taxonomy and distribution. Taxon, 63(5): 1092-1102.

Dietrich W (1977). The South American Species of Oenothera Sect. Oenothera (Raimannia, Renneria; Onagraceae). Annals of the Missouri Botanical Garden, 64(3): 425-626.

Dillon MO (1997): Checklist of Lomas de Mejia, Arequipa, Peru. http://www.sacha.org/envir/deserts/locals/lists/mejia.htm

Dillon MO (2018). New combinations in Belloa J. Rémy and new diagnoses for Andean Lucilia Cass. and Mniodes (A.Gray) Benth. (Gnaphalieae, Asteraceae). Arnaldoa, 25(1): 51-74.

Dillon MO, Leiva S, Zapata M, Lezama P & Quipuscoa V (2011). Floristic Checklist of the Peruvian Lomas Formations. Arnaldoa, 18(1): 7-32.

Duke J (1961). Preliminary Revision of the Genus Drymaria. Annals of the Missouri Botanical Garden, 48(3), 173-268.

Eriksen B (1993). A Revision of Monnina Subg. Pterocarya (Polygalaceae) in Northwestern South America. Annals of the Missouri Botanical Garden, 80(1): 191-207.

Evrard C & van Hove C (2004). Taxonomy of the American Azolla Species (Azollaceae): A Critical Review. Systematics and Geography of Plants, 74(2): 301-318.

Faundez YL & Macaya BJ (1998). Presencia de Lophopappus foliosus Rusby (Asteraceae) en Chile y antecedentes taxonomicos sobre el genero Lophopappus Rusby. Noticiario Mensual de! Museo Nacional de Historia Natural, 332: 3-6.

Fernandez Ardiles CR & Quipuscoa V (2021). A tiny genus of Argentina and Peru: Leptoglossis Benth. Solanaceae Seminar Online.

Fernandez Monteiro JA (2012). Functional morphology and productivity of a tussock grassland in the Bolivian Altiplano. PhD Thesis, Universität Basel.

Flora Argentina (2018). Flora Argentina y del Cono Sur. Instituto de Botánica Darwinion. Buenos Aires, Argentina.

Flora of North America (1993ff.). eFloras.org. Missouri Botanical Garden.

Flora of Peru (1936ff.). Macbride J.F. et al. Fieldiana Botany (1936 -1995). Field Museum of Natural History. Chicago.

Flores Fuentes RA (2016). Caracterización florística de las formaciones de vegetación de la Región de Arica y Parinacota. Thesis, Universidad de Chile. Santiago, Chile.

Freire SE, Monti C, Bayón ND & Migoya MA (2018). Taxonomic Studies in Pseudognaphalium Kirp. (Asteraceae, Gnaphalieae) from Peru. Systematic Botany, 43(1): 325-343.

Funk VA (1997). Werneria s.l. (Compositae: Senecioneae) in Ecuador. In: R. Valencia & H. Balslev (eds), Estudios sobre diversidad y ecologia de plantas. Memorias del II Congreso Ecuatoriano de Botánica. Univ. Catól. Quito, 16-20: 25-35.

Galán de Mera A, Linares Perea E (2012). La Vegetacion de la Region Arequipa. Libreria Junior. Arequipa, Peru.

Galán de Mera A, Linares Perea E, Campos de la Cruz J, Vicente Oreallan JA (2009). Nuevas observaciones sobre la vegetacion del sur del Peru. Del desierto Pacifico al altiplano. Acta Botanica Malacitana, 34.

Galán de Mera A, Linares Perea E, Montoya Quino J, Vicente Orellana JA (2019). Prosopis andicola (Algarobia, Caesalpinioideae, Leguminosae), a new combination and rank, and P. calderensis, a new species for mesquite populations from Southern Peru. Phytotaxa, 414(1).

Galán de Mera A, Linares Perea E, Montoya Quino J, Torres Marquina I, Vicente Oreallan JA (2020). Acerca de Sambucus peruviana Kunth (Viburnaceae). La identidad de una planta medicinal andina. Arnaldoa, 27(2): 553-560.

Gomez-Sosa E (2005). Taxonomic novelties in Astragalus (Leguminosae) for South America. Novon, 15(4): 542-547.

Gonzáles P (2021). How many Salpichroa species there are, and where are they? Solanaceae Seminars Online.

Gonzáles P et al. (2022). Viola ornata and Viola longibracteolata (Violaceae, subgen. Neoandinium), two rare, new rosulate species from southern Peru. Phytotaxa, 571(1): 52-64.

Gonzáles P, Cano A & Müller J (2019). An unusual new record of Baccharis (Asteraceae) from the Peruvian Andes and its relation with the northern limit of the dry puna. Acta Botanica Mexicana, 126: e1393.

GrassBase (2002ff.): Clayton WD, Vorontsova MS, Harman KT & Williamson H - The Online World Grass Flora. https://www.kew.org/data/grasses-db/index.htm

Gutierrez DG, Muñoz-Schick M, Grossi MA, Rodriguez-Cravero JF, Morales V, Moreira-Muñoz A (2016). The genus Stevia (Eupatorieae, Asteraceae) in Chile: a taxonomical and morphological analysis. Phytotaxa, 282(1): 001-018.

Hassler M (2024). World Plants. Synonymic Checklist and Distribution of the World Flora. www.worldplants.de

Heim E (2014): Flora of Arequipa, Peru. Edition 1. Books on Demand, Norderstedt (D).

Heim E (2021): Catalog of ferns, gymnosperms and flowering plants of the department of Arequipa, Peru. Edition 1. Books on Demand, Norderstedt (D).

Hellwig F (1988). Anmerkungen zu zwei Baccharis-Arten (Compositae-Astereae) aus Peru und Chile. Mitteilungen der Botanischen Staatssammlung München, 27: 99-109.

Hermann M & Heller J (editors) (1997). Andean roots and tubers: Ahipa, arracacha, maca and yacon. Promiting the conservation and use of neglected and underutilized crops, 21.

Hershkovitz MA (1991). Taxonomic notes on Cistanthe, Calandrinia, and Talinum (Portulacaceae). Phytologia, 70: 209-225.

Hill S (1982). A MONOGRAPH OF THE GENUS MALVASTRUM - III. Rhodora, 84(839): 317-409.

Hind DJN (2011). An Annotaded Preliminary Checklist of the Compositae of Bolivia. Royal Botanic Gardens, KEW.

Hind DJN (2022). A new genus, Rockhausenia (Compositae: Senecioneae: Senecioninae). Kew Bull.

Hofreiter A, Rodriguez EF (2006). The Alstroemeriaceae in Peru and neighbouring areas. Revista Peruana de Biologia, 13(1): 5-69.

Huertas de Schneider ML (2010). Systematics and evolution of the genus Palaua (Malvaceae). PhD Thesis, Goethe Universität Frankfurt am Main.

Huiet L, Li FW, Kao T-T, Prado J, Smith AR, Schuettpelz E & Pryer KM (2018). A worldwide phylogeny of Adiantum (Pteridaceae) reveals remarkable convergent evolution in leaf blade architecture. Taxon, 67(3): 488-502.

Hunziker AT & Subils R (1979). Salpiglossis, Leptoglossis and Reyesia (Solanaceae) a synoptical survey. Botanical Museum Leaflets, Harvard University, 27(1/2): 1-43.

Iamonico D, Montesinos-Tubée DB (2023). The Genus Paronychia (Caryophyllaceae) in South America: Nomenclatural Review and Taxonomic Notes with the Description of a New Species from North Peru. Plants, 12(5): 1064.

Ickert-Bond SM & Wojciechowski MF (2004). Phylogenetic Relationships in Ephedra (Gnetales): Evidence from Nuclear and Chloroplast DNA Sequence Data. Systematic Botany, 29(4): 834-849.

iNaturalist.org (2024). iNaturalist App.

INGEMMET (2014). Instituto Geológico Minero y Metalúrgico. Lima, Peru. Mapa Geológico 50,000. https://www.ingemmet.gob.pe/

Jepson eFlora (2024). The Jepson Herbarium. University of California, Berkeley.

Johnston IM (1928). Studies in the Boraginaceae - VII. Contributions Gray Herbarium.

Johnston IM (1947). Astragalus in Argentina, Bolivia and Chile. Journal of Arnold Arboretum, 28(3): 336-374.

Katinas L (2012). Revisión del género Perezia (Compositae). Boletín de la Sociedad Argentina de Botánica, 47(1-2): 159-261.

Katinas L, Sancho G, Vitali MS (2013). A revision of Lophopappus (Asteraceae, Nassauvieae). Phytotaxa, 103: 25-45.

Kerr MS (2004). A phylogenetic and biogeographic analysis of Sanguisorbeae (Rosaceae), with emphasis on the Pleistocene radiation of the high Andean genus Polylepis. PhD Thesis University of Maryland.

Kessler M, Schmidt-Lebuhn AN (2006). Taxonomical and distributional notes on Polylepis (Rosaceae). Org. Divers. Evol. 6, Electr. Suppl. 1: 1-10.

Knapp S, Särkinen T, Barboza GE (2023). A revision of the South American species of the Morelloid clade (Solanum L., Solanaceae). PhytoKeys, 231: 1-342.

Krapovickas AC (1974). Acaulimalva, nuevo género de Malváceas. Darwiniana, 19(1): 9-39.

Krapovickas AC (1996). Sinopsis der genero Gaya (Malvaceae). Bonplandia, 9(1-2): 77.

Leiva Gonzales S, Mione T, Yacher L (2016). The genus Jaltomata Schltdl. (Solanaceae) from La Libertad Region and a new taxon from Northern Peru. Arnaldoa, 23(1): 21-98.

León B, Roque J, Ulloa Ulloa C, Pitman N, Jørgensen PM, Cano A (2006): El libro rojo de las plantas endémicas del Perú. Universidad Nacional Mayor De San Marcos, Lima

León JF, Cáceres Musaja C & Sulca Quispe L (2004): Flora y vegetación del departamento de Tacna. Ciencia & Desarrollo 8.

Liede-Schumann S & Meve U (2013). The Orthosiinae Revisited (Apocynaceae, Asclepiadoideae, Asclepiadeae). Annals of the Missouri Botanical Garden, 99(1): 44-81.

Lizarazu AM & Freire SE (2019). A Taxonomic Revision of Heterosperma (Asteraceae: Coreopsideae). Annals of the Missouri Botanical Garden, 104: 633-663.

Lleellish M, Odar J & Trinidad H (2015). Guía de Flora de las Lomas de Lima. Servicio Nacional Forestal y de Fauna Silvestre. Lima, Peru.

Lopez Laphitz RM, Ezcurra C & Vidal-Russel R (2015). Taxonomic revision of the South American genus Quinchamalium (Schoepfiaceae). Boletín de la Sociedad Argentina de Botánica, 50(2): 235-246.

Lozada-Gobilard S, Avila-Calero S, Ortuño T & Weigend M (2020). Taxonomical revision of the genus Hypseocharis in Peru and Bolivia. Revista Peruana de Biologia, 27(3): 383-394.

Magenta MAG, Pirani JR & Mondin CA (2010). Novos táxons e combinações de Viguiera Kunth (Asteraceae-Heliantheae) no Brasil. Rodriguésia, 61: 01-11.

Martínez Carretero E (2018). Ephedraceae Dumort. in Flora de Mendoza. Multequina, 27: Fascículo 10, FM 1-9.

Meerow AW, Gardner EM & Nakamura K (2020). Phylogenomics of the Andean Tetraploid Clade of the American Amaryllidaceae (Subfamily Amaryllidoideae): Unlocking a Polyploid Generic Radiation Abetted by Continental Geodynamics. Frontiers in Plant Science, 11.

Metzgar JS, Schneider H and Pryer (2007). Phylogeny and Divergence Time Estimates for the Fern Genus Azolla (Salviniaceae). Int. J. Plant Sci., 168(7): 1045-1053.

Miller JS and Levin RA (2023). Project Lycieae. Online: Miller Laboratory @ Amherst College.

Mione T, Leiva GS, Yacher L (2007). Five new species of Jaltomata (Solanaceae) from Cajamarca, Peru. Novon, 17(1): 49-58.

Mione T, Leiva GS, Yacher L, Cameron AM (2011). Jaltomata atiquipa (Solanaceae): a new species of southern Peru. Phytologia, 93(2): 203-207.

Molau U (1988). Scrophulariaceae Part I. Calceolarieae. Flora Neotropica, 47: 1-325.

Molinari-Novoa EA (2016). Review of the Trifolium amabile Complex in Peru, with the Description of a New Species. Scientifia, Article ID 5435781.

Montesinos-Tubée DB & Borsch T (2023). Molecular phylogenetics and morphology reveal the Plettkea lineage including several members of Arenaria and Pycnophyllopsis to be a clade of 21 South American species nested within Stellaria (Caryophyllaceae, Alsineae). Willdenowia, 53(3): 115-148.

Montesinos-Tubée DB & Gonzáles P (2020). Senecio Beltránii (Asteraceae, Senecioneae): a new caespitose species endemic to South Peru. Blumea, 65(2): 162 -166.

Montesinos-Tubée DB & Mondragón LP (2013). Flora y vegetación en tres localidades de una cuenca costeña: río Acarí, provincia de Caravelí (Arequipa, Perú). Zonas Áridas, 15(1): 11-30.

Montesinos-Tubée DB & Mondragón LP (2016). Adiciones a la flora de la cuenca del río Acarí, provincia de Caravelí, Arequipa, Perú. Chloris Chilensis, 19(2).

Montesinos-Tubée D & Teillier S (2022). Arenaria L. (Caryophyllaceae) en Chile. Gayana Bot., 79(2): 124-139.

Montesinos-Tubée DB & Zegarra Flores JA (2019). Plantas silvestres de Arequipa en áreas urbanas y rurales. Fondo Editorial Universidad Católica de Santa Maria. Arequipa.

Montesinos-Tubée DB (2012). Andean shrublands of Moquegua, South Peru: prepuna plant communities. Phytocoenologia, 42: 29-55.

Montesinos-Tubée DB (2015): Flora Moqueguana - Guia practica para la identificacion de plantas silvestres. Anglo American. Moquegua, Peru.

Montesinos-Tubée DB et al. (2019): Diversidad florística, comunidades vegetales y propuestas de conservación del monte ribereño en el río Chili (Arequipa, Perú). Arnaldoa, 26(1), 97-130.

Morales-Fierro V & Beltrán H (2021). Two new records of Senecio (Asteraceae, Senecioneae) for the Flora of Peru. Darwiniana, nueva serie, 9(1): 130-138.

Müller J (2006). Systematics of Baccharis (Compositae-Astereae) in Bolivia, including an Overview of the Genus. Mountain Research and Development, 14(3): 189-211.

Müller J (2013). World checklist of Baccharis L. (Compositae - Asteraceae). Draft der Friedrich-Schiller-Universität, Jena.

Muñuico JW, Dillon MO y Quipuscoa V (2024a). Jarava disticha (Poaceae, Pooideae, Stipeae), una nueva combinación nomenclatural y delimitación taxonómica con Jarava pachypus. Acta Botanica Mexicana, 131: e2292.

Muñuico JW, Dillon MO, Quipuscoa V (2024b). Dos nuevos registros para la flora agrostológica de Perú. Revista peruana de Biología, 31(1): e26851: 001-010.

Nesom GL (1993). Synopsis of Parastrephia (Asteraceae: Astereae). Phytologia, 75(5): 347-367.

NYBG Steere Herbarium (2020). World Flora Online. New York Botanical Garden, New York, USA.

Nylinder S & Anderberg AA (2015). Phylogeny of the Inuleae (Asteraceae) with special emphasis on the Inuleae-Plucheinae. Taxon, 64(1): 110-130.

O'Leary N & Moroni P (2014). Hierobotana Briq., an intriguing monotypic genus of tribe Verbeneae (Verbenaceae). Phytotaxa, 164(4): 286-290.

O'Leary N& Mulgura ME (2014). Synopsis of tribe Verbeneae Dumortier (Verbenaceae) in Peru. Phytotaxa, 163(3): 121-148.

O'Leary N, Yuan YW, Chemisquy A, Olmstead RG (2009). Reassignment of species of paraphyletic Junellia s.l. to thte new genus Mulguraea (Verbenaceae) and new circumscription of genus Junellia: molecular and morphological congruence. Systematic Botany, 34(4): 777-786.

O'Leary N, Lu-Irving P, Moroni P & Siedo S (2016). Taxonomic Revision of Aloysia (Verbenaceae, Lantaneae) In South America. Annals of the Missouri Botanical Garden, 101(3): 568-609.

O'Leary N, Frost LA, Mirra F & Moroni P (2021). Insights into the Taxonomy of Citharexylum (Verbenaceae): A Revision of the South American Taxa. Annals of the Missouri Botanical Garden, 106: 167-233.

Pacific Bulb Society (2020). Photographs and Information. www.pacificbulbsociety.org.

Pauca A & Quipuscoa V (2017). Catálogo de las cactáceas del departamento de Arequipa, Perú. Arnaldoa, 24(2): 447-496.

Pauca-Tanco GA & Quipuscoa V (2020). El género Cumulopuntia (Cactaceae, Opuntioideae) en el Departamento de Arequipa, Perú. Darwiniana, nueva serie, 8(1): 337-371.

Payne Willard W (1966). Notes on the Ragweeds of South America with the Description of two new species: Ambrosia pannosa and A. parvifolia (Compositae).

Pedersen TM (1972). Cyperus laetus Presl and Cyperus rigens Presl, two badly understood South American sedges, with notes on some related species. Darwiniana, 17: 527-547.

Peralta IE, Knapp S, Spooner DM (2005). New Species of Wild Tomatoes (Solanum Section Lycopersicon: Solanaceae) from Northern Peru. Systematic Botany, 30(2): 424-434.

Peterson PM & Sanchez Vega I (2007). Eragrostis (Poaceae: Chloridoideae: Eragrostideae: Eragrostidinae) of Peru. Annals of the Missouri Botanical Garden, 94(4): 745-790.

Peterson PM, Sanchez Vegea I, Romaschenko K, Giraldo-Canas D, Refulio Rodriguez NF (2018). Revision of Muhlenbergia (Poaceae, Chloridoideae, Cynodonteae, Muhlenbergiinae) in Peru: Classification, phylogeny, and a new species, M. romaschenkoi. PhytoKeys, 114: 123-206.

Peterson PM, Soreng RJ, Romaschenko K, Barberá P, Quintanar A & Aedo C (2019). New combinations and new names in American Cinnagrostis, Peyritschia, and Deschampsia, and three new genera: Greeneochloa, Laegaardia, and Paramochloa (Poeae, Poaceae). Phytoneuron 39: 1-23.

Pfanzelt S, von Hagen KB (2016). Morphological variation of Gentiana section Chondrophyllae in South America and taxonomic implications. Plant Systematics and Evolution, 302: 155-172.

POWO (2024). Plants of the World online. KEWScience (2019-2024).

Pteridophyte Phylogeny Group (PPG) (2016). A community-derived classification for extant lycophytes and ferns. J. Syst. Evol., 54: 563-603.

Pruski JF (2021). Studies of Neotropical Compositae–XV. The new genus Chaetacalia, retention of Aetheolaena, Culcitium, Haplosticha, and Iocenes, two new species of Senecio, and Lasiocephalus revisited again (Senecioneae: Senecioninae). Phytoneuron 65: 1-83.

Quipuscoa Silvestre V & Dillon MO (2018). Four new endemic species of Nolana (Solanaceae-Nolaneae) from Arequipa, Peru. Arnaldoa, 25(2): 295-322.

Quipuscoa Silvestre V & Dillon MO (2020). Three New Species of Mniodes A. Gray (Gnaphaliinae, Gnaphalieae, Asteraceae) from Bolivia and Peru and Nomenclatural Changes in the Lucilia-Group. Arnaldoa, 27(2): 375-404.

Quipuscoa Silvestre V & Huamantupa I (2010). Plantas vasculares de la Reserva Nacional de Salinas y Aguada Blanca, Arequipa-Peru. Lima: desco, PROFONANPE, SERNANP, ISBN: 978-612-4043-09-3.

Quipuscoa Silvestre V, Tejada Perez C, Pernandez Ardiles C, Pauca Tanco A, Durand Vera K, Dillon MO (2016). Diversity of vascular plants in Lomas de Yuta, Islay province, Arequipa, Peru, 2016. Arnaldoa, 23(2): 517 - 546.

Ravenna P (1988). Notes on Iridaceae VII. Phytologia, 64(4): 289.

Richardson AT (1977). Monograph of the genus Tiquilia (Coldenia, sensu lato), Boraginaceae: Ehretioideae. Rhodora, 79(820): 467-572.

Ritz CM, Reiker J, Charles G, Hoxey P, Hunt D, Lowry M, Stuppy W, Taylor N (2008). Molecular phylogeny and character evolution in trete-stemmed Andean oputias (Cactaceae-Opuntioideae). Molecular Phylogenetics and Evolution, 65: 668-681.

Rivas-Martínez S, Rivas-Sáenz S (2018). Worldwide bioclimatic classification system, 1996–2018. Phytosociological Research Center, Spain.

Rivera S, Leon B, Arakaki M (2024). Taxonomic updates of Cheilanthes Sw. (Pteridaceae) in Peru, with a conservation assessment for four endemics. Revista peruana de Biología, 31(1) e26205: 001-030.

Robinson H & Moore AJ (2004). New species and new combinations in Rhysolepis (Heliantheae: Asteraceae). Proceedings of the Biological Society of Washington, 117(3): 423-446.

Rodríguez EF & Sagástegui AA (2014). Notas sobre el génereo Stachys (Lamiaceae) en el Perú. Revista de Investigación Científica REBIOL 34(2): 83-89.

Rodriguez Diaz M (1998). Estudio de la Biodiversidad Cuenca del Cotahuasi: La Union - Arequipa

Flora medicinal. Asociacion Especializada para el Desarrollo - AEDES."

Ruhm J, Böhnert T, Mutke J, Liebert F, Montesinos-Tubée DB & Weigend M (2022). Two Sides of the Same Desert: Floristic Connectivity and Isolation Along the Hyperarid Coast and Precordillera in Peru and Chile. Frontiers in Ecology and Evolution, 10.

Ruhsatz B (2012). Vegetation and ecology of the high Andean peatlands of Bolivia. Phytocoenologia, 42(3-4): 133-179.

Saarela JM, Peterson PM & Refulio Rodriguez NF (2006). Bromus ayacuchensis (Poaceae: Pooideae: Bromeae), a new species from Peru, with a key to Bromus in Peru. SIDA, Contributions to Botany 22(2): 915-926.

Salomon L, Sklenar P, Freire S (2018). Synopsis of Senecio series Culcitium (Asteraceae: Senecioneae, Senecioninae) in the Andean region of South America. Phytotaxa, 340(1): 1-47.

Schaefer H & Acevedo-Rodriguez P (2021). Guide to genera of lianas and climbing plants in the Neotropics - Cucurbitaceae.

Schiavinato DJ, Gutierrez DG & Bartoli A (2017). Typifications and nomenclatural clarifications in South American Tagetes (Asteraceae, Tageteae). Phytotaxa, 326(3): 175-188.

Schiavinato DJ, Gutierrez DG & Bartoli A (2023). Tagetes dombeyi (Asteraceae, Tageteae), a new species from the Central Andes first collected in the 18th century. Rodriguésia, 74: e00162023.

Schlumpberger BO & Renner SS (2012). Molecular phylogenetics of Echinopsis (Cactaceae): Polyphyly at all levels and convergent evolution of pollination modes and growth forms. American Journal of Botany, 99(8): 1335-1349.

Schneider JV (2013). The Peruvian species of Cristaria (Malveae, Malvaceae): Taxonomic revision, chromosome counts, and breeding system. Phytotaxa, 110(1).

Schwarzer C (2007). Systematische Untersuchungen an den peruanischen Vertretern der Gattungen Pectocarya D.C. ex Meisn., Amsinckia Lehm., Plagiobothrys Fisch. & C.A.Mey. und Cryptantha Lehm. ex G.Don (Boraginaceae). Thesis, Universität Berlin.

Shaw JMH (2021). The identity of Oxalis megalorrhiza – a 300-year puzzle. CactusWorld, 39(1): 67-72.

Shaw JMH (2023). Oxalis OXALIDACEAE. In: Eggli U., Nyffeler R. (eds) Dicotyledons: Rosids. Illustrated Handbook of Succulent Plants. Springer.

Siedo SJ (2012). Four new species of the genus Aloysia Palau (Verbenaceae). Lundellia, 15: 35-46.

Simpson BB, Ulibarri EA (2006). A Synopsis of the Genus Hoffmannseggia (Leguminosae). Lundellia, 9: 7-33.

SINIA (2024). Sistema Nacional de Información Ambiental. Ministerio del Ambiente, Peru. https://sinia.minam.gob.pe/

Solbrig OT (1962). The South American Species of Erigeron. Contributions Gray Herbarium, 191: 3-79.

Solomon JC (1982). The Systematics and Evolution of Epilobium (Onagraceae) in South America. Annals of the Missouri Botanical Garden, 69(2): 239-335.

Squeo FA, Arancio G & Gutiérrez JR (2008). Catálogo de la Flora Vascular de la Región de Atacama. Libro Rojo de la Flora Nativa y de los Sitios Prioritarios para su Conservación, 6: 97-120.

Stevens PF (2017). Angiosperm Phylogeny Website. Version 14. http://www.mobot.org/MOBOT/research/APweb/

Svenson HK (1946). Vegetation of the Coast of Ecuador and Peru and Its Relation to that of the Galapagos Islands. II. Catalogue of Plants. American Journal of Botany, 33(6): 427-498.

Sylvester SP (2014). Bartsia lydiae, a new species of Bartsia sect. Laxae (Orobanchaceae) from the southern Peruvian Andes with a revised key to Bartsia sect. Laxae. Phytotaxa, 164(1): 41-46.

Sylvester S, Soreng R, Peterson PM & Sylvester MDPV (2016). An updated checklist and key to the open-panicled species of Poa L. (Poaceae) in Peru including three new species, Poa ramoniana, Poa tayacajaensis, and Poa urubambensis. PhytoKeys, 65: 57-90.

Talavera C, Pauca A, Fernandez C, Villasante F, Villegas L y Delgado A (2017). Flora de Lomas de Atiquipa. Universidad Nacional de San Agustín de Arequipa. ISBN: 978-612-4337-11-6.

Tate JA (2011). The status of Urocarpidium (Malvaceae): Insight from nuclear and plastid-based phylogenies. Taxon, 60(5): 1330-1338.

Taylor CM (1994). Revision of Tetragonia (Aizoaceae) in South America. Systematic Botany, 19(4): 575-589.

Testoni D, Linder HP (2017). Synoptic taxonomy of Cortaderia Stapf (Danthonioideae, Poaceae). PhytoKeys, 76: 39-69.

Timana M (2005). Systematic studies in Pycnophyllum and Pycnophyllopsis (Caryophyllaceae) of the high Andes. PhD Thesis, University of Texas.

Tropicos.org (2023). Database of all specimen data in Missouri Botanical Garden.

Trujillo D, Gonzales P, Trinidad H & Cano A (2016). The Andean genus Myrosmodes (Orchidaceae, Cranichideae) in Peru. Lankersteriana, 16(2): 129-151.

Tryon RM, Stolze RG (1989). Pteridophyta of Peru - Part I-VI. Fieldiana, Botany New Series 20 (1989), 22 (1989), 27 (1991), 29 (1992), 32 (1993), 34 (1994).

Urtubey E et al. (2016). New circumscription of the genus Gamochaeta (Asteraceae, Gnaphalieae) inferred from nuclear and plastid DNA sequences. Plant Systematics and Evolution, 302: 1047-1066.

van Ooststroom SJ (1934). A monograph of the genus Evolvulus. Mededeelingen van het Botanisch Museum en Herbarium van de Rijks Universiteit te Utrecht.

Villanueva-Almanza L, Landis JB, Koenig D, Ezcurra E (2021). Genetic and morphological differentiation in Washingtonia (Arecaceae): solving a century-old palm mystery. Botanical Journal of the Linnean Society, 196(4): 506-523.

Walter H und Breckle SW (1999). Vegetation und Klimazonen. Ulmer.

Watson L, Macfarlane TD & Dallwitz MJ (2024). The grass genera of the world.

Weigend M (2011). The genus Balbisia (Vivianiaceae, Geraniales) in Peru, Bolivia and northern Chile. Phytotaxa, 22: 47-56.

Weigend M, Cano Echeverria A & Rodriguez-Rodriguez EF (2005). New species and new records of the flora in Amotape-Huancabamba Zone: Endemics and biogeographic limits. Revista Peruana de Biologia, 12(2).

Whaley OQ, Orellana A, Pérez E, Tenorio M, Quinteros F, Mendoza M & Pecho O.(2010): Plantas y Vegetación de Ica, Perú. Un recurso para su restauración y conservación.

Zhou B, Tu T, Kong F, Wen J, Xu X (2018). Revised phylogeny and historical biogeography of the cosmopolitan aquatic plant genus Typha (Typhaceae). Sci. Rep., 8(1): 8813.

Appendix

Appendix 1: List of genera and their families

Aa (Orchidaceae)
Abutilon (Malvaceae)
Acaulimalva (Malvaceae)
Achyrocline (Asteraceae)
Aciachne (Poaceae)
Acmella (Asteraceae)
Adesmia (Fabaceae)
Adiantum (Pteridaceae)
Agave (Asparagaceae)
Ageratina (Asteraceae)
Agrostis (Poaceae)
Airampoa (Cactaceae)
Alchemilla (Rosaceae)
Aldama (Asteraceae)
Allionia (Nyctaginaceae)
Alnus (Betulaceae)
Aloe (Asphodelaceae)
Alonsoa (Scrophulariaceae)
Aloysia (Verbenaceae)
Alshehbazia (Brassicaceae)
Alstroemeria (Alstroemeriaceae)
Alternanthera (Amaranthaceae)
Amaranthus (Amaranthaceae)
Amauropelta (Thelypteridaceae)
Ambrosia (Asteraceae)
Ammannia (Lythraceae)
Amsinckia (Boraginaceae)
Andicolea (Asteraceae)
Andrachne (Phyllanthaceae)
Annona (Annonaceae)
Anredera (Basellaceae)
Aphyllocladus (Asteraceae)
Apium (Apiaceae)
Apodanthera (Cucurbitaceae)
Arcytophyllum (Rubiaceae)
Arenaria (Caryophyllaceae)
Argemone (Papaveraceae)
Argylia (Bignoniaceae)
Argyrochosma (Pteridaceae)
Aristeguietia (Asteraceae)
Aristida (Poaceae)
Armatocereus (Cactaceae)
Arracacia (Apiaceae)
Arundo (Poaceae)
Aschersoniodoxa (Brassicaceae)
Asclepias (Apocynaceae)
Asparagus (Asparagaceae)
Asplenium (Aspleniaceae)
Astragalus (Fabaceae)
Astrolepis (Pteridaceae)

Atriplex (Amaranthaceae)
Austrocylindropuntia (Cactaceae)
Avena (Poaceae)
Azolla (Salviniaceae)
Azorella (Apiaceae)
Baccharis (Asteraceae)
Bacopa (Plantaginaceae)
Balbisia (Francoaceae)
Berberis (Berberidaceae)
Bidens (Asteraceae)
Blechnum (Blechnaceae)
Boerhavia (Nyctaginaceae)
Bolboschoenus (Cyperaceae)
Bomarea (Alstroemeriaceae)
Bougainvillea (Nyctaginaceae)
Bouteloua (Poaceae)
Bowlesia (Apiaceae)
Brassica (Brassicaceae)
Bromus (Poaceae)
Browningia (Cactaceae)
Buddleja (Scrophulariaceae)
Bulnesia (Zygophyllaceae)
Burnellia (Asteraceae)
Caiophora (Loasaceae)
Calandrinia (Montiaceae)
Calceolaria (Calceolariaceae)
Calliandra (Fabaceae)
Callitriche (Plantaginaceae)
Calycera (Calyceraceae)
Camissonia (Onagraceae)
Canna (Cannaceae)
Cantua (Polemoniaceae)
Capsella (Brassicaceae)
Cardionema (Caryophyllaceae)
Cardiospermum (Sapindaceae)
Carex (Cyperaceae)
Carica (Caricaceae)
Castilleja (Orobanchaceae)
Cenchrus (Poaceae)
Centaurea (Asteraceae)
Centaurium (Gentianaceae)
Cerastium (Caryophyllaceae)
Ceratophyllum (Ceratophyllaceae)
Cestrum (Solanaceae)
Chaerophyllum (Apiaceae)
Chaetanthera (Asteraceae)
Cheilanthes (Pteridaceae)
Chenopodiastrum (Amaranthaceae)
Chenopodium (Amaranthaceae)
Chersodoma (Asteraceae)

Chionopappus (Asteraceae)
Chloris (Poaceae)
Chorizanthe (Polygonaceae)
Chuquiraga (Asteraceae)
Cicendia (Gentianaceae)
Cichorium (Asteraceae)
Cinnagrostis (Poaceae)
Cistanthe (Montiaceae)
Citharexylum (Verbenaceae)
Clematis (Ranunculaceae)
Cleome (Cleomaceae)
Clinanthus (Amaryllidaceae)
Clinopodium (Lamiaceae)
Colignonia (Nyctaginaceae)
Colletia (Rhamnaceae)
Colobanthus (Caryophyllaceae)
Colocasia (Araceae)
Commelina (Commelinaceae)
Commicarpus (Nyctaginaceae)
Conium (Apiaceae)
Convolvulus (Convolvulaceae)
Cordia (Boraginaceae)
Corryocactus (Cactaceae)
Cortaderia (Poaceae)
Cottea (Poaceae)
Cotula (Asteraceae)
Crassula (Crassulaceae)
Cremolobus (Brassicaceae)
Cressa (Convolvulaceae)
Cristaria (Malvaceae)
Crotalaria (Fabaceae)
Croton (Euphorbiaceae)
Cryptantha (Boraginaceae)
Cuatrecasasiella (Asteraceae)
Culcitium (Asteraceae)
Cumulopuntia (Cactaceae)
Cuscuta (Convolvulaceae)
Cyclanthera (Cucurbitaceae)
Cyclospermum (Apiaceae)
Cylindropuntia (Cactaceae)
Cymbalaria (Plantaginaceae)
Cynodon (Poaceae)
Cyperus (Cyperaceae)
Cystopteris (Cystopteridaceae)
Dalea (Fabaceae)
Datura (Solanaceae)
Daucus (Apiaceae)
Dayia (Polemoniaceae)
Dendrophthora (Santalaceae)
Deschampsia (Poaceae)

Descurainia (Brassicaceae)
Desmodium (Fabaceae)
Dichondra (Convolvulaceae)
Dicliptera (Acanthaceae)
Dictyophragmus (Brassicaceae)
Digitaria (Poaceae)
Diplachne (Poaceae)
Diplostephium (Asteraceae)
Diplotaxis (Brassicaceae)
Dipsacus (Caprifoliaceae)
Distichia (Juncaceae)
Distichlis (Poaceae)
Domeykoa (Apiaceae)
Draba (Brassicaceae)
Drymaria (Caryophyllaceae)
Dunalia (Solanaceae)
Dyschoriste (Acanthaceae)
Dysphania (Amaranthaceae)
Echeandia (Asparagaceae)
Echeveria (Crassulaceae)
Echinochloa (Poaceae)
Eclipta (Asteraceae)
Eleocharis (Cyperaceae)
Eleusine (Poaceae)
Elodea (Hydrocharitaceae)
Encelia (Asteraceae)
Enneapogon (Poaceae)
Ephedra (Ephedraceae)
Epilobium (Onagraceae)
Equisetum (Equisetaceae)
Eragrostis (Poaceae)
Eremocharis (Apiaceae)
Erigeron (Asteraceae)
Eriochloa (Poaceae)
Eriosyce (Cactaceae)
Erodium (Geraniaceae)
Erythranthe (Phrymaceae)
Erythrostemon (Fabaceae)
Escallonia (Escalloniaceae)
Eschscholzia (Papaveraceae)
Eudema (Brassicaceae)
Eulychnia (Cactaceae)
Euphorbia (Euphorbiaceae)
Euploca (Boraginaceae)
Evolvulus (Convolvulaceae)
Exhalimolobos (Brassicaceae)
Exodeconus (Solanaceae)
Fabiana (Solanaceae)
Facelis (Asteraceae)
Festuca (Poaceae)
Ficus (Moraceae)
Flaveria (Asteraceae)
Foeniculum (Apiaceae)
Frankenia (Frankeniaceae)

Froelichia (Amaranthaceae)
Fuertesimalva (Malvaceae)
Fumaria (Papaveraceae)
Funastrum (Apocynaceae)
Furcraea (Asparagaceae)
Galinsoga (Asteraceae)
Galium (Rubiaceae)
Galvezia (Plantaginaceae)
Gamochaeta (Asteraceae)
Gaya (Malvaceae)
Genista (Fabaceae)
Gentiana (Gentianaceae)
Gentianella (Gentianaceae)
Geoffroea (Fabaceae)
Geranium (Geraniaceae)
Gilia (Polemoniaceae)
Glandularia (Verbenaceae)
Gochnatia (Asteraceae)
Gomphocarpus (Apocynaceae)
Gomphrena (Amaranthaceae)
Grindelia (Asteraceae)
Guilleminea (Amaranthaceae)
Gynerium (Poaceae)
Gynoxys (Asteraceae)
Haageocereus (Cactaceae)
Halerpestes (Ranunculaceae)
Hedeoma (Lamiaceae)
Heiseria (Asteraceae)
Heliotropium (Boraginaceae)
Helogyne (Asteraceae)
Herissantia (Malvaceae)
Herniaria (Caryophyllaceae)
Hesperoxiphion (Iridaceae)
Heterosperma (Asteraceae)
Hieracium (Asteraceae)
Hoffmannseggia (Fabaceae)
Hordeum (Poaceae)
Hydrocotyle (Araliaceae)
Hypericum (Hypericaceae)
Hypertelis (Molluginaceae)
Hypochaeris (Asteraceae)
Hypseocharis (Geraniaceae)
Indigofera (Fabaceae)
Inga (Fabaceae)
Ipomoea (Convolvulaceae)
Isolepis (Cyperaceae)
Jaborosa (Solanaceae)
Jacquemontia (Convolvulaceae)
Jaltomata (Solanaceae)
Jarava (Poaceae)
Jatropha (Euphorbiaceae)
Jobinia (Apocynaceae)
Johnstonella (Boraginaceae)
Juncus (Juncaceae)

Junellia (Verbenaceae)
Jungia (Asteraceae)
Kageneckia (Rosaceae)
Kalanchoe (Crassulaceae)
Kallstroemia (Zygophyllaceae)
Koeleria (Poaceae)
Krameria (Krameriaceae)
Lactuca (Asteraceae)
Laennecia (Asteraceae)
Lantana (Verbenaceae)
Larrea (Zygophyllaceae)
Lemna (Araceae)
Leonotis (Lamiaceae)
Lepechinia (Lamiaceae)
Lepidium (Brassicaceae)
Leptoglossis (Solanaceae)
Leucaena (Fabaceae)
Leucheria (Asteraceae)
Ligaria (Loranthaceae)
Lilaeopsis (Apiaceae)
Limonium (Plumbaginaceae)
Limosella (Scrophulariaceae)
Linum (Linaceae)
Loasa (Loasaceae)
Lobelia (Campanulaceae)
Lobivia (Cactaceae)
Lobularia (Brassicaceae)
Lolium (Poaceae)
Lomanthus (Asteraceae)
Lonicera (Caprifoliaceae)
Lophopappus (Asteraceae)
Lorenzochloa (Poaceae)
Loxanthocereus (Cactaceae)
Lucilia (Asteraceae)
Ludwigia (Onagraceae)
Lupinus (Fabaceae)
Luzula (Juncaceae)
Lycianthes (Solanaceae)
Lycium (Solanaceae)
Lysimachia (Primulaceae)
Lysipomia (Campanulaceae)
Machaerophorus (Brassicaceae)
Malesherbia (Passifloraceae)
Malva (Malvaceae)
Malvastrum (Malvaceae)
Mancoa (Brassicaceae)
Marrubium (Lamiaceae)
Mastigostyla (Iridaceae)
Mathewsia (Brassicaceae)
Matricaria (Asteraceae)
Matthiola (Brassicaceae)
Matucana (Cactaceae)
Maytenus (Celastraceae)
Medicago (Fabaceae)

Melica (Poaceae)
Melilotus (Fabaceae)
Melocactus (Cactaceae)
Mentha (Lamiaceae)
Mentzelia (Loasaceae)
Mesembryanthemum (Aizoaceae)
Mesosphaerum (Lamiaceae)
Microchloa (Poaceae)
Microsteris (Polemoniaceae)
Mimosa (Fabaceae)
Minthostachys (Lamiaceae)
Mirabilis (Nyctaginaceae)
Misbrookea (Asteraceae)
Mniodes (Asteraceae)
Modiola (Malvaceae)
Mollugo (Molluginaceae)
Monnina (Polygalaceae)
Montia (Montiaceae)
Montiopsis (Montiaceae)
Mostacillastrum (Brassicaceae)
Muehlenbeckia (Polygonaceae)
Muhlenbergia (Poaceae)
Mulguraea (Verbenaceae)
Munnozia (Asteraceae)
Munroa (Poaceae)
Mutisia (Asteraceae)
Myoporum (Scrophulariaceae)
Myrcianthes (Myrtaceae)
Myrica (Myricaceae)
Myriophyllum (Haloragaceae)
Myriopteris (Pteridaceae)
Myrosmodes (Orchidaceae)
Nama (Boraginaceae)
Nasa (Loasaceae)
Nassella (Poaceae)
Nasturtium (Brassicaceae)
Neltuma (Fabaceae)
Neobartsia (Orobanchaceae)
Neoraimondia (Cactaceae)
Neuontobotrys (Brassicaceae)
Nicandra (Solanaceae)
Nicotiana (Solanaceae)
Nolana (Solanaceae)
Notholaena (Pteridaceae)
Nothoscordum (Amaryllidaceae)
Nototriche (Malvaceae)
Novenia (Asteraceae)
Nuttallanthus (Plantaginaceae)
Oenothera (Onagraceae)
Olsynium (Iridaceae)
Onoseris (Asteraceae)
Ophioglossum (Ophioglossaceae)
Ophryosporus (Asteraceae)
Opuntia (Cactaceae)

Oreocereus (Cactaceae)
Oriastrum (Asteraceae)
Oritrophium (Asteraceae)
Orobanche (Orobanchaceae)
Otholobium (Fabaceae)
Ourisia (Plantaginaceae)
Oxalis (Oxalidaceae)
Oxychloe (Juncaceae)
Oziroe (Asparagaceae)
Palaua (Malvaceae)
Pappostipa (Poaceae)
Paquirea (Asteraceae)
Parasenegalia (Fabaceae)
Paraserianthes (Fabaceae)
Parastrephia (Asteraceae)
Parietaria (Urticaceae)
Parkinsonia (Fabaceae)
Paronychia (Caryophyllaceae)
Pascalia (Asteraceae)
Pasithea (Asphodelaceae)
Paspalum (Poaceae)
Passiflora (Passifloraceae)
Patosia (Juncaceae)
Pectis (Asteraceae)
Pectocarya (Boraginaceae)
Pellaea (Pteridaceae)
Pentacyphus (Apocynaceae)
Peperomia (Piperaceae)
Perezia (Asteraceae)
Perityle (Asteraceae)
Persicaria (Polygonaceae)
Phacelia (Boraginaceae)
Phalaris (Poaceae)
Philibertia (Apocynaceae)
Philoglossa (Asteraceae)
Phragmites (Poaceae)
Phyla (Verbenaceae)
Phylloscirpus (Cyperaceae)
Physalis (Solanaceae)
Phytolacca (Phytolaccaceae)
Picradeniopsis (Asteraceae)
Picrosia (Asteraceae)
Piptochaetium (Poaceae)
Pitraea (Verbenaceae)
Pityrogramma (Pteridaceae)
Plagiobothrys (Boraginaceae)
Plantago (Plantaginaceae)
Plazia (Asteraceae)
Pleopeltis (Polypodiaceae)
Pluchea (Asteraceae)
Plumbago (Plumbaginaceae)
Poa (Poaceae)
Poissonia (Fabaceae)
Polyachyrus (Asteraceae)

Polycarpon (Caryophyllaceae)
Polygonum (Polygonaceae)
Polylepis (Rosaceae)
Polypogon (Poaceae)
Polystichum (Dryopteridaceae)
Pontederia (Pontederiaceae)
Porophyllum (Asteraceae)
Portulaca (Portulacaceae)
Presliophytum (Loasaceae)
Proustia (Asteraceae)
Pseudelephantopus (Asteraceae)
Pseudognaphalium (Asteraceae)
Pteridium (Dennstaedtiaceae)
Puya (Bromeliaceae)
Pycnophyllum (Caryophyllaceae)
Pyrolirion (Amaryllidaceae)
Quinchamalium (Schoepfiaceae)
Randia (Rubiaceae)
Ranunculus (Ranunculaceae)
Raphanus (Brassicaceae)
Rhodoscirpus (Cyperaceae)
Ribes (Grossulariaceae)
Richardia (Rubiaceae)
Ricinus (Euphorbiaceae)
Rockhausenia (Asteraceae)
Rorippa (Brassicaceae)
Rostraria (Poaceae)
Rubus (Rosaceae)
Rumex (Polygonaceae)
Ruppia (Ruppiaceae)
Salicornia (Amaranthaceae)
Salix (Salicaceae)
Salpichroa (Solanaceae)
Salsola (Amaranthaceae)
Salvia (Lamiaceae)
Sambucus (Viburnaceae)
Samolus (Primulaceae)
Sanguisorba (Rosaceae)
Sapindus (Sapindaceae)
Schinus (Anacardiaceae)
Schkuhria (Asteraceae)
Schoenoplectus (Cyperaceae)
Scutia (Rhamnaceae)
Sedum (Crassulaceae)
Senecio (Asteraceae)
Senna (Fabaceae)
Serjania (Sapindaceae)
Sesuvium (Aizoaceae)
Setaria (Poaceae)
Sicyos (Cucurbitaceae)
Sida (Malvaceae)
Sidastrum (Malvaceae)
Sigesbeckia (Asteraceae)
Silene (Caryophyllaceae)

Silybum (Asteraceae)
Siphocampylus (Campanulaceae)
Sisymbrium (Brassicaceae)
Sisyrinchium (Iridaceae)
Solanum (Solanaceae)
Soleirolia (Urticaceae)
Soliva (Asteraceae)
Sonchus (Asteraceae)
Sorghum (Poaceae)
Spananthe (Apiaceae)
Spartium (Fabaceae)
Spergularia (Caryophyllaceae)
Spilanthes (Asteraceae)
Sporobolus (Poaceae)
Stachys (Lamiaceae)
Stellaria (Caryophyllaceae)
Stevia (Asteraceae)
Stipa (Poaceae)
Stuckenia (Potamogetonaceae)
Suaeda (Amaranthaceae)
Symphyotrichum (Asteraceae)
Tagetes (Asteraceae)
Talinum (Talinaceae)
Tanacetum (Asteraceae)
Tara (Fabaceae)
Tarasa (Malvaceae)
Taraxacum (Asteraceae)
Tecoma (Bignoniaceae)
Tessaria (Asteraceae)
Tetraglochin (Rosaceae)
Tetragonia (Aizoaceae)
Thlaspi (Brassicaceae)
Tigridia (Iridaceae)
Tillandsia (Bromeliaceae)
Tiquilia (Boraginaceae)
Tragus (Poaceae)
Tribulus (Zygophyllaceae)

Trichocereus (Cactaceae)
Trifolium (Fabaceae)
Triglochin (Juncaginaceae)
Trihesperus (Asparagaceae)
Triodanis (Campanulaceae)
Tripogonella (Poaceae)
Trixis (Asteraceae)
Tropaeolum (Tropaeolaceae)
Typha (Typhaceae)
Urocarpidium (Malvaceae)
Urtica (Urticaceae)
Vachellia (Fabaceae)
Valeriana (Caprifoliaceae)
Vallesia (Apocynaceae)
Vallisneria (Hydrocharitaceae)
Vasconcellea (Caricaceae)
Verbena (Verbenaceae)
Veronica (Plantaginaceae)
Vicia (Fabaceae)
Vigna (Fabaceae)
Viguiera (Asteraceae)
Villanova (Asteraceae)
Viola (Violaceae)
Visnaga (Apiaceae)
Wahlenbergia (Campanulaceae)
Waltheria (Malvaceae)
Washingtonia (Arecaceae)
Weberbauera (Brassicaceae)
Weberbauerella (Fabaceae)
Weberbauerocereus (Cactaceae)
Werneria (Asteraceae)
Woodsia (Woodsiaceae)
Xanthium (Asteraceae)
Zameioscirpus (Cyperaceae)
Zannichellia (Potamogetonaceae)
Zantedeschia (Araceae)
Zinnia (Asteraceae)

Zygophyllum (Zygophyllaceae)
Tropaeolum (Tropaeolaceae)
Tunilla (Cactaceae)
Typha (Typhaceae)
Ullucus (Basellaceae)
Urocarpidium (Malvaceae)
Urtica (Urticaceae)
Vachellia (Fabaceae)
Valeriana (Caprifoliaceae)
Varronia (Boraginaceae)
Vasconcellea (Caricaceae)
Verbena (Verbenaceae)
Veronica (Plantaginaceae)
Viburnum (Adoxaceae)
Vicia (Fabaceae)
Vigna (Fabaceae)
Viguiera (Asteraceae)
Villanova (Asteraceae)
Viola (Violaceae)
Visnaga (Apiaceae)
Vulpia (Poaceae)
Wahlenbergia (Campanulaceae)
Waltheria (Malvaceae)
Washingtonia (Arecaceae)
Weberbauera (Brassicaceae)
Weberbauerella (Fabaceae)
Weberbauerocereus (Cactaceae)
Wedelia (Asteraceae)
Werneria (Asteraceae)
Woodsia (Woodsiaceae)
Xanthium (Asteraceae)
Xenophyllum (Asteraceae)
Zannichellia (Potamogetonaceae)
Zea (Poaceae)
Zinnia (Asteraceae)

Appendix 2: List of families and their orders

Acanthaceae (Lamiales)
Aizoaceae (Caryophyllales)
Alstroemeriaceae (Liliales)
Amaranthaceae (Caryophyllales)
Amaryllidaceae (Asparagales)
Anacardiaceae (Sapindales)
Annonaceae (Magnoliales)
Apiaceae (Apiales)
Apocynaceae (Gentianales)
Araceae (Alismatales)
Araliaceae (Apiales)
Arecaceae (Arecales)
Asparagaceae (Asparagales)
Asphodelaceae (Asparagales)
Aspleniaceae (Polypodiales)
Asteraceae (Asterales)
Basellaceae (Caryophyllales)
Berberidaceae (Ranunculales)
Betulaceae (Fagales)
Bignoniaceae (Lamiales)
Blechnaceae (Polypodiales)
Boraginaceae (Boraginales)
Brassicaceae (Brassicales)
Bromeliaceae (Poales)
Cactaceae (Caryophyllales)
Calceolariaceae (Lamiales)
Calyceraceae (Asterales)
Campanulaceae (Asterales)
Cannaceae (Zingiberales)
Caprifoliaceae (Dipsacales)
Caricaceae (Brassicales)
Caryophyllaceae (Caryophyllales)
Celastraceae (Celastrales)
Ceratophyllaceae (Ceratophyllales)
Cleomaceae (Brassicales)
Commelinaceae (Commelinales)
Convolvulaceae (Solanales)
Crassulaceae (Saxifragales)
Cucurbitaceae (Cucurbitales)

Cyperaceae (Poales)
Cystopteridaceae (Polypodiales)
Dennstaedtiaceae (Polypodiales)
Dryopteridaceae (Polypodiales)
Ephedraceae (Ephedrales)
Equisetaceae (Equisetales)
Escalloniaceae (Escalloniales)
Euphorbiaceae (Malpighiales)
Fabaceae (Fabales)
Francoaceae (Geraniales)
Frankeniaceae (Caryophyllales)
Gentianaceae (Gentianales)
Geraniaceae (Geraniales)
Grossulariaceae (Saxifragales)
Haloragaceae (Saxifragales)
Hydrocharitaceae (Alismatales)
Hypericaceae (Malpighiales)
Iridaceae (Asparagales)
Juncaceae (Poales)
Juncaginaceae (Alismatales)
Krameriaceae (Zygophyllales)
Lamiaceae (Lamiales)
Linaceae (Malpighiales)
Loasaceae (Cornales)
Loranthaceae (Santalales)
Lythraceae (Myrtales)
Malvaceae (Malvales)
Molluginaceae (Caryophyllales)
Montiaceae (Caryophyllales)
Moraceae (Rosales)
Myricaceae (Fagales)
Myrtaceae (Myrtales)
Nyctaginaceae (Caryophyllales)
Onagraceae (Myrtales)
Ophioglossaceae (Ophioglossales)
Orchidaceae (Asparagales)
Orobanchaceae (Lamiales)
Oxalidaceae (Oxalidales)
Papaveraceae (Ranunculales)

Passifloraceae (Malpighiales)
Phrymaceae (Lamiales)
Phyllanthaceae (Malpighiales)
Phytolaccaceae (Caryophyllales)
Piperaceae (Piperales)
Plantaginaceae (Lamiales)
Plumbaginaceae (Caryophyllales)
Poaceae (Poales)
Polemoniaceae (Ericales)
Polygalaceae (Fabales)
Polygonaceae (Caryophyllales)
Polypodiaceae (Polypodiales)
Pontederiaceae (Commelinales)
Portulacaceae (Caryophyllales)
Potamogetonaceae (Alismatales)
Primulaceae (Ericales)
Pteridaceae (Polypodiales)
Ranunculaceae (Ranunculales)
Rhamnaceae (Rosales)
Rosaceae (Rosales)
Rubiaceae (Gentianales)
Ruppiaceae (Alismatales)
Salicaceae (Malpighiales)
Salviniaceae (Salviniales)
Santalaceae (Santalales)
Sapindaceae (Sapindales)
Schoepfiaceae (Santalales)
Scrophulariaceae (Lamiales)
Solanaceae (Solanales)
Talinaceae (Caryophyllales)
Thelypteridaceae (Polypodiales)
Tropaeolaceae (Brassicales)
Typhaceae (Poales)
Urticaceae (Rosales)
Verbenaceae (Lamiales)
Viburnaceae (Dipsacales)
Violaceae (Malpighiales)
Woodsiaceae (Polypodiales)
Zygophyllaceae (Zygophyllales)

Appendix 3: Phylogenetic tree of the Flora of Arequipa

Subdivisios		Class	Clades				Order	Family
Pteridophyta (Ferns and Allies)		Equisetopsida (Horsetails)					Equisetales	Equisetaceae
		Psilotopsida					Ophioglossales	Ophioglossaceae
		Polypodiopsida (Ferns)					Polypodiales	Aspleniaceae, Blechnaceae, Cystopteridaceae, Dennstaedtiaceae, Dryopteridaceae, Polypodiaceae, Pteridaceae, Thelypteridaceae, Woodsiaceae
							Salviniales	Salviniaceae
Spermatophyta (Seed Plants)	Gymnospermae (Fruitless Seed Plants)	Gnetopsida					Ephedrales	Ephedraceae
		Pinopsida					Pinales	Araucariaceae, Cupressaceae
	Angiospermae (Flowering Plants)	Magnoliopsida	Magnoliidae				Magnoliales	Annonaceae, Magnoliaceae
							Laurales	Lauraceae
							Piperales	Piperaceae
			Monocotyledoneae				Alismatales	Araceae, Hydrocharitaceae, Juncaginaceae, Potamogetonaceae
							Liliales	Alstroemeriaceae
							Asparagales	Amaryllidaceae, Asparagaceae, Asphodelaceae, Iridaceae, Orchidaceae
				Commelinids			Arecales	Arecaceae
							Poales	Bromeliaceae, Cyperaceae, Juncaceae, Poaceae, Typhaceae
							Commelinales	Commelinaceae
			Eudicotiledoneae				Ranunculales	Berberidaceae, Papaveraceae, Ranunculaceae
							Proteales	Proteaceae
				Superrosids			Saxifragales	Crassulaceae, Grossulariaceae, Halogaraceae
					Rosids	Fabids	Zygophyllales	Krameriaceae, Zygophyllaceae
							Oxalidales	Oxalidaceae
							Malpighiales	Euphorbiaceae, Hypericaceae, Linaceae, Passifloraceae, Salicaceae, Violaceae
							Fabales	Fabaceae, Polygalaceae
							Rosales	Moraceae, Rhamnaceae, Rosaceae, Urticaceae
							Cucurbitales	Cucurbitaceae
							Fagales	Casuarinaceae, Myricaceae
						Malvids	Geraniales	Geraniaceae, Vivianiaceae
							Myrtales	Myrtaceae, Onagraceae
							Sapindales	Anacardiaceae, Meliaceae, Rutaceae, Sapindaceae
							Malvales	Malvaceae
							Brassicales	Brassicaceae, Caricaceae, Cleomaceae, Tropaeolaceae
				Superasterids			Santalales	Loranthaceae, Santalaceae, Schoepfiaceae
							Caryophyllales	Aizoaceae, Amaranthaceae, Basellaceae, Cactaceae, Caryophyllaceae, Frankeniaceae, Molluginaceae, Montiaceae, Nyctaginaceae, Phytolaccaceae, Plumbaginaceae, Polygonaceae, Portulacaceae, Talinaceae
					Asterids		Cornales	Loasaceae
							Ericales	Polemoniaceae, Primulaceae, Sapotaceae
						Lamids	Solanales	Convolvulaceae, Solanaceae
							Gentianales	Apocynaceae, Gentianaceae, Rubiaceae
							Lamiales	Acanthaceae, Bignoniaceae, Calceolariaceae, Lamiaceae, Oleaceae, Orobanchaceae, Phrymaceae, Plantaginaceae, Scrophulariaceae, Verbenaceae
							Boraginales	Boraginaceae, Hydrophyllaceae
						Campanulids	Asterales	Asteraceae, Calyceraceae, Campanulaceae
							Escalloniales	Escalloniaceae
							Apiales	Apiaceae, Araliaceae
							Dipsacales	Adoxaceae, Caprifoliaceae

Source: based on APG IV (2016), PPG I (2016), and Stevens (2017ff.)

Appendix 4: Synonyms of the accepted species

Aa gymnandra (Rchb.f.) Schltr. = Myrosmodes gymnandra (Rchb.f.) C.A.Vargas

Aa nervosa (Kraenzl.) Schltr. = Myrosmodes nervosa (Kraenzl.) Novoa, C.A.Vargas & Cisternas

Aa nubigena (Rchb.f.) Schltr. = Myrosmodes nubigena Rchb.f.

Abronia parviflora Kunth = Colignonia parviflora (Kunth) Choisy

Abutilon crispum (L.) Medik. = Herissantia crispa (L.) Brizicky

Abutilon erosum Schltdl. = Abutilon bivalve (Cav.) Dorr

Abutilon weberbaueri Ulbr. = Abutilon bivalve (Cav.) Dorr

Acacia macracantha Humb. & Bonpl. ex Willd. = Vachellia macracantha (Humb. & Bonpl. ex Willd.) Seigler & Ebinger

Acacia macracanthoides Bertero ex DC. = Vachellia macracantha (Humb. & Bonpl. ex Willd.) Seigler & Ebinger

Acacia pallida Humb. & Bonpl. ex Willd. = Neltuma pallida (Humb. & Bonpl. ex Willd.) C.E.Hughes & G.P.Lewis

Acacia visco Lorentz ex Griseb. = Parasenegalia visco (Lorentz ex Griseb.) Seigler & Ebinger

Acacia visite Griseb. = Parasenegalia visco (Lorentz ex Griseb.) Seigler & Ebinger

Acanthonychia ramosissima (Weinm.) Rohrb. = Cardionema ramosissima (Weinm.) A.Nelson & J.F.Macbr.

Acetosella corniculata (L.) Kuntze = Oxalis corniculata L.

Acetosella laxa (Hook. & Arn.) Kuntze = Oxalis laxa Hook. & Arn.

Acetosella megalorrhiza (Jacq.) Kuntze = Oxalis megalorrhiza Jacq.

Achyranthes halimifolia Lam. = Alternanthera halimifolia (Lam.) Standl. ex Pittier

Achyranthes porrigens Jacq. = Alternanthera porrigens (Jacq.) Kuntze

Achyrocline brittoniana Deble & Marchiori = Achyrocline ramosissima Britton

Achyrocline flavescens Griseb. = Achyrocline alata (Kunth) DC.

Achyrocline pterocaula DC. = Achyrocline alata (Kunth) DC.

Achyrocline rufescens (Kunth) DC. = Achyrocline alata (Kunth) DC.

Achyrophorus chillensis (Kunth) Sch.Bip. = Hypochaeris chillensis (Kunth) Britton

Achyrophorus elatus Wedd. = Hypochaeris elata Griseb.

Achyrophorus eriolaenus Sch.Bip. = Hypochaeris eriolaena (Sch.Bip.) Reiche

Achyrophorus meyenianus (Walp.) Walp. = Hypochaeris meyeniana (Walp.) Benth. & Hook.f. ex Griseb.

Achyrophorus taraxacifolius (Meyen & Walp.) Sch.Bip. = Hypochaeris taraxacoides Ball

Achyrophorus taraxacoides (Ball) Walp. = Hypochaeris taraxacoides Ball

Acmella fimbriata (Kunth) Cass. = Acmella ciliata (Kunth) Cass.

Acorellus laevigatus (L.) Palla = Cyperus laevigatus L.

Acrostichum niveum (Poir.) Desv. = Argyrochosma nivea (Poir.) Windham

Acrostichum sinuatum Lag. ex Sw. = Astrolepis sinuata (Lag. ex Sw.) D.M.Benham & Windham

Acrostichum trifoliatum L. = Pityrogramma trifoliata (L.) R.M.Tryon

Actinochloa simplex (Lag.) Roem. & Schult. = Bouteloua simplex Lag.

Adesmia clarenii (R.E.Fr.) Burkart = Adesmia miraflorensis J.Rémy

Adesmia nordenskioldii (R.E.Fr.) Burkart = Adesmia miraflorensis J.Rémy

Adesmia polyacantha Wedd. = Adesmia miraflorensis J.Rémy

Adesmia rupicola Wedd. = Adesmia spinosissima Meyen ex Vogel

Adesmia vicina J.F.Macbr. = Adesmia miraflorensis J.Rémy

Adipera bicapsularis (L.) Britton & Rose = Senna bicapsularis (L.) Roxb.

Aeschynomene muricata (Jacq.) Desv. = Adesmia muricata (Jacq.) DC.

Aethusa leptophylla (Pers.) Spreng. = Cyclospermum leptophyllum (Pers.) Sprague ex Britton & P.Wilson

Agathophytum ambrosioides (L.) Peterm. = Dysphania ambrosioides (L.) Mosyakin & Clemants

Agave cordillerensis Lodé & Pino = Agave americana L.

Ageratina azangaroensis (Sch.Bip. ex Wedd.) R.M.King & H.Rob. = Ageratina glechonophylla (Less.) R.M.King & H.Rob.

Ageratina remyana (Phil.) R.M.King & H.Rob. = Ageratina glechonophylla (Less.) R.M.King & H.Rob.

Agrostis eminens (J.Presl) Griseb. = Deschampsia eminens (J.Presl) Saarela

Agrostis glomerata (J.Presl) Kunth = Agrostis tolucensis Kunth

Agrostis hoffmannii Mez = Agrostis tolucensis Kunth

Agrostis indica L. = Sporobolus indicus (L.) R.Br.

Agrostis nana (J.Presl) Kunth = Agrostis breviculmis Hitchc.

Agrostis peruviana (P.Beauv.) Spreng. = Muhlenbergia peruviana (P.Beauv.) Steud.

Agrostis preslii Kunth = Cinnagrostis preslii (Kunth) P.M.Peterson, Soreng, Romasch. & Barberá

Agrostis punctata (L.) Lam. = Eriochloa punctata (L.) Ham.

Agrostis radiata L. = Chloris radiata (L.) Sw.

Agrostis rigescens J.Presl = Cinnagrostis rigescens (J.Presl) P.M.Peterson, Soreng, Romasch. & Barberá

Agrostis virginica L. = Sporobolus virginicus (L.) Kunth

Aira spicata L. = Koeleria spicata (L.) Barberá, Quintanar, Soreng & P.M.Peterson

Airampoa ayrampo (Haenke) Doweld = Airampoa soehrensii (Britton & Rose) Lodé

Alchemilla appendiculata Wedd. ex Murb. = Alchemilla diplophylla Diels

Alchemilla calchaquina Lillo = Alchemilla pinnata Ruiz & Pav.

Aleuritopteris sulfurea Fée = Notholaena sulphurea (Cav.) J.Sm.

Alguelagum lamiifolium (Benth.) Kuntze = Lepechinia lamiifolia (Benth.) Epling

Allionia elegans (Choisy) Kuntze = Mirabilis elegans (Choisy) Heimerl

Allionia expansa (Ruiz & Pav.) Kuntze = Mirabilis expansa (Ruiz & Pav.) Standl.

Allionia ovata Pursh = Mirabilis ovata (Ruiz & Pav.) F.Meigen

Allionia prostrata (Ruiz & Pav.) Kuntze = Mirabilis prostrata (Ruiz & Pav.) Heimerl

Allium andicola (Kunth) Regel = Nothoscordum andicola Kunth

Allium bivalve (L.) Kuntze = Nothoscordum bivalve (L.) Britton

Allium gracile Aiton = Nothoscordum gracile (Aiton) Stearn

Allocarya humilis (Ruiz & Pav.) Greene = Plagiobothrys humilis (Ruiz & Pav.) I.M.Johnst.

Allocarya linifolia (Lehm.) J.F.Macbr. = Plagiobothrys linifolius (Lehm.) I.M.Johnst.

Allocarya macbridei (I.M.Johnst.) Ferreyra = Plagiobothrys macbridei I.M.Johnst.

Allosorus esculentus (G.Forst.) C.Presl = Pteridium esculentum (G.Forst.) Cockayne

Allosorus myriophyllus Farw. = Myriopteris myriophylla J.Sm.

Allosorus pilosus (Goldm.) Farw. = Cheilanthes pilosa Goldm.

Allosorus pruinatus Farw. = Cheilanthes pruinata Kaulf.

Allosorus sagittatus (Cav.) C.Presl = Pellaea sagittata (Cav.) Link.

Allosorus ternifolius (Cav.) Kuntze = Pellaea ternifolia (Cav.) Link

Allosperma tuberosa (L.) Raf. = Commelina tuberosa L.

Alonsoa albiflora G.Nicholson = Alonsoa meridionalis (L.f.) Kuntze

Alonsoa caulialata Ruiz & Pav. = Alonsoa meridionalis (L.f.) Kuntze

Alonsoa grandiflora Voss = Alonsoa meridionalis (L.f.) Kuntze

Alonsoa parviflora (Kunth) G.Don = Alonsoa meridionalis (L.f.) Kuntze

Alonsoa peruviana López Guillén = Alonsoa meridionalis (L.f.) Kuntze

Alopecurus elongatus (Kunth) Poir. = Polypogon elongatus Kunth

Alopecurus interruptus Poir. = Polypogon interruptus Kunth

Alsine cuspidata (Willd. ex D.F.K.Schltdl.) Wooton & Standl. = Stellaria cuspidata Willd. ex D.F.K.Schltdl.

Alstroemeria dulcis Hook. = Bomarea dulcis (Hook.) Beauverd

Alstroemeria involucrosa (Herb.) Hunz. = Bomarea involucrosa (Herb.) Baker

Alstroemeria latifolia Ruiz & Pav. = Bomarea latifolia (Ruiz & Pav.) Herb.

Alstroemeria lineatiflora Ruiz & Pav. = Alstroemeria chorillensis Herb.

Alstroemeria ovata Cav. = Bomarea ovata (Cav.) Mirb.

Alstroemeria violacea Phil. = Alstroemeria chorillensis Herb.

Altensteinia gymnandra Rchb.f. = Myrosmodes gymnandra (Rchb.f.) C.A.Vargas

Altensteinia nervosa Kraenzl. = Myrosmodes nervosa (Kraenzl.) Novoa, C.A.Vargas & Cisternas

Altensteinia nubigena (Rchb.f.) Rchb.f. = Myrosmodes nubigena Rchb.f.

Altensteinia weddelliana Rchb.f. = Aa weddelliana (Rchb.f.) Schltr.

Alternanthera ciliata Poepp. ex Seub. = Alternanthera pungens Kunth

Alternanthera dolichocephala (Urb.) Urb. = Alternanthera halimifolia (Lam.) Standl. ex Pittier

Alternanthera echinata Sm. = Alternanthera pungens Kunth

Alternanthera fastigiata Suess. = Alternanthera porrigens (Jacq.) Kuntze

Alternanthera junciflora (Remy) I.M.Johnst. = Alternanthera porrigens (Jacq.) Kuntze

Alternanthera macrorrhiza Hauman = Alternanthera caracasana Kunth

Alternanthera mollendoana Suess. = Alternanthera pubiflora (Benth.) Kuntze

Alternanthera nigriceps Hook. = Alternanthera porrigens (Jacq.) Kuntze

Alternanthera parvifolia (Moq.) Fawc. & Rendle = Alternanthera caracasana Kunth

Alternanthera peploides (Humb. & Bonpl. ex Schult.) Urb. = Alternanthera caracasana Kunth

Alternanthera sericea Kunth = Alternanthera porrigens (Jacq.) Kuntze

Alternanthera villiflora Scheele = Alternanthera caracasana Kunth

Alternanthera williamsii (Standl.) Standl. = Alternanthera pubiflora (Benth.) Kuntze

Amaranthus caracasanus Kunth = Amaranthus spinosus L.

Amaranthus coracanus Mart. = Amaranthus spinosus L.

Amaranthus diacanthus Raf. = Amaranthus spinosus L.

Amaranthus fasciatus Roxb. = Amaranthus viridis L.

Amaranthus gracilis Desf. ex Poir. = Amaranthus viridis L.

Amaranthus polystachyus Willd. = Amaranthus viridis L.

Amaranthus tortuosus Hornem. = Amaranthus dubius Mart. ex Thell.

Amaryllis arvensis F.Dietr. = Pyrolirion arvense (F.Dietr.) Erhardt, Götz & Seybold

Ambrina ambrosioides (L.) Spach = Dysphania ambrosioides (L.) Mosyakin & Clemants

Ambrina incisa (Poir.) Moq. = Dysphania incisa (Poir.) ined.

Ambrosia orobanchifera Meyen = Ambrosia cumanensis Kunth

Ambrosia parvifolia W.W.Payne = Ambrosia dentata (Cabrera) M.O.Dillon

Ambrosia peruviana Willd. = Ambrosia cumanensis Kunth

Ambrosia tacorensis Meyen = Ambrosia artemisioides Meyen & Walp.

Amoria polymorpa (Poir.) C.Presl = Trifolium polymorphum Poir.

Amsinckia angustifolia Lehm. = Amsinckia calycina (Moris) Chater

Amsinckia aurantiaca (Brand) Brand = Amsinckia calycina (Moris) Chater

Amsinckia humifusa Walp. = Amsinckia calycina (Moris) Chater

Amsinckia parviflora Bernh. = Amsinckia calycina (Moris) Chater

Anacharis potamogeton (Bertero) Vict. = Elodea potamogeton (Bertero) Espinosa

Anarrhinum canadense L. = Nuttallanthus canadensis (L.) D.A.Sutton

Anatherostipa bomanii (Hauman) Peñail. = Lorenzochloa bomanii (Hauman) Romasch.

Anatherostipa mucronata (Griseb.) F.Rojas = Lorenzochloa mucronata (Griseb.) Romasch.

Anatherostipa obtusa (Nees & Meyen) Peñail. = Lorenzochloa obtusa (Nees & Meyen) Romasch.

Anatherostipa rigidiseta (Pilg.) Peñail. = Lorenzochloa rigidiseta (Pilg.) Romasch.

Anchusa linifolia Lehm. = Plagiobothrys linifolius (Lehm.) I.M.Johnst.

Andinocleome chilensis (DC.) Iltis ex E.M.McGinty & Roalson = Cleome chilensis DC.

Andinopuntia floccosa (Salm-Dyck) Guiggi = Austrocylindropuntia floccosa (Salm-Dyck) F.Ritter

Andrachne ciliatoglandulosa (Millsp.) Croizat = Andrachne microphylla (Lam.) Baill.

Anil humilis (Kunth) Kuntze = Indigofera humilis Kunth

Anisophyllum hypericifolium (L.) Haw. = Euphorbia hypericifolia L.

Anisophyllum lasiocarpum (Klotzsch) Klotzsch & Garcke = Euphorbia lasiocarpa Klotzsch

Anisophyllum meyenianum (Klotzsch) Klotzsch & Garcke = Euphorbia meyeniana Klotzsch

Anisophyllum serpens (Kunth) Klotzsch & Garcke = Euphorbia serpens Kunth

Anomocarpus pulvinatus (J.Rémy) Miers = Calycera pulvinata J.Rémy

Anotis thymifolia (Ruiz & Pav.) DC. = Arcytophyllum thymifolium (Ruiz & Pav.) Standl.

Anthericum caeruleum Ruiz & Pav. = Pasithea caerulea (Ruiz & Pav.) D.Don

Anthericum eccremorrhizum Ruiz & Pav. = Echeandia eccremorrhiza (Ruiz & Pav.) ined.

Anthericum glaucum Ruiz & Pav. = Trihesperus glaucus (Ruiz & Pav.) Herb.

Anthochloa lepidula Nees & Meyen = Poa lepidula (Nees & Meyen) Soreng & L.J.Gillespie

Antiphytum linifolium (Lehm.) A.DC. = Plagiobothrys linifolius (Lehm.) I.M.Johnst.

Antirrhinum canadense L. = Nuttallanthus canadensis (L.) D.A.Sutton

Apargia chillensis Kunth = Hypochaeris chillensis (Kunth) Britton

Aphanes pinnata (Ruiz & Pav.) Pers. = Alchemilla pinnata Ruiz & Pav.

Aphyllon tacnaense (Mattf.) A.C.Schneid. = Orobanche tacnaensis Mattf.

Aphyllon weberbaueri (Mattf.) A.C.Schneid. = Orobanche weberbaueri Mattf.

Apium laciniatum (DC.) Urb. = Cyclospermum laciniatum (DC.) Constance

Apium leptophyllum (Pers.) F.Muell. = Cyclospermum leptophyllum (Pers.) Sprague ex Britton & P.Wilson

Aplotheca interrupta (L.) Mart. ex Cham. = Froelichia interrupta (L.) Moq.

Apodanthera herrerae Harms = Apodanthera mandonii Cogn.

Apodanthera moqueguana Mart.Crov. = Apodanthera mandonii Cogn.

Arabis lanata Walp. = Neuontobotrys lanatus (Walp.) Al-Shehbaz

Arcytophyllum juniperifolium (Ruiz & Pav.) Standl. = Arcytophyllum thymifolium (Ruiz & Pav.) Standl.

Arcytophyllum laricifolium (Cav.) W.H.Lewis = Arcytophyllum thymifolium (Ruiz & Pav.) Standl.

Arcytophyllum weberbaueri K.Krause = Arcytophyllum thymifolium (Ruiz & Pav.) Standl.

Arenaria scopulorum Kunth = Arenaria digyna Willd. ex D.F.K.Schltdl.

Arequipa haynii (Otto ex Salm-Dyck) Krainz = Matucana haynii (Otto ex Salm-Dyck) Britton & Rose

Arequipa hempeliana (Gürke) Oehme = Oreocereus hempelianus (Gürke) D.R.Hunt

Arequipa leucotricha (Phil.) Britton & Rose = Oreocereus leucotrichus (Phil.) Wagenkn.

Arequipiopsis hempeliana (Gürke) Kreuz. & Buining = Oreocereus hempelianus (Gürke) D.R.Hunt

Argylia canescens D.Don = Argylia radiata (L.) D.Don

Argylia chrysantha Phil. = Argylia radiata (L.) D.Don

Argylia digitalina Phil. = Argylia radiata (L.) D.Don

Argylia eremophila Phil. = Argylia radiata (L.) D.Don

Argylia glabriuscula Phil. = Argylia radiata (L.) D.Don

Argylia puberula DC. = Argylia radiata (L.) D.Don

Argylia tenuifolia C.Presl = Argylia radiata (L.) D.Don

Argylia villosa Phil. = Argylia radiata (L.) D.Don

Argyrochosma flava (Hook.) M.Kessler & A.R.Sm. = Argyrochosma nivea (Poir.) Windham

Argyrochosma tenera (Gillies ex Hook.) M.Kessler & A.R.Sm. = Argyrochosma nivea (Poir.) Windham

Aristida bromoides Kunth = Aristida adscensionis L.

Aristida cardosoi Cout. = Aristida adscensionis L.

Aristida chaetophylla Steud. = Aristida adscensionis L.

Aristida corctata Kunth = Aristida adscensionis L.

Aristida debilis Mez = Aristida adscensionis L.

Aristida depressa Retz. = Aristida adscensionis L.

Aristida effusa Henrard = Aristida adscensionis L.

Aristida elatior Cav. = Aristida adscensionis L.

Aristida fasciculata Torr. = Aristida adscensionis L.

Aristida festucoides Poir. = Aristida adscensionis L.

Aristida heymannii Regel = Aristida adscensionis L.

Aristida humilis Kunth = Aristida adscensionis L.

Aristida interrupta Cav. = Aristida adscensionis L.

Aristida macrochloa Hochst. = Aristida adscensionis L.

Aristida maritima Steud. = Aristida adscensionis L.

Aristida modatica Steud. = Aristida adscensionis L.

Aristida nana Steud. = Aristida adscensionis L.

Aristida nigrescens J.Presl = Aristida adscensionis L.

Aristida peruviana Beetle = Aristida adscensionis L.

Aristida racemosa Biehler = Aristida adscensionis L.

Aristida schaffneri E.Fourn. = Aristida adscensionis L.

Aristida simplicissima Steud. = Aristida adscensionis L.

Aristida spicigera Trin. & Rupr. = Aristida adscensionis L.

Aristida strictiflora Trin. & Rupr. = Aristida adscensionis L.

Aristida submucronata Schumach. & Thonn. = Aristida adscensionis L.

Aristida vulpioides Hance = Aristida adscensionis L.

Armatocereus arboreus Rauh & Backeb. = Armatocereus matucanensis Backeb.

Armatocereus churinensis Rauh & Backeb. = Armatocereus matucanensis Backeb.

Armatocereus ghiesbreghtii (K.Schum.) F.Ritter = Armatocereus matucanensis Backeb.

Armatocereus oligogonus Rauh & Backeb. = Armatocereus matucanensis Backeb.

Armatocereus riomajensis Rauh & Backeb. = Armatocereus matucanensis Backeb.

Arthrocereus bylesianus (Andreae & Backeb.) Buxb. = Haageocereus bylesianus (Andreae & Backeb.) Lodé

Arundo rigida (Kunth) Poir. = Cinnagrostis rigida (Kunth) P.M.Peterson, Soreng, Romasch. & Barberá

Arundo sagittata (Aubl.) Pers. = Gynerium sagittatum (Aubl.) P.Beauv.

Asclepias clausa Jacq. = Funastrum clausum (Jacq.) Schltr.

Aspidium orbiculatum Desv. = Polystichum orbiculatum (Desv.) Gay

Aspidium rufum (Poir.) Mett. = Amauropelta rufa (Poir.) Salino & T.E.Almeida

Asplenium debile Fée = Asplenium gilliesii Hook.

Asplenium diaphanum (Bory) Lojac. = Cystopteris diaphana (Bory) Blasdell

Asplenium herbaceum Fée = Asplenium triphyllum C.Presl

Asplenium imbricatum Hook. & Grev. = Asplenium triphyllum C.Presl

Asplenium rhomboideum Brack. = Asplenium triphyllum C.Presl

Asplenium tenue C.Presl = Asplenium triphyllum C.Presl

Asplenium ternatum C.Presl = Asplenium triphyllum C.Presl

Aster canadensis (L.) E.H.L.Krause = Erigeron canadensis L.

Aster limnophilus (Sch.Bip.) Hemsl. & Pearson = Oritrophium limnophilum (Sch.Bip.) Cuatrec.

Aster squamatus (Spreng.) Hieron. = Symphyotrichum squamatum (Spreng.) G.L.Nesom

Asteriscium amplexicaule H.Wolff = Domeykoa amplexicaulis (H.Wolff) Mathias & Constance

Astragalus benthamianus Gillies = Astragalus garbancillo Cav.

Astragalus bolivianus Phil. = Astragalus arequipensis Vogel

Astragalus capitellus Britton = Astragalus micranthellus Wedd.

Astragalus colliculus Rusby = Astragalus minimus Vogel

Astragalus deminutivus (Phil.) I.M.Johnst. = Astragalus diminutivus (Phil.) Gómez-Sosa

Astragalus drepanophorus Griseb. = Astragalus arequipensis Vogel

Astragalus herzogii Ulbr. = Astragalus uniflorus DC.

Astragalus macrorrhynchus Ulbr. = Astragalus richii A.Gray

Astragalus mandonii Rusby = Astragalus garbancillo Cav.

Astragalus minor Clos = Astragalus garbancillo Cav.

Astragalus orbignyanus Wedd. = Astragalus arequipensis Vogel

Astragalus patancanus Ulbr. = Astragalus micranthellus Wedd.

Astragalus pilgeri J.F.Macbr. = Astragalus weddellianus (Kuntze) I.M.Johnst.

Astragalus punensis J.F.Macbr. = Astragalus weddellianus (Kuntze) I.M.Johnst.

Astragalus sinocarpus Rusby = Astragalus arequipensis Vogel

Astragalus unifultus L'Hér. = Astragalus garbancillo Cav.

Astragalus viciiformis Ulbr. = Astragalus triflorus (DC.) A.Gray

Astrephia chaerophylloides (Sm.) DC. = Valeriana chaerophylloides Sm.

Astrephia coarctata (Ruiz & Pav.) Dufr. = Valeriana coarctata Ruiz & Pav.

Astrephia interrupta (Ruiz & Pav.) Dufr. = Valeriana interrupta Ruiz & Pav.

Atriplex ambrosioides (L.) Crantz = Dysphania ambrosioides (L.) Mosyakin & Clemants

Atriplex andina R.E.Fr. = Atriplex myriophylla Phil.

Atriplex herzogii Standl. = Atriplex myriophylla Phil.

Atriplex lilloi Hauman = Atriplex myriophylla Phil.

Atriplex microphylla Phil. = Atriplex imbricata (Moq.) D.Dietr.

Atriplex rotundifolia (Moq.) Dombey ex Standl. = Atriplex espostoi Speg.

Atriplex serpyllifolia Herzog = Atriplex myriophylla Phil.

Atropa glandulosa Hook. = Salpichroa glandulosa (Hook.) Miers

Atropa hirsuta Meyen = Salpichroa hirsuta (Meyen) Miers

Atropa humifusa Gouan = Nolana humifusa (Gouan) I.M.Johnst.

Atropa physalodes L. = Nicandra physalodes (L.) Gaertn.

Atropa ramosissima Mathews ex Dunal = Salpichroa ramosissima Miers

Atropa spinosa Meyen = Dunalia spinosa (Meyen) Dammer

Austrocylindropuntia hirschii (Backeb.) E.F.Anderson = Austrocylindropuntia floccosa (Salm-Dyck) F.Ritter

Austrocylindropuntia lauliacoana F.Ritter = Austrocylindropuntia floccosa (Salm-Dyck) F.Ritter

Austrocylindropuntia machacana F.Ritter = Austrocylindropuntia floccosa (Salm-Dyck) F.Ritter

Austrocylindropuntia sphaerica (C.F.Först.) G.D.Rowley = Cumulopuntia sphaerica (C.F.Först.) E.F.Anderson

Azolla arbuscula Desv. = Azolla filiculoides Lam.

Azolla magellanica Willd. = Azolla filiculoides Lam.

Azolla squamosa Molina = Azolla filiculoides Lam.

Azorella columnaris H.Wolff = Azorella compacta Phil.

Azorella glabra Wedd. = Azorella diapensioides A.Gray

Azorella prismatoclada Domin = Azorella compacta Phil.

Azorella yareta Hauman = Azorella compacta Phil.

Baccharis absinthioides Hook. & Arn. = Tessaria absinthioides (Hook. & Arn.) DC.

Baccharis adscendens Pers. = Baccharis scandens Pers.

Baccharis alata Loudon = Baccharis genistelloides (Lam.) Pers.

Baccharis decurrens (Vell.) Stellf. = Baccharis genistelloides (Lam.) Pers.

Baccharis doniana Hook. & Arn. = Baccharis gnidiifolia Kunth

Baccharis frigida Kunth = Baccharis gnidiifolia Kunth

Baccharis graveolens Sch.Bip. = Andicolea graveolens (Sch.Bip.) Mayta & Molinari

Baccharis incarum Wedd. = Baccharis tola Phil.

Baccharis lejia Phil. = Baccharis tola Phil.

Baccharis lucida Meyen = Parastrephia lucida (Meyen) Cabrera

Baccharis petiolata DC. = Baccharis scandens Pers.

Baccharis procumbens Hieron. = Baccharis caespitosa Pers.

Baccharis quadrangularis Meyen = Parastrephia quadrangularis (Meyen) Cabrera

Baccharis reticulata (Ruiz & Pav.) Pers. = Baccharis genistelloides (Lam.) Pers.

Baccharis semiserrata DC. = Baccharis gnidiifolia Kunth

Baccharis sternbergiana Steud. = Baccharis gnidiifolia Kunth

Baccharis tafiensis Heering = Baccharis tola Phil.

Baccharis triptera Mart. = Baccharis genistelloides (Lam.) Pers.

Baccharis venosa (Ruiz & Pav.) Pers. = Baccharis genistelloides (Lam.) Pers.

Balbisia weberbaueri R.Knuth = Balbisia verticillata Cav.

Bargemontia gracillima I.M.Johnst. = Nolana gracillima (I.M.Johnst.) I.M.Johnst.

Bargemontia tarapacana (Phil.) I.M.Johnst. = Nolana tarapacana (Phil.) I.M.Johnst.

Bartsia bartsioides (Hook.) Edwin = Neobartsia bartsioides (Hook.) Uribe-Convers & Tank

Bartsia camporum Diels = Neobartsia camporum (Diels) Uribe-Convers & Tank

Bartsia diffusa Benth. = Neobartsia diffusa (Benth.) Uribe-Convers & Tank

Bartsia elongata Wedd. = Neobartsia elongata (Wedd.) Uribe-Convers & Tank

Bartsia pedicularoides Benth. = Neobartsia pedicularoides (Benth.) Uribe-Convers & Tank

Bartsia peruviana Walp. = Neobartsia peruviana (Walp.) Uribe-Convers & Tank

Bartsia serrata Molau = Neobartsia serrata (Molau) Uribe-Convers & Tank

Bartsia weberbaueri Diels = Neobartsia weberbaueri (Diels) Uribe-Convers & Tank

Basella diffusa Ruiz & Pav. ex Moq. = Anredera diffusa (Moq.) Sperling

Bastardia bivalvis (Cav.) Kunth ex Griseb. = Abutilon bivalve (Cav.) Dorr

Bastardia crispa (L.) A.St.-Hil. = Herissantia crispa (L.) Brizicky

Belloa caespititia (Wedd.) Cabrera = Mniodes caespititia (Wedd.) Quip. & M.O.Dillon

Belloa kunthiana (DC.) Anderb. & S.E.Freire = Lucilia kunthiana (DC.) Zardini

Belloa longifolia (Cuatrec. & Aristeg.) Sagást. & M.O.Dillon = Mniodes longifolia (Cuatrec. & Aristeg.) S.E.Freire, Chemisquy, Anderb. & Urtubey

Belloa piptolepis (Wedd.) Cabrera = Mniodes piptolepis (Wedd.) S.E.Freire, Chemisquy, Anderb. & Urtubey

Beloere crispa (L.) Shuttlew. = Herissantia crispa (L.) Brizicky

Berberis conferta Kunth = Berberis lutea Ruiz & Pav.

Berberis huanucensis (C.K.Schneid.) J.F.Macbr. = Berberis lutea Ruiz & Pav.

Berberis lobbiana (C.K.Schneid.) C.K.Schneid. = Berberis lutea Ruiz & Pav.

Berberis psiloclada (C.K.Schneid.) Ahrendt = Berberis lutea Ruiz & Pav.

Berberis spruceana (C.K.Schneid.) Ahrendt = Berberis lutea Ruiz & Pav.

Berberis virgata Ruiz & Pav. = Berberis lutea Ruiz & Pav.

Bermudiana chilensis (Hook.) Kuntze = Sisyrinchium chilense Hook.

Bermudiana jamesonii (Baker) Kuntze = Sisyrinchium jamesonii Baker

Bermudiana juncea (E.Mey. ex C.Presl) Kuntze = Olsynium junceum (E.Mey. ex C.Presl) Goldblatt

Bermudiana micrantha (Cav.) Kuntze = Sisyrinchium micranthum Cav.

Bidens abadiae DC. = Bidens pilosa L.

Bidens arenaria Gand. = Bidens pilosa L.

Bidens bimucronata Turcz. = Bidens pilosa L.

Bidens canescens Bertol. = Bidens triplinervia Kunth

Bidens chilensis DC. = Bidens pilosa L.

Bidens consolidifolia Turcz. = Bidens triplinervia Kunth

Bidens geraniifolia Brandegee = Bidens triplinervia Kunth

Bidens hirtella Kunth = Bidens triplinervia Kunth

Bidens hispida Kunth = Bidens pilosa L.

Bidens humilis Kunth = Bidens triplinervia Kunth

Bidens macrantha Griseb. = Bidens triplinervia Kunth

Bidens ododrata Cav. = Bidens pilosa L.

Bidens odorata Cav. = Bidens pilosa L.

Bidens paleacea Vis. = Bidens pilosa L.

Bidens pectinata Sch.Bip. = Bidens triplinervia Kunth

Bidens scandicina Kunth = Bidens pilosa L.

Bidens striata Schott ex Sweet = Bidens pilosa L.

Bidens viciosoi Pau = Bidens pilosa L.

Bidwillia glauca (Ruiz & Pav.) G.Don = Trihesperus glaucus (Ruiz & Pav.) Herb.

Bignonia fulva Cav. = Tecoma fulva (Cav.) G.Don

Bignonia radiata L. = Argylia radiata (L.) D.Don

Bignonia stans L. = Tecoma stans (L.) Juss. ex Kunth

Binghamia decumbens (Vaupel) Werderm. = Haageocereus decumbens (Vaupel) Backeb.

Binghamia platinospina (Werderm. & Backeb.) Werderm. = Haageocereus platinospinus (Werderm. & Backeb.) Backeb.

Biscutella chilensis Lag. ex DC. = Cremolobus chilensis (Lag. ex DC.) DC.

Blairia nodiflora (L.) Gaertn. = Phyla nodiflora (L.) Greene

Blechnum ciliatum M.Martens & Galeotti = Blechnum occidentale L.

Blechnum cunninghamii T.Moore = Blechnum occidentale L.

Blechnum drepanophyllum Trevis. = Blechnum occidentale L.

Blechnum falcatum Link = Blechnum occidentale L.

Blechnum rugosum T.Moore = Blechnum occidentale L.

Blechnum scaberulum Sodiro = Blechnum occidentale L.

Blechnum suburbicum Vell. = Blechnum occidentale L.

Blitum ambrosioides (L.) Beck = Dysphania ambrosioides (L.) Mosyakin & Clemants

Blumenbachia carduifolia Ball = Caiophora carduifolia C.Presl

Boerhavia bracteata T.Cooke = Boerhavia coccinea Mill.

Boerhavia chaerophylloides (Sm.) Willd. = Valeriana chaerophylloides Sm.

Boerhavia glandulosa Andersson = Boerhavia coccinea Mill.

Boerhavia hirsuta Willd. = Boerhavia coccinea Mill.

Boerhavia litoralis Kunth = Commicarpus tuberosus (Lam.) Standl.

Boerhavia marlothii Heimerl = Boerhavia coccinea Mill.

Boerhavia patula Dombey ex Vahl = Boerhavia coccinea Mill.

Boerhavia polymorpha Rich. = Boerhavia coccinea Mill.

Boerhavia ramulosa M.E.Jones = Boerhavia coccinea Mill.

Boerhavia sonorae Rose = Boerhavia coccinea Mill.

Boerhavia squamata Raf. = Boerhavia coccinea Mill.

Boerhavia tuberosa Lam. = Commicarpus tuberosus (Lam.) Standl.

Boerhavia viscosa Lag. & Rodr. = Boerhavia coccinea Mill.

Bogenhardia crispa (L.) Kearney = Herissantia crispa (L.) Brizicky

Bomarea alpicola Kraenzl. = Bomarea latifolia (Ruiz & Pav.) Herb.

Bomarea amoena (Herb.) M.Roem. = Bomarea ovata (Cav.) Mirb.

Bomarea biflora Vargas = Bomarea dulcis (Hook.) Beauverd

Bomarea calcensis Vargas = Bomarea dulcis (Hook.) Beauverd

Bomarea campanuliflora Killip = Bomarea dulcis (Hook.) Beauverd

Bomarea cuzcoensis Vargas = Bomarea dulcis (Hook.) Beauverd

Bomarea grandifolia (Kunth) Herb. = Bomarea latifolia (Ruiz & Pav.) Herb.

Bomarea macrocarpa (Ruiz & Pav.) Herb. = Bomarea ovata (Cav.) Mirb.

Bomarea maculata Killip ex Vargas = Bomarea involucrosa (Herb.) Baker

Bomarea nobilis Herb. = Bomarea ovata (Cav.) Mirb.

Bomarea phyllostachya Mast. ex Baker = Bomarea dulcis (Hook.) Beauverd

Bomarea puberula (Herb.) Kraenzl. ex Perkins = Bomarea dulcis (Hook.) Beauverd

Bomarea punctata Herb. = Bomarea ovata (Cav.) Mirb.

Bomarea subsessilis Killip = Bomarea ovata (Cav.) Mirb.

Bomarea tomentosa (Ruiz & Pav.) Herb. = Bomarea ovata (Cav.) Mirb.

Bomarea uniflora (M.Roem.) Killip = Bomarea dulcis (Hook.) Beauverd

Bomarea villosa M.Roem. = Bomarea ovata (Cav.) Mirb.

Borzicactus decumbens (Vaupel) Britton & Rose = Haageocereus decumbens (Vaupel) Backeb.

Borzicactus haynii (Otto ex Salm-Dyck) Kimnach = Matucana haynii (Otto ex Salm-Dyck) Britton & Rose

Borzicactus hempelianus (Gürke) Donald = Oreocereus hempelianus (Gürke) D.R.Hunt

Borzicactus jajoanus Backeb. = Loxanthocereus jajoanus (Backeb.) Backeb.

Borzicactus leucotrichus (Phil.) Kimnach = Oreocereus leucotrichus (Phil.) Wagenkn.

Borzicactus platinospinus (Werderm. & Backeb.) Borg = Haageocereus platinospinus (Werderm. & Backeb.) Backeb.

Borzicactus sextonianus (Backeb.) Kimnach = Loxanthocereus sextonianus (Backeb.) Backeb.

Bougainvillea patagonica Decne. = Bougainvillea spinosa (Cav.) Heimerl

Bougueria nubicola Decne. = Plantago nubicola (Decne.) Rahn

Boussingaultia diffusa (Moq.) Volkens = Anredera diffusa (Moq.) Sperling

Bouteloua brachyathera Phil. = Bouteloua simplex Lag.

Bouteloua humilis (Kunth) Hieron. = Bouteloua simplex Lag.

Bouteloua procumbens (P.Durand) Griffiths = Bouteloua simplex Lag.

Bouteloua prostrata Lag. = Bouteloua simplex Lag.

Bouteloua pusilla Vasey = Bouteloua simplex Lag.

Bouteloua tenuis (P.Beauv. ex Kunth) Griseb. = Bouteloua simplex Lag.

Bowlesia acutangula Benth. = Bowlesia lobata Ruiz & Pav.

Bowlesia acutiloba H.Wolff = Bowlesia lobata Ruiz & Pav.

Bowlesia cirrosa Phil. = Bowlesia tropaeolifolia Gillet & Hook.

Bowlesia diversifolia Meyen ex Walp. = Bowlesia tropaeolifolia Gillet & Hook.

Bowlesia flexilis Meyen = Bowlesia tropaeolifolia Gillet & Hook.

Bowlesia mandonii Rusby = Bowlesia tenella Meyen

Bowlesia pulchella Wedd. = Bowlesia tropaeolifolia Gillet & Hook.

Bowlesia rupestris H.Wolff = Bowlesia tropaeolifolia Gillet & Hook.

Bramia monnieri (L.) Drake = Bacopa monnieri (L.) Wettst.

Brandesia porrigens (Jacq.) Mart. = Alternanthera porrigens (Jacq.) Kuntze

Brandesia pubiflora Benth. = Alternanthera pubiflora (Benth.) Kuntze

Braya cachensis Speg. = Aschersoniodoxa cachensis (Speg.) Al-Shehbaz

Braya calycina Wedd. = Eudema calycinum (Desv.) Al-Shehbaz, Salariato, A.Cano & Zuloaga

Braya monimocalyx Gilg & Muschl. ex Hosseus = Eudema monimocalyx (Gilg & Muschl. ex Hosseus) Al-Shehbaz, Salariato, A.Cano & Zuloaga

Brayopsis arequipa Al-Shehbaz, A.Cano, M.A.Cueva & Salariato = Eudema arequipa (Al-Shehbaz, A.Cano, M.A.Cueva & Salariato) Al-Shehbaz, Salariato, A.Cano & Zuloaga

Brayopsis calycina (Desv.) Gilg & Muschl. = Eudema calycinum (Desv.) Al-Shehbaz, Salariato, A.Cano & Zuloaga

Brayopsis monimocalyx (Gilg & Muschl. ex Hosseus) O.E.Schulz = Eudema monimocalyx (Gilg & Muschl. ex Hosseus) Al-Shehbaz, Salariato, A.Cano & Zuloaga

Brayulinea densa (Willd. ex Schult.) Small = Guilleminea densa (Willd. ex Schult.) Moq.

Brizopyrum spicatum (L.) Hook. & Arn. = Distichlis spicata (L.) Greene

Bromidium rigescens (J.Presl) Nees & Meyen = Cinnagrostis rigescens (J.Presl) P.M.Peterson, Soreng, Romasch. & Barberá

Bromus barbatoides Beal = Bromus berteroanus Colla

Bromus bicuspis Nees ex Steud. = Bromus berteroanus Colla

Bromus brongniartii Kunth = Bromus catharticus Vahl

Bromus frigidus Ball = Bromus lanatus Kunth

Bromus haenkeanus (J.Presl) Kunth = Bromus catharticus Vahl

Bromus leyboldtii Phil. = Bromus berteroanus Colla

Bromus mathewsii Steud. = Bromus catharticus Vahl

Bromus oliganthus Pilg. = Bromus lanatus Kunth

Bromus preslii Kunth = Bromus catharticus Vahl

Bromus schraderi Kunth = Bromus catharticus Vahl

Bromus spicatus Nees = Tripogonella spicata (Nees) P.M.Peterson & Romasch.

Bromus trinii É.Desv. = Bromus berteroanus Colla

Bromus unioloides Kunth = Bromus catharticus Vahl

Bryantiella glutinosa (Phil.) J.M.Porter = Dayia glutinosa (Phil.) J.M.Porter

Buddleja buxifolia Kraenzl. = Buddleja coriacea J.Rémy

Buddleja rhododendroides Kraenzl. = Buddleja coriacea J.Rémy

Buddleja rondeletiiflora Benth. = Buddleja incana Ruiz & Pav.

Buddleja rugosa Kunth = Buddleja incana Ruiz & Pav.

Buddleja ususch Kraenzl. = Buddleja coriacea J.Rémy

Buddleja utilis Kraenzl. = Buddleja coriacea J.Rémy

Bulnesia macrocarpa Phil. = Bulnesia retama (Gillies ex Hook. & Arn.) Griseb.

Bystropogon mollis Kunth = Minthostachys mollis (Benth.) Griseb.

Bystropogon sidifolius L'Hér. = Mesosphaerum sidifolium (L'Hér.) Harley & J.F.B.Pastore

Bystropogon spicatus Benth. = Minthostachys spicata (Benth.) Epling

Cacabus flavus I.M.Johnst. = Exodeconus flavus (I.M.Johnst.) Axelius & D'Arcy

Cacabus inflatus (Ruiz & Pav.) Miers = Nolana inflata Ruiz & Pav.

Cacabus integrifolius Phil. = Exodeconus integrifolius (Phil.) Axelius

Cacabus pusillus Bitter = Exodeconus pusillus (Bitter) Axelius

Cacalia ruderalis (Jacq.) Sw. = Porophyllum ruderale (Jacq.) Cass.

Cactus aureus Meyen = Corryocactus aureus (Meyen) Hutchison

Caesalpinia miranda (Sandwith) J.F.Macbr. = Hoffmannseggia miranda Sandwith

Caesalpinia prostrata (Lag. ex DC.) J.F.Macbr. = Hoffmannseggia prostrata Lag. ex DC.

Caesalpinia spinosa (Molina) Kuntze = Tara spinosa (Molina) Britton & Rose

Caesalpinia viscosa (Ruiz & Pav.) J.F.Macbr. = Hoffmannseggia viscosa (Ruiz & Pav.) Hook. & Arn.

Caiophora cinerea Urb. & Gilg = Caiophora cirsiifolia C.Presl

Caiophora cymbifera Urb. & Gilg = Caiophora cirsiifolia C.Presl

Caiophora mandoniana Urb. & Gilg = Caiophora andina Urb. & Gilg

Caiophora pachylepis Urb. & Gilg = Caiophora cirsiifolia C.Presl

Caiophora pauciseta Killip = Caiophora carduifolia C.Presl

Caiophora preslii Urb. & Gilg = Caiophora cirsiifolia C.Presl

Caiophora sepiaria (Ruiz & Pav. ex G.Don) J.F.Macbr. = Caiophora cirsiifolia C.Presl

Calacinum fruticulosum (Walp.) J.F.Macbr. = Muehlenbeckia fruticulosa (Walp.) Standl.

Calacinum hastulatum (Sm.) J.F.Macbr. = Muehlenbeckia hastulata (Sm.) I.M.Johnst.

Calacinum tamnifolium (Kunth) J.F.Macbr. = Muehlenbeckia tamnifolia (Kunth) Meisn.

Calamagrostis breviaristata (Wedd.) Pilg. = Cinnagrostis breviaristata (Wedd.) P.M.Peterson, Soreng, Romasch. & Barberá

Calamagrostis brevifolia (J.Presl) Steud. = Cinnagrostis brevifolia (J.Presl) P.M.Peterson, Soreng, Romasch. & Barberá

Calamagrostis chrysantha (J.Presl) Steud. = Deschampsia chrysantha (J.Presl) Saarela

Calamagrostis curvula (Wedd.) Pilg. = Cinnagrostis curvula (Wedd.) P.M.Peterson, Soreng, Romasch. & Barberá

Calamagrostis eminens (J.Presl) Steud. = Deschampsia eminens (J.Presl) Saarela

Calamagrostis heterophylla (Wedd.) Pilg. = Cinnagrostis heterophylla (Wedd.) P.M.Peterson, Soreng, Romasch. & Barberá

Calamagrostis jamesonii Steud. = Cinnagrostis jamesonii (Steud.) P.M.Peterson, Soreng, Romasch. & Barberá

Calamagrostis minima (Pilg.) Tovar = Cinnagrostis minima (Pilg.) P.M.Peterson, Soreng, Romasch. & Barberá

Calamagrostis ovata (J.Presl) Steud. = Deschampsia ovata (J.Presl) Saarela

Calamagrostis preslii (Kunth) Hitchc. = Cinnagrostis preslii (Kunth) P.M.Peterson, Soreng, Romasch. & Barberá

Calamagrostis rigescens (J.Presl) Scribn. = Cinnagrostis rigescens (J.Presl) P.M.Peterson, Soreng, Romasch. & Barberá

Calamagrostis rigida (Kunth) Trin. ex Steud. = Cinnagrostis rigida (Kunth) P.M.Peterson, Soreng, Romasch. & Barberá

Calamagrostis setiflora (Wedd.) Pilg. = Cinnagrostis setiflora (Wedd.) P.M.Peterson, Soreng, Romasch. & Barberá

Calamagrostis spicigera (J.Presl) Steud. = Cinnagrostis spicigera (J.Presl) P.M.Peterson, Soreng, Romasch. & Barberá

Calamagrostis vicunarum (Wedd.) Pilg. = Cinnagrostis vicunarum (Wedd.) P.M.Peterson, Soreng, Romasch. & Barberá

Calamagrostis violacea (Wedd.) Hack. = Cinnagrostis violacea (Wedd.) P.M.Peterson, Soreng, Romasch. & Barberá

Calandrinia calycina Phil. = Cistanthe calycina (Phil.) Carolin ex Hershk.

Calandrinia caulescens Kunth = Calandrinia ciliata (Ruiz & Pav.) DC.

Calandrinia cumingii Hook. & Arn. = Montiopsis cumingii (Hook. & Arn.) D.I.Ford

Calandrinia lingulata (Ruiz & Pav.) DC. = Cistanthe lingulata (Ruiz & Pav.) Hershk.

Calandrinia menziesii (Hook.) Torr. & A.Gray = Calandrinia ciliata (Ruiz & Pav.) DC.

Calandrinia micrantha Schltdl. = Calandrinia ciliata (Ruiz & Pav.) DC.

Calandrinia muricata Rydb. = Calandrinia ciliata (Ruiz & Pav.) DC.

Calandrinia paniculata DC. = Cistanthe paniculata (DC.) Carolin ex Hershk.

Calandrinia phacosperma DC. = Calandrinia ciliata (Ruiz & Pav.) DC.

Calandrinia pulchella Lilja = Calandrinia ciliata (Ruiz & Pav.) DC.

Calandrinia stenophylla Rydb. = Calandrinia ciliata (Ruiz & Pav.) DC.

Calandrinia weberbaueri Diels = Cistanthe weberbaueri (Diels) Carolin ex Hershk.

Calceolaria anagalloides Kraenzl. = Calceolaria dichotoma Lam.

Calceolaria aurea Pennell = Calceolaria colquepatana Pennell

Calceolaria fiebrigiana Kraenzl. = Calceolaria plectranthifolia Walp.

Calceolaria gemelliflora Cav. = Calceolaria angustiflora Ruiz & Pav.

Calceolaria glutinosa Heer & Regel = Calceolaria chelidonioides Kunth

Calceolaria grandipinnata Edwin = Calceolaria chelidonioides Kunth

Calceolaria hieronymi Kraenzl. = Calceolaria plectranthifolia Walp.

Calceolaria lysimachioides Kraenzl. = Calceolaria dichotoma Lam.

Calceolaria meistantha Pennell = Calceolaria dichotoma Lam.

Calceolaria multiflora Cav. = Calceolaria angustiflora Ruiz & Pav.

Calceolaria obscura Pennell = Calceolaria chelidonioides Kunth

Calceolaria ovata Sm. = Calceolaria dichotoma Lam.

Calceolaria paucifolia Edwin & Wooden = Calceolaria plectranthifolia Walp.

Calceolaria ribesiifolia Rusby = Calceolaria lobata Cav.

Calceolaria verticillata Ruiz & Pav. = Calceolaria angustiflora Ruiz & Pav.

Caldasia andicola Lag. ex DC. = Chaerophyllum andicola (Kunth) K.F.Chung

Calliandra expansa Benth. = Calliandra taxifolia (Kunth) Benth.

Calliandra prostrata Benth. = Calliandra taxifolia (Kunth) Benth.

Calophanes repens Nees = Dyschoriste repens (Nees) Kuntze

Calymenia expansa (Ruiz & Pav.) Pers. = Mirabilis expansa (Ruiz & Pav.) Standl.

Calymenia ovata Pers. = Mirabilis ovata (Ruiz & Pav.) F.Meigen

Calymenia prostrata (Ruiz & Pav.) Pers. = Mirabilis prostrata (Ruiz & Pav.) Heimerl

Calyxhymenia expansa Ruiz & Pav. = Mirabilis expansa (Ruiz & Pav.) Standl.

Calyxhymenia ovata Ruiz & Pav. = Mirabilis ovata (Ruiz & Pav.) F.Meigen

Calyxhymenia prostrata Ruiz & Pav. = Mirabilis prostrata (Ruiz & Pav.) Heimerl

Camassia biflora Cocucci = Oziroe biflora (Ruiz & Pav.) Speta

Campanula perfoliata L. = Triodanis perfoliata (L.) Nieuwl.

Canna edulis Ker Gawl. = Canna indica L.

Cantua dependens Pers. = Cantua buxifolia Juss. ex Lam.

Cantua laciniata Poir. = Gilia laciniata Ruiz & Pav.

Capraria monnieria Roxb. = Bacopa monnieri (L.) Wettst.

Cardamindum minus (L.) Moench = Tropaeolum minus L.

Cardamine nana Müll.Berol. = Rorippa nana (Schltdl.) J.F.Macbr.

Cardionema multicaule DC. = Cardionema ramosissima (Weinm.) A.Nelson & J.F.Macbr.

Cardiospermum acuminatum Miq. = Cardiospermum halicacabum L.

Cardiospermum alatum Bremek. & Oberm. = Cardiospermum corindum L.

Cardiospermum canescens Wall. = Cardiospermum corindum L.

Cardiospermum clematideum A.Rich. = Cardiospermum corindum L.

Cardiospermum corycodes Kunze = Cardiospermum halicacabum L.

Cardiospermum ferrugineum A.Rich. = Cardiospermum corindum L.

Cardiospermum giganteum Barb.Rodr. = Cardiospermum corindum L.

Cardiospermum glabrum Schumach. & Thonn. = Cardiospermum halicacabum L.

Cardiospermum luridum Blume = Cardiospermum halicacabum L.

Cardiospermum molle Kunth = Cardiospermum corindum L.

Cardiospermum moniliferum Sw. ex Steud. = Cardiospermum halicacabum L.

Cardiospermum oblongum A.Rich. = Cardiospermum corindum L.

Cardiospermum palmeri Vasey & Rose = Cardiospermum corindum L.

Cardiospermum parviflorum Cambess. = Cardiospermum corindum L.

Cardiospermum villosum Macfad. = Cardiospermum corindum L.

Carex hermaphrodita Jacq. = Cyperus hermaphroditus (Jacq.) Standl.

Carex psammogaea Steud. = Carex melanocystis É.Desv.

Carica candicans A.Gray = Vasconcellea candicans (A.Gray) A.DC.

Caryochloa montevidensis Spreng. = Piptochaetium montevidense (Spreng.) Parodi

Cassebeera ternifolia (Cav.) Farw. = Pellaea ternifolia (Cav.) Link

Cassia bicapsularis L. = Senna bicapsularis (L.) Roxb.

Cassia birostris Dombey ex Vogel = Senna birostris (Dombey ex Vogel) H.S.Irwin & Barneby

Cassia brongniartii Gaudich. = Senna brongniartii (Gaudich.) H.S.Irwin & Barneby

Castelia cuneato-ovata Cav. = Pitraea cuneato-ovata (Cav.) Caro

Castilleja bracteata Edwin = Castilleja pumila (Benth.) Wedd.

Cathartocarpus bicapsularis (L.) Ham. = Senna bicapsularis (L.) Roxb.

Celastrus verticillatus Ruiz & Pav. = Maytenus verticillata (Ruiz & Pav.) DC.

Celosia interrupta Seub. = Froelichia interrupta (L.) Moq.

Cenchrus brevisetus E.Fourn. ex Hemsl. = Cenchrus echinatus L.

Cenchrus crinitus Mez = Cenchrus echinatus L.

Cenchrus humilis Hitchc. = Cenchrus spinifex Cav.

Cenchrus incertus M.A.Curtis = Cenchrus spinifex Cav.

Cenchrus insularis Scribn. = Cenchrus echinatus L.

Cenchrus microcephalus Nash ex Hitchc. & Chase = Cenchrus spinifex Cav.

Cenchrus parviceps Shinners = Cenchrus spinifex Cav.

Cenchrus parviflorus Poir. = Setaria parviflora (Poir.) Kerguélen

Cenchrus pauciflorus Benth. = Cenchrus spinifex Cav.

Cenchrus pungens Kunth = Cenchrus echinatus L.

Cenchrus roseus E.Fourn. ex Hemsl. = Cenchrus spinifex Cav.

Cenchrus strictus Chapm. = Cenchrus spinifex Cav.

Centranthus ruber (L.) DC. = Valeriana rubra L.

Ceratocephalus ciliatus (Kunth) Kuntze = Acmella ciliata (Kunth) Cass.

Ceratocephalus leiocarpus Kuntze = Spilanthes leiocarpa DC.

Ceratocephalus pilosus (L.) Rich. = Bidens pilosa L.

Ceratochloa cathartica (Vahl) Herter = Bromus catharticus Vahl

Cereus arequipensis Meyen = Neoraimondia arequipensis (Meyen) Backeb.

Cereus brachypetalus Vaupel = Corryocactus brachypetalus (Vaupel) Britton & Rose

Cereus brevistylus K.Schum. ex Vaupel = Corryocactus brevistylus (K.Schum. ex Vaupel) Britton & Rose

Cereus candelaris Meyen = Browningia candelaris (Meyen) Britton & Rose

Cereus cephalomacrostibas Werderm. & Backeb. = Weberbauerocereus cephalomacrostibas (Werderm. & Backeb.) F.Ritter

Cereus cuzcoensis (Britton & Rose) Werderm. = Trichocereus cuzcoensis Britton & Rose

Cereus decumbens Vaupel = Haageocereus decumbens (Vaupel) Backeb.

Cereus haynii (Otto ex Salm-Dyck) Croucher = Matucana haynii (Otto ex Salm-Dyck) Britton & Rose

Cereus macrogonus Salm-Dyck = Trichocereus macrogonus (Salm-Dyck) Riccob.

Cereus platinospinus Werderm. & Backeb. = Haageocereus platinospinus (Werderm. & Backeb.) Backeb.

Cereus weberbaueri K.Schum. ex Vaupel = Weberbauerocereus weberbaueri (K.Schum. ex Vaupel) Backeb.

Ceropteris trifoliata (L.) Kuhn = Pityrogramma trifoliata (L.) R.M.Tryon

Cestrum densiflorum Francey = Cestrum tomentosum L.f.

Cestrum foetidum Salisb. = Cestrum auriculatum L'Hér.

Cestrum granadense Willd. ex Roem. & Schult. = Cestrum tomentosum L.f.

Cestrum hediundinum Dunal = Cestrum auriculatum L'Hér.

Cestrum hirsutum Jacq. = Cestrum tomentosum L.f.

Cestrum lanatum M.Martens & Galeotti = Cestrum tomentosum L.f.

Cestrum lanuginosum Ruiz & Pav. = Cestrum tomentosum L.f.

Cestrum lasianthum Dunal = Cestrum auriculatum L'Hér.

Cestrum leptanthum Dunal = Cestrum auriculatum L'Hér.

Cestrum meridanum Pittier = Cestrum tomentosum L.f.

Cestrum neomiersianum Benítez = Cestrum tomentosum L.f.

Cestrum serratum Dunal = Cestrum auriculatum L'Hér.

Cestrum verbascifolium Zucc. ex Francey = Cestrum tomentosum L.f.

Chabraea daucifolia (D.Don) Wedd. = Leucheria daucifolia (D.Don) Crisci

Chaetachlaena odorata D.Don = Onoseris odorata Hook. & Arn.

Chaetanthera multiflora Bonpl. = Perezia multiflora (Bonpl.) Less.

Chaetanthera pinnatifida Bonpl. = Perezia pinnatifida (Bonpl.) Wedd.

Chaetanthera pungens Bonpl. = Perezia pungens (Bonpl.) Less.

Chaetanthera stuebelii Hieron. = Oriastrum stuebelii (Hieron.) A.M.R.Davies

Chaetochloa parviflora (Poir.) Scribn. = Setaria parviflora (Poir.) Kerguélen

Chaetocyperus albibracteatus (Nees & Meyen ex Kunth) Nees & Meyen = Eleocharis albibracteata Nees & Meyen ex Kunth

Chaetotropis elongata (Kunth) Björkman = Polypogon elongatus Kunth
Chamaefilix monanthes (L.) Farw. = Asplenium monanthes L.
Chamaesyce hirta (L.) Millsp. = Euphorbia hirta L.
Chamaesyce hypericifolia (L.) Millsp. = Euphorbia hypericifolia L.
Chamaesyce lasiocarpa (Klotzsch) Arthur = Euphorbia lasiocarpa Klotzsch
Chamaesyce meyeniana (Klotzsch) Croizat = Euphorbia meyeniana Klotzsch
Chamaesyce serpens (Kunth) Small = Euphorbia serpens Kunth
Cheilanthes andina Hook. = Cheilanthes pilosa Goldm.
Cheilanthes doradilla (Colla) Domin = Cheilanthes mollis (Kunze) C.Presl
Cheilanthes fasciculata Goldm. = Cheilanthes pruinata Kaulf.
Cheilanthes macleanii Hook. = Cheilanthes pilosa Goldm.
Cheilanthes mathewsii Kunze = Cheilanthes pruinata Kaulf.
Cheilanthes myriophylla Desv. = Myriopteris myriophylla J.Sm.
Cheilanthes saundersii Alston = Cheilanthes fractifera R.M.Tryon
Cheilanthes sinuata (Lag. ex Sw.) Domin = Astrolepis sinuata (Lag. ex Sw.) D.M.Benham & Windham
Cheilanthes sulphurea (Cav.) Mickel & Beitel = Notholaena sulphurea (Cav.) J.Sm.
Cheilanthes ternifolia (Cav.) T.Moore = Pellaea ternifolia (Cav.) Link
Chenopodium amboanum (Murr) Aellen = Dysphania ambrosioides (L.) Mosyakin & Clemants
Chenopodium ambrosioides L. = Dysphania ambrosioides (L.) Mosyakin & Clemants
Chenopodium bolivianum Murr = Chenopodium petiolare Kunth
Chenopodium hastatum Phil. = Chenopodium petiolare Kunth
Chenopodium incisum Poir. = Dysphania incisa (Poir.) ined.
Chenopodium paniculatum Hook. = Chenopodium petiolare Kunth
Chenopodium sparsiflorum Phil. = Chenopodium petiolare Kunth
Chersodoma diclina (Wedd.) Cabrera = Chersodoma antennaria (Wedd.) Cabrera
Chersodoma longipedicellata Ricardi & Martic. = Chersodoma arequipensis (Cuatrec.) Cuatrec.
Chloris alba J.Presl = Chloris virgata Sw.
Chloris caudata Trin. = Chloris virgata Sw.
Chloris compressa DC. = Chloris virgata Sw.
Chloris decora Nees ex Steud. = Chloris virgata Sw.
Chloris elegans Kunth = Chloris virgata Sw.
Chloris glaucescens Steud. = Chloris radiata (L.) Sw.
Chloris gracilis P.Durand = Chloris radiata (L.) Sw.
Chloris multiradiata Hochst. = Chloris virgata Sw.
Chloris notocoma Hochst. = Chloris virgata Sw.
Chloris penicillata (Vahl) Pers. = Chloris virgata Sw.
Chloris pubescens Lag. = Chloris virgata Sw.
Chloris scoparia Desf. = Chloris radiata (L.) Sw.
Chlorocharis geniculata (L.) Rikli = Eleocharis geniculata (L.) Roem. & Schult.
Chlorocyperus articulatus (L.) Rikli = Cyperus articulatus L.
Chlorocyperus laevigatus (L.) Palla = Cyperus laevigatus L.
Chondrosum simplex (Lag.) Kunth = Bouteloua simplex Lag.
Chrysochosma sulphurea Kümmerle = Notholaena sulphurea (Cav.) J.Sm.
Chucoa lanceolata (H.Beltrán & Ferreyra) G.Sancho, S.E.Freire & Katinas = Paquirea lanceolata (H.Beltrán & Ferreyra) Panero & S.E.Freire
Chymocarpus tuberosus (Ruiz & Pav.) Heynh. = Tropaeolum tuberosum Ruiz & Pav.
Cicendia poeppigii Griseb. = Cicendia quadrangularis (Lam.) Griseb.
Cicendia quitensis (Kunth) Griseb. = Cicendia quadrangularis (Lam.) Griseb.
Cieca suberosa (L.) Moench = Passiflora suberosa L.
Cincinalis esculenta (G.Forst.) Trevis. = Pteridium esculentum (G.Forst.) Cockayne
Cinna phleoides (Kunth) Kunth = Muhlenbergia coerulea (Griseb.) Mez
Cinna stricta (Kunth) Kunth = Muhlenbergia coerulea (Griseb.) Mez
Citharexylum kobuskianum Moldenke = Citharexylum flexuosum (Ruiz & Pav.) D.Don
Citharexylum megacanthum Rusby = Citharexylum weberbaueri Hayek

Citharexylum retusum D.Don = Citharexylum flexuosum (Ruiz & Pav.) D.Don
Clarionea ciliosa Phil. = Perezia ciliosa Reiche
Clarionea pinnatifida (Bonpl.) DC. = Perezia pinnatifida (Bonpl.) Wedd.
Claytonia alba (Ruiz & Pav.) Kuntze = Calandrinia alba (Ruiz & Pav.) DC.
Claytonia calycina Colenso = Cistanthe calycina (Phil.) Carolin ex Hershk.
Claytonia ciliata (Ruiz & Pav.) Kuntze = Calandrinia ciliata (Ruiz & Pav.) DC.
Claytonia cumingii (Hook. & Arn.) Kuntze = Montiopsis cumingii (Hook. & Arn.) D.I.Ford
Claytonia fontana (L.) R.J.Davis = Montia fontana L.
Claytonia lingulata (Ruiz & Pav.) Kuntze = Cistanthe lingulata (Ruiz & Pav.) Hershk.
Claytonia paniculata (DC.) Kuntze = Cistanthe paniculata (DC.) Carolin ex Hershk.
Cleistocactus aureus (Meyen) F.A.C.Weber = Corryocactus aureus (Meyen) Hutchison
Cleistocactus sextonianus (Backeb.) D.R.Hunt = Loxanthocereus sextonianus (Backeb.) Backeb.
Clematis cochabambensis Rusby = Clematis millefoliolata Eichler
Clematis parvifrons Ulbr. = Clematis millefoliolata Eichler
Clematis seemannii Kuntze = Clematis millefoliolata Eichler
Clitanthes humilis Herb. = Clinanthus humilis (Herb.) Meerow
Clomena peruviana P.Beauv. = Muhlenbergia peruviana (P.Beauv.) Steud.
Coburgia humilis (Herb.) Herb. = Clinanthus humilis (Herb.) Meerow
Coldenia conspicua I.M.Johnst. = Tiquilia conspicua (I.M.Johnst.) A.T.Richardson
Coldenia dichotoma (Ruiz & Pav.) Lehm. = Tiquilia dichotoma (Ruiz & Pav.) Pers.
Coldenia elongata Rusby = Tiquilia elongata (Rusby) A.T.Richardson
Coldenia ferreyrae I.M.Johnst. = Tiquilia ferreyrae (I.M.Johnst.) A.T.Richardson
Coldenia grandiflora Phil. = Tiquilia grandiflora (Phil.) A.T.Richardson
Coldenia litoralis Phil. = Tiquilia litoralis (Phil.) A.T.Richardson
Coldenia paronychioides Phil. = Tiquilia paronychioides (Phil.) A.T.Richardson
Coldenia simulans I.M.Johnst. = Tiquilia simulans (I.M.Johnst.) A.T.Richardson
Coleogeton striatus (Ruiz & Pav.) Les & R.R.Haynes = Stuckenia striata (Ruiz & Pav.) Holub
Colignonia acutifolia (Heimerl) Heimerl = Colignonia parviflora (Kunth) Choisy
Colignonia biumbellata Ball = Colignonia parviflora (Kunth) Choisy
Colignonia pubigera Heimerl = Colignonia parviflora (Kunth) Choisy
Colignonia weberbaueri Heimerl = Colignonia parviflora (Kunth) Choisy
Collania dulcis (Hook.) Herb. = Bomarea dulcis (Hook.) Beauverd
Collania involucrosa Herb. = Bomarea involucrosa (Herb.) Baker
Colletia assimilis N.E.Br. = Colletia spinosissima J.F.Gmel.
Colletia atrox Miers = Colletia spinosissima J.F.Gmel.
Colletia ferox Gillies & Hook. = Colletia spinosissima J.F.Gmel.
Colletia horrida Willd. = Colletia spinosissima J.F.Gmel.
Colletia infausta N.E.Br. = Colletia spinosissima J.F.Gmel.
Colletia intricata Miers = Colletia spinosissima J.F.Gmel.
Colletia invicta Miers = Colletia spinosissima J.F.Gmel.
Colletia polyacantha Humb. & Bonpl. ex Schult. = Colletia spinosissima J.F.Gmel.
Colletia spicata Humb. & Bonpl. ex Schult. = Scutia spicata (Humb. & Bonpl. ex Schult.) Weberb.
Colletia tenuicola Miers = Colletia spinosissima J.F.Gmel.
Colletia trifurcata N.E.Br. = Colletia spinosissima J.F.Gmel.
Collomia gracilis (Hook.) Douglas ex Benth. = Microsteris gracilis (Hook.) Greene
Colobanthus alatus Pax = Colobanthus quitensis (Kunth) Bartl.
Colobanthus aretioides Gillies ex Hook. = Colobanthus quitensis (Kunth) Bartl.
Colobanthus cherlerioides Hook.f. = Colobanthus quitensis (Kunth) Bartl.
Colobanthus crassifolius (d'Urv.) Hook.f. = Colobanthus quitensis (Kunth) Bartl.

Colobanthus saginoides Bartl. = Colobanthus quitensis (Kunth) Bartl.

Colutea triflora (DC.) Poir. = Astragalus triflorus (DC.) A.Gray

Commelina clandestina Mart. = Commelina tuberosa L.

Commelina coelestis Willd. = Commelina tuberosa L.

Commelina elliptica Kunth = Commelina tuberosa L.

Commelina fasciculata Ruiz & Pav. = Commelina tuberosa L.

Commelina graminifolia Kunth = Commelina tuberosa L.

Commelina hispida Ruiz & Pav. = Commelina tuberosa L.

Commelina intermedia Schltdl. = Commelina tuberosa L.

Commelina nervosa Ruiz & Pav. = Commelina tuberosa L.

Commelina reflexa Rusby = Commelina tuberosa L.

Commelina scapigera Kunth = Commelina tuberosa L.

Commelina variabilis Schltdl. = Commelina tuberosa L.

Commicarpus crassifolius Heimerl = Commicarpus tuberosus (Lam.) Standl.

Conanthus dichotomus (Ruiz & Pav.) Kuntze = Nama dichotoma (Ruiz & Pav.) Choisy

Convolvuloides purpurea (L.) Moench = Ipomoea purpurea (L.) Roth

Convolvulus carneus (Jacq.) Spreng. = Ipomoea carnea Jacq.

Convolvulus dumetorum Kunth = Ipomoea dumetorum Willd.

Convolvulus purpureus L. = Ipomoea purpurea (L.) Roth

Convolvulus unilateralis Roem. & Schult. = Jacquemontia unilateralis (Roem. & Schult.) O'Donell

Conyza artemisiifolia Meyen & Walp. = Laennecia artemisiifolia (Meyen & Walp.) G.L.Nesom

Conyza bonariensis (L.) Cronquist = Erigeron bonariensis L.

Conyza canadensis (L.) Cronquist = Erigeron canadensis L.

Conyza chingoyo Kunth = Pluchea chingoyo DC.

Conyza deserticola Phil. = Erigeron deserticolus (Phil.) comb.ined.

Conyza floribunda Kunth = Erigeron floribundus (Kunth) Sch.Bip.

Conyza kunthiana DC. = Lucilia kunthiana (DC.) Zardini

Conyza squamata Spreng. = Symphyotrichum squamatum (Spreng.) G.L.Nesom

Conyzanthus squamatus (Spreng.) Tamamsch. = Symphyotrichum squamatum (Spreng.) G.L.Nesom

Cooperia albicans (Herb.) Sprague = Pyrolirion albicans Herb.

Cordia flava (Andersson) Gürke = Cordia lutea Lam.

Cordia marchionica Drake = Cordia lutea Lam.

Cordia rotundifolia Ruiz & Pav. = Cordia lutea Lam.

Coreopsis fasciculata Wedd. = Burnellia fasciculata (Wedd.) Mesfin & D.J.Crawford

Coreopsis irmscheriana Bruns = Heiseria irmscheriana (Bruns) Mesfin

Coronopus didymus (L.) Sm. = Lepidium didymum L.

Corryocactus acervatus F.Ritter = Corryocactus aureus (Meyen) Hutchison

Corryocactus cuajonesensis F.Ritter = Corryocactus aureus (Meyen) Hutchison

Corryocactus krausii Backeb. = Corryocactus brevistylus (K.Schum. ex Vaupel) Britton & Rose

Corryocactus pachycladus Rauh & Backeb. = Corryocactus brevistylus (K.Schum. ex Vaupel) Britton & Rose

Corryocactus prostratus F.Ritter = Corryocactus quadrangularis (Rauh & Backeb.) F.Ritter ex D.R.Hunt, N.P.Taylor & G.J.Charles

Corryocactus puquiensis Rauh & Backeb. = Corryocactus brevistylus (K.Schum. ex Vaupel) Britton & Rose

Cosmea pilosa (L.) Spreng. = Bidens pilosa L.

Cotula mexicana (DC.) Cabrera = Soliva mexicana DC.

Cotula prostrata (L.) L. = Eclipta prostrata (L.) L.

Cotyledon peruviana Baker = Echeveria peruviana Meyen

Coursetia weberbaueri Harms = Poissonia weberbaueri (Harms) Lavin

Covillea divaricata (Cav.) Coville = Larrea divaricata Cav.

Cremolobus aphanopterus A.Gray = Cremolobus chilensis (Lag. ex DC.) DC.

Cremolobus humilis Muschl. = Cremolobus chilensis (Lag. ex DC.) DC.

Cremolobus parviflorus Wedd. = Cremolobus chilensis (Lag. ex DC.) DC.

Cremolobus pinnatifidus Hook. = Cremolobus chilensis (Lag. ex DC.) DC.

Cremolobus sinuatus Hook. = Cremolobus chilensis (Lag. ex DC.) DC.

Cremolobus weberbaueri Muschl. = Cremolobus chilensis (Lag. ex DC.) DC.

Cressa arenaria Willd. ex Schult. = Cressa truxillensis Kunth

Cressa depressa Goodd. = Cressa truxillensis Kunth

Cressa erecta Rydb. = Cressa truxillensis Kunth

Cressa insularis House = Cressa truxillensis Kunth

Cressa minima A.Heller = Cressa truxillensis Kunth

Cressa multiflora Willd. ex Schult. = Cressa truxillensis Kunth

Critesion muticum (J.Presl) Á.Löve = Hordeum muticum J.Presl

Crotalaria affinis DC. = Crotalaria incana L.

Crotalaria montana A.Rich. = Crotalaria incana L.

Crotalaria schimperi A.Rich. = Crotalaria incana L.

Croton microphyllus Lam. = Andrachne microphylla (Lam.) Baill.

Croton quitensis Spreng. = Croton alnifolius Lam.

Crypsinna stricta (Kunth) E.Fourn. = Muhlenbergia coerulea (Griseb.) Mez

Crypsis phleoides Kunth = Muhlenbergia coerulea (Griseb.) Mez

Crypsis stricta Kunth = Muhlenbergia coerulea (Griseb.) Mez

Crypsis virginica (L.) Nutt. = Sporobolus virginicus (L.) Kunth

Cryptantha cajabambensis Brand = Cryptantha peruviana I.M.Johnst.

Cryptantha latifolia I.M.Johnst. = Cryptantha granulosa (Ruiz & Pav.) I.M.Johnst.

Cryptantha macbridei I.M.Johnst. = Cryptantha limensis (A.DC.) I.M.Johnst.

Cryptantha parviflora (Phil.) Reiche = Johnstonella parviflora (Phil.) Hasenstab & M.G.Simpson

Cryptantha weberbaueri Brand = Cryptantha peruviana I.M.Johnst.

Cryptantha woitschachii Brand = Cryptantha limensis (A.DC.) I.M.Johnst.

Culcitium albifolium Zoellner = Culcitium oligocephalum Cabrera

Culcitium glaciale Meyen & Walp. = Culcitium humile DC.

Cumulopuntia alboareolata (F.Ritter) F.Ritter = Cumulopuntia zehnderi (Rauh & Backeb.) F.Ritter

Cumulopuntia crassicylindrica (Rauh & Backeb.) F.Ritter ex Eggli = Cumulopuntia sphaerica (C.F.Först.) E.F.Anderson

Cumulopuntia frigida F.Ritter = Cumulopuntia ignescens (Vaupel) F.Ritter

Cumulopuntia glomerata (Pfeiff.) Hoxey = Cumulopuntia ignescens (Vaupel) F.Ritter

Cumulopuntia hystrix F.Ritter = Cumulopuntia ignescens (Vaupel) F.Ritter

Cumulopuntia mistiensis (Backeb.) E.F.Anderson = Cumulopuntia corotilla (K.Schum. ex Vaupel) E.F.Anderson

Cumulopuntia multiareolata (F.Ritter) F.Ritter = Cumulopuntia sphaerica (C.F.Först.) E.F.Anderson

Cumulopuntia tortispina F.Ritter = Cumulopuntia ignescens (Vaupel) F.Ritter

Cumulopuntia tubercularis F.Ritter = Cumulopuntia sphaerica (C.F.Först.) E.F.Anderson

Cumulopuntia tumida F.Ritter = Cumulopuntia sphaerica (C.F.Först.) E.F.Anderson

Cumulopuntia unguispina (Backeb.) F.Ritter ex A.Pauca & Quip. = Cumulopuntia sphaerica (C.F.Först.) E.F.Anderson

Cyanopsis chingoyo Steud. = Pluchea chingoyo DC.

Cyclolepis denticulata J.Rémy = Aphyllocladus denticulatus (J.Rémy ex Gay) Cabrera

Cylindropuntia subulata (Muehlenpf.) F.M.Knuth = Austrocylindropuntia subulata (Muehlenpf.) Backeb.

Cynanchum clausum (Jacq.) Jacq. = Funastrum clausum (Jacq.) Schltr.

Cynanchum formosum N.E.Br. = Jobinia formosa (N.E.Br.) Liede & Meve

Cynanchum tiaratum Malme = Jobinia tiarata (Malme) Liede & Meve

Cynodon ciliaris (L.) Raspail = Eragrostis ciliaris (L.) R.Br.

Cynoglossum lineare Ruiz & Pav. = Pectocarya linearis (Ruiz & Pav.) DC.

Cypella cyrtophylla (I.M.Johnst.) Diels = Mastigostyla cyrtophylla I.M.Johnst.

Cypella pardalis Ravenna = Hesperoxiphion pardalis (Ravenna) Ravenna

Cyperus alpinus Liebm. = Cyperus hermaphroditus (Jacq.) Standl.

Cyperus andinus Palla ex Kük. = Cyperus seslerioides Kunth

Cyperus bakeri C.B.Clarke = Cyperus ochraceus Vahl

Cyperus bufonius (L.) Maack ex Trautv. = Juncus bufonius L.

Cyperus declinatus Moench = Cyperus eragrostis Lam.

Cyperus fistulosus Ehrenb. ex Boeckeler = Cyperus articulatus L.
Cyperus gymnos Schult. = Cyperus articulatus L.
Cyperus incompletus (Jacq.) Link = Cyperus hermaphroditus (Jacq.) Standl.
Cyperus interceptus Steud. = Cyperus articulatus L.
Cyperus longespicatus Boeckeler = Cyperus hermaphroditus (Jacq.) Standl.
Cyperus monandrus Roth = Cyperus eragrostis Lam.
Cyperus navicularis Steud. = Cyperus ochraceus Vahl
Cyperus nodosus Humb. & Bonpl. ex Willd. = Cyperus articulatus L.
Cyperus ochrocephalus Steud. = Cyperus eragrostis Lam.
Cyperus perpusillus Boeckeler = Cyperus seslerioides Kunth
Cyperus prionotropis Steud. = Cyperus eragrostis Lam.
Cyperus sertularinus Liebm. = Cyperus hermaphroditus (Jacq.) Standl.
Cyperus subambiguus Kük. = Cyperus hermaphroditus (Jacq.) Standl.
Cyperus subnodosus Nees & Meyen = Cyperus articulatus L.
Cyperus vegetus Willd. = Cyperus eragrostis Lam.
Cystopteris atrovirens C.Presl = Cystopteris diaphana (Bory) Blasdell
Cystopteris emarginatodenticulata Fomin = Cystopteris diaphana (Bory) Blasdell
Cystopteris emarginulata C.Presl = Cystopteris diaphana (Bory) Blasdell
Cystopteris nivalis (Pirotta) Pic.Serm. = Cystopteris diaphana (Bory) Blasdell
Cystopteris sempervirens T.Moore = Cystopteris diaphana (Bory) Blasdell
Cystopteris viridula (Desv.) Desv. = Cystopteris diaphana (Bory) Blasdell
Dalea calliantha Ulbr. = Dalea boliviana Britton
Dalea calocalyx Ulbr. = Dalea cylindrica Hook.
Dalea hofstenii R.E.Fr. = Dalea boliviana Britton
Dalea peruviana (J.F.Macbr.) J.F.Macbr. = Dalea exilis DC.
Dalea retusifolia Harms ex Kuntze = Dalea boliviana Britton
Dalea samancoensis Ulbr. = Dalea cylindrica Hook.
Dalea tapacariensis Harms ex Kuntze = Dalea boliviana Britton
Danthonia nardoides Phil. = Nassella nardoides (Phil.) Barkworth
Daucus australis Poepp. ex DC. = Daucus montanus Humb. & Bonpl. ex Schult.
Daucus toriloides DC. = Daucus montanus Humb. & Bonpl. ex Schult.
Dendrophthora mesembrianthemifolia Urb. = Dendrophthora mesembryanthemifolia Griseb.
Dendropogon usneoides (L.) Raf. = Tillandsia usneoides (L.) L.
Descurainia brachycarpa Reiche = Descurainia depressa (Phil.) Reiche
Descurainia gilgiana Muschl. = Descurainia athrocarpa (A.Gray) O.E.Schulz
Descurainia leptoclada Muschl. = Descurainia myriophylla (Kunth ex DC.) R.E.Fr.
Descurainia minutiflora (Phil.) Reiche = Descurainia stricta (Phil.) Reiche
Descurainia pflanzii Muschl. = Descurainia depressa (Phil.) Reiche
Descurainia pulcherrima Muschl. = Descurainia myriophylla (Kunth ex DC.) R.E.Fr.
Descurainia rubescens (Phil.) Reiche = Descurainia stricta (Phil.) Reiche
Descurainia titicacensis (Walp.) Lillo ex Hauman & Irigoyen = Descurainia myriophylla (Kunth ex DC.) R.E.Fr.
Descurainia urbaniana Muschl. = Descurainia athrocarpa (A.Gray) O.E.Schulz
Desmodium arenarium Kunth = Desmodium scorpiurus (Sw.) Poir.
Desmodium multicaule DC. = Desmodium scorpiurus (Sw.) Poir.
Desmodium parviflorum M.Martens & Galeotti = Desmodium scorpiurus (Sw.) Poir.
Desmodium pilosiusculum DC. = Desmodium uncinatum (Jacq.) DC.
Deyeuxia breviaristata Wedd. = Cinnagrostis breviaristata (Wedd.) P.M.Peterson, Soreng, Romasch. & Barberá
Deyeuxia brevifolia J.Presl = Cinnagrostis brevifolia (J.Presl) P.M.Peterson, Soreng, Romasch. & Barberá
Deyeuxia chrysantha J.Presl = Deschampsia chrysantha (J.Presl) Saarela
Deyeuxia curvula Wedd. = Cinnagrostis curvula (Wedd.) P.M.Peterson, Soreng, Romasch. & Barberá
Deyeuxia eminens J.Presl = Deschampsia eminens (J.Presl) Saarela
Deyeuxia filifolia Wedd. = Cinnagrostis filifolia (Wedd.) P.M.Peterson, Soreng, Romasch. & Barberá
Deyeuxia heterophylla Wedd. = Cinnagrostis heterophylla (Wedd.) P.M.Peterson, Soreng, Romasch. & Barberá
Deyeuxia jamesonii (Steud.) Munro ex Wedd. = Cinnagrostis jamesonii (Steud.) P.M.Peterson, Soreng, Romasch. & Barberá
Deyeuxia lagurus Wedd. = Cinnagrostis lagurus (Wedd.) P.M.Peterson, Soreng, Romasch. & Barberá
Deyeuxia minima (Pilg.) Rúgolo = Cinnagrostis minima (Pilg.) P.M.Peterson, Soreng, Romasch. & Barberá
Deyeuxia ovata J.Presl = Deschampsia ovata (J.Presl) Saarela
Deyeuxia rigescens (J.Presl) Türpe = Cinnagrostis rigescens (J.Presl) P.M.Peterson, Soreng, Romasch. & Barberá
Deyeuxia rigida Kunth = Cinnagrostis rigida (Kunth) P.M.Peterson, Soreng, Romasch. & Barberá
Deyeuxia setiflora Wedd. = Cinnagrostis setiflora (Wedd.) P.M.Peterson, Soreng, Romasch. & Barberá
Deyeuxia spicigera J.Presl = Cinnagrostis spicigera (J.Presl) P.M.Peterson, Soreng, Romasch. & Barberá
Deyeuxia vicunarum Wedd. = Cinnagrostis vicunarum (Wedd.) P.M.Peterson, Soreng, Romasch. & Barberá
Deyeuxia violacea Wedd. = Cinnagrostis violacea (Wedd.) P.M.Peterson, Soreng, Romasch. & Barberá
Diadesma rhombifolia (L.) Raf. = Sida rhombifolia L.
Diallobus bicapsularis (L.) Raf. = Senna bicapsularis (L.) Roxb.
Diapedium congestum (Kunth) Kuntze = Dicliptera congesta Kunth
Diapedium tomentosum (Vahl) Kuntze = Dicliptera tomentosa (Vahl) Nees
Diaphoranthema capillaris (Ruiz & Pav.) Beer = Tillandsia capillaris Ruiz & Pav.
Diaphoranthema recurvata (L.) Beer = Tillandsia recurvata (L.) L.
Diaphoranthema virescens (Ruiz & Pav.) Beer = Tillandsia virescens Ruiz & Pav.
Diatrema purpurea (L.) Raf. = Ipomoea purpurea (L.) Roth
Dicksonia montevidensis Spreng. = Woodsia montevidensis Hieron.
Dicliptera ruiziana Wassh. = Dicliptera tomentosa (Vahl) Nees
Dielsiochloa floribunda (Pilg.) Pilg. = Festuca floribunda (Pilg.) P.M.Peterson, Soreng & Romasch.
Digitaria vaginata (Sw.) Philippe = Paspalum vaginatum Sw.
Diplachne procera (Kunth) Spreng. = Festuca procera Kunth
Diplachne rigescens J.Presl = Festuca rigescens (J.Presl) Kunth
Diplachne spicata (Nees) Döll ex Benth. = Tripogonella spicata (Nees) P.M.Peterson & Romasch.
Diplandra potamogeton Bertero = Elodea potamogeton (Bertero) Espinosa
Diplostephium tacorense Hieron. = Diplostephium meyenii Wedd.
Dipyrena arequipensis (Botta) Ravenna = Mulguraea arequipensis (Botta) N.O'Leary & P.Peralta
Dissanthelium macusaniense (E.H.L.Krause) R.C.Foster & L.B.Sm. = Poa macusaniensis (E.H.L.Krause) Refulio
Distichia andina (Phil.) Benth. & Hook.f. = Oxychloe andina Phil.
Distichia clandestina (Phil.) Buchenau = Patosia clandestina (Phil.) Buchenau
Distichia tolimensis (Decne.) Buchenau = Distichia muscoides Nees & Meyen
Distichlis dentata Rydb. = Distichlis spicata (L.) Greene
Distichlis hirsuta Phil. = Distichlis spicata (L.) Greene
Distichlis hirta Phil. = Distichlis spicata (L.) Greene
Distichlis marginata Phil. = Distichlis spicata (L.) Greene
Distichlis misera Phil. = Distichlis humilis Phil.
Distichlis prostrata (Kunth) É.Desv. = Distichlis spicata (L.) Greene
Distichlis stricta (Torr.) Rydb. = Distichlis spicata (L.) Greene
Distichlis tenuifolia Phil. = Distichlis spicata (L.) Greene
Distichlis viridis Phil. = Distichlis spicata (L.) Greene
Dolia crassulifolia (Kunze ex Walp.) Kuntze = Nolana crassulifolia Kunze ex Walp.
Dolia laxa Miers = Nolana laxa (Miers) I.M.Johnst.
Dolia tarapacana Phil. = Nolana tarapacana (Phil.) I.M.Johnst.
Dolichos luteolus Jacq. = Vigna luteola (Jacq.) Benth.
Dortmanna decurrens (Cav.) Kuntze = Lobelia decurrens Cav.

Dorystigma squarrosum Miers = Jaborosa squarrosa (Miers) Hunz. & Barboza

Draba calycina Desv. = Eudema calycinum (Desv.) Al-Shehbaz, Salariato, A.Cano & Zuloaga

Drynaria macrocarpa (Willd.) Fée = Pleopeltis macrocarpa (Willd.) Kaulf.

Dryopteris glandulosolanosa C.Chr. = Amauropelta glandulosolanosa (C.Chr.) Salino & T.E.Almeida

Dryopteris rufa (Poir.) C.Chr. = Amauropelta rufa (Poir.) Salino & T.E.Almeida

Dumerilia axillaris Lag. ex DC. = Jungia axillaris Spreng.

Dunalia angustifolia Dammer = Dunalia spinosa (Meyen) Dammer

Dunalia inermis Borzì = Dunalia spinosa (Meyen) Dammer

Dunalia lycioides Miers = Dunalia spinosa (Meyen) Dammer

Dunalia ramiflora Miers = Dunalia spinosa (Meyen) Dammer

Dyctisperma urticifolius (Poir.) Raf. ex B.D.Jacks. = Rubus urticifolius Poir.

Dysosmia foetida (L.) M.Roem. = Passiflora foetida L.

Dysphania graveolens (W.A.Weber) Mosyakin & Clemants = Dysphania incisa (Poir.) ined.

Echinocactus aureus (Meyen) Pfeiff. = Corryocactus aureus (Meyen) Hutchison

Echinocactus haynii Otto ex Salm-Dyck = Matucana haynii (Otto ex Salm-Dyck) Britton & Rose

Echinocactus islayensis C.F.först. = Eriosyce islayensis (C.F.Först.) Katt.

Echinocactus leucotrichus Phil. = Oreocereus leucotrichus (Phil.) Wagenkn.

Echinopsis bylesiana (Andreae & Backeb.) Mayta = Haageocereus bylesianus (Andreae & Backeb.) Lodé

Echinopsis cephalomacrostibas (Werderm. & Backeb.) H.Friedrich & G.D.Rowley = Weberbauerocereus cephalomacrostibas (Werderm. & Backeb.) F.Ritter

Echinopsis chalaensis (Rauh & Backeb.) H.Friedrich & G.D.Rowley = Trichocereus chalaensis Rauh & Backeb.

Echinopsis cuzcoensis (Britton & Rose) H.Friedrich & G.D.Rowley = Trichocereus cuzcoensis Britton & Rose

Echinopsis decumbens (Vaupel) Mayta = Haageocereus decumbens (Vaupel) Backeb.

Echinopsis haynii (Otto ex Salm-Dyck) Molinari = Matucana haynii (Otto ex Salm-Dyck) Britton & Rose

Echinopsis hempeliana Gürke = Oreocereus hempelianus (Gürke) D.R.Hunt

Echinopsis leucotricha (Phil.) Anceschi & Magli = Oreocereus leucotrichus (Phil.) Wagenkn.

Echinopsis macrogona (Salm-Dyck) H.Friedrich & G.D.Rowley = Trichocereus macrogonus (Salm-Dyck) Riccob.

Echinopsis maximiliana Heyder ex A.Dietr. = Lobivia maximiliana (Heyder ex A.Dietr.) Backeb. ex Rausch

Echinopsis pampana (Britton & Rose) D.R.Hunt = Lobivia pampana Britton & Rose

Echinopsis platinospina (Werderm. & Backeb.) Anceschi & Magli = Haageocereus platinospinus (Werderm. & Backeb.) Backeb.

Echinopsis rauhii (Backeb.) Mayta = Weberbauerocereus rauhii Backeb.

Echinopsis sextoniana (Backeb.) Mayta = Loxanthocereus sextonianus (Backeb.) Backeb.

Echinopsis weberbaueri (K.Schum. ex Vaupel) Anceschi & Magli = Weberbauerocereus weberbaueri (K.Schum. ex Vaupel) Backeb.

Eclipta alba (L.) Hassk. = Eclipta prostrata (L.) L.

Eclipta angustata Umemoto & H.Koyama = Eclipta prostrata (L.) L.

Eclipta angustifolia C.Presl = Eclipta prostrata (L.) L.

Eclipta erecta L. = Eclipta prostrata (L.) L.

Eclipta flexuosa Raf. = Eclipta prostrata (L.) L.

Eclipta heterophylla Bartl. = Eclipta prostrata (L.) L.

Eclipta hirsuta Bartl. = Eclipta prostrata (L.) L.

Eclipta longifolia Schrad. = Eclipta prostrata (L.) L.

Eclipta marginata Boiss. = Eclipta prostrata (L.) L.

Eclipta nutans Raf. = Eclipta prostrata (L.) L.

Eclipta parviflora Wall. ex DC. = Eclipta prostrata (L.) L.

Eclipta patula Schrad. = Eclipta prostrata (L.) L.

Eclipta pumila Raf. = Eclipta prostrata (L.) L.

Eclipta punctata L. = Eclipta prostrata (L.) L.

Eclipta pusilla M.E.Jones = Eclipta prostrata (L.) L.

Eclipta simplex Raf. = Eclipta prostrata (L.) L.

Eclipta spicata Spreng. = Eclipta prostrata (L.) L.

Eclipta sulcata Raf. = Eclipta prostrata (L.) L.

Eclipta thermalis Bunge = Eclipta prostrata (L.) L.

Eclipta tinctoria Raf. = Eclipta prostrata (L.) L.

Eclipta zippeliana Blume = Eclipta prostrata (L.) L.

Ectemis lutea (Lam.) Raf. = Cordia lutea Lam.

Eichhornia crassipes (Mart.) Solms = Pontederia crassipes Mart.

Eleocharis brizantha Steud. = Eleocharis geniculata (L.) Roem. & Schult.

Eleocharis capitata (L.) R.Br. = Eleocharis geniculata (L.) Roem. & Schult.

Eleocharis densisquamata Steud. = Eleocharis geniculata (L.) Roem. & Schult.

Eleocharis dispar E.J.Hill = Eleocharis geniculata (L.) Roem. & Schult.

Eleocharis melanosperma Steud. = Eleocharis geniculata (L.) Roem. & Schult.

Eleocharis riparia Nees ex Spreng. = Eleocharis geniculata (L.) Roem. & Schult.

Eleocharis setacea R.Br. = Eleocharis geniculata (L.) Roem. & Schult.

Eleocharis valida Boeckeler = Eleocharis geniculata (L.) Roem. & Schult.

Eleogiton cernua (Vahl) A.Dietr. = Isolepis cernua (Vahl) Roem. & Schult.

Eleutheranthera prostrata (L.) Sch.Bip. = Eclipta prostrata (L.) L.

Elodea chilensis (Planch.) Casp. = Elodea potamogeton (Bertero) Espinosa

Elodea peruviensis H.St.John = Elodea potamogeton (Bertero) Espinosa

Elodea titicacana H.St.John = Elodea potamogeton (Bertero) Espinosa

Elytrospermum californicum C.A.Mey. = Schoenoplectus californicus (C.A.Mey.) Soják

Englerocharis calycina (Desv.) Baehni & J.F.Macbr. = Eudema calycinum (Desv.) Al-Shehbaz, Salariato, A.Cano & Zuloaga

Enneapogon borealis (Griseb.) Honda = Enneapogon desvauxii P.Beauv.

Enneapogon brachystachyus (Jaub. & Spach) Stapf = Enneapogon desvauxii P.Beauv.

Enneapogon phleioides Roem. & Schult. = Enneapogon desvauxii P.Beauv.

Enneapogon pusillus Rendle = Enneapogon desvauxii P.Beauv.

Enneapogon wrightii (S.Watson) Roshev. = Enneapogon desvauxii P.Beauv.

Enydria aquatica Vell. = Myriophyllum aquaticum (Vell.) Verdc.

Ephedra breana Phil. = Ephedra americana Humb. & Bonpl. ex Willd.

Ephedra humilis Wedd. = Ephedra rupestris Benth.

Ephedra peruviana Bertero ex Carr. = Ephedra americana Humb. & Bonpl. ex Willd.

Epicampes coerulea Griseb. = Muhlenbergia coerulea (Griseb.) Mez

Epicampes kunthiana Griseb. = Muhlenbergia coerulea (Griseb.) Mez

Epicampes phleoides (Kunth) Griseb. = Muhlenbergia coerulea (Griseb.) Mez

Epilobium aequinoctiale Sam. = Epilobium denticulatum Ruiz & Pav.

Epilobium assurgens Sam. = Epilobium denticulatum Ruiz & Pav.

Epilobium bolivianum Sam. = Epilobium denticulatum Ruiz & Pav.

Epilobium caesium Hausskn. = Epilobium denticulatum Ruiz & Pav.

Epilobium densifolium Kunze ex Hausskn. = Epilobium pedicellare C.Presl

Epilobium erosum Spach ex Hausskn. = Epilobium denticulatum Ruiz & Pav.

Epilobium euphrasioides Sieber ex Hausskn. = Epilobium denticulatum Ruiz & Pav.

Epilobium haenkeanum Hausskn. = Epilobium pedicellare C.Presl

Epilobium helodes H.Lév. = Epilobium denticulatum Ruiz & Pav.

Epilobium hirtum Sam. = Epilobium denticulatum Ruiz & Pav.

Epilobium peruvianum Hausskn. = Epilobium pedicellare C.Presl

Epilobium repandum Spach ex Hausskn. = Epilobium denticulatum Ruiz & Pav.

Equisetum chilense C.Presl = Equisetum bogotense Kunth

Equisetum rinihuense G.Kunkel = Equisetum bogotense Kunth

Equisetum stipulaceum Vaucher = Equisetum bogotense Kunth

Eragrostis alba J.Presl = Eragrostis mexicana (Hornem.) Link

Eragrostis brizoides Schult. = Eragrostis pectinacea (Michx.) Nees

Eragrostis chilensis (Moris) Nees = Eragrostis virescens J.Presl

Eragrostis cognata Steud. = Eragrostis pectinacea (Michx.) Nees

Eragrostis compta Link = Eragrostis ciliaris (L.) R.Br.

Eragrostis contracta Pilg. = Eragrostis lurida J.Presl

Eragrostis contristata Nees = Eragrostis lurida J.Presl

Eragrostis delicatula Trin. = Eragrostis virescens J.Presl

Eragrostis deserticola Phil. = Eragrostis peruviana (Jacq.) Trin.

Eragrostis diffusa Buckley = Eragrostis pectinacea (Michx.) Nees

Eragrostis glandulosa L.H.Harv. = Eragrostis mexicana (Hornem.) Link

Eragrostis leptantha Trin. = Eragrostis virescens J.Presl

Eragrostis limbata E.Fourn. = Eragrostis mexicana (Hornem.) Link

Eragrostis lobata Trin. = Eragrostis ciliaris (L.) R.Br.

Eragrostis orcuttiana Vasey = Eragrostis mexicana (Hornem.) Link

Eragrostis pauciflora Trin. ex E.Fourn. = Eragrostis mexicana (Hornem.) Link

Eragrostis pulchella Parl. = Eragrostis ciliaris (L.) R.Br.

Eragrostis scabra Phil. = Eragrostis virescens J.Presl

Eragrostis subatra Jedwabn. = Eragrostis nigricans (Kunth) Steud.

Eragrostis tristis Jedwabn. = Eragrostis nigricans (Kunth) Steud.

Eragrostis unionis Steud. = Eragrostis pectinacea (Michx.) Nees

Erdisia quadrangularis Rauh & Backeb. = Corryocactus quadrangularis (Rauh & Backeb.) F.Ritter ex D.R.Hunt, N.P.Taylor & G.J.Charles

Erdisia sextoniana Backeb. = Loxanthocereus sextonianus (Backeb.) Backeb.

Eremodraba schulzii Al-Shehbaz = Neuontobotrys schulzii (Al-Shehbaz) Al-Shehbaz

Ericoila sedifolia (Kunth) G.Don = Gentiana sedifolia Kunth

Erigeron albidus (Willd. ex Spreng.) A.Gray = Erigeron floribundus (Kunth) Sch.Bip.

Erigeron artemisiifolius Sch.Bip. = Laennecia artemisiifolia (Meyen & Walp.) G.L.Nesom

Erigeron artemisioides (Meyen & Walp.) Sch.Bip. = Laennecia artemisiifolia (Meyen & Walp.) G.L.Nesom

Erigeron bilbaoanus (J.Rémy) Cabrera = Erigeron floribundus (Kunth) Sch.Bip.

Erigeron brittonianus Rusby = Erigeron rosulatus Wedd.

Erigeron cinerascens Sch.Bip. = Erigeron deserticolus (Phil.) comb.ined.

Erigeron crispus Pourr. = Erigeron bonariensis L.

Erigeron limnophilus Sch.Bip. = Oritrophium limnophilum (Sch.Bip.) Cuatrec.

Erigeron linearifolius Cav. = Erigeron bonariensis L.

Erigeron myriocephalus (J.Rémy) Herter = Erigeron floribundus (Kunth) Sch.Bip.

Erigeron pulvinatus Wedd. = Erigeron rosulatus Wedd.

Erigeron sordidus Gillies ex Hook. & Arn. = Erigeron bonariensis L.

Erinus laciniatus L. = Glandularia laciniata (L.) Schnack & Covas

Eritrichium filagineum Phil. = Cryptantha filaginea (Phil.) Reiche

Eritrichium limense A.DC. = Cryptantha limensis (A.DC.) I.M.Johnst.

Eritrichium linifolium Wedd. = Plagiobothrys linifolius (Lehm.) I.M.Johnst.

Eritrichium parviflorum Phil. = Johnstonella parviflora (Phil.) Hasenstab & M.G.Simpson

Erosion ciliare (L.) Lunell = Eragrostis ciliaris (L.) R.Br.

Erythranthe andicola (Kunth) G.L.Nesom = Erythranthe glabrata (Kunth) G.L.Nesom

Erythranthe parviflora (Lindl.) G.L.Nesom = Erythranthe glabrata (Kunth) G.L.Nesom

Erythranthe pilosiuscula (Kunth) G.L.Nesom = Erythranthe glabrata (Kunth) G.L.Nesom

Escallonia coquimbensis J.Rémy = Escallonia angustifolia C.Presl

Escallonia graefiana Hosseus = Escallonia angustifolia C.Presl

Ethulia bidentis L. = Flaveria bidentis (L.) Kuntze

Eudema friesii O.E.Schulz = Alshehbazia friesii (O.E.Schulz) Salariato, Zuloaga & Al-Shehbaz

Eudema werdermannii O.E.Schulz = Alshehbazia werdermannii (O.E.Schulz) Salariato, Zuloaga & Al-Shehbaz

Eugenia ferreyrae McVaugh = Myrcianthes ferreyrae (McVaugh) McVaugh

Eupatorium ballii Oliv. = Aristeguietia ballii (Oliv.) R.M.King & H.Rob.

Eupatorium glechonophyllum Less. = Ageratina glechonophylla (Less.) R.M.King & H.Rob.

Eupatorium melissifolium Lam. = Stevia melissifolia Sch.Bip.

Eupatorium pentlandianum DC. = Ageratina pentlandiana (DC.) R.M.King & H.Rob.

Eupatorium sternbergianum DC. = Ageratina sternbergiana (DC.) R.M.King & H.Rob.

Eupatorium stramineum DC. = Helogyne straminea B.L.Rob.

Euphorbia begoniifolia Lehm. = Euphorbia serpens Kunth

Euphorbia boliviana Rusby = Euphorbia hypericifolia L.

Euphorbia capitata Lam. = Euphorbia hirta L.

Euphorbia chrysochaeta W.Fitzg. = Euphorbia hirta L.

Euphorbia flexicaulis Scheele = Euphorbia serpens Kunth

Euphorbia gemella Lag. = Euphorbia hirta L.

Euphorbia globulifera Kunth = Euphorbia hirta L.

Euphorbia glomerifera (Millsp.) L.C.Wheeler = Euphorbia hypericifolia L.

Euphorbia herniarioides Nutt. = Euphorbia serpens Kunth

Euphorbia inflexa Urb. & Ekman = Euphorbia serpens Kunth

Euphorbia microclada Urb. = Euphorbia serpens Kunth

Euphorbia minutiflora N.E.Br. = Euphorbia serpens Kunth

Euphorbia nodiflora Steud. = Euphorbia hirta L.

Euphorbia obliterata Jacq. = Euphorbia hirta L.

Euphorbia papilligera Boiss. = Euphorbia hypericifolia L.

Euphorbia pileoides Millsp. = Euphorbia serpens Kunth

Euxolus viridis (L.) Moq. = Amaranthus viridis L.

Evolvulus helianthemoides Meisn. = Evolvulus villosus Ruiz & Pav.

Evolvulus lanceolatus (Poir.) Sweet = Evolvulus villosus Ruiz & Pav.

Exacum quadrangulare (Lam.) Willd. = Cicendia quadrangularis (Lam.) Griseb.

Facelis weddelliana Beauverd = Facelis plumosa (Wedd.) Sch.Bip.

Fagelia angustiflora (Ruiz & Pav.) Kuntze = Calceolaria angustiflora Ruiz & Pav.

Fagelia cuneiformis (Ruiz & Pav.) Kuntze = Calceolaria cuneiformis Ruiz & Pav.

Fagelia lobata (Cav.) Kuntze = Calceolaria lobata Cav.

Fagelia pinnata (L.) Kuntze = Calceolaria pinnata L.

Fagelia pisacomensis (Meyen ex Walp.) Kuntze = Calceolaria pisacomensis Meyen ex Walp.

Fagelia plectranthifolia (Walp.) Kuntze = Calceolaria plectranthifolia Walp.

Fagelia tenuis (Benth.) Kuntze = Calceolaria tenuis Benth.

Fagonia chilensis Hook. & Arn. = Zygophyllum chilense (Hook. & Arn.) Christenh. & Byng

Fedia chaerophylloides (Sm.) Kunth = Valeriana chaerophylloides Sm.

Festuca deserticola Phil. = Festuca rigescens (J.Presl) Kunth

Festuca eriostoma Hack. = Festuca chrysophylla Phil.

Festuca fusca L. = Diplachne fusca (L.) P.Beauv. ex Roem. & Schult.

Festuca haenkei Kunth = Festuca rigescens (J.Presl) Kunth

Festuca laeteviridis Pilg. = Festuca dolichophylla J.Presl

Festuca meyenii (St.-Yves) E.B.Alexeev = Festuca chrysophylla Phil.

Festuca orthophylla Pilg. = Festuca chrysophylla Phil.

Festuca paupera Phil. = Festuca rigescens (J.Presl) Kunth

Festuca pflanzii Pilg. = Festuca dolichophylla J.Presl

Festuca scirpifolia (J.Presl) Kunth = Festuca dolichophylla J.Presl

Festuca sublimis Pilg. = Festuca procera Kunth

Feuilleea taxifolia (Kunth) Kuntze = Calliandra taxifolia (Kunth) Benth.

Flaveria capitata Juss. ex Sm. = Flaveria bidentis (L.) Kuntze

Flaveria chilensis J.F.Gmel. = Flaveria bidentis (L.) Kuntze

Flaveria contrayerba Pers. = Flaveria bidentis (L.) Kuntze

Flaveria peruviana J.F.Gmel. = Ophryosporus peruvianus (J.F.Gmel.) R.M.King & H.Rob.

Flaveria spicata Sm. = Flaveria bidentis (L.) Kuntze

Fortunatia acaulis (Baker) Guagl. & S.C.Arroyo = Oziroe acaulis (Baker) Speta

Fortunatia biflora (Ruiz & Pav.) J.F.Macbr. = Oziroe biflora (Ruiz & Pav.) Speta

Frankenia aspera Phil. = Frankenia chilensis C.Presl ex Schult. & Schult.f.

Frankenia campestris Schauer = Frankenia chilensis C.Presl ex Schult. & Schult.f.

Frankenia erecta Gay = Frankenia chilensis C.Presl ex Schult. & Schult.f.

Frankenia farinosa J.Rémy = Frankenia chilensis C.Presl ex Schult. & Schult.f.

Frankenia florida Phil. = Frankenia chilensis C.Presl ex Schult. & Schult.f.

Frankenia glabrata Phil. = Frankenia chilensis C.Presl ex Schult. & Schult.f.

Frankenia lignosa Rusby = Frankenia chilensis C.Presl ex Schult. & Schult.f.

Frankenia peruviana Schellenb. = Frankenia chilensis C.Presl ex Schult. & Schult.f.

Franseria arborescens (Mill.) Brandegee = Ambrosia arborescens Mill.

Franseria dentata Cabrera = Ambrosia dentata (Cabrera) M.O.Dillon

Freirea debilis (G.Forst.) Jarm. = Parietaria debilis G.Forst.

Froelichia alata S.Watson = Froelichia interrupta (L.) Moq.

Froelichia tomentosa (Mart.) Moq. = Froelichia interrupta (L.) Moq.

Funastrum apiculatum (Decne.) Schltr. = Funastrum clausum (Jacq.) Schltr.

Funastrum barbatum (Mart. ex E.Fourn.) Schltr. = Funastrum clausum (Jacq.) Schltr.

Funastrum crassifolium (Decne.) Schltr. = Funastrum clausum (Jacq.) Schltr.

Funastrum cumanense (Kunth) Schltr. = Funastrum clausum (Jacq.) Schltr.

Funastrum cuspidatum (E.Fourn.) Schltr. = Funastrum clausum (Jacq.) Schltr.

Funastrum filiforme (Jacq.) Schltr. = Funastrum clausum (Jacq.) Schltr.

Funastrum fragile Rusby = Funastrum clausum (Jacq.) Schltr.

Funastrum lanceolatum Rusby = Funastrum clausum (Jacq.) Schltr.

Funastrum lasianthum (Schltr.) Schltr. = Funastrum clausum (Jacq.) Schltr.

Funastrum pedunculatum (E.Fourn.) Schltr. = Funastrum clausum (Jacq.) Schltr.

Funastrum pubescens (Kunth) Schltr. = Funastrum clausum (Jacq.) Schltr.

Galega coerulea L.f. = Dalea coerulea (L.f.) Schinz & Thell.

Galinsoga acmella Steud. = Galinsoga parviflora Cav.

Galinsoga ciliata (Raf.) S.F.Blake = Galinsoga quadriradiata Ruiz & Pav.

Galinsoga hirsuta Baker = Galinsoga parviflora Cav.

Galinsoga hispida Benth. = Galinsoga quadriradiata Ruiz & Pav.

Galinsoga laciniata Retz. ex Hoffm. = Galinsoga parviflora Cav.

Galinsoga purpurea H.St.John & D.White = Galinsoga unxioides Griseb.

Galinsoga quinqueradiata Ruiz & Pav. = Galinsoga parviflora Cav.

Galinsoga semicalva (A.Gray) H.St.John & D.White = Galinsoga parviflora Cav.

Galium apiculatum Roem. & Schult. = Galium corymbosum Ruiz & Pav.

Galium hirsutum Ruiz & Pav. = Galium corymbosum Ruiz & Pav.

Galium involucratum Kunth = Galium corymbosum Ruiz & Pav.

Galium sessiliflorum Willd. = Galium corymbosum Ruiz & Pav.

Galliaria spitosa (L.) Nieuwl. = Amaranthus spinosus L.

Gamochaeta coarctata (Willd.) Kerguélen = Gamochaeta americana (Mill.) Wedd.

Gamochaeta guatemalensis (Gand.) Cabrera = Gamochaeta americana (Mill.) Wedd.

Gamochaeta irazuensis G.L.Nesom = Gamochaeta americana (Mill.) Wedd.

Gamochaeta rosacea (I.M.Johnst.) Anderb. = Gamochaeta purpurea (L.) Cabrera

Gamochaeta spicata (Klatt) Cabrera = Gamochaeta americana (Mill.) Wedd.

Gamochaeta weddelliana (Rusby) Anderb. = Gamochaeta capitata Wedd.

Gardenia rotundifolia (Ruiz & Pav.) F.Dietr. = Randia rotundifolia Ruiz & Pav.

Gayoides crispum (L.) Small = Herissantia crispa (L.) Brizicky

Geboscon bivalve (L.) House = Nothoscordum bivalve (L.) Britton

Gentiana casapaltensis (Ball) J.S.Pringle = Gentiana sedifolia Kunth

Gentiana cespitosa Willd. ex Schult. = Gentiana sedifolia Kunth

Gentiana microphylla Griseb. = Gentiana sedifolia Kunth

Gentiana potamophila Gilg = Gentianella potamophila (Gilg) Zarucchi

Gentiana quadrangularis Lam. = Cicendia quadrangularis (Lam.) Griseb.

Gentiana sandiensis Gilg = Gentianella sandiensis (Gilg) J.S.Pringle

Gentiana scarlatina Gilg = Gentianella scarlatina (Gilg) J.S.Pringle

Geranium bolivianum R.Knuth = Geranium fallax Steud.

Geranium chamaense Pittier = Geranium diffusum Kunth

Geranium commutatum Steud. = Geranium core-core Steud.

Geranium fiebrigianum R.Knuth = Geranium fallax Steud.

Geranium filipes Killip = Geranium fallax Steud.

Geranium fuscicaule R.Knuth = Geranium fallax Steud.

Geranium holm-nielsenii Halfd.-Niels. = Geranium diffusum Kunth

Geranium melanopotamicum Speg. = Geranium core-core Steud.

Geranium mollendinense R.Knuth = Geranium limae R.Knuth

Geranium multiflorum R.Knuth = Geranium limae R.Knuth

Geranium pallidifolium R.Knuth = Geranium sessiliflorum Cav.

Geranium rapulum A.St.-Hil. & Naudin = Geranium core-core Steud.

Geranium razuhillcaense R.Knuth = Geranium sessiliflorum Cav.

Geranium robustipes R.Knuth = Geranium diffusum Kunth

Geranium santacruzense R.Knuth = Geranium sessiliflorum Cav.

Geranium sodiroanum R.Knuth = Geranium diffusum Kunth

Geranium squamosum Phil. = Geranium core-core Steud.

Geranium subsericeum R.Knuth = Geranium core-core Steud.

Geranium titicacaense R.Knuth = Geranium fallax Steud.

Geranium totorense R.Knuth = Geranium fallax Steud.

Geranium weberbauerianum R.Knuth = Geranium sessiliflorum Cav.

Gerascanthus luteus (Lam.) Borhidi = Cordia lutea Lam.

Gilia alpina (Wedd.) Brand = Gilia laciniata Ruiz & Pav.

Gilia erecta Hieron. = Gilia laciniata Ruiz & Pav.

Gilia glutinosa Phil. = Dayia glutinosa (Phil.) J.M.Porter

Gilia gracilis Hook. = Microsteris gracilis (Hook.) Greene

Glandularia clavata (Ruiz & Pav.) Botta = Junellia clavata (Ruiz & Pav.) O'Leary & Múlgura

Glandularia fasciculata (Benth.) P.Jørg. = Junellia fasciculata (Benth.) N.O'Leary & P.Peralta

Glomeraria viridis (L.) Cav. = Amaranthus viridis L.

Gnaphalium americanum Mill. = Gamochaeta americana (Mill.) Wedd.

Gnaphalium capitatum Griseb. = Gamochaeta capitata Wedd.

Gnaphalium cheiranthemifolium Pers. = Pseudognaphalium cheiranthifolium (Lam.) Hilliard & B.L.Burtt

Gnaphalium dysodes Spreng. = Pseudognaphalium dysodes (Spreng.) S.E.Freire, N.Bayón & C.Monti

Gnaphalium elegans Kunth = Pseudognaphalium elegans Kartesz

Gnaphalium gaudichaudianum DC. = Pseudognaphalium gaudichaudianum (DC.) Anderb.

Gnaphalium kunthianum (DC.) Kuntze = Lucilia kunthiana (DC.) Zardini

Gnaphalium lacteum Meyen & Walp. = Pseudognaphalium lacteum (Meyen & Walp.) Anderb.

Gnaphalium lanuginosum Kunth = Pseudognaphalium lanuginosum (Kunth) Anderb.

Gnaphalium pensylvanicum Willd. = Gamochaeta pensylvanica (Willd.) Cabrera

Gnaphalium piptolepis (Wedd.) Griseb. = Mniodes piptolepis (Wedd.) S.E.Freire, Chemisquy, Anderb. & Urtubey

Gnaphalium purpureum L. = Gamochaeta purpurea (L.) Cabrera

Gnaphalium ramosissimum Sch.Bip. = Achyrocline ramosissima Britton

Gnaphalium tunariense Kuntze = Novenia tunariensis (Kuntze) S.E.Freire

Gnaphalium viravira Molina = Pseudognaphalium viravira (Molina) Anderb.

Gochnatia iserniana Cuatrec. = Gochnatia arequipensis Sandwith

Gochnatia lanceolata H.Beltrán & Ferreyra = Paquirea lanceolata (H.Beltrán & Ferreyra) Panero & S.E.Freire

Gochnatia tarapacana Phil. = Lophopappus tarapacanus (Phil.) Cabrera

Gomphrena conwayi Rusby = Gomphrena meyeniana Walp.

Gomphrena interrupta L. = Froelichia interrupta (L.) Moq.

Gomphrena pearcei Oliv. = Gomphrena umbellata J.Rémy

Gourliea decorticans Gillies ex Hook. & Arn. = Geoffroea decorticans (Gillies ex Hook. & Arn.) Burkart

Grabowskia disticha (Meyen) Walp. = Lycium distichum Meyen

Graminastrum macusaniense E.H.L.Krause = Poa macusaniensis (E.H.L.Krause) Refulio

Grammica acutiloba (Engelm.) Hadac & Chrtek = Cuscuta acutiloba Engelm.

Grammica foetida (Kunth) Hadac & Chrtek = Cuscuta foetida Kunth

Grammica grandiflora (Kunth) Hadac & Chrtek = Cuscuta grandiflora Kunth

Granadilla foetida (L.) Gaertn. = Passiflora foetida L.

Granadilla suberosa (L.) Gaertn. = Passiflora suberosa L.

Gratiola monnieri (L.) L. = Bacopa monnieri (L.) Wettst.

Grindelia montana Phil. = Grindelia glutinosa (Cav.) Mart.

Grindelia peruana Sch.Bip. = Grindelia glutinosa (Cav.) Mart.

Grindelia peruviana Sch.Bip. = Grindelia glutinosa (Cav.) Mart.

Grindelia virgata Nutt. = Grindelia glutinosa (Cav.) Mart.

Guilleminea illecebroides Kunth = Guilleminea densa (Willd. ex Schult.) Moq.

Gymnogramma nivea (Poir.) Mett. = Argyrochosma nivea (Poir.) Windham

Gymnogramma sinuata (Lag. ex Sw.) C.Presl = Astrolepis sinuata (Lag. ex Sw.) D.M.Benham & Windham

Gymnogramma trifoliata (L.) Desv. = Pityrogramma trifoliata (L.) R.M.Tryon

Gymnopogon radiatus (L.) Parodi = Chloris radiata (L.) Sw.

Gynerium jubatum Lemoine = Cortaderia jubata (Lemoine) Stapf

Gynopleura tenuifolia (D.Don) M.Roem. = Malesherbia tenuifolia D.Don

Haageocereus ambiguus Rauh & Backeb. = Haageocereus decumbens (Vaupel) Backeb.

Haageocereus australis Backeb. = Haageocereus decumbens (Vaupel) Backeb.

Haageocereus cephalomacrostibas (Werderm. & Backeb.) P.V.Heath = Weberbauerocereus cephalomacrostibas (Werderm. & Backeb.) F.Ritter

Haageocereus chalaensis F.Ritter = Haageocereus decumbens (Vaupel) Backeb.

Haageocereus familiaris (F.Ritter) Lodé = Haageocereus bylesianus (Andreae & Backeb.) Lodé

Haageocereus litoralis Rauh & Backeb. = Haageocereus decumbens (Vaupel) Backeb.

Haageocereus mamillatus Rauh & Backeb. = Haageocereus decumbens (Vaupel) Backeb.

Haageocereus multicolorispinus Buining = Haageocereus decumbens (Vaupel) Backeb.

Haageocereus ocona-camanensis Rauh & Backeb. = Haageocereus decumbens (Vaupel) Backeb.

Haageocereus pluriflorus Rauh & Backeb. = Haageocereus platinospinus (Werderm. & Backeb.) Backeb.

Haageocereus subtilispinus F.Ritter = Haageocereus decumbens (Vaupel) Backeb.

Haageocereus weberbaueri (K.Schum. ex Vaupel) D.R.Hunt = Weberbauerocereus weberbaueri (K.Schum. ex Vaupel) Backeb.

Halimolobos weddellii (E.Fourn.) O.E.Schulz = Exhalimolobos weddellii (E.Fourn.) Al-Shehbaz & C.D.Bailey

Hartmannia rosea (L'Hér. ex Aiton) G.Don = Oenothera rosea L'Hér. ex Aiton

Hedeoma adscendens Rusby = Hedeoma mandoniana Wedd.

Hedysarum muricatum Jacq. = Adesmia muricata (Jacq.) DC.

Hedysarum scorpiurus Sw. = Desmodium scorpiurus (Sw.) Poir.

Hedysarum uncinatum Jacq. = Desmodium uncinatum (Jacq.) DC.

Heleogiton pungens (Vahl) Rchb. = Schoenoplectus pungens (Vahl) Palla

Heleophylax americanus (Pers.) Schinz & Thell. = Schoenoplectus americanus (Pers.) Volkart

Heliotropium bangii Rusby = Heliotropium microstachyum Ruiz & Pav.

Heliotropium brachystachyum Gürke = Heliotropium microstachyum Ruiz & Pav.

Heliotropium corymbosum Ruiz & Pav. = Heliotropium arborescens L.

Heliotropium grandiflorum Donn = Heliotropium arborescens L.

Heliotropium odoratissimum Gaterau = Heliotropium arborescens L.

Heliotropium odoratum Moench = Heliotropium arborescens L.

Heliotropium odorum Salisb. = Heliotropium arborescens L.

Heliotropium peruvianum L. = Heliotropium arborescens L.

Heliotropium phaenocarpum Reiche = Heliotropium microstachyum Ruiz & Pav.

Heliotropium pilosum Ruiz & Pav. = Euploca pilosa (Ruiz & Pav.) Luebert

Heliotropium toratense I.M.Johnst. = Euploca toratensis (I.M.Johnst.) J.I.M.Melo

Helogyne fiebrigii Hieron. = Helogyne straminea B.L.Rob.

Helogyne weberbaueri B.L.Rob. = Helogyne straminea B.L.Rob.

Helopus obtusus (Nees & Meyen) Steud. = Lorenzochloa obtusa (Nees & Meyen) Romasch.

Helopus punctatus (L.) Nees = Eriochloa punctata (L.) Ham.

Helosciadium laciniatum DC. = Cyclospermum laciniatum (DC.) Constance

Helosciadium leptophyllum (Pers.) DC. = Cyclospermum leptophyllum (Pers.) Sprague ex Britton & P.Wilson

Hemionitis arequipensis (Maxon) Christenh. = Cheilanthes arequipensis (Maxon) R.M.Tryon & A.F.Tryon

Hemionitis bonariensis (Willd.) Christenh. = Myriopteris aurea (Poir.) Grusz & Windham

Hemionitis emperatricella Christenh. = Pellaea sagittata (Cav.) Link.

Hemionitis fractifera (R.M.Tryon) Christenh. = Cheilanthes fractifera R.M.Tryon

Hemionitis incarum (Maxon) Christenh. = Cheilanthes incarum Maxon

Hemionitis macleanii (Hook.) ined. = Cheilanthes pilosa Goldm.

Hemionitis mollis (Kunze) Christenh. = Cheilanthes mollis (Kunze) C.Presl

Hemionitis myriophylla (Desv.) Christenh. = Myriopteris myriophylla J.Sm.

Hemionitis nivea (Poir.) Christenh. = Argyrochosma nivea (Poir.) Windham

Hemionitis peruviana (Desv.) Christenh. = Cheilanthes peruviana T.Moore

Hemionitis pruinata (Kaulf.) Christenh. = Cheilanthes pruinata Kaulf.

Hemionitis sinuata (Lag.) Christenh. = Astrolepis sinuata (Lag. ex Sw.) D.M.Benham & Windham

Hemionitis sulphurea (Cav.) Christenh. = Notholaena sulphurea (Cav.) J.Sm.

Hemionitis ternifolia (Cav.) Christenh. = Pellaea ternifolia (Cav.) Link

Hemionitis trifoliata (L.) Kunth = Pityrogramma trifoliata (L.) R.M.Tryon

Hepetis ferruginea (Ruiz & Pav.) Mez = Puya ferruginea (Ruiz & Pav.) L.B.Sm.

Herissantia trichoda (A.Rich.) Fryxell = Herissantia crispa (L.) Brizicky

Herpestis monnieri (L.) Rothm. = Bacopa monnieri (L.) Wettst.

Hesperis amplexicaulis Kuntze = Neuontobotrys amplexicaulis (Kuntze) Al-Shehbaz

Hesperis athrocarpa Kuntze = Descurainia athrocarpa (A.Gray) O.E.Schulz

Hesperis calycina Kuntze = Eudema calycinum (Desv.) Al-Shehbaz, Salariato, A.Cano & Zuloaga

Hesperis gracilis Kuntze = Mostacillastrum gracile (Wedd.) Al-Shehbaz

Hesperis myriophylla Kuntze = Descurainia myriophylla (Kunth ex DC.) R.E.Fr.

Hesperis peruviana (DC.) Kuntze = Weberbauera peruviana (DC.) Al-Shehbaz

Hesperis weddellii Kuntze = Exhalimolobos weddellii (E.Fourn.) Al-Shehbaz & C.D.Bailey

Heterosperma depressum Griseb. = Heterosperma diversifolium Kunth

Heterosperma involucratum (Phil.) Reiche = Heterosperma nanum (Nutt.) Sherff

Heterosperma maritimum Kunth = Heterosperma ovatifolium Cav.

Heterosperma rhombifolium (Sch.Bip.) Griseb. = Heterosperma ovatifolium Cav.

Heterothalamus acaulis Wedd. ex R.E.Fr. = Baccharis acaulis (Wedd. ex R.E.Fr.) Cabrera

Heterothalamus boliviensis Wedd. = Baccharis boliviensis (Wedd.) Cabrera

Heterothrix gracilis O.E.Schulz = Mostacillastrum gracile (Wedd.) Al-Shehbaz

Hieracium chiclense Ball = Hieracium mandonii (Sch.Bip.) Britton

Hieracium eriosphaerophorum Zahn = Hieracium mandonii (Sch.Bip.) Britton

Hoffmannseggia gracilis (Ruiz & Pav.) Hook. & Arn. = Hoffmannseggia prostrata Lag. ex DC.

Hoffmannseggia pilosa G.Don = Hoffmannseggia prostrata Lag. ex DC.

Hoffmannseggia stipulata Sandwith = Hoffmannseggia miranda Sandwith

Hoffmannseggia ternata Phil. = Hoffmannseggia viscosa (Ruiz & Pav.) Hook. & Arn.

Homanthis multiflorus (Bonpl.) Kunth = Perezia multiflora (Bonpl.) Less.

Homanthis pinnatifidus (Bonpl.) Kunth = Perezia pinnatifida (Bonpl.) Wedd.

Homanthis pungens (Bonpl.) Kunth = Perezia pungens (Bonpl.) Less.

Homoianthus multiflorus (Bonpl.) D.Don = Perezia multiflora (Bonpl.) Less.

Homoianthus pinnatifidus (Bonpl.) D.Don = Perezia pinnatifida (Bonpl.) Wedd.

Homoianthus pungens (Bonpl.) Spreng. = Perezia pungens (Bonpl.) Less.

Hordeum andicola Griseb. = Hordeum muticum J.Presl

Hydrocotyle batrachioides DC. = Hydrocotyle ranunculoides L.f.

Hydrocotyle caffra Meisn. = Hydrocotyle umbellata L.

Hydrocotyle cymbalariifolia Muhl. ex Elliott = Hydrocotyle ranunculoides L.f.

Hydrocotyle fluitans DC. = Hydrocotyle umbellata L.

Hydrocotyle multiflora Ruiz & Pav. = Hydrocotyle bonariensis Comm. ex Lam.

Hydrocotyle natans Cirillo = Hydrocotyle ranunculoides L.f.

Hydrocotyle pelviformis Gand. = Hydrocotyle bonariensis Comm. ex Lam.

Hydrocotyle petiolaris DC. = Hydrocotyle bonariensis Comm. ex Lam.

Hydrocotyle polystachya A.Rich. = Hydrocotyle umbellata L.

Hydrocotyle scaposa Steud. = Hydrocotyle umbellata L.

Hydrocotyle tribotrys Ruiz & Pav. = Hydrocotyle bonariensis Comm. ex Lam.

Hydrocotyle yucatanensis Millsp. = Hydrocotyle bonariensis Comm. ex Lam.

Hydrolea dichotoma Ruiz & Pav. = Nama dichotoma (Ruiz & Pav.) Choisy

Hypochaeris brasiliensis (Less.) Benth. & Hook.f. ex Griseb. = Hypochaeris chillensis (Kunth) Britton

Hypochaeris cryptocephala (Sch.Bip.) Domke = Hypochaeris eriolaena (Sch.Bip.) Reiche

Hypochaeris ornata J.Kost. = Hypochaeris echegarayi Hieron.

Hypochaeris spinneri Beauverd = Hypochaeris eriolaena (Sch.Bip.) Reiche

Hypochaeris stenocephala (A.Gray ex Wedd.) Kuntze = Hypochaeris taraxacoides Ball

Hypochaeris tweediei (Hook. & Arn.) Cabrera = Hypochaeris chillensis (Kunth) Britton

Hypsela oligophylla (Wedd.) Benth. & Hook.f. ex Zahlbr. = Lobelia oligophylla (Wedd.) Lammers

Hypseocharis bilobata Killip = Hypseocharis pimpinellifolia J.Rémy

Hypseocharis corydalifolia R.Knuth = Hypseocharis pimpinellifolia J.Rémy

Hypseocharis malpasensis R.Knuth = Hypseocharis pimpinellifolia J.Rémy

Hypseocharis moschata R.Knuth = Hypseocharis pimpinellifolia J.Rémy

Hypseocharis pedicularifolia R.Knuth = Hypseocharis pimpinellifolia J.Rémy

Hypseocharis pilgeri R.Knuth = Hypseocharis pimpinellifolia J.Rémy

Hyptis sidifolia (L'Hér.) Briq. = Mesosphaerum sidifolium (L'Hér.) Harley & J.F.B.Pastore

Illecebrum densum Willd. ex Schult. = Guilleminea densa (Willd. ex Schult.) Moq.

Indigofera anil L. = Indigofera suffruticosa Mill.

Indigofera divaricata Jacq. = Indigofera suffruticosa Mill.

Indigofera micrantha Desv. = Indigofera suffruticosa Mill.

Inga taxifolia Kunth = Calliandra taxifolia (Kunth) Benth.

Inula glutinosa Pers. = Grindelia glutinosa (Cav.) Mart.

Ionoxalis latifolia (Kunth) Rose = Oxalis latifolia Kunth

Ipomoea affinis M.Martens & Galeotti = Ipomoea purpurea (L.) Roth

Ipomoea caesia Hoffmanns. = Ipomoea purpurea (L.) Roth

Ipomoea chilensis A.Braun & C.D.Bouché = Ipomoea dumetorum Willd.

Ipomoea diehlii M.E.Jones = Ipomoea purpurea (L.) Roth

Ipomoea discolor Jacq. = Ipomoea purpurea (L.) Roth

Ipomoea egregia House = Ipomoea plummerae A.Gray

Ipomoea glandulifera Ruiz & Pav. = Ipomoea purpurea (L.) Roth

Ipomoea glaucescens (Kunth) G.Don = Ipomoea dumetorum Willd.

Ipomoea hermanniae (L'Hér.) G.Don = Convolvulus hermanniae L'Hér.

Ipomoea hirsutula J.Jacq. = Ipomoea purpurea (L.) Roth

Ipomoea intermedia Schult. = Ipomoea purpurea (L.) Roth

Ipomoea minuta R.E.Fr. = Ipomoea plummerae A.Gray

Ipomoea oligantha Choisy = Ipomoea dumetorum Willd.

Ipomoea paposana Phil. = Ipomoea dumetorum Willd.

Ipomoea patens (A.Gray) House = Ipomoea plummerae A.Gray

Ipomoea pilosissima M.Martens & Galeotti = Ipomoea purpurea (L.) Roth

Ipomoea pulchella (Kunth) G.Don = Ipomoea dumetorum Willd.

Ipomoea serotina Roem. & Schult. = Ipomoea dumetorum Willd.

Islaya islayensis (C.F.Först.) Backeb. = Eriosyce islayensis (C.F.Först.) Katt.

Jaborosa crispa (Miers) Benth. & Hook.f. ex Griseb. = Jaborosa squarrosa (Miers) Hunz. & Barboza

Jaborosa decurrens Miers = Jaborosa squarrosa (Miers) Hunz. & Barboza

Jaborosa leiocalyx Dammer = Jaborosa squarrosa (Miers) Hunz. & Barboza

Jarava annua (Mez) Peñail. = Stipa annua Mez

Jarava chrysophylla (É.Desv.) Peñail. = Pappostipa chrysophylla (É.Desv.) Romasch.

Jarava eriostachya (Kunth) Peñail. = Jarava ichu Ruiz & Pav.

Jarava pachypus (Pilg.) Peñail. = Stipa pachypus Pilg.

Jarava plumosula (Nees ex Steud.) F.Rojas = Stipa plumosa Trin.

Jarava usitata Pers. = Jarava ichu Ruiz & Pav.

Jatropha rumicifolia Fern.Casas = Jatropha macrantha Müll.Arg.

Juncellus laevigatus (L.) C.B.Clarke = Cyperus laevigatus L.

Juncoides racemosa (Desv.) Kuntze = Luzula racemosa Desv.

Juncus brunneus Buchenau = Juncus ebracteatus E.Mey.

Juncus cespifolius Raf. = Juncus bufonius L.

Juncus chamissonis Kunth = Juncus imbricatus Laharpe

Juncus collinus Steud. = Juncus imbricatus Laharpe

Juncus prolifer Kunth = Juncus bufonius L.

Juncus pumilus Raf. = Juncus bufonius L.

Juncus spanianthus Steud. = Juncus imbricatus Laharpe

Juncus trinervis Liebm. = Juncus ebracteatus E.Mey.

Junellia arequipense (Botta) Botta = Mulguraea arequipensis (Botta) N.O'Leary & P.Peralta

Junellia hayekii Moldenke = Junellia minima (Meyen) Moldenke

Junellia lucanensis (Moldenke) N.O'Leary & P.Peralta = Junellia fasciculata (Benth.) N.O'Leary & P.Peralta

Junellia straguloides Moldenke = Junellia minima (Meyen) Moldenke

Jungia seleriana Muschl. = Jungia axillaris Spreng.

Jussiaea hexapetala Hook. & Arn. = Ludwigia hexapetala (Hook. & Arn.) Zardini, H.Y.Gu & P.H.Raven

Jussiaea octovalvis (Jacq.) Sw. = Ludwigia octovalvis (Jacq.) P.H.Raven

Jussiaea peploides Kunth = Ludwigia peploides (Kunth) P.H.Raven

Justicia tomentosa Vahl = Dicliptera tomentosa (Vahl) Nees

Kageneckia amygdalifolia C.Presl = Kageneckia lanceolata Ruiz & Pav.

Kageneckia glutinosa Kunth = Kageneckia lanceolata Ruiz & Pav.

Kalanchoe tuberosa H.Perrier = Kalanchoe delagoensis Eckl. & Zeyh.

Kardanoglyphos nana Schltdl. = Rorippa nana (Schltdl.) J.F.Macbr.

Kleinia ruderalis Jacq. = Porophyllum ruderale (Jacq.) Cass.

Koeleria trachyantha Phil. = Rostraria trachyantha (Phil.) Soreng

Krameria canescens Willd. = Krameria lappacea (Dombey) Burdet & B.B.Simpson

Krameria illuca Phil. = Krameria lappacea (Dombey) Burdet & B.B.Simpson

Krameria pentapetala Ruiz & Pav. = Krameria lappacea (Dombey) Burdet & B.B.Simpson

Krameria triandra Ruiz & Pav. = Krameria lappacea (Dombey) Burdet & B.B.Simpson

Krynitzkia linifolia (Lehm.) A.Gray = Plagiobothrys linifolius (Lehm.) I.M.Johnst.

Kyrstenia glechonophylla Greene = Ageratina glechonophylla (Less.) R.M.King & H.Rob.

Lachemilla diplophylla (Diels) Rothm. = Alchemilla diplophylla Diels

Lachemilla pinnata (Ruiz & Pav.) Rothm. = Alchemilla pinnata Ruiz & Pav.

Lamourouxia bartsioides Hook. = Neobartsia bartsioides (Hook.) Uribe-Convers & Tank

Landia lappacea Dombey = Krameria lappacea (Dombey) Burdet & B.B.Simpson

Lantana microphylla (Kunth) Mart. = Glandularia microphylla (Kunth) Cabrera

Lantana svensonii Moldenke = Lantana sprucei Hayek

Lantana zahlbruckneri Hayek = Lantana sprucei Hayek

Laretia compacta (Phil.) Reiche = Azorella compacta Phil.

Larrea monticellii Perrone & Caro = Larrea divaricata Cav.

Larrea viscosa Ruiz & Pav. = Hoffmannseggia viscosa (Ruiz & Pav.) Hook. & Arn.

Lasiospora eriolaena P.Candargy = Hypochaeris eriolaena (Sch.Bip.) Reiche

Ledocarpon meyenianum (Klotzsch) Steud. = Balbisia meyeniana Klotzsch

Legousia perfoliata (L.) Britton = Triodanis perfoliata (L.) Nieuwl.

Lemaireocereus matucanensis (Backeb.) W.T.Marshall = Armatocereus matucanensis Backeb.

Lemna abbreviata (Hegelm.) Hegelm. = Lemna minuta Kunth

Leontodon chillensis (Kunth) DC. = Hypochaeris chillensis (Kunth) Britton

Lepicystis macrocarpa (Willd.) Paxton & Lindl. = Pleopeltis macrocarpa (Willd.) Kaulf.

Lepidium americanum Vell. = Lepidium didymum L.

Lepidium anglicum Huds. = Lepidium didymum L.

Lepidium auritum Turcz. = Lepidium bipinnatifidum Desv.

Lepidium humboldtii DC. = Lepidium bipinnatifidum Desv.

Lepidium kalenbornii C.L.Hitchc. = Lepidium bipinnatifidum Desv.

Lepidium prostratum Savi = Lepidium didymum L.

Lepidium reticulatum Howell = Lepidium strictum (S.Watson) Rattan ex B.L.Rob.

Lepidium sectifolium Steud. = Lepidium bipinnatifidum Desv.

Lepidium walpersii J.F.Macbr. = Lepidium chichicara Desv.

Lepidophyllum lucidum (Meyen) Cabrera = Parastrephia lucida (Meyen) Cabrera

Lepidophyllum quadrangulare Benth. & Hook.f. = Parastrephia quadrangularis (Meyen) Cabrera

Lepisorus macrocarpa (Willd.) Ching ex R.D.Dixit = Pleopeltis macrocarpa (Willd.) Kaulf.

Leptilon bonariense (L.) Small = Erigeron bonariensis L.

Leptochloa fusca (L.) Kunth = Diplachne fusca (L.) P.Beauv. ex Roem. & Schult.

Leptochloa spicata (Nees) Scribn. = Tripogonella spicata (Nees) P.M.Peterson & Romasch.

Leptofeddea lomana Diels = Leptoglossis lomana (Diels) Hunz.

Leptoglossis ferreyraei Hunz. & Subils = Leptoglossis lomana (Diels) Hunz.

Leucothauma albicans (Herb.) Ravenna = Pyrolirion albicans Herb.

Leysera odorata Ruiz & Pav. ex DC. = Onoseris odorata Hook. & Arn.

Liabum lyratum A.Gray = Munnozia lyrata (A.Gray) H.Rob. & Brettell

Ligaria coronata Tiegh. = Ligaria cuneifolia (Ruiz & Pav.) Tiegh.

Ligaria emarginata Tiegh. = Ligaria cuneifolia (Ruiz & Pav.) Tiegh.

Ligaria lanceolata Tiegh. = Ligaria cuneifolia (Ruiz & Pav.) Tiegh.

Ligaria viscoides (Poepp. ex Eichler) Tiegh. = Ligaria cuneifolia (Ruiz & Pav.) Tiegh.

Lilaeopsis andina A.W.Hill = Lilaeopsis occidentalis J.M.Coult. & Rose

Lilaeopsis exigua Pérez-Mor. = Lilaeopsis occidentalis J.M.Coult. & Rose

Lilaeopsis hillii Pérez-Mor. = Lilaeopsis occidentalis J.M.Coult. & Rose

Lilaeopsis macloviana (Gand.) A.W.Hill = Lilaeopsis occidentalis J.M.Coult. & Rose

Lilaeopsis masonii Mathias & Constance = Lilaeopsis occidentalis J.M.Coult. & Rose

Lilaeopsis sinuata A.W.Hill = Lilaeopsis occidentalis J.M.Coult. & Rose

Limnochloa geniculata (L.) Nees = Eleocharis geniculata (L.) Roem. & Schult.

Limosella americana Glück = Limosella australis R.Br.

Limosella brachystema Raf. = Limosella australis R.Br.

Limosella coerulea Burch. = Limosella australis R.Br.

Limosella lineata Glück = Limosella australis R.Br.

Limosella maritima Raf. = Limosella australis R.Br.

Limosella minuta Dinter & Suess. = Limosella australis R.Br.

Limosella monticola Dinter = Limosella australis R.Br.

Limosella subulata E.Ives = Limosella australis R.Br.

Limosella tenella Quézel & Contandr. = Limosella aquatica L.

Linaria canadensis (L.) Chaz. = Nuttallanthus canadensis (L.) D.A.Sutton

Lippia nodiflora (L.) Michx. = Phyla nodiflora (L.) Greene

Lithospermum calycinum Moris = Amsinckia calycina (Moris) Chater

Lithospermum dichotomum Ruiz & Pav. = Tiquilia dichotoma (Ruiz & Pav.) Pers.

Lithospermum myosotoides Lehm. = Plagiobothrys myosotoides (Lehm.) Brand

Loasa chenopodiifolia Desr. = Nasa chenopodiifolia (Desr.) Weigend

Loasa incana Graham = Presliophytum incanum (Graham) Weigend

Loasa pentlandii Paxton = Caiophora pentlandii (Paxton) Paxton & Lindl.

Loasa rosulata Wedd. = Caiophora rosulata (Wedd.) Urb. & Gilg

Loasa urens Jacq. = Nasa urens (Jacq.) Weigend

Lobelia pratiana Gaudich. ex Lammers = Lobelia oligophylla (Wedd.) Lammers

Lobivia aureosenilis Kníže = Lobivia pampana Britton & Rose

Lobivia glaucescens F.Ritter = Lobivia pampana Britton & Rose

Lobivia mistiensis (Werderm. & Backeb.) Backeb. = Lobivia pampana Britton & Rose

Loeflingia ramosissima Weinm. = Cardionema ramosissima (Weinm.) A.Nelson & J.F.Macbr.

Lonchestigma squarrosum Dunal = Jaborosa squarrosa (Miers) Hunz. & Barboza

Lophopappus berberidifolius Cabrera = Lophopappus tarapacanus (Phil.) Cabrera

Loranthus cuneifolius Ruiz & Pav. = Ligaria cuneifolia (Ruiz & Pav.) Tiegh.

Lorentea sessiliflora Less. = Pectis sessiliflora Sch.Bip.

Loricaria graveolens (Sch.Bip.) Wedd. = Andicolea graveolens (Sch.Bip.) Mayta & Molinari

Loxanthocereus aticensis Rauh & Backeb. = Loxanthocereus sextonianus (Backeb.) Backeb.

Loxanthocereus camanaensis Rauh & Backeb. = Loxanthocereus sextonianus (Backeb.) Backeb.

Loxanthocereus gracilis (Akers & Buining) Backeb. = Loxanthocereus sextonianus (Backeb.) Backeb.

Loxanthocereus nanus Backeb. = Loxanthocereus sextonianus (Backeb.) Backeb.

Loxanthocereus puquiensis F.Ritter = Loxanthocereus sextonianus (Backeb.) Backeb.

Loxanthocereus riomajensis Rauh & Backeb. = Loxanthocereus jajoanus (Backeb.) Backeb.

Loxanthocereus splendens Backeb. = Loxanthocereus sextonianus (Backeb.) Backeb.

Loxanthocereus variabilis F.Ritter = Loxanthocereus sextonianus (Backeb.) Backeb.

Lucilia affinis Wedd. = Lucilia kunthiana (DC.) Zardini

Lucilia lehmannii Hieron. = Lucilia kunthiana (DC.) Zardini

Lucilia longifolia Cuatrec. & Aristeg. = Mniodes longifolia (Cuatrec. & Aristeg.) S.E.Freire, Chemisquy, Anderb. & Urtubey

Lucilia piptolepis Wedd. = Mniodes piptolepis (Wedd.) S.E.Freire, Chemisquy, Anderb. & Urtubey

Lucilia plumosa Wedd. = Facelis plumosa (Wedd.) Sch.Bip.

Lucilia pusilla Hieron. = Lucilia kunthiana (DC.) Zardini

Lucilia tunariensis K.Schum. = Novenia tunariensis (Kuntze) S.E.Freire

Lucilia violacea Wedd. = Lucilia kunthiana (DC.) Zardini

Luciliocline longifolia (Cuatrec. & Aristeg.) M.O.Dillon & Sagást. = Mniodes longifolia (Cuatrec. & Aristeg.) S.E.Freire, Chemisquy, Anderb. & Urtubey

Luciliocline piptolepis (Wedd.) M.O.Dillon & Sagást. = Mniodes piptolepis (Wedd.) S.E.Freire, Chemisquy, Anderb. & Urtubey

Luciliopsis argentina Cabrera = Cuatrecasasiella argentina (Cabrera) H.Rob.

Lugonia andina Ball = Pentacyphus andinus (Ball) Liede

Lupinaster amabilis (Kunth) C.Presl = Trifolium amabile Kunth

Lupinus punensis C.P.Sm. = Lupinus eriocladus Ulbr.

Luzula humilis Buchenau = Luzula racemosa Desv.
Luzula macusaniensis Steud. ex Buchenau = Luzula racemosa Desv.
Lycianthes candicans (Dunal) Hassl. = Lycianthes lycioides (L.) Hassl.
Lycianthes dombeyi (Dunal) Hassl. = Lycianthes lycioides (L.) Hassl.
Lycium confusum Dammer ex R.E.Fr. = Lycium americanum Jacq.
Lycium divaricatum Rusby = Lycium distichum Meyen
Lycium fragosum Miers = Lycium stenophyllum J.Rémy
Lycium johnstonii S.F.Blake = Lycium americanum Jacq.
Lycium nodosum Miers = Lycium americanum Jacq.
Lycium obtusum Willd. ex Roem. & Schult. = Lycium americanum Jacq.
Lycium oreophilum Wedd. = Lycium distichum Meyen
Lycium pruinosum Griseb. = Lycium americanum Jacq.
Lycium scabrum Nees = Lycium distichum Meyen
Lycium spathulifolium Britton = Lycium americanum Jacq.
Lycium subglabrum (Moldenke) Moldenke = Lycium americanum Jacq.
Lycium subtridentatum Dammer = Lycium americanum Jacq.
Lycopersicon chilense Dunal = Solanum chilense (Dunal) Reiche
Lycopersicon pennellii (Correll) D'Arcy = Solanum pennellii Correll
Lycopersicon peruvianum (L.) Mill. = Solanum peruvianum L.
Lycopersicon pimpinellifolium (L.) Mill. = Solanum pimpinellifolium L.
Lysimachia monnieri L. = Bacopa monnieri (L.) Wettst.
Malacocarpus islayensis (C.F.Först.) Britton & Rose = Eriosyce islayensis (C.F.Först.) Katt.
Malesherbia pulchra Phil. = Malesherbia tenuifolia D.Don
Maligia gracilis (Aiton) Raf. = Nothoscordum gracile (Aiton) Stearn
Mallostoma thymifolium (Ruiz & Pav.) B.D.Jacks. = Arcytophyllum thymifolium (Ruiz & Pav.) Standl.
Malva capitata Cav. = Tarasa capitata (Cav.) D.M.Bates
Malva chilensis A.Braun & C.D.Bouché = Fuertesimalva chilensis (A.Braun & C.D.Bouché) Fryxell
Malva coromandeliana L. = Malvastrum coromandelianum (L.) Garcke
Malva echinata C.Presl = Fuertesimalva echinata (C.Presl) Fryxell
Malva limensis L. = Fuertesimalva limensis (L.) Fryxell
Malva nubigena (Walp.) Wedd. = Acaulimalva nubigena (Walp.) Krapov.
Malva operculata Cav. = Tarasa operculata (Cav.) Krapov.
Malva peruviana L. = Fuertesimalva peruviana (L.) Fryxell
Malva rhizantha (A.Gray) Wedd. = Acaulimalva rhizantha (A.Gray) Krapov.
Malva rhombifolia (L.) E.H.L.Krause = Sida rhombifolia L.
Malva spinosa (L.) E.H.L.Krause = Sida spinosa L.
Malva tenella Cav. = Tarasa tenella (Cav.) Krapov.
Malvastrum anthemidifolium (Remy) A.Gray = Nototriche anthemidifolia (Remy) A.W.Hill
Malvastrum capitatum (Cav.) Griseb. = Tarasa capitata (Cav.) D.M.Bates
Malvastrum congestiflorum I.M.Johnst. = Tarasa congestiflora (I.M.Johnst.) Krapov.
Malvastrum englerianum Ulbr. = Acaulimalva engleriana (Ulbr.) Krapov.
Malvastrum limense (L.) Ball = Fuertesimalva limensis (L.) Fryxell
Malvastrum longirostre Wedd. = Nototriche longirostris (Wedd.) A.W.Hill
Malvastrum mandonianum Wedd. = Nototriche mandoniana (Wedd.) A.W.Hill
Malvastrum nototrichoides Hochr. = Tarasa nototrichoides (Hochr.) Krapov.
Malvastrum nubigenum (Walp.) Baker f. = Acaulimalva nubigena (Walp.) Krapov.
Malvastrum obcuneatum Baker f. = Nototriche obcuneata (Baker f.) A.W.Hill
Malvastrum operculatum (Cav.) Hochr. = Tarasa operculata (Cav.) Krapov.
Malvastrum orbignyanum Wedd. = Nototriche orbignyana (Wedd.) A.W.Hill
Malvastrum pediculariifolium (Meyen) A.Gray = Nototriche pediculariifolia (Meyen) A.W.Hill
Malvastrum pennellii Ulbr. = Fuertesimalva pennellii (Ulbr.) Fryxell
Malvastrum peruvianum (L.) A.Gray = Fuertesimalva peruviana (L.) Fryxell
Malvastrum rhizanthum A.Gray = Acaulimalva rhizantha (A.Gray) Krapov.
Malvastrum sandemanii Sandwith = Palaua sandemanii (Sandwith) Fryxell
Malvastrum tenellum (Cav.) Hieron. = Tarasa tenella (Cav.) Krapov.

Malveopsis anthemidifolia (Remy) Kuntze = Nototriche anthemidifolia (Remy) A.W.Hill
Malveopsis copelandii Hieron. = Nototriche orbignyana (Wedd.) A.W.Hill
Malveopsis coromandeliana (L.) Morong = Malvastrum coromandelianum (L.) Garcke
Malveopsis longirostris (Wedd.) Kuntze = Nototriche longirostris (Wedd.) A.W.Hill
Malveopsis mandoniana (Wedd.) Kuntze = Nototriche mandoniana (Wedd.) A.W.Hill
Malveopsis orbignyana (Wedd.) Kuntze = Nototriche orbignyana (Wedd.) A.W.Hill
Malveopsis pediculariifolia (Meyen) Kuntze = Nototriche pediculariifolia (Meyen) A.W.Hill
Malveopsis peruviana (L.) Kuntze = Fuertesimalva peruviana (L.) Fryxell
Malvinda spinosa (L.) Medik. = Sida spinosa L.
Mammillaria haynii (Otto ex Salm-Dyck) C.Ehrenb. = Matucana haynii (Otto ex Salm-Dyck) Britton & Rose
Mancoa minima Rollins = Mancoa hispida Wedd.
Marginaria macrocarpa (Willd.) B.K.Nayar & S.Kaur = Pleopeltis macrocarpa (Willd.) Kaulf.
Marginaria pycnocarpa (C.Chr.) Herter = Pleopeltis pycnocarpa (C.Chr.) A.R.Sm.
Margyricarpus cristatus Britton = Tetraglochin cristata (Britton) Rothm.
Marica micrantha (Cav.) Ker Gawl. = Sisyrinchium micranthum Cav.
Marilaunidium dichotomum (Ruiz & Pav.) Kuntze = Nama dichotoma (Ruiz & Pav.) Choisy
Mariscus hermaphroditus (Jacq.) Urb. = Cyperus hermaphroditus (Jacq.) Standl.
Mariscus tacnensis (Nees & Meyen) J.Raynal = Cyperus tacnensis Nees & Meyen
Marsea artemisiifolia Kuntze = Laennecia artemisiifolia (Meyen & Walp.) G.L.Nesom
Martrasia axillaris Lag. ex Steud. = Jungia axillaris Spreng.
Maytenus cuezzoi Legname = Maytenus verticillata (Ruiz & Pav.) DC.
Megastachya ciliaris (L.) P.Beauv. = Eragrostis ciliaris (L.) R.Br.
Megastachya nigricans (Kunth) Roem. & Schult. = Eragrostis nigricans (Kunth) Steud.
Meibomia scorpiurus (Sw.) Kuntze = Desmodium scorpiurus (Sw.) Poir.
Meibomia uncinata (Jacq.) Kuntze = Desmodium uncinatum (Jacq.) DC.
Meioperis suberosa (L.) Raf. = Passiflora suberosa L.
Melandrium mandonii Rohrb. = Silene mandonii (Rohrb.) Bocquet
Melica cajamarcensis Pilg. = Melica scabra Kunth
Melica majuscula Pilg. = Melica scabra Kunth
Melica pallida Kunth = Melica scabra Kunth
Melica pyrifera Hack. = Melica scabra Kunth
Melica weberbaueri Pilg. = Melica scabra Kunth
Melocactus fortalezensis Rauh & Backeb. = Melocactus peruvianus Vaupel
Melocactus huallancensis Rauh & Backeb. = Melocactus peruvianus Vaupel
Melocactus trujilloensis Rauh & Backeb. = Melocactus peruvianus Vaupel
Melocactus unguispinus Backeb. = Melocactus peruvianus Vaupel
Mentha mollis Benth. = Minthostachys mollis (Benth.) Griseb.
Mentzelia cordifolia Dombey ex Urb. & Gilg = Mentzelia scabra Kunth
Merope argentea Wedd. = Mniodes argentea (Wedd.) M.O.Dillon
Merope caespititia Wedd. = Mniodes caespititia (Wedd.) Quip. & M.O.Dillon
Merope kunthiana (DC.) Wedd. = Lucilia kunthiana (DC.) Zardini
Merope piptolepis Wedd. = Mniodes piptolepis (Wedd.) S.E.Freire, Chemisquy, Anderb. & Urtubey
Mesosphaerum graveolens Kuntze = Mesosphaerum sidifolium (L'Hér.) Harley & J.F.B.Pastore
Mesosphaerum polyanthum (Poit.) Kuntze = Mesosphaerum sidifolium (L'Hér.) Harley & J.F.B.Pastore
Meyenia weberbaueri Backeb. = Weberbauerocereus weberbaueri (K.Schum. ex Vaupel) Backeb.

Microcala quadrangularis (Lam.) Griseb. = Cicendia quadrangularis (Lam.) Griseb.

Microchloa abyssinica Hochst. ex A.Rich. = Microchloa kunthii Desv.

Microchloa setacea Kunth = Microchloa kunthii Desv.

Microdonta nana Nutt. = Heterosperma nanum (Nutt.) Sherff

Micromeria boliviana Benth. = Clinopodium bolivianum (Benth.) Kuntze

Micromeria gilliesii Benth. = Clinopodium gilliesii (Benth.) Kuntze

Mikania melissifolia (Lam.) Willd. = Stevia melissifolia Sch.Bip.

Milium punctatum L. = Eriochloa punctata (L.) Ham.

Milla andicola (Kunth) Baker = Nothoscordum andicola Kunth

Mimosa adhaerens Kunth = Mimosa albida Humb. & Bonpl. ex Willd.

Mimosa macracantha (Humb. & Bonpl. ex Willd.) Poir. = Vachellia macracantha (Humb. & Bonpl. ex Willd.) Seigler & Ebinger

Mimosa manzanilloana Rose = Mimosa albida Humb. & Bonpl. ex Willd.

Mimosa pallida Poir. = Neltuma pallida (Humb. & Bonpl. ex Willd.) C.E.Hughes & G.P.Lewis

Mimosa racemosa Schltdl. = Mimosa albida Humb. & Bonpl. ex Willd.

Mimulus glabratus Kunth = Erythranthe glabrata (Kunth) G.L.Nesom

Minthostachys tomentosa (Benth.) Epling = Minthostachys mollis (Benth.) Griseb.

Mirabilis arenaria Heimerl = Mirabilis elegans (Choisy) Heimerl

Mirabilis campestris (Griseb.) I.M.Johnst. = Mirabilis ovata (Ruiz & Pav.) F.Meigen

Mirabilis micrantha (Choisy) Heimerl = Mirabilis prostrata (Ruiz & Pav.) Heimerl

Mitostax pallida (Humb. & Bonpl. ex Willd.) Raf. = Neltuma pallida (Humb. & Bonpl. ex Willd.) C.E.Hughes & G.P.Lewis

Mniodes cerrateae Ferreyra = Mniodes caespititia (Wedd.) Quip. & M.O.Dillon

Mniodes kunthiana (DC.) S.E.Freire, Chemisquy, Anderb. & Urtubey = Lucilia kunthiana (DC.) Zardini

Mniodes schultzii (Wedd.) S.E.Freire, Chemisquy, Anderb. & Urtubey = Mniodes caespititia (Wedd.) Quip. & M.O.Dillon

Mniodes tunariensis (Kuntze) Hieron. ex Weberb. = Novenia tunariensis (Kuntze) S.E.Freire

Modiolastrum sandemanii (Sandwith) S.R.Hill & Fryxell = Palaua sandemanii (Sandwith) Fryxell

Molina salicifolia Ruiz & Pav. = Baccharis salicifolia (Ruiz & Pav.) Pers.

Molina scandens Ruiz & Pav. = Baccharis scandens Pers.

Mollugo arenaria Kunth = Mollugo verticillata L.

Mollugo chevalieri Hutch. & Dalziel = Mollugo verticillata L.

Mollugo costata Y.T.Chang & Wei = Mollugo verticillata L.

Mollugo dichotoma Schrank = Mollugo verticillata L.

Mollugo juncea Fenzl = Mollugo verticillata L.

Mollugo spergulifolia Willd. ex Steud. = Mollugo verticillata L.

Monachne punctata (L.) Nash = Eriochloa punctata (L.) Ham.

Monactineirma suberosa (L.) Bory = Passiflora suberosa L.

Monnina angustifolia DC. = Monnina pterocarpa Ruiz & Pav.

Monnina arenicola Ferreyra = Monnina macrostachya Ruiz & Pav.

Monnina chanduyensis Chodat = Monnina pterocarpa Ruiz & Pav.

Monnina lanceolata (Poir.) DC. = Monnina macrostachya Ruiz & Pav.

Monnina piurensis Ferreyra = Monnina pterocarpa Ruiz & Pav.

Monnina weberbaueri Chodat = Monnina macrostachya Ruiz & Pav.

Montiopsis boliviana Kuntze = Montiopsis cumingii (Hook. & Arn.) D.I.Ford

Morella pavonis (C.DC.) Parra-Os. = Myrica pavonis C.DC.

Muehlenbeckia chilensis Meisn. = Muehlenbeckia hastulata (Sm.) I.M.Johnst.

Muehlenbeckia dumosa I.M.Johnst. = Muehlenbeckia fruticulosa (Walp.) Standl.

Muehlenbeckia peruviana Meisn. = Muehlenbeckia tamnifolia (Kunth) Meisn.

Muehlenbeckia rupestris Wedd. = Muehlenbeckia fruticulosa (Walp.) Standl.

Muhlenbergia bourgaei E.Fourn. = Muhlenbergia peruviana (P.Beauv.) Steud.

Muhlenbergia breviculmis Swallen = Muhlenbergia ligularis (Hack.) Hitchc.

Muhlenbergia calcicola Swallen = Muhlenbergia ligularis (Hack.) Hitchc.

Muhlenbergia cleefii Laegaard = Muhlenbergia fastigiata (J.Presl) Henrard

Muhlenbergia minuscula H.Scholz = Muhlenbergia ligularis (Hack.) Hitchc.

Muhlenbergia nana Benth. = Muhlenbergia peruviana (P.Beauv.) Steud.

Muhlenbergia pulcherrima Scribn. ex Beal = Muhlenbergia peruviana (P.Beauv.) Steud.

Muhlenbergia pusilla Steud. = Muhlenbergia peruviana (P.Beauv.) Steud.

Murucuia peduncularis (Cav.) Spreng. = Passiflora peduncularis Cav.

Mutisia cneorifolia Decne. ex Wedd. = Mutisia orbignyana Wedd.

Mutisia neriifolia Decne. ex Wedd. = Mutisia orbignyana Wedd.

Myobroma uniflora (DC.) Steven = Astragalus uniflorus DC.

Myosotis granulosa Ruiz & Pav. = Cryptantha granulosa (Ruiz & Pav.) I.M.Johnst.

Myosotis humilis Ruiz & Pav. = Plagiobothrys humilis (Ruiz & Pav.) I.M.Johnst.

Myriophyllum brasiliense Cambess. = Myriophyllum aquaticum (Vell.) Verdc.

Myriophyllum elatinoides Gaudich. = Myriophyllum quitense Kunth

Myriophyllum mattogrossense Hoehne = Myriophyllum aquaticum (Vell.) Verdc.

Myriophyllum pallidum Rusby = Myriophyllum quitense Kunth

Myriophyllum proserpinacoides Gillies ex Hook. & Arn. = Myriophyllum aquaticum (Vell.) Verdc.

Myriophyllum ternatum Gaudich. = Myriophyllum quitense Kunth

Myriophyllum titikakense J.Rémy = Myriophyllum quitense Kunth

Myriophyllum viridescens Gillies ex Hook. & Arn. = Myriophyllum quitense Kunth

Myriopteris elegans J.Sm. = Myriopteris myriophylla J.Sm.

Myriopteris paleacea Fée = Myriopteris myriophylla J.Sm.

Myrosmodes cleefii Szlach., Mytnik & S.Nowak = Myrosmodes nubigena Rchb.f.

Myrosmodes weberbaueri (Schltr.) C.A.Vargas = Myrosmodes gymnandra (Rchb.f.) C.A.Vargas

Myrrhis andicola Kunth = Chaerophyllum andicola (Kunth) K.F.Chung

Myzorrhiza tacnaensis (Mattf.) Holub = Orobanche tacnaensis Mattf.

Myzorrhiza weberbaueri (Mattf.) Holub = Orobanche weberbaueri Mattf.

Napaea crispa (L.) Moench = Herissantia crispa (L.) Brizicky

Napaea rhombifolia (L.) Moench = Sida rhombifolia L.

Nassella ligularis (Griseb.) Barkworth & Torres = Nassella neesiana (Trin. & Rupr.) Barkworth

Nassella mucronata (Kunth) R.W.Pohl = Lorenzochloa mucronata (Griseb.) Romasch.

Nasturtium bipinnatifidum (Desv.) Kuntze = Lepidium bipinnatifidum Desv.

Nasturtium chichicara (Desv.) Kuntze = Lepidium chichicara Desv.

Nasturtium nanum Wedd. = Rorippa nana (Schltdl.) J.F.Macbr.

Navarretia glutinosa (Phil.) Kuntze = Dayia glutinosa (Phil.) J.M.Porter

Navarretia gracilis (Hook.) Kuntze = Microsteris gracilis (Hook.) Greene

Navarretia laciniata (Ruiz & Pav.) Kuntze = Gilia laciniata Ruiz & Pav.

Neobartsia fiebrigii (Diels) Uribe-Convers & Tank = Neobartsia elongata (Wedd.) Uribe-Convers & Tank

Neobotrydium ambrosioides (L.) M.L.Zhang & G.L.Chu = Dysphania ambrosioides (L.) Mosyakin & Clemants

Neobotrydium incisum (Poir.) M.L.Zhang & G.L.Chu = Dysphania incisa (Poir.) ined.

Neomolina gnidiifolia (Kunth) F.H.Hellw. = Baccharis gnidiifolia Kunth

Neoporteria islayensis (C.F.Först.) Donald & G.D.Rowley = Eriosyce islayensis (C.F.Först.) Katt.

Neoraimondia aticensis Rauh & Backeb. = Neoraimondia arequipensis (Meyen) Backeb.

Neoraimondia gigantea Backeb. = Neoraimondia arequipensis (Meyen) Backeb.

Neoraimondia macrostibas (K.Schum.) Britton & Rose = Neoraimondia arequipensis (Meyen) Backeb.

Neoraimondia roseiflora Backeb. = Neoraimondia arequipensis (Meyen) Backeb.

Neoschroetera divaricata (Cav.) Briq. = Larrea divaricata Cav.

Nepeta sidifolia J.F.Gmel. = Mesosphaerum sidifolium (L'Hér.) Harley & J.F.B.Pastore

Nephrodium rufum (Poir.) Desv. = Amauropelta rufa (Poir.) Salino & T.E.Almeida

Neuontobotrys grayanus (Baehni & J.F.Macbr.) Al-Shehbaz = Neuontobotrys amplexicaulis (Kuntze) Al-Shehbaz

Nicandra macrocalyx Bitter = Nicandra physalodes (L.) Gaertn.

Nicandra minor Fisch., C.A.Mey. & Avé-Lall. = Nicandra physalodes (L.) Gaertn.

Nicandra nana Bitter = Nicandra physalodes (L.) Gaertn.

Nicandra undulata Bitter = Nicandra physalodes (L.) Gaertn.

Nicandra violacea André ex H.Lemoine = Nicandra physalodes (L.) Gaertn.

Nicoraella bomanii (Hauman) Torres = Lorenzochloa bomanii (Hauman) Romasch.

Nicoraella mucronata (Griseb.) Torres = Lorenzochloa mucronata (Griseb.) Romasch.

Nicoraella obtusa (Nees & Meyen) Torres = Lorenzochloa obtusa (Nees & Meyen) Romasch.

Nicoraella rigidiseta (Pilg.) Torres = Lorenzochloa rigidiseta (Pilg.) Romasch.

Nicotia rustica (L.) Opiz = Nicotiana rustica L.

Nicotiana brachysolen Phil. = Nicotiana undulata Ruiz & Pav.

Nicotiana breviflora Poir. = Nicotiana undulata Ruiz & Pav.

Nicotiana chilensis Millán = Nicotiana undulata Ruiz & Pav.

Nicotiana pallida Salisb. = Nicotiana paniculata L.

Nicotiana pavonii Dunal = Nicotiana rustica L.

Nicotiana pulmonarioides Kunth = Nicotiana undulata Ruiz & Pav.

Nicotiana pumila Steud. = Nicotiana rustica L.

Nicotiana rugosa Mill. = Nicotiana rustica L.

Nicotiana scabra Lehm. = Nicotiana rustica L.

Nicotiana viridiflora Lehm. = Nicotiana paniculata L.

Nolana amplexicaulis Ferreyra = Nolana adansonii (Schult.) I.M.Johnst.

Nolana confinis (I.M.Johnst.) I.M.Johnst. = Nolana lycioides I.M.Johnst.

Nolana cordata (J.Rémy) Dunal = Nolana adansonii (Schult.) I.M.Johnst.

Nolana johnstonii Vargas = Nolana laxa (Miers) I.M.Johnst.

Nolana spergularioides Ferreyra = Nolana lycioides I.M.Johnst.

Nolana volcanica Ferreyra = Nolana lycioides I.M.Johnst.

Nolana weberbaueri I.M.Johnst. = Nolana laxa (Miers) I.M.Johnst.

Nolana weissiana Ferreyra = Nolana inflata Ruiz & Pav.

Nordenstamia longistyla (Greenm. & Cuatrec.) B.Nord. = Gynoxys longistyla (Greenm. & Cuatrec.) Cuatrec.

Nothites melissifolia DC. = Stevia melissifolia Sch.Bip.

Notholaena arequipensis Maxon = Cheilanthes arequipensis (Maxon) R.M.Tryon & A.F.Tryon

Notholaena argentea E.J.Lowe = Notholaena sulphurea (Cav.) J.Sm.

Notholaena aurea (Poir.) Desv. = Myriopteris aurea (Poir.) Grusz & Windham

Notholaena cretacea Liebm. = Notholaena sulphurea (Cav.) J.Sm.

Notholaena incarum (Maxon) comb.ined. = Cheilanthes incarum Maxon

Notholaena lepida (Phil.) Looser = Notholaena sulphurea (Cav.) J.Sm.

Notholaena mollis Kunze = Cheilanthes mollis (Kunze) C.Presl

Notholaena monosticha T.Moore = Notholaena sulphurea (Cav.) J.Sm.

Notholaena myriophylla J.Sm. = Myriopteris myriophylla J.Sm.

Notholaena nivea (Poir.) Desv. = Argyrochosma nivea (Poir.) Windham

Notholaena peruviana Desv. = Cheilanthes peruviana T.Moore

Notholaena pulveracea Kunze = Notholaena sulphurea (Cav.) J.Sm.

Notholaena sinuata (Lag. ex Sw.) Kaulf. = Astrolepis sinuata (Lag. ex Sw.) D.M.Benham & Windham

Notholaena ternifolia (Cav.) Keyserl. = Pellaea ternifolia (Cav.) Link

Nothoscordum striatum (Jacq.) Kunth = Nothoscordum bivalve (L.) Britton

Nototriche acuminata A.W.Hill = Nototriche argentea A.W.Hill

Nototriche aristata A.W.Hill = Nototriche longirostris (Wedd.) A.W.Hill

Nototriche bicolor Pax = Nototriche obcuneata (Baker f.) A.W.Hill

Nototriche borussica (Meyen) A.W.Hill = Nototriche meyenii Ulbr.

Nototriche candaravica Ulbr. = Nototriche turritella A.W.Hill

Nototriche cheilanthifolia Turcz. = Nototriche anthemidifolia (Remy) A.W.Hill

Nototriche discolor Turcz. = Nototriche anthemidifolia (Remy) A.W.Hill

Nototriche herrerae Ulbr. ex A.W.Hill = Nototriche sulphurea A.W.Hill

Nototriche incana Turcz. = Nototriche pediculariifolia (Meyen) A.W.Hill

Nototriche lobulata (Wedd.) Hochr. = Nototriche obcuneata (Baker f.) A.W.Hill

Nototriche pseudoglabra A.W.Hill = Nototriche anthemidifolia (Remy) A.W.Hill

Nototriche sericea A.W.Hill = Nototriche argentea A.W.Hill

Nototriche ticsanica Ulbr. = Nototriche obcuneata (Baker f.) A.W.Hill

Nototriche villosissima Hochr. = Nototriche sulphurea A.W.Hill

Novenia acaulis (Wedd. ex Benth. & Hook.f.) S.E.Freire & F.Hellwig = Novenia tunariensis (Kuntze) S.E.Freire

Nuttallanthus subandinus (Diels) D.A.Sutton = Nuttallanthus canadensis (L.) D.A.Sutton

Obione espostoi (Speg.) G.L.Chu = Atriplex espostoi Speg.

Obione imbricata Moq. = Atriplex imbricata (Moq.) D.Dietr.

Obione myriophylla (Phil.) Ulbr. = Atriplex myriophylla Phil.

Obione peruviana (Moq.) G.L.Chu = Atriplex peruviana Moq.

Odostelma pedunculata Raf. = Passiflora peduncularis Cav.

Oedipachne punctata (L.) Link = Eriochloa punctata (L.) Ham.

Oenothera dentata Cav. = Camissonia dentata (Cav.) Reiche

Oenothera kuntziana H.Lév. = Oenothera punae Kuntze

Oenothera lyrata H.Lév. = Oenothera rosea L'Hér. ex Aiton

Oenothera mandonii H.Lév. = Oenothera nana Griseb.

Oenothera octovalvis Jacq. = Ludwigia octovalvis (Jacq.) P.H.Raven

Oenothera psychrophila Ball = Oenothera rosea L'Hér. ex Aiton

Oenothera rubida Rusby = Oenothera sandiana Hassk.

Oenothera rubra Cav. = Oenothera rosea L'Hér. ex Aiton

Oenothera virgata Ruiz & Pav. = Oenothera rosea L'Hér. ex Aiton

Oenothera weberbaueri K.Krause = Oenothera sandiana Hassk.

Onoseris longipedicellata Muschl. = Onoseris odorata Hook. & Arn.

Onoseris parva Muschl. = Onoseris odorata Hook. & Arn.

Ophioglossum stipatum Colla = Ophioglossum crotalophoroides Walter

Ophioglossum tuberosum Hook. & Arn. = Ophioglossum crotalophoroides Walter

Ophryosporus origanoides Hieron. = Ophryosporus heptanthus (Sch.Bip. ex Wedd.) R.M.King & H.Rob.

Oplotheca interrupta Nutt. = Froelichia interrupta (L.) Moq.

Opuntia corotilla K.Schum. ex Vaupel = Cumulopuntia corotilla (K.Schum. ex Vaupel) E.F.Anderson

Opuntia depauperata Britton & Rose = Opuntia pubescens H.L.Wendl. ex Pfeiff.

Opuntia dimorpha C.F.Först. = Cumulopuntia dimorpha (C.F.Först.) A.Pauca & Quip.

Opuntia dumetorum A.Berger = Opuntia pubescens H.L.Wendl. ex Pfeiff.

Opuntia floccosa Salm-Dyck = Austrocylindropuntia floccosa (Salm-Dyck) F.Ritter

Opuntia hoffmannii Bravo = Opuntia pubescens H.L.Wendl. ex Pfeiff.

Opuntia ignescens Vaupel = Cumulopuntia ignescens (Vaupel) F.Ritter

Opuntia ignota Britton & Rose = Cumulopuntia ignota (Britton & Rose) F.Ritter ex D.R.Hunt

Opuntia infesta (F.Ritter) Iliff = Opuntia pubescens H.L.Wendl. ex Pfeiff.

Opuntia leptarthra F.A.C.Weber = Opuntia pubescens H.L.Wendl. ex Pfeiff.

Opuntia leucophaea Phil. = Cumulopuntia leucophaea (Phil.) Hoxey

Opuntia maxima Mill. = Opuntia ficus-indica (L.) Mill.

Opuntia nana (DC.) Vis. = Opuntia pubescens H.L.Wendl. ex Pfeiff.

Opuntia pascoensis Britton & Rose = Opuntia pubescens H.L.Wendl. ex Pfeiff.

Opuntia pestifer Britton & Rose = Opuntia pubescens H.L.Wendl. ex Pfeiff.

Opuntia pumila Rose = Opuntia pubescens H.L.Wendl. ex Pfeiff.

Opuntia soehrensii Britton & Rose = Airampoa soehrensii (Britton & Rose) Lodé

Opuntia sphaerica C.F.Först. = Cumulopuntia sphaerica (C.F.Först.) E.F.Anderson

Opuntia subulata (Muehlenpf.) Engelm. = Austrocylindropuntia subulata (Muehlenpf.) Backeb.

Opuntia zehnderi (Rauh & Backeb.) G.D.Rowley = Cumulopuntia zehnderi (Rauh & Backeb.) F.Ritter

Oreobroma acaule (Kunth) Rydb. = Calandrinia acaulis Kunth

Oreocereus australis (F.Ritter) A.E.Hoffm. = Oreocereus hempelianus (Gürke) D.R.Hunt

Oreocereus hendriksenianus (Backeb.) Backeb. = Oreocereus leucotrichus (Phil.) Wagenkn.

Oreocereus ritteri Cullmann = Oreocereus leucotrichus (Phil.) Wagenkn.

Oreocereus variicolor Backeb. = Oreocereus leucotrichus (Phil.) Wagenkn.

Oreomyrrhis andicola (Lag. ex DC.) Hook.f. = Chaerophyllum andicola (Kunth) K.F.Chung

Oreomyrrhis andina Hemsl. = Chaerophyllum andicola (Kunth) K.F.Chung

Oreophila meyeniana Walp. = Hypochaeris meyeniana (Walp.) Benth. & Hook.f. ex Griseb.

Oreophila taraxacifolia Meyen & Walp. = Hypochaeris taraxacoides Ball

Ornithogalum biflorum (Ruiz & Pav.) D.Don = Oziroe biflora (Ruiz & Pav.) Speta

Ornithogalum bivalve L. = Nothoscordum bivalve (L.) Britton

Oryzopsis montevidensis (Spreng.) Hauman & Van der Veken = Piptochaetium montevidense (Spreng.) Parodi

Oryzopsis mucronata (Griseb.) Parodi = Lorenzochloa mucronata (Griseb.) Romasch.

Oryzopsis rigidiseta Pilg. = Lorenzochloa rigidiseta (Pilg.) Romasch.

Ourisia nana Benth. = Ourisia muscosa Benth.

Oxalis adenocaulos Phil. = Oxalis laxa Hook. & Arn.

Oxalis alsinoides Walp. = Oxalis laxa Hook. & Arn.

Oxalis arborescens Perr. = Oxalis megalorrhiza Jacq.

Oxalis arequipensis R.Knuth = Oxalis pachyrrhiza Wedd.

Oxalis bicolor Savigny = Oxalis megalorrhiza Jacq.

Oxalis buchtienii Bruns = Oxalis sepalosa Diels

Oxalis bulbigera R.Knuth = Oxalis lomana Diels

Oxalis chosicensis R.Knuth = Oxalis laxa Hook. & Arn.

Oxalis cumingii Herb. = Oxalis laxa Hook. & Arn.

Oxalis deflexa Poepp. ex Progel = Oxalis laxa Hook. & Arn.

Oxalis dichotomiflora Steud. = Oxalis laxa Hook. & Arn.

Oxalis diffusa Boreau = Oxalis corniculata L.

Oxalis foliosa Blatt. = Oxalis corniculata L.

Oxalis geranioides R.Knuth = Oxalis laxa Hook. & Arn.

Oxalis juninensis R.Knuth = Oxalis pachyrrhiza Wedd.

Oxalis leoncitensis R.Knuth = Oxalis nubigena Walp.

Oxalis lomana var. glabrescens R.Knuth = Oxalis lomana Diels

Oxalis micrantha Bertero ex Savi = Oxalis laxa Hook. & Arn.

Oxalis ornata Phil. = Oxalis megalorrhiza Jacq.

Oxalis paniculata Reiche = Oxalis megalorrhiza Jacq.

Oxalis paposana Phil. = Oxalis megalorrhiza Jacq.

Oxalis parvula J.Rémy = Oxalis nubigena Walp.

Oxalis platycaulis Steud. = Oxalis laxa Hook. & Arn.

Oxalis pubescens Stokes = Oxalis corniculata L.

Oxalis pygmaea A.Gray = Oxalis nubigena Walp.

Oxalis radicosa A.Rich. = Oxalis corniculata L.

Oxalis recisa Noronha = Oxalis corniculata L.

Oxalis repens Thunb. = Oxalis corniculata L.

Oxalis reticulata Steud. = Oxalis megalorrhiza Jacq.

Oxalis rigida (Barnéoud) Lourteig = Oxalis laxa Hook. & Arn.

Oxalis rubrocincta Lindl. = Oxalis megalorrhiza Jacq.

Oxalis simulans Baker = Oxalis corniculata L.

Oxalis succulenta Barnéoud = Oxalis megalorrhiza Jacq.

Oxalis tarapacana Phil. = Oxalis megalorrhiza Jacq.

Oxalis tenuicaulis A.Cunn. = Oxalis corniculata L.

Oxalis uittienii J.T.Jansen = Oxalis corniculata L.

Oxalis urvillei A.Cunn. = Oxalis corniculata L.

Oxalis villosa M.Bieb. = Oxalis corniculata L.

Oxalis yauliensis R.Knuth = Oxalis nubigena Walp.

Oxybaphus elegans Choisy = Mirabilis elegans (Choisy) Heimerl

Oxybaphus expansus (Ruiz & Pav.) Vahl = Mirabilis expansa (Ruiz & Pav.) Standl.

Oxybaphus ovatus (Ruiz & Pav.) Vahl = Mirabilis ovata (Ruiz & Pav.) F.Meigen

Oxybaphus prostratus (Ruiz & Pav.) Vahl = Mirabilis prostrata (Ruiz & Pav.) Heimerl

Oxychloe clandestina (Phil.) Hauman = Patosia clandestina (Phil.) Buchenau

Oxydectes alnifolia (Lam.) Kuntze = Croton alnifolius Lam.

Oxydectes ruiziana (Müll.Arg.) Kuntze = Croton ruizianus Müll.Arg.

Oxystelma solanoides (Kunth) K.Schum. = Philibertia solanoides Kunth

Oziroe sessilis (R.E.Fr.) Ravenna = Oziroe acaulis (Baker) Speta

Palaua flexuosa Mast. = Palaua dissecta Benth.

Palaua geranioides Ulbr. = Palaua dissecta Benth.

Palaua lomageiton Ulbr. = Palaua moschata Cav.

Palaua prostrata G.Don = Palaua moschata Cav.

Palaua pusilla Ulbr. = Palaua dissecta Benth.

Panicum californicum Benth. = Digitaria californica (Benth.) Henrard

Papaya candicans (A.Gray) Kuntze = Vasconcellea candicans (A.Gray) A.DC.

Parietaria appendiculata Webb & Berthel. = Parietaria debilis G.Forst.

Parietaria australis (Nees) Blume = Parietaria debilis G.Forst.

Parietaria carnosula Blume = Parietaria debilis G.Forst.

Parietaria gracilis Lowe = Parietaria debilis G.Forst.

Parietaria humifusa Blume = Parietaria debilis G.Forst.

Parietaria laxiflora Engl. = Parietaria debilis G.Forst.

Parietaria scandens Engl. = Parietaria debilis G.Forst.

Parietaria squalida Hook.f. = Parietaria debilis G.Forst.

Parkinsonia inermis Spreng. = Parkinsonia aculeata L.

Parkinsonia spinosa Kunth = Parkinsonia aculeata L.

Parkinsonia thornberi M.E.Jones = Parkinsonia aculeata L.

Parodia islayensis (C.F.Först.) Borg = Eriosyce islayensis (C.F.Först.) Katt.

Parodiochloa trachyantha (Phil.) A.M.Molina = Rostraria trachyantha (Phil.) Soreng

Paronychia compacta Montesinos & Iamonico = Paronychia andina A.Gray

Paronychia microphylla var. arequepensis Chaudhri = Paronychia arequipensis (Chaudhri) Montesinos & Iamonico

Paronychia microphylla var. arequipensis Chaudhri = Paronychia arequipensis (Chaudhri) Montesinos & Iamonico

Paronychia ramosissima (Weinm.) DC. = Cardionema ramosissima (Weinm.) A.Nelson & J.F.Macbr.

Parosela boliviana (Britton) J.F.Macbr. = Dalea boliviana Britton

Parosela coerulea (L.f.) J.F.Macbr. = Dalea coerulea (L.f.) Schinz & Thell.

Parosela cylindrica (Hook.) J.F.Macbr. = Dalea cylindrica Hook.

Parosela exilis (DC.) J.F.Macbr. = Dalea exilis DC.

Parosela onobrychis (DC.) J.F.Macbr. = Dalea onobrychis DC.

Parosela pennellii J.F.Macbr. = Dalea pennellii (J.F.Macbr.) J.F.Macbr.

Paspalum cristatum Trin. = Paspalum flavum J.Presl

Paspalum depauperatum J.Presl = Paspalum candidum (Humb. & Bonpl. ex Flüggé) Kunth

Paspalum foliosum (Lag.) Kunth = Paspalum vaginatum Sw.

Paspalum inflatum A.Rich. = Paspalum vaginatum Sw.

Paspalum lineispatha Mez = Paspalum candidum (Humb. & Bonpl. ex Flüggé) Kunth

Paspalum littorale R.Br. = Paspalum vaginatum Sw.

Paspalum mollendense Mez = Paspalum flavum J.Presl

Paspalum mononeurum Steud. = Paspalum candidum (Humb. & Bonpl. ex Flüggé) Kunth

Paspalum punctatum (L.) Flüggé = Eriochloa punctata (L.) Ham.

Paspalum scabrum Scribn. = Paspalum candidum (Humb. & Bonpl. ex Flüggé) Kunth

Paspalum squamatum Steud. = Paspalum vaginatum Sw.

Paspalum vinosum Mez = Paspalum candidum (Humb. & Bonpl. ex Flüggé) Kunth

Passiflora gaudichaudiana Planch. = Passiflora peduncularis Cav.

Passiflora muralis Barb.Rodr. = Passiflora foetida L.

Passiflora thaumasiantha Harms = Passiflora trisecta Mast.

Passiflora variegata Mill. = Passiflora foetida L.
Patagonium melanthes (Phil.) Reiche = Adesmia melanthes Phil.
Patagonium miraflorense (Remy) Rusby = Adesmia miraflorensis J.Rémy
Patagonium muricatum (Jacq.) Kuntze = Adesmia muricata (Jacq.) DC.
Patagonium spinosissimum (Meyen ex Vogel) Kuntze = Adesmia spinosissima Meyen ex Vogel
Patosia tucumanensis Castillón = Patosia clandestina (Phil.) Buchenau
Pectis cryptocephala Sch.Bip. = Pectis sessiliflora Sch.Bip.
Pectis cryptopetala Sch.Bip. = Pectis sessiliflora Sch.Bip.
Pectis pinnata Lam. = Schkuhria pinnata (Lam.) Kuntze ex Thell.
Pectocarya lateriflora (Lam.) DC. = Pectocarya linearis (Ruiz & Pav.) DC.
Pellaea lanuginosa Fée = Pellaea ternifolia (Cav.) Link
Pellaea nivea (Poir.) Prantl = Argyrochosma nivea (Poir.) Windham
Pelta palustris (L.) Dulac = Zannichellia palustris L.
Pennellia gracilis (Wedd.) O.E.Schulz = Mostacillastrum gracile (Wedd.) Al-Shehbaz
Pennisetum parviflorum (Poir.) Trin. = Setaria parviflora (Poir.) Kerguélen
Pentacaena ramosissima (Weinm.) Hook. & Arn. = Cardionema ramosissima (Weinm.) A.Nelson & J.F.Macbr.
Pentacyphus boliviensis Schltr. = Pentacyphus andinus (Ball) Liede
Pentagonia perfoliata (L.) Kuntze = Triodanis perfoliata (L.) Nieuwl.
Peperomia anisophylla C.DC. = Peperomia galioides Kunth
Peperomia apoda Trel. = Peperomia galioides Kunth
Peperomia artatiflora Trel. = Peperomia galioides Kunth
Peperomia dendroides Trel. = Peperomia galioides Kunth
Peperomia dendromorphis Trel. = Peperomia galioides Kunth
Peperomia distractiflora Trel. = Peperomia galioides Kunth
Peperomia granata Trel. = Peperomia galioides Kunth
Peperomia longispica Trel. = Peperomia galioides Kunth
Peperomia muscisedens C.DC. = Peperomia galioides Kunth
Peperomia oblongifolia C.DC. = Peperomia galioides Kunth
Peperomia redondoana Trel. = Peperomia galioides Kunth
Peperomia subcorymbosa Sodiro = Peperomia galioides Kunth
Peperomia trullifolia Trel. = Peperomia galioides Kunth
Perdicium cacalioides Kunth = Trixis cacalioides D.Don
Pereskia subulata Muehlenpf. = Austrocylindropuntia subulata (Muehlenpf.) Backeb.
Pereskiopsis subulata (Muehlenpf.) Britton & Rose = Austrocylindropuntia subulata (Muehlenpf.) Backeb.
Perezia abbiattii Cabrera = Perezia ciliosa Reiche
Perezia aracencis J.Kost. = Perezia pungens (Bonpl.) Less.
Perezia atacamensis Reiche = Perezia pungens (Bonpl.) Less.
Perezia carduncelloides Griseb. = Perezia pungens (Bonpl.) Less.
Perezia ciliaris D.Don = Perezia pungens (Bonpl.) Less.
Perezia cirsiifolia Wedd. = Perezia pinnatifida (Bonpl.) Wedd.
Perezia coerulescens Wedd. = Perezia pinnatifida (Bonpl.) Wedd.
Perezia coriacea Tovar = Perezia pungens (Bonpl.) Less.
Perezia elongata Kuntze = Perezia pungens (Bonpl.) Less.
Perezia foliosa Rusby = Perezia pungens (Bonpl.) Less.
Perezia hunzikeri Cabrera = Perezia pungens (Bonpl.) Less.
Perezia integrifolia Wedd. = Perezia pinnatifida (Bonpl.) Wedd.
Perezia keshua Cabrera = Perezia pungens (Bonpl.) Less.
Perezia laurifolia Kuntze = Perezia pungens (Bonpl.) Less.
Perezia mandonii Rusby = Perezia pungens (Bonpl.) Less.
Perezia nitidifolia J.Kost. = Perezia pinnatifida (Bonpl.) Wedd.
Perezia nivalis Wedd. = Perezia pungens (Bonpl.) Less.
Perezia obtusisquama J.Kost. = Perezia pungens (Bonpl.) Less.
Perezia purpurata Wedd. = Perezia pungens (Bonpl.) Less.
Perezia pygmea Wedd. = Perezia pinnatifida (Bonpl.) Wedd.
Perezia scalpellifolia J.Kost. = Perezia pungens (Bonpl.) Less.
Perezia scalpellifolia var. parvifolia J.Kost. = Perezia pungens (Bonpl.) Less.
Perezia sublyrata Domke = Perezia pungens (Bonpl.) Less.
Perezia violacea Wedd. = Perezia pinnatifida (Bonpl.) Wedd.
Perezia weberbaueri Hieron. ex Weberbauer = Perezia pungens (Bonpl.) Less.
Perityle discoidea (Phil.) I.M.Johnst. = Perityle emoryi Torr.
Perityle grayi Rose = Perityle emoryi Torr.

Perityle nuda Torr. = Perityle emoryi Torr.
Persicaria opelousana (Riddell) Small = Persicaria hydropiperoides (Michx.) Small
Persicaria persicarioides (Kunth) Small = Persicaria hydropiperoides (Michx.) Small
Petroravenia friesii (O.E.Schulz) Al-Shehbaz = Alshehbazia friesii (O.E.Schulz) Salariato, Zuloaga & Al-Shehbaz
Petroravenia werdermannii (O.E.Schulz) Al-Shehbaz = Alshehbazia werdermannii (O.E.Schulz) Salariato, Zuloaga & Al-Shehbaz
Phaca diminutiva Phil. = Astragalus diminutivus (Phil.) Gómez-Sosa
Phaca richii (A.Gray) A.Gray = Astragalus richii A.Gray
Phaca triflora DC. = Astragalus triflorus (DC.) A.Gray
Phacelia foliosa Phil. = Phacelia pinnatifida Griseb. ex Wedd.
Phacelia nana Wedd. = Phacelia pinnatifida Griseb. ex Wedd.
Phacelia setigera Phil. = Phacelia pinnatifida Griseb. ex Wedd.
Phacelia sinuata Phil. = Phacelia pinnatifida Griseb. ex Wedd.
Phacelia viscosa Phil. = Phacelia pinnatifida Griseb. ex Wedd.
Pharbitis purpurea (L.) Bojer = Ipomoea purpurea (L.) Roth
Pharnaceum verticillatum (L.) Spreng. = Mollugo verticillata L.
Phaseolus luteolus (Jacq.) Gagnep. = Vigna luteola (Jacq.) Benth.
Phelloderma cuneato-ovata (Cav.) Miers = Pitraea cuneato-ovata (Cav.) Caro
Philibertella clausa (K.Schum.) Vail = Funastrum clausum (Jacq.) Schltr.
Philibertia clausa K.Schum. = Funastrum clausum (Jacq.) Schltr.
Philibertia hastata (Decne.) Schltr. = Philibertia solanoides Kunth
Philibertia obtusata Malme = Philibertia solanoides Kunth
Philibertia obtusiuscula Malme = Philibertia solanoides Kunth
Philibertia variifolia (Decne.) Schltr. = Philibertia solanoides Kunth
Philibertia weberbaueri Schltr. = Philibertia solanoides Kunth
Philippiamra calycina (Phil.) Hershk. = Cistanthe calycina (Phil.) Carolin ex Hershk.
Philippiamra celosioides (Phil.) Kuntze = Cistanthe celosioides (Phil.) Carolin ex Hershk.
Phlebodium macrocarpum (Willd.) J.Sm. = Pleopeltis macrocarpa (Willd.) Kaulf.
Phlox gracilis (Hook.) Greene = Microsteris gracilis (Hook.) Greene
Phrygilanthus cuneifolius Eichler = Ligaria cuneifolia (Ruiz & Pav.) Tiegh.
Phyllanthidea microphylla (Lam.) Didr. = Andrachne microphylla (Lam.) Baill.
Physaloides physalodes (L.) Britton ex Small & Vail = Nicandra physalodes (L.) Gaertn.
Physematium montevidensis (Spreng.) Shmakov = Woodsia montevidensis Hieron.
Physematium montividense (Spreng.) Shmakov = Woodsia montevidensis Hieron.
Phytolacca australis Phil. = Phytolacca bogotensis Kunth
Phytolacca micrantha H.Walter = Phytolacca bogotensis Kunth
Phytolacca parviflora Hauman = Phytolacca bogotensis Kunth
Picrosia australis A.D.Orb. = Picrosia longifolia D.Don
Picrosia runcinata Gillies ex Hook. & Arn. = Picrosia longifolia D.Don
Pilosella mandonii Sch.Bip. = Hieracium mandonii (Sch.Bip.) Britton
Pimpinella leptophylla Pers. = Cyclospermum leptophyllum (Pers.) Sprague ex Britton & P.Wilson
Pingraea alnifolia (Meyen & Walp.) F.H.Hellw. = Baccharis alnifolia Meyen & Walp.
Pingraea salicifolia (Ruiz & Pav.) F.H.Hellw. = Baccharis salicifolia (Ruiz & Pav.) Pers.
Pingraea scandens (Ruiz & Pav.) F.H.Hellw. = Baccharis scandens Pers.
Piper galioides (Kunth) Poir. = Peperomia galioides Kunth
Piptatherum obtusum Nees & Meyen = Lorenzochloa obtusa (Nees & Meyen) Romasch.
Piptatherum punctatum (L.) P.Beauv. = Eriochloa punctata (L.) Ham.
Piptochaetium depressum (Steud.) C.Peña = Piptochaetium montevidense (Spreng.) Parodi
Piptochaetium granulatum Phil. = Piptochaetium montevidense (Spreng.) Parodi

Piptochaetium humile Phil. = Piptochaetium montevidense (Spreng.) Parodi

Piptochaetium mexicanum (Hitchc.) Beetle = Nassella mexicana (Hitchc.) R.W.Pohl

Piptochaetium moelleri Phil. = Piptochaetium montevidense (Spreng.) Parodi

Piptochaetium mucronatum Griseb. = Lorenzochloa mucronata (Griseb.) Romasch.

Piptochaetium subnudum Phil. = Piptochaetium montevidense (Spreng.) Parodi

Piptochaetium tuberculatum É.Desv. = Piptochaetium montevidense (Spreng.) Parodi

Piptochaetium verrucosum Phil. = Piptochaetium montevidense (Spreng.) Parodi

Piqueria peruviana (J.F.Gmel.) B.L.Rob. = Ophryosporus peruvianus (J.F.Gmel.) R.M.King & H.Rob.

Piqueria pubescens Sm. = Ophryosporus pubescens (Sm.) R.M.King & H.Rob.

Pitcairnia ferruginea Ruiz & Pav. = Puya ferruginea (Ruiz & Pav.) L.B.Sm.

Pitraea chilensis Turcz. = Pitraea cuneato-ovata (Cav.) Caro

Placus floresiana (Sch.Bip. ex Miq.) Kuntze = Erigeron floribundus (Kunth) Sch.Bip.

Plagiobothrys pygmaeus (Kunth) I.M.Johnst. = Plagiobothrys linifolius (Lehm.) I.M.Johnst.

Plagiobothrys tinctorius A.Gray = Plagiobothrys myosotoides (Lehm.) Brand

Plantago hirsuta Ruiz & Pav. = Plantago limensis Pers.

Plantago humilior Pilg. = Plantago myosuros Lam.

Plantago nigritella Pilg. = Plantago myosuros Lam.

Plantago nitida Roem. & Schult. = Plantago limensis Pers.

Plantago occidentalis Decne. = Plantago myosuros Lam.

Plantago pseudomyosuros Pilg. = Plantago myosuros Lam.

Plantago purpusii Brandegee = Plantago tubulosa Decne.

Plantago taraxacoides (Speg.) Pilg. = Plantago myosuros Lam.

Platonia nodiflora (L.) Raf. = Phyla nodiflora (L.) Greene

Platyloma sagittatum (Cav.) J.Sm. = Pellaea sagittata (Cav.) Link.

Platyloma ternifolium (Cav.) J.Sm. = Pellaea ternifolia (Cav.) Link

Platyopuntia soehrensii (Britton & Rose) F.Ritter = Airampoa soehrensii (Britton & Rose) Lodé

Platystachys latifolia (Meyen) K.Koch = Tillandsia latifolia Meyen

Platystachys purpurea (Ruiz & Pav.) Beer = Tillandsia purpurea Ruiz & Pav.

Pleiarina humboldtiana (Willd.) Raf. = Salix humboldtiana Willd.

Pleopeltis bryopoda (Maxon) de la Sota = Pleopeltis pycnocarpa (C.Chr.) A.R.Sm.

Pleopeltis ensifolia Carmich. ex Hook. = Pleopeltis macrocarpa (Willd.) Kaulf.

Pleopeltis lanceolata Kaulf. = Pleopeltis macrocarpa (Willd.) Kaulf.

Pleopeltis lepidota (Willd. ex Schltdl.) C.Presl = Pleopeltis macrocarpa (Willd.) Kaulf.

Pleopeltis linearis Kaulf. = Pleopeltis macrocarpa (Willd.) Kaulf.

Pleopeltis marginalis Kaulf. = Pleopeltis macrocarpa (Willd.) Kaulf.

Plettkea weberbaueri (Muschl.) Mattf. = Stellaria weberbaueri (Muschl.) Montesinos & Borsch

Pluchea absinthioides (Hook. & Arn.) H.Rob. & Cuatrec. = Tessaria absinthioides (Hook. & Arn.) DC.

Plumbago glandulosa Willd. ex Roem. & Schult. = Plumbago caerulea Kunth

Plumbago humboldtiana Roem. & Schult. = Plumbago caerulea Kunth

Plumbago rhomboidea Hook. = Plumbago caerulea Kunth

Poa algida Trin. = Poa annua L.

Poa altoperuana R.Lara & Fern.Casas = Poa kurtzii R.E.Fr.

Poa asperiflora Hack. = Poa kurtzii R.E.Fr.

Poa atacamensis Parodi = Poa laetevirens R.E.Fr.

Poa candamoana Pilg. = Poa adusta J.Presl

Poa ciliaris L. = Eragrostis ciliaris (L.) R.Br.

Poa dumetorum Hack. = Poa horridula Pilg.

Poa fusca (L.) Desf. = Diplachne fusca (L.) P.Beauv. ex Roem. & Schult.

Poa lurida (J.Presl) Kunth = Eragrostis lurida J.Presl

Poa meyenii Nees & Meyen = Poa annua L.

Poa munozensis Hack. = Poa kurtzii R.E.Fr.

Poa nana Phil. = Poa laetevirens R.E.Fr.

Poa nigricans Kunth = Eragrostis nigricans (Kunth) Steud.

Poa ovata Tovar = Poa gymnantha Pilg.

Poa pectinacea Michx. = Eragrostis pectinacea (Michx.) Nees

Poa peruviana Jacq. = Eragrostis peruviana (Jacq.) Trin.

Poa pflanzii Pilg. = Poa kurtzii R.E.Fr.

Poa piifontii J.Fernandez Casas, J.Molero & A.Susanna = Poa horridula Pilg.

Poa pseudoaequigluma Tovar = Poa gymnantha Pilg.

Poa staffordiae Tovar = Poa spicigera Tovar

Poa virescens (J.Presl) Kunth = Eragrostis virescens J.Presl

Podosemum virginicum (L.) Link = Sporobolus virginicus (L.) Kunth

Poinciana spinosa Molina = Tara spinosa (Molina) Britton & Rose

Polemonium laciniatum (Ruiz & Pav.) Kuntze = Gilia laciniata Ruiz & Pav.

Polyachyrus echinopsoides DC. = Polyachyrus sphaerocephalus D.Don

Polyachyrus glandulosus Nutt. = Polyachyrus sphaerocephalus D.Don

Polyachyrus mollendoensis I.M.Johnst. = Polyachyrus annuus I.M.Johnst.

Polyachyrus niveus Lag. ex DC. = Polyachyrus sphaerocephalus D.Don

Polyachyrus tarapacanus Phil. = Polyachyrus sphaerocephalus D.Don

Polyachyrus uniflorus Sch.Bip. = Polyachyrus sphaerocephalus D.Don

Polyachyrus villosus Wedd. = Polyachyrus sphaerocephalus D.Don

Polygonum fruticulosum Walp. = Muehlenbeckia fruticulosa (Walp.) Standl.

Polygonum hastulatum (Sm.) Kuntze = Muehlenbeckia hastulata (Sm.) I.M.Johnst.

Polygonum hydropiperoides Michx. = Persicaria hydropiperoides (Michx.) Small

Polygonum tamnifolium Kunth = Muehlenbeckia tamnifolia (Kunth) Meisn.

Polylepis tenuiruga Bitter = Polylepis rugulosa Bitter

Polypodium diaphanum Bory = Cystopteris diaphana (Bory) Blasdell

Polypodium macrocarpum Willd. = Pleopeltis macrocarpa (Willd.) Kaulf.

Polypodium pycnocarpum C.Chr. = Pleopeltis pycnocarpa (C.Chr.) A.R.Sm.

Polypodium rufum Poir. = Amauropelta rufa (Poir.) Salino & T.E.Almeida

Polypogon affinis Brongn. = Polypogon elongatus Kunth

Polypogon brachyatherus Phil. = Polypogon interruptus Kunth

Polypogon breviaristatus Phil. = Polypogon interruptus Kunth

Polypogon inaequalis Trin. = Polypogon elongatus Kunth

Polypogon mexicanus Steud. = Polypogon elongatus Kunth

Polypogon microstachys Phil. = Polypogon interruptus Kunth

Polypogon tarapacanus Phil. = Polypogon interruptus Kunth

Polystichum polyphyllum C.Presl = Polystichum orbiculatum (Desv.) Gay

Polystichum pycnolepis T.Moore = Polystichum orbiculatum (Desv.) Gay

Polystichum pygmaeum Copel. = Polystichum orbiculatum (Desv.) Gay

Polystichum sodiroi Christ = Polystichum orbiculatum (Desv.) Gay

Pontederia elongata Balf. = Pontederia crassipes Mart.

Poponax macracantha (Humb. & Bonpl. ex Willd.) Killip = Vachellia macracantha (Humb. & Bonpl. ex Willd.) Seigler & Ebinger

Poponax macracanthoides (Bertero ex DC.) Britton & Rose = Vachellia macracantha (Humb. & Bonpl. ex Willd.) Seigler & Ebinger

Porophyllum ellipticum Cass. = Porophyllum ruderale (Jacq.) Cass.

Portulaca andicola Poelln. = Portulaca perennis R.E.Fr.

Portulaca arequipaensis Poelln. = Portulaca perennis R.E.Fr.

Portulaca officinarum Crantz = Portulaca oleracea L.

Portulaca portulacastrum L. = Sesuvium portulacastrum (L.) L.

Potamogeton striatus Ruiz & Pav. = Stuckenia striata (Ruiz & Pav.) Holub

Pourretia ferruginea (Ruiz & Pav.) Spreng. = Puya ferruginea (Ruiz & Pav.) L.B.Sm.

Pratia oligophylla Wedd. = Lobelia oligophylla (Wedd.) Lammers

Prismatocarpus perfoliatus (L.) Sweet = Triodanis perfoliata (L.) Nieuwl.

Priva cuneato-ovata (Cav.) Rusby = Pitraea cuneato-ovata (Cav.) Caro

Prosopis alba Griseb. = Neltuma alba (Griseb.) C.E.Hughes & G.P.Lewis

Prosopis calderensis A.Galán, E.Linares, J.Montoya & Vicente Orell. = Neltuma calderensis (A.Galán, E.Linares, J.Montoya & Vicente Orell.) C.E.Hughes & G.P.Lewis

Prosopis pallida (Humb. & Bonpl. ex Willd.) Kunth = Neltuma pallida (Humb. & Bonpl. ex Willd.) C.E.Hughes & G.P.Lewis

Proustia berberidifolia (Cabrera) Ferreyra = Lophopappus tarapacanus (Phil.) Cabrera

Proustia foliosa (Rusby) Ferreyra = Lophopappus foliosus Rusby

Pseudobaccharis acaulis (Wedd. ex R.E.Fr.) Cabrera = Baccharis acaulis (Wedd. ex R.E.Fr.) Cabrera

Pseudobaccharis boliviensis (Wedd.) Cabrera = Baccharis boliviensis (Wedd.) Cabrera

Pseudognaphalium austrobrasilicum Deble & Marchiori = Pseudognaphalium gaudichaudianum (DC.) Anderb.

Pseudognaphalium badium (Wedd.) Anderb. = Pseudognaphalium lacteum (Meyen & Walp.) Anderb.

Pseudognaphalium cabrerae (S.E.Freire) S.E.Freire, N.Bayón, C.M.Baeza, Giuliano & C.Monti = Pseudognaphalium gaudichaudianum (DC.) Anderb.

Pseudognaphalium dombeyanum (DC.) Anderb. = Pseudognaphalium dysodes (Spreng.) S.E.Freire, N.Bayón & C.Monti

Pseudognaphalium fastigiatum N.Bayón = Pseudognaphalium gaudichaudianum (DC.) Anderb.

Pseudognaphalium frigidum Anderb. = Pseudognaphalium lacteum (Meyen & Walp.) Anderb.

Pseudognaphalium graveolens (Kuntze) Anderb. = Pseudognaphalium dysodes (Spreng.) S.E.Freire, N.Bayón & C.Monti

Pseudognaphalium melanosphaeroides (Sch.Bip. ex Wedd.) Anderb. = Pseudognaphalium lanuginosum (Kunth) Anderb.

Pseudognaphalium mendocinum (Phil.) Deble & Marchiori = Pseudognaphalium gaudichaudianum (DC.) Anderb.

Pseudognaphalium pellitum (Kunth) Anderb. = Pseudognaphalium cheiranthifolium (Lam.) Hilliard & B.L.Burtt

Pseudognaphalium versatile (Rusby) Anderb. = Pseudognaphalium viravira (Molina) Anderb.

Psittacanthus cuneifolius (Ruiz & Pav.) G.Don = Ligaria cuneifolia (Ruiz & Pav.) Tiegh.

Psoralea pubescens Poir. = Otholobium pubescens (Poir.) J.W.Grimes

Pteris aurea Poir. = Myriopteris aurea (Poir.) Grusz & Windham

Pteris esculenta G.Forst. = Pteridium esculentum (G.Forst.) Cockayne

Pteris sagittata Cav. = Pellaea sagittata (Cav.) Link.

Pteris sulphurea Cav. = Notholaena sulphurea (Cav.) J.Sm.

Pteris ternifolia Cav. = Pellaea ternifolia (Cav.) Link

Pterocladis genistelloides Lamb. ex G.Don = Baccharis genistelloides (Lam.) Pers.

Pteromonnina macbridei (Chodat) B.Eriksen = Monnina macbridei Chodat

Pteromonnina macrostachya (Ruiz & Pav.) B.Eriksen = Monnina macrostachya Ruiz & Pav.

Pteromonnina pterocarpa (Ruiz & Pav.) B.Eriksen = Monnina pterocarpa Ruiz & Pav.

Pteromonnina ramosa (I.M.Johnst.) B.Eriksen = Monnina ramosa I.M.Johnst.

Ptilurus daucifolius D.Don = Leucheria daucifolia (D.Don) Crisci

Ptychotis leptophylla (Pers.) Penz. = Cyclospermum leptophyllum (Pers.) Sprague ex Britton & P.Wilson

Puya echinotricha André = Puya ferruginea (Ruiz & Pav.) L.B.Sm.

Puya grandiflora Hook. = Puya ferruginea (Ruiz & Pav.) L.B.Sm.

Pycnophyllopsis weberbaueri (Muschl.) Timaná = Stellaria weberbaueri (Muschl.) Montesinos & Borsch

Pycnophyllum aculeatum Muschl. = Pycnophyllum molle J.Rémy

Pycnophyllum filiforme Mattf. = Pycnophyllum tetrastichum J.Rémy

Pycnophyllum glomeratum Mattf. = Pycnophyllum tetrastichum J.Rémy

Pycnophyllum lechlerianum Rohrb. = Pycnophyllum tetrastichum J.Rémy

Pycnophyllum leptothamnum Mattf. = Pycnophyllum tetrastichum J.Rémy

Pycnophyllum markgrafianum Mattf. = Pycnophyllum molle J.Rémy

Pycnophyllum weberbaueri Muschl. = Stellaria weberbaueri (Muschl.) Montesinos & Borsch

Pycreus laevigatus (L.) Nees = Cyperus laevigatus L.

Pygmaeocereus bylesianus Andreae & Backeb. = Haageocereus bylesianus (Andreae & Backeb.) Lodé

Pyrolirion aurantiacum Lem. = Pyrolirion arvense (F.Dietr.) Erhardt, Götz & Seybold

Pyrolirion aureum Herb. = Pyrolirion arvense (F.Dietr.) Erhardt, Götz & Seybold

Pyrrosia macrocarpa (Willd.) Kaulf. = Pleopeltis macrocarpa (Willd.) Kaulf.

Quamoclit nationis Hook. = Ipomoea nationis (Hook.) G.Nicholson

Quinchamalium andinum Phil. = Quinchamalium chilense Molina

Quinchamalium bracteosum Phil. = Quinchamalium chilense Molina

Quinchamalium breviflorum C.Presl = Quinchamalium chilense Molina

Quinchamalium brevistaminatum Pilg. = Quinchamalium chilense Molina

Quinchamalium carnosum Phil. = Quinchamalium chilense Molina

Quinchamalium elegans C.Presl = Quinchamalium chilense Molina

Quinchamalium elongatum Pilg. = Quinchamalium chilense Molina

Quinchamalium ericoides Brongn. = Quinchamalium chilense Molina

Quinchamalium excrecens Phil. = Quinchamalium chilense Molina

Quinchamalium fruticulosum Steud. ex Miers = Quinchamalium chilense Molina

Quinchamalium gracile (Hook. & Arn.) Brongn. = Quinchamalium chilense Molina

Quinchamalium linarioides Phil. = Quinchamalium chilense Molina

Quinchamalium linifolium Meyen ex Walp. = Quinchamalium chilense Molina

Quinchamalium litorale Phil. = Quinchamalium chilense Molina

Quinchamalium lomae Pilg. = Quinchamalium chilense Molina

Quinchamalium majus Brongn. = Quinchamalium chilense Molina

Quinchamalium minutum Phil. = Quinchamalium chilense Molina

Quinchamalium parviflorum Phil. = Quinchamalium chilense Molina

Quinchamalium pratense Phil. = Quinchamalium chilense Molina

Quinchamalium procumbens Ruiz & Pav. = Quinchamalium chilense Molina

Quinchamalium purpureum Phil. = Quinchamalium chilense Molina

Quinchamalium rugosum Phil. = Quinchamalium chilense Molina

Quinchamalium tarapacanum Phil. = Quinchamalium chilense Molina

Quinchamalium tenue Steud. = Quinchamalium chilense Molina

Quinchamalium thesioides Phil. = Quinchamalium chilense Molina

Rabdochloa spicata (Nees) Stuck. = Tripogonella spicata (Nees) P.M.Peterson & Romasch.

Radicula nana Druce = Rorippa nana (Schltdl.) J.F.Macbr.

Raimannia punae (Kuntze) Sprague & L.Riley = Oenothera punae Kuntze

Raimundochloa trachyantha (Phil.) A.M.Molina = Rostraria trachyantha (Phil.) Soreng

Ranunculus uniflorus Phil. ex Reiche = Halerpestes uniflora (Phil. ex Reiche) Emadzade, Lehnebach, P.J.Lockh. & Hörandl

Rapuntium decurrens (Cav.) C.Presl = Lobelia decurrens Cav.

Rauvolfia flexuosa Ruiz & Pav. = Citharexylum flexuosum (Ruiz & Pav.) D.Don

Rauvolfia glabra Cav. = Vallesia glabra (Cav.) Link

Reigera maritima (L.) Opiz = Bolboschoenus maritimus (L.) Palla

Reimaria candida Humb. & Bonpl. ex Flüggé = Paspalum candidum (Humb. & Bonpl. ex Flüggé) Kunth

Relbunium corymbosum (Ruiz & Pav.) K.Schum. = Galium corymbosum Ruiz & Pav.

Relbunium hypocarpium (L.) Hemsl. = Galium hypocarpium (L.) Endl. ex Griseb.

Renealmia recurvata L. = Tillandsia recurvata (L.) L.

Renealmia usneoides L. = Tillandsia usneoides (L.) L.

Rhysolepis dilloniorum A.J.Moore & H.Rob. = Aldama dilloniorum (A.J.Moore & H.Rob.) E.E.Schill. & Panero

Rhysolepis helianthoides (Rich.) A.J.Moore & H.Rob. = Aldama helianthoides (Rich.) E.E.Schill. & Panero

Rhysolepis lanceolata (Britton) H.Rob. & A.J.Moore = Aldama lanceolata (Britton) E.E.Schill. & Panero

Richardsonia lomensis K.Krause = Richardia lomensis (K.Krause) Standl.
Rostkovia clandestina Phil. = Patosia clandestina (Phil.) Buchenau
Rothia pinnata (Lam.) Kuntze = Schkuhria pinnata (Lam.) Kuntze ex Thell.
Rubia corymbosa (Ruiz & Pav.) DC. = Galium corymbosum Ruiz & Pav.
Rubia hypocarpia (L.) DC. = Galium hypocarpium (L.) Endl. ex Griseb.
Rubus hassleri Chodat = Rubus urticifolius Poir.
Rubus trichomallus Schltdl. = Rubus urticifolius Poir.
Rumex chilensis Moris = Rumex cuneifolius Campd.
Rumex hastulatus Sm. = Muehlenbeckia hastulata (Sm.) I.M.Johnst.
Rumex montevidensis Spreng. = Rumex cuneifolius Campd.
Ruppia andina Phil. = Ruppia maritima L.
Ruppia obliqua Griseb. & Schenk = Ruppia maritima L.
Ruppia pectinata Rydb. = Ruppia maritima L.
Ruppia salina Schur = Ruppia maritima L.
Ruppia taquetii H.Lév. = Ruppia maritima L.
Ruppia zosteroides Lojac. = Ruppia maritima L.
Saccharum sagittatum Aubl. = Gynerium sagittatum (Aubl.) P.Beauv.
Sageretia spicata (Humb. & Bonpl. ex Schult.) Brongn. = Scutia spicata (Humb. & Bonpl. ex Schult.) Weberb.
Sagina quitensis Kunth = Colobanthus quitensis (Kunth) Bartl.
Salicornia copiapina Phil. = Salicornia neei Lag.
Salicornia corticosa (Meyen) Walp. ex Ung.-Sternb. = Salicornia neei Lag.
Salicornia gaudichaudiana Moq. = Salicornia neei Lag.
Salicornia peruviana Kunth = Salicornia neei Lag.
Salpichroa dilatata Dammer = Salpichroa ramosissima Miers
Salpichroa lehmannii Dammer = Salpichroa tristis Walp.
Salpichroa tristis var. lehmannii (Dammer) Keel = Salpichroa tristis Walp.
Salpichroa uncu Benoist = Salpichroa ramosissima Miers
Salpiglossis acutiloba I.M.Johnst. = Leptoglossis acutiloba (I.M.Johnst.) Hunz. & Subils
Salpiglossis albiflora I.M.Johnst. = Leptoglossis albiflora (I.M.Johnst.) Hunz. & Subils
Salpiglossis lomana (Diels) J.F.Macbr. = Leptoglossis lomana (Diels) Hunz.
Salvia adenoclada Briq. = Salvia striata Benth.
Salvia avicularis Briq. = Salvia haenkei Benth.
Salvia biflora Ruiz & Pav. = Salvia tubiflora Sm.
Salvia bridgesii Britton ex Rusby = Salvia haenkei Benth.
Salvia cupheifolia Kunth = Salvia oppositiflora Ruiz & Pav.
Salvia excisa Ruiz & Pav. = Salvia tubiflora Sm.
Salvia foliosa Benth. = Salvia rhombifolia Ruiz & Pav.
Salvia grata Vahl = Salvia oppositiflora Ruiz & Pav.
Salvia pilosa Vahl = Salvia rhombifolia Ruiz & Pav.
Salvia pseudoavicularis Briq. = Salvia haenkei Benth.
Salvia scrobiculata Meyen = Salvia tubiflora Sm.
Salvia strictiflora Hook. = Salvia oppositiflora Ruiz & Pav.
Salvia subtriflora Roem. & Schult. = Salvia tubiflora Sm.
Sambucus graveolens Willd. ex Schult. = Sambucus peruviana Kunth
Sambucus oreopola Donn.Sm. = Sambucus peruviana Kunth
Samolus floribundus Kunth = Samolus parviflorus Raf.
Samolus pyrolifolius Greene = Samolus parviflorus Raf.
Sanguinaria vaginata (Sw.) Bubani = Paspalum vaginatum Sw.
Sanvitalia helianthoides Rich. = Aldama helianthoides (Rich.) E.E.Schill. & Panero
Sapindus peruvianus Walp. = Sapindus saponaria L.
Sarcocornia neei (Lag.) M.A.Alonso & M.B.Crespo = Salicornia neei Lag.
Sarcogonum fruticulosum (Walp.) Rusby = Muehlenbeckia fruticulosa (Walp.) Standl.
Sarcogonum tamnifolium (Kunth) Rusby = Muehlenbeckia tamnifolia (Kunth) Meisn.
Sarcostemma andinum (Ball) R.Holm = Pentacyphus andinus (Ball) Liede
Sarcostemma clausum (Jacq.) Schult. = Funastrum clausum (Jacq.) Schltr.
Sarcostemma solanoides (Kunth) Decne. = Philibertia solanoides Kunth
Sassia latifolia (Kunth) Holub = Oxalis latifolia Kunth
Satureja boliviana (Benth.) Briq. = Clinopodium bolivianum (Benth.) Kuntze
Schedonorus lanatus (Kunth) Roem. & Schult. = Bromus lanatus Kunth
Schkuhria advena Thell. = Schkuhria pinnata (Lam.) Kuntze ex Thell.

Schkuhria anthemoidea (DC.) J.M.Coult. = Schkuhria pinnata (Lam.) Kuntze ex Thell.
Schkuhria glabrescens Gand. = Schkuhria pinnata (Lam.) Kuntze ex Thell.
Schkuhria hopkirkia A.Gray = Schkuhria pinnata (Lam.) Kuntze ex Thell.
Schkuhria isopappa Benth. = Schkuhria pinnata (Lam.) Kuntze ex Thell.
Schkuhria multiflora Hook. & Arn. = Picradeniopsis multiflora (Hook. & Arn.) B.G.Baldwin
Schkuhria octoaristata DC. = Schkuhria pinnata (Lam.) Kuntze ex Thell.
Schkuhria virgata DC. = Schkuhria pinnata (Lam.) Kuntze ex Thell.
Schoenoplectus cernuus (Vahl) Hayek = Isolepis cernua (Vahl) Roem. & Schult.
Schoenoplectus chamissoi (Nees) Soják = Schoenoplectus californicus (C.A.Mey.) Soják
Schoenoplectus chilensis (Nees & Meyen) Soják = Schoenoplectus americanus (Pers.) Volkart
Schoenoplectus maritimus (L.) Lye = Bolboschoenus maritimus (L.) Palla
Schoenoplectus olneyi (A.Gray) Palla = Schoenoplectus americanus (Pers.) Volkart
Schoenoplectus pseudotriqueter (Steud.) Soják = Schoenoplectus californicus (C.A.Mey.) Soják
Schoenoplectus riparius (Nees & Meyen) Palla = Schoenoplectus californicus (C.A.Mey.) Soják
Schoenoplectus tatora (Kunth) Palla = Schoenoplectus californicus (C.A.Mey.) Soják
Schroeterella divaricata (Cav.) Briq. = Larrea divaricata Cav.
Scilla acaulis Baker = Oziroe acaulis (Baker) Speta
Scilla biflora Ruiz & Pav. = Oziroe biflora (Ruiz & Pav.) Speta
Scirpus albibracteatus (Nees & Meyen ex Kunth) Kuntze = Eleocharis albibracteata Nees & Meyen ex Kunth
Scirpus americanus pers. = Schoenoplectus americanus (Pers.) Volkart
Scirpus asper J.Presl & C.Presl = Rhodoscirpus asper (J.Presl & C.Presl) Lév.-Bourret, Donadío & J.R.Starr
Scirpus californicus (C.A.Mey.) Steud. = Schoenoplectus californicus (C.A.Mey.) Soják
Scirpus cernuus Vahl = Isolepis cernua (Vahl) Roem. & Schult.
Scirpus deserticola Phil. = Phylloscirpus deserticola (Phil.) Dhooge & Goetgh.
Scirpus geniculatus L. = Eleocharis geniculata (L.) Roem. & Schult.
Scirpus maritimus L. = Bolboschoenus maritimus (L.) Palla
Scirpus pungens Vahl = Schoenoplectus pungens (Vahl) Palla
Scrophularia meridionalis L.f. = Alonsoa meridionalis (L.f.) Kuntze
Selinum leptophyllum (Pers.) E.H.L.Krause = Cyclospermum leptophyllum (Pers.) Sprague ex Britton & P.Wilson
Senecio abadianus DC. = Lomanthus abadianus (DC.) B.Nord. & Pelser
Senecio acaulis Phil. = Senecio breviscapus DC.
Senecio adenophylloides Sch.Bip. = Senecio rufescens DC.
Senecio antennaria Wedd. = Chersodoma antennaria (Wedd.) Cabrera
Senecio arequipensis Cuatrec. = Chersodoma arequipensis (Cuatrec.) Cuatrec.
Senecio armeriifolius Phil. = Senecio scorzonerifolius Meyen & Walp.
Senecio arnaldii Cabrera = Lomanthus arnaldii (Cabrera) B.Nord. & Pelser
Senecio austrorufescens Cuatrec. = Senecio rufescens DC.
Senecio candollei Wedd. = Culcitium humile DC.
Senecio canoi P.Gonzáles, Montesinos & Ed.Navarro = Senecio melanandrus (Wedd.) J.Calvo, A.Granda & V.A.Funk
Senecio chamaecephalus Wedd. = Senecio breviscapus DC.
Senecio discoloratus Cuatrec. = Senecio rhizomatus Rusby
Senecio erosus Wedd. = Senecio rhizomatus Rusby
Senecio flagellisectus Griseb. = Senecio rudbeckiifolius Meyen & Walp.
Senecio graveolens Wedd. = Senecio nutans Sch.Bip.
Senecio icaensis H.Beltrán & A.Galán = Lomanthus icaensis (H.Beltrán & A.Galán) B.Nord.
Senecio jodopappus Sch.Bip. = Chersodoma jodopappa (Sch.Bip.) Cabrera
Senecio juanisernii Cuatrec. = Chersodoma juanisernii (Cuatrec.) Cuatrec.
Senecio keshna Cabrera = Culcitium oligocephalum Cabrera
Senecio keshua Cabrera = Culcitium oligocephalum Cabrera

Senecio lomincola Cabrera = Lomanthus lomincola (Cabrera) B.Nord. & Pelser

Senecio longistylus Greenm. & Cuatrec. = Gynoxys longistyla (Greenm. & Cuatrec.) Cuatrec.

Senecio modestus Wedd. = Culcitium humile DC.

Senecio mollendoensis Cabrera = Lomanthus mollendoensis (Cabrera) B.Nord.

Senecio okopanus Cabrera = Lomanthus okopanus (Cabrera) B.Nord.

Senecio phylloleptus Cuatrec. = Senecio reicheanus Cabrera

Senecio psychrophilus Wedd. = Senecio nutans Sch.Bip.

Senecio pucapampaensis H.Beltrán = Senecio melanandrus (Wedd.) J.Calvo, A.Granda & V.A.Funk

Senecio pulviniformis Hieron. = Senecio humillimus Sch.Bip.

Senecio serratifolius (Meyen & Walp.) Cuatrec. = Culcitium serratifolium Meyen & Walp.

Senecio subcandidus A.Gray = Lomanthus subcandidus (A.Gray) B.Nord.

Senecio sykorae Montesinos = Senecio melanandrus (Wedd.) J.Calvo, A.Granda & V.A.Funk

Senecio tassaensis Montesinos = Senecio melanandrus (Wedd.) J.Calvo, A.Granda & V.A.Funk

Senecio tovarii Cabrera = Lomanthus tovarii (Cabrera) B.Nord. & Pelser

Senecio vallestris DC. = Senecio pentlandianus DC.

Senecio vittatus Reiche = Senecio ctenophyllus Phil.

Senecio weddellii Cabrera = Senecio rhizomatus Rusby

Senecio wernerioides Wedd. = Senecio breviscapus DC.

Senecio zoellneri Martic. & Quezada = Culcitium oligocephalum Cabrera

Senegalia visco (Lorentz ex Griseb.) Seigler & Ebinger = Parasenegalia visco (Lorentz ex Griseb.) Seigler & Ebinger

Seris odorata Kuntze = Onoseris odorata Hook. & Arn.

Setaria affinis Schult. = Setaria parviflora (Poir.) Kerguélen

Setaria brachytricha Mez ex R.A.W.Herrm. = Setaria parviflora (Poir.) Kerguélen

Setaria discolor Hack. = Setaria parviflora (Poir.) Kerguélen

Setaria flava (Nees) Kunth = Setaria parviflora (Poir.) Kerguélen

Setaria geniculata (Poir.) P.Beauv. = Setaria parviflora (Poir.) Kerguélen

Setaria gracilis Kunth = Setaria parviflora (Poir.) Kerguélen

Setaria imberbis (Poir.) Roem. & Schult. = Setaria parviflora (Poir.) Kerguélen

Setaria laevigata (Nutt.) Schult. = Setaria parviflora (Poir.) Kerguélen

Setaria penicillata J.Presl = Setaria parviflora (Poir.) Kerguélen

Setaria purpurascens Kunth = Setaria parviflora (Poir.) Kerguélen

Setaria stipiculmis Müll.Hal. = Setaria parviflora (Poir.) Kerguélen

Setaria streptobotrys E.Fourn. = Setaria parviflora (Poir.) Kerguélen

Setaria tenella Desv. = Setaria parviflora (Poir.) Kerguélen

Setaria vulpina (Willd.) P.Beauv. = Setaria parviflora (Poir.) Kerguélen

Sicyos bryoniifolius Moris = Sicyos baderoa Hook. & Arn.

Sida angustifolia Lam. = Sida spinosa L.

Sida anthemidifolia J.Rémy = Nototriche anthemidifolia (Remy) A.W.Hill

Sida bicolor Cav. = Sida spinosa L.

Sida bicuspidata J.F.Gmel. = Sida spinosa L.

Sida bivalvis Cav. = Abutilon bivalve (Cav.) Dorr

Sida crispa L. = Herissantia crispa (L.) Brizicky

Sida emarginata Willd. = Sida spinosa L.

Sida glandulosa Roxb. ex Wight & Arn. = Sida spinosa L.

Sida heterocarpa Engelm. ex A.Gray = Sida spinosa L.

Sida lomageiton Ulbr. = Sida oligandra K.Schum.

Sida lomana Bruns = Sida oligandra K.Schum.

Sida minor Macfad. = Sida spinosa L.

Sida modesta Phil. = Palaua modesta (Phil.) Reiche & Johow

Sida multifida Dombey ex Cav. = Cristaria multifida (Dombey ex Cav.) Cav.

Sida nubigena Walp. = Acaulimalva nubigena (Walp.) Krapov.

Sida paniculata L. = Sidastrum paniculatum (L.) Fryxell

Sida patuliloba R.E.Fr. = Sida oligandra K.Schum.

Sida pediculariifolia Meyen = Nototriche pediculariifolia (Meyen) A.W.Hill

Sida pimpinellifolia Mill. = Sida spinosa L.

Sida rupo Ulbr. = Sida jatrophoides L'Hér.

Sida scabra Thonn. = Sida spinosa L.

Sida subdistans A.St.-Hil. & Naudin = Sida spinosa L.

Sida tenuicaulis Hook.f. = Sida spinosa L.

Sieglingia spicata (Nees) Kuntze ex Stuck. = Tripogonella spicata (Nees) P.M.Peterson & Romasch.

Sigesbeckia bogotensis D.L.Schulz = Sigesbeckia jorullensis Kunth

Sigesbeckia cordifolia Kunth = Sigesbeckia jorullensis Kunth

Sigesbeckia melanolepis Sch.Bip. = Sigesbeckia jorullensis Kunth

Silvaea celosioides Phil. = Cistanthe celosioides (Phil.) Carolin ex Hershk.

Sison laciniatum L'Hér. ex DC. = Cyclospermum laciniatum (DC.) Constance

Sisymbrium amplexicaule A.Gray = Neuontobotrys amplexicaulis (Kuntze) Al-Shehbaz

Sisymbrium athrocarpum A.Gray = Descurainia athrocarpa (A.Gray) O.E.Schulz

Sisymbrium calycinum Wedd. = Eudema calycinum (Desv.) Al-Shehbaz, Salariato, A.Cano & Zuloaga

Sisymbrium depressum Phil. = Descurainia depressa (Phil.) Reiche

Sisymbrium dianthoides Phil. = Mostacillastrum dianthoides (Phil.) Al-Shehbaz

Sisymbrium ferreyrae Förther & Weigend = Mostacillastrum ferreyrae (Förther & Weigend) Al-Shehbaz

Sisymbrium gracile Wedd. = Mostacillastrum gracile (Wedd.) Al-Shehbaz

Sisymbrium lactucoides Förther & Weigend = Dictyophragmus lactucoides (Förther & Weigend) Al-Shehbaz

Sisymbrium lanatum (Walp.) O.E.Schulz = Neuontobotrys lanatus (Walp.) Al-Shehbaz

Sisymbrium morrisonii Al-Shehbaz = Mostacillastrum morrisonii (Al-Shehbaz) Al-Shehbaz

Sisymbrium myriophyllum Kunth ex DC. = Descurainia myriophylla (Kunth ex DC.) R.E.Fr.

Sisymbrium oleraceum O.E.Schulz = Mostacillastrum oleraceum (O.E.Schulz) Al-Shehbaz

Sisymbrium pectinifolium Al-Shehbaz = Mostacillastrum pectinifolium (Al-Shehbaz) Al-Shehbaz

Sisymbrium peruvianum DC. = Weberbauera peruviana (DC.) Al-Shehbaz

Sisymbrium strictum Phil. = Descurainia stricta (Phil.) Reiche

Sisymbrium weddellii E.Fourn. = Exhalimolobos weddellii (E.Fourn.) Al-Shehbaz & C.D.Bailey

Sisyrinchium berteroi Steud. ex Baker = Sisyrinchium chilense Hook.

Sisyrinchium dichroum Poepp. ex Klatt = Sisyrinchium micranthum Cav.

Sisyrinchium fimbriatum Dombey ex Klatt = Sisyrinchium micranthum Cav.

Sisyrinchium iridifolium Kunth = Sisyrinchium micranthum Cav.

Sisyrinchium junceum E.Mey. ex C.Presl = Olsynium junceum (E.Mey. ex C.Presl) Goldblatt

Sisyrinchium quinquevulnerum Dombey ex Klatt = Sisyrinchium chilense Hook.

Sisyrinchium uniflorum Gay ex Phil. = Sisyrinchium chilense Hook.

Sisyrinchium valdivianum Phil. = Sisyrinchium chilense Hook.

Sisyrinchium vulgare Herter = Sisyrinchium chilense Hook.

Solanum abbottianum Juz. = Solanum candolleanum Berthault

Solanum achacachense Cárdenas = Solanum candolleanum Berthault

Solanum alticola Bitter = Solanum boliviense Dunal

Solanum amabile Vargas = Solanum candolleanum Berthault

Solanum amarantoides Dunal = Solanum americanum Mill.

Solanum andersonii Ochoa = Solanum edmondstonii Hook.f.

Solanum apalophyllum Dunal = Solanum filiforme Ruiz & Pav.

Solanum arahuayum Ochoa = Solanum medians Bitter

Solanum bermejense Bitter = Solanum cochabambense Bitter

Solanum bukasovii Juz. ex Rybin = Solanum candolleanum Berthault

Solanum calvum Bitter = Solanum americanum Mill.

Solanum calygnaphalum Ruiz & Pav. = Solanum nitidum Ruiz & Pav.

Solanum chamaesarachidium Bitter = Solanum weddellii Phil.

Solanum chrysobotrys Walp. = Solanum peruvianum L.

Solanum coerulescens Bitter = Solanum pentlandii Dunal

Solanum commutatum Spreng. = Solanum peruvianum L.

Solanum crassipes Phil. = Solanum montanum L.

Solanum curtipes Bitter = Solanum americanum Mill.

Solanum cuzcoense Ochoa = Solanum candolleanum Berthault
Solanum decachondrum Bitter = Solanum cochabambense Bitter
Solanum decurrentilobum Cárdenas & Hawkes = Solanum boliviense Dunal
Solanum depexum Juz. = Solanum acaule Bitter
Solanum egranulatum Bitter = Solanum interandinum Bitter
Solanum ellipsifolium Cárdenas & Hawkes = Solanum boliviense Dunal
Solanum erythrocarpon G.Mey. = Solanum americanum Mill.
Solanum espinarense Vargas = Solanum candolleanum Berthault
Solanum excisirhombeum Bitter = Solanum grandidentatum Phil.
Solanum extuspellitum Bitter = Solanum cochabambense Bitter
Solanum ferreyrae Ugent = Solanum edmondstonii Hook.f.
Solanum gnaphaloides Pers. = Solanum nitidum Ruiz & Pav.
Solanum heteranthera Willd. = Solanum nitidum Ruiz & Pav.
Solanum inconspicuum Bitter = Solanum americanum Mill.
Solanum indecorum A.Rich. = Solanum americanum Mill.
Solanum insulae-solis Bitter = Solanum pentlandii Dunal
Solanum lopez-camarenae Ochoa = Solanum acroscopicum Ochoa
Solanum lyciiforme Dammer = Lycianthes lycioides (L.) Hassl.
Solanum lycioides L. = Lycianthes lycioides (L.) Hassl.
Solanum mandonii A.DC. = Solanum candolleanum Berthault
Solanum megistacrolobum Bitter = Solanum boliviense Dunal
Solanum mendocianum Gillies ex Nees = Solanum fragile Wedd.
Solanum microspermum Dunal = Solanum americanum Mill.
Solanum minutibaccatum Bitter = Solanum americanum Mill.
Solanum murphyi I.M.Johnst. = Solanum edmondstonii Hook.f.
Solanum myriadenium Bitter = Solanum grandidentatum Phil.
Solanum nodiflorum Jacq. = Solanum americanum Mill.
Solanum oleraceum Dunal = Solanum americanum Mill.
Solanum onagrifolium Bitter = Solanum interandinum Bitter
Solanum pachystylum Polgar = Solanum americanum Mill.
Solanum papilionaceum Dum.Cours. = Solanum americanum Mill.
Solanum parviflorum Nocca = Solanum corymbosum Jacq.
Solanum photeinocarpum Nakamura & Odash. = Solanum americanum Mill.
Solanum phyllanthum Cav. = Solanum montanum L.
Solanum pissisi (Phil.) Reiche = Solanum pimpinellifolium L.
Solanum probolospermum Bitter = Solanum cochabambense Bitter
Solanum pterocaulum Dunal = Solanum americanum Mill.
Solanum ptychanthum Dunal = Solanum americanum Mill.
Solanum punae Juz. = Solanum acaule Bitter
Solanum punoense Hawkes = Solanum candolleanum Berthault
Solanum purpureum Dunal = Solanum montanum L.
Solanum quadrangulare L.f. = Solanum americanum Mill.
Solanum rhamnoides Dunal = Solanum nitidum Ruiz & Pav.
Solanum rhopalostigma Bitter = Solanum montanum L.
Solanum ruderale F.Phil. = Solanum radicans L.f.
Solanum sandemanii Hawkes = Solanum medians Bitter
Solanum sarachioides Rusby = Solanum cochabambense Bitter
Solanum saxatile Ochoa = Solanum candolleanum Berthault
Solanum sciaphilum Bitter = Solanum americanum Mill.
Solanum senecioides Domb. ex Dunal = Solanum multifidum Lam.
Solanum soukupii Hawkes = Solanum candolleanum Berthault
Solanum strictum Zuccagni = Solanum americanum Mill.
Solanum tacnaense Ochoa = Solanum medians Bitter
Solanum tafallae J.F.Macbr. = Solanum multifidum Lam.
Solanum tarapacanum Phil. = Solanum grandidentatum Phil.
Solanum tenellum Bitter = Solanum americanum Mill.
Solanum tenuiflorum Steud. = Solanum americanum Mill.
Solanum tuberiferum Dunal = Solanum montanum L.
Solanum uyunense Cárdenas = Solanum acaule Bitter
Solanum velardei Ochoa = Solanum candolleanum Berthault
Solanum weberbaueri Bitter = Solanum medians Bitter
Solena rotundifolia (Ruiz & Pav.) D.Dietr. = Randia rotundifolia Ruiz & Pav.
Soliva minuta (L.f.) Sweet = Soliva mexicana DC.
Soliva pedicellata Ruiz & Pav. = Soliva mexicana DC.
Soliva pedunculata Ruiz & Pav. ex Steud. = Soliva mexicana DC.

Soliva pygmaea Kunth = Soliva mexicana DC.
Sophia myriophylla Rydb. = Descurainia myriophylla (Kunth ex DC.) R.E.Fr.
Specularia perfoliata (L.) A.DC. = Triodanis perfoliata (L.) Nieuwl.
Spergularia laciniata Baehni & J.F.Macbr. = Spergularia fasciculata Phil.
Spergularia stuebelii (Hieron.) I.M.Johnst. = Spergularia fasciculata Phil.
Sphacele lamiifolia Benth. = Lepechinia lamiifolia (Benth.) Epling
Sphacele meyenii (Walp.) J.F.Macbr. = Lepechinia meyenii (Walp.) Epling
Sphaeralcea sandemanii (Sandwith) J.F.Macbr. = Palaua sandemanii (Sandwith) Fryxell
Sphaeropuntia sphaerica (C.F.Först.) Guiggi = Cumulopuntia sphaerica (C.F.Först.) E.F.Anderson
Sphaeropuntia zehnderi (Rauh & Backeb.) Guiggi = Cumulopuntia zehnderi (Rauh & Backeb.) F.Ritter
Sphaerostigma dentatum (Cav.) Walp. = Camissonia dentata (Cav.) Reiche
Spicanta occidentalis (L.) Kuntze = Blechnum occidentale L.
Spilanthes ciliata Kunth = Acmella ciliata (Kunth) Cass.
Spilanthes macraei Hook. & Arn. = Spilanthes leiocarpa DC.
Sporobolus angustus Buckley = Sporobolus indicus (L.) R.Br.
Sporobolus confertus J.A.Schmidt = Sporobolus virginicus (L.) Kunth
Sporobolus exilis (Trin.) Balansa = Sporobolus indicus (L.) R.Br.
Sporobolus fastigiatus J.Presl = Muhlenbergia fastigiata (J.Presl) Henrard
Sporobolus geniculatus (Nees ex Steud.) Nees ex Aitch. = Sporobolus virginicus (L.) Kunth
Sporobolus ligularis Hack. = Muhlenbergia ligularis (Hack.) Hitchc.
Sporobolus littoralis (Lam.) Kunth = Sporobolus virginicus (L.) Kunth
Sporobolus orientalis Kunth = Sporobolus indicus (L.) R.Br.
Sporobolus tremulus (Trin.) Kunth = Sporobolus virginicus (L.) Kunth
Stachys meyenii Walp. = Lepechinia meyenii (Walp.) Epling
Stangea rhizantha (A.Gray) Killip = Valeriana rhizantha A.Gray
Stangea wandae Graebn. & Tessend. = Valeriana wandae (Graebn. & Tessend.) Christenh. & Byng
Stenolobium fulvum (Cav.) Sprague = Tecoma fulva (Cav.) G.Don
Stenolobium stans (L.) Seem. = Tecoma stans (L.) Juss. ex Kunth
Stenomesson humile (Herb.) Baker = Clinanthus humilis (Herb.) Meerow
Stenomesson incarum Kraenzl. = Clinanthus incarum (Kraenzl.) Meerow
Stenotheca mandonii (Sch.Bip.) Sennikov = Hieracium mandonii (Sch.Bip.) Britton
Stevia dodecachaeta DC. = Stevia melissifolia Sch.Bip.
Stichophyllum bryoides Phil. = Pycnophyllum bryoides (Phil.) Rohrb.
Stipa bicolor Vahl = Piptochaetium bicolor (Vahl) É.Desv.
Stipa bomanii Hauman = Lorenzochloa bomanii (Hauman) Romasch.
Stipa brachyphylla Hitchc. = Nassella brachyphylla (Hitchc.) Barkworth
Stipa chrysophylla É.Desv. = Pappostipa chrysophylla (É.Desv.) Romasch.
Stipa depauperata Pilg. = Nassella depauperata (Pilg.) Barkworth
Stipa disticha Hitchc. = Stipa pachypus Pilg.
Stipa ichu (Ruiz & Pav.) Kunth = Jarava ichu Ruiz & Pav.
Stipa inconspicua J.Presl = Nassella inconspicua (J.Presl) Barkworth
Stipa mexicana Hitchc. = Nassella mexicana (Hitchc.) R.W.Pohl
Stipa mucronata Kunth = Lorenzochloa mucronata (Griseb.) Romasch.
Stipa nardoides (Phil.) Hack. ex Hitchc. = Nassella nardoides (Phil.) Barkworth
Stipa neesiana Trin. & Rupr. = Nassella neesiana (Trin. & Rupr.) Barkworth
Stipa obtusa (Nees & Meyen) Hitchc. = Lorenzochloa obtusa (Nees & Meyen) Romasch.
Stipa pubiflora (Trin. & Rupr.) Muñoz-Schick = Nassella pubiflora (Trin. & Rupr.) É.Desv.
Stipa rigidiseta (Pilg.) Hitchc. = Lorenzochloa rigidiseta (Pilg.) Romasch.
Stipa smithii Hitchc. = Nassella smithii (Hitchc.) Barkworth
Streptanthus englerianus Muschl. = Dictyophragmus englerianus (Muschl.) O.E.Schulz
Stuckertiella capitata Beauverd = Gamochaeta capitata Wedd.
Stylagrostis chrysantha (J.Presl) Mez = Deschampsia chrysantha (J.Presl) Saarela
Stylagrostis eminens (J.Presl) Mez = Deschampsia eminens (J.Presl) Saarela
Stylagrostis ovata (J.Presl) Mez = Deschampsia ovata (J.Presl) Saarela

Stypandra caerulea (Ruiz & Pav.) R.Br. = Pasithea caerulea (Ruiz & Pav.) D.Don

Suaeda brevifolia Phil. = Suaeda foliosa Moq.

Suaeda tenuifolia Phil. = Suaeda foliosa Moq.

Tacsonia peduncularis (Cav.) Juss. = Passiflora peduncularis Cav.

Tagetes andina M.Ferraro = Tagetes multiflora Kunth

Tagetes cabrerae M.Ferraro = Tagetes terniflora Kunth

Tagetes congesta Hook. & Arn. = Tagetes filifolia Lag.

Tagetes dichotoma Turcz. = Tagetes filifolia Lag.

Tagetes erythrocephala Rusby = Tagetes multiflora Kunth

Tagetes foeniculacea (Cass.) Desf. = Tagetes filifolia Lag.

Tagetes fragrantissima Sessé & Moc. = Tagetes filifolia Lag.

Tagetes gigantea Carrière = Tagetes terniflora Kunth

Tagetes graveolens L'Hér. ex DC. = Tagetes terniflora Kunth

Tagetes multifida DC. = Tagetes filifolia Lag.

Tagetes pseudomicrantha Lillo = Tagetes filifolia Lag.

Tagetes pusilla Kunth = Tagetes filifolia Lag.

Tagetes scabra Brandegee = Tagetes filifolia Lag.

Tagetes silenoides Meyen & Walp. = Tagetes filifolia Lag.

Talinum album Ruiz & Pav. = Calandrinia alba (Ruiz & Pav.) DC.

Talinum ciliatum Ruiz & Pav. = Calandrinia ciliata (Ruiz & Pav.) DC.

Talinum lingulatum Ruiz & Pav. = Cistanthe lingulata (Ruiz & Pav.) Hershk.

Tandonia diffusa Moq. = Anredera diffusa (Moq.) Sperling

Tara tinctoria Molina = Tara spinosa (Molina) Britton & Rose

Tarasa hornschuchiana (Walp.) Krapov. = Tarasa capitata (Cav.) D.M.Bates

Tarasa joergensenii (I.M.Johnst.) Krapov. = Tarasa capitata (Cav.) D.M.Bates

Tarasa plumosa (C.Presl) Kearney = Tarasa operculata (Cav.) Krapov.

Tarasa rahmeri Phil. = Tarasa operculata (Cav.) Krapov.

Tecomaria fulva (Cav.) Seem. = Tecoma fulva (Cav.) G.Don

Telanthera caracasana (Kunth) Moq. = Alternanthera caracasana Kunth

Telanthera pubiflora (Benth.) Moq. = Alternanthera pubiflora (Benth.) Kuntze

Tephrocactus corotilla (K.Schum. ex Vaupel) Backeb. = Cumulopuntia corotilla (K.Schum. ex Vaupel) E.F.Anderson

Tephrocactus dimorphus (C.F.Först.) Backeb. = Cumulopuntia dimorpha (C.F.Först.) A.Pauca & Quip.

Tephrocactus floccosus (Salm-Dyck) Backeb. = Austrocylindropuntia floccosa (Salm-Dyck) F.Ritter

Tephrocactus ignescens (Vaupel) Backeb. = Cumulopuntia ignescens (Vaupel) F.Ritter

Tephrocactus ignotus (Britton & Rose) Backeb. = Cumulopuntia ignota (Britton & Rose) F.Ritter ex D.R.Hunt

Tephrocactus soehrensii (Britton & Rose) G.D.Rowley = Airampoa soehrensii (Britton & Rose) Lodé

Tephrocactus sphaericus (C.F.Först.) Backeb. = Cumulopuntia sphaerica (C.F.Först.) E.F.Anderson

Tephrocactus zehnderi Rauh & Backeb. = Cumulopuntia zehnderi (Rauh & Backeb.) F.Ritter

Tephrosia coerulea (L.f.) Pers. = Dalea coerulea (L.f.) Schinz & Thell.

Tessaria dentata Ruiz & Pav. = Tessaria integrifolia Ruiz & Pav.

Tetraglochin paucijugata (I.M.Johnst.) Rothm. = Tetraglochin cristata (Britton) Rothm.

Tetraglochin tragacantha Rothm. = Tetraglochin cristata (Britton) Rothm.

Tetragonia kuntzei Buchw. = Tetragonia microcarpa Phil.

Tetragonia philippii V.V.Byalt = Tetragonia ovata Phil.

Tetragonia pusilla Phil. = Tetragonia ovata Phil.

Tetragonia tenella I.M.Johnst. = Tetragonia microcarpa Phil.

Tetragonia trigona Phil. = Tetragonia microcarpa Phil.

Thelypteris glandulosolanosa (C.Chr.) R.M.Tryon = Amauropelta glandulosolanosa (C.Chr.) Salino & T.E.Almeida

Thelypteris rufa (Poir.) A.R.Sm. = Amauropelta rufa (Poir.) Salino & T.E.Almeida

Tillaea connata Ruiz & Pav. = Crassula connata (Ruiz & Pav.) A.Berger

Tillandsia azurea C.Presl = Tillandsia purpurea Ruiz & Pav.

Tillandsia chilensis Baker = Tillandsia paleacea C.Presl

Tillandsia crinita Willd. ex Beer = Tillandsia usneoides (L.) L.

Tillandsia dependens Hieron. ex Mez = Tillandsia capillaris Ruiz & Pav.

Tillandsia favillosa Mez = Tillandsia paleacea C.Presl

Tillandsia filiformis Lodd. ex Schult. & Schult.f. = Tillandsia usneoides (L.) L.

Tillandsia fusca Baker = Tillandsia paleacea C.Presl

Tillandsia hieronymi Mez = Tillandsia capillaris Ruiz & Pav.

Tillandsia lanata Mez = Tillandsia paleacea C.Presl

Tillandsia lichenoides Hieron. = Tillandsia virescens Ruiz & Pav.

Tillandsia longibracteata Meyen = Tillandsia purpurea Ruiz & Pav.

Tillandsia monostachya W.Bartram = Tillandsia recurvata (L.) L.

Tillandsia pauciflora Sessé & Moc. = Tillandsia recurvata (L.) L.

Tillandsia permutata A.Cast. = Tillandsia capillaris Ruiz & Pav.

Tillandsia propinqua Gay = Tillandsia virescens Ruiz & Pav.

Tillandsia pusilla Gillies ex Baker = Tillandsia virescens Ruiz & Pav.

Tillandsia scalarifolia Baker = Tillandsia paleacea C.Presl

Tillandsia scoparia Willd. ex Schult. & Schult.f. = Tillandsia purpurea Ruiz & Pav.

Tillandsia trichoides Kunth = Tillandsia usneoides (L.) L.

Tillandsia uniflora Kunth = Tillandsia recurvata (L.) L.

Tissa fasciculata (Phil.) Hieron. = Spergularia fasciculata Phil.

Tithymalus hinkleyorum (I.M.Johnst.) Soják = Euphorbia hinkleyorum I.M.Johnst.

Tournefortia lilloi I.M.Johnst. = Heliotropium lilloi (I.M.Johnst.) Luebert

Tournefortia microstachya (Ruiz & Pav.) Roem. & Schult. = Heliotropium microstachyum Ruiz & Pav.

Tournefortia undulata Ruiz & Pav. = Heliotropium lilloi (I.M.Johnst.) Luebert

Tragacantha arequipensis (Vogel) Kuntze = Astragalus arequipensis Vogel

Tragacantha garbancillo (Cav.) Kuntze = Astragalus garbancillo Cav.

Tragacantha micranthella (Wedd.) Kuntze = Astragalus micranthellus Wedd.

Tragacantha minima (Vogel) Kuntze = Astragalus minimus Vogel

Tragacantha pusilla (Vogel) Kuntze = Astragalus pusillus Vogel

Tragacantha triflora (DC.) Kuntze = Astragalus triflorus (DC.) A.Gray

Tragacantha uniflora (DC.) Kuntze = Astragalus uniflorus DC.

Tragacantha weddelliana Kuntze = Astragalus weddellianus (Kuntze) I.M.Johnst.

Tragus occidentalis Nees = Tragus berteronianus Schult.

Tragus tcheliensis Debeaux = Tragus berteronianus Schult.

Trichachne californica (Benth.) Chase ex Hitchc. = Digitaria californica (Benth.) Henrard

Trichocereus cephalomacrostibas (Werderm. & Backeb.) Backeb. = Weberbauerocereus cephalomacrostibas (Werderm. & Backeb.) F.Ritter

Trichocereus glaucus F.Ritter = Trichocereus chalaensis Rauh & Backeb.

Trichocereus knuthianus Backeb. = Trichocereus cuzcoensis Britton & Rose

Trichocereus tarmaensis Rauh & Backeb. = Trichocereus cuzcoensis Britton & Rose

Trichocereus tulhuayacensis Ochoa ex Backeb. = Trichocereus cuzcoensis Britton & Rose

Trichocereus uyupampensis Backeb. = Trichocereus chalaensis Rauh & Backeb.

Tricratus parviflorus (Kunth) F.Dietr. = Colignonia parviflora (Kunth) Choisy

Tricycla spinosa Cav. = Bougainvillea spinosa (Cav.) Heimerl

Triglochin atacamensis Phil. = Triglochin striata Ruiz & Pav.

Triglochin decipiens R.Br. = Triglochin striata Ruiz & Pav.

Triglochin densiflora Dombey ex Kunth = Triglochin striata Ruiz & Pav.

Triglochin filifolia Sieber ex Spreng. = Triglochin striata Ruiz & Pav.

Triglochin flaccida A.Cunn. = Triglochin striata Ruiz & Pav.

Triglochin floridana Gand. = Triglochin striata Ruiz & Pav.

Triglochin littoralis Phil. ex Micheli = Triglochin striata Ruiz & Pav.

Triglochin pumila Larrañaga = Triglochin striata Ruiz & Pav.

Triglochin pycnostachya Gand. = Triglochin striata Ruiz & Pav.

Triglochin triandra Michx. = Triglochin striata Ruiz & Pav.

Tripogon spicatus (Nees) Ekman = Tripogonella spicata (Nees) P.M.Peterson & Romasch.

Tripsilina foetida (L.) Raf. = Passiflora foetida L.

Trisetaria spicata (L.) Paunero = Koeleria spicata (L.) Barberá, Quintanar, Soreng & P.M.Peterson

Trisetum floribundum Pilg. = Festuca floribunda (Pilg.) P.M.Peterson, Soreng & Romasch.

Trisetum spicatum (L.) K.Richt. = Koeleria spicata (L.) Barberá, Quintanar, Soreng & P.M.Peterson

Trismeria trifoliata (L.) Diels = Pityrogramma trifoliata (L.) R.M.Tryon

Trixis hexantha S.Moore = Trixis cacalioides D.Don

Trixis papillosa D.Don = Trixis cacalioides D.Don

Trixis paradoxa Cass. = Trixis cacalioides D.Don

Trixis subparadoxa B.Herrera = Trixis cacalioides D.Don

Tropaeolum aduncum Sm. = Tropaeolum peregrinum L.

Tropaeolum canariense Markham = Tropaeolum peregrinum L.

Tropaeolum dentatifolium Stokes = Tropaeolum minus L.

Tropaeolum morreanum Klatt = Tropaeolum peregrinum L.

Tropaeolum rectangulum Buchenau = Tropaeolum seemannii Buchenau

Tropaeolum weberbaueri Loes. = Tropaeolum peregrinum L.

Trophaeum minus (L.) Kuntze = Tropaeolum minus L.

Trophaeum peregrinum (L.) Kuntze = Tropaeolum peregrinum L.

Trophaeum tuberosum (Ruiz & Pav.) Kuntze = Tropaeolum tuberosum Ruiz & Pav.

Trychinolepis hoppii B.L.Rob. = Ophryosporus hoppii (B.L.Rob.) R.M.King & H.Rob.

Tula adansonii Schult. = Nolana adansonii (Schult.) I.M.Johnst.

Tunilla soehrensii (Britton & Rose) D.R.Hunt & Iliff = Airampoa soehrensii (Britton & Rose) Lodé

Tupa decurrens (Cav.) G.Don = Lobelia decurrens Cav.

Typha abyssinica Rchb.f. ex Rohrb. = Typha domingensis Pers.

Typha aequalis Schnizl. = Typha domingensis Pers.

Typha angustata Bory & Chaub. = Typha domingensis Pers.

Typha australis Schumach. = Typha domingensis Pers.

Typha basedowii Graebn. = Typha domingensis Pers.

Typha bracteata Greene = Typha domingensis Pers.

Typha brownii Kunth = Typha domingensis Pers.

Typha gigantea Schur ex Kunth = Typha domingensis Pers.

Typha macranthelia Webb & Berthel. = Typha domingensis Pers.

Typha spiralis Raf. = Typha domingensis Pers.

Typha tenuifolia Kunth = Typha domingensis Pers.

Uniola spicata L. = Distichlis spicata (L.) Greene

Urachne inconspicua Trin. & Rupr. = Nassella inconspicua (J.Presl) Barkworth

Urachne obtusa (Nees & Meyen) Trin. & Rupr. = Lorenzochloa obtusa (Nees & Meyen) Romasch.

Urachne pubiflora Trin. & Rupr. = Nassella pubiflora (Trin. & Rupr.) É.Desv.

Uralepis fusca (L.) Steud. = Diplachne fusca (L.) P.Beauv. ex Roem. & Schult.

Urocarpidium chilense (A.Braun & C.D.Bouché) Krapov. = Fuertesimalva chilensis (A.Braun & C.D.Bouché) Fryxell

Urocarpidium corniculatum Krapov. = Fuertesimalva corniculata (Krapov.) Fryxell

Urocarpidium echinatum (C.Presl) Krapov. & Fryxell = Fuertesimalva echinata (C.Presl) Fryxell

Urocarpidium limense (L.) Krapov. = Fuertesimalva limensis (L.) Fryxell

Urocarpidium peruvianum (L.) Krapov. = Fuertesimalva peruviana (L.) Fryxell

Urtica andicola Wedd. = Urtica echinata Benth.

Urtica copeyana Killip = Urtica leptophylla Kunth

Urtica debilis (G.Forst.) Endl. = Parietaria debilis G.Forst.

Urtica nicaraguensis Liebm. = Urtica leptophylla Kunth

Vachellia lutea (Mill.) Speg. = Vachellia macracantha (Humb. & Bonpl. ex Willd.) Seigler & Ebinger

Valantia hypocarpia L. = Galium hypocarpium (L.) Endl. ex Griseb.

Valeriana altoandina Cabrera = Valeriana nivalis Wedd.

Valeriana armeriifolia Schltdl. = Valeriana coarctata Ruiz & Pav.

Valeriana brachiata Pers. = Valeriana pinnatifida Ruiz & Pav.

Valeriana calchaquina Borsini = Valeriana wandae (Graebn. & Tessend.) Christenh. & Byng

Valeriana interrupta var. elatior (Graebn.) Killip = Valeriana interrupta Ruiz & Pav.

Valeriana pygmaea Graebn. = Valeriana globularis A.Gray

Vallesia cymbifolia Ortega = Vallesia glabra (Cav.) Link

Vallesia dichotoma Ruiz & Pav. = Vallesia glabra (Cav.) Link

Vallesia inedita Guib. = Vallesia glabra (Cav.) Link

Vallesia punctata Spreng. = Vallesia glabra (Cav.) Link

Varasia sedifolia (Kunth) Soják = Gentiana sedifolia Kunth

Vasquezia oppositifolia (Lag.) S.F.Blake = Villanova oppositifolia Lag.

Velaea peruviana H.Wolff = Arracacia peruviana (H.Wolff) Constance

Verbena affinis M.Martens & Galeotti = Verbena litoralis Kunth

Verbena arequipensis Botta = Mulguraea arequipensis (Botta) N.O'Leary & P.Peralta

Verbena bangiana Moldenke = Verbena hispida Ruiz & Pav.

Verbena cajamarcensis Binder ex Molinari = Verbena litoralis Kunth

Verbena clavata Ruiz & Pav. = Junellia clavata (Ruiz & Pav.) O'Leary & Múlgura

Verbena fasciculata Benth. = Junellia fasciculata (Benth.) N.O'Leary & P.Peralta

Verbena glandulosa Moris = Verbena hispida Ruiz & Pav.

Verbena integrifolia Sessé & Moc. = Verbena litoralis Kunth

Verbena juniperina Lag. = Junellia juniperina (Lag.) Moldenke

Verbena laciniata (L.) Briq. = Glandularia laciniata (L.) Schnack & Covas

Verbena lanceolata Willd. ex Spreng. = Verbena litoralis Kunth

Verbena longifolia M.Martens & Galeotti = Verbena litoralis Kunth

Verbena microphylla Kunth = Glandularia microphylla (Kunth) Cabrera

Verbena minima Meyen = Junellia minima (Meyen) Moldenke

Verbena nodiflora L. = Phyla nodiflora (L.) Greene

Verbena nudiflora Nutt. ex Turcz. = Verbena litoralis Kunth

Verbena sedula Moldenke = Verbena litoralis Kunth

Verbesina prostrata L. = Eclipta prostrata (L.) L.

Vesicarex collumanthus Steyerm. = Carex collumanthus (Steyerm.) L.E.Mora

Vicia mathewsii A.Gray = Vicia andicola Kunth

Vigna acuminata Hayata = Vigna luteola (Jacq.) Benth.

Vigna brachystachys Benth. = Vigna luteola (Jacq.) Benth.

Vigna glabra Savi = Vigna luteola (Jacq.) Benth.

Vigna helicopus (E.Mey.) Walp. = Vigna luteola (Jacq.) Benth.

Vigna hirsuta Feay ex Alph.Wood = Vigna luteola (Jacq.) Benth.

Vigna longepedunculata Taub. = Vigna luteola (Jacq.) Benth.

Vigna repens (L.) Kuntze = Vigna luteola (Jacq.) Benth.

Vigna villosa Savi = Vigna luteola (Jacq.) Benth.

Viguiera lanceolata Britton = Aldama lanceolata (Britton) E.E.Schill. & Panero

Viguiera pazensis Rusby = Aldama helianthoides (Rich.) E.E.Schill. & Panero

Vilfa fastigiata (J.Presl) Nees = Muhlenbergia fastigiata (J.Presl) Henrard

Vilfa indica (L.) Trin. ex Steud. = Sporobolus indicus (L.) R.Br.

Vilfa virginica (L.) P.Beauv. = Sporobolus virginicus (L.) Kunth

Villanova dissecta (Hook.) DC. = Villanova oppositifolia Lag.

Waltheria americana L. = Waltheria indica L.

Waltheria angustifolia L. = Waltheria indica L.

Waltheria arborescens Cav. = Waltheria indica L.

Waltheria corchorifolia Pers. = Waltheria indica L.

Waltheria elliptica Cav. = Waltheria indica L.

Waltheria erioclada DC. = Waltheria indica L.

Waltheria laxa Thulin = Waltheria indica L.

Waltheria martii Colla = Waltheria indica L.

Waltheria paniculata Benth. = Waltheria indica L.

Waltheria prostrata K.Schum. = Waltheria indica L.

Waltheria reticulata Hook.f. = Waltheria ovata Cav.

Waltheria sericea Turcz. = Waltheria ovata Cav.

Weberbauerocereus horridispinus Rauh & Backeb. = Weberbauerocereus weberbaueri (K.Schum. ex Vaupel) Backeb.

Weberbauerocereus torataensis F.Ritter = Weberbauerocereus weberbaueri (K.Schum. ex Vaupel) Backeb.

Wedelia incarnata (L.) L. = Allionia incarnata L.

Wedeliella incarnata (L.) Cockerell = Allionia incarnata L.

Werneria apiculata Sch.Bip. = Rockhausenia apiculata (Sch.Bip.) D.J.N.Hind

Werneria aretioides Wedd. = Rockhausenia aretioides (Wedd.) D.J.N.Hind

Werneria decumbens Hieron. = Werneria weddellii Phil.

Werneria lorentziana Hieron. = Werneria poposa Phil.

Werneria melanandra Wedd. = Senecio melanandrus (Wedd.) J.Calvo, A.Granda & V.A.Funk

Werneria microphylla H.Beltrán & S.Leiva = Rockhausenia microphylla (H.Beltrán & S.Leiva) D.J.N.Hind

Werneria nubigena Kunth = Rockhausenia nubigena (Kunth) D.J.N.Hind

Werneria orbignyana Wedd. = Rockhausenia orbignyana (Wedd.) D.J.N.Hind

Werneria pectinata Lingelsh. = Rockhausenia pectinata (Lingelsh.) D.J.N.Hind

Werneria pinnatifida J.Rémy = Rockhausenia pinnatifida (J.Rémy) D.J.N.Hind

Werneria pygmaea Gillies ex Hook. & Arn. = Rockhausenia pygmaea (Gillies ex Hook. & Arn.) D.J.N.Hind

Werneria sedoides S.F.Blake = Werneria marcida S.F.Blake

Werneria solivifolia Sch.Bip. = Rockhausenia solivifolia (Sch.Bip.) D.J.N.Hind

Werneria spathulata Wedd. = Rockhausenia spathulata (Wedd.) D.J.N.Hind

Werneria strigosissima A.Gray = Misbrookea strigosissima (A.Gray) V.A.Funk

Witheringia montana Dunal = Solanum montanum L.

Witheringia multifida Dunal = Solanum multifidum Lam.

Woodsia crenata (Kunze) Hieron. = Woodsia montevidensis Hieron.

Woodsia euporolepis (Trevis.) Trevis. = Woodsia montevidensis Hieron.

Woodsia incisa Gillies ex Hook. & Grev. = Woodsia montevidensis Hieron.

Woodsia peruviana Hook. = Woodsia montevidensis Hieron.

Xanthoxalis laxa (Hook. & Arn.) Holub = Oxalis laxa Hook. & Arn.

Xenophyllum ciliolatum (A.Gray) V.A.Funk = Werneria ciliolata A.Gray

Xenophyllum dactylophyllum (Sch.Bip.) V.A.Funk = Werneria dactylophylla Sch.Bip.

Xenophyllum digitatum (Wedd.) V.A.Funk = Werneria digitata Wedd.

Xenophyllum esquilachense (Cuatrec.) V.A.Funk = Werneria esquilachensis Cuatrec.

Xenophyllum marcidum (S.F.Blake) V.A.Funk = Werneria marcida S.F.Blake

Xenophyllum poposum (Phil.) V.A.Funk = Werneria poposa Phil.

Xenophyllum staffordiae (Sandwith) V.A.Funk = Werneria staffordiae Sandwith

Xenophyllum weddellii (Phil.) V.A.Funk = Werneria weddellii Phil.

Xenopoma bolivianum (Benth.) Griseb. = Clinopodium bolivianum (Benth.) Kuntze

Xeraea umbellata (Remy) Kuntze = Gomphrena umbellata J.Rémy

Xylopleurum roseum (L'Hér. ex Aiton) Raim. = Oenothera rosea L'Hér. ex Aiton

Zappania nodiflora (L.) Lam. = Phyla nodiflora (L.) Greene

Zephyranthes albicans (Herb.) Baker = Pyrolirion albicans Herb.

Zinnia florida Salisb. = Zinnia peruviana (L.) L.

Zinnia fulva Schrank = Zinnia peruviana (L.) L.

Zinnia intermedia Engelm. = Zinnia peruviana (L.) L.

Zinnia leptopoda DC. = Zinnia peruviana (L.) L.

Zinnia linearifolia Moench = Zinnia peruviana (L.) L.

Zinnia lutea Gaertn. = Zinnia peruviana (L.) L.

Zinnia multiflora L. = Zinnia peruviana (L.) L.

Zinnia pauciflora L. = Zinnia peruviana (L.) L.

Zinnia revoluta Cav. = Zinnia peruviana (L.) L.

Zinnia tenuiflora Jacq. = Zinnia peruviana (L.) L.

Zygophyllum retama Gillies ex Hook. & Arn. = Bulnesia retama (Gillies ex Hook. & Arn.) Griseb.

Appendix 5: Endemic species of Arequipa

Family	Species
Acanthaceae	Dicliptera congesta Kunth
Alstroemeriaceae	Bomarea latifolia (Ruiz & Pav.) Herb.
Amaranthaceae	Alternanthera arequipensis Suess.
Apiaceae	Domeykoa amplexicaulis (H.Wolff) Mathias & Constance
Apiaceae	Eremocharis ferreyrae Mathias & Constance
Apiaceae	Eremocharis hutchisonii Mathias & Constance
Apocynaceae	Jobinia tiarata (Malme) Liede & Meve
Asteraceae	Aldama dilloniorum (A.J.Moore & H.Rob.) E.E.Schill. & Panero
Asteraceae	Ambrosia dentata (Cabrera) M.O.Dillon
Asteraceae	Ambrosia pannosa W.W.Payne
Asteraceae	Lomanthus mollendoensis (Cabrera) B.Nord.
Asteraceae	Lomanthus okopanus (Cabrera) B.Nord.
Asteraceae	Ophryosporus bipinnatifidus B.L.Rob.
Asteraceae	Paquirea lanceolata (H.Beltrán & Ferreyra) Panero & S.E.Freire
Asteraceae	Senecio acarinus Cabrera
Asteraceae	Senecio calcicola Meyen & Walp.
Asteraceae	Senecio crassilodix Cuatrec.
Asteraceae	Senecio smithianus Cabrera
Asteraceae	Stevia hoppii B.L.Rob.
Boraginaceae	Tiquilia ferreyrae (I.M.Johnst.) A.T.Richardson
Boraginaceae	Tiquilia hunteri A.T.Richardson
Brassicaceae	Dictyophragmus englerianus (Muschl.) O.E.Schulz
Brassicaceae	Dictyophragmus lactucoides (Förther & Weigend) Al-Shehbaz
Brassicaceae	Lepidium crassius (C.L.Hitchc.) Al-Shehbaz
Brassicaceae	Machaerophorus arequipa Al-Shehbaz, A.Cano, M.A.Cueva & Salariato
Brassicaceae	Machaerophorus laticarpus Al-Shehbaz, A.Cano, M.A.Cueva & Salariato
Brassicaceae	Mathewsia peruviana O.E.Schulz
Brassicaceae	Neuontobotrys camanaensis Al-Shehbaz & A.Cano
Brassicaceae	Neuontobotrys schulzii (Al-Shehbaz) Al-Shehbaz
Bromeliaceae	Puya cahuachensis A.Galán, J.Montoya, Vicente Orell. & E.Linares
Bromeliaceae	Puya colca-canyonese sp. nov. ined.
Bromeliaceae	Puya colcaensis Treviño, Quip. & Gouda
Cactaceae	Eulychnia ritteri Cullmann
Cactaceae	Haageocereus bylesianus (Andreae & Backeb.) Lodé
Convolvulaceae	Cuscuta cockerellii Yunck.
Convolvulaceae	Cuscuta hitchcockii Yunck.
Convolvulaceae	Evolvulus lanatus Helwig
Fabaceae	Astragalus neobarnebyanus Gómez-Sosa
Fabaceae	Hoffmannseggia arequipensis Ulibarri
Fabaceae	Lupinus arequipensis C.P.Sm.
Fabaceae	Lupinus hinkleyorum C.P.Sm.
Fabaceae	Lupinus mollendoensis Ulbr.
Fabaceae	Lupinus munzianus C.P.Sm.
Fabaceae	Lupinus saxatilis Ulbr.

Fabaceae	Weberbauerella brongniartioides Ulbr.
Iridaceae	Mastigostyla hoppii R.C.Foster
Iridaceae	Tigridia arequipensis Montesinos, A.Pauca & Revilla
Loasaceae	Presliophytum arequipense Weigend
Malvaceae	Abutilon arequipense Ulbr.
Malvaceae	Gaya atiquipana Krapov.
Malvaceae	Gaya mollendoensis Krapov.
Montiaceae	Cistanthe weberbaueri (Diels) Carolin ex Hershk.
Myrtaceae	Myrcianthes ferreyrae (McVaugh) McVaugh
Onagraceae	Oenothera verrucosa I.M.Johnst.
Orobanchaceae	Orobanche weberbaueri Mattf.
Passifloraceae	Malesherbia fatimae Weigend & H.Beltrán
Rubiaceae	Galium arequipicum Dempster
Solanaceae	Jaltomata diversa (J.F.Macbr.) Mione
Solanaceae	Nolana chancoana M.O.Dillon & Quip.
Violaceae	Viola weberbaueri W.Becker

Appendix 6: Species excluded for Arequipa

The following species were mentioned at least once for Arequipa but have been excluded in this paper. The reason for their exclusion is explained in the specific part of this catalog.

Acaulimalva weberbaueri (Ulbr.) Krapov. (Malvaceae)

No voucher. Endemic to Ancash.

Achyrocline peruviana M.O.Dillon & Sagást. (Asteraceae)

Species of N-Peru (Cajamarca and La Libertad). Mentioned for Arequipa (Salinas y Aguada Blanca) by Quipuscoa & Huamantupa (2010), probably a confusion with A. ramosissima.

Achyrocline saturejoides (Lam.) DC. (Asteraceae)

Species of the humid E-slopes of the Andes. No voucher, mentioned for RN Salinas y Aguada Blanca by Quipuscoa & Huamantupa (2010), probably a mistake.

Adiantum raddianum C.Presl (Pteridaceae)

Species of the E-slope of the Andes. Specimen Cano Bellido 2613 (HUSA) from Atiquipa is probably a confusion with A. subvolubile. No other vouchers or mentions.

Ageratina scopulorum (Wedd.) R.M.King & H.Rob. (Asteraceae)

Species of the E-slopes of the Andes. Specimen Hutchison 7232 (MO) from Chachani is labeled as A. azangaroensis (= A. glechnophylla), no other voucher, mentioned by Brako & Zarucchi (1993).

Aldama australis (S.F.Blake) E.E.Schill. & Panero (Asteraceae)

Galvão et al. (2017) restrict the species to Bolivia (cerrado).

Aldama peruviana (A.Gray) E.E.Schill. & Panero (Asteraceae)

Specimen P.C.Hutchison-7137 from Islay, 450 m (1964) is Viguiera weberbaueri. Species of C-Peru.

Aloysia scorodonioides (Kunth) Cham. (Verbenaceae)

Aloysia scorodonioides var. hypoleuca (Briq.) Moldenke has been mentioned for Arequipa. The specimens from Arequipa all correspond to A. arequipensis, described by Siedo (2012).

Aloysia spathulata (Hayek) Moldenke (Verbenaceae)

Aloysia spathulata (Hayek) Moldenke has been mentioned for Arequipa. The specimens from Arequipa all correspond to A. arequipensis, described by Siedo (2012).

Alstroemeria ligtu L. (Alstroemeriaceae)

Endemic to C- and S-Chile. The peruvian vouchers are probably A. ligtu var. lineatiflora (= A. chorillensis).

Alstroemeria paupercula Phil. (Alstroemeriaceae)

Hofreiter & Rodriguez (2006) synonymize A. paupercula with A. violacea and restrict it to Chile. The Peruvian species A. chorillensis is separated from A. violacea.

Alstroemeria pelegrina L. (Alstroemeriaceae)

In the herbarium of Ruiz & Pavón A. pelegrina is supposed to be from Peru. Perhaps A. pelegrina was mislabeled because it has never been found again in Peru (Hofreiter & Rodriguez, 2006).

Amaranthus deflexus L. (Amaranthaceae)

Native to southern South America. No voucher. Mentioned by Brako & Zarucchi (1993) for Arequipa, based on specimen Cook & Gilbert 43 in Flora of Peru, which probably is a confusion with A. viridis.

Arenaria serpens Kunth (Caryophyllaceae)

Montesinos-Tubée & Teillier (2022) state that A. serpylloides (= A. serpens) was erroneously synonymized with A. digyna by Macbride (1937) and Brako & Zarucchi (1993). The Peruvian specimens of A. serpens are most likely A. digyna.

Aschersoniodoxa pilosa Al-Shehbaz (Brassicaceae)

A narrow Peruvian endemic known only by the type collection of Cuzco (Al-Shehbaz Navarro & Cano, 2012, Brako & Zarucchi, 1993). Mentioned for Arequipa (Aguada Blanca) by Quipuscoa & Huamantupa (2010), confusion with Aschersoniodoxa cachensis?

Astragalus dillinghamii J.F.Macbr. (Fabaceae)

Known only from the type locality in Junín. Mentioned for RN Salinas y Aguada Blanca by Quipuscoa & Huamantupa (2010), probably a mistake.

Atriplex taltalensis I.M.Johnst. (Amaranthaceae)

According to Brignone, Denham & Pozner (2016) the species is endemic to N-Chile. Dillon et al. (2011) do not mention the species for the Peruvian lomas. The specimen Lopez 69 (MO) from Mollendo is probably a confusion with A. espostoi.

Azorella multifida (Ruiz & Pav.) Pers. (Apiaceae)

No voucher for Arequipa, but mentioned for RN Salinas y Aguada Blanca by Quipuscoa & Huamantupa (2010). The species grows in humid parts of the Andes.

Baccharis aphylla DC. (Asteraceae)

A species from E-Brazil and Bolivia, growing at 1000-2000 m. The specimen Vargas 19490 (CUZ) corresponds to the description of the species, but has connate stems and was found at 3900 m. There are no other vouchers for Peru.

Baccharis buxifolia Pers. (Asteraceae)

No voucher. Mentioned for Salinas y Aguada Blanca by Quipuscoa & Huamantupa (2010) and listed as Parastrephietalia-species for Arequipa by Galán de Mera & Linares (2012), confusion with B. tola? Müller (2013) synonymizes B. peruviana Cuatrec. with B. buxifolia and excludes the species for Bolivia (humid parts of C-Peru northwards).

Baccharis latifolia Pers. (Asteraceae)

A species of the humid Paramo and the Yungas. The vouchers for Arequipa are doubtful and may be B. alnifolia / B.scandens. Specimens of Galiano and van der Werfft not seen, van der Werfft collected a specimen of B. alnifolia near the site as his specimen for B. latifolia.

Baccharis sphaerocephala Hook. & Arn. (Asteraceae)

Endemic to C- and S-Chile. No voucher and no observation. Mentioned by Brako & Zarucchi (1993) for Arequipa.

Baccharis tricuneata (L.f.) Pers. (Asteraceae)

Repeatedly mentioned by several authors. Müller (2006) points out that these vouchers are a confusion with B. caespitosa and B. tola. B. tricuneate is restricted to Venezuela and Colombia.

Balbisia peduncularis (Lindl.) D.Don (Francoaceae)

Endemic to N-Chile (Weigend, 2011). No voucher. Mentioned by Brako & Zarucchi (1993) for Arequipa, probably confused with Balbisia verticillata. Endemic to Chile.

Besleria tetrangularis Ruiz ex Hanst. (Gesneriaceae)

No redord. Mentioned by Brako & Zarucchi (1993) for Arequipa. Restricted to the humid E-slopes of the Andes.

Browningia hertlingiana (Backeb.) Buxb. (Cactaceae)

Specimen Ritter 669 (ZSS) is mentioned for Arequipa by Pauca & Quipuscoa (2017), but the correct location is Ayacucho (pers. comm. E. Wyss from ZSS).

Bulbostylis juncoides (Vahl) Kük. ex Herter (Cyperaceae)

No voucher. Mentioned for RN Salinas y Aguada Blanca by Quipuscoa & Huamantupa (2010). Mainly distributed in the E-Andes, an occurrence in Arequipa is rather unlikely.

Caiophora chuquitensis (Meyen) Urb. & Gilg (Loasaceae)

No voucher. Species of E-Andes. C. horrida (syn. of C. chuquitensis) is mentioned by Brako & Zarucchi for Arequipa, based on specimen Schmidt (F) from Sumbay, now renamed as C. andina.

Caiophora contorta (Desr.) C.Presl (Loasaceae)

Either a synonym of C. cirsiifolia or a vicariant species of C. cirsiifolia in N-Peru and Ecuador. The available specimens and observations from Chile and Ecuador, and one from Cuzco, all fit the description of C. cirsiifolia, except for the longer calyx. Specimen Vargas 22866 (CUZ) from Chiguata was not seen.

Caiophora coronata (Gillies ex Arn.) Hook. & Arn. (Loasaceae)

No recent vouchers or mentions. A species of the S-Andes (Chile, Argentina, S-Bolivia). Mentioned by Brako & Zarucchi for Arequipa (mistaken for C. pentlandii?).

Calceolaria bicolor Ruiz & Pav. (Calceolariaceae)

Species of WC-Peru growing between 2000 and 4000 m, specimens FLSP 5805 (CUZ) and FLSP 1907 (MO) probably confusion (lomas, below 1000 m).

Calceolaria hispida Benth. (Calceolariaceae)

A species of C-and N-Peru. No voucher, mentioned by Brako & Zarucchi (1993) for Arequipa.

Calceolaria utricularioides Hook. ex Benth. (Calceolariaceae)

Species of N- and C-Peru. Mentioned for the Lomas de Yuta by Quipuscoa et al. (2016), probably a confusion with C. dichotoma.

Carex bonariensis Desf. ex Poir. (Cyperaceae)

Species of SE-South America. Specimen FLSP 234 (MO) from Islay. Occurrence in Arequipa rather unlikely.

Castilleja fissifolia L.f. (Orobanchaceae)

Species of the humid paramos. No voucher. Mentioned by Quipuscoa & Huamantupa (2010) for RN Salinas y Aguada Blanca.

Castilleja virgatoides Edwin (Orobanchaceae)

Specimen Vargas 19411 (CUZ) is C. profunda (bracts pink).

Cenchrus pilosus Kunth (Poaceae)

Species of northern South America to Ecuador and N-Peru. Mention of Acari by Montesinos & Mondragón (2016) is probably C. echinatus.

Cerastium nutans Raf. (Caryophyllaceae)

Specimen van der Werff 20797 (MO) is labeled with "cf." and is no longer listed on tropicos.org (2024).

Cerastium subspicatum Wedd. (Caryophyllaceae)

Specimen Montesinos 4564 (B) is labeled with "cf." and probably a confusion with C. peruvianum.

Cheilanthes scariosa (Sw.) C.Presl (Pteridaceae)

Species of the wet tropics. No specimen, the observation from Sogay in iNaturalist (2024) is dubious.

Chlidanthus fragrans Herb. (Amaryllidaceae)

Specimen Iltis 1546 (US) from Camaná and specimen Raimondi 11645 (F) from Caravelí both appear to be Pyrolirion arvense (1 flower, growing in lomas).

Chuquiraga jussieui J.F.Gmel. (Asteraceae)

No voucher. Mentioned for Arequipa by Brako & Zarucchi (1993). A species of the more humid areas of the Andes.

Cinnagrostis rupestris (Trin.) P.M.Peterson, Soreng, Romasch. & Barberá (Poaceae)

Species of the E-slopes of the Andes. Mentioned for RN Salinas y Aguada Blanca by Quipuscoa & Huamantupa (2010), no specimen.

Cleistocactus samaipatanus (Cárdenas) D.R.Hunt (Cactaceae)

No voucher. Mentioned for Arequipa by Brako & Zarucchi (1993). A species of the dry forests in Bolivia and W-Brazil.

Clinanthus recurvatus (Ruiz & Pav.) Meerow (Amaryllidaceae)

Species of C-Peru. The Chala specimens are probably C. incarus.

Commelina diffusa Burm.f. (Commelinaceae)

Species of the E-slopes of the Andes.

Cristaria dissecta Hook. & Arn. (Malvaceae)

Species of Chile and Argentina. Specimen van der Werff H. 20606 (MO), Schneider (2013) recognizes only C. multifida for Peru.

Croton callicarpifolius Vahl (Euphorbiaceae)

Species of the Ecuadorian dry forests. Specimen Ferreyra 13511 (MO) from Caravelí, not reconfirmed by Brako & Zarucchi (1993).

Croton spurcus Croizat (Euphorbiaceae)

The species is known only from the type locality in Lima. The specimen Ferreyra 11969 (MO) from Caraveli (not seen), collection date 1956, is probably misidentified.

Culcitium canescens Bonpl. (Asteraceae)

No voucher. The species has been confused with C. oligocephalum and C. nivale in the southern part of the Andes (H.Beltrán, pers. comm.).

Cuscuta acuta Engelm. (Convolvulaceae)

Specimen Dillon 3291 (FTG) from Islay was removed from the herbarium. No further observations. Species of coastal Ecuador, and perhaps N-Peru.

Cuscuta corymbosa Ruiz & Pav. (Convolvulaceae)

No voucher, mentioned by Brako & Zarucchi (1993) as C. corymbosa Juss. ex Yunck., a synonym of C. foetida.

Cyperus laetus J.Presl & C.Presl (Cyperaceae)

Species of humid regions. Specimen FLSP 2147 (MO, HUSA) from Lomas de Jesus is probably confusion with C. tacnensis. No observation and no mention.

Cyperus rigens J.Presl & C.Presl (Cyperaceae)

C. tacnensis was formerly included in C. rigens. The records from Arequipa are probably C. tacnensis.

Dalea kuntzei Harms ex Kuntze (Fabaceae)

Species from E-Bolivia. The specimens Vargas 8042 and 8057 (MO) are probably misidentified.

Datura innoxia Mill. (Solanaceae)

Specimen Ochoa 14903 (F) from Caravelí is D. stramonium.

Draba matthioloides Gilg ex O.E.Schulz (Brassicaceae)

Species of N-Peru. No voucher. Mention for RN Salinas y Aguada Blanca (Quipuscoa & Huamantupa, 2010) probably misidentified.

Draba pickeringii A.Gray (Brassicaceae)

Species from C-Peru. No voucher. Mention for RN Salinas y Aguada Blanca (Quipuscoa & Huamantupa, 2010).

Drymaria cordata (L.) Willd. ex Schult. (Caryophyllaceae)

Vargas 19297 (CUZ) from Pocsi has been redetermined to D. fasciculata.

Duranta triacantha Juss. (Verbenaceae)

Mentioned for Arequipa by Brako & Zarucchi (1993), the presumed underlying specimen C.R.Worth & J.L.Morrison 15700 (F) from Chala has been re-identified as Citharexylum flexuosus.

Eleocharis bonariensis Nees (Cyperaceae)

Species of southern South America. Not mentioned for Peru in Brako & Zarucchi (1993) and Flora of Peru (1936ff.). Specimen Eyerdam & Beetle 22101 (MO) from the city of Arequipa may be a confusion with E. albibracteata.

Eleocharis montevidensis Kunth (Cyperaceae)

No voucher, mentioned for Arequipa by Brako & Zarucchi (1993, based on specimen Solomon 2808 (MO), which was redetermined to E. albibracteata in 2001.

Encelia pilosiflora S.F.Blake (Asteraceae)

Unresolved species with one voucher from Lima, mentioned for Arequipa by Brako & Zarucchi (1993). No voucher. Most likely synonymous with E. canescens.

Ephedra breana Phil. (Ephedraceae)

Martínez Carretero (2018) treats Ephedra breana as synonymous with E. americana. Hassler (2024) accepts this synonymy, POWO (2024) treats them separately. The 2 species do not differ clearly in morphology and show little difference in the phylogenetic study of Ickert-Bond & Wojciechowski (2004). Here, the 2 species are treated as synonyms.

Erechtites hieraciifolius (L.) Raf. ex DC. (Asteraceae)

Not mentioned by Brako & Zarucchi (1993) and Dillon et al. (2011). The only entry for Arequipa, specimen Ferreyra 13971 (MO) from Caravelí is a misidentification (yellow flowers).

Erigeron incaicus Solbrig (Asteraceae)

Species known only from the type locality in Cajamarca. The specimen Linares-Perea 241 (CUZ) from Chivay does not correspond to the description of Solbrig (1962).

Erigeron primulifolius (Lam.) Greuter (Asteraceae)

Species of the E-slopes of the Andes. The specimen Vargas (19332) from Puquina is doubtful (no basal leaves visible, short peduncles).

Escallonia paniculata (Ruiz & Pav.) Schult. (Escalloniaceae)

Species of the humid E-Andes. Specimen Raimondi 11481 (MO) is mentioned for Arequipa in tropicos.org, this is the holotype of E. salicifolia.

Escallonia pendula (Ruiz & Pav.) Pers. (Escalloniaceae)

No voucher. Mentioned by Brako & Zarucchi (1993). A species of northern South America.

Euphorbia glyptosperma Engelm. (Euphorbiaceae)

Species of North America. Mentioned for Arequipa by Gutte et al. (1986), the only voucher from South America. Probably a misidentification.

Euphorbia heterophylla L. (Euphorbiaceae)

Species of the E-slopes of the Andes. Specimen Hutchison 1312 (MO) from Caravelí, no modern voucher.

Euphorbia laurifolia Juss. ex Lam. (Euphorbiaceae)

Species of the eastern humid slopes of the Andes, therefore not expected. The specimen C.Vargas 19366 (CUZ) from Packaichacra is not mentioned by Brako & Zarucchi (1993). Probably a confusion with E. apurimacensis.

Euphorbia portulacoides L. (Euphorbiaceae)

Not mentioned by Dillon et al. (2011) and Brako & Zarucchi (1993). Specimen Ferreyra 11545 (MO) from Atico is probably a misidentification.

Fabiana densa J.Rémy (Solanaceae)

No voucher. Mentioned for RN Salinas y Aguada Blanca by Quipuscoa & Huamantupa (2010). Brako & Zarucchi (1993) mentions it for Tacna, the corresponding specimen was redetermined as F. stephanii.

Festuca humilior Nees & Meyen (Poaceae)

Species of Bolivia and NW-Argentina. No voucher. Mentioned for RN Salinas y Aguada Blanca by Quipuscoa & Huamantupa (2010), occurrence possible.

Flaveria trinervia (Spreng.) C.Mohr (Asteraceae)

Species of North America. Mentioned by Brako & Zarucchi (1993), probably confused with F. bidentis.

Gaya pilosa K.Schum. (Malvaceae)

Restricted to S-Brazil, NE-Argentina and Paraguay. The specimens Worth & Morrison 15616 (F) and Dillon et al. 8808 (F) are probably Gaya atiquipana. Specimen Dillon & Dillon 3937 (F) is probably G. mollendoensis.

Geranium berteroanum Colla (Geraniaceae)

Species from C-Chile to Tierra del Fuego (Aedo, Navarro & Alarcon, 2005). There seems to be a confusion with G. fallax in the vouchers from S-Peru (Arequipa, Moquegua, Tacna).

Glandularia gynobasis (Wedd.) N.O'Leary & P.Peralta (Verbenaceae)

Mentioned for Arequipa by Brako & Zarucchi (1993) and Hassler (2024). O'Leary & Mulgura (2014) note that the species is known for Peru only from the type locality of Tacora, Tacna.

Grindelia brachystephana Griseb. (Asteraceae)

Bartoli & Tortosa (1999) classify the species as endemic to Argentina. The specimens from Arequipa of G. bergii Cabrera (= synonym of G. brachystephana) have all glaucous, ± glabrous leaves and are confusions with G. tarapacana.

Halerpestes cymbalaria (Pursh) Greene (Ranunculaceae)

Specimen Tutin 911 (not seen), confusion with H. uniflora?.

Heliotropium angiospermum Murray (Boraginaceae)

N- and E-South America southward to C-Peru. Mentioned by Johnston (1928) for Murillo Bay (province?, not found on map) and the city of Arequipa. No modern vouchers.

Heliotropium verdcourtii Craven (Boraginaceae)

No vouche. A species of humid South America. Mentioned for Arequipa by Brako & Zarucchi (1993).

Helogyne calocephala Mattf. (Asteraceae)

No voucher. Restricted to N-Peru (pers. comm. H. Beltrán). Mentioned for Arequipa by Brako & Zarucchi (1993).

Hesperomeles ferruginea (Juss. ex Pers.) Benth. (Rosaceae)

No specimens, mentioned for Atiquipa by Talavera et al. (2017), a convincing observation from Atiquipa on iNaturalist (2024). Not mentioned for the lomas of Peru by Dillon et al. (2011). The occurrence in Atiquipa seems to be the only population of S-Peru and N-Chile near the Pacific coast, all other occurrences grow on the E-slopes of the Andes. A cultivated species?

Hesperoxiphion herrerae (Diels ex R.C.Foster) Ravenna (Iridaceae)

Endemic to Cuzco. Specimen Vargas 19058 (CUZ) from Quequena (prov. Arequipa) in bad condition and probably mistaken.

Hierobotana inflata (Kunth) Briq. (Verbenaceae)

O'Leary & Moroni (2014) restrict this species to Ecuador. Confused in Flora of Peru (1936ff.) with Junellia fasciculata.

Hoffmannseggia glauca (Ortega) Eifert (Fabaceae)

No voucher, mentioned for S-Peru by Simpson & Ulibarri (2006).

Hypochaeris chondrilloides (A.Gray) Cabrera (Asteraceae)

Species of Argentina and Chile. Reported for Peru only once in Mejia (specimen Mejia group 1374 (MO)). The collector group collected many ornamental plants and is not reliable.

Juncus cordobensis Barros (Juncaceae)

Species from S Brazil and Argentina. Mentioned by Brako & Zarucchi (1993) for Arequipa. The holotype of the synonym J. arequipensis, Lopez 015 (MO) (not seen), described by Balslev (1983) is from the lomas of Mollendo. Not mentioned for the Lomas of Peru by Dillon et al. (2011). Specimen Stafford 472 (K) from Juliaca (Puno), identified by Balslev as A. arequipensis, is in my opinion J. stipulatus.

Juncus pallescens Lam. (Juncaceae)

Species of humid climates. No voucher, mentioned by Quipuscoa & Huamantupa (2010) for RN Salinas y Aguada Blanca, probably a confusion with J. ebracteatus.

Junellia occulta (Moldenke) N.O'Leary & P.Peralta (Verbenaceae)

No voucher. Mentioned for Arequipa by Brako & Zarucchi (1993). Endemic species to NW-Peru (O'Leary & Mulgura, 2014).

Knowltonia integrifolia (DC.) Christenh. & Byng (Ranunculaceae)

A species of the east slope of the Andes. No voucher. Mentioned by Brako & Zarucchi, confusion with Misbrookea strigosissima?

Lantana haughtii Moldenke (Verbenaceae)

Binder (2002) synonymizes L. haughtii with L. angustibraceata, a species of N-and C-Peru. Specimen Vargas 19390 from Packaichacra (not curated) is probably a mistake.

Lantana radicans Ruiz & Pav. (Verbenaceae)

Original material of Ruiz & Pavón from Atiquipa. Not mentioned for the lomas of Peru by Dillon et al. (2011).

Lantana svensonii Moldenke (Verbenaceae)

Binder (2002) synonymizes L. svensonii with L. sprucei.

Leptoglossis darcyana Hunz. & Subils (Solanaceae)

Endemic to Tacna, no vouchers for other localities.

Lomanthus yauyensis (Cabrera) B.Nord. & Pelser (Asteraceae)

No voucher. Mentioned for Arequipa by Brako & Zarucchi, probably a confusion. Endemic to C-Peru (Beltran & Galán de Mera, 1996).

Loxanthocereus hystrix Rauh & Backeb. (Cactaceae)

No voucher, mentioned for Caravelí by Pauca & Quipuscoa (2017), not mentioned for Arequipa by Ostolaza (2019).

Lupinus paniculatus Desr. (Fabaceae)

Mentioned once for Arequipa, no specimen, species of humid puna.

Lupinus paruroensis C.P.Sm. (Fabaceae)

Mentioned for Arequipa, no voucher, species of humid puna.

Lysimachia minima (L.) U.Manns & Anderb. (Primulaceae)

Specimen FLSP 574 (MO) from Atiquipa (delisted by tropicos.org in 2024), mentioned by Dillon et al. (2011) for the lomas of Peru.

Lysipomia laciniata A.DC. (Campanulaceae)

A species of the E-slopes of the Andes. No recorrd, mentioned for RN Salinas y Aguada Blanca by Quipuscoa & Huamantupa (2010).

Malesherbia ardens J.F.Macbr. (Passifloraceae)

One mention for Caylloma, probably confusion with M. fatimae. Endemic to Tacna and Moquegua (Beltran et al., 2018).

Malva sylvestris L. (Malvaceae)

Mentioned by Montesinos & Zegarra (2019), the picture is M. assurgentiflora.

Margyricarpus alatus Gillies ex Hook. & Arn. (Rosaceae)

Specimens Vargas 19498 and 9444 (both CUZ) have cristate fruits (= M. cristatus). M. alatus seems to be restricted to more humid climates.

Mniodes andina (A.Gray) Cuatrec. (Asteraceae)

No voucher. A species endemic to C-Peru (Ancash).

Monnina herbacea DC. (Polygalaceae)

Species of C-Ecuador to C-Peru (Eriksen, 1993). Specimen FLSP 1730 (HUSA) from Taimara probably a mistake.

Montiopsis modesta (Phil.) D.I.Ford (Montiaceae)

Species of C- and N-Chile. Specimen Vargas 7994 (CUZ) from Yura, not seen, is probably a mistake.

Mulguraea aspera (Gillies & Hook.) N.O'Leary & P.Peralta (Verbenaceae)

Species of Argentina. Specimens from Arequipa Vargas 7927 & 8008 (MO), both not seen, are probably a confusion with M. arequipensis, a species described in 2009 by O'Leary et al. (2009).

Myrcianthes minimifolia (McVaugh) McVaugh (Myrtaceae)

A species of NW-Argentina. Mentioned for Arequipa by Brako & Zarucchi (1993). Specimen Ferreyra 13423 and 13481 (MICH) from Caravelí may be Myrcianthes ferreyrae.

Nama jamaicensis L. (Boraginaceae)

A species of humid climates E of the Andes. Specimens Jaime León 263 + 52 (MO) for Atiquipa, a confusion with N. dichotomum?

Neltuma andicola (Burkart) C.E.Hughes & G.P.Lewis (Fabaceae)

Galán de Mera et al. (2019) separated P. andicola from P. laevigata var. andicola based on the geographic range and restricted P. andicola to Cuzco. The species from Arequipa was renamed P. calderensis. Specimen Pennell F.W. 13062 (F) from Tiabaya fits the description of P. andicola (cultivated?).

Neltuma chilensis (Molina) C.E.Hughes & G.P.Lewis (Fabaceae)

No voucher. Mentioned for S-Peru by Burkart (1976) and for Arequipa by Brako & Zarucchi. Several specimens from Cuzco. May be a confusion or a planted specimen.

Neobartsia crenoloba (Wedd.) Uribe-Convers & Tank (Orobanchaceae)

No voucher, mentioned for RN Salinas y Aguada Blanca (Quipuscoa & Huamantupa, 2010).

Neobartsia thiantha (Diels) Uribe-Convers & Tank (Orobanchaceae)

No voucher. Mentioned for Arequipa by Brako & Zarucchi (1993). Presumably endemic to Cuzco.

Neuontobotrys intricatissimus (Phil.) Al-Shehbaz (Brassicaceae)

The Peruvian specimens were renamed N. schulzii by Al-Shehbaz (1990b).

Nolana aplocaryoides (Gaudich.) I.M.Johnst. (Solanaceae)

No voucher. Species of N-Chile to Tacna.

Nolana jaffuelii I.M.Johnst. (Solanaceae)

Mentioned for Arequipa in Brako & Zarucchi (1993). Restricted to Chile, no modern Peruvian voucher (Quipuscoa & Dillon, 2018).

Nototriche foetida Ulbr. (Malvaceae)

Specimen van der Werff 20794 (MO) from Arequipa is confusedly labeled and has a well developped stem. No other mentions for Arequipa, species restricted to Tacna and Moquegua.

Oenothera nocturna Jacq. (Onagraceae)

Specimen Cardenas & Rodriguez 3 (F) from faldas del Misti was identified by Dietrich in 1974 as O. nocturna, in his revision of the genus (Dietrich, 1977) the species is mentioned for C-Peru at low altitudes.

Ophryosporus chilca Hieron. (Asteraceae)

Species of Ecuador and N-Peru. No voucher. Mentioned by Brako & Zarucchi (1993) for Arequipa (O. heptanthus?).

Oxalis arenaria Bertero (Oxalidaceae)

Mentioned for Peru in Brako & Zarucchi (1993) and Flora of Peru (1936ff.), the cited specimen Herrera 3632 has been reclassified as O. moqueguensis (syn. of O. pinguiculacea), the specimen Vargas 1182 is now filed as O. latifolia.

Oxalis dombeyi A.St.-Hil. (Oxalidaceae)

Species of northern South America, from Panama to C-Peru. Herbarium sepcimen Weigend & Förther 97/721 (F) from Caravelí is Oxalis peruviana Norlind (confusion with O. peruviana R.Knuth = O. dombeyi).

Oxybasis macrosperma (Hook.f.) S.Fuentes, Uotila & Borsch (Amaranthaceae)

Unresolved species. Mentioned by Montesinos & Zegarra (2019) for Tiabaya. The occurrence in Arequipa is rather unlikely.

Palaua malvifolia Cav. (Malvaceae)

Several specimens from Caravelí. Huertas (2010) notes that the species is restricted to Lambayeque, La Libertad and Lima and that the specimens for Arequipa are based on misidentified specimens of P. tomentosa.

Panicum miliaceum L. (Poaceae)

No voucher. Montesionos & Zegarra (2021) mention it as introduced for Arequipa, the corresponding pictures are ambiguous. An occurrence in Arequipa is possible.

Parastrephia lepidophylla (Wedd.) Cabrera (Asteraceae)

Species of Bolivia, N-Chile and NW-Argentina. Synonymized with P. quadrangularis by Nesom (1993), the synonymy is not accepted by POWO (2024) and Hassler (2024).

Paronychia microphylla Phil. (Caryophyllaceae)

P. microphylla var. arequepensis was renamed to P. arequipensis by Iamonico & Montesinos-Tubée (2023).

Passiflora serratodigitata L. (Passifloraceae)

Species of the E-Andes. Mentioned by Brako & Zarucchi (1993).

Peperomia inaequalifolia Ruiz & Pav. (Piperaceae)

Species from N- and C-Peru. Specimen Worth & Morrison 15670 from Atiquipa was identified as P. inaequalifolia by Hutchison in 1982 and re-identified as P. galioides by Mathieu in 2005.

Peperomia scutellifolia Ruiz & Pav. (Piperaceae)

Unresolved species, probably endemic to Ancash or a synonym of P. peruviana. Mentioned for Arequipa by Brako & Zarucchi (1993).

Phyla betulifolia (Kunth) Greene (Verbenaceae)

The species is restricted to the humid areas of South America. No voucher. Mentioned for Arequipa by McBride (1938ff.) and Brako & Zarucchi (1993), with the remark "Moldenke" in Flora of Peru (1936ff.).

Plantago hispidula Ruiz & Pav. (Plantaginaceae)

Restricted to N- & C-Chile. Specimen Vargas 8454 (CUZ) from Mollendo looks like P. limensis. No other vouchers and mentions.

Poa calycina (J.Presl) Kunth (Poaceae)

No specimen, two doubtful observations in iNaturalist (2024).

Poa fibrifera Pilg. (Poaceae)

Sylvester et al. (2016b) declare it endemic to N- and C-Peru and note that the species is often confused.

Poa lilloi Hack. (Poaceae)

Reports of Poa lilloi and P. supina for Peru are considered erroneous (Sylvester et al., 2016b).

Polypsecadium llatasii (Al-Shehbaz) Al-Shehbaz (Brassicaceae)

Specimen Vargas 10917 (CUZ) from Atiquipa was transferred to Mostacillastrum ferreyrae. No other vouchers or mentions.

Pseudognaphalium luteoalbum (L.) Hilliard & B.L.Burtt (Asteraceae)

Specimen Rose & Rose 18828 (NY) from 1914, not seen. Mentioned for Arquipa by Brako & Zarucchi (1993). No modern voucher and no mention.

Punotia lagopus (K.Schum.) D.R.Hunt (Cactaceae)

No voucher. Mentioned for Arequipa by Brako & Zarucchi (1993). Not present in Arequipa, all observations are likely misidentifications of A. floccosa (M. Lowry, pers. comm.).

Puya densiflora Harms

Several observations in iNaturalist (2024) and mentioned for Arequipa, no specimen. Observations from Arequipa are Puya colca-canyonensis.

Puya weberbaueri Mez (Bromeliaceae)

Often mentioned for Arequipa, probably all mentions are confusion with Puya colca-canyonensis.

Pyrolirion tubiflorum (L'Hér.) M.Roem. (Amaryllidaceae)

No voucher. Mentioned as a cultivated herb for Arequipa by Brako & Zarucchi (1993). A species of the E-slopes of the Andes.

Ranunculus flagelliformis Sm. (Ranunculaceae)

No voucher. Mentions for Arequipa, probably confusion with Halerpestes uniflora.

Rockhausenia caespitosa (Wedd.) D.J.N.Hind (Asteraceae)

No voucher. Mention for RN Salinas y Aguada Blanca by Quipuscoa & Huamantupa (2010), may be a confusion with W. apiculata. Probably not present in Arequipa.

Rockhausenia glaberrima (Phil.) D.J.N.Hind (Asteraceae)

Species restricted to Bolivia and N-Chile. The collections from S-Peru identified as W. glaberrima by Beltrán (2017) should be referred to W. orbignyana (Hind, 2022).

Salicornia andina Phil. (Amaranthaceae)

Alonso et al. (2017) rename the Peruvian clades from S. andina to S. cuscoensis.

Salpichroa weberbaueri Dammer (Solanaceae)

Species of C-Peru, excluded for Arequipa by Gonzáles (2021).

Senecio attenuatus Sch.Bip. (Asteraceae)

Restricted to Bolivia. Specimens Montesinos 5056, 5171, 5353 (all B) are Senecio reicheanus. Mentioned for Arequipa by Brako & Zarucchi (1993) and Quipuscoa & Huamantupa (2010), probably also misidentifications.

Senecio bolivarianus Cuatrec. (Asteraceae)

Species of C-Peru. Specimen van der Werff H. 20632 (MO) labeled as cf. (not seen).

Senecio ferreyrae Cabrera (Asteraceae)

Species from C-Peru. Specimen Arenas P. 85 (MO) from Arequipa (not seen) is probably a mistake.

Senecio flaccidifolius Wedd. (Asteraceae)

Specimen Vargas 22847 (CUZ) not curated. Mention Rodriguez Diaz (1998) listed as affinis.

Senecio laricifolius Kunth (Asteraceae)

Endemic to N-Peru. Mentioned for Arequipa by Brako & Zarucchi (1993). Several specimens from 1900 to 1950, no modern vouchers.

Senecio mathewsii Wedd. (Asteraceae)

Species of the humid C-Andes of Peru. No voucher, mentioned for RN Salinas y Aguada Blanca by Quipuscoa & Huamantupa (2010)

Senecio octophyllus Sch.Bip. & Rusby (Asteraceae)

Specimen Pennell 13296 (F) from Nevado Chachani is probably S. crassilodyx. Mentioned by Brako & Zarucchi (1993).

Senecio weberbaueri Cuatrec. (Asteraceae)

Only one collection, Weberbauer 6618, Rio Comas in Junin. Brako & Zarucchi (1993) have erroneously assigned this specimen to Arequipa (pers. comm. H. Bertran).

Sida decandra R.E.Fr. (Malvaceae)

Species of Ecudor and N-Peru. Specimen Dillon et al. 4833 (F) is probably Sida oligandra.

Sigesbeckia orientalis L. (Asteraceae)

Specimen Tovar 2720 (MO) from Atiquipa is probably S. jorullensis.

Solanum physalifolium Rusby (Solanaceae)

Species of the E-slopes of the Andes. Often confused with S. nitidibaccatum, a species from temperate South America. Specimen Ellenberg 77 from rio Chili is probably N. nitidibaccatum.

Spergularia media (L.) C.Presl (Caryophyllaceae)

No voucher. A dubious observation from Chiguata in iNaturalist, which may be a confusion with S. fasciculata.

Spergularia stenocarpa (Phil.) I.M.Johnst. (Caryophyllaceae)

Species of C-Chile. Specimen Weigend & Förther 97/725 (F) from Caravelí looks different from the type and is probably S. collina.

Spermacoce tenuior L. (Rubiaceae)

Species of N- and E-South America. Specimen FLSP 2093 (HUSA) from Lomas de Atiquipa and FLSP 2049 (CUZ, MO), both not seen. Probably confused with Richardia lomensis.

Stellaria pedunculosa (Wedd.) Montesinos & Borsch (Caryophyllaceae)

Specimens from Moquegua by Montesinos and a doubtful observation in iNaturalist (2024) with obtuse leaves from Arequipa.

Stevia mandonii Sch.Bip. (Asteraceae)

No voucher. Mentioned for Arequipa by Quipuscoa & Huamantupa (2010), not confirmed by Bedoya-Cuno, Dillon & Quipuscoa (2024).

Stevia ovata Willd. (Asteraceae)

No observation. Mentioned for Atiquipa by Talavera et al. (2017) and RN de Salinas y Aguada Blanca by Quipuscoa & Huamantupa (2010), not confirmed in Bedoya-Cuno, Dillon & Quipuscoa (2024).

Stevia puberula Hook. (Asteraceae)

No voucher. Mentioned for Arequipa by Brako & Zarucchi (1993), not confirmed by Bedoya-Cuno, Dillon & Quipuscoa (2024).

Tagetes laxa Cabrera (Asteraceae)

Species of Bolivia and NW-Argentina. Mentioned for RN Salinas y Aguada Blanca by Quipuscoa & Huamantupa (2010).

Tarasa rhombifolia Krapov. (Malvaceae)

Endemic to Cuzco. Specimen Werff 20531 (MO) from Yura is probably a mistake.

Tigridia pavonia (L.f.) Redouté (Iridaceae)

Species of Mexico to Honduras. Specimens Raimondi 11652 (USM) and Ferreyra 11929 (USM) were transferred to Tigridia raimondii a new species described by Ravenna (1988). The specimen Schmidt s.n. (F) from Chala corresponds to the description of T. raimondii. Montesinos-Tubée, Pauca & Revilla (2016) exclude the species for Arequipa. Still mentioned for the Lomas de Yuta by Quipuscoa et al. (2016) and for the lomas of Peru by Dillon et al. (2011). Either a cultivated/introduced species or confused with T. raimondii.

Tillandsia myosura Griseb. ex Baker (Bromeliaceae)

Species of Bolivia and NW-Argentina. No voucher. Mentioned for Arequipa by Brako & Zarucchi (1993).

Tillandsia oroyensis Mez (Bromeliaceae)

Species of C- and N-Peru. No voucher. Mentioned for Arequipa by Brako & Zarucchi (1993).

Tillandsia straminea Kunth (Bromeliaceae)

Species of Ecuador and N-Peru. Mentioned by Brako & Zarucchi (1993).

Tiquilia nesiotica (J.T.Howell) A.T.Richardson (Boraginaceae)

Endemic species of the Galapagos Islands. No voucher. Mentioned for Arequipa by Brako & Zarucchi (1993).

Urtica magellanica Juss. ex Poir. (Urticaceae)

Urtica magellanica, a species of Patagonia, is often erroneously employed for U. leptophylla and U. urens (Weigend et al., 2005).

Valeriana clematitis Kunth (Caprifoliaceae)

Specimen from La Unión (Vargas 19389 (CUZ), labeled as aff., confusion with V. pinnatifida? Species of the E-slopes of the Andes, no contemporary voucher.

Varronia curassavica Jacq. (Boraginaceae)

Specimens from Arequipa are no longer listed under this name in GBIF(2024) and tropicos.org (2024).

Varronia cylindristachya Ruiz & Pav. (Boraginaceae)

pecies of the wet tropics. Specimen Weberbauer 1530 (F) is a fragment of a leaf. Probably a confusion.

Varronia spinescens (L.) Borhidi (Boraginaceae)

No voucher. Mentioned by Brako & Zarucchi (1993), the species grows on the E-slopes of the Andes.

Verbena bonariensis L. (Verbenaceae)

Mentioned for Arequipa by Brako & Zarucchi (1993) with the comment "According to Yeo (1990), V. bonariensis probably does not occur in Peru". In the last 30 years, the V. bonariensis complex has been studied in detail by Binder (2002), Nesom (2010), O'Leary et al. (2014), and others. All of them excluded the species for Peru.

Viburnum tridentatum Killip & A.C.Sm. (Viburnaceae)

Unresolved species, known only from the type locality at Vitoc (Junín) and mentioned for Arequipa in Flora of Peru (1936ff.) and by Brako & Zarucchi (1993). McBride may have confused the type locality Vitoc (Junin) with Vitor (Arequipa). Probably a synonym of V. triphyllum. Not mentioned by Landis et al. (2021).

Villanova titicacensis (Meyen & Walp.) Walp. (Asteraceae)

No voucher, mentioned by Brako & Zarucchi (1993), probably confusion with V. oppositifolia.

Wedelia hoffmanniana Bruns (Asteraceae)

Unresolved name, no voucher, mentioned only once in 1923 for the type locality in Mejia, specimen Guenther & Buchtien 47 (F), not seen.

Appendix 7: Taxonomic key for the species of Arequipa.

The taxonomic key is an artificial key. Specific taxonomic terms have been omitted so that a basic knowledge of botany is sufficient to use the key. Wherever possible, only characters that can be seen in the field by eye or with a magnifying glass have been included.

The key was adapted from existing keys. If no keys were available or the information was contradictory, the key was made on the basis of the species descriptions and/or the type specimens.

The development of the key to the angiosperm families was a major challenge. Several taxonomic guides were consulted and adapted to the plant families found in Arequipa. Despite these efforts, some inconsistencies are to be expected in this part of the key.

The keys for the genera Cinnagrostis (Poaceae), Dalea and Lupinus (both Fabaceae), Nassella (Poaceae), Nolana (Solanaceae), and Nototriche (Malvaceae) are tentative and provisional because available information is scarce and often inconsistent.

The taxonomic key will also be published as a booklet for use in the field. It is designed to accompany the illustrated Flora of Arequipa, Edition 2, which will be published soon.

Key to divisions

1	Plants that reproduce by spores. → Key flowering plants p. 1	Pteridophyta
1*	Plants that reproduce by seeds. → Key flowering plants p. 4	Spermatophyta

Key ferns and allies (Pteridophyta)

	Key based on Jepson (2024) and Cano Bellido (2006)	
1	Herbs, aerial shots articulate, hollow. Leaves small, scale-like, not photosynthetically active. Leaves and lateral branches whorled. Sporangia in terminal strobili.	Equisetaceae
1*	Aquatic or terrestrial herbs, shots not articulate. Lamina green, photosynthetically active. Sporangia on the underside of leaves.	2
2	Floating on water, slightly emerging from the water surface. Pinnately branched. Leaves reddish-green.	Salviniaceae
2*	Terrestrial or epiphytic ferns.	3
3	Leaf divided into dissimilar sterile and fertile parts.	Ophioglossaceae
3*	Leaf not divided into dissimilar sterile and fertile parts, sporangia borne on underside of leaf blade.	4
4	Pinnae arranged in horizontal layers, 5-10 cm or more apart. Frond often >80 cm, sometimes up to 2 m, 2-pinnate-pinnatifid.	Dennstaedtiaceae
4	Distance between pinnae <5 cm. Frond <80 cm (except for Amauropelta, but pinnae are close together in this genus).	5
5	Sori borne along or near margins of leaf. Indusia absent or sori covered by reflexed or recurved leaf margins or leaf margins segment.	Pteridaceae
5*	Sori borne away from margins on underside of pinnae, or sporangia scattered along veins and not clustered into sori.	6
6	Sporangia scattered along veins, not clustered into distinct sori.	Pteridaceae
6*	Sporangia clustered in distinct sori.	7
7	Sori round or moon shaped.	8
7*	Sori linear.	12
8	Blade entire or 1-pinnatifid.	Polypodiaceae
8*	Blade 2-pinnatifid.	9
9	Tall fern, >30 cm.	10
9*	Small fern, <30 cm.	11
10	Blade herbaceous, pinnule obtuse to acute.	Thelypteridaceae
10*	Blade leathery, base of pinnule shaped like a thumbs-up sign, apex acuminate, slightly spiny.	Dryopteridaceae
11	Fragile fern of humid places.	Cystopteridaceae
11*	Robust fern on rocks, walls and cliffs.	Woodsiaceae
12	Sori along the central vein of the pinnula.	Blechnaceae
12*	Sori along the lateral veins of the pinnula.	Aspleniaceae

Aspleniaceae

Asplenium

1	Pinnae usually trifoliolate or 2-foliolate, or occasionally with 2 pairs of lateral pinnules.	A. triphyllum
1*	Pinnae monofoliate.	2
2	Pinnae incised to the main vein.	A. praemorsum
2*	Pinnae dentate or lobed, but not incised as far as the main vein.	3
3	More than 30 pairs of pinnae, ladder-like appearance.	A. monanthes
3*	Less than 30 pairs of pinnae.	4
4*	Pinnae rectangular. Pinna base-angle approx. 45°.	A. lorentzii
4	Pinnae rhombic. Pinna base-angle >45°.	5
5	Rachis flexuous, green.	A. gilliesii
5*	Rachis stiff, reddish or grayish-brown.	A. peruvianum
	Note: Asplenium gilliesii and A. peruvianum may be variants with intermediates or conspecific.	

Blechnaceae

1	Sori along the central vein of the pinnula.	Blechnum occidentale

Cystopteridaceae

1	Fragile fern of humid places.	Cystopteris diaphana

Dennstaedtiaceae

1	Pinnae arranged in horizontal layers, 5-10+ cm apart. Frond often >80 cm, sometimes up to 2 m, 2-pinnate-pinnatifid.	Pteridium esculentum

Dryopteridaceae

1	Blade leathery, base of pinnule shaped like a thumbs-up sign.	Polystichum orbiculatum

Equisetaceae

Equisetum

1	Small horsetail, stems 2-3 mm in ∅.	E. bogotense
1*	Giant horsetail, 2-4 m tall, stems up to 40 mm in ∅.	E. xylochaetum

Ophioglossaceae

1	Blade divided into dissimilar sterile and fertile parts, narrowly elliptic.	Ophioglossum crotalophoroides

Polypodiaceae

Pleopeltis

1	Blade entire, narrowly elliptic.	P. macrocarpa
1*	Blade pinnatifid to pinnatisect.	P. pycnocarpa

Pteridaceae

1	Sporangia scattered on underside of pinnula, pseudoindusia absent.	2
1*	Sporangia marginal, often covered by a revolute margin (pseudoindusium).	3
2	Blade 1-pinnate. Pinnules lobed, distinctly petiolate.	Astrolepis sinuata
2*	Blade 1-2-pinnate. Pinnules linear, all divided from the base of the pinna, appearing 2-3-foliate.	Pityrogramma trifoliata
3	Pinnules reniform or flabellate, herbaceous. Rachis blackish, almost glabrous, glossy.	Adiantum
3*	Pinnules deltoid, usually leathery. Rachis scaly or pubescent.	4
4	Underside of blade glabrous or farinose.	5
4*	Underside of blade pubescent, scaly or with exudate.	6
5	Underside of blade farinose, upper side glabrous. Sporangia along the veins in the outer third of the pinnula.	Argyrochosma nivea
5*	Both sides of blade glabrous.	Pellaea
6	Blade pentagonal, abaxially densely covered with white or yellow exudate.	Notholaena sulphurea
6*	Blade deltoid, abaxially with scales, hairs, or glands.	7
7	Blades 2-3-pinnate. 32 small and 16 large spores per sporangium.	Cheilanthes
7*	Blade 1- or 4-pinnate (magnifier!). 64 small and 32 large spores per sporangium.	Myriopteris
	Note: Whether the subfamily Cheilanthoidea (Agryochosma, Astrolepis, Cheilanthes, Myriopteris, Notholaena, Pellaea) consists of only a single genus Hemionitis or of 14 smaller genera is currently under debate	

Adiantum

1	Pinnules reniform, beneath sparsely hairy.	A. chilense
1*	Pinnules flabellate, glabrous.	A. subvolubile

Cheilanthes

1	Fronds with dense indument, scales on the lamina and axes, or if with hairs, these are scattered to dense on the adaxial surface of the blade.	2
1*	Fronds glabrous, glabrescent or with indument of hairs on the lamina and axes, sometimes scattered scales on axes.	4
2	Blade 2-3-pinnate in the basal pinnae. Ultimate segments ovate-triangular, flabellate.	C. incarum
2*	Blade 2-pinate-pinnatisect or 2-pinnate-pinnatifid. Ultimate segments oblong-triangular.	3
3	Blade surface above glabrous or occasionally with inconspicuous scales. Fertile segment margins not distinctive. Pseudoindusium hyaline.	C. arequipensis
3*	Blade surface above mostly with microscales. Fertile segment margins distinctive. Pseudoindusium membranous.	C. peruviana
4	Hairs of blade branched, sparse above, dense beneath.	C. mollis
4*	Hairs of blade simple, dense or absent.	5
5	Blade and axis with eglandular hairs.	C. fractifera
5*	Blade and axis glabrous or with glandular hairs.	6
6	Blade subleathery. Fronds glutinous, glandular hairs with dark yellow secretion.	C. pruinata
6*	Blade herbaceous to papery. Fronds glutinous without dark yellow secretion.	C. pilosa

Myriopters

1	Blade 4-pinnate (magnifier!).	M. myriophylla
2	Blade 1-pinnate, pinnula deeply lobed, sessile to very short petiolate.	M. aurea

Pellaea

1	Blade 1-pinnate. Pinna ternate or entire, sessile, or subsessile.	P. ternifolia
1*	Blade 2-pinnate. Pinnula deltoid, sagittate.	P. sagittata

Salviniaceae

1	Floating on water, slightly emerging from the water surface. Pinnately branched.	Azolla filiculoides

Thelypteridaceae

Amauropelta

1	Veins beneath sunken. Midvein of pinna with red globose glands.	A. glandulosolanosa
1*	Veins beneath risen. Midvein of pinna without red globose glands. Indusia smaller.	A. rufa
	Note: A. glandulosolanosa and A. rufa grow together, probably hybridize, and may be conspecific. The occurrence of A. rufa in Arequipa is doubtful.	

Woodsiaceae

1	Robust fern on rocks, walls and cliffs.	Woodsia montevidensis

Key seed plants (Spermatophyta)

1	Obligate aquatic plants, whole plant floating or submersed in water. Inflorescences sometimes rising above the water surface.	Subkey 1: Aquatic plants
1*	Terrestrial plants, most parts not floating or submersed in water, but sometimes growing in shallow water.	3
2	Middle-sized to tall trees, usually >5 m high.	Subkey 2: Trees >5 m
2*	Herbs, shrubs or small trees, usually <5 m tall.	4
3	Parasitic plants without chlorophyll (not green), and usually growing on other plants, or root parasites with chlorophyll.	Subkey 3: Parasitic plants
3*	Plants not parasitic.	5
4	Plants almost leafless, at least when flowering, or stem-succulents.	Subkey 4: Leafless plants
4*	Plants with leaves when in flower.	6
5	Vines and lianas, twining or climbing with tendrils or prickles on other plants.	Subkey 5: Vines and lianas
5*	Not climbing plants.	5
6	Leaves with parallel veins. Leaves entire, margins entire.	Subkey 6: Herbs with parallel-veined leaves
6*	Leaves with reticulate veins, sometimes veins parallel, but then the veins connected by veinlets. Leaves entire, incised or compound, margins entire or not.	7
7	Perianth absent, or with either petals or sepals, or with petals and sepals, but then petals and sepals not or only slightly different colored.	Subkey 7: Plants with reticulate veins and simple perianth
7*	Perianth clearly composed of petals and sepals, sometimes the outer calyx strongly modified. Flowers sometimes densely packed in heads or corolla composed of different layers.	Subkey 8: Plants with reticulate veins and composed perianth

Subkey 1: Aquatic plants

1	Plant floating on water surface, without rooting in soil.	2
1*	Plant submersed, usually rooting in soil. Inflorescences sometimes rising above the water surface.	3
2	Small elliptic to ovate leaves (=fronds) 2-3 mm in Ø. Minute, inconspicuous flowers.	Araceae (Lemna)
2*	Leaves taller than 5 mm in Ø, petiole inflated. Flowers showy, whitish-blue.	Pontederiaceae
3	Leaves deeply incised.	4
3*	Leaves entire, usually linear, sometimes ovate.	6
4	Leaves pinnate, lobes linear.	Haloragaceae
4*	Leaves forked or palmately incised.	5
5	Leaves whorled. Flowers inconspicuous.	Ceratophyllaceae
5*	Leaves alternate. Flowers white, arising above the water surface.	Ranunculaceae
6	Leaves elliptic, ovate to spatulate.	7
6*	Leaves linear, entirely submersed.	8
7	Leaves ovate to spatulate, forming a floating terminal rosette on the water surface.	Plantaginaceae (Callitriche)
7*	Leaves cordate, single leaves floating on the water surface. Flowers yellow.	Ranunculaceae (Halerpestes)
8	Inflorescence axillary.	Potamogetonaceae
8*	Inflorescence terminal.	9
9	Leaves <1mm wide. Stipules entirely fused to blade, sheath-like.	Ruppiaceae
9*	Leaves >5 mm wide.	Hydrocharitaceae
1	*Semi-aquatic plants of inundated areas and bofedales: Apiaceae (Lilaeopsis), Brassicaceae (Nasturtium), Cyperaceae (several genera), Juncaceae (several genera), Juncaginaceae (Triglochin), Poaceae (Agrostis, Phragmites), Plantaginaceae (Veronica), Rosaceae (Lachemilla), Salviniaceae (Azorella), Scrophulariaceae (Limosella), Typhaceae.*	

Subkey 2: Trees >5 m

1	Palm, single-stemmed, unbranched. Leaves palmate.	Arecaceae
1*	Branched trees or columnar cacti.	2
2	Columnar cacti. Leafless, spiny, succulent.	Cactaceae
2*	Branched trees or treelets.	3
3	Fruit a legume (pods opening on both sides). Leaves usually pinnate. Flowers bean- or acacia-like.	Fabaceae
3*	Not in above combination.	4
4	Leaves pinnate.	5
4*	Leaves entire.	8
5	Leaves villous-lanate, at least beneath.	Rosaceae (Polylepis)
5*	Leaves glabrous.	6
6	Fruits 2 cm in Ø.	Sapindaceae (Sapindus)
6*	Fruits <1 cm in Ø.	7
7	Inflorescence a hanging panicle. Flowers greenish-yellow. Berries pink-red.	Anacardiaceae
7*	Inflorescence an erect cyme. Flowers white, fragrant. Berries purple-black.	Viburnaceae
8	Flowers solitary or in clusters.	9
8*	Flowers in catkins or enclosed in an urn-shaped tissue (fig), inconspicuous.	14
9	Significantly more stamens than petals.	10
9*	Stamens as much as petals.	11
10	Stamens numerous, bundled.	Myrtaceae
10*	Stamens 2-3x more than petals.	Rosaceae
11	Fruit <2 cm in Ø. Leaves <3 cm wide, sessile or base attenuate.	12
11*	Fruit >5 cm in Ø. Leaves >5 cm wide, base obtuse, truncate or slightly cordate, petiole distinct.	13
12a	Leaves crowded, glaucous. Flowers in white. Capsule 3-valved, seeds orange.	Celstraceae
12b	Inflorescence terminal, 6+ flowered. Flowers yellow to orange.	Scrophulariaceae (Buddleja)
12c	Flowers axillary, in clusters of <5. Flowers white. Leaves glabrous heavily spotted with translucent dots.	Scrophulariaceae (Myoporum)
13	Flowers 3-merous. Fruits globose, 10-15 cm in Ø, green, flesh white with black seeds.	Annonaceae
13*	Flowers 4 or 5-merous. Fruits elongate 10-18 x 5 cm, yellow-green, flesh orange (papaya-like)	Caricaceae
14	Flowers enclosed in a fig.	Moraceae
14*	Flowers in catkins.	15
15	Leaves ovate, pinnately veined, margins serrate.	Betulaceae
15*	Leaves linear-lanceolate. Often growing in riparian forests.	16
16	Fruit a globose drupe covered with waxy and verrucous papillae.	Myricaceae
16*	Fruit a dark ovoid capsule, 4-5 mm long, with numerous very small and long hairy seeds.	Salicaceae
	Cultivated trees in forests or in fields near settlements (not included in the key): *Arecaceae (Phoenix), Cupressaceae (Cupressus), Myrtaceae (Eucalyptus), Pinaceae (Pinus), Sapotaceae (Pouteria) and others.*	

Subkey 3: Parasitic plants

1	Parasitic plants without chlorophyll (not green).	2
1*	Green plants, sometimes reddish-green, hemiparasites.	3
2	Vines, parasitizes stems and leaves with haustoria (special organs that connect the parasite to the host plant)	Convolvulaceae (Cuscuta)
2*	Terrestrial plants, root parasites.	Orobanchaceae (Orobanche)
3	Root parasites. The parasitic habit is not obvious and can be recognized when the roots of the plant are connected with haustoria in the roots of other plants.	4
3*	Epiphytic shrubs, usually parasitizing shrubs or trees.	6

4	Shrub up to 1 m tall. Flowers 4-merous, pink to red, zygomorphic. Leaves green, ovate, acuminate, margins entire.	Krameriaceae
4*	Herbs.	5
5	Leaves usually green, laminar, margins crenate. Corolla zygomorphic.	Orobanchaceae
5*	Leaves reddish-green or brownish-gray, linear, margins entire. Corolla actinomorphic.	Schoepfiaceae
6	Flowers showy, petals yellowish-red, 30 mm long.	Loranthaceae
6*	Flowers inconspicuous, tepals whitish, 1 mm long.	Santalaceae
	Note: The epiphytic Tillandsia (Bromeliaceae) are not parasites.	

Subkey 4: Leafless plants

1	Cacti. Leafless (some Opuntioideae with early deciduous leaves), spiny, succulent.	Cactaceae
1*	Herbs and shrubs.	2
2	Plant without chlorophyll, not green.	3
2*	Green herbs and shrubs.	4
3	Parasitic vines, leaf and stem parasites.	Convolvulaceae (Cuscuta)
3*	Erect plant, root parasites.	Orobanchaceae (Orobanche)
4	Stems articulate. Plant apparently leafless, leaves as small scales in the nodes.	5
4*	Stems not articulate.	7
5	Seeds exposed, covered by bracts, but not fruits (gymnosperms). Seeds covered by 6-9 fleshy, red bracts, cone-like.	Ephedraceae
5*	Seeds enclosed within a fruit.	6
6	Shrub up to 2 m tall. Stems terete, grayish-green, leafless after flowering. Flowers yellow. Fruits alate, lantern-like.	Zygophyllaceae (Bulnesia)
6*	Plant of saline habitats.	Amaranthaceae (Salicornia)
7	Thorny shrubs. Leaves early caducous, therefore often seeming leafless.	Rhamnaceae
7*	Herbs.	8
8	Bulbous herbs. Leaves appearing after flowering.	Amaryllidaceae (Clinanthus)
8*	Grass-like herbs. Leaves reduced to sheaths or absent.	9
9	Stem ± triangular in cross section.	Cyperaceae
9*	Stem round in cross section.	Juncaceae
	Cultivated near settlements and as fence: Euphorbia ingens (Euphorbiaceae), cactus-like, with milky sap (not included in the key).	

Subkey 5: Vines and lianas

1	Flowers zygomorphic.	2
1*	Flowers actinomorphic.	5
2	Leaves parallel-veined, in a resupinate position (turned upside-down). Flowers pink to red, in umbels.	Alstroemeriaceae (Bomarea)
2*	Leaves reticulate-veined.	3
3	Pea-like flowers, consisting of a banner, a keel and two lateral wings.	Fabaceae (Vicia)
3*	Flowers not pea-like.	4
4	Leaves opposite. Flowers in pairs.	Caprifoliaceae (Lonicera)
4*	Leaves alternate. Flowers solitary.	Tropaeolaceae
5	Plant climbing with prickles.	Rosaceae (Rubus)
5*	Plant climbing with tendrils or twining.	6
6	Flowers in racemes or cymes.	7
6*	Flowers solitary.	10
7	Flowers in axillary racemes. Petals outside red. Leaves cordate.	Basellaceae
7*	Flowers in cymes.	8
8	Leaves opposite.	Apocynaceae

8*	Leaves alternate.	9
9	Flowers without inner corolla.	Cucurbitaceae
9*	Flowers with inner corolla.	Sapindaceae
10	Plant with stinging hairs.	Loasaceae (Caiophora)
10*	Plant without stinging hairs.	11
11	Petals free or absent, then sepals free.	12
11*	Petals fused.	13
12	Tendrils borne from stipules. Showy flowers (Passionflower). Fruit indehiscent.	Passifloraceae (Passiflora)
12*	Tendrils borne from leaf-apex. Flowers white to yellow, petals absent. Fruit an aggregate of achenes.	Ranunculaceae (Clematis)
13	Plant twining.	Convolvulaceae
13*	Plant climbing with tendrils.	Cucurbitaceae
	Note: Some Mutisia sp. (Asteraceae) form tendrils from the leaf apex to hold themselves upright, they are not classified here as lianas or vines.	

Subkey 6: Herbs with parallel-veined leaves

1	Perianth lacking or consisting of bristles and hairs.	2
1*	Perianth of at least two 3-merous circles.	4
2	Stem round, with nodes, hollow between the nodes.	Poaceae
2*	Stem triangular, at least at base, without nodes, firm.	3
3	Inflorescence a cob.	Typhaceae
3*	Inflorescence consisting of 1 to several spikes, not cob-like.	Cyperaceae
4	Perianth strongly reduced, usually <5 mm, yellowish-green or brownish, often dry membranous.	5
4*	Perianth conspicuous, colored or white, usually >5mm.	6
5	Inflorescence a corymb.	Juncaceae
5*	Inflorescence a raceme.	Juncaginaceae
6	Stamens 1 or 6 but only 1 fertile.	7
6*	Stamens 3 or more.	8
7	Stamens 1.	Orchidaceae
7*	Stamens 6, but only 1 fertile, the fertile stamen petal-like. Flowers showy, petals red or yellow, spotted.	Cannaceae
8	Stamens 3. Ovary inferior.	Iridaceae
8*	Stamens 6. Ovary inferior or superior.	9
9	Flowers with 3 petals and 0-3 sepals, petals different to sepals.	10
9*	Flowers with 6 tepals, petals and sepals ± identical.	12
10	Flowers terminal, blue, enclosed in well-developed bracts. Petals shortly nailed.	Commelinaceae
10*	Flowers axillary or terminal, not enclosed by bracts, not blue.	11
11	Inflorescence a panicle. Leaves herbaceous, cauline, resupinate (upside-down).	Alstroemeriaceae (Alstroemeria)
11*	Inflorescence a spike or a panicle. Leaves usually gray scaled and usually in a basal rosette (Tillandsia usneoides hanging without rosette), firm, often armed with marginal spikes. Epiphytic or terrestrial herbs.	Bromeliaceae
12	Leaves in a basal rosettes, succulent, triangular to lanceolate, margins armed with spines.	13
12*	Leaves linear, margins not armed.	14
13	Inflorescence >3 m long. Flowers yellowish-white.	Asparagaceae (Agave, Furcraea)
13*	Inflorescence <1 m long. Flowers usually yellow, sometimes orange-red.	Asphodelaceae (Aloe)
14	Shrubs, much-branched. Flowers solitary in leaf-axils. Leaves	Asparagaceae (Asparagus)
14*	Bulbous herbs.	15
15	Flowers usually a terminal umbel on a long scape (Clianthus sometimes with solitary flowers).	Amaryllidaceae

15*	Inflorescence corymbose, racemose or paniculate.	16
16	Petals blue.	Asphodelaceae (Pasithea)
16*	Flowers white or yellowish-white.	Asparagaceae
	Note: Plantago (Plantaginaceae) also has leaves with parallel veins, but the secondary veins are reticulated.	

Subkey 7: Plants with reticulate veins and simple perianth

1	Inflorescence a cob, surrounded by a spathe. Leaves large, base sagittate.	Araceae
1*	Inflorescence not a cob, without spathe.	2
2	Fruit a trigonous-globose capsule.	3
2*	Fruit of different shape.	4
3	Plants with milky latex (aqueous in Ricinus).	Euphorbiaceae
3*	Plants lacking latex.	Phyllanthaceae
4	Inflorescence a long, thin spike.	Piperaceae
4*	Inflorescence not a long thin spike.	5
5	Leaves orbicular, peltate (leaves of Hydrocotyle ranunculoides are not peltate, but incised to the base of the petiole in the middle of the leaf).	Araliaceae
5*	Leaves not peltate.	6
6	Perianth sepal-like, white, green or brownish.	7
6*	Perianth petal-like, conspicuously colored or at least not entirely white, green or brownish.	14
7	Leaves opposite or verticillate.	8
7*	Leaves alternate.	12
8	Leaves petiolate.	9
8*	Leaves sessile.	10
9	Plant with stinging hairs.	Urticaceae (Urtica)
9*	Plant without stinging hairs.	Amaranthaceae
10	Leaves opposite.	11
10*	Leaves verticillate.	Molluginaceae
11	Flowers pedicellate.	Caryophyllaceae
11*	Flowers sessile in globose heads.	Amaranthaceae
12	Leaves sessile.	Caryophyllaceae
12*	Leaves petiolate.	13
13	Weak herbs. Stems break easily.	Urticaceae
13*	Strong herbs or shrubs. Stems do not break easily.	Amaranthaceae
14	Shrubs or treelets.	15
14*	Herbs.	16
15	Thorny shrubs. Leaves early caducous, therefore often seeming leafless. Flowers in clusters, campanulate.	Rhamnaceae
15*	Treelets with imparipinnate leaves.	Rosaceae (Polylepis)
16	Flowers zygomorphic.	Papaveraceae (Fumaria)
16*	Flowers actinomorphic.	17
17	Flowers orange or yellow, >2 cm in Ø.	Papaveraceae
17*	Flowers greenish-yellow, blue or whitish-pink, <2 cm in Ø.	18
18	Leaves succulent. Flowers usually 4-merous, sometimes 5-merous.	Aizoaceae
18*	Leaves not succulent. Flowers 5-merous.	19
19	Flowers greenish-yellow. Leaves orbicular or pinnatisect, sometimes plied like an accordion, not succulent.	Rosaceae (Lachemilla)
19*	Flowers white or pinkish-white. Leaves lanceolate to oblong.	20
20	Fruits nutlets. Stipules sheathing.	Polygonaceae
20*	Fruit berries. Stipules absent.	Phytolaccaceae

Subkey 8: Plants with reticulate veins and composed perianth

1	Composite inflorescence, called a head or capitulum, which may superficially look like a single flower.	2
1*	Inflorescence not composite.	3
2	Involucral bracts different from leaves.	Asteraceae
2*	Involucral bracts leaf-like.	Calyceraceae
3	Flowers in umbels.	4
3*	Flowers not in umbels.	5
4a	Herbs or large cushion plants. Inflorescence a double umbel, in cushion plants a single umbel.	Apiaceae
4b	Inflorescence a single umbel. Plant with milky latex.	Apocynaceae (Asclepias)
4c	Inflorescence a single umbel. Fruit a capsule.	Primulaceae
4d	Inflorescence a single umbel. Fruit a schizocarp.	Geraniaceae (Erodium)
5	Fruit an inflated balloon with prickles. Seeds achenes.	Apocynaceae (Gomphocarpus)
5*	Fruit not balloon-like	6
6	Pea-like flowers, consisting of a banner, a keel and two lateral wings.	7
6*	Flowers not pea-like.	8
7	Wings borne from corolla.	Fabaceae
7*	Wings borne from calyx.	Polygalaceae
8	Stamens fused to a tube around the pistil.	Malvaceae
8*	Stamens not fused and not forming a tube around the pistil.	9
9	Petals absent, but flowers with numerous petal-like stamens. Plant succulent.	Aizoaceae
9*	Petals present.	10
10	Flowers zygomorphic.	12
10*	Flowers actinomorphic.	20
11	Petals free.	12
11*	Petals fused.	13
12a	Flowers 4-merous. Herbs. Leaves alternate, palmately compound. Flowers spider-like.	Cleomaceae
12b	Flowers 4-merous. Shrubs. Leaves alternate, entire. Flowers red. Root-parasite.	Krameriaceae
12c	Flowers 5-merous. Leaves basal, usually rosulate. Fruit a 3-valved capsule.	Violaceae
13	Leaves either orbicular and rosulate or linear and verticillate in 3, always distinctly succulent.	Crassulaceae
13*	Leaves not succulent, if ± succulent, then not orbicular or verticillate.	14
14	Fruit an achene.	Caprifoliaceae
14*	Fruits a schizocarp or a capsule.	15
15	Fruits a schizocarp.	16
15*	Fruits a capsule.	17
16	Style arises from a 4-lobed superior ovary with 4 apparent locules (like nuts). Smell of etheric oil (mint odor). Leaf margins always serrate. Stem square. Flowers strongly zygomorphic.	Lamiaceae
16*	Style attenuate from the ovary, ovary 2 locular, sometimes becoming divided by a false septum in 4-5 locules. Aromatic leaves in herbs, non-aromatic in shrubs. Stems square or irregularly angled. Flowers less zygomorphic than Lamiaceae.	Verbenaceae
17	Leaves alternate.	18
17*	Leaves opposite or rosulate.	19
18a	Ovary inferior. Corolla 2-labiate, tubular, anthers fused in a column around the stigma.	Campanulaceae
18b	Hemiparasitic herb. Inflorescence with colorful margined bracts (pink, red, lavender). Corolla 2-labiate, tubular, usually inconspicuously colored.	Orobanchaceae (Castilleja)
18c	Herbs or shrubs (Galvezia) with tubular, ± 2-labiate, red flowers and mostly 2 or 5-8 stamens.	Plantaginaceae
19a	Herbs. Strongly 2-labiate flowers. Fruit a capsule with elastic dehiscence.	Acanthaceae

19b	Herbs or shrubs. Tubular flowers, trumpet-like, bright yellow or orange. Fruit an elongated capsule.	Bignoniaceae
19c	Herbs or subshrubs. Flowers with enlarged, ± inflated, slipper-shaped lower lip. Corolla yellow or red, often red spotted inside.	Calceolariaceae
19d	Hemiparasitic herb. Corolla 2-labiate, tubular, lip trilobed, helmet often discolored to lip.	Orobanchaceae (Neobartsia)
19e	Herbs of humid places. Leaves triangular, petiole ± sheathing the stem. Flowers yellow with red spots on inferior lip.	Phrymaceae
19f	Herbs with tubular, ± 2-labiate flowers and mostly 2 or 5-8 stamens. If rosulate, then leaves with parallel veins and spicate inflorescence (Plantago).	Plantaginaceae
19g	Herb. Flowers bright orange-red.	Scrophulariaceae (Alonsoa)
20	Small herbs with terete and succulent leaves.	21
20*	Herbs or shrubs, leaves not terete and succulent.	22
21a	Flowers white or pink. Leaf apex often pointed.	Caryophyllaceae
21b	Flowers yellow, 5-merous, star-like, or reddish-white, but then 4-merous. Leaf apex obtuse.	Crassulaceae
21c	Flowers pink, magenta or deep yellow. Plant usually with stipular hairs.	Portulacaceae
22	Shrubs >50 cm tall (treelets >5 m tall see subkey 3).	23
22*	Herbs or small shrubs <50 cm tall.	26
23	Spiny or thorny shrubs.	24
23*	Not armed shrubs.	25
24a	Leaves in whorls often spinulose. Spines trifid. Flowers yellow, 6-merous, with an inner corolla.	Berberidaceae
24c	Leaves clustered in axils. Flowers with conspicuously colored bracts around the corolla.	Nyctaginaceae
24d	Leaves opposite. Flowers solitary. Fruits orbicular, yellowish, pulp black.	Rubiaceae
24e	Leaves alternate. Flowers tubular.	Solanaceae
25a	Leaves Flowers white, windmill shaped. Plant with milky latex. Berries white, translucent.	Apocynaceea (Vallesia)
25b	Leaves entire. Inflorescence scorpioid, flowers bluish-white (Heliotropium) or inflorescence a cyme with showy yellow flowers (Cordia).	Boraginaceae
25c	Leaves alternate, entire, linear-lanceolate, margins entire. Inflorescence a terminal raceme. Flowers tubular, sepals reddish, petals white, disc red.	Escalloniaceae
25d	Leaves triparted to base. Flowers solitary, yellow.	Francoaceae
25e	Leaves alternate, 3-lobed, margins serrate. Inflorescence a terminal raceme.	Grossulariaceae
25f	Leaves alternate. Flowers solitary, axillary, tubular (in Lycianthes rotate).	Solanaceae
25g	Leaves opposite, blade deeply bifid in V-shape. Flowers yellow. Capsule white-hairy.	Zygophyllaceae (Larrea)
26	Flowers 4- or 6-merous.	27
26*	Flowers 5-merous.	28
27a	Leaves alternate. Flowers 4-merous, petals free. Ovary superior. Fruit a siliqua.	Brassicaceae
27b	Leaves alternate. Flowers 4-merous, petals free. Ovary superior. Fruit a capsule.	Lythraceae
27c	Leaves opposite. Flowers 4-merous, petals fused. Ovary superior. Fruit a capsule. Calyx amplified, furrowed.	Gentianaceae (Microcala)
27d	Leaves usually alternate. Flowers 4- or 6-merous, petals free. Ovary inferior. Fruit a capsule.	Onagraceae
27e	Leaves strictly opposite or verticillate. Flowers 4- or 6-merous, petals fused. Ovary inferior. Fruit a drupe.	Rubiaceae
28	Dense cushion plant at high altitudes.	Caryophyllaceae
28*	Erect, decumbent or mat-forming plant, but not cushion plant.	29
29	Petals fused.	30
29*	Petals free.	34
30	Leaves alternate or rosulate.	31
30	Leaves opposite.	33
31	Ovary inferior	Campanulaceae (Triodanis)

31*	Ovary superior	32
32a	Leaves 3-foliate.	Oxalidaceae
32b	Either a rosulate plant with a long, multi-branched inflorescence and lavender flowers, or an erect herb with a terminal spike of lavender flowers with a conspicuous glandular calyx.	Plumbaginaceae
32c	Glutinous, pubescent annual herbs. Leaves entire or pinnate, blade and segments linear.	Polemoniaceae
32d	Semi-aquatic plant. All leaves basal, spathulate. Flowers white.	Scrophulariaceae (Limosella)
32e	Not in the above combination of characteristics.	Solanaceae
33a	Fruit a wind dispersed achene. Leaves usually deeply incised or rosulate.	Caprifoliaceae (Valeriana)
33b	Fruit a capsule. Leaves entire. Flowers bluish-white or yellow with scarlet, never pink.	Gentianaceae
33c	Fruit a nut. Leaves entire. Flowers pink or white, if white, then with whitish bracts.	Nyctaginaceae
34	Fruit a schizocarp.	Geraniaceae
34*	Fruit a capsule.	35
35	Leaves opposite	36
35*	Leaves alternate.	37
36a	Leaves entire. Flowers white.	Caryophyllaceae
36b	Small shrub. Leaves ericoid due to revolute margins. Flowers white, petals crenulate.	Frankeniaceae
36c	Leaves entire, linear lanceolate. Flowers yellow to reddish.	Hypericaceae
36d	Leaves entire. Flowers blue or red.	Primulaceae (Lysimachia)
36e	Leaves pinnate. Flowers yellow. Fruit a warty or prickly capsule.	Zygophyllaceae
37	Ovary superior.	38
37*	Ovary inferior.	39
38a	Leaves not succulent, sometimes semi-opposite. Inflorescence 3-5-flowered. Flowers yellow.	Linaceae
38b	Leaves succulent, if not then an acaulescent herb (Calandrinia). Flowers white or pinkish-red.	Montiaceae
38c	Leaves not succulent. Inflorescence many flowered. Flowers yellow.	Talinaceae
39	Stamens numerous. Flowers with inner corolla (nectar scales).	Loasaceae
39*	Stamens 5. Flower without double corolla.	Onagraceae

Acanthaceae

1	Stem terete to slightly quadrangular. Corolla 5-merous, obscurely 2-labiate.	Dyschoriste repens
1*	Stem hexagonal. Flowers 2-labiate.	Dicliptera

Dicliptera

1	Leaves ovate, 1-1.5x as long as wide. Stems whitish-tomentose.	D. tomentosa
1*	Leaves elliptic, 2-3x as long as wide. Stems not whitish.	D. congesta

Aizoaceae

	Key based on Jepson (2024)	
1	No petal-like stamens, petals 4-5.	2
1*	Many petal-like stamens.	Mesembryanthemum
2	Flowers yellow to yellowish-green.	Tetragonia
2*	Flowers green outside, pink, red or purplish inside.	Sesuvium portulacastrum

Mesembryanthemum

1	Flowers red.	M. cordifolium
1*	Flowers white.	M. crystallinum

Tetragonia

	Key based on Taylor (1994)	
1	Leaves lanceolate to ovate or triangular-hastate, 2-8 cm wide. Fruit with distal horns.	T. tetragonoides
1*	Leaves oblanceolate to obovate, elliptic or lanceolate, 0.2-2 cm wide. Fruit with sides smooth and angles rounded to sharp or narrowly winged.	2
2	Fruit with both sides and angles ridged to tuberculate.	3
2*	Fruit with sides smooth and angles rounded to sharp or narrowly winged.	4
3	Fruit 3.5-7 mm long, glabrous to pilose, moderately tuberculate to usually longitudinally ridged.	T. macrocarpa
3*	Fruit 9-10 mm long, densely pilose-villose and tuberculate.	T. pedunculata
4	Fruit rhomboidal to ellipsoid, widest at middle. Oldest leaves 1 to 2x as long as youngest leaves.	T. microcarpa
4*	Fruit ovoid or obovoid, widest at base or apex. Oldest leaves 1 to >3x as long as youngest leaves.	5
5	Plants pilose, at least on young stems and fruit.	T. vestita
5*	Plants and fruit glabrous.	6
6	Fruits obovoid, widest above middle, 6-9 mm long. Peduncles 3-8 mm long.	T. crystallina
6*	Fruits ovoid, widest below middle, 4-5 mm long. Peduncles 5-20 mm long.	T. ovata
	Note: All species except T. tetragonoides are similar and difficult to separate. Further phylogenetic analysis is needed to define species boundaries.	

Alstroemeriaceae

	Key based on Hofreiter & Rodriguez (2006))	
1	Fruit a dry explosive capsule. Plant erect. Flowers strongly zygomorphic.	Alstroemeria chorillensis
1*	Fruit a leathery, slowly dehiscent capsule or a berry. Plants erect or twining. Flowers weakly zygomorphic or actinomorphic.	Bomarea
	Note: Hofreiter & Rodriguez (2006) propose to synonymize A. lineatiflora Ruiz & Pav. and the Peruvian population of A. violaceae Phil. with A. chorillensis.	

Bomarea

	Key based on Hofreiter & Rodriguez (2006))	
1	Outer tepals yellow-green.	B. involucrosa
1*	Outer tepals pink.	2
2	Leaves 4-8 cm wide. Restricted to the lomas.	B. latifolia
2*	Leaves <3 cm wide.	3
3	Inner tepals pink, with green tip. Ovary semiinferior.	B. dulcis
3*	Inner tepals yellow, with green tip and round dark spots. Ovary inferior.	B. ovata

Amaranthaceae

	Key based on Jepson (2024)	
1	Stem apparently leafless. Leaves connate ensheathing completely the internodes. Plant of coastal salt marshes.	Salicornia
1*	Leaves well developed, not connate.	2
2	Leaves alternate.	3
2*	Leaves opposite.	7
3	Leaves ± semicylindric or linear, ± fleshy.	5
3*	Leaves flattened, not or weakly fleshy.	6
4	Leaves green, linear, 1-1.5 cm long, apex acuminate, forming a rather firm spine, 1-1.5 mm long.	Salsola kali
4*	Leaves glaucous, semiterete, 5-8 mm long, apex obtuse or acute, without spine.	Suaeda foliosa
5	Two subtending bracts around flower, enlarged in fruit (fruiting bracteoles).	Atriplex
5*	No fruiting bracteoles present.	6
6	Flower unisexual, plants monoecious or dioecious. Perianth scarious or membranous.	Amaranthus
6*	Perianth herbaceous.	Chenopodium

7	Leaves in one or more stemless open rosettes, often withered during the flowering period.	Gomphrena
7*	Leaves along the stem.	8
8	Inflorescence lax and interrupted. Stems ascending or decumbent.	Froelichia interrupta
8*	Inflorescence dense. Stems prostrate or ascending to decumbent.	9
9	Perianth parts fused.	Guilleminea densa
9*	Perianth parts free.	Alternanthera

Alternanthera

	Key based on Agudelo (2008)	
1	Some of the inflorescences pedunculate.	2
1*	All inflorescences sessile.	5
2	Foliage purple. Flowers white, on a pedicel of 1-2 mm.	A. brasiliana
2*	Foliage green.	3
3	Flowers purple or pink, seldom whitish-pink. Spreading shrub, sparsely branched. Upper leaves usually <2cm long. Sepals without dorsal veins.	A. porrigens
3*	Flowers white. Ascending shrub, branched, all parts white-grayish pubescent. Upper leaves (except bracts) usually >2 cm long. Sepals with 3 dorsal veins.	4
4	Heads in clusters of 3, lateral ones pedunculate, terminal ones sessile.	A. pubiflora complex
4*	Heads partly solitary on peduncles of <12 mm, partly subsessile at the apices of the lateral branches.	A. arequipensis
5	Prostrate perennial, branched stems forming dense mats	6
5*	Procumbent to ascending herb.	7
6	Opposite leaves of ± equal size. Sepals mucronate, but without a spine.	A. caracasana
6*	Opposite leaves of distinct unequal size. Sepals spiny.	A. pungens
7	Leaves distinctly petiolate, slightly tomentose.	A. halimifolia
7*	Leaves sessile or petiole very short, furry tomentose.	A. albotomentosa
	Note: The A. pubiflora complex includes A. pubiflora (Benth.) Kuntze, A. albosquarrosa Suess. and A. mollendoana Suess., all accepted as separate species by POWO (2024), and sometimes synonymized by other authors. The morphological differences are rather small. Here we treat them as synonyms.	

Amaranthus

	Key based on Agudelo (2008)	
1	Plants armed with spines in leaf axils.	A. spinosus
1*	Plants unarmed in leaf axils.	2
2	Perianth with 3 sepals. Bracts <50% of perianth.	A. viridis
2*	Perianth with 4-5 sepals. Bracts >50% of perianth.	3
3	At least 1 of the bracts clearly elongated and pointed.	A. hybridus complex
3a	Bracts deep red or purple, >2x as long as the sepals. Terminal spike slender, elongate, recurved.	A. cruentus
3b	Bracts green, usually 2x as long as the sepals. Terminal spike short and thick, erect.	A. hybridus s.str.
3*	All bracts of the same shape.	A. dubius

Atriplex

	Key based on Brignone et al. (2016)	
1	Fruiting bracteoles fused only at base.	2
1*	Fruiting bracteoles fused at base and along margins to various extents.	3
2	Fruiting bracteoles suborbicular, with two dorsal, 3-5-toothed crests at base, margins dentate at apex	A. espostoi
2*	Fruiting bracteoles rhomboid, dorsally smooth, rarely tuberculate at base, margins entire at apex.	A. peruviana
3	Fruiting bracteoles laterally fused half of their length.	4
3*	Fruiting bracteoles laterally fused to a different extent, but never half of their length.	5

4	Leaves canescent, with entire margins. Plants combining both smooth and ornamented, crested and tuberculate fruiting bracteoles.	A. myriophylla
4*	Leaves green, with margins entire in the basal half, dentate in the apical half. Plants with uniform, smooth, fruiting bracteoles.	A. suberecta
5	Monoecious herbs. Leaves spreading, margins loosely dentate.	A. semibaccata
5*	Shrub, usually dioecious. Leaves imbricate, triangular, vertically adjacent to stem, margins entire.	A. imbricata

Chenopodium s.l. (Chenopodiastrum, Chenopodium and Dysphania)

1	Leaves sticky glandular, odorous.	2
1*	Leaves glabrous or farinose.	3
2	Leaves up to 12 cm long. Flowers sessile.	Dysphania ambrosioides
2*	Leaves up to 4 cm long. Flowers pedicellate.	Dysphania incisa
3	Leaves glossy above, not farinose, margins irregularly dentate throughout.	Chenopodiastrum murale
3*	Leaves farinose, margins entire or with few teeth.	4
4	Leaves variable, even on the same plant, rhombic-ovate to lanceolate.	Chenopodium album
4*	Leaves variable, even on the same plant, usually deltoid and distinctly hastate-lobate at base.	Chenopodium petiolare

Gomphrena

1	Inflorescence unbranched, head-like.	G. meyeniana
1*	Inflorescence branched, umbellate.	G. umbellata

Salicornia

1	Stems 2-5 cm high. Plant of high altitude salt marshes.	S. cuscoensis
1*	Stems >20 cm high. Plant of coastal salt marshes	S. neei

Amaryllidaceae

	Key based on Flora of Peru (1936ff.), Meerow et al. (2020), Pacific Bulb Society (2020).	
1	Ovary superior.	Nothoscordum
1*	6-parted flowers with undifferentiated tepals and inferior ovaries, borne in pseudoumbels atop a leafless stalk known as a scape.	2
2	Flowers with 3 spoon shaped stigmas, yellow.	Pyrolirion
2*	Flowers orange or red.	Clinanthus

Clinanthus

1	Plant <10 cm tall.	C. humilis
1*	Plant >20 cm tall.	C. incarum

Nothoscordum

1	Leaves 4-12 mm wide. Umbel of 10-20 flowers, flowers pinkish-white.	N. gracile
1*	Leaves max. 4 mm wide. Umbel of 3-10 flowers.	2
2	Pedicels subequal, about 1 cm long, flowers pinkish-white.	N. andicola
2*	Pedicels mostly unequal, the longer often 2 cm long, flowers white with brown or green veins.	N. bivalve

Pyrolirion

1	Flowers white	P. albicans
1*	Flowers yellow	P. arvense

Anacardiaceae

1	Tree. Leaves pinnate. Inflorescence a panicle. Fruits red to pink.	Schinus molle

Annonaceae

1	Tree. Leaves ovate-lanceolate. Fruit a globular or heart-shaped berry, 10-15 cm in in Ø, green, pulp white, seeds black.	Annona cherimola

Apiaceae

	Key based on Flora of Peru (1936 ff.)	
1	Dense cushion plant of high altitudes.	Azorella
1*	Erect or trailing plant, not forming dense cushions.	2
2	Umbels all simple.	3
2	Umbels regularly compound (occasionally the lateral umbels simple).	8
3	Aquatic plant generally submersed. Leaves of septate linear phyllodes.	Lilaeopsis occidentalis
3*	Terrestrial plant.	4
4	Stipules present at base of petiole.	5
4*	Stipules lacking, the petioles sheathing at base.	6
5	Leaves and inflorescence stellate-pubescent.	Bowlesia
5*	Leaves and inflorescence glabrous.	Spananthe peruviana
6	Plant acaulescent or subcaulescent, generally hirsute (very variable species).	Chaerophyllum andicola
6*	Plants prominently caulescent, glabrous and glaucous.	7
7	Annuals. Petal apex merely acute, not prominently inflexed.	Domeykoa
7*	Perennials. Petal apex long inflexed.	Eremocharis
8	Fruits armed with bristles.	Daucus montanus
8*	Fruits unarmed.	9
9	Flowers yellow, greenish or purplish.	10
9*	Flowers white.	11
10	Leaflets filiform. Fennel odor.	Foeniculum vulgare
10*	Leaflets broad, serrate.	Arracacia
11	Stem purple-spotted.	Conium maculatum
11*	Stem green without red spots.	12
12	Primary umbels with up to 150 rays.	Visnaga daucoides
12*	Primary umbels with <20 rays.	13
13	Leaflets broad, triangular.	Apium graveolens
13*	Leaflets linear or filiform.	Cyclospermum

Arracacia

	Key based on the descriptions of Hermann & Heller (1997)	
1	Slender herb. Umbellule with 6-10 linear involucral bractlets longer than flowers. Flowers reddish-brown.	A. peruviana
1*	Stout herb. Umbellule without involucre. Flowers purple or greenish.	A. xanthorrhiza

Azorella

1	Cushions up to 1 m high and several meters in Ø.	A. compacta
1*	Cushions <10 cm high.	A. diapensioides

Bowlesia

	Key based on Flora of Peru (1936ff.)	
1	Calyx and involucre evident. Ovaries, fruits and stems stellate-pubescent, soft to touch.	2

1*	Calyx and involucre obsolete or very inconspicuous. Ovaries and fruits armed with glochids. Stems with hard stellate processes. Harsh to touch.	3
2	Petals stellate-pubescent dorsally. Lobes of basal leaves subequal, broadest near middle.	B. tropaeolifolia
2*	Petals usually glabrous. Central leaf lobe usually markedly longer than others, broadest at base.	B. lobata
3	Leaves divided nearly to base, leaflets pinnately parted.	B. palmata
3*	Leaves not lobed more than halfway to base, leaflets few toothed to entire.	4
4	Stems rather stout, armed with usually 4-rayed processes.	B. sodiroana
4*	Stems filiform, subglabrate.	B. tenella

Cyclospermum

1	Leaflets filiform, several times longer than broad. Fertile flowers >10 in each umbellet.	C. leptophyllum
1*	Leaflets linear, not several times longer than broad. Fertile flowers <10 in each umbellet.	C. laciniatum

Domeykoa

1	Basal leaves spinulose-dentate. Flowers yellowish-white.	D. amplexicaulis
1*	Basal leaves deeply 3-5-lobed or parted. Flowers pink to purplish.	D. saniculifolia

Eremocharis

1	Branches spinescent, the older ones with exfoliating bark. Growing above 3000 m.	E. hutchisonii
1*	Branches slender and flexuous, not spinescent, without prominently exfoliating bark. Growing below 2000 m.	2
2	Leaves ovate-deltoid, deeply trilobed. Umbels compact, 4-7 mm in Ø, with 20-30 flowers.	E. ferreyrae
2*	Leaves linear, ternate. Umbels lax, 7-10 mm in Ø, 5-10 flowered.	E. piscoensis

Apocynaceae

1	Leaves alternate. Shrub with white fruits.	Vallesia glabra
1*	Leaves opposite.	2
2	Flowers red-orange	Asclepias curassavica
2*	Flowers white or yellow cream, occasionally with brown spots or a purple center.	3
3	Fruit globose, like a balloon, armed with prickles. Shrub.	Gomphocarpus physocarpus
3*	Fruit not globose. Vines or low scramblers.	4
4	Flowers cream to yellow, with purple to brown spots or center.	5
4*	Flowers pale green to white.	6
5	Leaves subcordate at base. Suberect or low scrambler. Inflorescence few-flowered. Corolla incised to one third, center purple.	Pentacyphus andinus
5*	Leaves cordate at base, petiole sulcate. Vine. Inflorescence strictly umbelliform. Corolla not incised, spotted with purple.	Philibertia solanoides
6	Leaves gray-tomentose, veins pinnate, petiole terete.	Funastrum clausum
6*	Leaves green, veins palmate.	Jobinia

Jobinia

	Key based on Liede-Schumann & Meve (2013)	
1	Leaves cordate to semicordate at base. Corona bright white, nearly completely fused, plicate and emarginate.	J. formosa
1*	Leaves truncate to acute at base. Corona not plicate and only fused for about half its length and crowned by 5 long, slenderly lanceolate, free and erect lobules.	J. tiarata
	Note: The differentiation between J. formosa and J. tiarana is problematic. There are populations with lobed petals and cordate leaves.	

Araceae

1	Floating aquatic plant. Duckweed.	Lemna minuta
1*	Terrestrial plant.	2

2	Leaves leathery, glossy. Spathe white, conical.	Zantedeschia aethiopica
2*	Leaves leathery, not glossy. Spathe white, constricted.	Colocasia esculenta

Araliaceae

Hydrocotyle

1	Leaves not peltate, incised up to the middle into 3 lobes.	H. ranunculoides
1*	Leaves peltate, margins crenate, not incised or occasionally incised to one third of blade.	2
2	Inflorescence a compound umbel, branched.	H. bonariensis
2*	Inflorescence a simple umbel, unbranched.	H. umbellata

Arecaceae

1	Stout palm with prickly petioles and palmate leaves.	Washingtonia filifera

Asparagaceae

1	Leaves succulent, >50 cm long, margins prickly.	2
1*	Leaves not succulent, <50 cm long, margins not prickly.	3
2	Leaves bluish-gray. Flowers upright.	Agave americana
2*	Leaves green. Flowers pendulous.	Furcraea andina
3	Leaves needles, <2 cm long. Stem much-branched.	Asparagus officinalis
3*	Leaves linear, >10 cm long. Stem unbranched or sparsely branched.	4
4	Filaments subulate, broadest at base.	Oziroe
4*	Filaments filiform or club-shaped.	5
5	Filaments filiform. Tepals not overlapping. Anthers red-orange.	Echeandia eccremorrhiza
5*	Filaments club-shaped. Tepals broad, overlapping. Anthers yellow-orange.	Trihesperus glaucus

Oziroe

1	Inflorescence apparently sessile, peduncle subterranean.	O. acaulis
1*	Inflorescence pedunculate.	O. biflora

Asphodelaceae

1	Succulent perennial. Leaves armed with prickles. Flowers yellow in a many flowered, dense panicle.	Aloe vera
1*	Bulbous plant, up to 1 m tall. Leaves linear, keeled. Flowers blue in a loose panicle.	Pasithea caerulea

Asteraceae

	Key based on Hind (2011).	
1	Plants often with axillary spines or thorns, these not formed from branch or branchlet apexes. Corolla in- and outside and all other floral parts bearing simple, uniseriate, eglandular, 3-celled hairs.	Tribe Barnadesieae
1*	Stems lacking spines or thorns, or if stems apparently spiny these formed from branch and branchlet apexes. Corollas and other floral parts lacking such hairs, but often with other types of hairs or glabrous.	2
2	Corollas all ligulate, ligules equally 5-toothed at apex. Plants often with latex (no latex in Hypochaeris and Hieracium).	Tribe Cichorieae
2*	Corollas not ligulate, or if ligulate, then ligules with 4 or less apical teeth, or teeth unequal.	3
3	Involucral bracts uniseriate, usually cohering by overlapping margins, or partly or wholly connate. Pappus present.	4
3*	Involucral bracts imbricate in 2 or more series, free or connate, if uniseriate then not cohering or pappus absent or heads unisexual.	7

4	Plants lacking oil glands. Achenes never black when mature.	5
4*	Plants with oil glands on leaves and phyllaries. Achenes black when mature.	6
5	Involucre cylindrical or calyx-shaped.	Tribe Senecioneae
5*	Involucre cup-shaped to globose.	Tribe Perityleae
6	Involucre cylindrical.	Tribe Tageteae
6*	Involucre cup-shaped to globose.	Tribe Bahieae
7	Heads with all or only outer florets 2-labiate.	8
7*	Heads lacking 2-labiate florets.	10
8	Involucre 1-2-seriate.	Tribe Nassauvieae
8*	Involucre multiseriate.	9
9	Flowers pure white or orange-red. Shrubby vines or herbs.	Tribe Mutisieae
9*	Flowers whitish-pink, cream, pink or purple. Shrubs or herbs.	Tribe Onoserideae
10	Woody, tomentose shrubs. Leaves alternate. Involucral bracts 4-seriate. Florets isomorphic, tubular, lobes curled, >1/2 as long as tube.	Tribe Gochnatieae
10*	Not in the above combination of characteristics.	11
11	Heads arranged spirally in racemes. Flowers isomorphic, corolla tubular with lobes >1/3 of tube length. Small, stoloniferous herbs.	Tribe Vernonieae
11*	Not in the above combination of characteristics.	12
12	All leaves opposite, entire, often with 3 prominent longitudinal veins.	13
12*	Not in the above combination of characteristics.	15
13	Heads discoid, disc florets isomorphic. Corolla tubular, with lobes <1/3 of tube length.	Tribe Eupatorieae
13*	Heads radiate.	14
14	Petiole alate. Ray florets longer than Ø of disc, yellow. Disc florets brownish-red.	Tribe Liabeae
14*	Petiole not alate. Ray florets shorter than Ø of disc, yellow or white. Disc florets yellow.	Tribe Millerieae
15	Involucral bracts spiny and heads discoid.	Tribe Cardueae
15*	Not in the above combination of characteristics.	16
16	Involucral bracts papery, translucent, with strengthening tissue and plant woolly or glandular woolly.	Tribe Gnaphalieae
16*	Not in the above combination of characteristics.	17
17	Unarmed shrubs or treelets. All leaves alternate, leathery. Flowers in terminal corymbs, corolla pink.	Tribe Inuleae
17*	Not in the above combination of characteristics.	18
18	Leaves pinnate, opposite to alternate. Heads radiate, involucral bracts 2-seriate. Pappus in form of awns.	Tribe Coreopsideae
18*	Not in the above combination of characteristics.	19
19	Involucral bracts multiseriate, usually scarious with brown margins.	Tribe Antehmideae
19	Not in the above combination of characteristics.	20
20	Leaves frequently opposite and scabrid pubescent. Achenes black, usually glabrous or sometimes bristly. Pappus of scales, hairs or awns or absent, but never of capillary bristles, or if bristles are present, these are plumose.	Tribe Heliantheae
20*	Leaves usually alternate, but never scabrid pubescent. Achenes never black, very frequently bristly, rarely glabrous. Pappus usually of capillary barbellate bristles, or absent, rarely of small scales and then falling rapidly.	Tribe Asteraceae

Tribe Anthemideae

1	Prostrate herbs, <10 cm tall. Heads usually discoid, sometimes with inconspicuous ray florets.	2
1*	Erect herbs, >20 cm tall. Heads radiate, ray-florets prominent, white.	3
2	Heads on slender ± leafless peduncle, clearly exceeding leaves. Achenes not spiny at apex and not winged.	Cotula
2*	Heads subsessile or on short and thick peduncle. Leaves longer than peduncle or enveloping the sessile head. Achenes winged, with spiny apex.	Soliva
3	Leaves pinnatifid.	Tanacetum parthenium
3*	Leaves pinnate, lobes linear.	Matricaria chamomilla

Tribe Astereae

1	Heads discoid.	2
1*	Heads radiate, ray florets in some Erigeron sp. minute.	6
2	Florets yellow.	3
2*	Florets white.	4
3	Acaulescent plant, rosettes forming small colonies. Doubtful presence in AQ.	Novenia tunariensis
3*	Erect shrub.	Parastrephia
4	Woody shrubs.	Baccharis
4*	Herbs.	5
5	Florets uniform.	Erigeron
5*	Radial florets filiform, central florets tubiform.	Laennecia artemisiifolia
6	Ray florets yellow.	Grindelia
6*	Ray florets white.	7
7	Shrubs.	Diplostephium
7*	Herbs.	8
8	Stem and leaves hispid to densely pilose.	Erigeron
8*	Stem and leaves glabrous or with a few hairs on upper stem.	9
9	Heads solitary, on long peduncles. Rosulate herb in bogs and humid places.	Oritrophium limnophilum
9*	Heads in a branched inflorescence. Erect herbs.	Symphyotrichum squamatum

Tribe Bahieae

1	Herbs. Leaves linear. Heads solitary, discoid, disc florets yellow. Phyllaries 7-10, hirsutulous. Ray florets absent. Disc florets 15-30.	Picradeniopsis multiflora
	Herbs. Leaves linear. Phyllaries 4-6, usually glabrous. Ray florets 1-2 or absent. Disc florets 2-6.	Schkuhria pinnata

Tribe Bardadesieae

1	Intricately branched shrub with yellow stem spines. Bracts chestnut-reddish, florets yellow.	Chuquiraga spinosa

Tribe Cardueae

1	Flowers red-purple. Leaves green with distinct white along the vein, spiny margin.	Silybum marianum
1*	Flowers yellow. Leaves gray-green loosely villous, entire.	Centaurea melitensis

Tribe Cichorieae

1	Plant <30 cm tall, usually rosulate.	2
1*	Erect herb, >40 cm tall.	4
2	Plant with milky sap.	Taraxacum officinale s.l.
2*	Plants without milky sap.	3
3	Plant glabrescent.	Hypochaeris
3*	All parts hirsute.	Hieracium mandonii
4	Flowers white.	Picrosia longifolia
4*	Flowers yellow or blue.	5
5	Flowers blue.	Cichorium intybus
5*	Flowers yellow.	6
6	Leaf blades usually bended vertically. Heads axillary, solitary.	Lactuca serriola
6*	Leaf blades horizontal. Heads terminal, in clusters.	Sonchus

Tribe Coreopsideae

	Key based on Hind (2011).	2
1	Perennial shrub. Achenes lacking retrorsely barbed awns. Heads usually with 7-9 ray florets.	Burnellia fasciculata
1*	Annual herbs. Achenes (at least central) with pappus of retrorsely barbed awns. Heads usually with max. 8 ray florets.	2
2	Peripheral achenes obovate, without awns, disc achenes elongate, mostly beaked and with awns.	Heterosperma
2*	Peripheral and disc achenes mostly alike, all with similar pappus.	Bidens

Tribe Eupatorieae

1	Leaves linear to spatulate, >5x as long as wide, margins entire, sometimes pinnatifid, then lobes linear.	2
1*	Middle cauline leaves lanceolate to ovate., <5x as long as wide, margins serrate.	3
2	Heads with >20 florets.	Aristeguietia ballii
2*	Heads with <20 florets.	Helogyne
3	Heads with >20 florets.	Ageratina
3*	Heads with <10 florets.	4
4	Florets distinctly lobed. Lobes 2-5 mm long.	Stevia
	Florets tubular with small lobes. Lobes ca. 1 mm long.	Ophryosporus

Tribe Gnaphalieae

1	Leaves arranged in 2 lines, imbricate on stem. Plant erect.	Andicolea graveolens
1*	Leaves not in two lines.	2
2	White arachnoid filaments spanned between the leaf tips. Cushion plant of high-altitude bogs.	Cuatrecasasiella argentina
2*	No filaments spanned between the leaf tips.	3
3	Plant entirely grayish-white, tomentose.	4
3*	Plant green or grayish-green, at least on upper side of leaves.	7
4	Plant <10 cm tall.	5
4*	Plant >20 cm tall.	6
5	Decumbent herb.	Gamochaeta (humilis)
5*	Cushion plant or cespitose herb.	Lucilia / Mniodes
6	Heads in corymbs.	Achyrocline
6*	Heads in interrupted spikes.	Gamochaeta
7	Leaves oblong, lanceolate or spatulate.	8
7*	Leaves linear.	9
8	Heads terminal.	Pseudognaphalium
8*	Heads in interrupted spikes	Gamochaeta
9	Heads in dense, terminal clusters.	Gamochaeta (capitata)
9*	Heads axillary in clusters of 1-3.	Facelis plumosa

Tribe Gochnatieae

1	Woody, tomentose shrub.	Gochnatia arequipensis

Tribe Heliantheae

1	Plant armed with 2-3-furcate spines in leaf axils.	Xanthium spinosum
1*	Plant not armed with spines.	2
2	Heads in dense terminal racemes.	Ambrosia
2*	Heads solitary or in loose clusters.	3

3	Disc floret white. Herbs, often rooting at nodes.	4
3*	Disc florets yellow or purplish-brown.	5
4	Ray heads.	Eclipta prostrata
4*	Discoid heads.	Spilanthes leiocarpa
5	Ray florets red, pink or orange.	Zinnia peruviana
5*	Ray florets yellow.	6
6	Herb with aboveground stolons rooting at nodes.	Acmella ciliata
6*	Herb without stolons.	7
7	Leaves linear.	Pascalia glauca
7*	Leaves ovate to lanceolate.	8
8	Plant <30 cm tall, herb of the lomas.	Heiseria irmscheriana
8*	Plant >50 cm tall.	9
9	Leaves deltoid to ovate, grayish tomentose, margins entire.	Encelia canescens
9*	Leaves ovate, green, margins serrate.	10
10	Heads with ray florets up to 9 cm in Ø. Lower cauline leaves nearly sessile. Stem glabrous.	Viguiera weberbaueri
10*	Heads with ray florets <5cm in Ø. Lower cauline leaves short petiolate. Stem hirsute.	Aldama

Tribe Inuleae

1	Trees or erect shrubs. Branches spreading at an angle of <45°. Leaves oblanceolate, base attenuate.	Tessaria
1*	Decumbent shrubs. Branches spreading at an angle of >45°. Leaves broadly ovate, base truncate.	Pluchea chingoyo

Tribe Liabeae

1	Shrub. Ray florets >40.	Chionopappus benthamii
1*	Herbaceous plant. Ray florets <30.	2
2	Leaves broadly ovate, green beneath.	Philoglossa peruviana
2*	Leaves lyrate, deeply notched, white beneath.	Munnozia lyrata

Tribe Millerieae

1	Ray florets yellow.	Sigesbeckia jorullensis
1*	Ray florets white or pink.	Galinsoga

Tribe Mutisieae

1	Shrub or lianas, >30 cm tall. Leaves generally pinnately compound or with an apical tendril. Flowers yellow to red.	Mutisia
1*	Herb, 5-10 cm tall. Flowers white.	2
2	Leaves ovate.	Chaetanthera peruviana
2*	Leaves linear, terete.	Oriastrum stuebelii

Tribe Nassauvieae

1	Shrub, woody throughout.	2
1*	Perennial herb, sometimes with woody base.	5
2	Inflorescence with distal axils ending in spines.	Proustia cuneifolia
2*	Stem and branches not thorny, leaves sometimes spiny-dentate.	3
3	Florets yellow.	Trixis cacalioides
3*	Florets white to pinkish-white.	4
4	Leaves linear to obovate.	Lophopappus

4*	Leaves orbicular, pedately lobed.	Jungia axillaris
5	Heads in dense, globular pseudoheads (clusters without involucral bracts). Florets isomorphic.	Polyachyrus
5*	Heads solitary or in corymbs.	6
6	Leaves 2-pinnatifid, not spinulose.	Leucheria daucifolia
6*	Leaves lanceolate to oblanceolate, broadly dentate to pinnatifid, teeth often spinulose.	Perezia

Tribe Onoserideae

1	Herbs. Heads radiate, ray florets pink.	Onoseris
1*	Shrubs. Heads discoid or with some long-lobed, marginally arranged florets with long, ray-like florets.	2
2	Heads discoid. Leaves lanceolate, 4-6 cm long, ± glabrous. Florets creamy-white.	Paquirea lanceolata
2*	Heads with some ray-like florets. Leaves obovate to oblanceolate, <4 cm long. Disc florets pink to purple.	3
3	Branches whitish-tomentose. Leaves widely spaced.	Aphyllocladus denticulatus
3*	Branches dark gray with prominent leaf-scars. Leaves spirally arranged, imbricate.	Plazia daphnoides

Tribe Perityleae

1	Ray florets yellow.	Villanova oppositifolia
1*	Ray florets white.	Perityle emoryi

Tribe Senecioneae

1	Erect shrubs >1.5 m tall. Leaves lanceolate to ovate, >5 cm long.	Gynoxys
1*	Herbs or shrubs, <1 m tall. Leaves usually <5 cm long.	2
2	Plants dioecious. Usually much-branched shrubs with white-tomentose, entire leaves (C. antennaria is a herb).	Chersodoma
2*	Plants bisexual.	3
3	Heads hemispherical, distinctly wider than long.	4
3*	Heads campanulate or cylindrical, as long or longer than wide. Heads discoid or radiate, erect or slightly nodding.	5
4	Heads discoid, nodding. Herbs with distinct stem.	Culcitium
4*	Heads radiate. Acaulescent herb. Leaves and involucre covered with 3-5 mm long strigose hairs.	Misbrookea strigosissima
5	Erect herbs or shrubs, >20 cm tall (Werneria ciliolata sometimes <20 cm). Florets yellow or brownish-orange.	6
5*	Prostrate or rosulate herbs or dwarf shrubs, <15 cm tall. Florets yellow, white or purple.	8
6	Heads radiate, ray florets 5-veined. Heads in lax, corymbiform cymes. Leaves laminar. Blade white-tomentose beneath (not on L. mollendonensis).	Lomanthus
6*	Not in above combination.	7
7	Plant with broccoli- or clubmoss-like appearance. Leaves imbricate.	Werneria
7*	Not in above combination.	Senecio
8	Heads radiate. Ray florets white, sometimes purplish below.	9
8*	Not in above combination.	10
9	Acaulescent herbs. Leaves rosulate.	Rockhausenia
9*	Herbs or shrubs with distinct stem. Leaves imbricate.	Werneria
10	Florets yellow.	Senecio
10*	Florets white, purple or orange.	11
11	Leaves hirsute.	Senecio
11*	Leaves glabrous.	12
12	Florets purplish-orange. Heads radiate, ray florets short and inconspicuous.	Werneria
12*	Florets purplish-white or purple. Heads discoid.	Rockhausenia

Tribe Tageteae

1	Leaves pinnate.	Tagetes
1*	Leaves entire.	2
2	Leaves glaucous. Florets purplish-brown.	Porophyllum ruderale
2*	Leaves green. Florets yellow.	3
3	Plants >30 cm, stem purplish, leaf 3-veined, margins serrate, inflorescence a scorpioid glomerule.	Flaveria bidentis
3*	Plants <30 cm, stem green, leaf linear-lanceolate, margins entire, inflorescence solitary.	Pectis sessiliflora

Tribe Vernonieae

1	Corolla white. Perennial stoloniferous herb. Inflorescence racemose spicate, heads arranged spirally.	Pseudelephantopus spiralis

Achyrocline

	Key based on Flora of Peru (1936ff.)	
1	Leaves decurrent. Stem winged.	A. alata
1*	Leaves not decurrent. Stem not winged, <50 cm tall. Leaves linear-lanceolate, 20-30 x 1-3 mm.	A. ramosissima

Ageratina

1	Corolla purplish	A. pentlandiana
1*	Corolla predominantly white	2
2	Subshrub, up to 1 m tall. Heads in cluster of >10.	A. sternbergiana
2*	Subshrub, up to 75 cm tall. Heads in clusters of <6. Florets pinkish-white.	A. glechonophylla

Aldama

1	Growing in Lomas de Atiquipa (0-500 m).	A. dilloniorum
1*	Growing above 1500 m.	2
2	Erect to procumbent subshrub, strongly branched all over the stem.	A. helianthoides
2*	Smaller than A. helianthoides but erect and branched from the ground only.	A. lanceolata

Ambrosia

	Key based on Payne Willard (1966).	
1	Annual herb	A. artemisiifolia
1*	Shrub or perennial herb	2
2	Fruiting involucres less than 5 mm long, with 10 or fewer straight, reduced, or vestigial spines in 1-2 whorls below the beak, or without spines.	A. cumanensis
2*	Fruiting involucres more than 5 mm long, with more than 10 spines scattered over the surface.	3
3	Plant >1 m, leaves >10 cm in Ø.	A. arborescens
3*	Plant <1 m, leaves <10 cm in Ø.	4
4	Leaves entire, unlobed or with few lobe-like marginal teeth.	A. dentata
4*	Leaves 1-2 pinnately lobed.	5
5	Leaf segments 1.5 mm broad or less at widest point, blade hairy, but not tomentose. Spines of fruiting involucre easily broken from body.	A. artemisioides
5*	Leaf segments more than 1.5 mm broad at widest point, blades tomentose. Spines of fruiting involucre firmly attached to body.	A. pannosa

Baccharis

	Key based on Hellwig (1988), Müller (2006)	
1	Leafless shrub with winged stem.	B. genistelloides
1*	Shrub with leaves, stem terete.	2

2	Shrub <0.8 m tall.	3
2*	Shrub >1 m tall.	5
3	Rhizomatous herb. Small rosette with linear leaves. Head solitary, ± acaule.	B. acaulis
3*	Small shrub.	4
4	Low creeping cushion-shrub, up to 15 cm tall.	B. caespitosa
4*	Erect shrub up to 80 cm tall. Leaves spatulate, with up to 3 teeth on each side.	B. tola
5	Leaves linear, needle-like.	B. boliviensis
5*	Leaves ovate, elliptic or lanceolate, not needle-like.	6
6	Leaves ovate or elliptic, <4x longer than wide.	7
6*	Leaves lanceolate or elliptic, >4x longer than wide.	8
7	Branching sparse, branches often almost horizontal. Leaf broadly ovate, leaf margins with at least one tooth in lower quarter, teeth coarsely, broadly, directed outwardly. Grows from the coast to mid-elevation. Riparian, wet and saline areas.	B. scandens
7*	Branching loosely, branches ascending obliquely. Leaf elliptic, with entire margins in lower quarter, teeth small, narrow, pointing forward. Grows from medium to high altitudes. Riparian, roadsides.	B. alnifolia
8	Inflorescence with up to 10 heads. Leaves 6-50x longer than wide.	B. gnidiifolia
8*	Inflorescence a pyramidal or corymbiform panicle, many heads (>50). Leaves 4-15x longer than wide.	B. salicifolia

Bidens

1	Ray florets white or absent (see also B. exigua).	B. pilosa
1*	Ray florets yellow.	2
2	Heads >2 cm in Ø (incl. ray). Ray florets 1-2 cm long	3
2*	Heads <2 cm in Ø (incl. ray), ray florets may be absent. Ray florets <1 cm long.	4
3	At least some heads with more than 5 ray florets. Annual, suberect herb.	B. andicola
3*	All heads with 5 ray florets. Perennial, decumbent and trailing herb.	B. triplinervia
4	Leaf segments oblong to linear lanceolate.	B. exigua
4*	Leaf segments filiform.	B. pseudocosmos

Chersodoma

	Key based on Dillon & Sagastegui (1996)	
1	Erect shrub. Heads subsessile.	C. jodopappa
1*	Subshrubs or perennial herb. Heads pedunculate.	2
2	Perennial herbs. Involucral bracts ca. 18. Florets 50-200 per head.	C. antennaria
2*	Subshrub from a woody caudex. Involucral bracts 8-15. Florets 10-45 per head.	3
3	Inflorescence often weakly cymose, peduncles 5-40 mm long, 4-5 bracteolate.	C. arequipensis
3*	Inflorescence solitary, peduncles 3-9 mm long, 1-2 bracteolate.	C. juanisernii

Cotula

1	Annual herb, strigose to soft hairy. Leaves 2-3-pinnate. Heads 3-6 mm in Ø.	C. australis
1*	Perennial herb, glabrous. Leaves entire or irregularly dentate. Heads 6-15 mm in Ø.	C. coronopifolia

Culcitium

1	Heads in loose cymes, not solitary. Leaves basal and cauline.	C. oligocephalum
1*	Heads solitary. Leaves all basal, rosulate. Bract-like leave on peduncle present.	2
2	Leaves canescent, margins ± entire. Florets yellow. Growing in dry areas.	C. humile
2*	Leaves green, margins serrate. Florets reddish-purple. Growing in wet areas (bofedales).	C. serratifolium

Diplostephium

	Key based on Hind (2011).	
1	Leaves 20-30 mm long, light green, tomentose. Ray florets 5-6 mm long.	D. meyenii
1*	Leaves 8-13 mm long, dark green, tomentose. Ray florets 11-12 mm long.	D. cinereum
	Note: Diplostephium tacorense Hieron. is an unresolved species, probably synonym of D. meyenii.	

Erigeron

	Key based on Jepson (2024)	
1	Erect plant, branches with cauline leaves. Heads discoid, disciform, or inconspicuously radiate, ray florets absent or barely exceeding involucre (0-3 mm).	2
1*	All leaves in a basal rosette, without cauline leaves. Heads conspicuously radiate, rays exceeding to much-exceeding involucre.	6
2	Central stem much-branched. Heads 10+ per branch.	3
2*	Central stem poorly branched. Heads 1-10 per branch.	5
3	Central stem often overtopped by distal branches in age, plant ± gray hairy. Secondary venation of cauline leaves not visible. Heads in raceme-like clusters, slightly glutinous.	E. bonariensis
3*	Central stem generally exceeding distal branches in age, plant green, glabrous to hairy. Secondary venation of cauline leaves visible. Heads in panicle-like clusters.	4
4	At least some cauline leaves pinnately lobed. Stem reddish in mature plants. Ray florets 0-0.5 mm, disc florets generally 5-lobed.	E. floribundus
4*	Cauline leaves entire or sparsely serrate, apical margins ciliate. Stem green. Ray florets 0.5-3 mm, disc florets generally 4-lobed.	E. canadensis
5	Heads terminal and axillary, 1-3 heads per branch. Cauline leaves ovate, base ± amplexicaule.	E. pazensis
5*	Heads terminal, in loose corymbs of 1-10. Cauline leaves linear-oblanceolate, base sessile.	E. deserticolus
6	Heads pedunculate.	E. lanceolatus
6*	Heads short pedunculate, apparently sessile. Mat forming cushion-plant.	E. rosulatus

Galinsoga

1	Heads sessile or subsessile.	G. unxioides
1*	Heads pedunculate.	2
2	Stem and leaves hirsute.	G. quadriradiata
2*	Stem and leaves glabrous or slightly pubescent.	G. parviflora

Gamochaeta

	Key based on Urtubey et al. (2016), Jepson (2024)	
1	Plants <10 cm tall.	G. humilis
1*	Plants 10-50 cm tall.	2
2	Leaves discolorous.	3
2*	Leaves concolorous	4
3	Heads in a spike, interrupted when old.	G. americana
3*	Heads in terminal clusters.	G. capitata
4	Cauline leaves spatulate	G. pensylvanica
4*	Cauline leaves oblanceolate	G. purpurea

Grindelia

	Key based on Bartoli & Tortosa (1999), Flora of Peru (1936ff.).	
1	Lower leaves 3-13 cm long, herbaceous green, growing at elevations >2500 m. Head >2.5 cm in Ø, ray florets 36-45.	G. boliviana
1*	Leaves all <8 cm long. Glabrous, glandular shrubs.	2
2	Head	G. glutinosa
2*	Shrub of the Andes, 1 m tall, growing at elevations >2000 m. Leaves glaucous. Head <2.5 cm in Ø, ray florets <25, outer involucral bracts glabrous.	G. tarapacana

Gynoxys

1	Heads all discoid. Leaves green above, grayish-white beneath, rugose.	G. longifolia
1*	Heads radiate, at least some. Leaves canescent, smooth.	G. longistyla

Helogyne

1	Leaves linear, needle-like, margins entire. Involucral bracts pointed, straw-colored.	H. straminea
1*	Leaves spatulate to elliptical, apex acute to obtuse, at least some of them parted or dentate.	2
2	Leaves elliptic, entire, margins sometimes dentate. Heads usually solitary.	H. hutchisonii
2*	Leaves linear, often divided in 3 linear lobes. Heads in clusters of more than 1.	H. apaloidea

Heterosperma

	Key based on Lizarazu & Freire (2019)	
1	Leaf blade ovate.	H. ovatifolium
1*	Leaf blade pinnatisect (entire in the upper part of the plant)	2
2	Plants up to 10 cm tall.	H. nanum
2*	Plants >10 cm tall.	3
3	Outer phyllaries obovate.	H. ferreyrae
3*	Outer phyllaries linear to lanceolate.	H. diversifolium

Hypochaeris

1	Caulescent herb. Flowers yellow, several pedunculate heads on a branched and leafy stem.	2
1*	Acaulescent herb. Flowers white or yellow, heads solitary on a short stem or sessile.	3
2	Budding heads always erect.	H. chillensis
2*	Budding heads typically nutant.	H. elata
3	Corolla white.	4
3*	Corolla yellow (sometimes white in H. sessiliflora).	5
4	Involucre campanulate. Peduncle short, max. 2 cm.	H. echegarayi
4*	Involucre cylindric. Peduncle ± long, >2 cm.	H. taraxacoides
5	Leaves entire (often leafless when flowering). Phyllaries lanate, whitish.	H. eriolaena
5*	Leaves pinnatipartite to pinnatisect. Phyllaries green, glabrous or hirsute, but not lanate.	H. meyeniana

Lomanthus

1	Heads in corymbiform cymes of 20-40+.	2
1*	Heads solitary or in cymes of 2-15.	3
2	Leaves pinnatisect.	L. arnaldii
2*	Leaves entire, margins coarsely dentate-serrate.	L. mollendoensis
3	Ray florets whitish-yellow. Heads in cymes of 8-13.	L. subcandidus
3*	Ray florets deep yellow. Heads solitary or in cymes usually with <8 heads.	4
4	Leaves pinnatisect or deeply incised halfway to midrib.	5
4*	Leaves entire, margins dentate or lobed, usually not incised halfway to midrib.	7
5	Leaves pinnatisect.	L. icaensis
5*	Leaf margins dentate or lobed halfway to midrib.	6
6	Erect suffrutescent herb, usually without basal leaves.	L. abadianus
6*	Prostrate herb with many basal leaves, these long petiolate.	L. okopanus
7	Coastal herb (lomas), growing below 1000 m. Cauline leaves sessile, auriculate, blade green above	L. lomincola
7*	Andean herb, growing above 1500 m. Cauline leaves petiolate, blade bluish-green above.	L. tovarii

Lophopappus

	Key based on Katinas et al.(2013).	
1	Leaves linear-lanceolate, pungent at the apex.	L. foliosus
1*	Leaves oblanceolate or obovate-spatulate, obtuse at the apex.	2
2	Heads solitary, with 2-labiate corollas. 6-7 flowered.	L. cuneatus
2*	Heads 2-4 in dense cymes, with tubular-funnelform corollas. 7-11-flowered.	L. tarapacanus
	Note: Katinas et al.(2013) synonymize Lophopappus berberidifolius Cabrera with L. tarapacanus, which is followed here.	

Lucilia /Mniodes

	Key based on Dillon (2018) and Quipuscoa & Dillon (2020)	
1	Cespitose herb, leaves >5 mm long.	2
1*	Cushion plant, leaves <5 mm long.	4
2	Leaves 20-50 mm long.	Mniodes longifolia
2*	Leaves 5-20 mm long.	3
3	Leaves linear, pleated or canaliculate, 5-15 mm long, 0.8-1 mm wide, lower surface arachnoid-tomentose, glabrescent, upper surface densely silvery-tomentose.	Lucilia conoidea
3*	Leaves spatulate, plane to falcate, 10-20 mm long, 1-2 mm wide, both surfaces tomentose.	Lucilia kunthiana
4	Plants forming dense cushions.	5
4*	Cespitose herb, decumbent, forming loose mats.	6
5	Leaves densely imbricate, appressed on stem. Internodes not obvious on stem.	Mniodes coarctata
5*	Leaves spreading, not appressed on stem. Internodes on stem conspicuous.	Mniodes caespititia
6	Leaves densely lanate on both sides.	Mniodes piptolepis
6*	Leaves below lanate-tomentose, above glabrescent in age.	Mniodes argentea
	Note: The genus Lucilia is sometimes integrated in Mniodes. Lucilia conoidea is synonymized with L. kunthiana by some authors.	

Mutisia

	Key based on Flora of Peru (1936ff.).	
1	Leaves without tendrils.	2
1*	Leaves with terminal tendril.	3
2	Leaves entire, apex acute.	M. orbignyana
2*	Leaves lobed, apex pointed.	M. comptoniifolia
3	Leaves pinnate, 9-14 pairs of leaflets.	M. acuminata
3*	Leaves entire.	4
4	Leaf margins denticulate.	M. lanigera
4*	Leaf margins essentially entire.	M. arequipensis

Onoseris

1	Herb, up to 10-15 cm tall. Ray florets <10, involucral bracts not recurved.	O. minima
1*	Suffrutescent herb, stems usually >10 cm long. Ray florets usually >10.	2
2	Leaves 10-12 mm long, ovate, apex obtuse, both sides of blade white-tomentose.	O. humboldtiana
2*	Leaves usually >15 mm, lanceolate, upper surface of blade ± glabrescent, green. Involucral bracts recurved.	O. odorata

Ophryosporus

1	Leaves pinnatifid.	O. bipinnatifidus
1*	Leaves ovate to lanceolate.	2
2	Shrub up to 1 m tall. Leaves triangular-elliptic, <3 cm long, apex obtuse to acute, margins crenate.	O. pubescens
2*	Shrub >1 m tall. Leaves lanceolate, >3cm, apex acute to acuminate, margins serrate.	3
3	Plant yellowish-green. Inflorescence >30 cm long.	O. peruvianus
3*	Plant dark green. Inflorescence <20 cm long.	4
4	Heads in dense panicles. Leaf base acute.	O. hoppii
4*	Heads in lax panicles. Leaf base obtuse.	O. heptanthus

Parastrephia

	Key based on Nesom (1993).	
1	Leaves 4-6 mm long, not adnate to the stem, ascending to spreading recurving. Central flowers 13-28, peripheral flowers 7-16.	P. lucida
1*	Leaves 2-5 mm long, straight and tightly appressed to the stem, the basal portion adnate. Central flowers 3-10, peripheral flowers 3-9.	P. quadrangularis

Perezia

	Key based on Katinas (2012)	
1	Heads sessile to subsessile. Corolla blue, white, reddish or yellow.	P. pinnatifida
1*	Heads on peduncles >3 cm.	2
2	Heads in dense corymbs. Ray florets white.	P. multiflora
2*	Heads solitary or few flowered corymbose-racemes. Ray florets white, blue or pink.	3
3	Involucral bracts, imbricate, adpressed. Peduncle with linear bracts. Basal leaves entire.	P. ciliosa
3*	Involucral bracts lax and disorganized. Peduncle leavy or with ovate bracts. Basal leaves entire or parted.	P. pungens

Polyachyrus

	Key based on Katinas (2012)	
1	Annual herb. Leaves green. Heads in loose, hemispherical pseudo-heads. Growing below 1000 m.	P. annuus
1*	Perennial herb. Leaves whitish tomentose. Heads in dense, orbicular pseudo-heads. Growing above 2000 m.	P. sphaerocephalus

Pseudognaphalium

	Key based on Freire et al. (2018)	
1	Stems prostrate, branched.	P. lacteum
1*	Stems erect to ascending, branched or unbranched.	2
2	Leaves concolorous or slightly discolorous.	3
2*	Leaves conspicuously discolorous.	5
3	Stem and leaves protruding woolly.	P. lanuginosum
3*	Stems and leaves adherent woolly.	4
4	Heads usually arranged in dense terminal clusters. Stem leaves linear or linear-oblong.	P. viravira
4*	Heads arranged in corymbs. Stem leaves linear-lanceolate or lanceolate.	P. cheiranthifolium
5	Stem leaves lanceolate to elliptic, apex acute. Stems densely whitish-woolly.	P. elegans
5*	Stem leaves linear or linear-lanceolate to lanceolate, apex acute and attenuate. Stems lanuginose to whitish-woolly.	6
6	Stem leaves linear (1–8 mm wide).	P. gaudichaudianum
6*	Stem leaves linear-lanceolate to lanceolate (10–25 mm wide).	P. dysodes

Rockhausenia

	Key based on Calvo et al. (2020)	
1	Heads discoid. Leaves distinctly pinnate.	2
1*	Heads radiate. Leaves entire or denticulate.	3
2	Leaves <3 cm long, 4-6-lobed.	R. solivifolia
2*	Leaves >6 cm long, with 12+ lobes.	R. pinnatifida
3	Leaf margins denticulate or entire and ciliate.	4
3*	Leaf margins entire, not ciliate.	6
4	Leaf blade entire and ciliate.	R. pectinata
4*	Leaf blade denticulate, not ciliate.	5
5	Leaf blade 3-8 mm long. Plant forming mats.	R. aretioides
5*	Leaf blade 9-90 cm long. Plant forming clumps.	R. orbignyana
6	Leaves pseudopetiolate. Leaf blade oblanceolate, elliptic or spatulate.	R. spathulata
6*	Leaves sessile, extending into a sheath-like base. Leaf blade linear, narrowly oblanceolate or narrowly elliptic.	7
7	Leaf apex mucronate, at least in young leaves.	R. microphylla
7*	Leaf apex acute to obtuse, lacking an arista.	8
8	Involucral bracts 20-33, 15-30 mm long. Leaves usually >8 cm long and >3 mm wide. Ray florets >2 cm long.	R. nubigena

8*	Involucral bracts 11-14, 3-10 mm long. Leaves <8 cm long and <3 mm wide. Ray florets <1.2 cm long.	9
9	Leaves leathery, flat in cross section, shiny and conspicuously veined.	R. apiculata
9*	Leaves fleshy, elliptical to terete in cross section, matte with barely visible veins.	R. pygmaea
	Rockhausenia apiculata is synonymous with R. pygmaea by some authors.	

Senecio

	Key based on the descriptions by Cabrera (1985) and Montesinos-Tubée & Gonzales (2020)	
1	Spiny shrubs or subshrubs.	2
1*	Unarmed herbs or shrubs.	3
2	Intricately branched shrub, resembling Adesmia. Lateral branches ending in spines.	S. spinosus
2*	Subshrub with trifurcate leaves, leaf apex spiny.	S. trifurcifolius
3	Florets white, pinkish-white or purple. Heads discoid.	4
3*	Florets yellow. Heads radiate or discoid.	5
4	Heads sessile to subsessile. Leaves linear-oblong to spatulate, 0.5-1.5 cm long.	S. melanandrus
4*	Heads petiolate. Leaves linear, 4-8 cm long.	S. scorzonerifolius
5	Basal leaves tufted, linear-lanceolate, 20-35 cm long, 0.5-1.5 cm wide.	S. comosus
5*	Basal leaves <15 cm long.	6
6	Shrubs of the lomas, growing below 1000 m.	7
6*	Shrubs or herbs of elevations above 1000 m (Senecio vulgaris can be found in coastal areas).	9
7	Heads discoid. Leaves 10 cm long, linear, sessile, base obtuse. Heads 10-20 in terminal cymes.	S. acarinus
7*	Heads radiate. Leaves 2-3 cm long. Heads 3-10 in terminal cymes.	8
8	Leaves linear-lanceolate to ovate, sessile, base hastate to amplexicaul.	S. smithianus
8*	Leaves linear, sessile, base obtuse.	S. calcicola
9	Heads radiate.	10
9*	Heads discoid.	13
10	Acaulescent herb. Leaves in basal rosettes, margins dentate to lobed.	S. breviscapus
10*	Erect shrubs or herbs.	11
11a	Leaves 4-13 cm long, pseudopetiolate, not hastate. Heads numerous, in dense cymes. Stems reddish-purple.	S. rudbeckiifolius
11b	Leaves 2-7 cm long, sessile, not hastate. Heads numerous, in dense cymes. Stems brownish.	S. clivicola
11*	Leaves 1-4 cm long, sessile, base hastate to auriculate. Heads 1-10, in lax cymes.	12
12	Leaves pinnatifid, auricles at base denticulate. Stems purplish-green.	S. ctenophyllus
12*	Leaves linear, margins entire, base hastate. Stems glaucous.	S. yurensis
13	Plants white-tomentose, grayish-white in appearance.	14
13*	Plants glabrous or pubescent, green to glaucous in appearance.	15
14	Erect to decumbent shrub, >10 cm tall.	S. crassilodix
14*	Mat-forming dwarf shrub, >5 cm tall.	S. evacoides
15	Heads axillary. Aromatic shrub, glabrous. Heads nodding, yellow, orange or whitish-pink.	S. nutans
15*	Heads terminal, in cymes or solitary.	16
16	Heads solitary.	17
16*	Heads in terminal cymes of more than 1 head.	22
17	Leaves deeply lobed to pinnatisect, green.	18
17*	Leaves entire, glaucous-green. Heads almost sessile.	19
18	Leaves pinnatisect, with 5-12 pairs of segments. Heads on long peduncle, 15-30 mm in Ø. Involucral bracts 16-24.	S. jarae
18*	Leaves deeply dentate or lobed, with 2-4 pairs of segments. Heads on short peduncle, <15 mm in Ø. Involucral bracts 12-15.	S. pinnatilobatus
19	Involucre <5 mm long.	20
19*	Involucre >5 mm long.	21
20	Leaves entire, glabrous and fleshy. Involucral bracts 8. Florets 15-18.	S. humillimus

20*	Leaves entire or dentate, glabrous or lanuginose. Involucral bracts 13. Florets >20.	S. vegetus
21	Plants ± glabrous.	S. algens complex
21a	Pedicel 0-2 mm long. Involucre 10 mm in Ø. Plant glabrous, leaves glabrous.	S. gamolepis
21b	Pedicel 2-4 mm long. Involucre 5-7 mm in Ø. Plant glabrous, leaves with marginal trichomes.	S. anastasioi
21c	Pedicel 8-12 mm long. Involucre 8-12 mm in Ø. Plant entirely glabrous.	S. algens s.str.
21d	Pedicel 12-18 mm long. Involucre 7-9 mm in Ø. Plant with few papillose trichomes.	S. beltranii
21*	Plants tomentose, at least on the underside of leaves.	S. moqueguensis
22	Leaves pinnately lobbed to pinnatisect, incised more than halfway to main vein.	S. vulgaris
22*	Leaves entire, margins dentate or serrulate.	23
23	Heads nodding.	24
23*	Heads erect, in corymbiform cymes.	26
24	Herbs. Leaves long petiolate. Heads 2-3 cm in Ø.	S. rhizomatus
24*	Shrubs. Leaves sessile. Heads <1 cm in Ø.	25
25	Leaf margins strongly dentate.	S. adenophyllus
25*	Leaf margins entire or subentire.	S. rufescens
26	Small herb. Leaves elliptic, laminar, dark green.	S. violifolius
26*	Shrubs. Leaves linear.	27
27	Leaves conspicuously dentate.	S. reicheanus
27*	Leaves entire or minutely dentate.	28
28	Upper leaves hastate.	S. hastatifolius
28*	Leaves not hastate.	29
29	Leaves <2 cm long, crowded in leaf axils.	S. chachaniensis
29*	Leaves 2.5-6 cm long, alternate.	S. pentlandianus

Soliva

1	Heads on short and thick petiole, not woolly.	S. mexicana
1*	Heads ± sessile, woolly.	S. stolonifera

Sonchus

1	Leaves spiny, base auriculate, auricles round, spirally adpressed. Flowers <2 cm in Ø.	S. asper
1*	Leaves softly dentate, base auriculate, auricles not spirally adpressed. Flowers 2.5 cm in Ø.	S. oleraceus

Stevia

	Key based on Bedoya-Cuno et al. (2024).	
1	Basal and cauline leaves sessile to amplexicaul. Growing below 1000 m (lomas).	S. melissifolia
1*	Basal leaves petiolate, petiole 0.3-0.8 mm long. Growing above 1500 m.	2
2	Outer involucral bracts covered with simple trichomes, interspersed with sessile glands.	S. cuzcoensis
2*	Outer involucral bracts covered with stipitate glands, sometimes interspersed with simple trichomes.	3
3	Stems, leaves and peduncles integrally covered with stipitate glands, sometimes the leaves with simple trichomes.	S. weberbaueri
3*	Stem and peduncles on upper half with stipitate glands, lower part with simple trichomes.	4
4	Inflorescence densely corymbiform, sessile to subsessile.	S. hoppii
4*	Inflorescence laxly corymbiform, peduncle 4-8 mm long.	S. herrerae
	Note: Gutierrez et al. (2016) show that Stevia-species have a wide morphological variation, and they recognize only one species for Chile. Maybe the number of Peruvian Stevia species is also overestimated.	

Tagetes

	Key based on Schiavinato et al. (2023).	
1	Pinnae linear, 1 mm wide, margins entire. Involucres covered with punctiform glands, apex of phyllaries truncated, bearing a small tooth. Plants with anise-like odor.	T. filifolia

1*	Pinnae >1 mm wide, margins dentate. Involucres covered with linear to elliptic glands, apex of phyllaries acute or obtuse but not truncated. Plants odorless or with marigold-like odor.	2
2	Heads dimorphic, most heads multiflorous and some heads uniflorous. Plant up to 0.8 m tall. Leaves 5-15 cm long. 5 involucral bracts. Ray florets usually yellow.	T. terniflora
2*	Heads all multiflorous.	3
3	Plant 1-2 m tall. Leaves 3-28 cm long. 3-4 involucral bracts. Ray florets usually white.	T. minuta
3*	Plant <50 cm. Ray florets yellow to orange.	4
4	Leaf segments 2.5-5.5 mm wide. Involucre cylindrical, upper half partly or entirely purplish, lower half greenish. Pappus scales subulate, broader at base, margins barbellate, lateral projections much shorter than the width of the scale base.	T. dombeyi
4*	Leaf segments 1-2 mm wide. Involucre entirely purplish. Pappus scales oblanceolate, widest part in the upper 2/5, margins plumose, lateral projections longer than the scale base.	T. multiflora
	Note: Tagetes minuta probably does not occur in Peru. It is often confused with T. terniflora. Perhaps it is cultivated here and there and has escaped from cultivation.	

Tessaria

1	Slender tree, up to 12 m tall. Leaf margins entire.	T. integrifolia
1*	Small tree or shrub. Leaf margins dentate.	T. absinthioides
	Note: Both species hybridize.	

Werneria

	Key based on Calvo & Moreira-Muñoz (2020).	
1	Leaves entire, undivided at the apex.	2
1*	Leaves notched, forked, or finger-like at the apex.	4
2	Ray corollas not surpassing the involucre, yellow.	W. ciliolata
2*	Ray corollas conspicuously surpassing the involucre, white.	3
3	Involucral bracts 13-15. Ray florets 14-21 with corollas 7.4-12.9 mm long, disc florets 35-48.	W. marcida
3*	Involucral bracts 8-12. Ray florets 9-15 with corollas 5.9-8 mm long, disc florets 24-32.	W. weddellii
4	Heads disciform.	W. esquilachensis
4*	Heads radiate.	5
5	Ray corollas not surpassing the involucre, pale yellow. Heads ± nodding. Stems 30-60 cm tall.	W. staffordiae
5*	Ray corollas conspicuously longer than the involucre, white. Heads erect. Stems 2-24 cm tall.	6
6	Leaves finger-like at the apex, at least 9-divided.	W. dactylophylla
6*	Leaves 3-notched or 3-forked at the apex.	7
7	Leaves 3-notched at the apex. Disc florets 7-18.	W. poposa
7*	Leaves 3-forked. Disc florets 49-63.	W. digitata
	Note: Hind (2022) moved all former species of Werneria into a new genus Rockhausenia and transferred the former genus Xenophyllum to Werneria.	

Basellaceae

1	Twining vine. Leaves succulent. Inflorescences an axillary raceme.	Anredera diffusa

Berberidaceae

Berberis

1	Unarmed or armed shrub, up to 3 m tall, wood yellow. Leaves entire to spiny.	B. lutea
1*	Armed shrub. Leaves spiny-dentate.	B. saxicola
	Note: B. lutea and B. saxicola may be conspecific.	

Betulaceae

1	Deciduous shrub or tree, up to 30 m tall. Inflorescence a many-flowered catkin. Infructescence cone-like.	Alnus acuminata

Bignoniaceae

1	Perennial herbs. Leaves 2-pinnatifid.	Argylia radiata
1*	Shrubs. Leaves pinnate or entire.	Tecoma

Tecoma

2	Corolla yellow	T. stans
2*	Corolla bicolorous, inside yellow, outside reddish.	T. fulva

Boraginaceae

Incl. Cordiaceae, Ehretiaceae, Hydrophyllaceae, Namaceae.

	Key based on Flora of Peru (1936ff.) and Schwarzer (2007)	
1	Herbs, dichotomously branching.	Nama dichotoma
1*	Herbs or shrubs, not dichotomously branching	2
2	Leaves pinnatifid to 1-pinnate (3-foliate).	Phacelia
2*	Leaves entire.	3
3	Ovary slightly lobed if at all, sometimes 2-4-grooved laterally.	4
3*	Ovary deeply 4-parted, the fruit consisting of 4 (or fewer) nutlets.	7
4	Style entire, stigma 1. Inflorescences or its branches scorpioid or its flowers solitary.	5
4*	Style 2-parted or forked.	6
5	Anthers apically glabrous. Herbs or shrubs.	Heliotropium
5*	Anthers with pubescent or papillose apex. Subshrubs.	Euploca
6	Much-branched prostrate herbs or sub-shrubs. Flowers white, blue or lilac.	Tiquila
6*	Erect shrub. Flowers yellow.	Cordia lutea
7	Corolla yellow.	Amsinckia calycina
7*	Corolla white at least in large part.	8
8	Calyx in fruiting star-shaped, in nodding position.	Pectocarya
8*	Calyx in fruiting cup-shaped, in upright position.	9
9	Rosulate or semi-rosulate herb.	Plagiobothrys
9*	Plant not rosulate.	10
10	Nutlets thin-margined, dissimilar. Spikes not bracted or only at base.	Johnstonella parviflora
10*	Nutlets obtusely if at all margined, ± homomorphic. Spikes bracted except C. peruviana.	Cryptantha

Cryptantha

	Key based on Flora of Peru (1936ff.) and Schwarzer (2007)	
1	Corolla showy, to 7 mm wide. Style scarcely or not exserted.	C. granulosa
1*	Corolla 0.5-2 mm wide. Style longer than nutlets.	2
2	Spikes not bracted.	C. peruviana
2*	Spikes bracted.	3
3	Calyx in fruit 3-4 mm long, narrow lobes suberect to tips.	C. filaginea
3*	Calyx in fruit 5-6 mm long, broad lobes spreading at tips.	C. limensis

Euploca

1	Leaves elliptic-lanceolate, densely pilose, up to 4 cm long.	E. pilosa
1*	Leaves lanceolate, not pilose, up to 6 cm long.	E. toratensis

Heliotropium

1	Shrubs, >0.5 m tall.	2
1*	Herbs or subshrubs, <0.5m tall.	4

2	Leaves linear-lanceolate, 1 cm wide.	H. krauseanum
2*	Leaves ovate-lanceolate, >2 cm wide.	3
3	Flowers pink to blue. Fruit dry.	H. arborescens
3*	Flowers white. Fruit fleshy, white.	H. lilloi
4	Leaves linear-lanceolate, lateral veins faint, ± succulent, glaucous.	H. curassavicum
4*	Leaves obovate, lateral veins conspicuous, grayish-green.	H. microstachyum

Pectocarya

	Key based on Flora of Peru (1936ff.) and Schwarzer (2007)	
1	Decumbent to ascending herb. Nutlets subulately fringed dorsally as marginally. Calyx adaxially glabrous.	P. anomala
1*	Ascending to erect herb. Nutlets smooth dorsally. Calyx adaxially puberulent.	P. linearis

Phacelia

1	Leaves pinnatifid.	P. pinnatifida
1*	Leaves 3-foliate.	P. secunda

Plagiobothrys

	Key based on Flora of Peru (1936ff.) and Schwarzer (2007)	
1	Plant rosulate. Leaves mostly in a basal rosette.	P. myosotoides
1*	Plant semi-rosulate. Lower leaves opposite, even the earliest.	2
2*	Perennial herbs, rooting at lower nodes.	P. humilis
2	Annual herb.	3
3	Branched from the base, but not creeping and rooting at nodes. Stems up to 4 cm tall.	P. macbridei
3*	Branched from the base, decumbent and creeping, rooting at nodes. Stems up to 15 cm tall.	P. linifolius

Tiquilia

	Key based on Richardson (1977)	
1	Corolla persistent, stamens exserted 3-8 mm. Nutlet hemi-ovoid, ventral surface broadly flattened.	2
1*	Corolla deciduous, stamens included or slightly exserted. Nutlet not hemi-ovoid, ventral surface not broadly flattened.	5
2	Leaf blade deeply plicate, usually marginally crenate.	3
2*	Leaf blade not deeply plicate, marginally entire.	4
3	Corolla 8.5-13 mm long.	T. grandiflora
3*	Corolla 5-6 mm long.	T. dichotoma
4	Corolla 7.5-10 mm long.	T. ferreyrae
4*	Corolla 4.5-5.5 mm long.	T. simulans
5	Nutlet ovoid.	T. paronychioides
5*	Nutlet spheroid.	6
6	Leaf margins entire. Corolla lilac, sky-blue or milky-white.	7
6*	Leaf margins crenate. Corolla blue.	8
7	Corolla 4.5-8 mm long. Style attachment apical.	T. litoralis
7*	Corolla 7-13 mm long. Style attachment gynobasic.	T. conspicua
8	Leaves 10-23 mm long with 3-4 pairs of lateral veins. Corolla 5.5-12 mm long.	T. elongata
8*	Leaves 5-6 mm long with 2-3 pairs of lateral veins. Corolla 5.5-6.5 mm long.	T. hunteri

Brassicaceae

	Key based on Flora of Peru (1936ff.), Bailey (2007) and Al-Shehbaz (2006a)	
1	Fruit indehiscent, divided into 2 winged parts on a gynophore (stalk of ovary).	Cremolobus chilensis
1*	Fruit not divided into 2 parts, rarely with a gynophore.	2
2	Fruit with seedless terminal segment like a beak, beak usually longer than 2 seeds.	3

2*	Fruit without seedless terminal segment, if beaked, then beak <1 seed.	5
3	Hirsute herb. Fruit thick, >5 mm in Ø. usually white or purple with blue veins, sometimes yellowish.	Raphanus raphanistrum
3*	Glabrous herb. Fruit thin, <4 mm in Ø. Petals deep yellow.	4
4	Basal leaves >5 cm wide. Cauline leaves clasping.	Brassica rapa
4*	Basal leaves <3.5 cm wide. Cauline leaves sessile, if present.	Diplotaxis muralis
5	Fruit laterally compressed, the septum (partition that separates the locules of a fruit) ± linear.	6
5*	Fruits not laterally compressed, the septum broad.	9
6	Fruits heart-shaped or round to obovate, generally winged.	7
6*	Fruits triangular or elliptic to oblong, not winged.	8
7	Fruit >10 mm in Ø.	Thlaspi arvense
7*	Fruits <6 mm in Ø.	Lepidium
8	Fruits triangular.	Capsella bursa-pastoris
8*	Fruits elliptic to oblong.	Mancoa
9	Fruits short, thick, 1-4x longer than broad.	10
9*	Fruits elongate-linear, >5x longer than broad.	17
10	Inflorescence ± sessile to short pedunculate. Plant of high elevations.	11
10*	Inflorescence distinctly pedunculate.	13
11	Leaves herbaceous, pinnate.	Rorippa nana
11*	Leaves fleshy, entire.	12
12	Leaves linear, margins entire to lobbed or dentate.	Alshehbazia
12*	Leaves obovate-spatulate, purplish-gray.	Aschersoniodoxa cachensis
13	Leaves in rosettes, stems sparsely leafy or leafless.	14
13*	Much-branched herb, stems with many cauline leaves.	15
14	Flowers arranged in an inflorescence with more than 1 flower.	Draba
14*	Flowers solitary, pedicellate.	Eudema
15	Fruit spheroid.	Lobularia maritima
15*	Fruit cylindric.	16
16	Annual or perennial herb growing above 2500 m. Cauline leaves dentate.	Weberbauera
16*	Perennial herb growing below 2000 m. Cauline leaves deeply lobbed with narrowly linear lobes.	Mathewsia
17	Submersed aquatic plant or plant of wet areas.	Nasturtium officinale
17*	Terrestrial plant.	18
18	Leaves 2-pinnately parted, often dissected, the divisions fine. Moistened seeds mucilaginous.	Descurainia
18*	Leaves simple to variously divided but not 2-pinnately dissected, or the moistened seeds not mucilaginous.	19
19	Basal leaves not developed, cauline leaves petiolate, not auriculate at base, finely pinnatisect into narrowly linear lobes, usually grooved abaxially and densely pubescent along groves, fleshy, entire.	Machaerophorus
19*	Leaves not shaped as mentioned above.	20
20	Plant ± gray. Fruit with 2-3 horns.	Matthiola incana
20*	Plant green. Fruit without horns.	21
21	Flowers pure yellow (see also Mostacillastrum oleraceum with yellowish-white flowers and Neuontobotrys camanaensis with yellow flowers.	Sisymbrium
21*	Flowers white, pink or lilac (or yellowish-white, see remark above).	22
22	Stems decumbent to ascending.	Weberbauera
22*	Stems usually erect.	23
23	Stems solitary, branched at top, 40-90 cm tall. All parts densely pubescent, gray-green in appearance. Leaves auriculate to amplexicaule, oblong, irregularly dentate.	Exhalimolobos weddellii
23*	Not in the above combination of characteristics.	24

24	Annual herbs, glabrous throughout, usually branched from the base. Cauline leaves sessile, auriculate. Fruits <3 cm long. Growing below 1000 m on the coast or in lomas.	Dictyophragmus
24*	Not in the above combination of characteristics.	25
25	Plants glabrous or with simple trichomes, sometimes with stalked, 2-rayed (forked) trichomes. Fruit valves usually 1-4-veined, fruits 3-10 cm long, neither strongly falcate nor strongly recurved, fruiting pedicels divaricate to ascending.	Mostacillastrum
25*	Plants with dendritic trichomes at least on leaves and sepals, rarely glabrescent. Fruit valves not veined. Fruits to 0.5-3 cm long, often strongly falcate and/or strongly recurved, fruiting pedicels strongly reflexed or recurved, rarely divaricate and straight.	Neuontobotrys

Alshehbazia

	Key based on Al-Shehbaz et al. (2023)	
1	Leaves entire, adaxially densely pubescent, rarely glabrescent or only ciliate. Sepals free. Wetlands.	A. werdermannii
1*	Leaves lobed or rarely coarsely dentate, adaxially sparsely pubescent. Sepals connate. Dry puna.	A. friesii

Descurainia

	Key based on Al-Shehbaz (2012)	
1	Infructescence congested capitate at least along lateral branches. Sepals and petals persistent after fruit dehiscence.	D. athrocarpa
1*	Infructescence corymbose to elongated raceme. Sepals and petals usually caducous before or during fruit development.	2
2	Fruits 4-9 mm long. Perennial herb. Stem prostrate to decumbent from a caudex.	D. depressa
2*	Fruits 9-18 mm long. Annual herbs.	3
3	Fruits and young ovaries glabrous. Plants usually single stemmed.	D. myriophylla
3*	Fruits and young ovaries pubescent. Plants usually several- to many-stemmed.	D. stricta

Dictyophragmus

	Key based on Al-Shehbaz (2006b)	
1	Seeds broadly winged all around. Fruit 3-4 mm wide.	D. englerianus
1*	Seeds wingless. Fruit 1.5-2 mm wide.	D. lactucoides

Draba

	Key based on Al-Shehbaz (2006b) and Al-Shehbaz (2018)	
1	Petals blue.	D. werffii
1*	Petals white.	2
2	Perennial herb, much-branched. Peduncles <4 cm.	D. lapaziana
2*	Annual herb, rosulate. Peduncles >6 cm.	D. soratensis

Eudema

	Key based on Al-Shehbaz et al. (2023)	
1	Sepals persistent well after fruit dehiscence. Leaves above densely pubescent, below sparsely so.	E. monimocalyx
1*	Sepals caducous shortly after flowering. Leaves glabrous or sparsely pilose above, glabrous below.	2
2	Petals purple. Seeds reticulate. Leaves longer than to equaling fruit plus fruiting pedicel.	E. arequipa
2*	Petals white. Seeds smooth. Leaves shorter than fruit plus fruiting pedicels.	E. calycinum

Lepidium

	Key based on Al-Shehbaz (2010)	
1	At least some cauline leaves auriculate, sagittate or amplexicaul.	2
1*	Cauline leaves not auriculate, petiolate or sessile, sometimes absent.	4
2	Stamens 4, petals 2-3 mm. Apical notch of fruit absent.	L. crassius
2*	Stamens 2, petals 0.3-1.5 mm. Apical notch of fruit 0.1-1 mm.	3

3	Middle and upper cauline leaves pinnatifid or pinnatisect.	L. bipinnatifidum
3*	Middle and upper cauline leaves entire or dentate.	L. chichicara
4	Annuals or rarely biennials.	5
4*	Perennials with ± woody base or thickened root.	7
5	Fruits spectacle-like.	L. didymum
5*	Fruits ovate to elliptic.	6
6	Basal leaves pinnatifid.	L. strictum
6*	Basal leaves entire, serrate to dentate.	L. virginicum
7	Stems 30-150 cm long, erect. Basal leaves 10-30 cm long, margins dentate. Stamens 6.	L. latifolium
7*	Stems 4-25 cm long, decumbent to ascending. Basal leaves 1.7-11 cm long. Stamens 4.	L. werffii

Machaerophorus

	Key based on Al-Shehbaz et al. (2018)	
1	Fruit terete, cylindrical, torulose. Petals white. Seeds uniseriate.	M. arequipa
1*	Fruit strongly flattened, oblong-lanceolate to lanceolate, not torulose. Petals yellow. Seeds biseriate.	M. laticarpus

Mancoa

	Key based on Bailey et al. (2007)	
1	Stems and fruit valves pubescent.	M. hispida
1*	Stems and fruit valves glabrous.	M. laevis

Mathewsia

1	Flowers yellow, pedicels widely spreading, 3-5 mm long.	M. densifolia
1*	Flowers (yellowish-)white, pedicels ascending, 6-10 mm long.	M. peruviana

Mostacillastrum

	Key based on Al-Shehbaz (2006a)	
1	Uppermost leaves sessile, sagittate. All leaves linear to narrowly linear-lanceolate, entire.	M. dianthoides
1*	All leaves petiolate. Neither auriculate nor sagittate.	2
2	Basal and middle cauline leaves entire or dentate.	3
2*	Basal and middle cauline leaves pinnately lobed to pinnatisect.	5
3	Plants annual. Seeds 0.8-1.1 mm long, coarsely reticulate, with 150-200 muri on each side.	M. ferreyrae
3*	Plant perennial. Seeds 1.3-1.7 mm long, minutely reticulate, with 600-700 muri on each side.	4
4	Plants glabrous. Flowers yellowish-white. Fruits 3-4.5 cm long.	M. oleraceum
4*	Plants densely pubescent. Flowers white. Fruits 5-6.5 cm long.	M. morrisonii
5	Leaves pectinate, upper side sulcate. Sepals 1.8-2.5 mm long, petals 2.5-3.5 mm long.	M. pectinifolium
5*	Leaves pinnatifid to pinnatisect, upper side flat. Sepals 3-4 mm long, petals 4-5.5 mm long.	M. gracile

Neuontobotrys

	Key based on Al-Shehbaz (2004b)	
1	Upper cauline leaves linear, 0.5-2 mm wide.	N. schulzii
1*	Upper cauline leaves linear, >5 mm wide.	2
2	Middle and upper cauline leaves ± entire, glabrous.	N. amplexicaulis
2*	Middle and upper cauline leaves dentate or lobed, often densely pubescent, rarely glabrescent.	3
3	Flowers yellow.	N. camanaensis
3*	Flowers white.	N. lanatus

Sisymbrium

	Key based on Jepson (2024).	
1	Fruit narrowly awl-shaped, 1-1.4 cm long, adpressed to rachis. Seeds 10-20.	S. officinale
1*	Fruit narrowly linear, 2-10 cm, not adpressed to rachis. Seeds 30–140.	2
2	Pedicels narrower than fruit. Fruit 2-4 cm long.	S. irio
2*	Pedicels ± as thick as fruit. Fruit up to 10 cm long.	S. orientale

Weberbauera

	Key based on Al-Shehbaz (1990) and Al-Shehbaz & Montesinos-Tubée (2009)	
1	Fruit 1-3x longer than broad. Petals pale yellow. Plant ± fleshy.	W. ayacuchoensis
1*	Fruit more than 4x longer than broad. Petals white to pinkish. Plant herbaceous.	2
2	Basal leaves filiform to linear.	W. minutipila
2*	Basal leaves not filiform or linear.	3
3	Fruit 5-7 mm long.	W. arequipa
3*	Fruit 10-13 mm long.	W. peruviana

Bromeliaceae

1	Leaves usually spinose. Seeds winged or with caudate appendage.	Puya
1*	Leaves entire. Seeds plumose.	Tillandsia

Puya

1	Plants in flower >3 m tall.	2
1*	Plant in flower usually <2 m tall.	3
2	Plant up to 3 m in vegetative growth, leaves borne on stout stem. Inflorescence 9-10 m long. Flowers bluish-white.	P. raimondii
2*	Leaves rosulate, plant stemless in vegetative growth. Inflorescence 2-3 m long. Flowers greenish-yellow.	P. hoxeyi
3	Peduncle and inflorescence branches grayish-white.	4
3*	Peduncle and inflorescence branches reddish, pinkish or ferruginous.	6
4	Inflorescence densely branched, 20-35 cm long. Stout herb.	P. colca-canyonense sp. nov.
4*	Inflorescence laxly branched, >40 cm long. Slender herb.	5
5	Inflorescence dense, branches 4-8 cm long, 1-2 cm apart.	P. cahuachensis
5*	Inflorescence lax, branches 12-24 cm long, 4-6 cm apart.	P. colcaensis
6	Inflorescence lax, few-flowered, branches and bracts ferruginous.	P. ferruginea
6*	Inflorescence dense, many-flowered, branches and bracts pinkish-green.	P. cylindrica

Tillandsia

1	Plants usually terrestrial. Leaf blade at base >1 cm wide, in T. paleacea ± narrower.	2
1*	Plants epiphytic or growing on rocks. Leaf blade at base <0.5 cm wide.	4
2	Inflorescence a single terminal spike.	T. paleacea
2*	Inflorescence compound.	3
3	Plant rootless. Spikelets dense, 6-12-flowered. Peduncle and bracts pink-red, petals pink.	T. latifolia
3*	Plant rooting. Spikelets laxly 7-flowered. Peduncle green, bracts pink, petals white with blue apex.	T. purpurea
4	Plants 1-5-flowered.	T. recurvata
4*	Plants 1-2-flowered.	5
5	Plants hanging from trees and rocks in strands up to 3 m long.	T. usneoides
5*	Plants small, epiphytic or lithophytic.	6
6	Blade >2 mm wide.	T. hirta
6*	Blade <2 mm wide.	7
7	Petals uniformly colored.	T. capillaris
7*	Petals often spotted.	T. virescens

Cactaceae

	Key based on Pauca & Quipuscoa (2017)	
1	Deciduous leaves on young stems, areoles with glochids. Seeds with aril.	Subfam. Opuntioideae
1*	Leafless, areoles without glochids. Seeds without aril.	Subfam. Cactoideae

Subfamily Opuntioideae

	Key based on Pauca & Quipuscoa (2017)	
1	Stem flat, sometimes cylindrical, but then <3cm in Ø.	2
1*	Stem terete, globular, when cylindrical, then >3cm in Ø.	3
2	Plant <15 cm tall, prostrate. Style green. Growing above 3300 m.	Airampoa soehrensii
2*	Plant >15 cm tall, prostrate. Style not green. Usually growing below 3000 m.	Opuntia
3	Spines covered with a papyraceous sheath.	Cylindropuntia
3*	Spines not covered with a papyraceous sheath.	4
4	Stem with defined growth, regularly articulated.	Cumulopuntia
4*	Stem with undefined growth, not regularly articulated.	Austrocylindropuntia

Subfamily Cactoideae

	Key based on Pauca & Quipuscoa (2017)	
1	Fruits glabrous or covered by scales.	2
1*	Fruits hairy and scaly, sometimes bristly or spiny.	3
2	Tree-like cactus. Flowers open at night, white, no woolly and bristly swelling at the top (= cephalium).	Browningia candelaris
2*	Globular cactus. daytime open flowers, pink, with woolly and bristly swelling at the top (only in province Caravelí).	Melocactus peruvianus
3	Fruits hairy, scaly, and bristly or spiny.	4
3*	Fruits only hairy and scaly, not bristly or spiny.	8
4	Flowering terminal, flower buds at top of stem. Fruits dry, hairy and bristly.	Eriosyce islayensis
4*	Flowering axillary, flower buds lateral on stem. Fruit fleshy.	5
5	Tree cactus, stems up to 10 m tall and 40 cm in Ø. Areoles grow continuously, forming cylindrical buds.	Neoraimondia arequipensis
5*	Shrubby cactus, stems <7 m tall and <20 cm in Ø. Areoles not forming cylindrical buds.	6
6	Stems articulate.	Armatocereus
6*	Stems not articulate.	7
7	Stem basally branched.	Corryocactus
7*	Stem laterally branched.	Eulychnia ritteri
8	Globular, cylindrical or decumbent cactus, <90 cm tall.	9
8*	Columnar cactus, erect or ascending, >1 m tall.	13
9	Daytime open flowers, red or orange, sometimes pink.	10
9*	Flowers open at night, white.	Haageocereus
10	Globular cactus, up to 20 cm tall.	11
10*	Stems first erect, then decumbent, up to 40 cm long.	12
11	Flowering lateral.	Lobivia
11*	Flowering terminal.	Matucana haynii
12	Flowering terminal, flower buds at top of stem. Fruits dry, dehiscent.	Oreocereus
12*	Flowering axillary, flower buds lateral on stem. Fruit fleshy, indehiscent.	Loxanthocereus
13	Plants with abundant and long, white hairs.	Oreocereus
13*	Plants with few hairs, hairs not white.	14
14	Flowers 15-25 cm long. Fruit covered with long, black or brown hairs.	Trichocereus
14*	Flowers 9-12 cm long. Fruit covered with short, whitish or whitish-brown hairs.	Weberbauerocereus

Armatocereus

1	Stems usually unbranched and solitary, up to 7 m tall, 8-10 ribs. Fruit with white spines.	A. procerus
1*	Stem much-branched, up to 3 m tall, 7-9 ribs. Fruit with gray-violet spines.	A. matucanensis
	Note: A. matucanensis is often split. For Arequipa, the most commonly mentioned species is A. riomajensis, which is synonymized with A. matucanensis by POWO (2024).	

Austrocylindropuntia

1	Upright, tree-like cactus with many branches.	A. subulata
1*	Plant forming loose to dense cushions. The silvery hairs project well above the cushion (compare Punotia lagopus).	A. floccosa

Corryocactus

1	Stem 6-20 cm in Ø, erect, 1.5-5 m tall, branched from the base.	2
1*	Stem 3-5 cm in Ø, erect to decumbent, 0.1-0.5 m tall, up to 2 m long when decumbent, generally unbranched.	3
2	Stem 6-10 cm in Ø. Flowers orange. Growing in the lomas (0-1000 m).	C. brachypetalus
2*	Stem 10-20 cm in Ø. Longest spine often pointing down. Flowers yellow. Growing above 2000 m.	C. brevistylus
3	Plant without subterranean stems. Stem with 4-5 ribs. Flowers yellow to brilliant red.	C. quadrangularis
3*	Plants with subterranean stems. Stem with 5-8 ribs. Flowers usually yellowish-orange.	4
4	Growing above 1500 m.	C. aureus
4*	Growing below 1000 m (lomas).	C. dillonii

Cumulopuntia

	Key based on Pauca-Tanco & Quipuscoa (2020)	
1	Plants form dense cushion-like clusters. Stem segments with <32 areoles. Spines and areoles usually apically concentrated. Growing above 3000 m.	2
1*	Loosely clustered or solitary cacti. Stem segments with >32 areoles. Growing below 3200 m.	5
2	Root fibrous. Areoles at apex >4 mm in Ø. Longest spine up to 8 cm long, curved. Style apically thickened.	C. zehnderi
2*	Root napiform. areoles at apex <4 mm in Ø. Style basally thickened.	3
3	Stem segments with 20-32 areoles, ± uniformly distributed. Tepals yellowish or white.	C. corotilla
3*	Stem segments with 8-18 areoles, apically concentrated. Tepals deep yellow, orange or red.	4
4	Flowers red or orange-red. Fruit cylindrical.	C. ignescens
4*	Flowers yellow or orange. Fruit globose or subglobose.	C. ignota
5	Stem segments with 16-25 areolas.	C. dimorpha
5*	Stem segments with 28-60 areolas	6
6	Spines present on all the areoles, apical areoles with 12-22 spines. Spines curved and appressed on stems.	C. sphaerica complex
6*	Spines present from the middle part of stem segments, apical areoles with up to 11 spines. Spines usually straight, sometimes curved, not appressed on stems.	C. leucophaea
	Note: The taxonomy of Cumulopuntia is not yet settled. The most recent study of the Arequipa species was made by Pauca-Tanco & Quipuscoa (2020), who described 12 species. The Caryophyllales Network (2015ff.) accepts only 6 of them: C. corotilla (syn. C. mistiensis), C. ignescens (syn. C. glomerata), C. ignota, C. leucophaea, C. sphaerica (syn. C. crassicylindrica, C. dimorpha, C. multiareolata, C. tumida, C. unguispina) and C. zehnderi. POWO (2024) does the same but accepts C. dimorpha and synonymizes C. ignota with C. sphaerica. Here we include both disputed species (C. ignota and C. dimorpha).	

Cylindropuntia

1	Widely branched, large shrub, 1.5-2.5 m tall. Stem segments firmly attached. Flowers reddish-pink. Spines up to 3 cm long.	C. imbricata
1*	Densely branched shrub, up to 0.6 m tall. Stem segments easily detached. Flowers yellow-green. Spines up to 6 cm long.	C. tunicata

Haageocereus

1	Small cactus, subterranean stem <10 cm tall. Root napiform.	H. bylesianus
1*	Columnar cactus, decumbent, stems >50 cm long.	2
2	Stems decumbent, almost whole stem on the ground, 20 ribs. Growing in coastal areas.	H. decumbens
2*	Stems ascending, at least 1/3 of stem erect, 13-15 ribs. Growing above 1000 m.	H. platinospinus

Lobivia

1	Flowers 3-4 cm long. Spines ± erect.	L. maximiliana
1*	Flowers up to 6 cm long. Spines conspicuously curved.	L. pampana

Loxanthocereus

1	Erect to slightly decumbent columnar cactus. Flowers lateral, red.	L. jajoanus
1*	Prostrate columnar cactus. Flowers lateral, red.	L. sextonianus

Opuntia

1	Plant decumbent to ascending, <50 cm tall. Stem unequal, terete or flattened, segments up to 15 cm long.	O. pubescens
1*	Plant usually erect, >1 m tall. Stem usually flattened.	2
2	Plant to 5 m tall. Stem segments up to 50 cm long, with thin spines. Flowers red or yellow. Fruits pale green to red.	O. ficus-indica
2*	Plant to 2 m tall. Stem segments up to 25 cm long, with strong spines. Flowers bright yellow. Fruits purple when ripe, pulp purple.	O. stricta

Oreocereus

1	Flowering terminal, flower buds at top of stem. Fruits dry, dehiscent.	O. hempelianus
1*	Plants with abundant and long, white hairs.	O. leucotrichus

Trichocereus

1	Flowers up to 25 cm long. Stem often glaucous. Spines 1-4 cm long. Growing above 2000 m.	T. macrogonus
1*	Flowers up to 17 cm long.	2
2	Spines 2-3 cm long. Growing in lomas.	T. chalaensis
2*	Spines up to 12 cm long. Growing above 2500 m.	T. cuzcoensis
	Note: The taxonomy of Trichocereus is not yet settled. The most recent study of the Arequipa species was made by Pauca-Tanco & Quipuscoa (2020), who described 3 species, T. chalensis, T. cuzcoensis and T. schoenii. POWO (2024) synonymizes T. schoenii with T. macrogonus.	

Weberbauerocereus

1	Old plants with usually a single basal stem, 4-6 m tall. Spines 4-7 cm long.	W. rauhii
1*	Old plant with several basal stems, 2-4 m tall.	2
2	Plant up to 2 m tall. Spines up to 12 cm long. Growing below 1000 m.	W. cephalomacrostibas
2*	Plant up to 4 m tall. Spines up to 6 cm long. Growing above 1500 m.	W. weberbaueri

Calceolariaceae

Calceolaria

	Key based on Molau (1988).	
1	Leaves in general >5 cm long, decussate.	2
1*	Most leaves <5 cm long (in C. angustiflora and C. rugulosa some leaves are >5 cm long, combined with ternately arranged leaves on the main stem).	7
2	Leaves pinnatifid to pinnate or deeply lobed almost to the central vein.	3
2*	Leaves entire or shallowly lobed.	5
3	Leaves 4-15 x 3-12 cm, pinnatisect to pinnate with 4+ pairs of pinnae, pinnae pinnatifid to pinnatisect. Growing in coastal areas and in lomas.	C. pinnata

3*	Leaves lobed to pinnatifid, up to 2-4 pairs of lobes, lobes serrate to dentate. Growing above (1000-) 1500 m.	4
4	Plant up to 2 m tall. Leaves 5-25 x 4-19 cm long, 2-4 pairs of broad lobes. Petioles up to 12 cm long, conspicuously connate across the nodes with 2-11 mm wide wings.	C. chelidonioides
4*	Plant up to 0.5 m tall. Leaves 3-16 x 1-16 cm long, 3 pairs of lobes. Petioles up to 5 cm long, connate across the nodes with 0.3-5 mm wide wings.	C. tenuis
5	Cauline leaves sessile, basal leaves with an alate petiole.	C. plectranthifolia
5*	Leaves petiolate.	6
6	Leaves suborbicular to deltoid, palmately lobed up to 1/3 of radius. Corolla with spotted throat.	C. lobata
6*	Leaves ovate, margins serrate to slightly lobed. Corolla with a large purple blotch inside the throat.	C. rhacodes
7	Flowers orange to brownish-red.	C. pisacomensis
7*	Flowers pale to deep yellow.	8
8	Leaves ternate, at least at the main axis (in C. rugulosa sometimes decussate).	9
8*	Leaves decussate.	11
9	Leaves usually linear.	C. ajugoides
9*	Leaves ovate or lanceolate.	10
10	Leaves lanceolate, 0.7-2 cm wide, blade smooth. Corolla 10-18 mm long.	C. angustiflora
10*	Leaves ovate, 2-3.5 cm wide, blades rugulose or rugose. Corolla up to 10 mm long.	C. rugulosa
11	Stem dichotomously branched. Grows in the lomas, below 2000 m.	C. dichotoma
11*	Stem laterally branched. Grows above 2000 m.	12
12	Leaf-margins seeming entire. Leaves not fasciculate.	C. engleriana
12*	Leaf-margins lobulate, serrate or crenate. Leaves fasciculate.	13
13	Leaves cuneate to ± rounded at base. Upper leaf surface hirsute, lower surface puberulous. Blade lanceolate with 3-5 teeth on each side, entire in proximal third. Corolla whitish- to lemon-yellow.	C. cuneiformis
13*	Leaves truncate or cordate at base.	14
14	Leaves lobulate, with 3-9 lobes on each side.	C. inamoena
14*	Leaves serrate or crenulate, with 3-10 teeth on each side.	15
15	Glands on lower surface of leaves stalked, upper surface puberulous or pilose. Corolla golden yellow.	C. colquepatana
15*	Glands on lower surface of leaves orange, sessile, upper surface tomentose or short-hirsute, whitish. Corolla pale yellow.	C. parvifolia

Calyceraceae

Calycera

1	Rosulate herb. Leaves spatulate, crenate. Inflorescence with 7 involucral bracts, flowers white.	Calycera pulvinata

Campanulaceae

1	Minute, short-lived plant, stem <3 mm. Only present in March-Mai at altitudes above 4000 m.	Lysipomia mitsyae
1*	Plant >1 cm.	2
2	Flowers actinomorphic.	3
2*	Flowers zygomorphic.	4
3	Annual, erect herb of the lomas. Corolla divided to the middle, lilac.	Triodanis perfoliata
3*	Perennial, prostrate subshrub of high altitudes. Corolla campanulate, white to blue.	Wahlenbergia peruviana
4	Flowers white or blue to purple.	Lobelia
4*	Flowers orange to red.	Siphocampylus candollei

Lobelia

1	Prostrate cespitose herb, rooting at nodes. Corolla white, tinted pinkish-lavender, <1 cm long. Usually growing in wet soil.	L. oligophylla
1*	Erect herbs, >50 cm tall. Corolla blue to purple.	L. decurrens

Cannaceae

1	Tuberous, perennial herb, up to 1.5 m tall. Stem glabrous. Leaves large, stalk sheathing.	Canna indica

Caprifoliaceae

1	Twining vine climbing up to 10 m in trees.	Lonicera japonica
1*	Herbs, not climbing.	2
2	Tall biennial herb with prickles on stem and leaves.	Dipsacus sativus
2*	Unarmed herb up to 1 m tall.	Valeriana

Valeriana

	Key based on Flora of Peru (1936ff.)	
1	Plant caulescent, leaves erect, stem >1 cm.	2
1*	Plant ± acaule, leaves lying flat on the ground. Stem <1 cm.	9
2	Flowers in dense, globose or subglobose heads, terminal or in an interrupted spike.	3
2*	Flowers in cymes, loose clusters or panicles.	5
3	Leaves ciliolate.	V. coarctata
3*	Leaves not ciliolate.	4
4	Leaf blades >4 mm wide, abruptly or subabruptly tapering to the petiole.	V. nivalis
4*	Leaf blades <4 mm wide, gradually tapering to the petiole.	V. globularis
5	Lax, diffuse, annual herb. Leaves pinnate and pinnatifid. Looks like Apiaceae.	V. chaerophylloides
5*	Stout, erect herb.	6
6	Leaves entire. Flowers zygomorphic, deep reddish-pink.	V. rubra
6*	Leaves pinnate. Flowers ± actinomorphic, whitish-pink.	7
7	Basal and lower cauline leaves pinnatisect, not incised to the axis.	V. pinnatifida
7*	Basal and lower cauline leaves pinnate, incised to the axis.	8
8	Terminal leaflet much larger than lateral ones. Leaflets broad.	V. radicata
8*	Terminal leaflet not much larger than lateral ones. Leaflets narrow. Stem appears interrupted by the nodes.	V. interrupta
9	Plant loosely cespitose, stoloniferous. Leaves acuminate.	V. wandae
9*	Plant densely cespitose, without stolons. Leaves subrotund.	V. rhizantha

Caricaceae

1	Shrub or tree, up to 8 m tall, usually branching from base. Leaves broad ovoid to orbicular, base cordate.	Vasconcellea candicans
1*	Tree with broad stem. Leaves lanceolate, base obtuse.	Carica augusti
	Note: Vasconcellea pubescens A.DC. is often cultivated in Arequipa.	

Caryophyllaceae

	Key based on Flora of Peru (1936ff.) and Jepson (2024).	
1	Sepals connate, forming a tube.	Silene
1*	Sepals free or slightly connate, not tubular.	2
2	Fruits capsular. Flowers usually pedunculate and petals white, plants green.	3
2*	Fruits not capsular (fruit usually a small, bladdery, 1-seeded dry fruit = utricle). Flowers usually in axillary, sessile glomeruli, or large, yellowish-brown cushion plants.	10
3	Leaves ovate to orbicular, <2x as long as wide, base often cordate and clasping. Stem slender, with long internodes. Petals parted at least half of their length (variable genus with no distinctive characteristics).	Drymaria
3*	Leaves longer than wide, not orbicular.	4

4	Plants without stipules.	5
4*	Plants with stipules.	9
5	Petals absent.	Colobanthus quitensis
5*	Petals present	6
6	Petals notched or parted to base in 2-segments.	7
6*	Petals entire.	8
7	Plants glandular pubescent on leaves, stem and inflorescence.	Cerastium
7*	Plants glabrous or pubescent, but not glandular.	Stellaria
8	Procumbent to prostrate herb, forming lax mats, but not cushions.	Arenaria
8*	Prostrate herb forming dense cushions.	Stellaria
9	Leaves oblong, broader than linear. Plants glabrous. Petals inconspicuous, distinctly shorter than sepals.	Polycarpon tetraphyllum
9*	Leaves linear. Plants pubescent. Petals longer or as long as sepals.	Spergularia
10	Shrubs forming large cushions. Leaves yellowish-brown. Flowers solitary, terminal.	Pycnophyllum
10*	Plants prostrate, if forming cushions, then leaves green. Flowers axillary, if terminal, then in glomeruli.	11
11	Stipules <1 mm. Annual herb, Portulaca-like, but inflorescence in axillary glomeruli.	Herniaria sp.
11*	Stipules >1 mm. Perennial herbs.	12
12	Leaves linear, spine-tipped at apex.	Cardionema ramosissima
12*	Leaves elliptic to obovate, apex mucronate, but not spinulose.	Paronychia

Arenaria

	Key based on Montesinos-Tubée & Teillier (2022).	
1	Stamens 5. Petals shorter than sepals.	A. orbignyana
1*	Stamens 10. Petals longer than sepals.	2
2	Herb forming solitary cushions 4-6 cm in Ø.	A. acaulis
2*	Cespitose herb, not forming cushions.	3
3	Leaves oblanceolate, not succulent, margins ciliate.	A. digyna
3*	Leaves ovate-triangular, distinctly fleshy, glabrous. Growing in salt marshes.	A. rivularis

Cerastium

1	Leaves oblong, pubescent to glabrous. Low hemispherical herb, stems procumbent.	C. behmianum
1*	Leaves spatulate, glandular. Cespitose herb, stems usually erect.	C. peruvianum

Drymaria

	Key based on descriptions of Duke (1961).	
1	Inflorescence (sub)capitulate, pedicels <3 mm. Petals exceeding the sepals. Restricted to coastal lomas.	D. paposana
1*	Inflorescence cymose, pedicels usually >3 mm. Petals as long as sepals.	2
2	Inflorescence a lax cyme, flowers pedicellate. Leaves usually wider than long. Growing between 0-3000 m..	D. divaricata
2*	Inflorescence dichotomously branched. Leaves usually longer than wide. Growing above 1500 m.	D. fasciculata

Paronychia

	Key based on Chaudhri (1968) and Iamonico & Montesinos-Tubée (2023).	
1	Erect to decumbent shrub, often 1 m tall. Internodes long.	P. arequipensis
1*	Prostrate herbs or shrubs. Internodes short.	2
2	Flowers densely clustered into terminal glomeruli. Leaves 2.5-4 mm long, stipules larger than leaves.	P. andina
2*	Flowers clustered axillary. Stipules shorter or equaling the leaves.	3
3	Leaves 5-9 mm long. Internodes 5-9 mm.	P. muschleri
3*	Leaves 1.8-2.2 mm long. Internodes <0.5 mm.	P. ubinensis

Pycnophyllum

	Key based on descriptions of Timana (2005).	
1	Shoots terete. Forming large cushions up to 3 m in Ø. Petals always present, apex obtuse to truncate.	P. molle
1*	Leaves opposite, therefore shoots with square appearance.	2
2	Leaves spirally arranged. Forming small cushions <50 cm in Ø. Petals small, apically divided.	P. bryoides
2*	Leaves not spirally arranged. Forming large cushions >1 m in Ø. Petals absent.	P. tetrastichum

Silene

1	Flowers solitary on short, stout pedicel. Pedicel elongating up to 10 cm in fruit, often bitten off by herbivores.	S. mandonii
1*	Flowers in racemes or cymes.	2
2	Inflorescences one-sided racemes.	S. gallica
2*	Inflorescences few flowered cymes.	S. genovevae

Spergularia

1	Plants <5 cm tall. Flowers solitary, peduncle <1 cm.	S. andina
1*	Plants >5 cm tall. Flowers in few-flowered cymes.	2
2	Leaves strongly congested. Stipules as long as the leaves, deeply laciniate. Restricted to the lomas.	S. congestifolia
2*	Leaves not congested. Stipules 1 mm long or, if longer, then only laciniate at the tip.	3
3	Stipules 1 mm long. Leaves spreading, apex not mucronate. Restricted to lomas.	S. collina
3*	Stipules as long as the leaves. Leaves terete, apex mucronate. Growing from 0-3000 m.	S. fasciculata

Stellaria

1	Petals entire. Leaves imbricate.	S. weberbaueri
1*	Petals notched or parted to base in 2-segments. Leaves spreading.	2
2	Petals longer than the calyx. Petiole of lower leaves longer than blade.	S. cuspidata
2*	Petals shorter than the calyx. Petiole of lower leaves shorter than blade.	3
3	Leaves flat.	S. media
3*	Leaves undulate.	S. weddellii

Celastraceae

1	Shrub up to 4 m tall. Leaves >3 cm, oblong lanceolate, leathery, veins obvious. Flowers green, pedicellate, in axillary, 1-5-flowered, sessile umbels.	Maytenus sp.

Ceratophyllaceae

1	Submersed aquatic plant, root absent, but leafy branches modified as rhizoids. Leaves dichotomously divided into linear segments.	Ceratophyllum demersum

Cleomaceae

1	Annual herb, unarmed. Leaves palmately compound, 3-5 leaflets. Flowers white, 1 cm long.	Cleome chilensis

Commelinaceae

1	Prostrate herb up to 40 cm tall. Tuberous roots. Leaf-sheath hispid. Bracts well developed. Petals clearly overlapping and only shortly nailed, blue.	Commelina tuberosa

Convolvulaceae

	Key based on Jepson (2024)	
1	Plants non-green, parasitic, without leaves.	Cuscuta
1*	Plants green, photosynthetic, with leaves.	2

2	Styles 2, free or united only at base. Small-leaved creeping herb or erect herbs.	3
2*	Style 1. Tall or elongate plants, often climbing.	5
3	Ovary deeply lobed, fruit bicarpellate. Small-leaved creeping herb.	Dichondra
3*	Ovary not lobed, fruit not carpellate.	4
4	Styles bifid, stigmas long.	Evolvulus
4*	Styles entire, stigmas globose.	Cressa truxillensis
5	Corolla long tubular, tube at least as long as lobes.	Ipomoea
5*	Corolla funnel-shaped, but not forming a long tube.	6
6	Flowers white to pinkish-white, slightly tubular. Herbaceous, trailing perennial. Leaf base sagittate to cordate.	Convolvulus
6*	Flowers entirely blue, funnel-shaped. Suffrutescent perennial. Leaves ovate to narrowly elliptic, base truncate, rounded or slightly cordate.	Jacquemontia unilateralis

Convolvulus

1	Plant ± glabrous. Petals white, usually with pinkish stripes.	C. arvensis
1*	Plant velvety villous. Petals white, obtuse, usually cuspidate.	C. hermanniae

Cuscuta

	Key based on Flora of Peru (1936ff.)	
1	Styles rather stout, more or less subulate and tapering into the ovary.	2
1*	Styles slender, usually little thicker at base than at apex.	4
2	Scales (leaves) not developed. Flowers 4-6 mm long.	C. grandiflora
2*	Scales (leaves) developed.	3
3	Flowers 4-7 mm long, subsessile or sessile.	C. foetida
3*	Flowers 2-3 mm long, pedicellate.	C. acutiloba
4	Calyx lobes imbricate at base. Flowers 4-6 mm long, in dense clusters, white.	C. cockerellii
4*	Calyx lobes free to base. Flowers 2-4 mm long.	C. hitchcockii

Dichondra

1	Plant green, pubescent on the underside of blade, but not sericeous.	D. microcalyx
1*	Plant sericeous throughout.	2
2	Leaves reniform, attenuate to petiole.	D. argentea
2*	Leaves orbicular, not decurrent to petiole.	D. sericea

Evolvulus

	Key based on van Oostroom (1934)	
	Leaves oblong-lanceolate, lanceolate to linear, blade densely sericeous with closely adpressed, shining hairs.	E. argyreus
	Leaves elliptic or oblong.	2
1	Leaves 5-10 x 2-6 mm, blade lanate on both sides.	E. lanatus
	Leaves 10-24 x 5-14 mm, blade ± densely villose-tomentose on both sides, glabrescent in age.	E. villosus

Ipomoea

1	Shrubs, generally not climbing.	2
1*	Climbing herbs.	3
2	Tall shrub, up to 4 m tall. Leaves entire, >5 cm long. Flowers whitish-pink (sometimes climbing in shade).	I. carnea
2*	Small shrub, <50 cm. Leaves divided into 5-7 linear segments, <5 cm long. Flowers purple, tube white.	I. plummerae
3	Leaves divided into 5-7 lanceolate segments. Flowers pink.	I. cairica
3*	Leaves entire or slightly lobed.	4
4	Flowers white with greenish strips, tube long.	I. alba
4*	Flowers pink to red.	5

5	Flowers scarlet.	I. nationis
5*	Flowers pink or purple.	6
	Leaves up to 18 x 16 cm. Flowers 1-5, corolla 4 cm in Ø.	I. purpurea
	Leaves up to 11 x 6 cm. Flowers solitary, corolla 2 cm in Ø.	I. dumetorum

Crassulaceae

	Key based on Pinto et al. (2019)	
1	Small plants, <10 cm tall. Leaves <1 cm long.	2
1*	Erect herbs, >20 cm tall. Leaves >2 cm long.	3
2	Annual, ± erect herb, stems solitary, often growing in colonies. Leaves opposite, connate at base.	Crassula connata
2*	Perennial, cespitose herb, forming mats. Leaves spirally imbricate.	Sedum ignescens
3	Perennial erect herb, forming rosettes. Leaves 3-8 cm long.	Echeveria
3*	Perennial erect herb. Leaves verticillate or opposite, terete, 1-13 x 0.5 cm, margins with propagules.	Kalanchoe delagoensis

Echeveria

1	Rosettes on a succulent stem.	E. vulcanicola
1*	Rosettes acaule.	E. peruviana

Cucurbitaceae

	Key based on Schaefer & Acevedo-Rodríguez (2021)	
1	Tendrils 2-4-fid (check several tendrils).	Sicyos
1*	Tendrils mostly 1-2-fid.	2
2	Fruit spiny, pale green. Leaves obtusely lobed.	Cyclanthera mathewsii
2*	Fruit smooth, dark green, whitish-dotted. Leaves deeply lobed.	Apodanthera mandonii

Sicyos

1	Leaves ± glabrous.	S. baderoa
1*	Leaves short-hirsutulous.	S. kunthii

Cyperaceae

	Key based on Jepson (2024)	
1	Flower and fruit enclosed in sac-like structure. Flowers unisexual.	Carex
1*	Flower and fruit not enclosed in sac-like structure. Flowers bisexual or some staminate.	2
2	Flower bracts 2-ranked. Spikelets generally flat.	Cyperus
2*	Flower bracts spiraled. Spikelets not flat.	3
3	Flower stems generally leafless.	4
3*	Flower stems with cauline leaves.	6
4	Inflorescence bractless, with 1 spikelet. Leaves 2, bladeless or tooth-like.	Eleocharis
4*	Inflorescence bracteate, with >1 spikelet. Leaves 1-3, generally some clearly bladed.	5
5	Flower bract with >= 3 veins. Stem 10-40 cm tall.	Isolepis cernua
5*	Flower bract with 1 vein, at least in distal-most part of spikelet. Stem 0.5-4 m tall.	Schoenoplectus
6	Small plant, <10 cm tall, forming mats or cushions in bofedales.	7
6*	Erect herb, >30 cm tall.	8
7	Inflorescence a head with several spikelets.	Phylloscirpus deserticola
7*	Inflorescence a single terminal spikelet.	Zameioscirpus muticus
8	Inflorescence a simple cluster or umbel, spikelets in groups of 3-10.	Bolboschoenus maritimus
8*	Inflorescence a compound umbel, spikelets in groups of 20+.	Rhodoscirpus asper

Carex

1	Leaves 1 mm wide, filiform. Inflorescence a single, terminal spikelet.	C. vallis-pulchrae
1*	Leaves wider than 2 mm. Inflorescence with 3-7 spikelets.	2
2	Elongated culms present, inflorescence in terminal hemispherical clusters.	3
2*	Elongated culms lacking, inflorescence hidden in basal leaves.	4
3	Runners 1-2 cm long.	C. melanocystis
3*	Runners long.	C. ecuadorica
4	Fruit yellowish-brown to bright yellow at maturity. Leaves <2cm long.	C. collumanthus
4*	Fruit yellowish-brown to brown at maturity. Leaves 2-4 cm long.	C. humahuacaensis

Cyperus

1	Culm 1, stolons with tubers. Fruit seldom maturing.	C. rotundus
1*	Culms generally >1, without stolons and tubers. Fruit maturing.	2
2	Inflorescence appearing lateral, spikes digitate.	C. laevigatus
2	Inflorescence an umbel, simple or compound.	3
3	Umbel compound (compare more than one plant, C. ochraceus often has simple umbels).	4
3*	Umbel simple.	7
4	Involucral bracts >10. Semi-aquatic plant, often >1.5 m tall.	C. alternifolius
4*	Involucral bracts <10. Terrestrial plant, <1.5 m.	5
5	Culms terete, occasionally trigonous for apical 1/3, pith-filled with conspicuous transverse septa 5-50 mm apart.	C. articulatus
5*	Culms trigonous.	6
6	Spikelets greenish, grayish, or stramineous. Leaves green.	C. ochraceus
6*	Spikelets reddish-brown. Leaves glaucous.	C. tacnensis
7	Spikes ovoid to oblong cylindric.	C. hermaphroditus
7*	Spikes globular.	8
8	Rays absent, spikes sessile. Involucral bracts horizontal to reflexed, parallel to culm.	C. seslerioides
8*	Rays present. Involucral bracts horizontal to ascending, not reflexed.	9
9	Plant annual, rhizome absent. Rays 1-5, involucral bracts 2-4.	C. difformis
9*	Plant perennial rhizomatous. Rays 3-10, involucral bracts 4-8.	C. eragrostis

Eleocharis

1	Tufted herb, up to 45 cm tall, rhizomes and stolons absent.	E. geniculata
1*	Extensively creeping, ligneous rootstock. Culms 2-15 cm long.	E. albibracteata

Schoenoplectus

	Key based on Jepson (2024)	
1	Inflorescence panicle-like. Leaf blades absent.	S. californicus
1*	Inflorescence head-like. Leaf blades present.	2
2	Distal leaf blade <1.5× sheath. Inflorescence bract 1-6 cm long. Spikelets 2-20.	S. americanus
2*	Distal leaf blade 2-5× sheath. Inflorescence bract 1-20 cm long. Spikelets 1-5.	S. pungens

Ephedraceae

Ephedra

	Key based on Martínez Carretero (2018)	
1	Shrub or creeping sub-shrub, >1 m tall. Seeds covered by 6-9 fleshy, pink to red bracts.	E. americana
1*	Dwarf shrub up to 60 cm tall. Female cone with 3-6 fleshy, red or white bracts.	E. rupestris
	Note: E. breana Phil. is considered as a synonym of E. americana, as suggested by Martínez Carretero (2018).	

Escalloniaceae

Escallonia

1	Shrub with elongate, spreading branches. Leaves ovate, 1 cm long. Flowers solitary, axillary.	E. myrtilloides
1	Inflorescence a many-flowered panicle with 1 to several flowers per inflorescence branch.	E. angustifolia
	Note: E. salicifolia Mattf: Allopatric with E. angustifolia and morphologically almost indistinguishable. The type is from Arequipa, but the species is probably restricted to Bolivia.	

Euphorbiaceae

1	Flowers in terminal inflorescence.	2
1*	Flowers solitary, axillary.	Euphorbia
2	Leaves palmately lobed.	3
2*	Leaves entire.	Croton
3	Shrub, much-branched. Leaves palmately 3-lobed. Petals bright red, >1 cm long.	Jatropha macrantha
3	Herb, usually single stemmed. Leaves palmately 7-11-lobed. Flowers apetalous, <2 cm in Ø.	Ricinus communis

Croton

1	Leaves elliptic, base and apex usually obtuse, stellate tomentose, soon glabrate, petiole 1-2 cm long. Inflorescence 10-20 cm long, dense, male flowers with 12-20 stamens. Capsule glabrescent.	C. alnifolius
1*	Leaves ovate to orbicular, apex acuminate, base truncate to cordate, whitish-gray villous. Raceme <10 cm, few-flowered, male flowers with 30-40 stamens. Capsule tomentose.	C. ruizianus

Euphorbia

1	Shrubs >50 cm tall. Leaves lanceolate, petiolate.	E. apurimacensis
1*	Herbs or subshrubs, <50 cm tall.	2
2	Inflorescence subtended by a sessile, deltoid bract clearly larger than the inflorescence. Robust herbs, usually ± erect.	3
2*	Inflorescence without such bract. Delicate herbs, usually prostrate.	4
3	Leaves irregularly serrate-dentate. Decumbent subshrub.	E. hinkleyorum
3*	Leaves entire. Erect herb.	E. peplus
4	Flowers (cyathia) solitary.	5
4*	Flowers in clusters of several or in umbels.	6
5	Flowers >5 mm in Ø. Appendage (petal-like bract) dentate, white. Leaf-apex acute.	E. peruviana
5*	Flowers <3mm in Ø. Appendage entire, white to pinkish. Leaf-apex emarginate.	E. serpens
6	Flowers in clusters of several, ± sessile. Leaves orbicular.	E. meyeniana
6*	Flowers in umbels. Leaves ovate or elongate.	7
7	Umbels ± sessile. Leaves broadly ovate, hairy.	E. hirta
7*	Umbels pedunculate. Leaves narrowly ovate, ± glabrous to puberulent.	8
8	Fruit densely hairy.	E. lasiocarpa
8*	Fruit ± glabrous	E. hypericifolia

Fabaceae

1	Flower radial, calyx, corolla generally inconspicuous, petals free or fused, lobes not overlapped in bud, stamens 10 or generally many, often long-exserted.	Subfam. Mimosoideae
1*	Flower generally bilateral, less often ± radial, calyx, corolla generally conspicuous, petals overlapped in bud, free or 2 lowermost ± fused, stamens 10 (or 5, 6, 7, 9), generally ± included	2
2	Flower ± radial to ± bilateral, sepals fused only at very base, leaf 1- or 2-pinnate.	Subfam. Caesalpinoideae
2*	Flower bilateral, sepals free to ± entirely fused, leaf simple, 1-pinnate, or palmate.	Subfam. Papilionoideae

Subfamily Caesalpinoideae

	Key based on Flora of Peru (1936ff.), Jepson (2024), Simpson & Ulibarri (2006).	
1	Leaves 1-pinnate	Senna
1*	Leaves 2-pinnate (Parkinsonia appearing 1-pinnate)	2
2	Plant armed with spines or prickles.	3
2*	Plant not armed with prickles or spines.	4
3	Pinnae 2-4, sessile, appearing as several leaves, c. 30 cm long. Leaflets 30-60, small, elliptic. Bark green.	Parkinsonia aculeata
3*	Leaves up to 10 cm long, with 2-3 pairs of pinnae, each with 8 pairs of subsessile, elliptic leaflets. Bark gray.	Tara spinosa
4	Shrub >1 m. Showy flowers, petals yellow with far protruding, red stamens.	Erythrostemon gilliesii
4*	Shrub <1m. Flowers yellow or orange.	Hoffmannseggia

Subfamily Mimosoideae

	Key based on Flora of Peru (1936ff.), Jepson (2024), NYBG (2020).	
1	Leaflets 3-4 per pinna.	Mimosa albida
1*	Leaflets >4 per pinna.	2
2	Plant armed with spines.	3
2*	Plant not armed with prickles and spines.	4
3	Inflorescences globular.	Vachellia
3*	Inflorescences elongate.	Neltuma
4	Decumbent shrub. Pods erect, thin, with large margins.	Calliandra taxifolia
4*	Tree. Pods pendulous, flat.	5
5	Inflorescences not globular.	6
5*	Inflorescences globular.	7
6	Leaves 2-pinnate. Inflorescence a bottlebrush, greenish-white. Pods thin, dry.	Paraserianthes lophantha
6*	Leaves 1-pinnate, rachis alate. Inflorescence straggly looking, flowers with long stamens, white. Pod thick, green, with thickened edge and white, sugary pulp around the seeds.	Inga feuillei
7	Inflorescence yellow. Leaves 2-pinnate 15-20 cm long, 5-16 pairs of pinnae, leaflets 14-48 pairs per pinna.	Parasenegalia visco
7*	Inflorescence white. Leaves 2-pinnate 5-9 cm long, 4-8 pairs of pinnae, leaflets 5-15 pairs.	Leucaena leucocephala

Subfamily Papilionoideae

1	Tree >5m. Leaves imparipinnate. Flowers clustered, yellow to orange.	Geoffroea decorticans
1*	Herb or shrub, <5 m (Otholobium pubescens can grow tall as a tree, but is usually a shrub).	2
2	Pods divided into 1-seeded segments.	3
2*	Pods not divided, sometimes slightly constricted.	5
3	Leaflets glandular-punctate.	Weberbauerella
3*	Leaflets not glandular punctate.	4
4	Leaves with >3 leaflets. Flowers yellow.	Adesmia
4*	Leaves 3-foliate. Flowers white, pink or purple.	Desmodium
5	Herbs, usually twining and climbing.	6
5*	Herbs or shrubs, not twining or climbing.	7
6	Climbs with tendrils on leaves. Leaves pinnate. Flowers white or purple.	Vicia
6*	Climbing through twining stems. Leaves 3-foliate. Flowers yellow or greenish-yellow.	Vigna luteola
7	Shrub, rush-like appearance. Leaves simple, early caducous.	Spartium junceum
7*	Leaves compound.	8
8	Leaves palmately compound with more than 3 leaflets.	Lupinus
8*	Leaves 3-foliate or pinnate.	9

Euphorbiaceae

9	Leaves 3-foliate.	10
9*	Leaves pinnate with more than 3 leaflets.	15
10	Shrub or tree, clearly woody in older parts.	11
10*	Herbs or subshrubs, if woody at all, only on lower parts.	12
11	Shrub. Leaves 3-foliate, 0.5-2 x 0.2-1.5 cm. Flowers yellow.	Genista monspessulana
11*	Shrub or tree. Leaves pinnately 3-foliate, 3-5 x 1-1.5 cm. Flowers blue to purple, rarely white.	Otholobium pubescens
12	Inflorescence a head, globular.	Trifolium
12*	Inflorescence a raceme, cylindrical to elongate.	13
13	Pods terete, slightly inflated.	Crotalaria incana
13*	Pods orbicular, sickle-shaped or spiraling.	14
14	Pods sickle-shaped or spiraling. Leaves 3-foliate, leaf-like stipule.	Medicago
14*	Pods orbicular to ovoid. Leaves 3-foliate, stipule linear.	Melilotus
15	Flowers red or pale red.	16
15*	Flowers not red.	17
16	Inflorescence a dense, many-flowered raceme, flowers subsessile to sessile. Leaves glabrous.	Indigofera
16*	Inflorescence a lax, few-flowered raceme, flowers clearly pedicellate. Leaves silky pubescent.	Poissonia weberbaueri
17	Acaule to erect herbs, most leaves basal. Leaves hirsute to pubescent. Calyx swollen.	Astragalus
17*	Erect herbs or shrubs, most leaves cauline. Leaves glabrous to puberulent, leaflets all strongly opposite. Calyx not swollen.	Dalea

Adesmia

	Key based in part on Ulibarri (1986).	
1	Annual of perennial herbs.	A. muricata
1*	Shrubs, often ± spinescent.	2
2	Entire plant, except flowers, glutinous.	A. verrucosa
2*	Plants sometimes glandular but not glutinous.	3
3	Leaves subsessile, the leaflets minute.	A. spinosissima
3*	Leaves well petiolate, the leaflets 3-4 mm long.	4
4	Leaflets 6-10 pairs, puberulent. Petals glabrous.	A. miraflorensis
4*	Leaflets 3-4 pairs, sericeous.	5
5	Leaflets 3 pairs, 1-5 mm long. Flowers 8-9 mm long (presumed hybrid A. spinosissima x augusti).	A. melanthes
5*	Leaflets 3-4 pairs, 5-11 mm long. Flowers 7-8 mm long.	A. augusti

Astragalus

	Key based on Flora of Peru (1936ff.), Johnston (1947) and Gomez-Sosa (2005).	
1	Flowers 10 mm long or longer.	2
1*	Flowers <8 mm.	4
2	Plants diffuse. Stem slender, elongated to >10 cm. Leaflets ashy-sericeous.	A. richii
2*	Plants stout, stem tufted erect.	3
3	Herb or subshrub, 10-80 cm tall. Leaves with 10-15 pairs of leaflets. Flowers in pedunculate inflorescence.	A. garbancillo
3*	Depressed herb, stem 1-3 cm long. Leaves with 3-12 pairs of leaflets, glabrous. Flowers usually solitary, 18 mm long.	A. uniflorus
4	Herbs of the lomas, growing below 1000 m.	5
4*	Herbs or shrubs growing above 2000 m.	6
5	Delicate annual, suberect to erect. Leaves with 4-6 pairs of leaflets. Inflorescence 1-3 flowered, flowers 5 mm long.	A. triflorus
5*	Stout prostrate perennial. Leaves with 7-8 pairs of leaflets. Inflorescence 1-5 flowered.	A. neobarnebyanus
6	Inflorescence pedunculate, usually more than 2 per inflorescence.	7
6*	Flowers ± sessile, solitary or in pairs.	9

7	Flowers 6-7 mm long.	8
7*	Flowers 3-4 mm long. Leaves with 6-8 pairs of leaflets.	A. micranthellus
8	Leaves with 7-14 pairs of leaflets.	A. arequipensis
8*	Leaves with 2-4 pairs of leaflets. Leaflets linear-oblong, 2-5 mm long, widely spaced.	A. confinis
9	Leaflets <2 mm long.	10
9*	Leaflets 2-5 mm long.	12
10	Plants silvery-strigose. Leaves with 5-8 pairs of leaflets. Flowers 8 mm long, corolla pale blue.	A. diminutivus
10*	Plants green, ± strigilose.	11
11	Leaves 1-2 cm long ± ashy-strigillose or glabrate above, 5-6 pairs of leaflets. Flowers 5 mm long.	A. minimus
11*	Leaves 1-1.5 cm long, glabrous, 3-6 pairs of leaflets. Flowers 7 mm long.	A. weddellianus
12	Leaflets 2 mm long.	A. peruvianus
12*	Leaflets 3-5 mm long.	A. pusillus

Dalea

	A tentative key, there is contradictory information about the morphology and distribution of the species.	
1	Openly branched shrubs, 1.5-3 m tall.	D. coerulea
1*	Herbs.	2
2	Leaflets up to 15 x 5 mm. Subdecumbent herbs. Inflorescence 2-8 cm long.	D. cylindrica
2*	Leaflets <10 mm long and <3 mm wide.	3
3	Erect herbs. Peduncle 4x longer than the leaves.	D. onobrychis
3*	Prostrate to ascending herbs, 4-5 pairs of leaflets, leaflets 6-10 x 3 mm. Flowers mulberry purple, banner yellowish-white.	D. pennellii
3*	Prostrate to ascending herbs, 5-8 pairs of leaflets, leaflets 3-5 x 1.5-3 mm. Flowers blue or purple.	4
4	Stems 10-45 cm long	D. boliviana
4*	Stems >50 cm long.	D. exilis
	Note: Dalea onobrychis may not occur in Arequipa.	

Desmodium

1	Leaves green, no shiny central part. Pods not deeply incised.	D. scorpiurus
1*	Leaflets with shiny central part. Pods deeply incised.	D. uncinatum

Hoffmannseggia

1	Subshrubs with woody branches. Flowers borne on leafy branches. Fruits dehiscent by simple spreading of the valves, valves occasionally in-rolling after dehiscence.	H. viscosa
1*	Perennial herbs, occasionally woody at base (H. miranda). Flowers borne on unbranched flowering stalks arising from the base of the plant. Fruits indehiscent or dehiscent with each valve twisting around itself longitudinally.	2
2	Herbs without woody base. Flowers bright yellow, with yellow (occasionally red), glandular clavate trichomes on the petal claws less than 0.5 mm long.	H. prostrata
2*	Stem bases woody with stems often branching basally. Flowers yellow orange, with purple or yellow filiform trichomes on the petal claws 0.5-3.0 mm long.	3
3	Leaflets subeliptical, obovate, or subfalcate in outline, without conspicuous veins, soft, (4)6-9(10) pair per pinna. Trichomes on the petal yellow.	H. arequipensis
3*	Leaflets ovate, obovate, or elliptical in outline, prominently veined, leathery, 3-5(9) pair per pinna. Trichomes on the petal purple.	H. miranda

Indigofera

1	Procumbent herbs. Leaves whitish-strigose. Flowers orange-red.	I. humilis
1*	Erect shrub. Leaves glabrescent. Flowers whitish-red.	I. suffruticosa

Euphorbiaceae

Lupinus

	Key based on Flora of Peru (1936ff.) and on the type specimens.	
1	Flowers entirely yellow.	L. cuzcensis
1*	Flowers purplish-blue, sometimes with reddish-yellow keel.	2
2	Densely congested perennial 5-8 cm tall. Peduncles almost not developed, racemes 2 cm long, concealed among the leaves.	L. ananeanus
2*	Herbs or shrubs, usually >10 cm tall. Peduncles well developed, racemes as long as or longer than the leaves.	3
3	Plants of the lomas, growing below 1000 m.	4
3*	Plants growing above 2000 m.	5
4	Annual herb of the lomas, <10 cm tall. Raceme with few flowers, lax.	L. mollendoensis
4*	Subshrub, 60-90 cm tall. Raceme dense.	L. arequipensis
5	Shrub, much-branched, up to 3 m tall.	L. proculaustrinus
5*	Herbs or shrubs <1 m tall.	6
6	Leaves glabrous to sparsely sericeous, appearing green.	7
6*	Leaves densely villous or sericeous, appearing whitish-gray.	9
7	Petiole >10 cm long. Peduncle >15 cm long. Leaflets 50 x 7 mm, folded and appearing linear.	L. misticola
7*	Petiole <10 cm long. Peduncle <10 cm long. Leaflets <40 mm long.	8
8	Stipules 12-18 mm long. Low shrub, 20-40 cm tall, prostrate-ascending, branchlets and petiole loosely pubescent.	L. ballianus
8*	Stipules 7-10 mm long. Subshrub forming low clumps, 20-30 cm tall, branches minutely appressed sericeous.	L. saxatilis
9	Leaflets <20 mm long, petiole 1-1.5 cm long. Peduncle short.	L. eriocladus
9*	Leaflets >20 mm long, petiole >1.5 cm long. Peduncle at least 3 cm long.	10
10	Stems, petioles and leaves densely villous with spreading, brownish hairs up to 8 mm long.	11
10*	Stems and petioles with silky tomentum. Leaflets sericeous with adpressed whitish hairs.	12
11	Peduncle 5 cm long. Leaflets 20-25 mm long.	L. hinkleyorum
11*	Peduncle 15 cm long. Leaflets 20-50 mm long.	L. munzianus
12	Petioles 1.5-3 cm long. Leaflets 25 mm long, 6 mm wide.	L. tarapacensis
12*	Petioles 4-6 cm long. Leaflets 45 mm long, 10 mm wide.	L. tomentosus
	Note: The genus Lupinus needs further research. Beside the key in Flora of Peru (1936ff.), no modern taxonomic treatment of the genus exists for S-Peru. I would not be surprised if L. ballianus and L. saxatilis respectively L. hinkleyorum and L. munzianus become synonyms in the future. Beside that, I have seen two species in Arequipa that do not fit in any of the circumscribed species above. One of them could be the low altitude species L. arequipensis, growing at 4000 m.	

Medicago

1	Erect herbs. Flowers purple to whitish-blue. Often cultivated.	M. sativa
1*	Prostrate to ascending herbs. Flowers yellow.	2
2	Pods discoid, coiled in 1.5-2.5 spirals, 15 spines or tubercles in each row.	M. polymorpha
2*	Pods sickle-shaped, spiraling less than 180°.	M. lupulina

Melilotus

1	Herb up to 2.5 m tall. Flowers white.	M. albus
1*	Herb up to 0.5 m tall. Flowers yellow.	M. indicus

Neltuma

1	Leaves all 1-pinnate. Flowers yellow. Shrub.	N. calderensis
1*	Leaves 1-2 pinnate. Flowers white or yellow.	2
2	Leaves 2-pinnate with 3 pairs of secondary leaf axis. Flowers white. Shrub.	N. pallida
2*	Leaves 1- and 2-pinnate. Flowers yellow. Tree, trunk in age reaching 1 m in Ø.	N. alba
	The record from N. alba may be planted.	

Senna

1	Leaflets 1-2 pairs	S. brongniartii
1*	Leaflets >2 pairs	2
2	Leaflets 3-5 pairs	S. bicapsularis
2*	Leaflets >5 pairs	3
3	Inflorescence a loose raceme. Flowers long pedicellate, yellow. Leaves up to 20 cm long.	S. birostris
3*	Inflorescence a dense raceme. Flowers ± sessile, yellow. Leaves up to 50 cm long.	S. didymobotrya

Trifolium

1	Decumbent to erect herb. A pair of reduced leaves at the base of the inflorescence. Flowers pink to reddish.	T. pratense
1*	Prostrate herb. Inflorescence long pedunculate, without leaf at the base. Flowers white to pink.	2
2	Apex of leaflets distinctly emarginate.	T. polymorphum
2*	Apex of leaflets acute or obtuse (rarely slightly emarginate in T. repens).	3
3	Prostrate and rooting at the nodes. Leaflets glabrous and glossy beneath.	T. repens
3*	Prostrate, but not rooting at the nodes. Leaflets pubescent and not glossy beneath.	T. amabile

Vachellia

1	Leaves mostly with >11 pairs of pinnae	V. macracantha
1*	Leaves with <8 pairs of pinnae	V. karroo

Vicia

1	Flowers subsessile, solitary. Tendril simple.	V. lomensis
1*	Flowers pedunculate, rarely solitary.	2
2	Raceme 5-6 flowered, flowers 1 cm long. Leaflets in 4-6 pairs, oblong, usually cuneate, tendril bifurcate.	V. andicola
2*	Raceme 1-5 flowered, flowers <1cm long. Leaflets in 1-3 pairs, narrow, usually cuspidately acute, tendril simple.	V. graminea

Weberbauerella

	Key based on Orellana-Garcia et al. (2023).	
1	Ovary pubescent to villous. Leaves with 8-18 leaflet pairs, Leaflets >1cm long. Endemic to the lomas of Camaná and Islay.	W. brongniartioides
1*	Ovary glabrous. Leaves with 14-34 leaflet pairs. Leaflets <1 cm long. Endemic to Ica and N-Arequipa (Caravelí).	W. raimondiana

Francoaceae

Balbisia

	Key based on Weigend (2011)	
1	Leaflets narrowly elliptical-acuminate, margins recurved, lamina sericeous, whitish in the living state, gray when dried. Calyx densely pubescent to sericeous, usually not flushed with red.	B. verticillata
1*	Leaflets narrowly obovate, margins flat, lamina subglabrous to finely pubescent, glaucous in the living state, fresh green when dried. Calyx finely puberulent, abaxially flushed with red.	B. meyeniana
	Note: Weigend (2011) states that B. meyeniana appears to be restricted to Moquegua and Tacna and has not so far been found further north. However, an occurrence in southern Arequipa is likely.	

Frankeniaceae

1	Low shrub. Leaves 0.5 mm long, margins revolute. Corolla white, petals crenulate at apex. Growing near the coast.	Frankenia chilensis

Gentianaceae

1	Branched herbs. Leaves basal and cauline.	2
1*	Unbranched herbs. Leaves rosulate, basal. Peduncle sometimes with bract-like leaves.	3
2	Corolla pink, 5-merous.	Centaurium erythraea
2*	Corolla yellow, 4-merous.	Cicendia quadrangularis
3	Corolla lobes partly connate, whitish-blue.	Gentiana sedifolia
3*	Corolla lobes not connate, white, whitish-violet or yellowish-scarlet.	Gentianella

Gentianella

1	Flowers whitish-violet.	G. potamophila
1*	Flowers scarlet or white.	2
2	Flowers white, tube sometimes yellowish.	G. sandiensis
2*	Flowers scarlet and yellow.	G. scarlatina

Geraniaceae

	Key based on Jepson (2024)	
1	Fruit a capsule without an extended beak. Plants glabrous.	Hypseocharis pimpinellifolia
1*	Fruit with an extended beak. Plants hairy.	2
2	Fertile stamens generally 10. Leaves palmately lobed to divided.	Geranium
2*	Fertile stamens generally 5. Leaves crenate to shallowly ± palmately lobed or generally pinnately lobed to compound.	Erodium

Erodium

1	Basal and lower cauline leaves entire, slightly lobbed.	E. malacoides
1*	Basal and lower cauline leaves pinnately compound, leaflets lobed to deeply incised.	2
2	Leaf with <5 pairs of leaflets.	E. geoides
2*	Leaf with >5 pairs of leaflets.	3
3	Leaflets petiolate, margins dentate. Petals 13-15 mm long.	E. moschatum
3*	Leaflets sessile, deeply incised up to the main vein. Petals 5-8 mm long.	E. cicutarium

Geranium

	Key based on Aedo et al. (2005)	
1	Plant acaulescent, hairs always eglandular. Petals slightly emarginate.	G. sessiliflorum
1*	Plant caulescent, hairs glandular or eglandular.	2
2	Peduncles 1-flowered, solitary.	G. diffusum
2*	Peduncles at least in part 2-flowered.	3
3	Peduncles with glandular hairs.	G. fallax
3*	Peduncles without glandular hairs.	4
4	Pedicels with hairs appressed backwards. Petals emarginate.	G. core-core
4*	Pedicels with protruding hairs.	G. limae

Grossulariaceae

Ribes

	Key based on Weigend & Andrade-Galán (2022) and Flora of Peru (1936ff.)	
1	Leaves up to 3 cm wide, usually with one pair of lateral lobes.	R. brachybotrys
1*	Leaves up to 2 cm wide, usually unlobed, some of them with obscurely developed lateral lobes.	R. ovalifolium

Haloragaceae

Myriophyllum

1	Dioecious aquatic herb. Segments of leaves 5-8 mm long. Flowers solitary in axils of leaves.	M. aquaticum
1*	Monoecious aquatic herb. Segments of leaves up to 15 mm long. Flowers in spikes with entire bracts.	M. quitense

Hydrocharitaceae

1	Submersed aquatic plant, stems long. Leaves <2 cm long, crowded in whorls of 3-4.	Elodea potamogeton
	Submersed aquatic plant, stoloniferous, stems short. Leaves >5 cm long. Peduncle screwed.	Vallisneria sp.

Hypericaceae

1	Herb, 10-60 cm tall. Leaves sessile, linear, 3 cm long. Flowers yellow to reddish.	Hypericum silenoides

Iridaceae

	Key based on Flora of Peru (1936ff.)	
1	Segments of the flowers similar or not strongly dissimilar. Style branches alternate with the anthers.	2
1*	Segments of the flowers definitely dissimilar in size or shape. Style branches opposite the anthers.	3
2	Stem terete or acaulescent plant. Leaves glaucous. Stem terete, Juncus-like. Flowers pink, sometimes white, ± nodding.	Olsynium junceum
2*	Stem flattened or winged. Leaves green.	Sisyrinchium
3	Style branches not winged, deeply divided.	Tigridia
3*	Style branches winged-lobed.	4
4	Corolla yellow-orange.	Hesperoxiphion pardalis
4*	Corolla blue to violet.	Mastigostyla

Mastigostyla

1	Spathes not subtended by cauline leaves, the plants obviously caulescent.	M. cyrtophylla
1*	Spathes subtended by cauline leaves, the plants appearing acaulescent.	M. hoppii

Sisyrinchium

1	Stem terete. Flowers yellow. Fruits oblong.	S. jamesonii
1*	Stem flattened. Flowers not pure yellow. Fruits globose.	2
2	Flowers lilac. Fruits erect when ripe.	S. chilense
2*	Flowers yellowish-white with purple center. Fruits pendulous when ripe.	S. micranthum

Tigridia

	Key based on Montesinos et al. (2016)	
1	Basal leaves 12-70 cm long, 1.5-9 mm wide. Flower white to pale purple. Grows above 2000 m.	T. arequipensis
1*	Basal leaves 15-30 cm long, 14-18 mm wide. Flowers white with basal section purple. Lomas below 1000 m.	T. raimondii

Juncaceae

	Key based on Flora of Peru (1936ff.)	
1	Cushion plants of high elevations.	2
1*	Loosely leafy or not cushion-forming plants.	4
2	Leaves closely imbricate, erect, pointed.	Distichia muscoides

2*	Leaves divaricate, long caudate.	3
3	Leaves >2 cm long, spreading vertically.	Oxychloe andina
3*	Leaves up to 1 cm long, spreading horizontally.	Patosia clandestina
4	Leaves more or less ciliate-pubescent, seeds 3.	Luzula racemosa
4*	Leaves glabrous, seeds many.	Juncus

Juncus

	Key based on Flora of Peru (1936ff.)	
1	Flowers in ± sessile, globular clusters.	J. ebracteatus
1*	Inflorescence branched, flowers pedicellate, clusters not globular.	2
2	Plant <20 cm tall.	3
2*	Plant >30 cm tall.	4
3	Strongly branched, decumbent annual in disturbed areas (resistant to trampling).	J. bufonius
3*	Unbranched, stiffly erect perennial in bofedales.	J. stipulatus
4	Leaves well developed.	J. imbricatus
4*	Leaves undeveloped, sheath-like.	J. balticus

Juncaginaceae

1	Emergent, perennial aquatic herb. Grass-like, bases not bulbous. Inflorescence racemose.	Triglochin striata

Krameriaceae

1	Shrub, up to 1 m tall. Flowers 4-merous, petals unequal, pink to red.	Krameria lappacea

Lamiaceae

1	Flowers clustered in axillary and terminal, dense heads of >8 flowers.	2
1*	Flowers axillary, solitary or in clustered of <7 flowers.	5
2	Flowers tubular, 25 mm long, corolla orange-red.	Leonotis nepetifolia
2*	Flowers <10 mm long, corolla white or pink.	3
3*	Flowers distinctly zygomorphic. Leaves rugose, gray tomentose. Flowers white.	Marrubium vulgare
3	Flowers ± actinomorphic to slightly zygomorphic. Leaves non-rugose, prominently veined, minty.	4
4	Flowers pink. Leaves glabrous to hispidulous.	Mentha aquatica
4*	Flowers white. Leaves densely pubescent.	Minthostachys
5	Leaves <20 mm long.	6
5*	Leaves >20 mm long.	7
6	Subshrub up to 15 cm tall. Leaves ovate-obovate, nearly orbicular, ± sessile, appearing imbricate, margins crenate.	Hedeoma mandoniana
6*	Shrub up to 2 m tall. Leaves oblong, shortly petiolate, more than 3x longer as broad, margins entire.	Clinopodium
7	Calyx 2-3-lobed.	Salvia
7*	Calyx 5-lobed.	8
8	Flowers pink to red.	Stachys
8*	Flowers white to bluish-purple.	9
9	Flowers sessile or on a shortly pedunculate, lax panicle.	Lepechinia
9*	Inflorescence a head on a 10-30 mm long, filiform peduncle.	Mesosphaerum sidifolium

Clinopodium

1	Corolla tube 5-8 mm long, clearly longer than the calyx.	C. bolivianum
1*	Corolla tube 1-2 mm long, equaling the calyx.	C. gilliesii

Lepechinia

1	Erect shrub, up to 2 m tall. Flowers 15-30 mm long, red to purple.	L. lamiifolia
1*	Procumbent-ascending perennial herb, 20-40 cm tall. Flowers 3.5-4.5 mm long, white.	L. meyenii

Minthostachys

1	Inflorescence axillary, in sessile or short-pedunculate heads of 8-10 flowers.	M. mollis
1*	Inflorescence axillary, in dense, many-flowered heads.	M. spicata

Salvia

	Key based on Flora of Peru (1936ff.)	
1	Flowers opposite in racemes.	2
1*	Flowers 3-6 or more in spicate verticils.	4
2	Leaves glabrous or nearly. Corolla red. 2000-4000 m.	S. oppositiflora
2*	Leaves softly pubescent beneath	3
3	Leaves obtuse at base, mostly <25 mm wide Corolla red. 0-3000 m.	S. striata
3*	Leaves cordate at base, mostly >25 mm wide. Corolla red. 0-3000 m.	S. tubiflora
4	Corolla red. 2500-3000 m.	S. haenkei
4*	Corolla dark blue, sometimes rather white.	5
5	Inflorescence dense, spike not interrupted, many-flowered whorls.	S. hispanica
	Spike interrupted, whorls 1-6-flowered.	6
6	Calyx glandular. Plant glandular-hispid. 0-4000 m.	S. rhombifolia
6*	Calyx eglandular. Plant subglabrous. 0-1500 m.	S. paposana

Stachys

	Key based on Rodríguez & Sagástegui (2014).	
1	Calyx 3-5 mm long in flower, twice as long in fruit. Corolla tube 3.5-5 mm long, ± inserted in calyx.	S. arvensis
1*	Calyx in fruit as long as in flower. Corolla tube 5-8 mm long, slightly longer than calyx.	2
2	Calyx 3-6.5 mm long. Corolla tube 5-7 mm long, longer than lower lip. Leaf base subcordate.	S. herrerae
2*	Calyx 7-8 mm long. Corolla tube 6-8 mm long, shorter than lower lip. Leaf base cordate.	S. aperta

Linaceae

1	Branched perennial subshrub. Leaves linear, 1 cm long, lower subopposite, upper alternate, gradually reduced. Flowers typically opposite, subsessile to petiolate, petals yellow.	Linum prostratum
	Note: In the lomas grows Linum prostratum var. parvum, annual, habit smaller than Linum prostratum, with 2 brown globose glands at the attenuate leaf-base.	

Loasaceae

	Key based on Flora of Peru (1936ff.)	
1	Fertile stamens not in fascicles.	Mentzelia scabra
1*	Fertile stamens clustered in 5 fascicles.	2
2	Flowers usually >3 cm in Ø, petals salmon-pink to deep-orange or red. Fruit usually twisted.	Caiophora
2*	Flowers <2.5 cm in Ø, petals white or orange-yellow. Fruit not twisted.	3
3	Shrubs with lignified aerial parts.	Presliophytum
3*	Herbs, aerial parts not lignified.	4
4	Leaves opposite, ± deeply palmately lobed. Flowers yellow-orange, with red center.	Loasa nitida
4*	Leaves alternate, pinnate or shallowly lobed. Flowers white or pale yellow, center white or yellow.	Nasa

Caiophora

	Key based on Ackermann (2011)	

Loranthaceae

1	Acaulescent, rosulate plant. Capsule up to 16 mm long.	C. rosulata
1*	Plants with well-developed aerial shoot, shoots erect, decumbent or winding. Capsule >20 mm long.	2
2	Perennial, winding herb up to 5 m tall, without basal rosette.	3
2*	Not climbing herbs.	4
3	Flowers red, sometimes orange.	C. cirsiifolia
3*	Flowers yellow to pale orange.	C. carduifolia
4	Petals free. Leaves pinnate, narrowly ovate in outline, densely covered with stinging hairs, green beneath.	C. pentlandii
4*	Petals usually connate. Leaves pinnatifid, ovate to lanceolate in outline	5
5	Leaves scarcely hairy, whitish beneath, only 1 pair of lobes free.	C. deserticola
5*	Leaves densely hairy, usually more than 1 pair of lobes free.	C. andina

Nasa

1	Flowers white with yellow center, petals spreading radial. Leaves shallowly lobed.	N. chenopodiifolia
1*	Flowers yellow with white center, petals reflexed. Leaves pinnate.	N. urens

Presliophytum

	Key based on Acuña & Weigend (2017)	
1	Sepals almost as long as wide. Petals subequal in size to sepals.	P. arequipense
1*	Sepals ca. 1.5-2× as long as wide. Petals at least 1.5× as long as sepals	P. incanum

Loranthaceae

1	Hemiparasitic shrub, growing on trees, cacti and shrubs. Flowers yellowish-red.	Ligaria cuneifolia

Lythraceae

1	Erect herb, up to 60 cm tall. Leaves opposite, sessile, lanceolate-linear, >6 cm long. Flowers 4-merous, sessile, axillary, in small clusters. Petals dark pink.	Ammannia coccinea

Malvaceae

	Key based on Fryxell (1997)	
1	Acaulescent or cespitose perennial herbs, no more than a few cm tall.	2
1*	Erect herbs, subshrubs or shrubs, sometimes trailing but not acaulescent or cespitose.	3
2	Involucel present. Stipules free. Pedicels axillary, free. Leaves simple.	Acaulimalva
2*	Involucel absent. Stipules adnate to the petiole forming a broad laminar sheath. Pedicels adnate to the petiole, at least in part. Leaves often complexly divided.	Nototriche
3	Flowers in axillary scorpioid cymes.	4
3*	Flowers solitary or fasciculate in the leaf axils or otherwise disposed, never in scorpioid cymes	5
4	Calyx bearing stipitate stellate hairs, the stipe purplish.	Tarasa
4*	Calyx bearing sessile stellate hairs.	Fuertesimalva
5	Involucel (involucre of bractlets) present (Malvastrum early caducous).	6
5*	Involucel absent.	9
6	Inflorescence axillary, pedunculate, with >20 flowers. Flowers ± sessile, densely clustered.	Waltheria
6*	Flowers solitary, sometimes congested toward branch tips.	7
7	Flowers yellow. Leaves ovate, distinctive pinnately veined.	Malvastrum coromandelianum
7*	Flowers white, pink or red.	8
8	Flowers red to orange-red. Petals not emarginate.	Modiola caroliniana
8*	Flowers white or pink. Petals emarginate.	Malva
9	Mericarps >30, irregularly arranged in several series, 1-seeded. Herbs or occasionally shrubs of the lomas, growing below 1000 m.	Palaua
9*	Mericarps 3-30, in single whorl around central column, 1-7 seeded.	10

10	Lower leaves petiolate, upper leaves immediately below inflorescence sessile and amplexicaule. Fruit subglobose, fragile-walled, lantern-like. Corolla white or yellow.	Herissantia crispa
10*	All leaves manifestly petiolate (in Sidastrum upper leaves subpetioate).	11
11	Flowers and fruits solitary, pendulous due to carpocrater (expanded distal end of pedicel).	Cristaria multifida
11*	Carpocrater lacking.	12
12	Inflorescence forming in age a many-flowered (>30), terminal panicle. Flowers reddish-purple, petals reflexed, 5 mm long, pedicels elongate.	Sidastrum paniculatum
12*	Inflorescence few-flowered (1-5), not paniculate.	13
13	Leaves incised, sometimes lanceolate-elliptic but base not cordate.	Sida
13*	Leaves entire, broadly ovate, base cordate, margins crenate-serrate.	14
14	Flowers solitary.	Abutilon
14*	Flowers in fascicles of 2-5.	Gaya

Abutilon

1	Corolla violet, soon reflexed, darker toward base within.	A. arequipense
1*	Petals dark yellow, becoming whitish-orange after flowering.	A. bivalve

Acaulimalva

1	Petals violet, at least in some parts.	A. nubigena
1*	Petals white, sometimes with pale pink tips.	2
2	Corolla 6-10 mm long.	A. rhizantha
2*	Corolla 15-35 mm long.	A. engleriana

Fuertesimalva (incl. Urocarpidium)

	Key based on Fryxell (1996)	
1	Leaves subglabrous above, stellate pubescent beneath. Flowers usually white to pale pink. Pistils 15, mericarps reticulate, at least laterally.	2
1*	Leaves with stellate hairs on both sides. Flowers usually deep pink or lavender, rarely whitish. Pistils 7-12, mericarps rugulose.	3
2	Leaves ovate-rhombic, indistinctly 3-5 lobed. Stipules membranous and persistent. Peduncle up to 10 cm long.	Urocarpidium albiflorus
2*	Leaves divided or very deeply lobbed, often glabrous, a bristle terminating each tooth. Stipules narrowly lanceolate, 13 mm long. Peduncle 2-6 cm long.	F. chilensis
3	Inflorescence 2-6-flowered.	4
3*	Inflorescence usually with more than 6 flowers.	5
4	Peduncle 1-2 cm. Inflorescence 2-3-flowered.	F. corniculata
4*	Peduncle >3 cm. Inflorescence 2-6-flowered.	F. echinata
5	Leaves bright green. Petals 7-8 mm long. Plant sparsely branched.	F. pennellii
5*	Leaves grayish-green. Petals 3-4 mm long. Plant much-branched.	6
6	Plants growing above 1500 m. Mericarps lacking an appendix (= endoglossum).	F. limensis
6*	Plants growing below 1000 m. Mericarps with small appendix.	F. peruviana
	Note: F. echinata and F. corniculata are very similar. It is possible that the two species have been mistaken for each other. F. pennellii is an unresolved species.	

Gaya

1	Inflorescence 3-flowered. Petals 10 mm long. Growing above 1500 m.	G. weberbaueri
1*	Flowers solitary. Petals 15-20 mm long.	2
2	Leaves velutinous without long hairs. Petals 15-20 mm. Mericarps 12-14, ± isometric (4.5 x 4 mm). Grows in the lomas below 1000 m in the north of Arequipa (prov. Caravelí).	G. atiquipana
2*	Leaves velutinous with long hairs on the veins. Petals 15 mm. Mericarps 10-11, longer than wide (6.5 x 4 mm). Grows in the lomas below 1000 m in the south of Arequipa (prov. Islay).	G. mollendoensis
	Note: The taxonomy of the Gaya species at medium altitudes in southern Peru is ambiguous. As G. weberbaueri is mentioned several times, we will use this name. The 3 species of Gaya are similar and may be subspecies or varieties of one and the same species.	

Malva

1	Shrub up to 4 m tall. Flowers showy, petals 2.5-4.5 cm long, rose to deep purple.	M. assurgentiflora
1*	Annual herb, trailing or ascending. Flowers white, petals 4-5 mm long.	M. parviflora

Nototriche

	Little information available. Key based on Flora of Peru (1936ff.), which Macbride himself considers suggestive.	
1	Dense cespitose cushion plant composed of small rosettes (<1 cm in Ø) borne on a highly branched subterranean ligneous caudex.	N. azorella
1*	Rosettes solitary or clustered, not forming a dense cushion.	2
2	Rosettes loose. Leaves distinctly petiolate.	3
2*	Rosettes dense. Petioles not clearly visible or absent.	9
3	Leaves orbicular to reniform, with palmate proliferations on the apex.	N. obcuneata
3*	Leaves 1-3-pinnate.	4
4	Petals 5-6 mm long. Growing in saline habitats.	N. salina
4*	Petals >8 mm long.	5
5	Leaves glabrous at least beneath.	6
5*	Leaves densely sericeous to stellate-tomentose.	7
6	Calyx tube glabrous.	N. anthemidifolia complex
6a	Corolla 10-12 mm long.	N. anthemidifolia s.str.
6b	Corolla 18-20 mm long. Leaves palmately lobed, segments multilobed.	N. argyllioides
6c	Corolla 15 mm long. Leaves 2-pinnate.	N. longirostris
6*	Calyx tube finely stellate.	N. mandoniana
7	All parts densely sericeous.	N. argentea
7*	Leaves stellate-tomentose	8
8	Leaves stellate sulphury-tomentose. Corolla 20-35 mm long, pale blue.	N. sulphurea
8*	Leaves ashy stellate-tomentose. Corolla 15 mm long, cream with brown stripes and red brown tip.	N. pediculariifolia
9	Flowers white to pale purple, at least inside the corolla, sometimes with dark purple stripes.	10
9*	Flowers dark blue, deep purple or reddish-brown.	12
10	Rosettes turret-like.	N. turritella
10*	Rosettes not or only slightly domed.	11
11	Calyx 6-9 mm long, corolla 8-15 mm long. Leaves whitish-green.	N. meyenii
11*	Calyx 12 mm long, corolla 20 mm long. Leaves blackish beneath.	N. nigrescens
12	Corolla dark blue to deep purple, petals overlapping. Calyx lobes entire.	N. orbignyana
12*	Corolla reddish-brown, greenish striped, petals not overlapping. Calyx lobes trifid.	N. staffordiae
	Other species for Arequipa: *Nototriche digitulifolia A.W.Hill: Similar to N. orbignyana, but primary leaf-lobes digituliform or 5-7. Andean III: 4000-4500 m. Peru. Arequipa: doubtful. No recent observation for Arequipa.* *Nototriche pedatiloba A.W.Hill: Similar to N. obcuneata, but leaves incised up to half and flowers pinkish-white. Andean III: 4000- >4500 m. Grasslands. S-Peru. Arequipa: doubtful. No recent record.* *Nototriche pellicea A.W.Hill: Similar to N. nigrescens but a denser habit, calyx 7-7.5 mm long and leaves not blackish beneath. Andean III: 4000- >4500 m. Grasslands. S-Peru. Arequipa: doubtful. Doubtful herbarium specimen from Chivay and several doubtful observations.* *Nototriche sepaliloba Hochr: Scarce information about this species. Type from Huancavelica. No herbarium specimen but mentioned for Arequipa.*	

Palaua

	Key based on Huertas (2010)	
1	All leaves (also basal leaves) deeply divided.	2
1*	Basal leaves slightly lobed to max. 1/3 or entire, sometimes cauline leaves divided.	3
2	Flower 7-25 mm long, in loose racemose inflorescences.	P. dissecta complex
2a	Subshrub. Petals 22-25 mm. Calyx 10-18 mm long. Grows only in lomas of Camaná.	P. camanensis

2b	Herb to subshrub. Petals 10-30 mm. Calyx 7-11 mm. Broader distribution, from Caravelí to Islay.	P. mollendoensis
2c	Herb. Petals 7-16 mm. Calyx 4-7 mm. Leaves (2-)pinnatifid, lobes 5-7(-9), basal lobes forming a sagittate base. Lima to Tacna.	P. dissecta s.str.
2d	Herb. Petals 7-22 mm. Calyx 5-7 mm. Leaves trilobed to trisect, basal lobes forming a truncate base. Arequipa to Tacna.	P. weberbaueri
2*	Flowers 5-8 mm long, in dense corymbiform inflorescences.	P. guentheri
3	Calyx 3-lobed.	P. trisepala
3*	Calyx 4-lobed.	4
4	Corolla inconspicuous, ± as long than the calyx, petals <5mm.	5
4*	Corolla conspicuous, much longer than calyx, petals >6 mm.	6
5	Prostrate perennial subshrub. Branches red or dark brown.	P. modesta
5*	Erect or ascending annual herb. Branches light green.	P. inconspicua
6	Leaves dimorphic: basal leaves elliptical to ovate to slightly trilobed, margins crenate or serrate, cauline leaves subsessile, trilobed, margins entire.	P. sandemanii
6*	Leaves not dimorphic.	7
7	Leaf blades orbicular, rhombiform or deltoid, papery, grayish-green. Calyx <10 mm long, calyx base laxly tomentose or hispid, shallowly winged, venation of calyx lobes inconspicuous or with one dark midvein.	P. tomentosa
7*	Leaf blades ovate, deltoid or elliptic, (sub)leathery, grayish to yellowish. Calyx usually >10 mm long, base velutinous, conspicuously winged, venation of calyx lobes striate.	8
8	Leaf blade ovate to deltoid, as long as wide.	P. moschata
8*	Leaf blade elliptic to ovate, twice as long as wide.	P. velutina

Sida

1	Flowers pink or violet. Petiole without spines.	2
1*	Flowers yellow. Petiole often with 1-2 tubercles or small spines just below the base.	3
2	Leaves palmately 7-9-lobed, lobes acute. Underside of blade purple.	S. jatrophoides
2*	Leaves palmately 3-5-lobed, lobes obtuse.	S. oligandra
3	Petiole often with 1-2 tubercles or small spines just below the base. Leaves elliptic-obovate. Pedicel 0.5 cm long.	S. spinosa
3*	Petiole without spines. Leaves rhomboidal. Pedicel 1-2.5 cm long.	S. rhombifolia

Tarasa

	Key based on Krapovickas (1954)	
1	Leaves ± entire, rhomboid.	T. congestiflora
1*	Leaves lobed or incised, triangular to ovate.	2
2	Corolla entirely pink to violet.	3
2*	Corolla white, sometimes with purple base.	4
3	Inflorescence a thyrse, flowers widely spaced on the peduncle.	T. thyrsoidea
3*	Flowers solitary or densely clustered, often distinctly scorpioid.	T. capitata
4	Shrubby habit. Leaves canescent stellate, lobed but not deeply incised.	T. operculata
4*	Small herb. Leaves hirsute, green, deeply incised.	5
5	Flowers >1cm in Ø, petals 7 mm long, longer than calyx.	T. tenuis
5*	Flowers <1cm in Ø, petals <5 mm, as long as the calyx.	6
6	Plant decumbent to ascending, <30 cm tall. Leaves <4 cm long. Peduncle fused to petiole.	T. nototrichoides
6*	Plant erect, up to 60 cm tall. Leaves >5 cm long. Peduncle free.	T. tenella
	Note: Tarasa hornschuchiana (Walp.) Krapov. has broader sepals and the same distribution and ecology as T. capitata. T. hornschuchiana is considered here to be a synonym of T. capitata.	

Waltheria

1	Leaves ovate-elliptic, at least 3x as long as wide. Inflorescences head-like, hemispherical.	W. indica
1*	Leaves ovate-rotund, 1-2x as long as wide. Inflorescences elongate, with several interrupted clusters of flowers.	W. ovata

Molluginaceae

	Key based on Jepson (2024)	
1	Stem erect. Leaves linear to spatulate, 1-5 mm wide, glaucous. Inflorescence pedunculate, pedicels 3-11 mm. Stamens 5.	Hypertelis cerviana
1*	Stem prostrate to ascending. Leaves linear or elliptic to wide-obovate, 0.5–15 mm wide, not glaucous. Inflorescence ± sessile, pedicels 3-30 mm. Stamens 3.	Mollugo verticillata

Montiaceae

1	Leaves with dilated membranous or scarious base, stem-clasping. Petals united at the base.	Montia fontana
1*	Leaves not dilated at base and not stem-clasping. Petals free.	2
2	Flowers in inflorescence of 2+ flowers, terminal. Cauline leaves lanceolate or spatulate, not needle-like, glabrous. Petals usually dark pink.	Cistanthe
2*	Flowers solitary, axillary, sometimes acaulescent, or flowers appearing in terminal clusters, but each subtended by a leaf and therefore axillary.	3
3	Plant acaulescent and if not, cauline leaves lanceolate or spatulate. Calyx ± glabrous.	Calandrinia
3*	Cauline leaves needle-like, hirsute. Petals pinkish-white. Calx hirsute.	Montiopsis cumingii

Calandrinia

1	Acaulescent plant.	2
1*	Stem developed.	3
2	Stamens 5-10. Capsule usually membranous at maturity. Leaves usually without trichomes.	C. acaulis
2*	Stamens 12-20. Capsule usually indurate at maturity. Leaves usually with trichomes.	C. carolinii
3	Flowers white	C. alba
3*	Flowers pink.	C. ciliata
	Note: Own observations show that plants with 10 and 12-20 stamens grow almost together in the same place. The species status of C. carolinii is questionable.	

Cistanthe

	Key based on	
1	Petals <2 mm long. Prostrate to ascending herb.	2
1*	Petals >4 mm long. Erect herb.	3
2	Flowers in dense glomeruli with >30 flowers. Dwarf-shrub growth.	C. celosioides
2*	Flowers in clusters of <10 flowers, inflorescence with conspicuous imbricate bracts.	C. calycina
3	Leaves orbicular to broadly obovate.	C. paniculata
3*	Leaves oblanceolate.	4
4	Leaves narrow, 3-4 mm wide.	C. lingulata
4*	Leaves broader, >10 mm wide.	C. weberbaueri
	Note: C. lingulata and C. weberbaueri intergrade, they may be local forms of the same species.	

Moraceae

1	Shrub with 3-5 lobed leaves. Figs axillary, solitary.	Ficus carica

Myricaceae

1	Dioecious shrub of riparian forests, similar to Salix but with globose drupes.	Myrica pavonis

Myrtaceae

1	Shrub of the lomas. Leaves opposite, entire, ovate. Flowers white.	Myrcianthes ferreyrae

Nyctaginaceae

	Key based on Flora of Peru (1936ff.)	
1	Flowers white, 3 mm long, tube minute, so that the petals appear free. Inflorescence a many-flowered, dense umbel, not bracteate. Uppermost cauline leaves sometimes white.	Colignonia parviflora
1*	Flowers pink to purple, in Mirabilis sometimes white, but then petals > 1 cm. Petals distinctly connate.	2
2	Flowers surrounded by a calyx-like involucre.	3
2*	Flowers not involucrate.	5
3	Thorny, much-branched shrub, 2-4 m tall.	Bougainvillea spinosa
3*	Unarmed herb or shrub.	4
4	Involucre 3-parted, involucral bracts orbicular, base free. Stem glandular-puberulent to villous.	Allionia incarnata
4*	Involucre calyx-like, usually 5-parted.	Mirabilis
5	Fruit 3-5-ribbed with trichomes. Corolla bell-shaped.	Boerhavia coccinea
5*	Fruit terete, 10-ribbed with large viscid gland. Corolla funnel-shaped.	Commicarpus tuberosus

Mirabilis

	Key based on Flora of Peru (1936ff.)	
1	Corolla 3-6 cm long.	M. jalapa
1*	Corolla <2 cm long.	2
2	Fruit constricted at base, conspicuously 5-sulcate. Involucre strongly enlarged in age. Petals clearly notched.	M. ovata
2*	Fruit not constricted at base, ± smooth. Involucre not enlarged in age.	3
3	Leaves thin, mostly truncate or subcordate at base, commonly acuminate or long-acuminate, with long and slender petioles. Petals not notched. Stamens 3, fruit shorter than the involucre.	M. prostrata
3*	Leaves ± thick, mostly rounded to acute at the base, broadly rounded to obtuse at apex (uppermost leaves acute), petiole stout and often very short.	4
4	Stamens 5. Fruit subglobose, much exceeding the involucre.	M. intercedens
4*	Stamens normally 3. Fruit shorter than the involucre.	5
5	Involucre 6-10 mm long. Petals 15 mm long, stamens exserted. Stem thick and ± hollow.	M. elegans
5*	Involucre 3.5-5 mm long. Petals 6 mm long, slightly notched, stamens not exserted. Stem not hollow.	M. expansa

Onagraceae

	Key based on Jepson (2024)	
1	Sepals persistent after flower	Ludwigia
1*	Sepals deciduous after flower	2
2	Seeds hair-tufted	Epilobium
2*	Seeds not hair-tufted	3
3	Stigma 4-lobed. Leaves entire. Petals >6 mm.	Oenothera
3*	Stigma hemispheric. Leaves linear, dentate. Petals <6 mm, yellow.	Camissonia dentata

Epilobium

	Key based on Solomon (1982)	
1	Low, densely cespitose, 10 cm tall. Leaves all opposite or alternate in the inflorescence, margins subentire. Flowers erect at flowering.	E. fragile
1*	Robust, erect or ascendent, >20 cm tall.	2
2	Leaves mostly alternate, opposite only near the base, margins clearly dentate. Flowers erect.	E. pedicellare
2*	Leaves mostly opposite, alternate above and in the inflorescence, margins weakly dentate. Flowers nodding.	E. denticulatum

Ludwigia

1	Robust, erect shrub. Petals 4.	L. octovalvis
1*	Herb or subshrub, floating on water or creeping on mud, rooting on nodes. Petals 5-6.	2
2	Stem decumbent or creeping, erect and woody with age, not succulent. Upper leaves acute at apex.	L. hexapetala
2*	Stem decumbent or creeping, ascendent, succulent. Upper leaves obtuse at apex.	L. peploides

Oenothera

	Key based on Dietrich (1977)	
1	Flowers pink. Perennial weed, branched from the ground.	O. rosea
1*	Flowers yellow, often aging orange or orange-brown.	2
2	High mountain plants of max. 15 cm high, either flowering in the rosette or forming short, prostrate side branches, capsules 1-2 cm long.	3
2*	Plant taller than 15 cm, erect or prostrate, but if prostrate, then the capsules more than 2 cm long.	4
3	Plants forming a rosette only, rarely with short side branches. Capsules standing out at right angles from the stem.	O. nana
3*	Plants with prostrate side branches 5-20 cm long. Capsule ± appressed to the stem.	O. punae
4	Annual or biannual plants, 50-100 cm tall. Capsule urn-shaped, gradually narrowed toward the tip, erect, 4-9 mm thick at the base and with the subtending leaf fused to it.	5
4*	Annual plants, mostly less than 30 cm tall. Capsule ± cylindrical, at most tapering toward the apex, erect or standing out from the stem, 1.5-4 mm thick, usually only indistinctly fused with the subtending leaf if at all.	6
5	Plants not forming a rosette, much-branched. Leaves 5-10 x 1-3 cm. Floral tube 2.5-5.5 cm long. Capsule 4-7 mm thick.	O. sandiana
5*	Plants forming a rosette, unbranched or somewhat obliquely branched. Leaves 6-20 x 1-3 cm. Floral tube 1.2-2 cm long. Capsule 6-9 mm thick.	O. peruana
6	Plants mostly unbranched. Leaves remotely and bluntly serrate. Floral tube 0.6-1.1 cm long. Seeds 1.5-1.7 mm long, dark brown to almost black. Plant of middle elevations (2000-3000 m).	O. verrucosa
6*	Plants branched, leaves sinuate-toothed. Floral tube 1-3 cm long. Seeds 1-1.3 mm long, brown. Plant of the lomas (0-1000 m).	O. arequipensis

Orchidaceae

	Key based on Trujillo et al. (2016)	
1	Peduncle elongate, thin and completely enclosed by loose to tight-fitting, cylindrical sheaths. Rachis of the spike and ovary sparsely pilose, floral bracts longer than the flowers, dorsal sepal and petals free from column, lip calceolate with involute and lacerate margin.	Aa weddelliana
1*	Peduncle acrescent, thick and completely enclosed by imbricate infundibuliform sheaths. Rachis of the spike and ovary glabrous, floral bracts same size or little shorter than the flower, dorsal sepal and petals adnate to back of column above base, lip cucullate with the margins irregularly erose or with moniliform hairs	Myrosmodes

Myrosmodes

	Key based on Trujillo et al. (2016)	
1	Floral bract with the upper margin crenulate. Petals with the upper margin lacerate-fimbriate. Lip shallowly 3-lobed in front (or with a conspicuous midlobe).	M. nubigena
1*	Floral bract with the margin entire, undulate or somewhat irregular. Petals entire or erose. Lip simple or with a short midlobe.	2
2	Lip with a short subquadrate midlobe.	M. gymnandra
2*	Lip simple without midlobe.	M. nervosa
	Note: The presence of M. nubigena in Arequipa is likely, but not confirmed.	

Orobanchaceae

1	Parasitic plant without chlorophyll.	Orobanche
1*	Green plants.	2
2	Leaves alternate. Bracts different color from leaves, tipped with bright red or lavender.	Castilleja
2.	Leaves opposite. Bracts same color as leaves.	Neobartsia

Castilleja

1	Bracts lavender. Leaves pinnate.	C. profunda
1*	Bracts red. Leaves entire.	2
2	Erect herb. Spike elongate.	C. arvensis
2*	Cespitose herb. Spike short or absent.	C. pumila

Neobartsia

	Key based pro parte on Sylvester (2014). Provisional key, the available information is contradictory.	
1	Corolla purple, red or whitish-purple, without predominant yellow parts.	2
1*	Corolla yellow, helmet often tipped with brown or reddish-brown.	5
2	Twining vine.	N. weberbaueri
2*	Erect or decumbent herb.	3
3	Corolla purplish-white, helmet purple-brown.	N. pedicularoides
3*	Helmet basically red, lip yellow or red.	4
4	Inflorescence dense, subspicate, upper bracts prominent, lanceolate to subulate, acute or acuminate. Lip red.	N. bartsioides
4*	Inflorescence loose, the bracts smaller than leaves, crenate, dentate or entire. Lip bright yellow.	N. camporum
5	Leaf margins deeply serrate, leaves acuminate. Corolla pale yellow, margins of helmet brownish.	N. serrata
5*	Leaf margins crenate to shallowly serrate. Corolla deep yellow.	6
6	Corolla yellow, helmet brownish. Leaves obtuse.	N. peruviana
6*	Corolla uniformly yellow.	7
7	Inflorescence a dense raceme.	N. diffusa
7*	Inflorescence an interrupted, elongated raceme.	N. elongata

Orobanche

	Key based on Flora of Peru (1936ff.)	
1	Flowers 2 cm long, sessile (lower subsessile. Growing above 2000 m.	O. tacnaensis
1*	Flowers 2.5 cm long, long-pedicellate (except upper. Growing below 1000 m.	O. weberbaueri

Oxalidaceae

Oxalis

	Key based in part on the descriptions of Shaw (2023).	
1	Flowers white to pink.	2
1*	Flowers yellow.	3
2	Flowers pure white.	O. nubigena
2*	Flowers pink.	O. latifolia
3	Stem long, creeping or decumbent spreading, often rooting at the nodes.	O. corniculata
3*	Stem developed to acaulescent, not creeping or decumbent spreading.	4
4	Stem clearly >10 cm long.	5
4*	Stem up to 10 cm long or acaulescent.	6
5	Procurement, irregularly branched stems, 30-60 cm long. Petioles up to 20 cm long, leaves not succulent. Flowers up to 15 mm in Ø.	O. peruviana
5*	Erect subshrub, stems to 40 cm long. Petioles <5 cm long, leaves succulent. Flowers 20-30 mm in Ø.	O. sepalosa
6	Plants setulose-villous.	O. laxa

6*	Plants glabrous or sparsely pilose.	7
7	Flowers distinctly smaller in Ø than the 3-foliate leaves.	8
7*	Flowers nearly equal to larger in Ø than the 3-foliate leaves.	9
8	Rhizome small, bulbiform. Growing in the lomas.	O. lomana
8*	Rhizome or caudex becoming stout, often producing a scarred stem. Sepals hastate, margins undulate. Growing above 2000 m.	O. pachyrrhiza
9	Petiole to 8 cm long, glabrous. Leaflets crystalline beneath. Peduncle to 10 cm long, flowers 18 mm in Ø.	O. megalorrhiza
9*	Petiole to 17 cm long, tomentose. Leaflets pubescent beneath. Peduncle to 22 cm long, flowers 35-40 mm in Ø.	O. mollendoensis

Papaveraceae

1	Plant spiny. Leaves generally cauline, dentate or deeply pinnate-lobed, prickly. Flowers yellow.	Argemone
1*	Plant unarmed. Leaves basal and cauline dissected.	2
2	Flowers radial, deep orange to yellow-orange.	Eschscholzia californica
2*	Flowers biradial, white with purple tips.	Fumaria capreolata

Argemone

1	Flowers bright yellow. Largest spines of capsule 5-7 mm long.	A. mexicana
1*	Flowers pale yellow. Largest spines of capsule 10-15 mm long.	A. subfusiformis

Passifloraceae

1	Erect shrub.	Malesherbia
1*	Liana with tendrils.	Passiflora

Malesherbia

	Key based on Beltrán et al. (2018)	
1	Leaves entire, margins slightly undulate, not glandular.	M. fatimae
1*	Leaves pinnatifid to pinnatisect.	2
2	Flowers reddish-yellow or reddish-white.	3
2*	Flowers white.	4
3	Leaves pinnatisect, densely tomentose. Tube yellowish-red, petals yellow.	M. tenuifolia
3*	Leaves pinnatisect, with thick, pinkish glandular hairs on the margins. Corolla reddish-white.	M. haemantha
4	Receptacle obconical, stamens exserted.	M. arequipensis
4*	Receptacle tubular, stamens not exserted.	M. angustisecta

Passiflora

1	Leaves 3-foliate or 3-lobed to below the middle, margins serrate. Flowers pure white. Fruit >3 cm in Ø.	2
1*	Leaves entire or 3-lobed not beyond half to mid-rib, margins ± entire. Flowers green or pale purplish-white. Fruit <2 cm in Ø.	3
2	Leaves deeply lobed, cordate. Stalk with anthers and style symmetrical in the center of the flower.	P. peduncularis
2*	Leaves 3-foliate, leaflets petiolulate. Stalk with anthers and style asymmetrical, on one side of the flower.	P. trisecta
3	Bracts 2-4 pinnatifid, segments filiform, surrounding the fruit. Flowers 2-5 cm in Ø. Sepals and petals white, filaments purplish, at least at base.	P. foetida
3*	Bracts linear-lanceolate, entire, not surrounding the fruit. Flowers 0.8-1.5 cm in Ø. Sepals green and petals absent, filaments white.	P. suberosa
	Note: Often cultivated, but probably not adventive are Passiflora tripartita var. mollissima (Kunth) Holm-Niels. & P.Jorg. and Passiflora pinnatistipula Cav.	

Phrymaceae

1	Prostrate herb. Leaves deltoid to orbicular, dentate, palmately veined. Flowers zygomorphic, yellow with orange blotches, calyx connate, conspicuously compressed at the sides.	Erythranthe glabrata

Phyllanthaceae

1	Annual herb up to 30 cm tall. Leaves orbicular, glandular, long petiolate. Flowers green. Fruit 3-lobed, tuberculate.	Andrachne microphylla

Phytolaccaceae

1	Herb up to 2 m tall, stems violet. Leaves broadly lanceolate to elliptic. Inflorescence a dense raceme. Sepals 5, white to red. Berries green-brown becoming dark brown.	Phytolacca bogotensis

Piperaceae

Peperomia

1	Plant erect to ascending, stems clearly defined. Leaves linear to ovate.	P. galioides
1*	Plant ± acaulescent. Leaves ovate to orbicular, peltate.	P. peruviana

Plantaginaceae

1	Plant aquatic, submersed. Terminal leaves floating on water surface in a rosette-like pattern.	Callitriche heteropoda
1*	Plants terrestrial, sometimes growing in shallow water or in moody soil, but without floating.	2
2	Corolla dry, membranous, translucent, persistent in fruit. Leaves generally basal, veins generally ± parallel.	Plantago
2*	Corolla of petal-like texture, not membranous or translucent, generally not persistent in fruit. Leaves basal or basal and cauline, veins generally pinnate or palmate.	3
3	Shrub with opposite leaves in whorls of 3. Flowers red.	Galvezia elisensii
3*	Herbs with opposite or alternate leaves. Flowers white, blue or purple.	4
4	Prostrate or mat-forming herbs of moist places, <3 cm tall. Leaves ± succulent. Flowers white, without spur.	5
4*	Erect or climbing herbs, >5 cm tall. Leaves not succulent. Flowers blue to lilac or purple, with or without spur.	6
5	Prostrate herb with long runners, growing below 1000 m. Leaves sessile. Flowers pinkish-white with purple center.	Bacopa monnieri
5*	Mat-forming herb growing above 3500 m. Leaves long petiolate. Flowers white with yellow center.	Ourisia muscosa
6	Leaves opposite. Flowers weakly zygomorphic, without spur.	Veronica
6*	Leaves alternate, at least on flowering stems. Flowers highly zygomorphic, with spur.	7
7	Procumbent to trailing herb. Leaves reniform, cordate at base, margins with 5-11 teeth.	Cymbalaria muralis
7*	Erect herb. Leaves linear-lanceolate.	Nuttallanthus canadensis

Plantago

1	Inflorescence 1-2 flowered, ± sessile. Flowers unisexual. Shrub of the high mountains, often forming dense mats. Leaves rigid, lanceolate, coarsely dentate.	P. tubulosa
1*	Inflorescence many-flowered, peduncle >1 cm.	2
2	Leaves broadly elliptic to ± cordate, narrowed abruptly to petiole.	P. major
2*	Leaves lanceolate or linear, gradually narrowing to petiole.	3
3	Leaves >5 mm wide.	4
3*	Leaves <5mm wide.	6
4	Spike <3 cm long, ovate. Peduncle much longer than the spike.	P. lanceolata
4*	Spike >4 cm long, linear. Peduncle ± equaling the length of the spike.	5
5	Perennial herb, primary root usually disappearing. Leaves 8-30 x 3.5-6 cm.	P. australis

5*	Annual herb, with taproot. Leaves 4-10 x 0.5-2 cm.	P. myosuros
6	Shrubs, forming dense clumps with a branching, decumbent to ascending stem. Leaves narrowly linear 2-5 x 0.1 cm, silky pilose. Peduncle 10-15 cm long, spike 1 cm long. Fruit 2-seeded.	P. sericea
6*	Herbs, ± acaulescent.	7
7	Leaves linear-lanceolate, 3-11 x max. 0.5 cm, margins slightly undulate or remotely denticulate. Peduncle 4-19 cm long, spike cylindric, 0.5-6 cm long. Fruit 2-seeded.	P. limensis
7*	Leaves linear, margins entire, not undulate	8
8	Leaves > 5 cm long. Peduncle longer than the leaves. Spikes elongated. Corolla lobes acute.	P. linearis
8*	Leaves 1-5 cm long. Peduncle 1-2.5 cm long. Spike 0.5-1 cm long, globular. Corolla lobes obtuse. Fruit 1-seeded.	P. nubicola

Veronica

1	Flowers in axillary, many flowered racemes.	V. anagallis-aquatica
1*	Flowers solitary in leaf-axils.	2
2	Flowers long pedicelled, pedicel elongated in fruit, several times as long as fruit. Leaves coarsely dentate.	V. persica
2*	Flowers and fruits ± sessile or short pedicelled, pedicel shorter or as long as fruit.	3
3	Herb with one or several erect stems. Corolla white.	V. peregrina
3*	Ascending herb, rooting at nodes. Corolla grayish-white with violet stripes.	V. serpyllifolia

Plumbaginaceae

1	Rosulate herb, all leaves basal. Adventive in the vicinity of the city of Arequipa.	Limonium hyblaeum
	Erect herb, all leaves cauline.	Plumbago caerulea

Poaceae

	Key based on Flora of North America (1993ff.)	
1	Spikelets almost always with 2 florets, the lower florets always sterile or staminate, frequently reduced to lemmas, occasionally missing, the upper florets bisexual, staminate, or sterile.	2
1*	Spikelets either with other than 2 florets or, if with 2, the lower floret bisexual.	3
2	Glumes stiff, leathery to indurate, often subequal, at least 1 and usually both exceeding the upper floret.	Sorghum halepense
2*	Glumes flexible, membranous, the lower glumes usually shorter than the upper glumes, sometimes missing, the upper glumes usually subequal to or exceeded by the upper floret.	Tribe Paniceae
3	Spikelets with 1 floret. Lemmas terminating in a 3-branched awn, lateral branches sometimes reduced.	Aristida adscensionis
3*	Spikelets with more than 1 floret or, if only 1, the lemma not terminating in a 3-branched awn.	4
4	Spikelets with 1 sexual floret.	5
4*	Spikelets usually with more than 1 sexual floret.	6
5	Culms 2-19 cm tall; plants of cold or damp habitats, not rhizomatous; sheaths of the flag leaves closed for at least 1/2 their length; caryopses exposed at maturity.	Tribe Poeae
5*	Culms 5-300 cm tall; plants usually of warm or dry habitats, often rhizomatous; sheaths of the flag leaves open to the base; caryopses not exposed at maturity.	Tribe Cynodonteae
6	Lemmas unawned, flabellate or with (5)7-15 awnlike teeth.	Tribe Pappophoreae
6*	Lemmas awned or unawned, lanceolate, rectangular, or ovate, apices entire, mucronate, bilobed, or bifid, occasionally 4-lobed or 4-5-toothed, sometimes erose.	7
7	Cauline leaf sheaths closed for 1/2 their length or more; glumes usually exceeded by the distal florets, sometimes greatly so.	8
7*	Cauline leaf sheaths open for at least 1/2 their length; glumes exceeding or exceeded by the distal florets.	10

8	Spikelets 15-30 mm long, usually torpedo-shaped when green. Lemmas usually awned, often bilobed or bifid, veins convergent distally.	Bromus
8*	Spikelets 0.7-60 mm long. Lemmas often unawned, not both bilobed/bifid and with convergent veins. Ovary apices usually glabrous.	9
9	Lemma veins (4)5-15, usually prominent, parallel distally.	Melica scabra
9*	Lemmas veins 1-9, often inconspicuous, usually convergent distally.	Tribe Poeae
10	Spikelets with 1 floret. Lemmas with terminal awn, the junction of the awn and lemma conspicuous.	Tribe Stipeae
10*	Spikelets with 1-60 florets. Lemmas unawned or awned, awns basal to terminal, if terminal or subterminal, the lemma-awn junction not conspicuous.	11
11	Ligules, at least of the flag leaves, of hairs, a ciliate ridge or membrane bearing cilia longer than the basal ridge or membrane. Leaves usually hairy on either side of the ligule.	12
11*	Ligules membranous, if ciliate, the cilia shorter than the membranous base. Leaves usually glabrous on either side of the ligule.	13
12	Cushion-forming grass, 20-30 cm tall. Leaves pungent (vegetable hedgehogs) or perennial bunchgrass several meters tall with razor-edged leaves (Pampas grass).	Cortaderia
12*	Not in the above combination of characteristics.	Tribe Cynodonteae
13	Inflorescences panicles or unilateral racemes, not spikelike. Spikelets solitary, the lowest 0-4 florets in a spikelet sterile or staminate, one distal floret bisexual or unisexual.	Tribe Poeae
13*	Inflorescences panicles, racemes, or spikes. Spikelets sometimes in pairs or triplets, sterile florets, if any, distal to the bisexual or pistillate florets.	14
14	Lemmas with 1-3 or 9-11 conspicuous veins. Sheaths open.	Tribe Cynodonteae
14*	Lemmas with 3-15 often inconspicuous veins, if with 3 conspicuous veins, the sheaths closed.	15
15	Inflorescences spikes or spikelike. Spikelets 1-5+ per node, at least 1 spikelet sessile or subsessile.	16
15*	Inflorescences panicles, with no sessile spikelets.	17
16	Inflorescences with 3 spikelets at a node, the middle spikelet sessile and the lateral ones reduced. Glumes and lemmas awned.	Hordeum muticum
16*	Not in the above combination of characteristics.	Tribe Poeae
17	Rachillas hairy, hairs 2-3 mm long. Lemmas leathery	Tribe Stipeae
17*	Rachillas glabrous to hairy, hairs shorter than 2 mm. Lemmas membranous to leathery.	18
18	Leaves to 2 cm wide, usually not conspicuously distichous. Culms <2 m tall, usually less than 1 cm thick.	Tribe Poeae
18*	Leaves 2-10 cm wide, often conspicuously distichous. Culms 2-10(15) m tall, often more than 1 cm thick.	19
19	Lower cauline blades disarticulating, upper cauline blades forming a flat, fan-shaped arrangement.	Gynerium sagittatum
19*	Lower cauline blades persistent, upper cauline blades not forming a flat, fan-shaped arrangement.	Tribe Arundineae

Tribe Arundineae

1	Perennial reed, up to 10 m tall, with hollow stems 2-3 cm in Ø. All parts distinctly bigger than Phragmites australis.	Arundo donax
1*	Stout reed, commonly 2-4 m high, producing stolons and rhizomes, often forming large colonies.	Phragmites australis

Tribe Cynodonteae

1	Inflorescence with 1 terminal branch per panicle, branch arcuate. Spikelets laterally compressed, pectinate.	Bouteloua simplex
1*	Inflorescence a spike, or a panicle with more than 1 branch.	2
2	Upper glume 5-7-veined, with 5-7 longitudinal rows of spinelike projections.	Tragus
2*	Glumes without spinelike projections.	3
3	Leaves conspicuously distichous, upper leaf sheaths strongly overlapping, upper blades stiff, glabrous, often spreading horizontally, equal to or exceeding the panicles (difference from Sporobolus virginicus).	Distichlis
3*	Leaves not as mentioned above.	4

4	Inflorescence a panicle with 1-30 spikelike branches, these usually borne digitately, occasionally in more than 1 whorl or racemose in Diplachne, sometimes with a few isolated branches below the primary whorl.	5
4*	Inflorescence a raceme, the branches not borne digitately.	9
5	Panicles with a solitary, slender, spikelike branch. Rachis curved or falcate.	Microchloa kunthii
5*	Panicle with more than 1 branch.	6
6	Spikelets with at least 2 bisexual florets. Florets unawned.	7
6*	Spikelets with only 1 bisexual floret, sometimes with 1 or more sterile florets distal to the bisexual floret. Florets unawned or awned.	8
7	Panicle with up to 35 spikelike branches in several whorls. Spikelets with 6-20 florets.	Diplachne fusca
7*	Panicle with 4-10 spikelike branches in a terminal whorl, sometimes with a few isolated branches below the primary whorl.	Eleusine indica
8	Spikelets usually without sterile or modified florets. Florets unawned.	Cynodon dactylon
8*	Spikelets with 1 or more sterile florets distal to the bisexual floret. Bisexual florets awned.	Chloris
9	Spikelets 1-flowered.	10
9*	Spikelets few to many flowered.	11
10	Ligules membranous, 1-15 mm long.	Muhlenbergia
10*	Ligules a short fringe of hairs or a hairy rim.	Sporobolus
11	Decumbent, stoloniferous herb. Inflorescence embraced at base by subtending leaf.	Munroa decumbens
11*	Erect herbs.	12
12	Inflorescence a panicle. Ligule a fringe of hairs.	Eragrostis
12*	Inflorescence a one-sided raceme. Ligule a ciliolate membrane.	Tripogonella spicata

Tribe Paniceae

	Key based on Flora of North America (1993ff.)	
1	Spikelets conspicuously hairy, hairs 1.5-5 mm long.	Digitaria californica
1*	Spikelets not hairy.	2
2	Most spikelets or groups of spikelets subtended by 1 to many, distinct to ± connate, stiff bristles or bracts.	3
2*	Spikelets not subtended by stiff bristles.	4
3	At least some bristles fused for more than 1/2 their length	Cenchrus
3*	Bristles free, not fused.	Setaria parviflora
4	Spikelets not subtended by a cuplike callus.	Echinochloa
4*	Spikelets subtended by a cuplike callus.	Eriochloa

Tribe Pappophoreae

1	Florets 6-10 per spikelet: Lemmas with both awns and awned teeth, not forming a pappuslike crown.	Cottea pappophoroides
1*	Florets 3-6 spikelet. Lemmas awned but without awned teeth, the awns forming a pappuslike crown.	Enneapogon desvauxii

Tribe Poeae

	Key based on Flora of Peru (1936ff.) and Peterson et al. (2019)	
1	Spikelets with 2 staminate, neuter, or rudimentary lemmas unlike and below the fertile lemma, no sterile or rudimentary florets above.	Phalaris canariensis
1*	Spikelets without sterile lemmas below the perfect floret, or these rarely present and like the fertile ones.	2
2	Spikelets sessile, placed edgewise on each node of a continuous rachis.	Lolium perenne
2*	Spikelets pedicelled, in open or contracted, sometimes spikelike panicles.	3
3	Spikelets 1-flowered.	4
3*	Spikelets 2-many-flowered.	7
4	Rachilla jointed below the glumes, the spikelets deciduous.	Polypogon
4*	Rachilla jointed above the glumes.	5
5	Rachilla not prolonged.	Agrostis
5*	Rachilla prolonged behind the palea.	6

6	Florets sessile. Ligules decurrent or not, 0.2-4 mm long, commonly truncate to obtuse.	Cinnagrostis
6*	Florets with a stipe between the upper glume and the callus of the floret. Ligules slightly to strongly decurrent, usually elongate, 4-20 mm long, acuminate.	Deschampsia
7	Glumes as long as the lowest floret, usually as long as the spikelet. Lemmas awned from the back.	8
7*	Glumes shorter than the lowest floret. Lemmas awnless or awned from the tip or from a bifid apex.	10
8	Awn short, inserted near the apex of the lemma. Annual.	Rostraria trachyantha
8*	Awn inserted near the base or toward the middle of the lemma.	9
9	Annual.	Avena sterilis
9*	Perennial.	Koeleria spicata
10	Lemmas awnless, obtuse or acute.	Poa
10*	Lemma awned or acuminate, awn terminal.	Festuca

Tribe Stipeae

	Key based on the genera descriptions by Watson et al. (2024) and in part on Flora of Peru (1936ff.)	
1	Spikelets on short peduncles among the leaves.	Aciachne pulvinata
1*	Spikelets in exserted panicles.	2
2	Fruit terete, 2-several times as long as wide. Awn persistent or tardily deciduous. Lemma without a crown of hairs at apex.	3
2*	Fruit 1-2 times as long as wide. Awn readily deciduous, ± eccentrically attached. Lemma with a crown of hairs at apex.	6
3	Awn not geniculate.	Lorenzochloa
3*	Awn geniculate.	4
4	Awn hairless. Lemma with pappus-like, silky hairs in the upper part.	Jarava ichu
4*	Awns plumose. Lemma hairy, not pappus-like.	5
5	Plant stoloniferous. Lemma apex bifid.	Pappostipa chrysophylla
5*	Plant without stolon. Lemma apex not bifid.	Stipa
6	Lemma entirely enclosing the palea. Palea keel-less, consistently nerveless.	Nassella
6*	Lemma not quite enclosing the palea, fitting into the paleal groove. Palea 2-keeled, 2 nerved.	Piptochaetium
	Note: The genus delimitation in the tripe Stipeae is still under discussion. Here we follow POWO (2024), who accepts Jarava as a homogeneous section with nine species. The other species of Jarava s.l. are listed as Stipa sp.	

Agrostis

1	Sprawling, stoloniferous grass.	A. stolonifera
1*	Cespitose grass.	2
2	Plant 4-10 cm tall.	A. tolucensis
2*	Plant 8-32 cm tall.	A. breviculmis

Bromus

	Key based on Saarela et al. (2006)	
1	Lemma awns geniculate and twisted. Plants short-lived annuals.	B. berteroanus
1*	Lemma awns straight and not twisted. Plants short-lived to long-lived annuals, biennials, or perennials.	2
2	Spikelets generally laterally compressed. Lemmas laterally compressed and keeled.	3
2*	Spikelets terete or dorsally flattened. Lemmas terete or dorsally flattened.	4
3	Lemma awns 12-18(-25) cm long. Plants short-lived annuals found only along the coastal desert of Arequipa. Panicle branches nodding, drooping and flexuous.	B. striatus
3*	Lemma awns 0.5-6 mm long. Plants long-lived annuals, biennials, or perennials, plants more wide-ranging. Panicle branches usually stiffly erect and ascending, rarely nodding, drooping or flexuous.	B. catharticus
4	Culms 5-30(-40) cm tall. Leaf blades generally 1-2.5 mm wide. Spikelets densely villous.	B. villosissimus
4*	Culms 30-60 cm tall. Leaf blades generally 2-12 mm wide. Pedicels hairy, lemmas villous.	B. lanatus

Cenchrus

1	Stoloniferous perennial, stolons much-branched and appressed to the ground. Inflorescence short, almost enclosed in the uppermost leaf sheath, bristles not obvious.	C. clandestinus
1*	Annual grass, erect. Inflorescence not enclosed in the leaf sheath, apparently bristly.	2
2	Fascicles having more than 1 whorl of flattened inner bristles, these originating at irregular intervals throughout the body of the cupule, sometimes subtended by terete outer bristles.	C. spinifex
2*	Fascicles having 1 whorl of fused, flattened inner bristles, subtended by free, terete, outer bristles.	3
3	Outer bristles half as long as the inner, flattened bristles.	C. echinatus
3*	Outer bristles equaling or slightly exceeding the inner, flattened bristles.	C. brownii

Chloris

1	Inflorescence green, sometimes reddish-green. Inflorescence branches ascending, open after flowering.	C. radiata
1*	Inflorescence silvery, due to silky hairs on the upper lemma margins. Inflorescence branches erect and not opening.	C. virgata
	Note: C. radiata and C. spicata are similar and probably often confused.	

Cinnagrostis

	Key based on Flora of Peru (1936ff.) and descriptions from GrassBase (2002ff.)	
1	Lemma 5-awned, one dorsal and 4 apical awns. Basal leaves involute-filiform, cauline leaves flat or loosely involute, all soft and lax.	C. heterophylla
1*	Lemma 1-awned. Leaf-blades all alike.	2
2	Culms <5 cm long. Blades <2 cm long. Panicles <1.5 cm long (see also C. preslii, sometimes this combination is present in small specimens).	3
2*	Culms >5 cm long. Blades >2 cm long. Panicles >1.5 cm long.	4
3	Panicle ovate, subtended by an inflated leaf-sheath. Ligule 0.5-1.2 mm long, eciliate. Cespitose.	C. lagurus
3*	Panicle linear. Ligule 0.3 mm long, truncate. Cushion forming.	C. minima
4	Cushion forming or stoloniferous grass.	5
4*	Cespitose grass, without stolons and usually not cushion forming.	6
5	Cushion forming grass. Leaves distinctly curved, 1-4 cm long, rough on both sides. Awn 1.5 mm long.	C. curvula
5*	Stoloniferous grass. Blades slightly curved, 3-9 cm long, margins pubescent, not rough. Awn 4-5 mm long.	C. spicigera
6	Panicle open, oblong. Ligule 2-4.5 mm long. Awn 1-2 mm long.	C. breviaristata
6*	Panicle spiciform, linear. Ligule 0.5-2 mm long. Awn 2-6 mm long.	7
7	Blades 4-10 cm long, 1-3 mm wide. Panicle 4-10 cm long. Awn 2-3.5 mm long. Rhachilla extension pilose.	C. rigescens
7*	Blades 1-4 cm long, 1 mm wide. Panicle 1-6 cm long. Awn 4-6 mm long. Rhachilla extension glabrous or hairy.	8
8	Blades conduplicate, surface glabrous. Panicle 1.5-2.5 cm long. Awn 4-4.5 mm long.	C. preslii
8*	Blades curved or flexuous filiform, surface slightly rough. Panicle 2-6 cm long. Awn 5.5-6 mm long.	C. vicunarum
	Species with doubtful occurrence in Arequipa: *Cinnagrostis brevifolia (J.Presl) P.M.Peterson, Soreng, Romasch. & Barberá* *Cinnagrostis filifolia (Wedd.) P.M.Peterson, Soreng, Romasch. & Barberá* *Cinnagrostis jamesonii (Steud.) P.M.Peterson, Soreng, Romasch. & Barberá* *Cinnagrostis rigida (Kunth) P.M.Peterson, Soreng, Romasch. & Barberá* *Cinnagrostis setiflora (Wedd.) P.M.Peterson, Soreng, Romasch. & Barberá* *Cinnagrostis violacea (Wedd.) P.M.Peterson, Soreng, Romasch. & Barberá* *All of them mentioned for Arequipa. No voucher. The presence is probable due to the distribution of the species.*	

Cortaderia

1	Cushion-forming grass, 20-30 cm tall. Leaves pungent (vegetable hedgehogs).	C. echinata
1*	Perennial bunchgrass. Long, razor-edged leaves. Inflorescence a purplish to white panicle, several meters tall (Pampas grass).	C. jubata
	Note: Barker et al. (2003) postulate that all South American species of Cortaderia sect. Cortaderia (Pampas grass) belong to C. selloana (dioecious, East) and C. jubata (apomictic, West).	

Deschampsia

	Key based on Flora of Peru (1936ff.)	
1	Panicles open, lower branches naked for 2 cm at the base.	D. eminens
1*	Panicles short and dense.	2
2	Panicle oblong, 5-10 cm long.	D. chrysantha
2*	Panicle oval or ovate, 2-4 cm long.	D. ovata

Distichlis

1	Perennial, rhizomatous herb growing above 3500 m. Panicle with 1-3 spikelets.	D. humilis
1*	Perennial, rhizomatous herb growing in coastal areas. Panicle with 2-20 spikelets.	D. spicata

Echinochloa

1	Panicle branches 0.7-2 cm long, without secondary branches. Spikelets 2-3 mm long, unawned.	E. colonum
1*	Panicle branches 1-14 cm long, usually rebranched, the secondary branches often short and inconspicuous. Spikelets 2.5-5 mm long, awned or unawned.	E. crus-pavonis

Eragrostis

	Key based on Peterson & Sanchez Vega (2007)	
1	Annuals, cespitose, stoloniferous or mat-forming to geniculate, without innovations at the basal nodes.	2
1*	Plants perennial or biennial, cespitose, forming innovations at the basal nodes.	8
2	Palea keels prominently ciliate, the cilia 0.2-0.8 mm long.	3
2*	Palea keels smooth or scabrous, the scabridities less than 0.2 mm long.	4
3	Spikelets 1.8-3.2 mm long. Palea keels tuberculate at base, ciliate. Stamens 2.	E. ciliaris
3*	Spikelets 3-4.5 mm long. Palea keels soft and silky, not tuberculate at base. Stamens 3.	E. peruviana
4	Spikelets 5-60 flowered. Warty glands present on the keels of lemma and usually on the keels of glumes and blade-margins.	E. cilianensis
4*	Spikelets 2-20 flowered. Warty glands absent on the keels of lemma and glumes, and blade-margins.	5
5	Caryopses without a ventral groove.	E. pectinacea
5*	Caryopses with a shallow or deep ventral groove.	6
6	Spikelets arranged in glomeruli, 2-4(-6)-flowered.	E. nigricans
6*	Spikelets not arranged in glomeruli, 5-15-flowered.	7
7	Spikelets linear to linear-lanceolate, 0.7-1.4 mm wide.	E. virescens
7*	Spikelets ovate to oblong in outline, 1.5-2.4 mm wide.	E. mexicana
8	Panicles 0.5-5 cm wide, open, not contracted, the primary branches ascending and spreading to divaricate.	E. lurida
8*	Panicles 0.2-2 cm wide, contracted, densely flowered, cylindrical and narrowly spicate to spiciform, the primary branches ascending and tightly appressed.	9
9	Panicle rachis glabrous. Culm nodes glabrous or occasionally with a tuft of hairs, the hairs <1 mm long. Blades glabrous, scabrous, with a few scattered hairs or short pilose but not densely silky.	E. attenuata
9*	Panicle rachis densely pilose, the hairs not rigid. Culm nodes silky pilose, the hairs up to 3 mm long. Blades densely silky pilose above and below.	E. weberbaueri

Poaceae

Eriochloa

	Key based on Flora of Peru (1936ff.)	
1	Spikelets 3 mm long. Plants annual.	E. procera
1*	Spikelets 4-6 mm long. Plants perennial.	E. punctata

Festuca

	Key based on Flora of Peru (1936ff.)	
1	Panicle shorter than basal leaves, embraced at base by subtending leaf.	F. floribunda
1*	Panicle longer than the basal leaves.	2
2	Lemma distinctly awned, awn 5-15 mm. Annual, culms solitary or cespitose.	F. myuros
2*	Awn <2 mm. Perennial, cespitose	3
3	Culms 5-12 cm long. Panicle 2-4 cm long. Spikelets with 3-4 fertile florets.	F. peruviana
3*	Culms >15 cm long. Panicle >4 cm long. Spikelets with 2-7 fertile florets.	4
4	Tussok rhomboidal, often forming circular colonies with bare soil in the center. Culms 30-65 cm long. Spikelets 7-10 mm long.	F. chrysophylla
4*	Cespitose, habit not as above.	5
5	Panicle elliptic, nodding. Awn 0-1 mm long. Lemma 1-2.5 mm long. Rhachilla extension 3-4 mm long. Apical sterile florets rudimentary.	F. dolichophylla
5*	Panicle straight.	6
6	Panicle oblong to ovate. Culms geniculate ascending, 80-120 cm long, 2-3-noded.	F. procera
6*	Panicle contracted, linear. Culms erect, 15-25 cm long, 1-noded.	F. rigescens

Lorenzochloa

	Key based on Flora of Peru (1936ff.)	
1	Awn 35-60 mm long.	2
1*	Awn <20 mm long.	3
2	Culms 20-40 cm long. Ligule 4-6 mm long. Blades filiform, convolute, 10-17 cm x 0.5 mm, surface ribbed.	L. bomanii
2*	Culms 60-100 cm long. Ligule 0.5-1 mm long. Blades ± flat, 10-20 cm x 1-3 mm, surface sparsely hairy on both sides.	L. mucronata
3	Glumes obtuse, 3 mm long.	L. obtusa
3*	Glumes acute, 5 mm long	L. rigidiseta

Muhlenbergia

	Key based on Peterson et al. (2018)	
1	Plants annual. Florets awned.	M. peruviana
1*	Plants perennial. Florets not awned	2
2	Culms 50-100 cm tall. Panicles contracted and spike-like, 0.6-2 cm wide, plumbeous to reddish-purple.	M. coerulea
2*	Culms 2-12 cm tall. Panicles loosely contracted, not spike-like.	3
3	Rhizomatous herb. Leaf blades tightly involute, 2-8.5 mm long, arcuate, apex often pungent.	M. fastigiata
3*	Plants not rhizomatous. Leaf blades usually flat, 3-25 mm long, straight, not pungent.	M. ligularis

Nassella

1	Awns >30 mm long. Fruit 8 mm long, pubescent only on the lower part. Floret callus pungent.	N. neesiana
1*	Awns <20 mm long Fruit <5 mm long.	2
2	Culms >50 cm long. Awns flexuous, not bigeniculate.	N. pubiflora
2*	Culms <50 cm long. Awns bigeniculate.	3
3	Culms 20-40 cm long. Ligule an eciliate membrane. Panicle 7-15 cm long. Grows above 2500 m.	N. depauperata
3*	Culms 8-10 cm long. Ligule a fringe of hairs. Panicle 2-3 cm long. Grows below 2000 m.	N. nardoides

Species with doubtful occurrence in Arequipa: *Nassella asplundii Hitchc.* *Nassella brachyphylla (Hitchc.) Barkworth* *Nassella inconspicua (J.Presl) Barkworth* *Nassella mexicana (Hitchc.) R.W.Pohl* *Nassella smithii (Hitchc.) Barkworth* *All of them growing above 2500. No voucher, mentioned for Arequipa. Occurrence is likely.*	

Paspalum

	Key based on Flora of Peru (1936ff.)	
1	Inflorescence usually with a terminal pair of branches. Rachis not winged. Plants perennial, stoloniferous or creeping.	P. vaginatum
1*	Inflorescence with more than 2 branches. Rachis winged. Plants annual, erect to decumbent, sometimes rooting at lower nodes.	2
2	Inflorescence much-branched. Second glume developed.	P. flavum
2*	Inflorescence with few branches. Both second and first glume suppressed.	P. candidum

Piptochaetium

	Key based on descriptions from GrassBase (2002ff.)	
1	Spikelet 10-12 mm long. Lemma 5-6.5 mm long. Awn 40-50 mm long.	P. bicolor
1*	Spikelet 3-4 mm long. Lemma 1.5-2 mm long. Awn 7-8 mm long.	P. montevidense

Poa

	Key for the Poas with open panicle based on Sylvester et al. (2016), other species based on descriptions from GrassBase (2002ff.)	
1	Panicle open.	2
1*	Panicle contracted or spiciform.	7
2	Plants annual. Longest anthers of proximal florets 0.2-1.5 mm long.	P. annua
2*	Plants perennial. Longest anthers of proximal florets 1.6-3 mm long.	3
3	Lemmas glabrous, smooth or scabrous.	4
3*	Lemmas, at least of the upper florets, pubescent or villous in their lower half (rarely scabrous-pubescent in P. kurtzii).	5
4	Lemmas surface completely smooth. Ligules 2-3.5 mm long. Leaf blades conspicuously folded.	P. gilgiana
4*	Lemmas slightly to strikingly scabrous between and on veins. Ligules 8-15 mm long.	P. pearsonii
5	Plants 60-150 cm tall. Leaf blades flat, 5-10 mm wide.	P. horridula
5*	Plants usually <60 cm tall. Leaf blades involute to narrowly convolute, 0.5-2 mm wide.	6
6	Leaf blade densely scabrous beneath. Ligules 5-8 mm long, acute.	P. kurtzii
6*	Leaf blade glabrous to scaberulous beneath with prickles or hooks usually restricted to the leaf margin. Ligules 0.5-3 mm long, truncate	P. adusta
7	Lemma translucent or papery, apex dentate or emarginate.	8
7*	Lemma membranous, apex obtuse or acute.	9
8	Spikelets with 3-8 fertile florets. Fertile lemma flabellate, translucent, apex emarginate. Glumes translucent. Perennials, cushion-forming, culms 3-12 cm long.	P. lepidula
8*	Spikelets with 2 fertile florets. Fertile lemma ovate, papery, apex dentate. Glumes membranous, with translucent margins. Annual plants, culms 3-7 cm long.	P. macusaniensis
9	Culms 30-40 cm long, erect.	P. marshallii
9*	Culms <25 cm long, erect or ascending.	10
10	Culms all >10 cm long.	11
10*	Culms, at least some of them, >10 cm long.	14
11	Spikelets with 3-4 fertile florets. Dwarf grass, culms 1-4 cm long.	P. humillima
11*	Spikelets with 2 fertile florets. Culms 4-7 cm long	12
12	Fertile spikelets 3-3.3 mm long. Apex of upper glume and lemma acute. Lemma 2.5-2.7 mm long.	P. brevis
12*	Fertile spikelets 3.6-4.5 mm long. Apex of upper glume and lemma obtuse. Lemma 3.5-4 mm long.	13

13	Panicle linear, 1.5-2 cm long and 0.3-0.4 cm wide. Spikelets 3.5-4 mm long. Lower glume 3.6-4 mm long.	P. aequigluma
13*	Panicle ovate, 2-2.5 cm long and 0.5-0.7 cm wide. Spikelets 4-4.5 mm long. Lower glume 2.5-2.8 mm long.	P. perligulata
14	Ligules 4-8 mm long. Culms 13-25 cm long. Blades puberulous, hairy above, margins scabrous.	P. gymnantha
14*	Ligules <4 mm long. Culms <15 cm long. Blades glabrous, margins smooth.	15
15	Spikelet with 3-4 fertile florets. Ligule 1-2 mm long, erose, obtuse. Lemma 2.5 mm long.	P. laetevirens
15*	Spikelet with 2 fertile florets. Ligule 2.5-3.5 mm long, erose, truncate. Lemma 4-4.5 mm long.	P. spicigera

Polypogon

	Key based on Flora of Peru (1936ff.) and Jepson (2024)	
1	Glumes awnless.	P. viridis
1*	Glumes awned.	2
2	Annual herb. Inflorescence dense.	P. monspeliensis
2*	Perennial herb. Inflorescence lobed or interrupted.	3
3	Culms decumbent, 30-80 cm long. Leaves 4-6 mm wide.	P. interruptus
3*	Culms erect to geniculately ascending, 60-100 cm long. Leaves 4-15 mm wide.	P. elongatus

Sporobolus

1	Panicle 20-35 cm long. Spikelets 4-10 mm long.	S. indicus
1*	Panicle 3-10 cm long. Spikelets 1-3 mm long. Leaves usually conspicuously distichous. Found on beaches.	S. virginicus

Stipa

	Key based in part on Muñuico et al. (2024)	
1	Awn glabrous or scabrous. Plant 20-150 cm tall.	S. pachypus
1*	Awn pilose or plumose.	2
2	Annual plant, 20-30 cm tall. Awn 10-15 mm long, hirtellous.	S. annua
2*	Perennial plant, 30-300 cm tall. Awn 30-40 mm long, plumose.	S. plumosa
	Note: Based on morphological differences, Muñuico et al. (2024) split Stipa pachypus into Jarava disticha, distributed in the lomas of Ica and Arequipa, and J. pachypus, endemic to the lomas of S-Arequipa.	

Tragus

1	Upper glumes with 5 longitudinal rows of spinelike projections, 5-veined.	T. berteronianus
1*	Upper glumes with (5)6-7 longitudinal rows of spinelike projections, 7-veined.	T. racemosus

Polemoniaceae

	Key based on Flora of Peru (1936ff.)	
1	Perennial shrubs with woody stems.	Cantua
1*	Plants herbaceous, usually annuals.	2
2	Leaves entire.	Microsteris gracilis
2*	Leaves dissected or (2-)pinnatifid.	3
3	Leaves linear, once dissected, <1 mm wide.	Dayia glutinosa
3*	Leaves pinnatisect, dissected 1-4 times. Rachis and segments 2 mm wide.	Gilia

Cantua

1	Leaves pinnately divided. Capsule shorter than calyx.	C. volcanica
1*	Leaves simple. Capsule longer than calyx.	2
2	Flowers pink to purple, sometimes yellow or white. Corolla-lobes >8 mm, spread ± 90°, often bilobate. Leaves usually >2 cm long, elliptic.	C. buxifolia
2*	Flowers reddish-orange. Corolla-lobes <8 mm, spread ± 45°. Leaves usually <2 cm long, lanceolate.	C. candelilla

Gilia

1	Erect herb, up to 20 cm tall. Leaf-rachis and -segments 1-2 mm wide. Growing above 1000 m.	G. laciniata
1*	Decumbent herb, up to 10 cm tall. Some of the leaf-rachis and -segments >2 mm wide. Growing below 500 m.	G. lomensis

Polygalaceae

Monnina

	Key based on Flora of Peru (1936ff.) and Eriksen (1993)	
1	Perennial shrub.	M. salicifolia
1*	Annual herbs.	2
2	Leaves linear.	3
2*	At least the lower leaves ovate-lanceolate.	4
3	Unbranched or sparsely branched distally. Leaves not at all revolute. Fruit glabrescent.	M. macbridei
3*	Much-branched from the base. Leaves revolute. Fruit puberulous.	M. ramosa
4	Plant <0.5 m tall. Stipules absent. Fruit 2 x 2 mm, irregularly winged.	M. macrostachya
4*	Plant 0.5-2.5 m tall. Stipules present. Fruit 4-7 mm in Ø, always with a broad, entire wing.	M. pterocarpa

Polygonaceae

1	Annual plant, throughout pilose, much-branched. Basal leaves lanceolate, cauline leaves and involucre uncinate.	Chorizanthe commissuralis
1	Plant ± glabrous.	2
2	Shrub or vine.	Muehlenbeckia
2*	Herb.	3
3	Prostrate annual herb, usually on pathways. Flowers white.	Polygonum aviculare
3*	Erect herb.	4
4	Inflorescence a dense spike of pink flowers.	Persicaria hydropiperoides
4*	Inflorescence in sessile glomeruli or racemes. Flowers inconspicuous.	Rumex

Muehlenbeckia

1	Flowers short pedicellate, few or numerous at the nodes. Leaves cuneately narrowed at base.	M. fruticulosa
1*	Flowers in simple or paniculate racemes. Leaves cordate or hastate lobate at base.	2
2	Leaves ± hastate-lobate, not cordate at base. Raceme simple.	M. hastulata
2*	Leaves cordate at base, Raceme paniculate.	M. tamnifolia

Rumex

1	Basal leaves hastate at base	R. acetosella
1*	Basal leaves cordate or cuneate at base	2
2	Plant <25 cm tall, stoloniferous herb.	R. cuneifolius
2*	Plant >25 cm, without stolon	3
3	Perigon lobes entire	4
3*	Perigone lobes toothed	5
4	Bracts along the entire length of the inflorescence.	R. conglomeratus
4*	Bracts only in the lower part of the inflorescence.	R. crispus
5	Inflorescence branched, branches forming angle of >60° with 1. order stem. Flowers 10-20 per whorls. Bracts throughout the inflorescence.	R. pulcher
5*	Inflorescence branched, branches forming angle of <60° with 1. order stem. Flowers 15-40 per whorls. Bracts only in the lower part of the inflorescence.	R. obtusifolius

Pontederiaceae

1	Free floating aquatic plant with swollen petiole and mauve and blue flowers.	Pontederia crassipes

Portulacaceae

Portulaca

1	Plant completely covered by whit-pilose stem hairs, the minute leaves hidden.	P. nivea
1*	Plant more or less pilose or glabrous, the leaves not minute.	2
2	Leaves terete or narrowly linear.	3
2*	Leaves flat.	4
3	Plants distinctly perennial, with thick caudex and stems. Stem pubescence short. Leaves 5-10 mm long, linear-spatulate.	P. perennis
3*	Plants evidently annual. Pubescence in leaf-axils long. Leaves 6-20 mm long, narrowly linear.	P. pilosa
4	Plant glabrous.	P. oleracea
4*	Plant pilose.	P. tingoensis

Potamogetonaceae

1	Inflorescence pedunculate. Leaves 5-20 cm long and 1-6 mm wide.	Stuckenia striata
1*	Inflorescence ± sessile. Leaves 1-8 cm long and 1 mm wide.	Zannichellia palustris

Primulaceae

1	Leaves 1.5-15 cm long, base decurrent. Flowers white.	Samolus parviflorus
1*	Leaves <2.5 cm long, base not decurrent. Flowers reddish-orange or purplish-blue.	Lysimachia arvensis

Ranunculaceae

1	Shrubby vines.	Clematis
1*	Aquatic plants.	2
2	Flowers white. Leaves all submersed, dissected with long, linear leaves.	Ranunculus trichophyllus
2*	Flowers yellow. Leaves floating, entire.	Halerpestes uniflora

Clematis

1	Leaflets or their divisions linear-filiform.	C. millefoliolata
1*	Leaflets or their divisions at least 10 mm wide.	C. peruviana

Rhamnaceae

1	Shrub, ± leafless, leaves minute, margins serrate. Flowers whitish-red, pendent. Fruits finally capsular.	Colletia spinosissima
1*	Shrub with ovate leaves, 2.5 m long, margins entire. Flowers yellow, ± sessile. Fruits drupiform.	Scutia spicata

Rosaceae

1	Woody tree or shrubs.	2
1*	Scrambling and thorny lianas.	Rubus
1**	Herbs or aquatic plants.	4
2	Leaves pinnate or 3-foliate.	Polylepis
2*	Leaves entire or spiniform.	3
3	Leaves spiniform, sometimes with 1 or 2 leaflets. Small shrub, fruit margins bristly.	Tetraglochin cristata
3*	Leaves oblong, elliptic or ovate. Fruits woody, star shaped.	Kageneckia lanceolata

5	Herbs or aquatic plants, forming stolons. Petals greenish-yellow.	Alchemilla
5*	Rosulate herb, no stolons. Leaves pinnate, bluish-green. Flowers apetalous, each flower subtended by a papery bract, greenish or white to red-purple, stigma red, stamens white or pink.	Sanguisorba minor

Alchemilla

1	Aquatic plant, floating. Leaves folded like accordion.	A. diplophylla
1*	Rosulate herb, stolon-forming. Leaves pinnate, pale green. Petals greenish-yellow, stigma and stamens yellow.	A. pinnata

Polylepis

	Key based on Boza Espinoza & Kessler (2022)	
1	Imparipinnate with 3-6 pairs of lateral leaflets.	P. microphylla
1*	Imparipinnate with 1 pair of lateral leaflets.	2
2	Lower leaflet surfaces densely tomentose with a dense underlying layer of very short, white pannose hairs. Widespread in the department of Arequipa and S-Peru.	P. rugulosa
2*	Lower leaflet surfaces densely pilose, villous or tomentose without pannose hairs. Found only in prov. Caravelí and adjacent Apurimac and Ayacucho.	P. fjeldsaoi

Rubus

1	Stems tomentose, glaucous when young. Flowers pale pink.	R. ulmifolius
1*	Stems densely hirsute, densely reddish-setose. Flowers white.	R. urticifolius

Rubiaceae

1	Trees or shrubs with lignified branches.	2
1*	Annual or perennial herbs.	3
2	Woody tree or shrub up to 3 m tall. Apex of lateral branches armed with thorns. Leaves orbicular to obovate.	Randia rotundifolia
2*	Small shrub, 20-30 cm tall, unarmed. Leaves linear-lanceolate, ericaceous.	Arcytophyllum thymifolium
3	Leaves verticillate	Galium
3*	Leaves opposite, decussate, clustered at tip of stem, but not verticillate.	Richardia lomensis

Galium

1	Stem up to 1 m long. Stem and leaves retrosely scabrous (plant sticks to clothing).	2
1*	Stem <50 cm long. Stem and leaves not retrosely scabrous.	3
2	Leaves in whorls of 6-8, 1-3 cm long. Fruit dry, green, uncinate.	G. aparine
2*	Leaves in whorls of 4, 0.5-1.5 cm long. Fruit orange-red, fleshy.	G. hypocarpium
3	Leaves and stem with long scattered hairs. Growing in coastal areas below 600 m.	G. arequipicum
3*	Leaves and stem glabrate to hispid. Growing above 1700 m.	G. corymbosum

Ruppiaceae

1	Aquatic plant, with floating leaves just below the water's surface.	Ruppia maritima

Salicaceae

1	Tree up to 10 m tall. Leaves deciduous, linear, acuminate, serrulate. Catkins appearing with the leaves.	Salix humboldtiana

Santalaceae

1	Parasitic shrub. Stems succulent, reddish-brown, articulate. Leaves opposite, red like the stem, oblanceolate, succulent. Berries white.	Dendrophthora mesembryanthemifolia

Sapindaceae

1	Small tree with symmetrical crown. Leaves pinnate with 6-13 leaflets.	Sapindus saponaria
1*	Climbing vines.	2
2	Fruits inflated, lantern-like.	Cardiospermum
2*	Fruits schizocarps with broadly-winged mericarps.	Serjania sp.

Cardiospermum

	Key based on Flora of Peru (1936ff.)	
1	Terminal leaflets usually distinctly petiolulate. Seed scar (hilum) typically small.	C. corindum
1*	Terminal leaflets usually decurrently petiolulate if at all. Seed scar typically large.	C. halicacabum
	Note: Both species are similar and some authors synonymize them as C. halicacabum.	

Schoepfiaceae

1	Perennial low shrub. Branched from the base. Leaves linear, fleshy, green to reddish. Inflorescence terminal, clusters of more than 20 dark orange flowers, 5-merous.	Quinchamalium chilense

Scrophulariaceae

1	Shrub or tree >2 m tall.	2
1*	Herbs or subshrubs, <1 m tall.	3
2	Flowers orange or yellow. Leaves canescent or tomentose beneath.	Buddleja
2*	Flowers white. Leaves translucently dotted, glabrous on both sides.	Myoporum laetum
3	Shrub. Flowers red.	Alonsoa meridionalis
3*	Flowers white. Semi-aquatic herb.	Limosella

Buddleja

1	Leaves 7-15 cm long, acute, petiole 1-2 cm long.	B. incana
1*	Leaves 1-4 cm long, obtuse, petiole 0.3-0.4 cm long.	B. coriacea

Limosella

1	Leaves flat, spatulate, long petiolate.	L. aquatica
1*	Leaves cylindrical, linear, without differentiated blade.	L. australis

Solanaceae

	Key based on Hunziker & Subils (1979), Jepson (2024)	
1	Inflorescence (2-)3 to many-flowered, generally bearing pedicels and peduncles (some Jaltomata without pedicel, but many flowers, or pedicellate and pedunculate, but only one flower).	2
1*	Inflorescence with solitary flowers bearing only pedicels and no peduncles.	6
2	Corolla a long tube with inserted stamens.	3
2*	Corolla flat or short-campanulate, stamens protruding from corolla.	5
3	Plant <30 cm tall. Leaves max. 3 cm long. Short-lived ephemeral herbs of the lomas.	Leptoglossis
3*	Plant >50 cm tall. Leaves more than 5 cm long (observe main leaf at Cestrum).	4
4	Fruit a berry. Plants mostly woody. Leaves paired, a normal main leaf accompanied by a much smaller secondary leaf.	Cestrum
4*	Fruit a dry capsule. Plants mostly herbaceous.	Nicotiana
5	Corolla flat, anther dehiscence by apical pores.	Solanum
5*	Corolla slightly campanulate, anther dehiscence longitudinal.	Jaltomata
6	Fruit prickly, erect. Leaves lobed and dentate.	Datura stramonium
6*	Fruit not prickly.	7
7	Perennial shrub.	8
7*	Herbs, mostly ephemeral.	12
8	Leaves linear to cylindrical, succulent.	Fabiana stephanii

8*	Leaves not linear or cylindrical, succulent or not.	9
9	Plant armed with spines.	10
9*	Plant not armed.	11
10	Stamens protruding the short corolla-tube.	Lycium
10*	Stamens inserted in a rather long corolla-tube.	Dunalia spinosa
11	Tall shrub, >1m. Flowers blue-purple.	Lycianthes lycioides
11*	Short shrub, <50 cm. Flowers white or yellow.	Salpichroa
12	Calyx lobes longer than the tube, winged toward base	13
12*	Calyx lobes shorter than the tube, not winged at base	15
13	Decumbent herb of the lomas. Calyx not enclosing fruit as a whole.	Exodeconus
13*	Erect herb of disturbed areas. Calyx encloses entire fruit.	14
14	Flowers yellow with deep brown center. Calyx inflated in fruit, sepals entirely connate.	Physalis peruviana
14*	Flowers blue to violet. Calyx inflated in fruit, sepals not entirely connate.	Nicandra
15	Acaulescent, rosulate herb growing above 3500 m. Leaves lanceolate, deeply lobed, leathery.	Jaborosa squarrosa
15*	Decumbent herb or shrub growing below 3000 m. Leaves mostly entire.	Nolana

Cestrum

1	Glabrous shrub. Secondary leaf auriculate.	C. auriculatum
1*	Tometose shrub. Secondary leaf protruding.	C. tomentosum

Exodeconus

	Key based on Axelius (1994)	
1	Flowers yellow.	E. flavus
1*	Flowers white, violet, bluish, greenish or a combination of these.	2
2	Leaves broadly ovate to suborbicular. Flowers 1.2-1.4 cm long. Calyx in fruit ribbed by protruding transverse veins.	E. integrifolius
2*	Leaves lanceolate to rhomboid. Flowers 1.5-2.0 cm long. Calyx in fruit thin, smooth.	E. pusillus

Jaltomata

	Key based on Mione et al. (2007), Mione et al. (2011), Leiva et al. (2016)	
1	Small herbaceous plant, 0.5-0.7 m tall. Flowers solitary, corolla purple, with a ring of green maculae at the base.	J. quipuscoae
1*	Shrub, up to 1.5 m tall. Corolla white, with a ring of green maculae at the base.	2
2	Inflorescence with peduncle and flowers pedicellate.	J. atiquipa
2*	Inflorescence sessile, without peduncle, but flowers pedicellate.	J. diversa

Leptoglossis

	Key based on Hunziker & Subils (1979)	
1	Calyx with 7-10 lobes. Flowers ± white.	L. albiflora
1*	Calyx with 5 lobes. Flowers yellow to purple.	2
2	Corolla usually purplish. Calyx with a few capitate hairs. Stems glabrous. Growing above 1500 m.	L. acutiloba
2*	Corolla usually cream or yellow. Calyx with a dense pubescence of bifurcate hairs. Stems hairy. Growing in the lomas below 1000 m.	L. lomana
	Note: L. ferreyraei Hunz. & Subils is treated here as synonym of L. lomana, and L. acutiloba is reestablished (in POWO a synonym of L. lomana), as proposed by Fernandez & Quipuscoa (2021).	

Lycium

	Key based on Miller and Levin (2023)	
1	Leaves glabrous, 0.5-6 cm long and 0.5-3 cm wide. Corolla white to violet, with dark purple venation, tube 5-8 mm long, lobes 2.4-4.5 mm long (at least half as long as the corolla tube). Berries red.	L. americanum

Solanaceae

1*	Leaves pubescent, <2 cm long and <0.7 cm wide. Corolla white to pale purple or pale blue, tube >10 mm long, lobes <2.5 mm long (clearly less than half as long as the corolla tube). Berries red, orange, or blackish.	2
2	Corolla slightly pubescent toward base. Berries orange, red, 3-7 seeds per locule.	L. distichum
2*	Corolla glabrate, but lobes usually ciliate. Berries 1-5 seeds per locule, color unknown.	L. stenophyllum

Nicandra

1	Flowers 3.5-5 cm in Ø, corolla pale blue and white with purple spots at the base. Leaves ovate, dentate and waved.	N. physalodes
1*	Flowers 1.5-2 cm in Ø, corolla pale blue and white without spots at the base. Leaves ovate, margins regularly dentate with 6-7 blunt teeth on each side.	N. yacheriana

Nicotiana

1	Corolla pink or whitish-pink. Shrub 3-5 m tall, whole plant tomentose. Leaves 30 cm long, petiole alate. Stamens well-exserted, ascending.	N. tomentosa
1*	Corolla white, yellow or greenish.	2
2	Perennial, spindly shrub 1-6 m high, glaucous, glabrous. Corolla yellow to slightly greenish.	N. glauca
2*	Annual or perennial, <3 m tall. Pubescent or Corolla whitish- or greenish-yellow.	3
3	Corolla 2-5 cm long, at least 3x as long as the calyx.	4
3*	Corolla <2 cm, less than 2x as long as calyx.	6
4	Stamens exserted. Corolla greenish-yellow. Leaves orbicular to ovate, long petiolate, cordate with closed sinus.	N. benavidesii
4*	Stamens inserted in corolla.	5
5	Annual. Corolla uniformly greenish-yellow.	N. paniculata
5*	Perennial. Corolla greenish-yellow with dark green corolla lobes.	N. knightiana
6	Leaves oblong, undulate, cauline leaves nearly sessile.	N. undulata
6*	Leaves broadly ovate, cauline leaves petiolate.	N. rustica

Nolana

	Alternative key by Quipuscoa & Dillon (2018), with all subspecies and local variants included.	
1	Plants with a persistent basal rosette.	2
1*	Plants without persistent basal leaves.	3
2	Margins of basal leaves strictly entire. Calyx 5 mm wide.	N. scaposa
2*	Margins of basal leaves dentate to crenate. Calyx inflated, globular, 10 mm wide.	N. inflata
3	Inflorescence a terminal, 3-branched, weak scorpioid cyme.	N. tricotiflora
3*	Flowers solitary or in dense cluster, but not scorpioid.	4
4	Calyx inflated.	N. chancoana
4*	Calyx not inflated.	5
5	Leaves clearly petiolate	6
5*	Leaves indistinctly petiolate or sessile.	9
6	Calyx >8 mm wide.	N. humifusa
6*	Calyx <7 mm wide.	7
7	Leaves slightly cordate or truncate at base. Perennial shrub with brownish stems.	N. adansonii
7*	Leaves attenuate to acute at base. Annual subshrubs.	8
8	Leaves 6-10 mm wide, leathery. Flowers pale blue.	N. arenicola
8*	Leaves 10-20 mm wide, fleshy. Flowers bluish-violet.	N. laxa
9	Leaves 5-12 mm wide. Flowers lavender, inner throat purple.	N. callae
9*	Leaves <4 mm wide (in N. lycioides sometimes wider, but margins revolute).	10
10	Corolla white.	11
10*	Corolla lavender, purple or blue.	12
11	Corolla 3-6 mm wide. Inner tube sometimes with bluish marks. Sandy soils between 100 and 300 m.	N. arequipensis
11*	Corolla 10-12 mm wide. Tube >15 mm long, yellowish. Known only from Quicacha (Caravelí).	N. quicachaensis

12	Corolla tube >15 mm long.	13
12*	Corolla tube <10 mm long.	14
13	Densely lanuginose to arachnoid subshrub. Leaves oblong, 5-8 mm long, 1-2 mm wide. Restricted to Punta Bombon (Islay).	N. bombonensis
13*	Not lanuginose or arachnoid. Leaves elliptic, 8-15 mm long, 2-3.5 mm wide. Restricted to Chapi (Arequipa).	N. chapiensis
14	Leaves terete, fleshy.	15
14*	Leaves not terete, but with revolute margins.	16
15	Erect herb. Corolla >10 mm wide, lavender with white throat.	N. crassulifolia
15*	Shrub growing near the ocean. Corolla <5 mm wide, lavender with dark purple throat.	N. tarapacana
16	Leaves 8-25 mm long.	N. gracillima
16*	Leaves 5-9 mm long.	N. lycioides

Salpichroa

	Key based on Gonzáles (2021).	
1	Flowers 1-1.5 cm long.	2
1*	Flowers >3 cm long.	3
2	Corolla greenish-yellow to deep yellow, tube with glandular trichomes on the outside.	S. tristis
2*	Corolla whitish-yellow, tube broadest at base, glabrous outside, lobes reflexed at anthesis.	S. ramosissima
3	Flowers 3-4 cm long, hirsute outside in upper half, lobes mucronate, anthers exserted.	S. hirsuta
3*	Flowers 5 cm long, ± glabrous outside, anthers not exserted.	S. glandulosa

Solanum

	Key based on Peralta et al. (2005)	
1	Flowers yellow.	Clade Tomato
1*	Flowers not yellow.	2
2	Low, tuber bearing herbs. Leaves pinnate. Corolla >2 cm in Ø, blue to purple (in S. acaule sometimes corolla 1.5 cm in Ø and white).	Clade Petota
2*	Not in above combination.	3
3	Corolla pentagonal to rotate (not lobed). Leaves pinnate or deeply incised.	Clade Regmandra
3*	Not in above combination.	4
4	Annuals to short-lived perennials. Leaves entire, margins entire to broadly dentate. Flowers white or purplish. Fruits small juicy berries.	Clade Morelloid
4*	Not in above combination.	5
5	Stem and leaves silvery-white and with prickles to 1 cm long. Leaves with white margin. Fruit globular, 3-4 cm in Ø, pale or deep yellow.	S. marginatum
5*	Stem and leaves not prickly.	6
6	Inflorescence unbranched, 1-3-flowered. Flowers white, calyx glabrescent. Berry bright orange-red.	S. pseudocapsicum
6*	Inflorescence branched, 10-50-flowered.	7
7	Shrub up to 1.5 m tall. Flowers lilac to purple, calyx grayish pubescent. Inflorescence 10-20-flowered. Berries dark red when ripe.	S. nitidum
7*	Shrub 1.5-3 m tall. Flowers white to pale purple.	8
8	Upright shrub up to 3 m tall. Stems strong, densely strigose. Inflorescence highly branched, 10-50-flowered. Fruits green.	S. basendopogon
8*	Scandent shrub, 1.5-3 m tall. Inflorescence branched, 10-20-flowered. Fruits whitish-green with brown stripes.	S. filiforme

Solanum clade Morelloid

	Key based on Knapp et al. (2023)	
1	Plants with glandular trichomes, sticky to touch.	2
1*	Plants without glandular trichomes, not sticky to touch.	4
2	Inflorescence unbranched.	S. nitidibaccatum
2*	Inflorescence forked or several times branching.	3

3	Plants with woody rootstock. Calyx lobes 2-3 mm long, tips obtuse. Growing in grasslands.	S. fragile
3*	Plants lacking a woody rootstock. Calyx lobes 1-1.5 mm long, tips acute. Growing in disturbed areas.	S. grandidentatum
4	Leaves pinnatisect, divided halfway or more to the midrib.	5
4*	Leaves simple, the margins dentate or not.	6
5	Leaves deeply 5-lobed. Stem strongly angled to winged. Mature berries orange or orange-yellow.	S. radicans
5*	Leaves deeply lobed. Fruiting calyx partly to completely enclosing the berry.	S. weddellii
6	Mature berries red. Leaves with entire margins	S. corymbosum
6*	Mature berries green or purple. Leaves usually broadly dentate, at least in proximal half of blade.	7
7	Inflorescence unbranched. Fruits deep purple.	S. americanum
7*	Inflorescence several times branched or forked.	8
8	Inflorescence several times branched. Petals usually dark violet. Leaves strongly dentate along the margin.	S. pentlandii
8*	Inflorescence forked. Petals white to pale violet, in S. cochabambense sometimes dark violet. Leaves entire or slightly dentate on proximal half of blade.	9
9	Scandent shrub, stems up to 3+ m long. Leaves usually entire. Inflorescence several times branched, 10-80-flowered.	S. cochabambense
9*	Subwoody herb, up to 1.5 m tall. Leaves slightly dentate on proximal half of blade. Inflorescence forked or branched, 10-20-flowered.	10
10	Leaves elliptic to ovate, 3.2-14 cm long and 1.5-6 cm wide. Corolla white with yellow-green center. Fruits grayish-green at maturity.	S. arequipense
10*	Leaves narrowly elliptic, 1.8-8 cm long and 0.8-4 cm wide. Corolla white, often violet or with violet stripes. Fruits green with purple blotches at maturity.	S. interandinum

Solanum clade Petota

1	Plant rosulate, acaulescent to semierect.	2
1*	Plant markedly erect.	3
2*	Leaves pinnate, with 4-6 pairs of leaflets, subequal except for the mot proximal 1-2 pairs.	S. acaule
2	Leaves with 0-4 pairs of leaflets and a terminal leaflet much larger than the laterals.	S. boliviense
3	Flowers long pedicellate, nodding.	S. medians
3*	Flowers not nodding.	4
4	Leaves pinnate, with 4-5 pairs of leaflets, subequal except for the mot proximal pair, which is smaller.	S. acroscopicum
4*	Leaves pinnate, with 2-6 pairs of leaflets, subequal except for the mot distal pair, which is larger.	S. candolleanum
	Note: This section includes Solanum tuberosum, the cultivated potato. The key does not include cultivated plants.	

Solanum clade Regmandra

1	Leaf blade decurrent on the petiole and the stem. Corolla blue, lilac to purple.	S. paposanum
1*	Leaves not decurrent. Corolla usually white, sometimes lilac.	2
2*	Inflorescence much-branched with 8-23 flowers. Leaves 1-3-pinnate.	S. multifidum
2	Inflorescence not branched or 1-branched, 1-11-flowered. Leaves entire, margins deeply lobed.	3
3	Leaves narrowly ovate, 4-5-lobed nearly to midvein, the lobes sometimes lobed. Restricted to coastal Ica and N-Arequipa (prov. Caravelí).	S. edmondstonii
3*	Leaves broadly ovate, 2-3-lobed halfway to midrib, lobes entire. Stem arising from a subterranean spheroidal swollen caudex (tuber-like).	S. montanum

Solanum clade Tomato

1	Inflorescence 20-50-flowered. Stems and leaves densely grayish pubescent, 5-7 pairs of leaflets. Anther tubes straight.	S. chilense
1*	Inflorescence 8-30-flowered. Stems and leaves usually not grayish pubescent, 2-4 pairs of leaflets. Anther tubes usually ± curved (± straight in S. pimpinellifolium).	2

2	Fruit bright red. Calyx lobes in fruit strongly reflexed and parallel to fruiting pedicel. Corollas stellate, divided >¾.	S. pimpinellifolium
2*	Fruit greenish-white, sometime with dark green or purple stripe from apex.	3
3	Leaves with almost orbicular leaflets, terminal leaflet wider than long, but not larger than the laterals. Secondary leaflets absent.	S. pennellii
3*	Leaflets broadly to narrowly elliptic, terminal leaflet larger than the laterals. Secondary leaflets occasionally present on lower lateral leaflets.	3
4	Growing at mid to high elevations above 1500 m. Calyx often longer than fruit.	S. corneliomulleri
4*	Growing in lomas and coastal areas below 800 m.	S. peruvianum
	Note: The species of Solanum clade Tomato are similar to each other and inbred. Therefore, identification to species level is challenging and requires combined morphological characters. S. chilense, S. corneliomulleri and S. peruvianum are sometimes subsumed as S. peruvianum s.l.	

Talinaceae

1	Perennial from a thick, fleshy root. Stems stout, up to 1 m high, reddish, little branched, or simple. Leaves numerous, thick and fleshy. Flowers on pedicels 1.5 cm long, petals bright yellow.	Talinum spathulatum
	Note: T. spathulatum is usually synonymized with Talinum paniculatum. Here, T. spathulatum is listed as a separate species due to the obvious morphological differences. T. paniculatum s.str. does not occur in Arequipa.	

Tropaeolaceae

Tropaeolum

1	Leaves lobed.	2
1*	Leaves entire, not lobed.	4
2	Petals orange-red, lobes not spread, entire.	T. tuberosum
2*	Petals yellow-orange, lobes spread, incised.	3
3	Petals incised halfway or more. Plants sparsely pilose. Leaf-lobes acute.	T. seemannii
3*	Petals incised marginally. Plants glabrous. Leaf-lobes obtuse.	T. peregrinum
4	Petals 3 cm long.	T. majus
4*	Petals 2 cm long. All parts distinctly smaller than by T. majus.	5
5	Central vein of peltate leaf ending in a mucronate tooth. Leaves and petiole glabrous.	T. minus
5*	Central vein of peltate leaf not mucronate. Leaves and petiole pubescent.	T. ferreyrae

Typhaceae

1	Perennial herb in shallow water. Leaves linear. Inflorescence a cigar-like spike.	Typha domingensis

Urticaceae

1	Leaves and stems with stinging hairs. Leaves opposite.	Urtica
1*	Leaves without stinging hairs. Leaves alternate.	2
2	Erect plant, leaves >1 cm	Parietaria debilis
2*	Decumbent plant, leaves orbicular to reniform, <0.5 cm in ∅.	Soleirolia soleirolii

Urtica

	Key based on Flora of Peru (1936ff.)	
1	Leaves flabellate-incised, the segments lobed.	U. flabellata
1*	Leaves ovate to ovate-lanceolate, or cordate.	2
2	Decumbent herb of high altitudes.	U. trichantha
2*	Erect herb.	3
3	Leaves thick, rugose-bullate.	U. echinata
3*	Leaves thin, ± flat.	4
4	Annual herb. Leaves incised-dentate. Flowers in glomeruli.	U. urens
4*	Perennial herb. Leaves coarsely dentate or serrate. Flowers ± elongate spikes.	U. leptophylla
	Note: The name Urtica magellanica Juss. ex Poir., a species of Patagonia, is often erroneously employed for U. leptophylla and U. urens	

Verbenaceae

	Key based on O'Leary & Mulgura (2014).	
1	Annual to perennial herbs or subshrubs.	2
1*	Shrubs, >1 m.	6
2	Rhizome bearing a tuber.	Pitraea cuneato-ovata
2*	Rhizome without tuber.	3
3	Corolla ± 2-lipped, nutlets 2, calyx 2-4 toothed.	Phyla nodiflora
3*	Corolla radial or slightly zygomorphic, nutlets 4, calyx 4-5-toothed.	4
4	Inflorescence unbranched.	Junellia, Mulguraea
4*	Inflorescence branched.	5
5	Style long, more than 3x longer than ovary, mature calyx usually longer than fruit.	Glandularia
5*	Style brief, less than 3x longer than ovary, mature calyx shorter than fruit.	Verbena
6	Fruits dry.	Aloysia arequipensis
6*	Fruits fleshy.	7
7	Stem without spines (sometimes with epidermis-borne prickles).	Lantana
7*	Stem with prominent spines.	Citharexylum

Citharexylum

1	Inflorescence with 12-20 flowers. Leaves papery or membranous.	C. flexuosum
1*	Inflorescence with 1-6 flowers. Leaves leathery.	C. weberbaueri
	Note: C. weberbaueri and C. flexuosum are morphologically similar, occur in similar habitats and are probably often confused.	

Glandularia

	Key based on O'Leary & Mulgura (2014).	
1	Stumpy suffruticose plants, <10 cm tall	G. microphylla
1*	Erect suffruticose plants, >10 cm tall	G. laciniata
	Note: Glandularia is treated under Verbena by POWO (2024).	

Junellia, Mulguraea

	Key based on O'Leary & Mulgura (2014).	
1	Anther connective tissue not surpassing tube.	M. arequipensis
1*	Anther connective tissue surpassing tube.	2
2	Cushion plants of high altitude.	J. minima
2*	Erect herbs.	3
3	Prickling plants	J. juniperina
3*	Plants not prickly	4
4	Leaf blades trisected from the base, leaves terete.	J. clavata
4*	Leaf blades 3-5 sected or 3-5-parted, never from the base. Leaves flat.	J. fasciculata

Lantana

	Key based on Binder (2002).	
1	Flowers white to whitish-violet, sometimes with yellow throat. Fruits fleshy, whitish when ripe.	L. sprucei
1*	Flowers yellow. Fruits dry, brown when ripe.	2
2	Dwarf, decumbent shrub, up to 30 cm high, creeping with 1 m long branches.	L. reptans
2*	Erect shrub, up to 2.5 m tall.	L. scabiosiflora

Verbena

	Key based on O'Leary & Mulgura (2014).	
1	Plants with dense hispid hairs.	V. hispida
1*	Plants subglabrous with strigose or villous pubescence, not hispid.	2

2	Inflorescence subconical. Leaf base slightly amplexicaule.	V. glabrata
2*	Inflorescence cylindrical. Leaf base cuneate.	V. litoralis

Violaceae

Viola

	Key based on Gonzáles et al. (2022).	
1	Leaf blade entire, green.	2
1*	Leaf blade crenate, dentate or pinnatifid, brownish to pinkish.	3
2	Flowers white. Leaf blade puberulent to glabrous, shiny, margins not ciliate.	V. micranthella
2*	Flowers yellow. Leaf blade sericeous, not shiny, margins ciliate.	V. weberbaueri
3	Plant laxly imbricate, stipules present. Petals brownish.	V. granulosa
3*	Plant densely imbricate, stipules absent.	4
4	Leaf blade scarcely wider than pseudopetiole, margin 5-crenate. Petals white, outside violet.	V. longibracteolata
4*	Leaf blade 2× wider than pseudopetiole, margin 11-crenate. Petals white on both sides.	V. ornata

Viburnaceae

1	Shrubs or trees. Leaves opposite, pinnate. Flowers in terminal compound cymes. Berries purple.	Sambucus peruviana

Zygophyllaceae

1	Flowers pink. Leaves 3-foliate.	Zygophyllum chilense
1*	Flowers yellow. Leaves pinnate or bifid.	2
2	Shrubs, >1 m tall.	3
2*	Stems prostrate, decumbent, <0.5 m tall. Leaves pinnate.	4
3	Branches angular, slightly alate, leafy. Fruit hairy. Leaves bifid.	Larrea divaricata
3*	Branches terete, usually leafless. Fruit alate, glabrous.	Bulnesia retama
4	Fruits with several 4-7 mm long prickles.	Tribulus terrestris
4*	Fruits warty, without prickles.	Kallstroemia parviflora

This catalog aims to enhance understanding of the flora and vegetation of the department of Arequipa, Peru, and to serve as a practical resource for researchers, conservationists, and nature enthusiasts. By expanding our knowledge, we hope to inspire appreciation for Arequipa's natural heritage and underscore the importance of preserving its biodiversity.

The catalog contains 1,350 wild-growing ferns and flowering plants. Each species has a brief morphological description, details of its ecological and geographic distribution, information on its human use, and notes on its taxonomy. The appendix includes a comprehensive taxonomic key to all species discussed and a list of commonly used synonyms.

PERIODENSYSTEM UND 2D-ATOMAUFBAU

EIN NEUES PERIODENSYSTEM DER ELEMENTE!

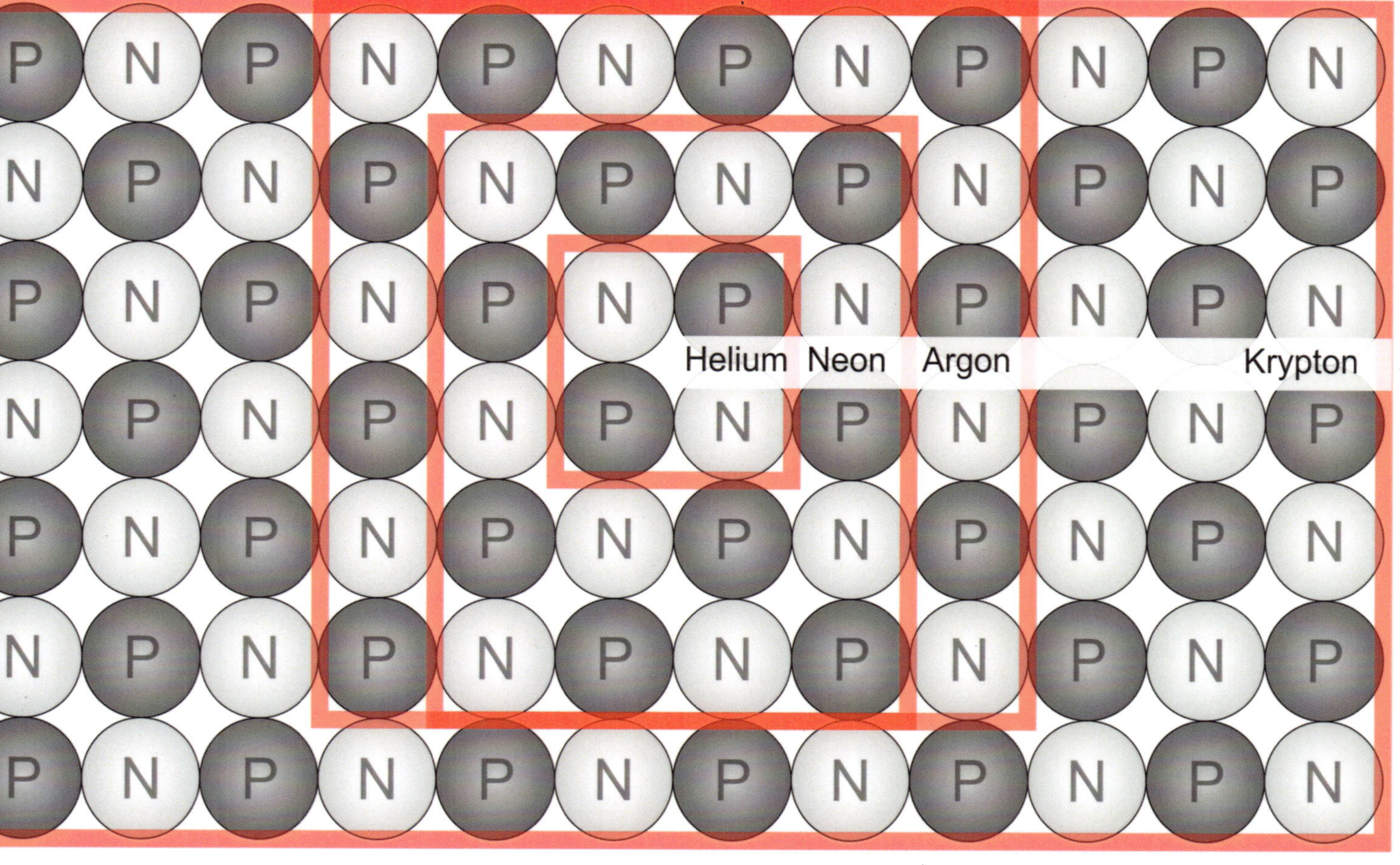

HELMUT M. ALBERT